AF539888

Climate Change and Agroforestry

Climate Change and Agroforestry

Adaptation, Mitigation and Livelihood Security

Editors

C.B. Pandey

Mahesh Kumar Gaur

R.K. Goyal

NEW INDIA PUBLISHING AGENCY

New Delhi – 110 034

NEW INDIA PUBLISHING AGENCY
101, Vikas Surya Plaza, CU Block, LSC Market
Pitam Pura, New Delhi 110 034, India
Phone: + 91 (11)27 34 17 17 Fax: + 91(11) 27 34 16 16
Email: info@nipabooks.com
Web: www.nipabooks.com

Feedback at feedbacks@nipabooks.com

© Publisher, 2018

ISBN: 978-93-86546-06-7

All rights reserved, no part of this publication may be reproduced, stored in a retrieval system or transmitted in any form or by any means, electronic, mechanical, photocopying, recording or otherwise without the prior written permission of the publisher or the copyright holder.

This book contains information obtained from authentic and highly regarded sources. Reasonable efforts have been made to publish reliable data and information, but the author/s, editor/s and publisher cannot assume responsibility for the validity of all materials or the consequences of their use. The author/s, editor/s and publisher have attempted to trace and acknowledge the copyright holders of all material reproduced in this publication and apologize to copyright holders if permission and acknowledgements to publish in this form have not been taken. If any copyright material has not been acknowledged please write and let us know so we may rectify it, in subsequent reprints.

Trademark notice: Presentations, logos (the way they are written/presented) in this book are under the trademarks of the publisher and hence, if copied/resembled the copier will be prosecuted under the law.

Composed, Designed and Printed in India

भारतीय कृषि अनुसंधान परिषद
कृषि अनुसंधान भवन-II, पूसा, नई दिल्ली 110 012
INDIAN COUNCIL OF AGRICULTURAL RESEARCH
KRISHI ANUSANDHAN BHAVAN-II, PUSA, NEW DELHI 110 012

डा. नरेन्द्र सिंह राठौड़
उप महानिदेशक (कृषि शिक्षा)
Dr. Narendra Singh Rathore
Deputy Director General (Agril. Edn.)

Phone : 011-25841760 (O)
Fax : 011-25843932
E-mail : ddgedn@gmail.com
nsrdsr@gmail.com
Website: www.icar.org.in

Foreword

Climate change, resulted from global warming, has started to show its impact world over in the form of storm, flood, drought and crop yields decline. It has been predicted to have a huge burden in future as the world needs to feed 9 billion populations by the turn of the 21st century, when global temperature rise is predicted to increase by 1.4 to 5.8 °C with significant inter-and intra-annual and regional variations. Rise in temperature is causing extremes in precipitation and droughts, and rises in sea level and melting of polar ice caps. Six of 10 countries, predicted the most vulnerable to climate change, lie in the Asia Pacific. Bangladesh tops the list followed by India, Nepal, the Phillippines, Afghanisthan and Myanmar. In India the warming may be more pronounced in its northern parts. The extremes in maximum and minimum temperatures are expected to increase under changing climate; few places are expected to get more rain while others may remain dry. States other than Punjab and Rajasthan in the north-west and Tamil Nadu in the south, which may experience a slight decrease, are projected to receive 20% more rainfall from summer monsoon. Number of rainy days may decline, but intensity is expected to rise at most parts of our country resulting in decline in gross per capita water availability from 1820 m^3 yr^{-1} in 2001 to as low as 1140 m^3yr^{-1} in 2050.

Indian agriculture by and large is monsoon dependant; hence any change in it is expected to affect crop production badly. Studies have predicted 4-5 million tons loss of wheat production for every 1°C rise, and a ton ha^{-1} loss in rice for each 2 °C rise in temperature. Decline in wheat yields is likely to be experienced on Indo-Gangetic basin, but rice yield decline may be visible in the states like Jharkhand, Odisha and Chaattisgarh. In Rajasthan each 2 °C rise in temperature is estimated to reduce pearl millet by 10-15%. If maximum and minimum temperature rise occur by 3°C and 3.5 °C, respectively, then soybean yields in MP will decline by 5%. Agriculture will be worst affected in coastal part of

India due to inundation and salinization owing to sea level rise. Overall, productivity of most crops is likely to decline only a little by 2020, but it will register 10-40% loss from the second half of the 21st century.

Climate change adaptation and mitigation is argued to be achieved by resorting to climate-smart practices at the field and farm scale, promoting diversity within farming systems and land uses across landscapes, and managing land use interactions to achieve synergies among a range of objectives. Agroforestry is one of the practices, which sequesters greater atmospheric carbon, reduces greenhouse gas emissions like N_2O by increasing nitrogen (N) use efficiency and uptake of excess N, and CH_4 by improving fodder quality in silvopstoral system; and makes nutrient and hydrological cycling agile due to increased SOC. Deep rooted trees allow better access to nutrients and water during droughts; increases soil porosity, reduces runoff and increases soil cover which can improve water infiltration and retention in the soil profile and thereby reduces moisture stress in low rainfall years, and during excessive soil moisture period, tree-based systems can maintain aerated soil conditions by pumping out excess water more rapidly than conventional agriculture. Thus agroforestry offers greater reduced risk and economic stability under climate change.

The editors have done a commendable job of bringing the contributions on climate change adaptation, mitigation and livelihood security with agroforestry into one publication. I am sure this book will serve as a repository of information in the field of agroforestry, which can be used by planner, researchers and undergraduate and post graduate students pursuing studies in the field of forestry/ agroforestry.

(N. S. Rathore)

Preface

Natural change in climate is slow and takes millions of years; and it is known to have made our planet hospitable to live. But, ruthless deforestation, industrialization, urbanization and intensive agriculture have led greenhouse gas (CO_2, N_2O and CH_4 mainly) emissions exacerbately high which increased ambient temperature of the globe by 0.74 °C within a short span of about two century, and it is forecasted to increase further by 1.4 to 5.8 °C by the end of 21st century. This increase in temperature has thrown climate out of gear as evident from occurrence of more than half of the mean annual rainfall within 24-yr in the Indian Thar desert and droughts in other parts of the Indian sub-continent same year. The climate change is not limited to one country or a continent. It is occurring across the globe as evident from droughts in Texas and flooding along the Missouri River in the United States and along the Red River in Canada. Climate change drives many stressors and interacts with many non-climatic stressors which make it difficult to forecast outcomes in any general way other than existing threats to agriculture. Increase in CO_2 on the one hand increases photosynthetic efficiency hence crop production particularly of C_4 crops, on the other hand rise in temperature affects reproductive biology of crops by reducing pollen viability and making spikelets sterile with ultimate result in crop yields reduction. Moreover, the decline in the crop yields is more in the area where temperature has already reached close to the physiological maxima. In India, warming during winter season is predicted to reduce production of *rabi* crops like wheat, mustard and chickpea on the Indo-Gangetic plain, while decreasing summer rainfall may decrease rice yields in Chattisgarh, Jharkhand and eastern part of MP. In horticulture sector, farmers in Himachal Pradesh are shifting their orchards on higher elevations to realize same fruit yields in the climate change regime.

Global warming is expected to accelerate many microbial processes too in different ecosystems which may impinge upon C and N cycles. Litter decomposition patterns may change and increased soil temperature may lead to autotrophic CO_2 loss from soil caused by root respiration, root exudation and fine root turnover. Fungal and bacterial diseases are predicted to increase in areas where precipitation increases leading to outbreak of new pest and diseases. Moreover, overall agricultural productivity for entire world is projected to decline between 3 to 16% by 2080, and it would be more (10-25%) in developing nations as they have temperature that is already close to or above crop physiological maxima. Rich countries, however, may experience increase in crop yields by 8%. India could see a drop in crop yields by 30-40%.

Several techniques have been advocated to mitigate and adapt the climate change. Agroforestry is one of them. Agroforestry generates more favourable microclimate, enhances biodiversity, reduces wind velocity, improves soil fertility, creates diversification in production, increases resource-use efficiency, reduces nutrient runoff, and decreases levels of erosion. All these together can be capitalized for climate change mitigation and adaptation services. Incorporation of trees into conventional agricultural operations results in new C being sequestered, which is greater compared to conventional agriculture. According to IPCC soil organic C (SOC) levels are assumed to stabilize at a new steady state after 20-years. Soils under agroforests, however, may accumulate carbon beyond this period. Stable carbon isotope analysis has proved that tree-derived SOC contributes more than 50% SOC beneath trees, which have greater mean residence time (45-yrs). Fertilization, liming, tillage, and livestock management in traditional agriculture are reported to cause N_2O and CH_4 emissions. Agroforestry, however, reduces N_2O emissions by eliminating nitrogen application on the areas occupied by trees and uptake of excess N by trees. Safety-net feature of agroforestry tree roots have the potential to take up excess N that would otherwise be available for N_2O emissions. Recently some studies have demonstrated that silvopsture may offer several options for reducing CH_4 and N_2O emissions, due to increase in nutrient use efficiency and decline in N fertilizer inputs. Improvement of fodder quality like crude protein, acid detergent fiber and total crude protein levels under tree shadings has been found to reduce CH_4 emissions from rumens.

Agroforestry increases a high level of diversity within agricultural lands which support numerous ecological and production services that bring resilience to the impact of climate change mitigation and adaptation. Climate change risk management is difficult in annual cropping systems due to increasing uncertainty of inter-annual variability in rainfall and temperature. Mixing of woody trees with crops, forage and livestock operations provides greater resilience to the inter-annual variability through crop diversification and increased resource use efficiency. Deep rooted trees allow better access to nutrients and water during droughts and when appropriately integrated into annual cropping systems and extract from different resource pools that would otherwise be lost from systems. Agroforestry increases soil porosity, reduces runoff and increases soil cover, which improve water infiltration and reduces moisture stress in low rainfall years. During periods of excessive soil moisture, tree based systems keep soils aerated by pumping out excess water and offer an economic return.

The book contains 36 chapters mainly on agroforestry practices found in India and its role in climate change mitigation and adaptation. The editors express their sincere thanks to contributors of the chapters and M/S New India Publishing Agency, New Delhi for timely publication of the book. The chapters have been compiled on contributor's individual capacity and they do not reflect any view or opinion of the editors and organizations to which they belong.

C.B. Pandey
Mahesh Kumar Gaur
R.K. Goyal

Contents

1

Climate Smart Agriculture and Carbon Sequestration

Abhishek Raj[1], *M. K. Jhariya*[2] *and S. S. Bargali*[3]

[1]*Department of Forestry, College of Agriculture, I.G.K.V., Raipur-492012 India*
[2]*Department of Farm Forestry, Sarguja University, Ambikapur-497001 India*
[3]*Department of Botany, Kumaun University, Nainital Uttarakhand-263001 India*

Introduction

The major environmental issue of global nature is increasing green house effect, which is warming earth. Enhanced green house effect is mainly due to increased emission of carbon dioxide, methane, nitrous oxide and CFCs. Because of the increasing concentrations of these greenhouse gases, there is much concern about future changes in our climate and direct or indirect effect on agriculture (Arora *et al.* 2011). Increases in atmospheric concentrations of green house gases and additional anthropogenic perturbations of the climate system affect the spatial and temporal distribution of weather regimes which contribute to shaping the diversity of agricultural activities throughout the world (Havlik *et al.* 2015). Consequences of changing climatic attributes such as rainfall, moisture regimes, temperature etc. affect qualitative and quantitative production of agricultural crops as well as livestock population by altering pasture quality and yield (Thornton *et al.* 2009). In the context of climate change mitigation, now a days adoption of climate smart agriculture work as an effective practice.

Climate smart agriculture is defined by three objectives: firstly, increasing agricultural productivity to support increased incomes, food security and development; secondly, increasing adaptive capacity at multiple levels (from farm to nation); and thirdly, decreasing greenhouse gas emissions and increasing carbon sinks (Campbell *et al.* 2014). As per Stoorvogel *et al.* (Stoorvogel *et al.* 2015), the Food and Agricultural Organization (FAO) defines climate-smart agriculture (CSA) as "agriculture that sustainably increases productivity, resilience (adaptation), reduces/removes GHGs (mitigation), and enhances achievement of national food security and development goals". Moreover, CSA has opportunity to maintain productivity to strengthen farmer's livelihood and enhance carbon sequestration potential by reducing GHG emissions from agriculture's sector. Different biomes have different capacity to store carbon which is ranged from 3 Gt in croplands to 212 Gt in tropical forests (Watson *et al.* 2000). The total carbon stocks in whole biomes are 2,477 Gt C in which the vegetation comprises 466 Gt C (18.8%) and soils comprise 2,011 Gt C (81.1%). Therefore, soils part comprises more carbon stocks as compared to vegetations. Among the soil this value ranges from 100 Gt C (Temperate forest) to 471 Gt C (Boreal forests). The value of total carbon stocks (Gt C) in decreasing order is as follow: boreal forest (559) > tropical forest (428)> tropical savannas (330) > temperate grassland (304) > wetlands (240) > deserts (199) > temperate forest (159) > croplands (131) > tundra (127) (Watson *et al.* 2000).

Agroforestry, an ecologically and environmentally sustainable land use system, offer great promise to sequester carbon (Parihar 2016). Along with agroforestry, another environmental friendly approach such as cultivation of biofuel species like *Jatropha curcus* is also effective in term of greenhouse gas reduction. As per Firdaus *et al.* (2010) biomass of a 32 month old *Jatropha curcas* on one hectare plantation sequestered total 13.0 Mg C ha^{-1} in biomass. The total C content of forests has been estimated at 638 Gt for 2005, which is more than the amount of carbon in the entire atmosphere (FAO 2007; Rawat 2010). Similarly, the estimated biomass carbon is 2.74 Tg and the soil carbon is 3.48 Tg in a natural forest area of Kolli hills, part of the Eastern Ghats of Tamil Nadu, India (Ramachandran *et al.* 2007). Wright *et al.* (2001) estimated that the goal of assimilating 3.3 Pg C $year^{-1}$ would require 670-760 M ha area of improved maize cultivation, whereas this goal can be achieved by adoption of 460 M ha under agroforestry practices. This chapter reviews the global scenario of climate change, vegetation shift and emergence of infectious plant disease under changing climate, climate change effects on biodiversity and mitigation strategies such as adoption of climate smart agriculture and other eco-friendly farming practices as agroforestry.

Global climate situation

Globally, carbon dioxide (70%) became the main cause of GHG emissions resulted primarily from burning of fossil fuel (petroleum), while the other sources are methane and nitrous oxide added due to deforestation and agricultural activities (Yohannes and Mebratu 2009). The different sector contributes emission of green house gases which is depicted in Table 1 (IPCC 2007; Smith *et al.* 2008). Tropical deforestation provides a significant contribution to anthropogenic increases in atmospheric CO_2 concentration that may lead to global warming (Jhariya and Raj 2014). Climate change, driven by fossil fuel combustion and deforestation, is a becoming threat to lives and livelihoods in every part of the world at this time (Ackerman 2009). A graphical representation (Fig. 1 and Fig. 2) is shown below for global projection of CO_2 emission from 1751 to 2010 (CDIAC 2013) and rise in global temperature from 1880 to 2014 (GISS 2015).

Table 1. Contribution of different sectors to greenhouse gas emissions

Sectors	GHG Emission %		
Energy	26		
Industry	19		
Forestry	17		
Agriculture	14	Nitrous oxide from soil	38%
		Enteric fermentation	32%
		Biomass burning	12%
		Rice production	11%
		Manure management	7%
Transport	13		
Building	8		
Waste	3		

Source: (IPCC 2007; Smith *et al.* 2008)

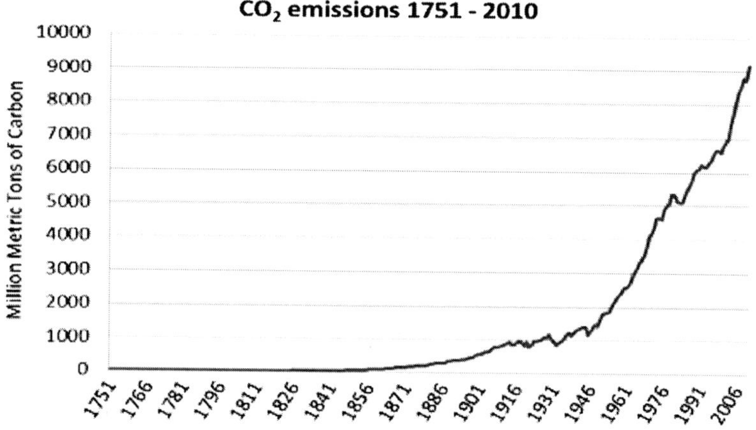

Fig. 1. Global projection of CO_2 emission (1751-2010) (*Source*: CDIAC 2013)

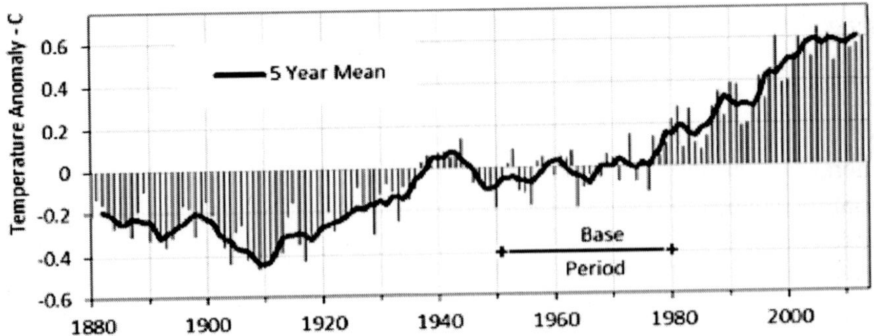

Fig. 2. Rise in global temperature from 1880 to 2014 (*Source*: Emanuel Woodward 1987; Prentice *et al.* 1992).

The Inter-Governmental Panel on Climate Change has projected the temperature increase to be between 1.1 °C and 6.4 °C by the end of the 21st Century (IPCC 2007). As per IPCC (2013) climate is changing globally at a pace unprecedented in human history and this change is directly related to anthropogenic GHG emissions. This continued high emission of GHGs would degrade the value of biodiversity, ecosystem services and economic development (IPCC 2014). Agriculture, forestry and other land use (AFOLU) are responsible for a quarter of all anthropogenic GHG emissions (Smith *et al.* 2014). Forest biomass burning also influence the emission of CO_2 in environment, which causes greenhouse effects and adversely affect the ecosystems (Jhariya *et al.* 2014; Jhariya *et al.* 2014; Kittur *et al.* 2014). Badarinath and Vadrevu (2011) reported that 2,414 km^2 area has been estimated to be burnt annually in forested area including closed broad-leaf deciduous forest, closed needle-leaf evergreen forest, closed to open broad-leaf evergreen/semi-deciduous forest, closed/open mixed broadleaf/needle-leaf forest, mosaic forest/grassland/shrub-land and open broad-leaf deciduous forest in India. The CO_2 emissions averaged across seven years were ~6.34 CO_2 Tg yr^{-1} from biomass burning of these forest types. The latest assessment report by IPCC (2013) states with 95% confidence that human influence is the main cause of the observed warming in the atmosphere and oceans and other indicators of climate change and that continued emissions of GHGs will cause further warming and changes in the components of the climate system. Therefore, it is observed that crystal clear impact of global warming on warming of atmosphere and ocean and melting snow and ice that leads to rise in sea level (IPCC 2014).

Climate change and vegetation shift

Each and every species has a range of climates for their survival and reproduction but climate change relocate and shift their distribution. Rise in $CO_{2,}$ climate change can affect structure and composition of vegetation and stand community. As per Kelly and Goulden (2008) change in climate would be expected to shift plant distribution as species expand in newly favorable areas and decline in increasingly hostile location. Effects of climate change is studied in term of general species distribution and populations (persistence, acclimation, genetic variability, dispersal, fragmentation, plant/animal interaction, species richness, conservation), potential response of vegetation (ecotonal shift- area, Physiography- changes in the composition, structural changes), Phenology, growth and productivity, and landscape (Theurillat and Guisan 2001). It is generally agreed that in many areas climatic conditions shape the ranges and distribution of species and thus the composition of biomes (Emanuel Woodward 1987; Prentice *et al.* 1992). In altitudinal studies, ACIA (ACIA 2004) has reported that each zone represent a definite type of vegetation. Shifting in current and future types of vegetation from north to south and from lower to higher elevation is the result of global warming. Under the range shifting, cold-adapted species from mountains at higher altitudes nowhere to move in warmer temperatures and lower-elevation species shift up to higher elevations (NRC 2008). Therefore, this type of climatic impacts has been observed in many species in many regions over long period of time.

Impact of climate change on agriculture

Climate change is the most important global environmental challenge which is facing by all living organism including humans and disturb natural ecosystems, agriculture and health (Painkra *et al.* 2016). The long lasting and ever-changing climate regimes such as temperature, solar radiation and precipitation will have an effect on crop productivity, livestock and agriculture. Climate change will also have an economic impact on agriculture, including changes in farm profitability, prices, supply, demand, trade and regional comparative advantages. Agriculture is sensitive to short-term changes in weather and to seasonal, annual and long-term variations in climate. Thus, the climate elements which affect the plant growth and development, hence the agriculture as a whole are carbon dioxide concentration, temperature, radiation, precipitation and humidity (Khan *et al.* 2009).

As we know, Indian agriculture is monsoon dependent and by the end of this century monsoon is expected to intensify with climate change resulting various consequences like rise in precipitation (Christensen *et al.* 2007) and leads to substantial regional variations in rainfall (Baettig *et al.* 2007). Effects of climate variability and occurrence of extreme weather pattern are seen in various tropics as in South Asia, South East Asia, South pacific, tropical highland, lowland, West Africa and American tropics. Among the tropics, South Asia has most productive agricultural land in Indo-Gangetic Plains (IGP), providing staple grain for 400 million people, primarily through a rice-wheat cropping system. But over the past few decades, the yield of rice and wheat in this highly intensive system have stagnated and even declined (Ladha *et al*. 2003). In this context, FAO estimates that South Asia will need to increase cereal production by almost 50% over the next three decades to meet future food demand but the current scenario of agricultural production and regional growth in population result the region will have an estimated 22 million tons cereal deficit by 2030 (Padgham 2009). As per Ortiz *et al.* (2008) approximately 50% of potentially wheat productive areas of the IGP could be reclassified as under the heat-stressed short-season production mega-environment by the end of 2050. Similarly, in every 1°C rise in night temperatures, rice yields are also expected to go down up to 10% (Peng *et al.* 2004). Climate change may already be contributing to productivity decline in the IGP due to decreased solar radiation and increased a minimum temperature that leads to suppress rice yields by decreasing photosynthesis and increasing respiration losses (Pathak *et al.* 2003). Moreover, Asian rice production also faces a long-term threat from sea-level rise. The major river deltas of South and Southeast Asia, along with the Nile, are among the most vulnerable in terms of lost agricultural lands and one meter sea level rise would threaten highly productive rice cultivation in the Mekong Delta, resulting in a 10% loss of GDP (Dasgupta *et al.* 2007). Mall and Singh (2000) observed that small changes in the growing season temperature over the years appeared to be the key aspect of weather affecting yearly wheat yield fluctuations.

Agriculture is backbone of developing countries and plays a crucial role in economic growth and development. Agriculture not only takes care of nation's food requirement for growing 9 billion population of the world by 2050 but it also improves the household food and nutrition security, food quality, environmental protection, resource sustainability, shift in food consumption pattern and mitigating climate change through sequestration of atmospheric carbon. Besides adding soil carbon, technology of sustainable resource management can be beneficial to farmers because they can enhance the production at reduced cost. Total private profits are estimated at US$105 billion for Africa, $274 billion for Latin America and $1.4 trillion for Asia by 2030 (CSAS 2012). Agriculture typically plays a larger role in developing economies than in the developed world (Guiteras 2007). As per FAO (2006) Indian agriculture contributes nearly 20% of GDP and provides 52% of employment as compared to 1% of GDP and 2% of employment for the US, with the majority of agricultural workers drawn from poorer segments of the population. But extreme weather conditions, such as floods, droughts, heat and cold-waves, flash floods, cyclones and hailstorm are direct hazards to crops. More subtle fluctuation in weather during critical phases of crop development can also have substantial impact on yields (Mall *et al.* 2006 b). Due to climate change it is observed that Indian economy has much been affected by drought than floods, that kept India's total GDP constant (between 2-5% decline in total GDP), despite significant economic diversification away from agriculture (Gadgil and Gadgil 2006).

Emergence of infectious plant disease under climate change

Climate change can lead to the disease emergence through gradual changes in climate and a high frequency of unusual weather events. Several works are reported in regard of consequences of global warming on incidence and severity of plant disease and influence the further co-evolution of plants and their pathogens (Burdon *et al.* 2006; Garrett *et al.* 2006; Crowl *et al.* 2008; Eastburn *et al.* 2011). As per Oerke *et al.* (1994) damage by disease and insect pests resulted in a 42% loss in the eight most important food and cash crops. Moreover, the pace of elevated temperature and CO_2 concentration has resulted important rice diseases namely blast (*Pyricularia oryzae*) and sheath blight (*Rhizoctonia solani*) and higher threat of potato late blight (*Phytophthora infestans*) (Gautam *et al.* 2013).

Temperature and moisture govern the rate of reproduction of many pathogens (Caffarra *et al.* 2012). It can also influence the interaction between pest and pathogen and ultimately cause emergence of infectious disease, which alter the physiological, morphological and anatomical characteristics of plants community. The species found in tropics are relatively more sensitive to alteration in temperature because of narrow temperature growth range behavior and these species are also currently living very close to their optimal temperature conditions (Ghini *et al.* 2011a). Similarly, risk of vector-borne disease at the local and regional level is limited by the climatic requirements of vectors (Malmstrom *et al.* 2011). Jones (2009) reported that climate change may regulate the host plant and insect-vector populations by spreading of plant viruses. Sometimes these extents of changing are beneficial for introduction of natural predator for minimizing disease by attacking their host pathogen. Introduction of the new resistance plant variety, mixed cropping, natural predator, controlling climatic regimes and a good scientific management practices like IPM (Integrated Pests Management), INP (Integrated Nutrient Management), IWM (Integrated Weed Management) etc. are beneficial tools for controlling the emerging disease in plant community, which help in setup of food production.

Role of climate smart agriculture for food security

By 2050, approximately 70% more food will have to be required to feed growing populations, particularly in developing countries (UN 2009; FAO 2011). Climate change presents a profound challenge to food security and social development. The impact of climate change on agriculture could result in problems with food security and may threaten the livelihood activities upon which much of the population depends. The FAO estimated that 1.02 billion people went hungry in 2009, the highest ever level of world hunger, mainly as a result of declining investment in agriculture (Anonymous 2010). Under major impact of climate change, temperature rise is likely to result in reduced food production leads to facing food insecurity. Yield of major cereal crops (rice, wheat, maize, sorghum) in the tropics and subtropics are expected to decline with a temperature increase as small as 1°C, such as could occur by around 2030 (Padgham 2009). South Asian agriculture is highly depends on monsoon and distribution and timing of monsoon precipitation can be highly variable. As per Mall *et al.* (2006 a) under extreme cases, up to 60% of annual rainfall can occur within a period of several days, resulting in severe flooding, high crop and livestock loss, and reduced groundwater recharge. Similarly, drought in India caused by a midsummer break in the monsoon reduced national cereal output by 18% and India's gross domestic product by 3% in 2002. The country's food grains production during 2002–03 had slumped to 174 Mt, due to widespread drought, from the record level of 212 Mt in 2001-02 (Mall *et al.* 2006 a).

Agriculture is already causing increased conversion of lands and placing greater pressure on biological diversity and natural resource functions than ever before (MEA 2005; IAASTD, 2009). As climate change causes temperature to rise and precipitation patterns to change, more weather extremes will potentially reduce global food production (IPCC 2007; Nelson *et al.* 2010). But now, these consequences of climate change can be solved by adopting climate smart agriculture. Climate smart agriculture addresses food security and lead to higher productivity by different practices that are consistent in smallholder systems. In the context of higher productivity and food security in smallholder agricultural production, Neufeldt *et al.* (2011) has given stressed on following practices of CSA comprises agroforestry (alley cropping, boundary plantation, woodlots, fruit tree garden), crop management (crop diversification, resistant varieties, incorporation of legumes, crop rotation, improves post-harvest techniques), livestock management (protein bank fodder production, grazing management, grassland management, livestock health and improvement), soil and water conservation and energy production system as biogas generation. Therefore, CSA has the potential to increase sustainable productivity, increase the resilience of farming systems to climate impacts and mitigate climate change through reduction of GHGs and carbon sequestration FAO 2010).

Mitigation and adaptation strategy

Strategies like reducing emission of GHGs from agriculture and deforestation, increased carbon sequestration through afforestation and agroforestry and substitution of fossil fuels could account 20 60% of the mitigation effort for climate change by 2030 (Rose *et al.* 2012). There is an increasing concern which has instigated a scheme under the Kyoto protocol to reduce carbon emissions which can be traded and offset by carbon sequesters (Harper *et al.* 2007; UNFCCC 2009). As per FAO (2013 b), forests are home to nearly 60 million indigenous people providing a variety of ecosystem services (both tangible and intangible) comprises food, fuel, timber, water, carbon sequestration and biodiversity. An estimate says 2.4 billion people cook using wood fuel which is major source of primary energy in developing regions (FAO 2013c). But this ecosystem services are badly threatens by climate change. Forest plays a lead role for maintaining the CO_2 level, carbon sequestration and protects the ecosystem from global warming. Likewise, agroforestry plays an important role in climate change mitigation through long term carbon sequestration i.e. storing of atmospheric carbon in vegetation and soil, produces wood and also contributes to farmer's income (Sudha *et al.* 2007). Similarly, Nair *et al.* (2011) summarized that the potential of agroforestry system in term of carbon storage varied from 0.3 to 15.2 Mg C ha^{-1} yr^{-1}; the highest being in the humid tropics receiving high rainfall. Therefore, various practices in forest, grassland and cropland help to sequester carbon. Protection and conservation of natural forest enhance the carbon sequestration potential as it store atmospheric carbon itself and reduce emission of GHGs. Reforestation and afforestation activity increase the carbon stock in vegetation (above and below ground) and enhance the soil organic carbon (SOC) to maintain soil fertility and health. In grassland ecosystem, certain management practices like optimization of stocking rates, incorporation of improved pasture species and legumes in degraded land, application of inorganic fertilizers and manure and introduction of earthworms are able to enhance above and below ground biomass by capturing and reducing emission of GHGs. Likewise, practices like conservation agriculture, agroforestry, application of mulching and biochar, introduction of cover crops and improved crop varieties, application of organic and inorganic fertilizers and residue management practices in cropland ecosystem can enhance the carbon storage capacity in both plant and soil (CSAS 2012).

Role of agroforestry in carbon mitigation

Indeed, agroforestry maintains overall farm productivity by incorporation of herbaceous food crops with perennial trees and livestock on the same piece of land that improve the socioeconomic condition of people. Agroforestry plays a major role in enhancement of overall farm productivity, soil enrichment through litter fall, maintaining environmental services such as climate change mitigation (carbon sequestration), phyto-remediation, watershed protection and biodiversity conservation. The carbon sequestration capacity depends upon tree species, their growing condition and management practices under agroforestry system (Jhariya *et al.* 2015). The carbon sequestration and their allocation may vary from species to species and in different components of the same tree species. In this context, Prasad *et al.* (2010) has reported carbon storage value in different tree species was in the order of *Eucalyptus tereticornis* = *Azadirachta indica* = *Acacia nilotica* = *Butea monosperma* >*Albizia procera* = *Dalbergia sissoo* >*Emblica officinalis* = *Anogeissus pendula*. Similarly, carbon storage value in tree components wise varies in order of branch = stem > root > foliage > stem bark = branch bark. Moreover, *Albizia procera* was found to be the most efficient in capturing C (128 kg C tree^{-1}) and removing CO_2 from the atmosphere (47 kg tree^{-1} yr^{-1}), while *Anogeissus pendula* was the least with corresponding values of carbon (8 kg C tree^{-1}) and CO_2 (3 kg tree^{-1} year^{-1}), respectively. Also by including trees in agricultural production systems, agroforestry can arguably increase the amount of C stored in lands devoted to agriculture while still allowing for the growing of food crops (Kursten 2000; Singh and Jhariya, 2016).

As per Izaurralde *et al.* (2001) around 2 billion ha of degraded land exists globally, of which 1.5 billion ha is located within tropics. Therefore, practice of afforestation and agroforestry have sequestered 8.7×109 Mg C yr^{-1} in the tropical and 4.9 × 109 Mg C yr^{-1} in the temperate above-ground C pools (Dixon and Turner 1991). In this context, IPCC (2000) has reported that 630 million ha of unproductive croplands and grasslands could be converted under agroforestry system globally; with sequestration potential varies from 391,000 Mg C yr^{-1} by 2010 to 586,000 Mg C yr^{-1} by 2040. Although there are various literatures (Table 2) on carbon storage capacity by different agroforestry models which varies from region to region and literatures (Table 3) on soil carbon-sequestration potential under agroforestry systems. Therefore, soil organic carbon pool (1550 Pg) and soil inorganic carbon pool (950 Pg) at 1-m depth are two distinct components of world soils that constitute the third largest global C pool. Other pools include the oceanic (38,400 Pg), geologic/fossil fuel (4500 Pg), biotic (620 Pg) and atmospheric (750 Pg) (Lal 2004a).

Table 2. Carbon storage capacity as per agroforestry model in different regions of India

Agroforestry model	Carbon storage capacity	Region	Source
Silvopastoral system (5 years)	9.5–19.7 t C ha^{-1}	Semiarid	83
Silvopastoral system (aged 6 years)	1.5–18.5 t C ha^{-1}	Northwestern India	84
Agrisilviculture system (aged 8 years)	4.7–13.0 t C ha^{-1}	Arid region	85
Agrisilviculture system (aged 11 years)	26.0 t C ha^{-1}	Semiarid region	86
Block plantation (aged 6 years)	24.1–31.1 t C ha^{-1}	Central India	87
Poplar block plantation	330,510 t	Punjab (Rupnagar district)	88
Eucalyptus bund plantation	59,361 t		
P. deltoides + wheat boundary plantation	4.66 t C ha^{-1}	Tarai region of central Himalaya	89
Populus deltoides 'G-48' + wheat	18.53 t C ha^{-1}		
Silvopasture	31.71 t C ha^{-1}	Himachal Pradesh	90
Agrisilviculture	13.37 t C ha^{-1}		
Agri-horticulture	12.28 t C ha^{-1}		
Hortipastoral	17.16 t C ha^{-1}		
Agrihorti silviculture	18.81 t C ha^{-1}		
Natural grassland	19.2 t C ha^{-1}		
Agroforestry woodlots (aged 8.8 years)	6.53 Mg ha^{-1} y^{-1}	Kerala, India	91
Agroforestry woodlots (aged 4 years)	12.04 Mg ha^{-1} y^{-1}	Puerto Rico	92
Silvopastoralism (aged 5 years)	6.55 Mg ha^{-1} y^{-1}	Kerala, India	91
Silvopastoralism (aged 6 years)	1.37 Mg ha^{-1} y^{-1}	Kurukshetra, India	84
Silvopasture(aged 11 years)	1.11 Mg ha^{-1} y^{-1}	W Oregon, USA	93
Agrisilviculture(aged 5 years)	1.26 Mg ha^{-1} y^{-1}	Chattisgarh, Central India	94
Fodder bank (aged 7.5 years)	0.29 Mg ha^{-1} y^{-1}	S_gou, Mali, W African Sahel	95
Indonesian homegardens (aged 13.4 years)	8.00 Mg ha^{-1} y^{-1}	Sumatra	96

Table 3. Soil carbon-sequestration potential under agroforestry systems

Agroforestry system/species	Region	Soil depth (cm)	Soil C (Mg ha^{-1})	Author
Mixed stands, Eucalyptus + Casuarina, Casuarina + *Leucaena* and Eucalyptus + *Leucaena* (aged 4 years)	Puerto Rico	0–40	61.9, 56.6, and 61.7	Parrotta (1999)
Agrisilviculture (*Gmelina arborea* +eight field crops) (aged 5 years)	Chhattisgarh, Central India	0–60	27.4	Swamy and Puri (2005)
Agroforest (home and outfield gardens)	Ipet_-Embera, Panama	0–40	45.0 –2.3	Kirby and Potvin (2007)
Tree-based pastures: slash pine (Pinuselliottii) + bahiagrass (Paspalum notatum) (aged 8-40 years)	Florida, USA	0–125	6.9 to 24.2	Haile *et al.* (2008)
Alley cropping: hybrid poplar + wheat,soybeans (Glycine max.), and maize rotation (aged 13 years)	South Canada	0–40	1.25	Oelbermann *et al.* (2006)

Role of Agriculture in carbon mitigation

Since over one third of arable land is in agriculture globally (World Bank 2015a) and agricultural systems would be a major component of using soils as a sink. Estimates of the global sequestration potential of agricultural soils are typically made for sequestration on an annual basis and range from 0.4 to 1.2 Gt yr^{-1} (Lal 2004 a). As per Hansen *et al.* (2013) an optimistic soil carbon accrual rate could sequester at least 10 % of the current annual emissions of 8-10 Gt yr^{-1}. As we know, agricultural activities have been a principal source of CO_2 emissions, along with those of CH_4 and N_2O. But a good and properly managed agriculture can be a solution to numerous environmental problems such as global warming. Therefore, improving input use efficiency, minimizing soil erosion, conserving energy and growing aerobic rice (rather than continuously flooded paddy), integrated nutrient management (INM), and integrated pest management (IPM) are among some of the promising options of reducing emissions. Similarly, in agriculture crop residue (shoot, leaves, cobs, husks etc) is a principal source of biomass which impacts on soil carbon dynamics, efficiency to stabilizing atmospheric concentration of CO_2 and improving agronomic/food production. Stimulation of CH_4 emissions, immobilization of available N, suppression of rice growth and accumulation of toxic materials are major problem arises due to short-term effects of cereal residues (wheat straw) incorporation into paddy field (Singh *et al.* 2005; Singh *et al.* 2008) but this problem can be reduce by the application of cereal residues before optimum time of rice transplanting which can help in reducing the adverse effect on rice growth and minimize the emission of CH_4 and N_2O (Ma *et al.* 2007). Therefore, return of crop residues, application of farm yard manure, green manure and balanced fertilizers can increase the soil organic carbon (SOC) content (Chaudhury *et al.* 2016). Compared to traditional tillage techniques, conservation tillage can play a key role in decreasing CO_2 emissions and in increasing the pool of C sequestered in the soil (Deen and Kataki 2003; Freibauer *et al.* 2004; Lal 2004 b; Lal 2004 c; Baker *et al.* 2007). As per Lal (2005) the amount of crop residues produced in 2001 was estimated at ~0.5 x 10^9 Mg yr^{-1} in the USA and ~ 4 x 10^9 Mg yr^{-1} in the world. Similarly, rice and rice-based cropping systems produce ~0.6 x 10^9 Mg yr^{-1} of crop residues in the tropics (Table 4) (Singh *et al.* 2005). Carbon sequestration potential under different land management practices showed that agroforestry plays an inevitable role in soil carbon sequestration potential in all regions (Africa, Asia, Europe, middle America, North America, Oceania, Russia and South America) as compared to other sole cropping systems (CSAS 2012). Asian region has great diversity in soil leads to highest soil carbon sequestration potential (2416.434 Mt C). That's why we stress upon agroforestry systems in which scientific management practices are able to increase carbon inputs into soil and possibly reduce GHGs emission. Moreover, soil carbon sequestration potential in agroforestry systems is followed by cover crops (1009.402 Mt C) and rice (516.843 Mt C) in Asian region (CSAS 2012).

Table 4. Crop residues production in the rice and rice-based cropping systems in the tropics and the world

Region	Residue produced (10^6 Mg yr^{-1})
Asia	166
Africa	39
South America	55
Sub-Total Tropics	260
World Total	604

Source: Singh *et al.* 2005

Conclusion

Climate change has emerged as the major challenge today and expected to impact on agricultural productivity by extreme weather i.e. unusual pattern of rainfall and temperature that leads to emergence of disease causing insect and pathogen and shifting crop patterns. There is a need to adopt climate smart agriculture i.e. transform agricultural sectors, including crop and livestock production, forestry, fishery etc in climate smart way in order to address the challenges of global warming and current food security. Therefore, climate smart agriculture has immense potential to perform carbon sequestration which varies widely, depending on the specific farming practices, type of intercropping in different land use management. Moreover, integration of better crops with other tree component and management practices could efficiently reduce and mitigate climate change worldwide.

References

ACIA (2004). Impacts of a Warming Arctic: Arctic Climate Impact Assessment. Cambridge University Press, Cambridge, UK.

Ackerman F (2009). Financing the Climate Mitigation and Adaptation Measures in Developing Countries, Stockholm Environment Institute, Working Paper WP-US-0910, pp.1-17.

Anonymous (2010). The Future of World Food Security. Rome, Italy: International Fund for Agricultural Development.

Arora VPS, Bargali SS, Rawat JS (2011). Climate change: challenges, impacts, androle of biotechnology in mitigation and adaptation. Progressive Agriculture 11:8-15.

Badarinath KVS, Vadrevu KP (2011). Carbon dioxide emissions from forest biomass burning in India. Global Environmental Research 15:45-52.

Baettig MB, Wild M, Imboden DM (2007). A climate change index: Where climate change may be most prominent in the 21st century. Geophysical Research Letters 34(1): L01705.

Baker JM, Ochsner TE, Venterea RT, Griffis TJ (2007). Tillage and soil carbon sequestration- What do we really know? Agric Ecosyst Environ 118:1-5.

Burdon JJ, Thrall PH, Ericson AL (2006). The current and future dynamics of disease in plant communities. Annu Rev Phytopathol 44:19-39.

Caffarra A, Rinaldi M, Eccel E, Rossi V, Pertot I (2012). Modeling the impact of climate change on the interaction between grapevine and its pests and pathogens: European grapevine moth and powdery mildew. Agric Ecosyst Environ 148:89-101.

Campbell BM, Thornton P, Zougmore R, Asten PV, Lipper L (2014). Sustainable intensification: What is its role in climate smart agriculture? Current Opinion in Environmental Sustainability 8:39–43.

Carbon Sequestration in Agricultural Soils (2012). Agriculture and Rural Development. Report no. 67395-GLB, The World Bank, pp. 43-50.

CDIAC(2013) Carbon Dioxide Information Analysis Centre. http://cdiac.ornl.gov/trends/emis/overview_2010.html

Chaudhury S, Bhattacharyya T, Wani SP, Pal DK, Sahrawat KL, Nimje A, Chandran P, Venugopalan MV, Telpande B (2016). Land use and cropping effects on carbon in black soils of semi-arid tropical India. Current Science 110(9):1692-1698.

Christensen JH, Hewitson B, Busuioc A, Chen A, Gao X, Held I, Jones R (2007). Regional climate projections. In: Solomon S, Qin D, Manning M, Chen Z, Marquis M, Averyt KB, Tignor M, Miller HL (eds) Climate Change 2007: The Physical Science Basis. Contribution of Working Group I to the Fourth Assessment Report of the Intergovernmental Panel on Climate Change, Cambridge University Press, Cambridge.

Crowl TA, Crist TO, Parmenter RR, Belovsky G, Lugo AE (2008). The spread of invasive species and infectious disease as drivers of ecosystem change. Frontiers Ecol Environ 6:238-246.

Dasgupta S, Laplante B, Meisner C, Wheeler D, Yan J (2007). The impact of sea level rise on developing countries: a comparative analysis. World Bank Policy Research Working Paper 4136.

Deen W, Kataki PK (2003). Carbon sequestration in a long-term conventional versus conservation tillage experiment. Soil Till Res 74:143-150.

Dixon RK, Turner DP (1991). The global carbon cycle and climate change: responses and feedback from below-ground systems. Environmental Pollution 73:245–261.

Eastburn DM, McElrone AJ, Bilgin DD (2011). Influence of atmospheric and climatic change on plant–pathogen interactions. Plant Pathol 60:54-69.

Emanuel WR, Shugart HH, Stevenson MP (1985). Climatic change and the broad scale distribution of terrestrial ecosystems complexes. Climatic Change 7: 29-43.

FAO (2006). FAO Statistical Yearbook 2005-2006. Food and Agricultural Organization.

FAO (2007) State of the World's Forests. FAO of United Nations, Rome.

FAO (2010). Climate-Smart Agriculture- Policies, Practices and Financing for Food Security, Adaptation and Mitigation. Food and Agriculture Organization of the United Nations, Rome.

FAO (2011). http://www.fao.org/news/story/en/item/35571/icode

FAO (2013 b). Climate Change Guidelines for Forest Managers. FAO Forestry Paper No. 172. Rome, Italy.

FAO (2014c). State of The World's Forest Enhancing-The Socio-Economic Benefits From Forests. Rome, Italy.

Firdaus MS, Hanif AHM, Safiee AS, Ismail MR (2010). Carbon sequestration potential in soil and biomass of *Jatropha curcas*. 19th World Congress of Soil Science, Soil Solutions for a Changing World. Brisbane, Australia. pp. 62-65.

Freibauer A, Rounsevell MDA, Smith P, Verhagen J (2004). Carbon sequestration in the agricultural soils of Europe. Geoderma 122:1-23.

Gadgil, S, Gadgil S (2006). The Indian monsoon, GDP and agriculture. Economic and Political Weekly, pp. 4887–95.

Garrett KA, Dendy SP, Frank EE, Rouse MN, Travers SE (2006). Climate change effects on plant disease: genomes to ecosystems. Annu Rev Phytopathol 44: 489-509.

Gautam HR, Bhardwaj ML, Kumar R (2013). Climate change and its impact on plant diseases. Current Science 105(12):1685-1691.

Gera M, Mohan G, Bisht NS, Gera N (2006). Carbon Sequestration potential under agroforestry in Rupnagar district of Punjab. Indian Forester 132(5): 543-555.

Ghini R, Bettiol W, Hamada E (2011 a). Diseases in tropical and plantation crops as affected by climate changes: Current knowledge and perspectives. Plant Pathol, 60:122-132.

GISS (2015). Surface Temperature Analysis, NASA, accessed January 25, 2015~ Global temperature, 1800- 2006, ProcessTrends.com, accessed October 27, 2009.

Guiteras R (2007). The Impact of Climate Change on Indian Agriculture. Job Market Paper. Draft, pp.1-53.

Haile SG, Nair PKR, Nair VD (2008). Carbon storage of different soil-size fractions in Florida silvo-pastoral systems. J Environ Qual 37:1789-1797.

Hansen J, Kharecha P, Sato M, Masson-Delmotte V, Ackerman F, Beerling DJ, Hearty PJ, Hoegh-Guldberg O, Hsu SL, Parmesan C, Rockstrom J, Rohling EJ, Sachs J, Smith P, Steffen K, Van Susteren L, Von Schuckmann K, Zachos JC (2013). In: Añel JA (ed) Assessing Dangerous Climate Change: Required Reduction of Carbon Emissions to Protect Young People, Future Generations and Nature. *PLoS ONE* 8(12): e81648.

Harper RJ, Beck AC, Ritson P, Hill MJ, Mitchell CD, Barrett DJ, Smettem KRJ, Mann SS (2007). The potential of greenhouse sinks to underwrite improved land management. Ecological Engineering 29(4): 329-341.

Havlík P, Valin H, Gusti M, Schmid E, Leclère D, Forsell D, Herrero M, Khabarov N, Mosnier A, Cantele M, Obersteiner M (2015). Climate change impacts and mitigation in the developing world: an integrated assessment of the agriculture and forestry sectors. Policy Research Working Paper 7477, pp.1-54.

IAASTD (2009). International Assessment of Agricultural Knowledge, Science and Technology for Development: Global Report. Island Press, Washington, DC.

IPCC (2000). Land use, Land-use Change, and Forestry. A Special Report of the IPCC. Cambridge University Press, Cambridge, UK, pp. 375.

IPCC (2007). Climate Change 2007: Synthesis Report. Contribution of Working Groups I, II and III to the Fourth Assessment Report of the Intergovernmental Panel on Climate Change. Geneva.

IPCC (2013). Climate Change 2013: The Physical Science Basis. Contribution of Working Group I to the Fifth Assessment Report of the Intergovernmental Panel on Climate Change. Cambridge University Press, Cambridge, UK and New York, USA.

IPCC (2014). In: Pachauri RK, Meyer LA (eds) Climate Change 2014: Synthesis Report. Contribution of Working Groups I, II and III to the Fifth Assessment Report of the Intergovernmental Panel on Climate Change. IPCC, Geneva, Switzerland, 151 pp.

Izaurralde RC, Rosenberg N, Lal R (2001). Mitigation of climatic change by soil carbon sequestration: issues of science, monitoring and degraded lands. Advances in Agronomy, 70:1–75.

Jhariya MK, Raj A (2014). Effects of wildfires on flora, fauna and physico-chemical properties of soil-An overview. Journal of Applied and Natural Science 6 (2):887–897.

Jhariya MK, Bargali, SS, Raj A (2015). Possibilities and Perspectives of Agroforestry in Chhattisgarh. In: Zlatic Miodrag (ed) Precious Forests- Precious Earth. InTech E-Publishing Inc, pp. 238-257.

Jhariya MK, Bargali SS, Swamy SL, Kittur B, Bargali K, Pawar GV (2014). Impact of forest fire on biomass and carbon storage pattern of tropical deciduous forests in Bhoramdeo wildlife sanctuary, Chhattisgarh. International Journal of Ecology and Environmental Science 40(1):57-74.

Jhariya MK (2014). Effect of forest fire on microbial biomass, storage and sequestration of carbon in a tropical deciduous forest of Chhattisgarh. Ph.D.Thesis, I.G.K.V., Raipur (C.G.), pp. 259.

Jones RAC (2009). Plant virus emergence and evolution: Origins, new encounter scenarios, factors driving emergence, effects of changing world conditions, and prospects for control. Virus Res. 141:113–130.

Kaur B, Gupta SR, Singh G (2002). Carbon storage and nitrogen cycling in silvo-pastoral systems on a sodic soil in northwestern India. Agroforest Syst 54: 21-29.

Kelly AE, Goulden ML (2008) Rapid shifts in plant distribution with recent climate change. *PNAS* 105(33):11823-11826.

Khan SA, Kumar K, Hussain MZ, Kalra N (2009). Climate change, climate variability and Indian agriculture: impacts vulnerability and adaptation strategies. Environmental Science and Engineering, pp. 19-38

Kirby KR, Potvin C (2007). Variation in carbon storage among tree species: Implications for the management of a small-scale carbon sink project. For Ecol Manage, 246:208–221.

Kittur B, Swamy SL, Bargali SS, Jhariya MK (2014). Wildland fires and moist deciduous forests of Chhattisgarh, India: divergent component assessment. Journal of Forestry Research 25(4):857-866.

Kumar BM, George SJ, Jamaludheen V, Suresh TK (1998 a). Comparison of biomass production, tree allometry and nutrient use efficiency of multipurpose trees grown in wood lot and silvopastoral experiments in Kerala, India. For Ecol Manage 112:145-163.

Kursten E (2000). Fuelwood production in agroforestry systems for sustainable land use and CO_2 mitigation. Ecological Engineering 16:69-72.

Ladha JK, Dawe D, Pathak H, Padre AT, Yadav RL, Singh B, Singh Y (2003). How extensive are yield declines in long-term rice-wheat experiments in Asia? Field Crops Research 81: 159-80.

Lal, R. 2004. Soil carbon sequestration impacts on global climate change and food security. Science, 304 (5677): 1623-7

Lal R (2004a). Soil carbon sequestrian impacts on global climate change and food security. Science 204:1623-1627.

Lal R (2004 b). Soil carbon sequestration to mitigate climate change. Geoderma 123:1-22.

Lal R (2004 c). Soil carbon sequestration impacts on global climate change and food security. Science 304:1623-1627.

Lal R (2005). World crop residues production and implication of its use as a biofuel. Env Intl 31:575-586.

Ma J, Li XL, Xu H, Han Y, Cai ZC, Yagi K (2007). Effects of nitrogen fertilizer and wheat straw application on CH_4 and N_2O emissions from a paddy rice field. Aus J Soil Res 45:359-367.

Mall RK, Singh KK (2000). Climate variability and wheat yield progress in Punjab using the CERES-wheat and WTGROWS models. Vayu Mandal 30(3-4): 35-41.

Mall RK, Gupta A, Singh R, Singh RS, Rathore LS (2006 a). Water resources and climate change: An Indian perspective. Current Science 90(12):1610-1626.

Mall RK, Singh R, Gupta A, Srinivasan G, Rathore LS (2006 b). Impact of climate change on Indian agriculture: A review. Climatic Change 78:445-78.

Malmstrom CM, Melcher U, Bosque-Pérezc NA (2011). The expanding field of plant virus ecology: Historical foundations, knowledge gaps, and research directions. Virus Res 159:84-94.

MEA (2005) Millennium Ecosystem Assessment. Ecosystems and Human Well-being: Synthesis. Island Press, Washington, DC.

Nair PKR, Vimala DN, Kumar BM, Showalter JM (2011). Carbon sequestration in agroforestry systems. Advances in Agronomy 108: 237–307.

Nelson GC, Rosegrant MW, Palazzo A, Gray I, Ingersoll C, Robertson R, Tokgoz S, Zhu T, Sulser T, Ringler C, Msangi S,You L (2010). Food Security, Farming, and Climate Change to 2050: Scenarios, Results, and Policy Options. IFPRI, Washington, DC.

Neufeldt H, Kristjanson P, Thorlakson T, Gassner A, Norton-Griffiths M, Place F, Langford K (2011). ICRAF Policy Brief 12: Making climate-smart agriculture work for the poor. Nairobi, Kenya. World Agroforestry Centre (ICRAF).

NRC (2008) Ecological Impacts of Climate Change. The National Academic Press, Washington, DC. pp.1-53.

NRCAF (2005). Annual Report. National Research Centre for Agroforestry, Jhansi.

Oelbermann M, Voroney RP, Gordon AM, Kass DCL, Schl_nvoigt AM, Thevathasan NV (2006). Carbon input, soil carbon pools, turnover and residue stabilization efficiency in tropical and temperate agroforestry systems. Agroforest Syst 68:27-36.

Oerke EC, Dehne HW, Schönbeck F, Weber A (1994). Crop production and crop protection. Amsterdam, The Netherlands: Elsevier Science, B.V.

Ortiz R, Sayre KD, Govaerts B, Gupta R, Subbarao GV, Ban T, Hodson D, Dixon JM, Ortiz-Monasterio JI, Reynolds M (2008). Climate change: Can wheat beat the heat? Agriculture, Ecosystems and Environment 126:46–58.

Padgham (2009) Agricultural development under a changing climate: opportunities and challenges for adaptation: Agriculture and rural development and environment departments, Joint Departmental Discussion Paper- Issue 1, pp.1-198.

Painkra GP, Bhagat PK, Jhariya MK, Yadav DK (2016). Beekeeping for poverty alleviation and livelihood security in Chhattisgarh, India. In: Narain Sarju, Rawat Sudhir Kumar (ed) Innovative Technology for Sustainable Agriculture Development. Biotech Books, New Delhi, India, pp.429-453.

Parihaar RS (2016). Carbon Stock and Carbon Sequestration Potential of different Land-use Systems in Hills and Bhabhar belt of Kumaun Himalaya. Ph.D thesis, Kumaun University, Nainital, pp. 330.

Parrotta JA (1999). Productivity, nutrient cycling and succession in single- and mixed-species stands of *Casuarina equisetifolia*, *Eucalyptus robusta* and *Leucaena leucocephala* in Puerto Rico. For Ecol Manage 124:45-77.

Pathak H, Ladha JK, Aggarwal PK, Peng S, Das S, Singh Y, Singh B (2003). Trends of climatic potential and on-farm yields of rice and wheat in the Indo-Gangetic Plains. Field Crops Research 80:223–34.

Peng S, Huang J, Sheehy JE, Laza RC, Visperas RM, Zhong X, Centeno GS, Khush GS, Cassman KG (2004) Rice yields decline with higher night temperature from global warming. Proceedings National Academy of Sciences 101(27):9971–9975.

Prasad R, Saroj NK, Newaj R, Venkatesh A, Dhyani SK, Dhanai CS (2010). Atmospheric carbon capturing potential of some agroforestry trees for mitigation of warming effect and climate change. Indian Journal of Agroforestry 12(2): 37–41.

Prentice IC, Cramer W, Harrison SP, Leemans R, Monserud RA, Solomon AM (1992). A global biome model based on plant physiology and dominance, soil properties and climate. Journal of Biogeography 19:117-134.

Rai P, Yadav RS, Solanki KR, Rao GR, Singh R (2001). Growth and pruned biomass production of multipurpose tree species in silvi-pastoral system on degraded lands in semi-arid region of Uttar Pradesh, India. Forest Tree and Livelihood, 11:347-364.

Ramachandran A, Jayakumar S, Haroon RM, Bhaskaran A, Arockiasamy DI (2007). Carbon sequestration: estimation of carbon stock in natural forests using geospatial technology in the eastern ghats of Tamil Nadu, India. Current Science 92(3): 323-331.

Rawat VRS (2010). Reducing emissions from deforestation in developing countries (REDD) and REDD plus under the UNFCC negotiations, Research note. Indian Forester 136(1):129-133.

Rose SK, Ahammad H, Eickhout B, Fisher B, Kurosawa A, Rao S, Riahi K, van Vuuren DP (2012). Land-based mitigation in climate stabilization. Energy Economics, pp. 365-380.

Roshetko M, Delaney M, Hairiah K, Purnomosidhi P (2002). Carbon stocks in Indonesian homegarden systems: Can smallholder systems be targeted for increased carbon storage? Am J Alt Agr 17:125-137.

Sharrow SH, Ismail S (2004). Carbon and nitrogen storage in agroforests, tree plantations, and pastures in western Oregon, USA. Agroforest Syst 60:123-130.

Singh B, Shan YH, Johnson-beeebout SE, Singh Y, Buresh RJ (2008) Crop residue management for lowland rice-based cropping systems in Asia. Adv Agron 98:118-199.

Singh Y, Singh B, Timsina J (2005). Crop residue management for nutrient cycling and improving soil productivity in rice-based cropping systems in the tropics. Adv Agron 85: 269-407.

Singh G (2005) Carbon sequestration under an agri-silvicultural system in the arid region. Indian Forester131(4) 543-552.

Singh NR, Jhariya MK (2016). Agroforestry and Agrihorticulture for Higher Income and Resource Conservation. In: Narain Sarju, Rawat Sudhir Kumar (ed) Innovative Technology for Sustainable Agriculture Development. Biotech Books, New Delhi, India, pp.125-145.

Smith P, Martino D, Cai Z (2008). Greenhouse Gas Mitigation in Agriculture. Philosophical Transactions of the Royal Society 363:789–813.

Smith, P, Bustamante, M, Ahammad, H, Clark, H, Don, H, Elsiddig, EA, Haberl, H, Harper, R, House, J, Jafari, M, Masera, O, Mbow, C, Ravindranath, NH, Rice, CW, Aba, CR, Romanovskaya, A, Sperling, F and Tubiello, F. 2014. Agriculture, Forestry and Other Land Use (AFOLU). Climate Change 2014: Mitigation of Climate Change. Contribution of Working Group III to the Fifth Assessment Report of the Intergovernmental Panel on

Climate Change. Edenhofer, O., R. Pichs-Madruga, Y. Sokona, E. Farahani, S. Kadner, K. Seyboth, A. Adler, I. Baum, S. Brunner, P. Eickemeier, B. Kriemann, J. Savolainen, S. Schlömer, C. von Stechow, T. Zwickel and J.C. Minx, Cambridge, United Kingdom and New York, NY, USA.

Stoorvogel JJ, Hendriks C, Claessens L (2015). When is agriculture climate-smart? A call for proper soil management. Conference paper: 7th Congress of the European Society for Soil Conservation "Agro ecological assessment and functional-environmental optimization of soils and terrestrial ecosystems" Moscow, Russian Federation.

Sudha P, Ramprasad V, Nagendra MDV, Kulkarni, HD, Ravindranath NH (2007). Development of an agroforestry carbon sequestration project in Khammam district, India. Mitigation and Adaptation Strategies for Climate Change, 12: 1131-1152.

Swamy SL, Puri S, Singh AK (2003). Growth, biomass, carbon storage and nutrient distribution in *Gmelina arborea* Roxb. stands on red lateritic soils in central India. Bio resource Technology 90:109-126.

Swamy SL, Puri S (2005). Biomass production and C-sequestration of Gmelina arborea in plantation and agroforestry system in India. Agroforest Syst 64: 181-195.

Takimoto A, Nair PKR, Nair VD (2008 b) Carbon stock and sequestration potential of traditional and improved agroforestry systems in the West African Sahel. Agri Eco Environ 125: 159-166.

Theurillat J, Guisan A (2001). Potential impact of climate change on vegetation in the European ALPS: a review. Climatic Change 50:77-109.

Thornton PK, van de Steeg J, Notenbaert A, Herrero M (2009). The impacts of climate change on livestock and livestock systems in developing countries: A review of what we know and what we need to know. Agricultural Systems 101:113-127.

UN (2009). The Millennium Development Goals Report 2009. United Nations, New York.

UNFCCC (2009). Land Use, Land-Use Change and Forestry, United Nations.

Verma KS, Kumar S, Bhardwaj DR (2008). Soil organic carbon stocks and carbon sequestration potential of agroforestry systems in H.P. Himalaya Region of India. Journal of Tree Sciences 27(1):14-27.

Watson RT, Noble IR, Bolin B, Ravindranath NH, Verardo DJ, Dokken, DJ (eds). (2000). Land Use, Land-Use Change, and Forestry. Intergovernmental Panel on Climate Change.Cambridge University Press, Cambridge.

Woodward FI (1987) Climate and plant distribution. Cambridge University Press, Cambridge.

World Bank (2015a) Agricultural land (% of land area). Available at http://data.worldbank.org/indicator/AG.LND.AGRI.ZS/countries?display=graph (verified 16 September 2015).

Wright DG, Mullen RW, Thomason WE, Raun WR (2001). Estimated land area increase of agricultural ecosystems to sequester excess atmospheric carbon dioxide. Commun. Soil Sci Plant Anal 32:1803-1812.

Yadava AK (2010). Biomass production and carbon sequestration in different agroforestry systems in Tarai region of central Himalaya. Indian Forester 136(2): 234-242.

Yohannes M, Mebratu K (2009). Local innovation in climate-change adaptation by Ethiopian pastoralists: PROLINNOVA–Ethiopia and Pastoralist Forum Ethiopia (PFE), Final report. Addis Ababa, Ethiopia.

2

Perceived Weather Anomalies and Climate Change: Community Based Adaptation in Sustaining Natural Resources and Livelihoods by Adi and Monpa Communities On Arunachal Pradesh, Eastern Himalaya

***Ranjay K. Singh*[1,2,3] *and Arun Agrawal*[3]**

[1]*College of Horticulture and Forestry, Central Agricultural University, Pashighat-791102. A.P.*
[2]*Central Soil Salinity Research Institute, Indian Council of Agricultural Research, Karnal-132001, Haryana, India*
[3]*School of Natural Resources andEnvironment, University of Michigan Ann Arbor, MI-48109, USA*

Introduction

Globally, now scholars accepted that the climate is changing and it has implications on human culture, social systems, economic stress and ecological problems (Ford *et al.* 2010). The IPCC advocates to search and develop local solutions for climate change adaptations (IPCC 1996; IPCC, 2001), however, it does not recognize the breadth and strength of centuries-tested community knowledge/traditional knowledge systems (Salick *et al.* 2009) which are available with the people of developing world, and has been for adaptation practices and strategies to extreme weather and climate change (Turner and Clifton 2009; Salick and Ross 2009). It is well recognized that with fragile ecosystems, poor economic status and with less robust technologies required to adapt climate change, marginal and poor people of developing countries are likely to be more vulnerable with the changing climate than the industrialized countries (Tompkin and Adger 2004; Agrawal, 2008).

Local observations of weather anomalies and environmental changes are well documented globally (Turner and Clifton 2009; Salick *et al.* 2009; Ford and Pearce 2010; Singh *et al.* 2011). Changes and anomalies observed in every community and farmers society include unpredictable weather, erratic rainfall, increasing temperature, prolonging rainfall period, changes in life cycle of plant species and disappearance/reduction in plant and animal population (Byg and Salick, 2009; Ford and Pearce 2010; Singh *et al.* 2011; Turner and Singh, 2011). The sensitivity of local community with different level and types of environmental risk compelled them to evolve a specific knowledge system and location specific adaptation skill in order to meet out challenges posed by climate (Lantz and Turner 2003; Ford and Smith 2004; Salick and Ross 2009; Singh *et al.* 2011).

The study of Reid *et al.* (2009) states that Glacier melt in the Himalayas is projected to increase flooding, rock avalanches from destabilized slopes, and to affect water resources within the next two to three decades. Other study indicated that effect of rising temperature and thereby melting Himalayan glaciers on Siang River passing through northeast India (specially Arunachal Pradesh), could be highly vulnerable to human and natural resources both (Himalaya Initiative News 2010). The abrupt changes in local weather and thereby flood in Siang river in September, 2010 has severely affected to the local communities and natural systems in upper as well as lower valley of Pasighat could be a recent phenomenon of climate change in Arunachal Pradesh (Financial Express 2010). Study of Bhattacharyya *et al.* (2007) on an exploratory pollen and carbon isotopic from a 1 m deep sediment profile at the Paradise Lake (4176 m asl) near Sela Pass (West Kameng district) of Arunachal Pradesh have indicated about changes in vegetation vis-à-vis climatic both during late Holocene in alpine ecosystem. The traditional communities- who are already experiencing weather anomalies and inter annual variations in climatic features (Singh and Padung 2010; Singh *et al.* 2011) in Arunachal Pradesh, would become more vulnerable due to impact of climate change on indigenous biodiversity, other natural resources as projected by various scholars.

Government of Arunachal Pradesh realized and considers that agriculture, forest and fisheries are likely to be severely hit due to climate change (The Sentinel 2011). Considering the problems to be aroused in future, state government State Action Plan on Climate Change (SAPCC). Being part of prepared Himalayan States, Arunachal Pradesh has become partner in passing and adapting declaration on climate change (The India Post 2009). Policy makers of state advocate having mission for sustainable habitat management and using locally available plant resources for development in organic manner (Sinha, 2009).

In response to environmental risk, tribal communities of Arunachal Pradesh living in varying fragile and harsh ecosystems (Bujarbarua and Baruah, 2009) have been using community based traditional knowledge in adaptations to secure livelihoods (Singh and Sureja 2006; Singh and Adi Women 2010; Singh *et al.* 2011). These adaptation practices vary from one social system to another and from one ecosystem to another on account of vast diversity in socio-ecological systems (Singh and Sureja 2006; Singh and Adi Women 2010), thus are highly specialized to a particular micro-ecosystem and social system. These adaptations have been in use by local communities to different levels of perceived sensitivity and vulnerability in a particular ecosystem (Singh and Bhowmik, 2008; Singh *et al.* 2011). But the question is that whether these CBAPs would be able to bear any future significant weather anomalies and climate changes? Or they would be vanish out? However, it is assumed that centuries tested few CBAPs may still remain in existences which are robust to absorb high level of environmental risks. In present research article, efforts were made to explore community based adaptation practices which are in use at large scale among the Adi and Monpa tribal communities. These CBAPs are evolved in response to harsh and risk prone ecosystems and variable climates in Arunachal Pradesh. This study was aimed to assess the people's perception and knowledge on weather anomalies and climate change among Adi (living in subtropical ecosystem) and Monpa (living in sub-temperate to temperate ecosystems) communities in Arunachal Pradesh, India; and to explore traditional knowledge of the communities and their adaptation practices.

Area and design of the study

Arunachal Pradesh is considered as one of the mega biodiversity centre (Myer 2000) and abode of cultural diversity (Singh and Srivastava, 2009). The subsistence economy of its peoples, all of whom depend largely on its forests and diverse ecosystems. The people of Arunachal Pradesh, represented by 26 major tribes and 110 ethnic groups, have developed unique bonds with the nature - expressed through Biocultural Diversity. The climate of Arunachal is governed by the Himalayan system and the altitudinal differences. The climate here is highly hot and humid at the lower altitudes and in the valleys wrapped by marshy thick forest particularly in the eastern region, while it becomes too cold in the higher altitudes. Average temperature during the winter months range from 15 to 21 °C and 22 to 30 °C during monsoon.

Study was carried out in East Siang and West Kameng districts (Fig. 1). The East Siang district occupies an area of 4,005 KM^2 and has a population of 87,430 (as of 2011 census). District lies between 27° 43' and 29°20' North latitudes and 94° 42' and 95° 35' East latitudes, and is located at 155 m sea level

(SL). It receives about 4168. 2 mm rainfall yearly. District has subtropical types of climate. East Siang shares its boundary with Upper Siang district in north, Dhemaji district (of Assam State) in south, Dibang Valley district in east and West Siang district in west. Siang is the biggest river of district originates in Tibet and joins the river Brahmaputra in plains of Assam. The total population of district is 87, 397 (2011 census).

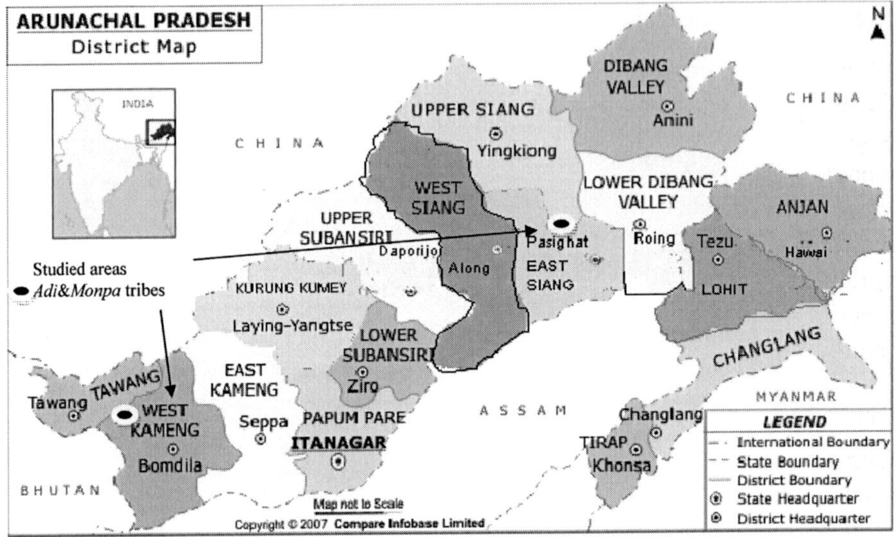

Fig. 1. Localization of study areas
Source: http://www. imd. gov. in/section/nhac/distforecast/arunachal-pradesh. htm

The East Siang district is mostly populated by the Adi tribe which comprises of a large number of ethnic groups. These groups can be divided into various subgroups such as the Minyongs, Padams, Shimongs, Milangs, Pasis, Karkos, Ashings, Pangis, Tangmas and Boris. The social relationships are determined on the basis of kinship or locality. Adis are totally dependent n agricultural and forest resources for their food and livelihood security. Women play central role in securing foods and livelihoods, while male hunts and perform heavy physical tasks.

West Kameng lies approximately between 91° 30' to 92° 40' East longitudes and 26° 54' to 28° 01' North latitudes. It covers an area of 7422 km^2 accounts for 8. 86 per cent of the total area of the state. The district shares an international border with Tibet in the north, Bhutan in the west, Tawang district in the northwest, and East Kameng district in the east. The southern border is shared with Sonitpur district and Darrang district of Assam. The topography is mostly mountainous. Much of West Kameng area is covered with the Himalayas. In

West Kameng there are three principal mountain chains - part of Sela range, Bomdila range and Chaku range. The altitude of Sela range varies from 4628 to 4573 m and Sela pass is 4181 m asl. West Kameng district experiences an arid tundra or a cool temperate climate in the north. Snow fall occurs from mid-November to February. Tenga, Bichom and Dirang Chu are the main rivers flowing through the district. All these rivers are tributaries of the river Kameng which flows through Bhalukpong circle of the district and joins the river Brahmaputra in plains of Assam.

West Kameng comprises five major tribes: Monpa (which makes up 78% of the district's population and includes Dirang, Bhut, Lish, and Kalaktang Monpa), Miji (Sajolang), Sherdukpen, Aka (Hrusso), and Khowa (Bugun). Minority tribes include Takpa, Lishipa, Chugpa, and Bupta. Most of the inhabitants are Buddhist. West Kameng district is having a population of 56,421 comprising of 30966 males and 25455 (Census 2011). Monpa and other tribes too of this district are agriculture, animal and forest depended communities for their and livelihood security. Except heavy physical work, Monpa women contribute in every activity required for foods and livelihood security.

Framework of study

Climate change could impact on various scales on rural livelihoods and food security. Though, few community's CBAPs could exhibit lower level of vulnerability owing to lower climate sensitivity and higher adaptive capacity (O'Brien *et al.* 2004; Agrawal 2008). We followed the framework of exposure sensitivity and adaptive capacity of communities continuously influenced by social and biophysical conditions suggested by Glantz (1996), Adger (2003a,b) and Ford and Smith (2004).

Defining community based adaptation practices (CBAPs)

We defined traditional ecological knowledge as the degree to which a person hold environmental knowledge about his/her surroundings environment and ecology, and he/she apply it in practical form to adapt to any weather anomalies/ or extreme climates. We defined CBAPs as the practices on food and livelihoods evolved traditionally and if known and adopted by majority of village people (at wider scale) of a particular community (publically domained knowledge and practices) in response to reduce, moderate or adjust the expected or actual risks associated to drought, flood or abrupt weather variability.

The CBAPs widely known and practiced by Adi and Monpa tribes are presented in this article. Five notable CBAP applied in agriculture and foods are presented in form of case studies. In first phase of study, the CBAPs were explored from both the communities. The traditional practices which were important as adaptive measures to perceived weather anomalies and climate changes, and varied from one social system to another, are summarized and presented in table form.

Defining micro-ecosystem

The index developed by Singh (2002) on identifying micro-ecosystems was applied during course of case studies (for case number one, two and three). Purpose of applying this index was to specify adaptive practices suited to a particular micro-ecosystem.

Sampling procedure

We adopted explorative and qualitative approach of carrying out this study (Ford and Smith 2004). Before conducting study, we carried out number of projects and activities relating to community-environmental research listed in Table 1. Study also contains data based on different types of meetings such as village workshops and regional and national seminars organized on the subject under study. Though, exclusive data were taken as observations from community members about their perception and CBAPs.

In order to collect data, East Siang (subtropical) and West Kameng (sub-temperate to temperate) districts were selected purposively. Adi from East Siang and Monpa from West Kameng districts (Fig. 1) were chosen as two tribal communities for this study. From East Siang district, twenty two (22) villages namely- Balek, Boleng, Dalung, Gune, Kebang, Mebo, Miram, Mirbuk, Mirmir, Ngopok, Pangin, Poglek, Rasam, Runne, Sibut, Sido, Sille-Oyan, Siluk Basti, Vijari, Yabgo and Yagrung, while seven (7) Monpa living villages namely-Chhung, Dirang, Lish, Namsu, Ramakamp, Thembang and Yang were selected from West Kameng district. Minyong, Padam, Pasi, and Pangi were ethnic groups of Adi tribe identified and living in sampled villages from East Siang district. Listed villages were selected purposively based on the variability of ecosystems, people dependency on various natural resources, degree of ethnicity and their exposure to various social and environmental issues.

From each village of East Siang, 30 respondents with 15 male and 15 female, thus total 660 from 22 villages (330 male and 330 female) were sampled randomly from a voter list provided by Village Panchayat Secretary. Further, we sampled 4 hunters from each Adi village totaling 88 in order to learn about their perception on climate change and hunting. From each of 7 Monpa dominating selected villages (West Kameng district), 36 respondents with 18 male and 18 female, thus total 252 (126 male and 126 female) from 7 villages were sampled with the same procedure. From each Monpa village, a sample of 2 hunters, thus total 14 was taken for studying hunting and climate relation. While making samples, help of Gaon Burha was taken in each village. Sampling

Table 1. Community- environmental research projects implemented as components of adaptation strategies to perceived weather anomalies and climate change

Carried out mitigation and adaptive approaches	Participants	Objectives	Communities	Follow-up actions
Establishing village traditional knowledge bank (VTKB)	Kebang,village level committee, elders and entire community members	To develop archives and digital data base of community knowledge systems in order to enhance conservation, cope-up environmental problems and promote livelihoods	Adi tribe	Community members are continuing their the work and preserving practices
Establishing community knowledge garden	Interested community members and clan	To domesticate and conserve threatened plants species, develop sustainable harvesting protocols, preserve related traditional knowledge, enhance climate change adaptive capacity and promote livelihood options	Monpa and Adi tribes	People have domesticated number of common and threatened plants species at individual and clan lands
Organizing biodiversity contests (BDC)	Kebang institution, elder men and women, school teachers and children	To create awareness about values of indigenous biodiversity and related knowledge systems in changing climate, to check knowledge and biodiversity erosion, select and create location specific plant based livelihoods as adaptive practices to cope-up with extreme weather and changing climate	Padam, Pasi, Minyong and Pangi ethnic groups of Adi tribe	Few grassroots people started in making knowledge networking with various institutions and community members on learning and conservation of indigenous resources
Organizing recipe contests (RC)	Rural women, members of Reglep (indigenous institution) and women who participate in barter system	To create awareness about values of indigenous biodiversity and related knowledge systems in changing climate, to check knowledge and biodiversity erosion, select and create location specific food based micro-enterprises as adaptive practices	Padam, Pasi, Minyong and Pangi ethnic groups of Adi tribe	Rural women are conserving food plants which are important for extreme weather, few knowledge networking of homogenous groups for livelihood adaptation
Organizing micro-enterprise development contests	Women who perform traditional food processing and weaving technologies, and interested women who participate in barter system	To screen best traditional food, weaving and handicrafts technologies, values addition, financial support through micro-venture funds and empowering women to appropriate livelihoods to cope-up with environmental and economic problems	Adi and Monpa tribes	Few plant based foods and handcrafts, weaving and handloom technologies are selected for adaptation practices
Organizing groups discussions contests	Elder men and women of community, school teachers and college students	To create awareness about values of indigenous biodiversity and related knowledge systems in changing climate, to check knowledge and biodiversity erosion, create knowledge network and human resources development of younger generation to become stewards of solving future climate problems in locality	Adi and Monpa tribes	Community members are energized and aware about potential of their local practices. They are paying more attention in exploring and using local practices
Organizing community workshops	Elders and younger men and women, farmers, hunters, healers	Exploring best indigenous/traditional practices and adaptive technology/strategies from community members, inculcate interest among them about conserving local natural resources in order to cope-up with extreme weather/changing climate, screen best	Adi and Monpa tribes	Community members are organizing location and problem specific meetings and workshops in order to explore,

was made of respondents with age varying from 25-85 years . Keeping diverse age of sample was driven with a notion of observing variations in the data set of adaptations. Summary for personal profile of sampled populations is presented in Table 2.

Table 2. Identified ethnic groups of Adi and Monpa communities and their major socioeconomic attributes

Tribe	Ethnic groups	Identified members in total sample	Mean age	Experience (in mean yrs) on environmental risk management	Types of livelihoods[1]	Degree of dependency on natural resources[2]
Adi (660)	Pasi	245	45.6±16.3	22.9±9.6	Diversified	Medium to high
	Padam	185	51.4±12.6	24.1±10.2	Diversified	Medium to high
	Minyong	120	50.5±11.5	21.8±8.6	Diversified	Medium to high
	Pangi	80	52.8±14.7	28.3±11.7	Diversified	Very high
Monpa (252)	Monpa	252	44.9±10.9	25.6±9.4	Diversified	High tovery high

Note

[1]Diversified livelihood ranged upto 5 components, as: agriculture, forest, animal husbandry, hunting, weaving/handlooms, making handicrafts, fishing and local marketing of bioroesurces

[2] Degree of dependency on natural resources is defined as:

(a) Very High= When livelihood and food security of a person is exclusively (100%) depend on natural resources

(b) High= When livelihood and food security of a person depend more than 75 % on natural resources

(c) Medium to high= When livelihood and food security of a person depend from 50- 75 % on natural resources

Protocol of data collection

Village workshops: methodological learning with communities

In East Siang district total 12 village workshops were organized on sensitization, awareness creation, recording data on weather anomalies, climate change and adaptation practices of Adi community and listening from elders in group. Objective was also to consult with cultural and community leaders and traditional knowledge holders, to listen their past experiences and present role in making local strategies relating to weather anomalies and climate change, and obtain their feedback on future planning of adaptations. While organizing village workshops, some time more than one village was clubbed together based on distance of one village from another. In similar fashion, a total of 4 village workshops were organized in West Kameng district. The range of participation in these workshops varied from 15-40 (with mean value 26.5), and thus over 600 community members of both the districts participated in total 16 workshops.

With the collaborative projects on community and environmental research, the cultural leaders of Adi and Monpa tribes were mobilized in order to develop and pursue certain 'protocols/norms' as adaptation practice for weather anomalies. These norms were framed in Kebang (customary institution) meeting for sustainable harvesting and management of wild resources. The objectives of this exercise were also to take help of these leaders as resource persons in projects implemented before and during this study. To develop protocols and frame rule of sustainable harvesting of plant species-especially about, dekang (*Gymnocladus burmanicus*) in changing socio-environmental scenario, customary chief of Bane Kebang (village level customary institution) and Bango Kebang (customary institution at regional level) of Adi tribe have pursued members of villages in a regional workshops held at Yagrung village (East Siang district). More than 125 Adi members have participated in this workshop. In similar fashion, two workshops were held at Dirang (West Kameng district) in association with Gaon Burhas and Chhopa (customary institution) with a participation of about 60 members to educate the community members, develop protocols and norms for sustainable conservation and adaptive practices on Paisang tree (*Quercus griffithii*) and local agrobiodiversity.

Focus groups discussions

The focus group discussions (FGDs) were held among both the communities to understand complex practices such as understanding those are given in case studies; listing various micro-ecosystems and their access for hunting and harvesting plant resources during drought (community managed mountain forests); cross verifying adaptations on cropping and farming systems practices, preparing seasonal calendar; matrix analysis of bioresources used at village level; historical contexts of subjects, and understanding about adaptation strategies of women- especially about the role of informal institutions. Total 14 FGDs were organized with Adi community while 4 FGDs with Monpa community with a participation of 6-12 members (average 7 members in each FGD) in each FGD. Total participation was 144 members in 18 FGDs. Few other tools of participatory rural appraisal (PRA such as transect walk and trend analysis were applied to observe adaptation practices in fields, and status and use pattern of resources by each community.

Case study approach

The notable adaptation practices were studied through case study approach. Use of this approach was to have detailed examinations and explore complete process and methods of a particular adaptation practice within its real-life context (Ford *et al.* 2010). This method was applied to go in-depth about process and system of wide scale adaptation or outstanding individual contribution in adaptation. For example we studied late Mrs. Pem Dolma of Dirang West

Kameng district as part of case study who died in 2005 after a few months of her interview and participant observations with her.

Personal interviews

Data were taken mostly through personal interview method using interview schedule containing open-ended questions. Though, very scanty data on climatic variables Arunachal Pradesh are available in literature or online, however, best efforts were made to review literature available on it and complemented to first hand observations wherever possible in discussions. Before final application, the interview schedules were pilot tested in non-sampled areas and questions have further been redefined and refined to make them effective and get reliable and valid data. Along with interview, audio and video aids were used complementarily to comprehends data and improve its credibility. Interview lasted from 1. 30 hrs to 4. 0 hrs. In case of case study, it took 2-4 days, and these have been conducted during their effective seasons through participant observations. First author (RKS) has lived in studied areas from October 2002 to December 2008, thus participated in most of the agricultural, festivals and cultural practices, hunting, food gathering and processing systems and local marketing of bioreosurces by Adi and Monpa to get first hand data during effective seasons through participant observations.

Experiment on effect of low temperature and ultraviolet radiation on yeast population in selected traditional fermented foods and beverages of Adi tribe

To find out scientific rationale about perception of *Adi* women on abnormalities in fermentation of traditional foods and beverages-caused by weather anomalies, a laboratory experiment was conducted. Objective was to know the effect of low temperature and ultraviolet radiation on the population of yeast in selected traditional fermented foods and beverages of Adi tribe. Aseptically collected fermented food samples *viz*., esing enging (sugar beet base), nongin (taker - local rice base), kala apong (traditional alcoholic beverage prepared from siye (traditional yeast table) and boiled rice of amkel variety- local rice base), ngosing (ribi, gali, ngoying–indigenous fish species base), and namdung (*Pyrillaocimodes*- local oil seed base) were incubated under different conditions of low temperature and ultraviolet radiation (from germicidal UV lamp). Yeast population was enumerated by standard plate count on YEPD agar and observations of 10 replicates were accounted.

Social verification and validation of explored adaptive practices through regional and national seminars

One regional (2005) and three national (2007, 2008 and 2009) seminars were organized at Pasighat where maximum participation (225, 120, 150 and 80, respectively) of state tribes (total 26) was ensured. Calling participation of large population of tribes also other than Adi and Monpa was aimed to develop reciprocal learning network with members having common interest on certain subjects of adaptive practices. In these four seminars, we further discussed and verified explored traditional practices through personal discussions with resource persons of studied populations. We further made aware other communities' members on current problems and future plan of adaptive practices for extreme weather and perceived climate changes which could be managed through local practices. Explored practices were summarised and project on screen bilingually in these seminars to get feedback of resource persons, and improvement in data-set was made accordingly. SWOT (strength, weakness, opportunity and threats) analysis of CBAPs was made during these seminars in which scientist other than communities members have also participated.

Ethical norms followed under study

The recent work of Pearce *et al.* (2009) advocated scientists to have ethical collaborations while working on community and environmental changes issues, though we followed such ethics from 2003 itself. Any activity of projects and proposal of organizing meetings and workshops for community-environmental issues, was properly communicated to Gaon Burhas, cultural leaders, Anchal Samithi Members (president of village panchayats) and elders before being their final implementation. This communication helped to get optimum support of community participation during operation of various projects (Table 1) and also to develop credible rapport with community for learning and researches on adaptation practices. Through this collaborative process, the feedback and suggested modification/inputs were further incorporated in future scheduled activities of a project. After taking the data and reports of each meeting and workshops, reports were prepared bilingually (in local dialect and English), and has been sent to community members of studied areas for reading and social verification of facts to be reported further. Prior informed consent (PIC) was obtained from each Gaon Burha of both the studied communities to use and publish data recorded from each village. For individual creativity- such as of Mrs. Pem Dolma, PIC was obtained from a person concerned.

Data were categorized and analysed using SPSS package available from University of Micihigan, Ann Arbor, USA in February, 2011. Looking to the nature of study, descriptive statistics such as frequency and percentage were run in SPSS in order to synthesis data and draw inference from them.

Perception of Adi and Monpa communities about weather anomalies and climate change

The results revealed that Adi and Monpa communities both perceive that normal rainfall pattern has declined in last 15- 20 years and it has become erratic (Fig. 2). Though, few respondents (Adi 28. 4 and Monpa 21. 7 %) perceived that there is no change in weather or climate. Majority of respondents from both the communities expressed that start-up of seasons and their end is also not usual and normal. These communities experienced that due to weather anomalies and sometimes abrupt changes in climate, changes in local ecology and landscapes are being noticed (Table 3). For example, communities observed that hazardous events have caused in making agricultural soil relatively poor in fertility, due to more landslides now agricultural lands needs more labour for controlling soil erosion, community managed small river tributaries and forest streams have become more devastating, and thus these events are contributing more threat to indigenous flora and faunas. Due to abrupt changes and weather anomalies, few most important species such as dekang (*Gymnocladus burmanicus*), mirangmose(*G. assamicus*), chandu (*Cordyceps sinensis*) and emo (*Aconitum ferox*) are threatened (Table 3).

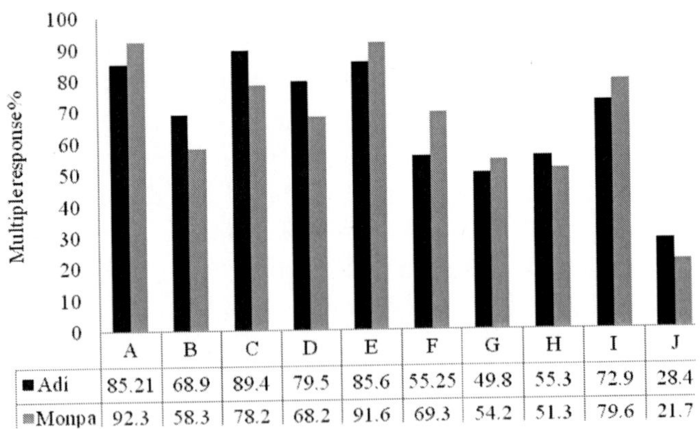

	A	B	C	D	E	F	G	H	I	J
■Adi	85.21	68.9	89.4	79.5	85.6	55.25	49.8	55.3	72.9	28.4
■Monpa	92.3	58.3	78.2	68.2	91.6	69.3	54.2	51.3	79.6	21.7

A= Normal rainfall has declined over last 15-20 years
B= Onset of rainfall has become late
C= Onset of rainfall has become erratic
D= Effective rains period became unpredictable
E= Weather has become warmer
F= Duration of cold is decreased
G= Snowfall at the mountain peaks are decreased
H= Frequency of storms are increased
I= Seasonal changes and inter-annual variations
J= There is no change in weather or climate

Fig. 2. Community perception about weather anomalies and climate change

Table 3. Ecological changes to landscape observed by communities

Aspects of ecological changes	Adi	Monpa
More hazardous events caused in poor fertility of soils due to either drought or heavy rains	76.4	68.8
Frequent land slide due to excessive rains in few hours	79.3	70.4
Frequent land slide causing to make agricultural forest fields more labor demanding job	89.4	78.6
Frequent land slide posing threats to local flora and fauna	48.4	42.9
Community rivers and regional rivers have become more devastating	65.8	59.3
The aquatic biodiversity are declining due to rapid course changes and silts in rivers and lakes	44.3	38.9
Some of keystone species used in food and ethnomedicinal webs are at threat (like *Gymnocladus assamicus, G. Burmanicus, Cordycepssinensis* and *Aconiume ferox*	65.9	42.7
Grazing lands are become unproductive gradually	39.4	78.9

It has been observed from both the communities that weather anomalies and climate change are not all of sudden phenomenon. These are the resultants and compounded impacts of number of factors including destruction of local ecology, erosion of ecological ethics among younger generation, gradual loss of socio-cultural respects and world views of local ecology (kincentric philosophy of community members) and loss of traditional ways of people survival (Fig. 3).

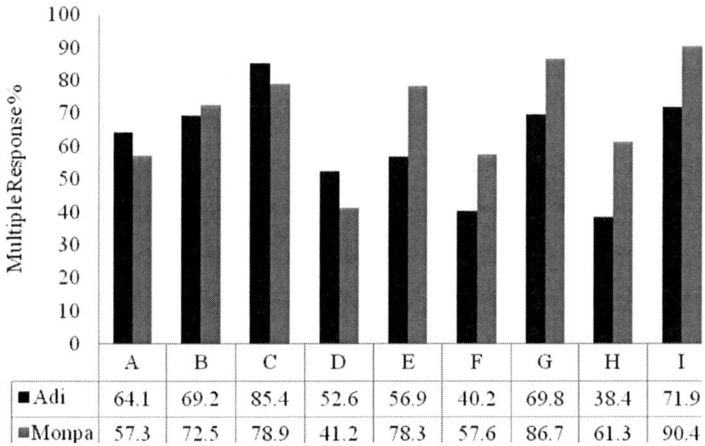

	A	B	C	D	E	F	G	H	I
■Adi	64.1	69.2	85.4	52.6	56.9	40.2	69.8	38.4	71.9
■Monpa	57.3	72.5	78.9	41.2	78.3	57.6	86.7	61.3	90.4

A= Destruction of ecology through commercial crops; B= Dishonoring mountain and community forest resources; C= Erosion of eco-cultural ethics among younger generation; D= Inappropriate government policies; E= Materialistic culture which cause harm to nature; F= Privatization of communal resources; G= Opting homogenous livelihood options instead of diversified; H= Polluting sacred mountain (*Lu, Mani,* etc.) and sacred forests; I= Replacing traditional ways of people survival strategies

Fig. 3. Perception of *Monpa* and *Adi* communities about causes of weather anomalies and perceived climate change

Learning with traditional hunters indicated that they feel changes in seasonal and climatic pattern. Now, they do not get usual hunt as they were able to harvest about 20-30 years back (Table 4). They are facing number of constraints in hunting, and thus their subsistence livelihood and eco-cultural identity are liable to become abnormal.

Table 4. Perception of hunters about hunting in weather anomalies and extreme conditions

Perception	Adi	Monpa
During excessive rain/flood, some animals change their natural track in forest	75.3	63.2
During excessive rain/flood, the trained dogs team get more trouble and problem in forest to chase animals	84.5	69.4
It's risky during the excessive rain to have hunting	96.4	86.1
Problem of setting hunting points in the forest	89.3	77.5
Landslide blocks path of migration of animals, sometimes could cause accidents to hunters	62. 8	43.8
Dynamics of grazing and habitation ground of wild animals during drought period gets disturbed	79. 6	55.8
Need to spent more time and energy for planning and hunting of animals	86.8	50.5
Elders cannot become part of hunting team during excessive rain- who are master of hunting strategy	70.2	65.3
Due to lack of elders' participation, the amount of hunts from forest become meager	66.8	59. 3

* Multiple response percentage
Total hunters from Adiwere 88 and Monpa 14 (to tal 102)

Social capitals of Adi and Monpa women

Both the communities have combined social and informal institutional adaptation strategies in order to cop-up with weather anomalies and climatic problems (Table 5). Activities varied from one to another in terms of percentage. In response to the mountain terrain and risk associated to livelihoods, special cultural institutions- called Mila (of Monpa) and Reglep (of Adi) led by women have been evolved. These two women's institutions help in performing task collectively. The task could be collection of forest resources for food and fire wood, harvesting and carrying loads from community forests and jhum lands to home. The group dynamics of both institutions enable women to play more effective roles during the period of drought and flood than the normal weather conditions. The other objective of this group is also to reduce excess drudgery to be caused during drought and or flood while performing physical task of harvesting and carrying food and other resources.

Table 5. Community based adaptation strategies of Monpa and Adi communities to cope-up with extreme weather and sustain livelihood

Perception	Adi	Monpa
Forming group (Reglep among Adi tribe and Mila among Monpa tribe) to make traditional dresses for sale and income generation	91.2	84.5
Collective collection of plant resources from morang (community forest) and homegardens (cane, bamboo and creepers) for making handicrafts for sale and income generation	89.5	78.4
Collective food processing from surplus food resources preserved in last season to add traditional values and sale in local market for income	76.9	70.5
Collective harvesting and collection of ethnobotanicals from community forest and homegardens to sale in local market and share the benefits	69.8	74.5
Forming self help groups and establishing micro-enterprises (making bamboo shoots, pickle from forest products) and earnings	80.2	76.8
Forming self help groups and collective purchasing of culturally important items such as cloths, handicrafts, beads, etc. and selling for collective income	67.8	59.8
Forming Reglep/Mila and collective raring of chickens, ducks and pigs for self foods and selling for income	59.3	55.8
Forming Reglep/Mila and promoting barter system to exchange foods and other culturally important items for promoting livelihoods	68.5	72.5

* Multiple response percentage

These institutions also perform in marketing of traditional bioreosurces. One of the objectives of these institutions is also to monitor plant populations of wild resources which are in community forest, and reducing probability of overexploitation of food resources that are used specially during drought. These resources specially include Anke (*Castanopsis kurzii*), Belang (*Artocarpus heterophyllus*) and Tasat (*Arenga obtusifolia*), straw berry, wild kiwi fruits, wild plum, *etc.* These wild resources are stored after processing, and are again re-processed in special traditional manner to use them during drought or flood.

Sustainability of these plant biodiversity is enhanced through harvesting in group. Individual harvesting of indigenous biodiversity from community forest is discouraged by women of both the communities.

Traditional knowledge based cropping and farming systems

Result revealed that Adi and Monpa farmers both have developed location specific adaptive indigenous cropping systems in order to cope up with harsh ecosystems and environmental risks (Table 6). Each adaptive practice of cropping system varies in percentage of use by farmers of both the communities. Interestingly, few components notably as Bathua (*Chenapodeum album*) with Phaphda (*Fagopyrum esculentum*) and finger millet with amaran thus and field pea are unique systems. These cropping systems do not exist even in formal agronomic practices, though Monpa farmers use these models as adaptive cropping system for rainfed agroecosystem. Similarly, Adi farmers have very sound and sustainable adaptations in agroforestry notably with Ogjok (*Bauhinia variegata*), bamboo species, Toko-patta (*Levistona jenkinsiana*) and citrus species to manage their crop field from wind and water erosion and landslides.

Table 6. Traditional knowledge based adaptation practices of cropping systems among Adi and Monpa tribes

S. N.	Traditional knowledge and adaptation practices of farming systems	Adaptive %	
		Adi	Monpa
1.0	**Relay cropping systems**		
1.1	Phaphda (*Fagopyrum esculentum*) followed by maize	00.0	95.0
1.2	Maize followed by paddy	48.0	00.0
2.0	**Micro-ecosystem based cropping systems**		
2.1	Barley + field pea; wheat + field pea	00.0	85.5
2.2	Finger millet (*Elcusine coracocna*) + *Amaranthus* species + field pea	00.0	57.8
2.3	Finger millet in selective patches	44.1	52.3
2.4	Foxtail millet (*Setaria italica*) in selective patches	38.2	47.2
2.3	Soybean + maize + cucurbit crops	00.0	48.5
2.4	Low land paddy followed by maize crop	00.0	30.5
2.5	Rainfed rice (in jhum field)* + beans + cucurbit crops + ethno-botanicals; rainfed rice + beans + beans	85.5	00.0
2.6	Rainfed rice (in jhum field) + soybean + solanaceous species + ethnobotanicals	82.8	00.0
2.7	Low land paddy	45.8	00.0
3.0	**Micro ecosystem based inter-cropping**		
3.1	Bathua (*Chenapodeum album*) with phaphda	00.0	45.5
3.2	Phaphda with rajma bean (*Phaseolus vulgaris*) and or field pea in gravelly and sloppy soils	00.0	50.5
4.0	**Planting living fence species on crops' boundary to avoid soil and water erosions**		
4.1	Ogjok (*Bauhinia variegata*), bamboo species and toko-patta (*Levistona jenkisiana*) and citrus species	90.7	00.0
4.2	Timbur (*Zanthoxylum americanum*), bamboo species, local peach, plum and almond on crops' boundary to avoid soil and water erosions	00.0	72.5
5.0	Traditional organic practices of crop and pest and disease management practices which are ecologically viable and economically feasible	95.2	96.6
6.0	Adding more organic matter in soil and applying mulch	79.6	98.5
7. 0	Intercropping with perennial woody tree species in between the crops	70.8	12.5
8.0	**Maintaining community seed banks of indigenous crops species and varieties**		
9.0	Selection of patches in land slopes according to nature of crops and their demand in relation to moisture and fertility gradient	925.9	85.7
10.0	Maintaining traditional water harvesting methods and sustain watershed in mountains	71.4	75.7
11.0	Using bamboo and wood made traditional agricultural tools and ensuring less exposure of soil to water and wind erosion	95.5	92.9

* Jhum field: Slash and burn agriculture system followed by *Adi* tribe

In order to reduce risk of crop failure and meet out other household needs, farmers of both the communities experiment informally with different adaptive diversified farming systems (Table 7). These farming systems do not include components only of agriculture and animals, but non-agricultural components are also equally adapted depending on types of ecosystems. The component of reported adaptive farming system carry gender specific role of both the communities. Each component has different level of environmental risk bearing capacity and becomes crucial during specific climatic sock- such as drought, erratic rainfall and other natural disasters.

Table 7. Adaptive diversified farming systems of Adi and Monpa communities

Components of diversified farming systems	Adi	Monpa
Crops with piggery, yak (*Bos...*), sheep and poultry	70.3	00.0
Crops with piggery, mithun (*Bos frontalis*), poultry, and duck	00.0	85.4
Agriculture + horticulture- one (growing crops along with indigenous fruits species)	50.4	55.4
Agriculture + horticulture- two (growing crops along with indigenous species of vegetables	65.9	79.8
Agriculture + microenterprise with silkworm raring	38.4	00.0
Agriculture + traditional weaving of cloths	85.4	55.5
Agriculture + marketing of wild resources	45.8	25.7
Agriculture + traditional hunting	40.5	20.4
Agriculture + laboring	07.0	30.2
Agriculture + traditional fishing	35.7	10.4
Agriculture + making handicrafts	65.8	35.6
Agriculture + making fishing net and tools (using indigenous plant species	12.6	03.8

Market driven adaptive practicesof Adi women

Adi women rely on terrestrial and aquatic biodiversity, and using them during food crisis- especially during drought and floods. We could explore 27 indigenous practices of Adi women which are important and traded during abnormal weather (Table 8). During drought, local wild resources are harvested from various ecological edges across the landscapes and altitudes. Women do such adaptive practices of combining biroesurces from two different micro-ecosystems in order to diversify products and reduce dependency on cultivated crops. These bioresources traded in local market become costly during abnormal weather than normal one (Table 8). But, these food products ensures supply of food resources for subsistence survival during extreme weather to those living nearby town, and has limited access of food resources due to blockage of road and other means of transports in between Assam State and Arunachal Pradesh. Local trading of wild products enable Adi women to fulfill their other needs which otherwise are affected due to drought or flood.

Table 8. Non-agricultural wild products used in food systems and local marketing by Adi women during normal and extreme weather/climate

Ingredient/Food	Availability in local market	Peak availability	Mean price (in Rs) per Kg Normal weather/ climate	Abnormal weather/ climate
A. Processed plant products				
E-kung (fermented bamboo shoot, fresh)	Year round	Aug – Sept	42.5±2.9	66.7±4.6
E-peng (fermented bamboo shoot, sliced fresh)	Year round	Aug – Sept	34.6±1.8	56.5±4.7
E-yup (fermented bamboo shoot, dried)	Year round	Sept – Oct	62.7±3.8	84.8±3.9
Peron-namsing (fermented soybean)	Year round	Jan – Feb	41.9±2.9	54.6±3.1
B.Processed animals prodcuts				
Ega-tapum (silk worm)	May – July	April – May	120.8±2.9	205.0±4.5
Engo (smoked fish)	Year round	June – July	252.4±2.5	404.80±5.4
Kebung (smoked squirrel)	Jan – April	March – April	250.5.0±4.5	350.4±5.8
Ngotar (sun dried fish)	Year round	April – May	150.8±3.7	305.6±3.9
Pumnger (caterpillars of *Delonix* spp.)	May – Aug	May – June	42.8±3.2	65.5±3.8
Tari (*Aspongopus najus*)	Dec-Feb	Nov – Dec	60.6±4.2	124.8±4.0
Tasum (smoked prawn)	Year round	June – July	82.0±4.7	154.5±5.9
C.Vegetables				
Akshap (*Mussenda roxburghii*)	Year round	May-Sep	43.0±3.5	64.6±4.2
Bangko (*Solanam spirale*)	Year round	May-Sep	45.0±2.7	75.60±3.8
Dhekia saag (*Diplazium esculentum*)	May-July	June-July	55.4±3.0	85.4±4.5
Gam oying (*Glochidion multilocular*)	April-Aug	May-July	44.5±3.9	69.8±3.9
Kopii (*Solanum torvum*)	April-Sep	May-July	60.5±1.9	84.7±5.4
Kopiir (*Solanum khasianum*)	April-Sep	May-July	62.8±2.7	92.7±6.2
Marsang (*Spilanthes acmella*)	Year round	March – April	35.5±4.0	59.4±2.1
Onger (*Zanthoxylum rhetsa*)	Year round	April – May	65.6±3.0	90.4±1.9
Ongin (*Clerodendrum colebrookianum*)	Year round	Nov – Dec	61.8±4.2	85.9±3.3
Tapar (indigenous mushroom)	April– July	June – July	22.3±3.8	34.5±3.7
Tapil (*Phoebe cooperians*)	Sept – Oct	Sept – Oct	65.4±4.1	119.4±4.5
D. Fruits				
Tadar (*Nephelium lappaceum*)	May – June	May – June	52.7±3.0	74.6±4.2
Anke (*Castanopsis kurzii*)	July – Aug	July – Aug	63.5±2.0	104.5±6.0
Sirang (*Castanopsis indica*)	July – Aug	July – Aug	60.0±1.8	119.5±3.5
Bureng (*Quercus latifolia*)	July – Sept	July – Aug	35.0±1.9	64.6±5.8
Belang (*Artocarpus heterophyllus*)	May-Sep	June-July	20.4±2.5	35.8±3.7

Note: US$ 1.0= Rs. 44.8

Data are taken from October 2002 to December 2008. Observations were taken from local market from Adi women during normal weather and extreme weather and disaster. Extreme weather and disaster were considered as drought and flood in Pasighat Valley caused by Siang River, and road and communication damaged at border area of Arunachal and Assam by which trade of food from north India and Assam is stopped.

In years 2000 in Pasighat valley flood was caused by Siang River originated in Tibet. Flood of 2003 in Pasighat valley was caused by erratic and heavy rainfall continuously for 45 days (personal observations).

Year 2002 and 2005 were almost drought years. Weather of 2006 and 2007 was very erratic like in beginning of April no rain but in June and July heavy rains
Other periods were of normal weather

Case studies of some notable CBAPs

First case study: Paisang (Quercus griffithi) and conservation of indigenous crops varieties for food security

Conservation of Paisang (*Quercux griffithi*) by *Monpa* tribe over a long period of time in west Kameng district has been possible because of its close association with the culture, agriculture, social institutions and compatibility of use in fragile ecosystem. Most Monpa farmers live in rainfed ecosystem and their economy is subsistence. Rainfed attributes of local ecosystem does not allow majority (92%) of Monpa farmers to use hybrid and improved varieties of crops and applying any inorganic fertilizer. Farmers depend on dry leaves of Paisangto use as mulch and compost in agriculture. Though, pine leaves are also used, however, more importance is given to Paisang.

Paisang is a deciduous woody perennial tree found in the sloppy hilly terrains. The leaf fall in this tree starts from last week of January and continues up to last week of February. Traditional importance of Paisang has inculcated effort of Monpa farmers to conserve it in the form of community forest, and managed collectively. Dry leaves of Paisang and pine trees as mulch material to conserve soil and water, and cultivate more than 48 indigenous crops and local varieties including, wheat and barley as a staple food crop.

The seeds of local varieties of maize (fenthina-dwarf variety, duration 3 months, thinasheru- tall variety, duration 5 months, and baklangboo-medium tall variety sown in festival Lohsar (January to February, duration 4 months) are spread in the fallow land and then ploughed using the bullock drawn local plough. Fenthina is grown in most fertile soils near kitchen gardens, thinasheru is grown in flat and main agricultural land where soil is brown with undulating land. Baklangboo is cultivated in gentle slope and jhum land. Less quantity of dry leaves of Paisang are used in black soil than in light textured and undulating lands. Improvement of soil fertility and total biomass from the local varieties of maize was recorded 30 to 40 per cent more in the land where Paisang leaves were applied as compared to the fields without Paisang practice.

If maize is grown after using the dry leaves of Paisang as natural mulch, then there is a better opportunity to increase cropping intensity by diversifying cropping systems. Such cropping systems are maize with indigenous varieties of black gram, soybean and rajmabean as mixed crops. Paisang leaves are also an integral part in the sole cropping of local wheat, barley and other about 35 landraces and varieties. In last 20 years, due to reduction in rainfall and warmer weather as perceived by Monpa, they have started adding more leaves and compost prepared from paisang in order to sustain soil moisture and crops' yield.

In rainfed agroecosystems of Dirang, we explored total 41 adaptive indigenous agrobiodiversity from Mrs. Pem Dolma (60) (Table 9). These crops species requires either no water or very least use of water and are cultivated using Paisang leaves. Most of these crops species and varieties are cultivated on natural rains, and are disease and insect pests tolerant. Though, productivity of each species is comparatively less than improved and hybrid varieties, but they provide insurance of food security without any failure during the environmental stresses.

Table 9. Indigenous agrobiodiversity of rainfed agroecosystems conserved by late Mrs. Pem Dolma

Local name	Botanical name	Special attributes
Ada	*Zingiber officinale*	Grown with least use of water
Amaranthus	Amaranthus *caudatus*	Grown without water
Bathua	*Chenopodium album*	Grown without water
Bong with awns	*Hordeum vulgare*	Drought tolerant
Bong without awns	*Hordeum vulgare*	Drought and disease tolerant
Broomsa peela	*Cucurbita moschata*	Disease tolerant and cultivated with least use of water
Broomsa saphed	*Benincasa hispida*	Disease tolerant and cultivated with least use of water
Bundagmo bada	*Panicum psilopodium* var. *coloratum*	Drought and disease tolerant
Bundagmo chhota	*Panicum psilopodium* var. *psilopodium*	Drought and disease tolerant
Chong	*Allium cepa*	Disease tolerant and can be cultivated with least use of water
Chukandar	*Beta vulgaris*	Disease tolerant and can be cultivated with lest use of water
Haldi	*Curcuma longa*	Disease tolerant and can be cultivated with least use of water
Kaibandu	Cucurbitaceous vegetable	Drought tolerant
Katili chaulai	*Amaranthus spinosus*	Drought tolerant
Lai Saag	*Brassica* sp.	Disease tolerant and can be cultivated with least use of water
Lal saag	*Amaranthus tricolor*	Drought and disease tolerant
Lau	*Lagenaria siceraria*	Drought and disease tolerant
Lee bada	*Glycine max*	Drought and disease tolerant
Lee choota	*Glycine max*	Drought and disease tolerant
Leme	*Brassica* sp.	Disease tolerant and can be cultivated with least use of water
Local gehun	Triticum aestivum	Drought and disease tolerant
Mandua	*Elcusine coracocna*	Drought and disease tolerant
Mann bada	*Allium* sp.	Disease tolerant and can be cultivated with lest use of water

(Contd.)

Mann Chhota	*Allium* sp.	Disease tolerant and can be cultivated with least use of water
Manthong	*Cucumis sativus*	Drought and disease tolerant
Matar	*Pisum sativum*	Drought and disease tolerant
Mongoicha	Solanum nigrum	Drought tolerant
Monpa lehsun lamm		Drought and disease tolerant
Penche	*Brassica* sp.	Disease tolerant and can be cultivated with least use of water
Phaphda meetha	*Polygonum fagopyrum*	Drought and disease tolerant
Phaphda teeta	*Polygonum fagopyrum*	Drought and disease tolerant
Rajma	*Phaseolus vulgaris*	Drought and disease tolerant
Sakarkand	*Ipomoea batatas*	Drought and disease tolerant
Sem	*Lablab purpureus*	Drought and disease tolerant
Solu	*Capsicum* sp.	Drought and disease tolerant
Suran	*Amorphophallus campanulatus*	Drought and disease tolerant
Taktak	Spinacia oleracea	Disease tolerant and can be cultivated with least use of water
Tamatar	*Solanum lycopersicum*	Drought and disease tolerant
Timbur	*Zanthoxylum americanum*	Drought and disease tolerant
Ush	*Coriandrum sativum*	Disease tolerant and can be cultivated with least use of water

The second case study: Ecosystem diversity and keystone biodiversity: climate change may threat to hunting culture based adaptation practices

It could be learned from Adi tribe that Dekang (*Gymnocladus burmanicus* C. E. Parkinson) and Emo (*Aconitum ferox*) are two keystone species that are found in two different ecosystems. Dekang is found at alleviation from 160 to 350 meters SL in subtropical climate (Singh *et al.* 2009), while Emo is found beyond 1200 meters SL. The pods of Dekang are used to trap deer and boar during hunting, while Emo is used in making poison to put as paste on arrow for hunting. Importance of these two keystone species have made them culturally most valuable and are being bartered in between ethnic groups (Padam, Pasi, Minyong and Pangi) of Ad itribe. Hunters make their hunting hut in the community forest nearby Dekang trees from the months of November till December. Due to the foul smell of Dekang's pods, dropped nearby the trees, deer and boars get attracted there and are being hunted by the Adihunters. Meat of deer and boar are integral part of food system of Adi tribe during festivals (Solung, Etar and Aran) as well as for the whole year.

The dung of deer and boar surrounding the Dekang tree helps in natural growing of indigenous mushroom species which are used as food and collected by the Adi women from April to May. Mushroom species are consumed, and in case of surplus amount, these are sold in the local market for the income generation also. Elder hunters revealed that 40 years back the flowering period of Dekang was last week of April to middle of May, but since last 10 years the flowering of Dekang is being noticed gradually pre-ponning in last week of March to middle of April. The average productivity of pods from a matured enough tree of Dekang in last seven years (from 2003-2009) recorded 13. 1±1. 3 kg. The elder hunters remember that about 30 years back the pod productivity of Dekang was on an average 20-25 kg. tree^{-1} as remembered by Mr. Tate Jamoh (65), a great hunter from Sibut village (June, 2007) .

It means after a 30 years of duration, productivity is reduced by 5 kg. Due to this change, the number of deers and boars are less attracted towards the Dekang tree and thus affecting the total meat to be hunted for the year round use specially during festivals and extreme weathers. A single hunter while staying of 15-20 days in the community forest may hunt 3-4 deer and 4-6 boar.. Forty years back, number of hunted deer and boar was average 5-7 and 10-12, respectively around Dekang trees.

In recent last 10 years, it is being noticed by the hunters that during November and December which are considered most dry period, the unusual and frequent erratic rainfall affect the quality of pods- by providing environment for fungal infection on pods dropped near the trees. The weather anomalies compounded with fungal infection could thus be affecting foul smell of Dekang pods. Therefore, this minimizes attraction of wild animals towards Dekang tree. Due to reduction in fruits dropping per tree and thereby reduction in the attraction of deer and boar population around Dekang, the frequency of hut is increasing gradually by the hunters in order to increase amount of hunts. As a result, the conflict may also occurs among the hunters about ownership of hunting points and hunting deers and boars, as reported by Mr. Litin Jamoh (50) from Yagrung village (May 2008).

In March 2008 it was noted that out of 22 trees, 10 trees were affected from the trunk borers and fungal infection, which might affect the total productivity of a Dekang tree. These biological problems are being noticed by the hunters too in last 6-7 years. Though, hunters use to rub the affected part of trunk with cow-dung and organic soil to remove the infection and trunk borers, the problem remains there and thus can affect survival of Dekang trees. Looking to biocultural and environmental values of Dekang tree, in one of the recent historical action, Bane Kebang (village level) and Bango Kebang (regional level) have passed a protocol and norms for its sustainable use and conservation in community forest and homegardens (Singh and Mukherjee 2008).

The rhizome of Emo (Aconitum ferox) is used in hunting the wild animals by the Adi hunters. The Emo is an integral part of hunting, food chain and livelihood of Adi tribe. The ecosystem where Emo is found is attributed with snow cover, without much vegetation, steep slope and rugged mountainous terrain at around

1200 1600 m asl. The cultural group Padam holds right to collect Emo from Dishang mountain and barter it with other neighboring sub-tribes. The Emo is collected during November to January (winter season) from particular niches of Dishang. Collectors target particular niches covered 2-3 cm cover of snow and ice and dig out its rhizomes. The rhizomes are processed collectively by elder hunters of a village to make it poisonous and use in hunting. The perception of Emo collector revealed that, in last 10 years the snow fall at the peak of Dishang Mountain is reduced, and thus consequently the plant population of plant availability and its productivity is being noticed reducing. Emo collectors now need more travel in Dishang mountain and spent more time in searching and digging out the rhizomes as reported by them.

Thus, we found that Dekang and Emo naturally grown in two different ecosystems are interconnected to adaptation of hunting culture and food security of Adi in mountain ecosystems.

The third case study: adaptations on culturally and nutritionally rich traditional food preservation

Projects on series of recipe contests of traditional foods prepared from indigenous biodiversity organized among Adi women brought an attention about some changes in fermented foods and beverages due to perceived weather anomalies. Adi women experienced that in comparison to 20 years back, now fermentation and preservation practices on legumes, cereals and fishes needs more care and frequent monitoring to avoid food from spoiling. Majority over 67 percent women reported that methods of fermentation and preservation which were used earlier now need refinement. For example, 20 years back, women use to ferment their bamboo shoots after keeping it in Chonga (cylinder of green bamboo stem) and placing them in water streams of community forest. Chonga was left for over 6 months to one year in order to get desired taste and flavor of fermented bamboo shoots, but now this practice is no more in existence. Women mentioned the reasons that it is because of erosion of cultural knowledge among younger generation (91. 2%) and abrupt changes in rainfall pattern and unpredictable weather (67. 4%).

Women experienced that now sometimes fermented traditional foods (prepared from legumes, tuber, rice, oil seed, leaves and fishes) and traditional alcoholic beverages (Apong prepared from amkel variety of rice, tapioca and sugar beet) take either less time in summer or more time in winter than the actual required and prescribed time by the elder women required for these foods. Further, to confirm the effect of weather anomalies on these foods, we carried out laboratory experiments on selected fermented foods and beverages presented below:

Effect of low temperature and ultraviolet radiation on the population of yeast in selected traditional fermented foods and beverages of Adi tribe

Experiment showed that concentration of yeast differs significantly among all the traditional food products that range from 5-8 log CFU gm^{-1}. The general trend of yeast load in the fermented food products was as Namdung (Pyrillaocimodes- local oil seed base), Ngosing (ribi, gali, ngoying – indigenous fish species base), Esingenging (sugar beet base), Nongin (a base prepared from taker local variety of rice), Kala Apong (traditional alcoholic beverage prepared from Siye (traditional yeast table) and boiled local rice variety called Amkel). Yeast cells were adversely affected by low temperature (6°C) when incubated for 6 weeks for Kala Apong followed by Nongin, and Esing enging; while it proved to be stimulatory for Namdung and Ngosing. A decline in yeast population was to the tune of 79 per cent on an average. Exposure to ultraviolet radiation for 5 min indicated maximum reduction of yeast population (up to 47%) and drastically affected the colony characters in Namdung and Ngosing among the fermented food products. Yeast, despite its mesophilic nature, positively responded to low temperature but exhibited population decline to a greater degree when exposed to UV ray for 5 and 15 min in Namdung and Ngosing.

Looking to abrupt changes in temperature and erratic rainfall, Adi women now manipulate in media culture- Siye (yeast tablet) required for preparation of Apongs, and appropriate changes in place and types of utensils required for fermentation. While, for fermenting legumes, oil seed, tubers and fish base products, women are still trying to formulate or retune some local adaptive practices. Few selected Adi women have also now been trained to follow proper retuned protocol of food fermentation and beverage preparation along with safety guidelines.

Plant and animal behaviour impacted directly or indirectly from climate change, and adaptation

Communities including farmers and hunters are experienced in observing the changes in local ecology and landscapes including few changes in indigenous plants species and behaviors of animals. For example abundance and availability of *Gymnocladus assamicus*, *G. burmanicus* and *Aconitum ferox*, which are part of food web and cultural identity of these communities, are affected due to perceived weather anomalies and environmental variability. Few species such as *G. burmanicus* got affected by stem borer and fungus. As Tompkins and Adger (2004) reported that climate change may leads to some of the hidden problems including epidemic, disease and other health issues, therefore, fish death in Siang River may be caused by compounded impacts of climate and local ecology.

Hunters stated that they are no more comfortable in easy hunt of wild animals. They have noticed many changes in forest ecosystems due to increasing landslides caused by erratic rainfall and other environmental variability, thus they are facing problems in hunting. The hunting practices of securing foods by both the communities living in mountain ecosystem are highly localized, and any significant climatic changes may further affect sustainability of plant and animals species which are already facing threats from various anthropogenic factors.

As Adi and Monpa both have delineated their concerns that changes in weather is caused by various social and ecological factors. The resilience of varying micro-ecosystems where jhum lands have been maintained by local communities for cultivating number of rainfed indigenous crops, are now facing threats due to a number of reasons including knowledge erosion, destruction in ecology through commercial crops and government policies (Fig. 4). The large scale farmers are converting their jhum lands into commercial crops field- like cultivating orange, pineapple, apple, kiwi fruits and number of exotic vegetables. This change has direct and indirect impact on a particular micro-ecosystem. This may affect adaptation practices being used by marginal farmers in varying micro-ecosystems. Such policies, as study indicates, sometimes could hamper the sustainability of entire ecosystem (Baletti 2011).

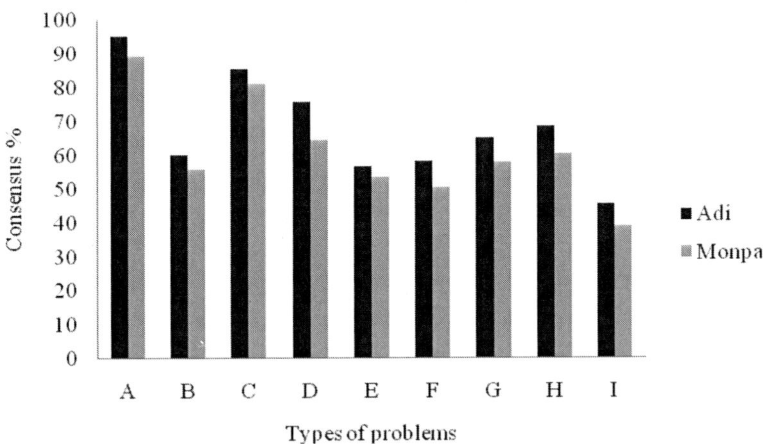

A= Erosion of traditional knowledge systems with modern practices; B= Weakening of social networks; C= Disappearance of indigenous institutions; D= Erosion in ecological ethics among younger population; E= Increasing rate of mono-cultivation of crops through subsidies given by government; F= Decreasing rate of diversity in livelihoods systems; G= Increased social mobility/ migration; H= Commercialization of resources

Fig. 4. Consensus percentage of *Adi* and *Monpa* communities about various problems that affect adaptation practices

Explored CBAPs do not necessarily mean that they are adapted in order to cope-up with extreme weather or climate change scenario after seeing the ongoing global debates on it; rather CBAPs are developed from time immemorial in order to survive under risk prone, fragile and harsh ecosystems of Arunachal Pradesh. As Agrawal (2008) and Byg and Salick (2009) reported that human adaptation to environmental risks, including those with climate change are especially necessary that are already ecologically stressed. Adi and Monp*a* both have been found to have adaptations for environmental risk and perceived climate changes through conserving location specific indigenous plants species and varieties, conserving community forests and integrate them with agriculture, diversifying farming systems and resource dependency, developing collective risk management strategies and creating social networks to spread risk. The other studies also revealed the similar adaptation practices from the world around adapted by vulnerable people (Pearce *et al.* 2010a; Ford and Pearce 2010; Singh *et al.* 2011).

Deciding level of vulnerability is a subject of debate. Vulnerability caused by climate change could be of different level, and low to medium level of vulnerability are of many times managed by local communities of Arunachal Pradesh (Singh *et al.* 2011). Perceived weather anomalies and changing climate in the regions have affected community and natural resources at different scale and time, accordingly the traditional farmers and local community have adapted their way of life and means of their livelihoods. Deaurden (2004), Eakin (2005), and Reid *et al.* (2009) have reported that CBAPs are important for immediate applications to reduce the percolation of climate change impact.

Study of O' Brien *et al.* (2004) shows that few parts of India have highest while few regions have lowest adaptive capacities to vulnerability of climate change, though this study does not clearly indicate the trend of vulnerability in Arunachal Pradesh. The other works of Chungand Ramanathan (2006) and Kumar *et al.* (2006) showed that most significant potential impacts of climate change on India are changes in the monsoon pattern with expected increases in mean monsoon intensity and variability. Most of the Indian population- especially marginal and poor including Arunachal Pradesh, is dependent on agricultural and forest resources for their livelihoods and food security, and any changes in climatic factors will make these communities more vulnerable (IPCC 1996, 2001; Panda 2009). The climate change- like increase of drought events has already been increasing in many parts of India including northeast India (Kumar *et al.* 1992) of which Arunachal Pradesh is a part.

Using leaves of *Quercus griffithi* to sustain crops in Dirang region is one of the classical model of adaptation in rainfed ecosystem. Though, this adaptation is very specific to a particular micro-ecosystem and is governed by ecological, social and cultural factors, however, it contributes in food security and resilience of socio-ecological system (Singh *et al.* 2006). The question is that whether such adaptive practices, which are contributing to food security and carbon sequestration both could be a part of REDD^{+}. Since a large portion of forest of Arunachal Pradesh comes under community control and management, and as forests in Arunachal Pradesh leads to carbon storage to the tune of 1497 million tones (Sinha 2009), thus cumulative impact of carbon sequestration by forests in Arunachal Pradesh is most significant in India. The community forests, which also contribute in carbon sequestration other than adaptations, provide direct livelihoods to 26 tribes of state and other environmental services. The healthy forests, which are under community control can also become a part of climate change adaptation programme under REDD^{+} schemes with proper supervision and monitoring systems (Singh and Padung 2010) provided forest resources are not going to be governed centrally (Phelps *et al.* 2010). Such CBAPs attributed with conservation of large scale plant species collectively, could have synergistic contribution in adaptation and carbon sequestration both (Olsson and Jerneck, 2010).

As Adi women indicated that they face problem in making fermented foods and traditional beverages through use of their older adaptive practices, and it is because of climatic variability and weather anomalies as perceived by them. Sometimes Adi foods get spoiled on certain specific period, which once upon a time was perfect time. These perceived changes in weather pattern affect food storability, processing, consumption pattern and even the selling in local market for income generation. This could even affect demand and supply chain of bioresources used in making fermented foods and beverages, and ultimately it may pose threat to cultural food habits and conservation dynamics. Our experiment too have led to the knowledge that oil and protein based (fishes) fermented food like Namdung (*Pyrillaocimodes*- local oil seed base) and Ngosing (Ribi, Gali and Ngoying – indigenous fish species base), respectively may be more sensitive than carbohydrate based traditional fermented foods (esing enging-sugar beet base, Nongin- local rice base and Kala apong--local rice based alcoholic beverage) to climate change and solar UV-B radiations. Thus, promoting and preserving biodiversity in Arunachal Pradesh is a challenge (Bhowmik *et al.* 2008).

Rainfed crops and uncultivated plants which are part of adaptive cropping and framing systems, of both the communities, are highly localized and contextual to a social system. The differences in such CBAPs for such systems were observed on account of perceived variability in climates, diversity in ecosystems and social systems, cultural variability and food habits. Adi lives in subtropical ecosystem while Monpa lives in sub-temperate to temperate ecosystems and the typology of environmental risks of these ecosystems vary from each other (Bujarbarua and Baruah 2009). According to variance in climatic features, both the communities have different CBAPs. Practicing with jhum land itself is century tested and locally evolved adaptation practice in order to sustain in harsh and fragile ecosystems (Ramakrishnan 2007). Though, it does not mean that components of diversified cropping systems practiced through indigenous plants varieties and farming systems diversified through varying components are always non-vulnerable to high climatic impact, these can also be affected if high level of climate changes happen.

The crops species which are components of adaptations in rainfed ecosystem are having potential to cope-up with environmental variability and their genes responsible for making crops tolerant to drought can be deducted and used with further breeding programmes. Therefore, these crops species needs special conservation for adaptation packages, hence deserve to be deposited to plant gene bank of India. Initiative was taken and over 38 rainfed adaptive crops species explored from Dirang agro-ecosystems has been submitted in the name of Monpa community to the National Bureau of Plant Genetic Resources, New Delhi (Singh 2004). Objective was to assist *ex-situ* conservation of rainfed crops species, and also assisting for further research to develop crops varieties adapted to drought, and disease and pests stresses.

Though, we lack any vulnerability study in Arunachal Pradesh, but few examples can justify our argument. The examples could be floods of year 2000, 2003 and most recent case of a sever flood of year September 2010 in East Siang river basin and Pasighat valley. In the flood of 2010, several houses were washed away in Mossing village in the upper Siang district and large area of agricultural fields and forest lands were devastated, and thus many farmers and local got affected (Financial Express 2010). The tributaries and catchment of Siang River usually affect Pasighat valley and thereby livelihoods of Adi. According to CISMHE (2009), the annual water inflows show maximum in 2003, when the erratic rainfall was more as experienced by Adi. The analysis of maximum and minimum recent temperatures of 2009 for the catchments of Siang River (from Pangin to Kaying, Tato and Pugging ecological points) indicated that this was as 33. 63 and 17. 1, 30. 66 and 14. 1, 26. 17 and 11. 45, and 32. 31 and 16. 48, respectively (CISMHE 2009). Such variability in temperature in mountain ecosystem from where Siang flows makes different micro-ecosystems. These micro-ecosystems are the landscape of agricultural fields and community forest of Adi community and might be affected and regulated by extreme weather. The similar situation may arise with Monpa also, but with different types of problems relating to rainfed ecosystem.

Social capitals and informal institutions were found to be effective in adaptation and securing food security during environmental risks, even in some extreme cases such as during drought, earthquake and moderate level of flood. For example, frequent landslides are common phenomenon in ecosystems of Monpa community. These landslides are, though, managed by state and central government agencies, however, social capital of Monpa community play pivotal roles in those areas where government agencies can reach immediately. Historical analysis with Adi elder women indicated that, an earthquake of 1951 has caused large-scale damaged of agricultural fields through landslides and flood in lower areas of Pasighat valley as well as in West Siang district too (some of the studied Adi women have native place in West Siang district) through the blockage of water streams and small river tributaries. The Adi community during this period faced severe food crisis. Though, this ecological crisis have been managed by government supports through providing foods and shelters, but adoption of social capital for sharing surplus food with each others, forming Reglep (informal institution) to mange agricultural field and collect food from community forests was equally effective. The male folk could contribute in controlling land erosion of village boundary and jhum fields through collective labor applied in stone pitching, replanting bamboo gardens, Toko-patta (*Levistona jenkinsiana*) gardens and Ogjok (*Bauhinia variegata*) as living fence on the boundaries.

The resources use scale and adaptation practices vary during extreme climate. For instance, in case of drought, the adaptation practices of accessing food resources become larger and energy consuming job. As an example, we found that Adi male and female both access and harvest certain community forests during drought for food resources (Table 10) from far away of ecological edges, and need more travel and time to hunt and collect food plants. During flood, especially when wetland paddy and availability of local fishes are affected in lower valley of mountain, then plant and animal based wild food resources become more important. The effect of flood on availability of bioreosurces used in foods at lower and middle levels of elevations is caused through landslides and severs soil erosion. During moderate level of floods, Adi women who live at higher alleviation could have chance to harvest plant and aquatic bioresources from water streams located in jhum lands and community forest where flood does not affect much. The bioresources harvested by this approach is used for own as well as traded the surplus amount in local market as they become important during abnormal weather (Table 8). Variations in price of these food resources depend on the degree of impact of abnormal weather.

Table 10. Adicommunity managed mountain forest ecosystems famous for hunting major wild animals and harvesting major plant resources during drought

Name of mountain/ valley	Wild animals	Wild plants
Dirap mountain	Sibi, Wild boar, deer, monkey, bear, etc.	Tapil, anke, bureng, belang
Billong mountain	Boar, deer, pigeon, cavil cat, kebungs (squirrels), etc.	Tapil, anke, bureng, belang, tadar (indigenous mushrooms species)
Partung valley	Deer, swan, birds, Kebungs (squirrels species) etc.	Tapil, anke, bureng, belang, tadar (indigenous mushrooms species)
Sirki Kosing	Bear, monkey, wild boar, porcupine, peacock, fowl, hornbill, etc.	Tapil, taan, tapang, tapil,anke, bureng, belang
Yagrung mountain	Wild boar, deer, monkey, bear, etc.	Tapil, anke, bureng, belang, tadar (indigenous mushrooms species)

Note: These four stated community forest located at higher alleviations in Adi living areas are rarely accessed in normal weather. These community forests are located much away from villages, situated at higher stream of mountain and need hours foot journey. These are accessed mostly during drought.

At the time of climatic problems such as landslides, flood and drought, when supply of food resources from Assam State and other parts of India is totally stopped, and Arunachal get disconnected due to road blockage or landslides then market value of locally available food resources are hiked automatically due to higher demand by tribal and non-tribal both the communities. For example, the price of potato imported from north India used to be Rs. 10-12 (US$ 0.23-0.27) per kg in 2003, but in the same year during June and July, rate of potato has reached to Rs. 45-55 a kg (US$1. 05-1.24). The reason was that Pasighat valley got totally disconnected from rest part of India due to sever flood caused by erratic rainfall of 45 days.

Our findings indicated that most of the CBAPs were very genders specific, though women across the communities were at par than men for most of the adaptations (Fig. 5). Further, an exclusive analysis of women according to the age groups indicated that elder women were found to be higher in adaptation practices than middle and younger women (Fig. 6). Agrawal (2008) have reported that community based adaptations are governed by a number of factors such as age, gender, social institutions and other factors. Hence, this showed that CBAPs are gender specific and women got more roles to play among the tribal communities to minimize environmental risks and secure foods and livelihood security.

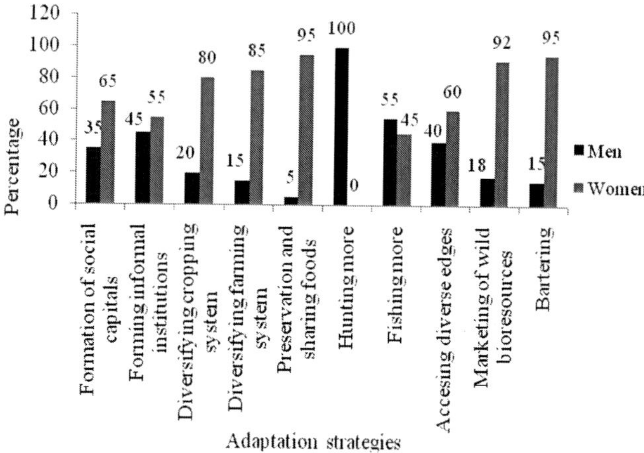

Fig. 5. Contribution percentage of male and female in adaptation strategy

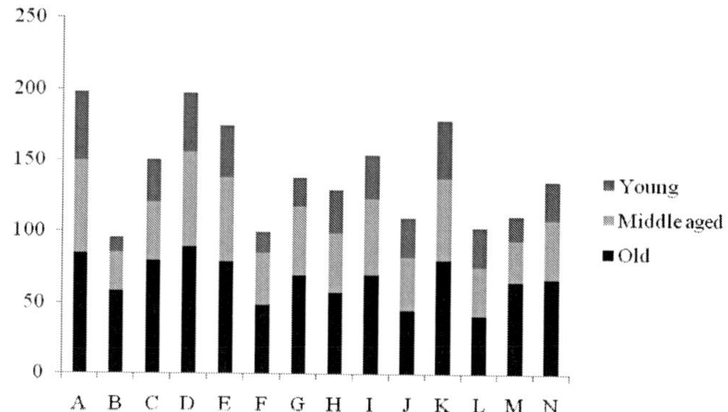

Age groups: Total women 492; Young (n= 130) = 25-40 yrs, middle aged ((n= 195) = 41-56 yrs and old aged (n= 167) = 57 and above

A= Increasingmore rainfed crops in jhum land

B= Domesticationof wild plant resources in homegarden

C= Diversifying food resources from various ecosystems

D= Preservation of surplus food resources through traditional practices

E= Strengthening bond with informal social institutions for sharing food resources

F= Sharing of food and related biocultural information through barter system

G= Collective responsibilities and plan to bear risk in collection and storage of food resources

H= Individual and collective conservation of seeds of plants for maintaining stock to grow next year

I= More fishing in community river

J= Asking males of family to hunt more meat from community forest for preservation

K= More harvesting of ethnobotanicals and food resources from community forest

L= Increase percentage on less preferred food resources

M= Teaching to young family members forsustainable use of food resources

N= Contingencyand long-term planning (at least for 6 months) for managing risk of foods

Fig. 6. Adaptation practiccs of women according to their age

Adaptation practices are affected by number of social, economic and ecological factors as explained by Adi and Monpa both (Fig. 3 and 4). Hence, in future in case of any sever environmental risk, the climatic change would be compounded with these factors, and community may become more vulnerable. As IPCC (1996, 2001) advocates to have some local solutions for adaptation, and for all the environmental problems, the policy makers and government do not have practices/technologies and solutions. Hence, locally available CBAPs could be rich source for it (Tompkins and Adger 2004), though some of ABAPs would require refinement and validation at low level, some at medium level and some at higher level. Because, all the CBAPs do not carry equal risk bearing capacity. Our SWOT analysis of explored CBAPs from both communities and discussed in workshops and seminars indicated that, apart from strengths and opportunity, there are also certain weaknesses and threats with each CBAPs (Fig. 7). The major weaknesses and threats were identified as CBAPs are highly specific to micro-ecosystems, more gender specific and in case of high climatic variability, few adaptive practices would be vanished out. Hence, strengths and opportunities with CBAPs are to be observed accordingly in order to use and promote them for future.

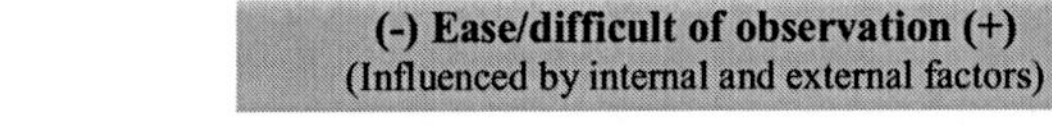

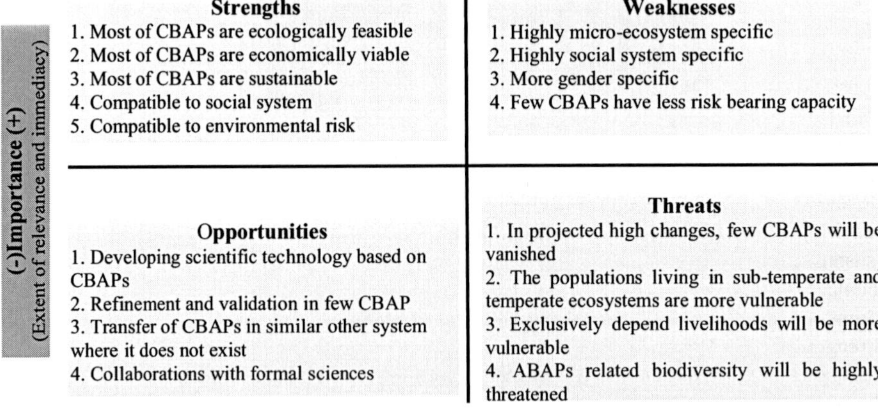

Fig. 7. SWOT (strength, weakness, opportunity and threat) dimension of community based adaptation practices (CBAPs)

According to level of perceived changes in local weather, communities under study have opined number of ways and means for adaptation through which weather anomalies and climate changes can be managed (Table 11), however, looking to fragility of ecosystems, sound planning and packages for Arunachal Pradesh need to be ready for coping up with any high and unpredictable changes in climate. Simultaneously, communities have also to be ready and willing to get networked with varying environmental institutions in order to have social learning, participate in co-management practices of adaptations (Tompkins and Adger 2004).

Farmers and local communities of state are having subsistence agriculture and other organic adaptation practices (Singh and Sureja 2006; Singh *et al.* 2006; Singh *et al.* 2010; Singh *et al.* 2011) in their survival strategies and thereby assumed that they contribute relatively less carbon emission than any developed state of India. However, perceiving climate change is a global challenge, and state government has framed policies on climate change adaptations and in carrying out a number of activities in lowering carbon emissions (Sinha 2009). This is a healthy political signal. Government is emphasizing to have sustainable adaptive measures and practices to conserve state forests, water bodies and mountains in order to be ready with any significant climatic stresses (Sinha 2009). However, the policies and activities on climate change adaptations need to be very specific and goal oriented to a particular problem/subject of climate change (Barnett 2010).

Table 11. Opinion ofAdi andMonpa members about the ways and means through which weather anomalies and climate change can be managed

Ways and means	Adi	Monpa
Improving conservation of plants and animals (wild and domesticated indigenous both) resources	70.8	56.8
Conserving water resources through community rivers, lakes, ponds and other traditional structures	89.4	79.5
Strengthening and promoting community efforts in local pitching, Gambian structures and land resources conservation practices	70.4	82.6
Developing integration and collaboration between community and environmental conservation agencies for exploring locally feasible and sustainable practices to combat with environmental problems	56.8	62.4
Inculcating interest among younger people to rely on traditional values of life	66.8	58.9
Maintaining eco-cultural practices through kincentric ecology	59.7	65.7
Strengthening socio-ecological network between peoples of different ecological edges	87.3	78.7
Strengthening and sustain social capitals	74.8	80.3
Promoting local practices and innovations	85.6	72.4
Maintaining existed belief and taboos associated to environments	57.4	68.2
Empowering local women since they contribute in environmental management	44.3	57.2

* Multiple percentages

Conclusion

From foregoing study it has been concluded that communities are aware about the weather anomalies and climate changes. Though community do not use the exact term of climate change, but they define climatic variability through correlating historical facts and types of problems correlated with environment, ecology and local livelihoods. They define and perceive climatic variability or anomalies through the notations and experiences on plants and animal behaviors. Women got more knowledge and skill to develop community based adaptive practices than men through various social, ecological and institutional processes. Adaptation practices vary from one social system to another based on types of micro-ecosystems and typology of environmental risks. Apart from, different level of capacity to absorb socks of environmental risk, the adaptation practices for food and livelihood security of both the communities is very much gender and age specific.

Not necessarily all the adaptive practices are able and would be applicable to cope-up with high environmental risks, however, few adaptive practices got logically considerable capacity such as use of Paisang tree leaves, cultivating rainfed crops species and varieties, and using plant and animals species from varying ecosystem for food security. These adaptations may be strengthened and promoted. Adi and Monpa have subsistence economy, living in fragile and harsh ecosystems; they share their resources collectively and have communal institutions in managing local forest resources. The community forest could be a part of REDD$^+$ with proper incentives and monitoring with community members (Singh and Padung 2010). The sound practices and social capitals of community can be used in future adaptations planning and policies. This will be important to state government who is trying to develop sound planning and polices on climate change (Sinha 2009) provided these policies are honoring and integrating notable CBAPs into package and ensuring to implement in participatory mode. Further, vulnerable communities needs the locations specific capacity building training to be ready for varying level of climatic changes.

Moreover, community based adaptive practices specially shown in case studies provide lessons and insights to scientists for future work on those practices for their refinement, up-gradation, quantifications (Ford *et al.* 2010), and more importantly collaborations between scientists and local communities to work on co-production of adaptive knowledge (Pearce *et al.* 2009). This is required for not only to cope-up future environmental risks but inclusive sustainable development too. As Weather head *et al.* (2010) reported that connecting indigenous knowledge and science is useful and valuable, but it would be more imperative for the scientists to use insights from community based adaptive practices to develop hypothesis, devise models and have the practical applications. Scientific models and measurement are becoming more and more accurate to record variability and climatic changes and predictions. Nevertheless, climate change is also a social, economic and cultural phenomenon, and to have sustainable adaptations, the perception, experiences and local knowledge of communities are much relevant depending upon their strength and capacity to deal with types of climatic problems (Byg and Salick 2009).

Acknowledgements

This article is dedicated to my most adorable son Mr. Anshumaan Singh (3. 2 years) and my lovely wife Mrs. Anamika Singh for their moral support, and who scarified my absence and responsibilities for them while being working on this paper at School of Natural Resources and Environment (SNRE), University of Michigan, Ann Arbor, USA. Author (RKS) express his gratitude to all the community members of Adi and Monpa communities who welcomed, cooperated and provided opportunity to learn with them. The help of Mrs. Orik Ralen, Dr. Egul Padung, Late Mrs Pem Dolma, Mr. Darge Tsering and Mr. Lobsang are gratefully acknowledged. The experimental analysis of fermented foods and beverages ofAditribe by Dr. S.N. Bhowmik- being a co-investigator of a joint project with RKS on traditional foods of Adi tribe is thankfully acknowledged. Most of the research works reported under this study was sponsored by National Innovation Foundation Ahmedabad, Central Agricultural University, Imphal and Northeastern Regional Council Shillong, India. Author RKS acknowledge Fulbright Commission in India, New Delhi for awarding Fulbright Fellowship and getting opportunity to work at SNRE, UM, USA.

References

Adger WN (2003a). Social aspects of adaptive capacity to climate change. In: Haque S, Smith J, Klein RTJ (eds) Climate Change, Adaptive Capacity and Development. Imperial College Press, London, pp. 29-50.

Adger WN (2003b). Social capital, collective action and adaptation to climate change. Economic Geography 79:387-404.

Agrawal A (2008). Role of local institutions in adaptation to climate change. Paper presented in conference on social dimensions of climate change. Social Development Department, The World Bank Washington DC.

Baletti B (2011). Soya scrutiny: A partnership to encourage sustainable farming in Brazil may not be as green as it seems. Nature, 472:5-6.

Barnett J (2010). Adapting to climate change: Three key challenges for research and policy–an editorial essay. WIREs Climate Change 1:314-317.

Bhattacharyya A, Sharmal J, Shah SK, Chaudhary V (2007). Climatic changes during the last 1800 yrs BP from paradise lake, Sela Pass, Arunachal Pradesh, Northeast Himalaya. Current Science 93(7):983-987.

Bhowmik SN, Singh RK, Bhardwaj R (2008). Variation in yeast population in some traditional fermented foods and beverages of *Adi* Tribe of Arunachal Pradesh. In: Singh RK (ed), National seminar-cum-workshop on Enhancing the traditional knowledge and grassroots innovations, exploration, consensus on prior informed consent and knowledge networking among various stakeholders, 29-30th March, College of Horticulture and Forestry, Pasighat, India.

Bujarbarua PS, Baruah S (2009). Vulnerability of fragile forest ecosystem of North East India in context with the global climate change: An ecological projection. Climate Change: Global Risks, Challenges and Decisions. IOP Conf. Series: Earth and Environmental Science 6 (2009) 072016:1-4.

Byg A, Salick J (2009). Local perspectives on a global phenomenon—Climate change in Eastern Tibetan villages. Global Environmental Change 19(2):156-166.

Chung CE, Ramanathan V (2006). Weakening of north Indian SST gadeinets and the monsoon rainfall in India and Sahel. Journal of Climate 19:2036-2045.

CISMHE (2009) Hydrometeorology. Environmental Impact Assessment – Hydrometeorology.

Climate Himalaya Initiative News (2010) Brahmaputra- The dying river. Climate Himalaya Initiative News.

Eakin H (2005). Institutional change, climatic risk, and rural vulnerability: Cases from Central Mexico. World Development 33(11):1923-1938.

Financial Express (2010). Flood of Pasighat Arunachal Pradesh, The Daily News paper, Financial Express, 11th September, 2010.

Ford JD, Keskitalo ECH, Smith T, Pearce T, Berrang-Ford L, Duerden F, Smit B (2010). Case study and analogue methodologies in climate change vulnerability research. *WIREs* Climate Change, 1:314-317.

Ford JD, Pearce T (2010). What we know, do not know and need to know about climate change vulnerability in western Canadian Arctic: A systematic literature review. Environmental Research Letter 5:1-9.

Ford JD, Smith B (2004). A framework for assessing the vulnerability in Canadian arctic to risk associated with climate change. Arctic 57:389-400.

Glantz M (1996). Currents of change: EI Ninos impact on climate and society. Cambridge University Press, Cambridge.

ICUN (2005). Medicinal plant conservation. The medicinal plant specialist group of the IUCN Species Survival Commission.

IPCC (1996). The regional impact of climate change: An assessment of vulnerability. Special report on regional impact of climate change.

IPCC (2001). Climate change: The scientific basis. Contribution of working group I to the third assessment report of the IPCC.

Inergovernment Panel on Climate Change (IPCC) (2007). Climate change 2007. The physical science basis. Summary for the policy makers. Contribution of Working Group I to the Fourth Assessment report of IPCC.

Kumar KK, Rajagopalan B, Hoerling M, Bates G, Cane M (2006). Unraveling the mystery of Indian monsoon failure during EI Nino. Science 314:115-119.

Kumar KR, Pant GB, Parthasarathy B, Sontakke NA (1992). Spatial and sub-seasonal pattern of long-term trends of Indian summer monsoon rainfall. International Journal of Climatology 12(3):257-268.

Lantz T, Turner NJ (2003). Traditional phenological knowledge of Aboriginal Peoples in British Columbia. J Ethnobiol 23(2):263-286.

Myers N, Mittermeier RA, Mittermeier CG, Da Fonseca GAB, Kent J (2000). Biodiversity hotspots for conservation priorities. Nature403:853-858.

NJC Hydropower Limited (2011). Baseline setting for physic-chemical aspects, EIA study for Nyamjangchhu hydroelectric project

O' Brien KL, Leinchenko R, Kelkar U,Venema H, Andahl G, Tompkins H (2004). Mapping vulnerability to multiple stresses: Climate change and globalization in India. Global Environmental Change, 4 (4):303-313.

Olsson L, Jerneck A (2010). Farmers fighting climate change—from victims to agents in subsistence livelihoods. *WIREs* Climate Change 1:363-373.

Panda A (2009). Assessing vulnerability to climate change in India. Economic and Political Weekly, XLIV(16):105-107.

Pearce T, Ford JD, Laidler GJ, Smith B, Duerden F, Allarut M, Andrachuk M, Baryluk S, Andrew Dialla A, Elee P, Goose A, Ikummaq T, Joamie E, Kataoyak F, Loring E, Meakin S, Nickels S, Shappa K, Shirley J, Wandel J (2009). Community collaboration and climate change research in the Canadian Arctic. Polar Research 28:10-27.

Pearce T, Smith B, Duerden D, Ford J, Goose A, Kataoyak F (2010). Inuit vulnerability and adaptive capacity to climate change in Ulukhaktok, Northwest territories, Canada. Polar Record 46:157-177.

Phelps J, Webb EL, Agrawal A (2010). Does REDD+ threaten to recentralize forest governance?. Science, 328: pp. 312-313.

Rai SK, Chakesang KC (2009). Impact of climate change on traditional agricultural practices: Cases from eastern Himalayas of India.

Ramakrishnan PS (2007). Sustainable mountain development: The Himalayan tragedy. Current Science 92(3): 308-316.

Reid H, Alam M, Berger R, Cannon T, Huq S, Milligan A (2009). Community-based adaptation to climate change: An overview. Participatory Learning and Action (PLA), Community based adaptation to climate change issue, 60: IIED, London, UK. pp. 13-38.

Salick J Fang Z, Byg A (2009). Eastern Himalayan alpine plant ecology, Tibetan ethnobotany and climate change. Global Environmental Change1:137-139.

Salick J, Ross N (2009). Traditional peoples and climate change: Introduction. Global Environ Change 19:137-139.

Singh RK (2002). Performance of soybean production technology in varying micro-farming situations assessed by resource-rich and resource-poor farmers. Ph. D. Dissertation, Jawaharlal Nehru Krishi Vishwa Vidyalaya, Jablpur, MP, India.

Singh RK (2004). ARIS-1455, E-200440006Z04, Indian collection numbers (IC No. 427030 to 427054) for the submitted folk varieties from Dirang, West Kameng district, Arunachal Pradesh, India (NBPGR, New Delhi), DIR,PA,D, 151, 4224.

Singh RK, AdiWomen (2010). Biocultural knowledge systems of tribes of eastern Himalayas. NISACIR, CSIR, New Delhi.

Singh RK, Sureja AK (2006). Community knowledge and sustainable natural resources management: Learning from *Monpa* tribe of Arunachal Pradesh. *T. D:* The Journal Transdisciplinary Research in Southern Africa2(1):73-102.

Singh RK, Bhowmik SN, Pandey CB (2011). Biocultural diversity, climate change and livelihood security of *Adi* community: Grassroots conservators of eastern Himalaya Arunachal Pradesh. Indian Journal of Traditional Knowledge, 10(1):39-56.

Singh RK, Mukherjee TK (2008). Community Based Sustainable Natural Resources Management and Development in Northeast India. Proceedings of national seminar held at College of Horticulture and Forestry, Central Agricultural University, Pasighat on 26-27th April.

Singh RK, Pretty J, Sarah P (2010). Traditional knowledge and biocultural diversity: Learning from tribal communities for sustainable development in northeast India. Journal of Environmental Planning and Management 53 (4):511-533.

Singh RK, Singh D, Sureja AK (2006). Community knowledge and biodiversity conservation by *Monpa* tribe. Indian Journal of Traditional Knowledge, 5(4):513-518.

Singh RK, Srivastava RC (2009). Biocultural Knowledge and *Adi* community: conservation and sustainability in biodiversity hotspot of Arunachal Pradesh. Current Science 96(7): 883-884.

Singh RK, Srivastava RC, Mukherjee TK (2009). Culturally important *dekang* (*Gymnocladus burmanicus* C. E. Parkinson): An addition to the flora of India from Arunachal Pradesh. Indian Journal of Traditional Knowledge 8(4):482-484.

Sinha GN (2009). Indian Himalayas: Glaciers, climate change and livelihoods. Paper presented in The Himalayan Chief Ministers Conclave, organized by Department of Environment, Science and Technology, Government of Himachal Pradesh and LEAD India, New Delhi, from 29-30 October, 2009.

The India Post (2009). Himalayan States pass declaration on climate change, Daily News, The India Post, Oct 30, 2009. URL:http://theindiapost. com/2009/10/30/himalayan-states-pass-declaration-on-climate-change/), 01-04-2011.

The Sentinel (2011). State Action Plan on Climate change soon. http://www. sentinelassam. com/arunachal/story. php?sec=2&subsec=7&id=62467&dtP=2011-02-01&ppr=1, The Daily, January, 31, 2011.

Tompkins EL, Adger WN (2004). Does adaptive management of natural resources enhance resilience to climate change? Ecology and Society 9(2):10

Turner II BL, Kasperson RE, Matson PA, McCarthy JJ, Corell RW, Christensen L, Eckley N, Kasperson JX, Luers A, Martello, ML, Polsky C, Pulsipher A, Schiller A (2003). A framework for vulnerability analysis in sustainability science. Proceedings of National Academy of Sciences, 100 (14):8074-8079.

Turner N, Clifton H (2009). It is so different today: Climate change and indigenous lifeways in British Columbia, Canada. Global Environ Change19:180-190.

Turner NJ, Singh RK (2011). Traditional knowledge in disaster prediction/ forecasting, management and climate change. Indian Journal of Traditional Knowledge 10(1):3-8.

3

Agroforestry: A Sustainable Land use System for Livelihood Security and Climate Change Mitigation

P. Saikia[1], Amit Kumar[1] and M. L. Khan[2]

[1]*School of Natural Resource Management, Central University of Jharkhand Brambe-835205, Ranchi, Jharkhand, India*
[2]*Department of Botany, Dr. Harisingh Gour Central University Sagar - 470003, Madhya Pradesh, India*

Introduction

Agroforestry is the intentional combination of agriculture and forestry technologies to create integrated, diverse, productive, profitable, and sustainable land-use systems (Rietveld 1995). It creates complex systems with impacts ranging from the site or practice level up to the landscape and beyond (Ellis, 2004). It is one of the most conspicuous land use systems that consists of annual and perennial plants, which are often integrated with livestock. It provides ecosystem services and reduces anthropological impacts on natural forests, arrests soil degradation, enhance soil fertility in many situations and improve farm resilience (Thangataa and Hildebrand 2012; Tewari *et al.* 2013). Agroforestry is a land management and farming system that are not only capable of fulfilling household needs but also maintaining and improving environmental quality. Itplays a vital role in achieving integrated rural and urban development. The changing climate has potential impacts on ecosystem goods and services by means of increased variability with greater risk of extreme weather events, such as prolonged drought, storms and floods (Lindner *et al.* 2010). Agroecosystem and forests are the ecosystems which are the most adversely affected by climate change by means of prevalence of pests, diseases, invasive species, species endangerment and high levels of food insecurity. The adoption of agroforestry reduce the impacts climate change by increasing tree cover outside forests, enhancing forest carbon stocks, conserving biodiversity, reducing risks and damage intensity, maintaining health and vitality, and scaling up multiple benefits (Zoysa and Inoue, 2014). Tree-based farming systems store carbon in soils and woody biomass, and they may also reduce greenhouse gas emissions from soils and often considered a cost-effective strategy for climate change mitigation (Verchot *et al.* 2007; Smith and Olesen, 2010). Trees have an important role to play not only in climate change mitigation but also in reducing vulnerability to climate-related risks. The value, role and contributions of agroforestry and the protection of endemic habitats, in the light of current global environmental challenges, cannot be overemphasized (Maathai, 2012). Agroforestry systems range from subsistence livestock silvo-pastoral systems to homegardens, on-farm timber production, all types of tree crops integrated with other crops and biomass plantations within a wide diversity of biophysical conditions and socio-ecological characteristics (Zomer *et al.* 2009). Agroforestry like most other natural resource management science, is characterized by high complexity of structure and function of which we have limited understanding (Sanchez 1995; Nair, 1998). Most of the agroforestry researches are focused on its potential to conservation of crop diversity, biodiversity, its carbon sequestration potential, soil fertility maintenance, biomass estimation, food and livelihood security, climate change adaptation and mitigation and its socio-economic prospects (Jensen 1993; Mercer and Miller 1998; Verchot *et al.* 2007; Zomer *et al.* 2009; Saikia and Khan, 2014). The use of geoinformatics in regular monitoring, potential site identification, precision farming is well established and thereby enhance the capability in decision making to achieve environmental protection and agricultural production goals. Therefore, this chapter emphasized to summarize the contribution of agroforestry in conservation of biodiversity, livelihood security, climate change mitigation and adaptation, soil quality maintenance etc. The role of geoinformatics technology in monitoring agroforestry system to achieve environmental sustainability and enhanced agricultural production to achieve livelihood security is also discussed.

Potential of agroforestry in conservation of biodiversity

Agroforestry systems practiced by rural poor are means for the environmental services such as biodiversity conservation, watershed protection and carbon sequestration. Traditional agroforestry systems significantly contribute to the conservation of biodiversity through *ex situ* conservation of tree species, reduction of pressure on remnant forests and the provision of suitable habitat for a number of animal and plant species including various rare endangered species like *Aquilariam alaccensis* Lam., *Livistona jenkinsiana* Griff. *Acorus calamus* L. etc. (Atta-Krah *et al.* 2004; Acharya 2006; McNeely and Schroth 2006; Saikia *et al.* 2012). The tree component of agroecosystems is particularly valued for specific roles including that of host plant to insects yielding marketable products such as silk (Singh *et al.*1994), lac products (Jaiswal *et al.* 2002) and honey (Dwivedi, 2001). It also provides shade, shelter, energy, food, fodder and many other goods and services that enable the agroecosystems to prosper (Leakey and Tchoundjeu 2001; McNeely and Schroth, 2006). Agroforestry systems contribute to biodiversity conservation by providing habitat for pollinators and seed dispersers that facilitate gene flow in other tree species (Slocum and Horvitz, 2000).

The greatest threat to biodiversity is land cover change-induced habitat destruction (Chapin *et al.* 2000), reinforcing the need for accurate maps of forest extent and change. Conservation planning is a complex task in which there is considerable uncertainty and often-competing objectives (Guikema and Milke 1999). It is becoming increasingly recognized that most data on biodiversity relates to small areas, while management and conservation activities typically operate at coarser, landscape scales and scaling between levels is difficult (Innes and Koch 1998; Griffiths *et al.* 2000; Negandra 2001). Farm level management decisions are mostly determined by the knowledge of the interactions among climatic and edaphic condition of the area, characteristics of crops and animals, technology, socio-economic factors and the institutional context including agricultural education, government policy and social customs (Oerke *et al.* 1994).

Livelihood security in agroforestry systems

Agroforestry systems play an important role in enhancing the productivity of lands to meet the demand of ever-growing human and livestock population. It has both productive and protective potential and provide opportunities for employment generation in rural areas. Dhyani *et al.* (2003, 2005) have highlighted the role of agroforestry to meet the subsistence needs of poor families and providing a platform for greater and sustained livelihood of the society. The major contributions made by agroforestry to the economy can be seen in terms of income, and employment generation. It is possible through combining food crops (fruits, vegetable, legumes, pulses, citrus fruits and edible medicines), timber crops and other economic crops with diverse products and benefits. Trees on agroforestry systems are important source of income and contributing to food security in difficult time. Multipurpose trees of traditional agroforestry systems are important for rural food security and income generation, and also for ensuring social and cultural stability to some extent (Boffa, 1999).

Role of agroforestry systems in climate change mitigation and adaptation

Vegetation have ability that can absorb and store CO_2, accumulating carbon in biomass of different parts thus contributes in maintaining the stability of climate.

It is depended on ecological system including wild species, the density of the vegetation, topography and other environmental factors (Ogawa *et al.* 1965; Senpaseuth *et al.* 2009). It is established that climate change will worsen the food security situation, with reduced yields, increased pest and disease attacks and extreme natural phenomena, such as floods and droughts (Kaimowitz, 2003). Agroforestry has a particular potential role in mitigation of atmospheric accumulation of greenhouse gases (IPCC 2000). Improving soil nitrogen through fertilization of crops and pastures increases N_2O emissions from soils and sometimes decreases the soil CH_4 sink (Steudler *et al.* 1989; Mosier and Delgado, 1997). High input of nitrogen and soil compaction can result in the reduction of sink strength of soils for CH_4 (Hansen *et al.* 1993; Palm *et al.* 2002). In agroforestry systems, where leguminous crops are managed to contribute nitrogen, there is little information on the amounts of N_2O produced or the effect on CH_4 consumption. Improved organic matter and flooding management in irrigated rice can decrease CH_4 emission from paddies (Wassman *et al.* 2000; Jain *et al.* 2000). Agroforestry systems are playing the greatest role in maintaining the resource base and, thus helping in building climatic resilient agriculture (Dhyani and Handa, 2014). Swaminathan (1983) has pointed out that biodiversity is the feed stock for a climate resilient agriculture. Agroforestry systems can potentially help farmers to adapt climate change mitigation through carbon sequestration (Luedeling *et al.* 2011). The destruction of forest resources by 20 percent results in the loss of carbon storage in xylems (Office of Environmental Policy and Planning 2000).

Soil quality management in agroforestry systems

Agroforestry systems are playing an important role in optimizing nutrient cycling, organic matter production and reducing a need for the external input of fertilizers (Handa *et al.* 2016). It has importance as a carbon sequestration strategy because of carbon storage potential in its multiple plant species and soil. A number of improved farming practices can increase the sustainability of farming systems and contribute to reducing farmers' vulnerability to climate variability while sequestering carbon from the atmosphere. Agroforestry systems can also have an indirect benefit on carbon sequestration when it help to decrease pressure on natural forests, which are the largest sinks of terrestrial carbon. The total carbon storage capacity of an agroforestry system depends on the growth and nature of the tree species, and varies from region to region (Newaj and Dhyani 2008). Tree species ameliorate soil by adding both above and below ground biomass into the soil system. However, variations do exist in the inherent capacity of different tree species in rehabilitating degraded lands. Tree species improved moisture retention capacity of soil as compared to the control. Protection of soils directly against erosive forces of raindrop and surface run off by improving physical and hydrological parameters of soil have been reported in many studies in India (Grewal and Abrol 1986; Deb *et al.* 2005). The adverse influences of widespread soil erosion on soil degradation, agricultural production, water quality, hydrological systems, and environments, have long been recognized as severe problems for human sustainability (Lal, 1998). However, estimation of soil erosion loss is often difficult due to the complex interplay of many factors, such as climate, land cover, soil, topography, and human activities. In addition to the biophysical parameters, social, economic, and political components also influence soil erosion (Ananda and Herath, 2003). Accurate and timely estimation of soil erosion loss or evaluation of soil erosion risk has become an urgent task (Lu *et al.* 2004).

Geoinformatics application in agroforestry research

The ever increasing population pressure has a significantly influenced land degradation, ecosystem resilience and sustainable soil and water use, which have substantial social and economic impacts. This led to the need to consistent monitoring of land surface dynamics in spatio-temporal framework within the context of sustainable land use. The satellite remote sensing provides cost-effective and feasible means of acquiring the necessary information about the environment condition within the spatial and temporal scales (Foody, 2003) and plays a major role in the provision of environmental indicators that may inform sustainable development and associated decision-making (Schultink 1992; Rao 2001; Chen, 2002). Although agroforestry, like most natural-resource management sciences, is characterized by high complexity of which we have limited understanding and data (Sanchez 1995; Nair 1998), the science and application of agroforestry can be greatly enhanced through the use of geoinformatics technology. The potential of remote sensing provides information pertaining to spatial extent of agroforestry, tree species diversity in agricultural land, land cover heterogeneity at the scale of landscape (Gould, 2000; Kerr *et al.* 2001; Oindo and Skidmore, 2002). A geospatial analysis of remote sensing derived datasets offer the relationship of tree cover, population density and climatic conditions within agricultural land. The multi-spectral and multi-spatial resolution satellite data enables in mapping and monitoring of local to global scale agroforestry methods and practices (Zomer *et al.* 2009). The spatio-temporal monitoring of agroforestry regions estimates above-ground carbon capture (Lu *et al.* 2002; Samaniego *et al.* 2009; Wang *et al.* 2011). Soil erosion estimation is being done with reasonable costs and better accuracy in larger areas (Millward and Mersey, 1999; Wang *et al.* 2003). The high resolution satellite images aid in assessment of trees species, diversity and quantification at parcel level, whereas the moderate resolution and high temporal resolution satellite offer near real time crop condition status at regional level and thus contribute in effective planning and implementations of agroforestry regime (Jeyaseelan and Kumar 2008; Arockraj *et al.* 2015). Using various spatial thematic layers like soil type, slope, land use/ land cover, groundwater potential, geomorphology in geospatial environment, suitable optimal locations for agroforestry practices can be identified to address community issues, such as water quality and wildlife habitat (Bentrup and Kellerman, 2003). The aspects of precision farming, tree and crop diversity, measures to arrest soil erosion can easily be determined using geoinformatics techniques and effective ground based implementation. However, a larger scale perspective and a multi-scale planning process is often required for community-driven goals to improve agroforestry practices (Rietveld and Francis, 2000).

Conclusion

Agroforestry has a high employment-generation potential in India. It offers opportunities for the improvement of the livelihood of poor people through provision of economic and environmental security. Non-timber forest products have been recognized as important resources for both sustainable livelihood and ecosystem conservation purposes. Rural communities have promoted conservation of biodiversity in their subsistence agricultural production systems and they are not only depend on wild plants as sources of food, medicine and fodder, but also developed methods of resource management, which may be important for the conservation of some of the world's important species. Agroforestry offers the potential to develop synergies between efforts to mitigate climate change, conserving rare plant species and to help vulnerable populations to adapt against the negative consequences of climate change. The geoinformatics technology offered an efficient opportunity in assessment of various aspects of agroforestry and contribute in policy making for sustainable development. We recommend following for future research:

1. Traditional agroforestry practices support species richness and provides evidence as biodiversity reservoirs which merit more research and development attention.
2. Optimum densities of plants to be maintained on agroforestry systems according to the farmers' socioeconomic conditions and the relative importance of plants in farmer livelihoods.
3. In agroforestry systems, care should be taken to avoid a monoculture in order to assure additional level of stability and resilience and minimize the chance of pest and disease outbreak.
4. In agroforestry systems management of more numbers of rare endangered plants with economic benefits may improve ecological structure of

agroforestry systems and ultimately, strengthen the conservation of the rare endangered species.

5. It is essential that research efforts using geoinformatics technology on these important cropping systems are intensified, so that future scaling-up of agroforestry can be rooted in robust scientific findings.

References

Acharya KP (2006). Linking trees on farms with biodiversity conservation in subsistence farming systems in Nepal. Biodivers Conserv 15:631–646.

Ananda J, Herath G (2003). Soil erosion in developing countries: a socio-economic appraisal. Journal of Environmental Management 68:343–353.

Arockraj S, Kumar A, Hoda N, Jeyaseelan AT (2015). Quantification and identification of tree species in open mixed forests using high resolution Quick Bird satellite imagery. Journal of Tropical Forestry and Environment 5(2):40-53.

Atta-Krah K, Kindt R, Skilton JN, Amaral W (2004) Managing biological and genetic diversity in tropical agroforestry. Agroforest Syst 61:183–194.

Bentrup G, Kellerman T (2003). Agroforestry and GIS: achieving land productivity and environmental protection. In: Proc. of the 8th North American Agroforestry Conference. Corvallis, OR. pp. 15-25.

Boffa JM (1999). Agroforestry parklands in sub-Saharan Africa. Food and Agriculture Organization of the United Nations, Rome.

Chapin FE, Zavaleta V, Eviner R, Naylor RL, Vitousek PM, Reynolds HL, Hooper DU, Lavorel S, Sala OE, Hobbie SE, Mack MC, Díaz S (2000). Consequences of changing biodiversity. Nature 405(6783):234-242.

Chen XW (2002). Using remote sensing and GIS to analyse land cover change and its impacts on regional sustainable development. International Journal of Remote Sensing 23:107–124.

Deb S, Tangjang S, Arunachalam A, Arunachalam K (2005). Role of litter, fine roots and microbial biomass in soil C and N budget in traditionally managed agroforestry systems. In: Bhatt BP, Bujarbaruah KM (eds.), Agroforestry in Northeast India: Opportunities and Challenges. ICAR Research Complex for NEH region, Umiam, pp. 491–506.

Dhyani SK, Handa AK (2014) Agroforestry in India and its Potential for Ecosystem Services. In: Dagar JC, Singh AK and Arunachalam A (eds), Agroforestry Systemsin India: Livelihood Security & Ecosystem Services. Advances in Agroforestry 10:345-365.

Dhyani SK, Sharda V, Sharma AR (2005). Agroforestry for sustainable management of soil, water and environmental quality: Looking back to think ahead. Range Management and Agroforestry 26(1):71-83.

Dhyani SK, Sharda VN, Sharma AR (2003). Agroforestry for water resources conservation: issues, challenges and strategies. In: Pathak PS, Ram N (eds) Agroforestry: Potentials and Opportunities.Jodhpur, India.

Dwivedi MK (2001). Apiculture in Bihar and Jharkhand: A study of costs and margins. Agricultural Marketing 44(1):12-14.

Ellis EA, Bentrup G, Schoeneberger MM (2004). Computer-based tools for decision support in agroforestry: Current state and future needs. Agroforestry Systems 61:401–421.

Foody GM (2003). Remote sensing of tropical forest environments: towards the monitoring of environmental resources for sustainable development. Int J Remote Sensing 24(20):4035–4046.

Gould W (2000). Remote sensing of vegetation, plant species richness, and regional biodiversity hotspots. Ecological Applications 10:1861 1870.

Grewal SS, Aborl IP (1986). Agroforestry on alkali soils: effect of some management practices on soil initial growth, biomass accumulation and chemical composition of selected tree species. Agrofor Syst 4:221–232.

Griffiths GH, Lee J, Eversham BC (2000). Landscape pattern and species richness: regional scale analysis from remote sensing. International Journal of Remote Sensing 21:2685–2704.

Guikema S, Milke M (1999). Quantitative decision tools for conservation programme planning: practice, theory and potential. Environmental Conservation 26:179–189.

Handa AK, Toky OP, Dhyani SK, Chavan SB (2016). Innovative agroforestry for livelihood security in India. World Agriculture 7-16.

Hansen S, Maechlum JE, Bakken LR (1993). N_2O and CH_4 fluxes in soils influenced by fertilization and tractor traffic. Soil Biol Biochem 25:621-630.

Innes JL, Koch B (1998). Forest biodiversity and its assessment by remote sensing. Global Ecology and Biogeography 7:397–419.

International Panel on Climate Change (IPCC) (2000). IPCC Special Report on Land Use, Land Use Change and Forestry. Summary for Policy Makers. Geneva, Switzerland.

Jain MC, Kumar K, Wassmann R, Mitra S, Singh SD, Singh JP, Singh R, Yadav AK, Gupta S (2000). Methane emissions from irrigated rice fields in Northern India (New Delhi). Nutr Cycl Agroecosys 58:75–83.

Jaiswal AK, Sharma KK, Kumar KK, Bhattacharya A (2002). Household's survey for assessing utilisation of conventional lac host trees for lac cultivation. New Agriculturist 13:13-17.

Jensen M (1993) Soil conditions, vegetation structure and biomass of a Javanese homegarden. Agrofor Syst 24:171-186.

Jeyaseelan AT, Kumar A (2008). Jharkhand Agricultural Information System - Satellite (IRS P6-AWiFS) based Crop Condition Monitoring during June, July, August, September and October 2008. Technical Project Report, JSAC-JAIS-2008/1-5, JSAC, Ranchi, Jharkhand, India.

Kaimowitz D (2003). Forest Law Enforcement and Rural Livelihoods. CIFOR, Bogor, Indonesia.

Kerr JT, Southwood TRE, Cihlar J (2001) Remotely sensed habitat diversity predicts butterfly species richness and community similarity in Canada. In: Proceedings of the National Academy of Sciences (USA) 98:11365–11370.

Lal R (1998). Soil erosion impact on agronomic productivity and environment quality: critical reviews. Plant Sciences 17:319–464.

Leakey RRB, Tchoundjeu Z (2001). Diversification of tree crops: domestication of companion crops for poverty reduction and environmental services. Exp Agric 37:279–296.

Lindner M, Maroschek M, Netherer S, Kremer A, Barbati A, Garcia-Gonzaloa J, Seidl R, Delzond S, Corona P, Kolström M, Lexer MJ, Marchetti M (2010). Climate change impacts, adaptive capacity, and vulnerability of European forest ecosystems. Forest Ecology and Management, 259(4):698–709.

Lu D, Li G, Valladares GS, Batistella M (2004). Mapping soil erosion risk in rondo nia, brazilianamazonia: Using RUSLE, remote sensing and GIS. Land degradation & development 15:499–512.

Lu D, Mausel P, Brondizio E, Moran E (2002). Assessment of atmospheric correction methods for Landsat TM data applicable to Amazon basin LBA research. Int J Remote Sens 23:2651–2671.

Luedeling E, Sileshi G, Beedy TD, Dietz J (2011). Carbon sequestration potential of agroforestry systems in Africa. In: Kumar BM and Nair PKR (eds) Carbon Sequestration Potential of Agroforestry Systems: Opportunities and Challenges.Advances in Agroforestry 8:61-83.

Maathai W (2012). Agroforestry, Climate Change and Habitat Protection. Agroforestry - The Future of Global Land Use. Advances in Agroforestry 9:3-6.

McNeely JA, Schroth G (2006). Agroforestry and biodiversity conservation—traditional practices, present dynamics, and lessons for the future. Biodivers Conserv 15:549–554.

Mercer DE, Miller RP (1998). Socioeconomic research in agroforestry: progress, prospects, priorities. Agroforestry Systems 38:177-193.

Millward AA, Mersey JE (1999). Adapting the RUSLE to model soil erosion potential in a mountainous tropical watershed. Catena 38:109-129.

Mosier AR, Delgado JA (1997). Methane and nitrous oxide fluxes in grasslands in western Puerto Rico.Chemosphere 35:2059–2082.

Nair PKR (1998) Directions in tropical agroforestry research: past, present and future. Agroforest Syst38:223–245.

Negandra H (2001). Using remote sensing to assess biodiversity. International Journal of Remote Sensing 22:237–240.

Newaj R, Dhyani SK (2008). Agroforestry for carbon sequestration: Scope and present status. Indian Journal of Agroforestry 10:1-9.

Oerke C, Dehne HW, Schoenbeck F, Weber A (1994). Crop production and crop protection: estimated losses in major food and cash crops. Elesvier, Amserdam, pp. 830.

Office of Environmental Policy and Planning (OEPP) (2000). Thailand's national greenhouse gas inventory 1994. Ministry of Science, Technology and Environment, Bangkok.

Ogawa H, Yoda K, Ogini K, Kira T (1965). Comparative ecological study on three main type of forest vegetation in Thailand. Nature and Life in Southeast Asia 4:49-80.

Oindo BO, Skidmore AK (2002). Inter-annual variability of NDVI and species richness in Kenya. International Journal of Remote Sensing 23:285-298.

Palm CA, Alegre JC, Arevalo L, Mutuo PK, Mosier AR, Coe R (2002). Nitrous oxide and methane fluxes insix different land use systems. Global Biogeochem Cycles 16:1073.

Rao DP (2001). A remote sensing-based integrated approach for sustainable development of land water resources. IEEE Transactions on Systems Man and Cybernetics 31:207-215.

Rietveld WJ (1995). Agroforestry: a maverick science and practice. In: Rietveld WJ (ed) Proceedings of Agroforestry and Sustainable Systems Symposium. Fort Collins, CO. General Technical Report RM-GTR-261. USDA, Forest Service, Rocky Mountain Forest and Range Experiment Station.

Rietveld WJ, Francis CA (2000). The future of agroforestry in the USA. In: Garrett HE, Rietveld WJ and Fisher RF (eds) North American Agroforestry: An Integrated Science and Practice. American Society of Agronomy, Madison, WI, pp. 387-402.

Saikia P, Choudhury BI, Khan ML (2012). Floristic composition and plant utilization pattern in homegardens of Upper Assam, India. Tropical Ecology 53(1):105-118.

Saikia P, Khan ML (2014). Homegardens of upper Assam, northeast India: a typical example of on farm conservation of Agarwood (*Aquilaria malaccensis* Lam.). International Journal of Biodiversity Science, Ecosystem Services and Management10(4):262-269.

Samaniego L, Schulz K (2009). Supervised classification of agricultural land cover using a modified k-NN technique (MNN) and Landsat remote sensing imagery. Remote Sens1: 875–895.

Sanchez PA (1995). Science in agroforestry. Agroforest Syst 30:5–55.

Schultink G (1992). Integrated remote-sensing, spatial information-systems, and applied models in resource assessment, economic-development, and policy analysis. Photogrammetric Engineering and Remote Sensing 58:1229–1237.

Senpaseuth P, Navanugraha C, Pattanakiat S (2009). The Estimation of carbon storage in dry evergreen and dry dipterocarp forests in Sang Khom District, Nong Khai Province, Thailand. Environment and Natural Resources Journal 7(2):1-11.

Singh MP, Dayal N, Singh BS (1994). Importance of genetic conservation of tasar host plants in agroforestry programme in Chhotanagpur region of Bihar. Journal of Palynology 30:157-163.

Slocum MG, Horvitz CC (2000). Seed arrival under different genera of trees in a neotropical pasture. Plant Ecol 149:51–62.

Smith P, Olesen JE (2010). Synergies between the mitigation of, and adaptation to, climate change in agriculture. J Agric Sci 148:543-552.

Steudler PA, Bowden RD, Mellilo JM, Aber JD (1989). Influence of nitrogen fertilization on methane uptakein temperate forest soils. Nature 341:314–316.

Swaminathan MS (1983). Genetic conservation: microbesto man. Presidential address to the 15th International Congress on Genetics. In: Genetics: new frontiers,Vol. 1. Oxford & IBH Publishing Co., New Delhi,India.

Tewari JC, Ram M, Roy MM, Dagar JC (2013). Livelihood improvements and climate change adaptations through agroforestry in hot arid environments. In: Dagar JC, Singh AK, Arunachalam A (eds) Agroforestry Systems in India: Livelihood Security and Ecosystem Services, Advances in Agroforestry. Springer New Delhi, India, Volume 10, pp.155-183.

Thangataa PH, Hildebrand PE (2012). Carbon stock and sequestration potential of agroforestry systems in smallholder agroecosystems of sub-Saharan Africa: mechanisms for 'reducing emissions from deforestation and forest degradation' (REDD+). Agric Ecosyst Environ 158: 172-183.

Verchot LV, Noordwijk MV, Kandji S, Tomich T, Ong C, Albrecht A, Mackensen J, Bantilan C, Anupama KV, Palm C (2007). Climate change: linking adaptation and mitigation through agroforestry. Mitigation Adaptation Strategies Global Change 12:901-918.

Wang G, Gertner G, Fang S, Anderson AB (2003). Mapping multiple variables for predicting soil loss by geostatistical methods with TM images and a slope map. Photogrammetric Engineering and Remote Sensing 69:889–898.

Wang G, Zhang M, Gertner GZ, Oyana T, McRoberts RE, Ge H (2011). Uncertainties of mapping aboveground forest carbon due to plot locations using national forest inventory plot and remotely sensed data. Scand J For Res 26:360–373.

Wassmann R, Lantin RS, Neue HU (2000). Methane emissions from major rice ecosystems in Asia. Nutr Cycl Agroecosys 58:1–398.

Zomer RJ, Trabucco A, Coe R, Place F (2009). Trees on farm: analysis of global extent and geographical patterns of agroforestry. ICRAF Working Paper - World Agroforestry Centre No. 89, pp. 63.

Zoysa MD, Inoue M (2014). Climate Change Impacts, Agroforestry Adaptation and Policy Environment in Sri Lanka. Open Journal of Forestry 4(5): 439-456.

4

Livelihood Opportunities through Natural Vegetation for Adaptation to Climate Change

Suresh Kumar

Central Arid Zone Research Institute, Jodhpur- 342 003

Introduction

It is well known that, over 50% of requirement of food for the entire humanity is obtained from just three crops- rice, maize and wheat though 150 crops are commercialized globally. Ethnobotanically, over 7000 plants species are cultivated or harvested from the wild which represent immense agro-biodiversity. Most of these species are under exploited but have the potential for contributing to food security, health (nutritional/medicinal), income generation and environmental services. Examples include little millets which provide calcium and iron and vitamins like niacin and sulphur containing amino acids, having more fiber than rice and wheat. Some other species which have in the recent past contributed to income generation in Indian arid zone include ber, anola, karonda and gonda. Likewise the perennial grasses in this area are not only source of nutritious fodder but also the grains e.g. gramana (*Panicum antidotale*) are consumed during famine. They have acquired importance being alternative sources of income as also alternate life support system during crisis. They often have neutraceutical, medicinal uses or often multiple uses. Demand for traditional food in large multi-ethnic cities and metros have brought these species into sharp focus, Hence, these are being viewed now as potential candidate for diversification in the cropping system, conservation of bio-diversity, enhancing the limits of green revolution even during climate change and finally as provider of molecules for curing various ailments of men and animals.

Prospects of natural vegetation

These nature species can contribute to food security and nutrition as also increasing income and satisfying the need of hidden hunger due to vitamins and mineral deficiency. These are also part of cultural heritage of local people. They have specific agro-ecological requirements and are adapted to grow on marginal land on their own. Hence, they do not have formal seeds supply systems but they do have traditional uses and traditional supply system in localized areas. These are collected from wild and produced traditionally with almost no external inputs. Because of little attention on their regeneration and protection, their population may be under threat of disappearance. Some examples of this type from Indian Arid Zone are *Moringa concanensis, Ziziphus truncata, Neurada procumbens* and *Cullen plicata.*

These plants can be grown in home gardens or grown along traditional crops as also at their margins and can also be used for diversification. They can also support, to a large extent, agro-pastoral systems and community seed systems. By virtue of their being evolved and sustained in a highly variable arid agro climates, these can withstand the harshness of projected climate change in these areas. In contrast 'pampered' crops will most likely either produce very less or will completely perish under such situations.

Plants for life support system

In a survey conducted in four districts namely Jodhpur, Bikaner, Banner and Jaisalmer, 51 plants were found to have people use for their survival and sustenance. Stems of 28 species are used in furniture, cartwheels, tools, stilts, windows, doors, toys and utensils, fencing, thatching, ropes, fuel, fodder, brooms, matchsticks, musical instruments, incense, deodorants, toothbrushes, dying leather, worshiping, magico-religious and explosives. Leaves of 17 taxa are used as fodder, vegetable, preparing drinks, mosquito repellent, pesticides, perfumes *etc.* Some 13 taxa are used as whole plant for fuel, fodder, food, fencing, thatching and even salt making (*Lycium barbarum*). Fruits of 8 taxa are used as vegetable, pickles or eaten raw. Seed of 10 taxa are used as pickles, delicacies vegetables, scarcity loads, pesticides and magico-religious beliefs. Roots of 6 taxa are used in making spoon, dying of leather. Gum of 4 taxa has neutraceauticals value while other 4 taxa have uses in stuffing pillows and other uses. These species have multiple uses. *Prosopis cineraria* and *Calotropis procera* each had 17 uses followed by 16 uses of *Capparis decidua*, 13 of *Ziziphus nummularia*, 9 of *Calligonum polygonoides*, *Prosopis juliflora,* 7 each of *Tecomella undulata*, *Acacia nilotica*, *Acacia senegal* and *Leptadenia pyrotechnica* (Kumar and Parveen 2004).

Medicinal plants sourced from natural habitats

Some 46 species found in high trade are collected from wastelands (Ved and Goraya 2007). Although these don't require immediate attention as these are abundantly available, their domestication and cultivation is essential to have uniform quality of the produce. These include the following:

Abrus precatorius (Rati), *Achyranthes aspera* (Aparnarg), *Aerva lanata* (Cheroola), *Andrographis paniculata* (Kalrnegh), *Boerhavia diffusa* (Punarnava), *Citrullus colocynthis* (Indrayan), *Convolvulus microphyllus* (Shankhapushpi), *Cynodon dactylon* (Ourva), *Cyperus rotundus* (Nagar motha), *Ocimum americanum* (Ban tulsi), *Peganum harmala* (Harmal), *Pluchea lanceolata* (Rasna), *Sida cordifolia* (Bala), *Sisymbrium irio* (Khubkalan), *Tephrosia purpurea* (Sarpankha), *Vetiveria zizanioides* (Lavancha) and *Clerodendrum phlomoides* (Arni).

Medicinal plants for diversification

Of the highly traded tropical medicinal plants sourced from the forest, following occur and can be grown in arid and drier semi arid zone (NMPB 2006):

Acacia nilotica (Babool), *Aegle marmelos* (Bael), *Asparagus racemosus* (shatavari), *Boswellia serrata, Emblica officinalis* (Arnla), *Gymnema sylvestre* (Gudmar), *Lannea coromandelica, Mucuna pruriens* (Kaunch), *Rauvolfia serpentina* (Sarpagandha) and *Commiphora wightii* (Gugal). Besides, following also can be grown in Indian arid and drier semi arid zone: *Adhatoda zeylanica* (Adusa), *Aloe vera* (Kurnari), *Cassia angustifolia* (Sonamukhi), *Catharanthus roseus* (Sadabahar), *Indigofera tinctoria* (Ali), *Jatropha curcas* (Jamalgota), *Lawsonia inermis* (Henna), *Lepidium sativum* (Halim), *Ocimum basilicum* (Sweet basil), *Vitex negundo* (Nirgundi), *Withania somnifera* (Ashwagandha) and *Ziziphus jujuba* (Ber).

Prioritized medicinal plants

Kumar *et al.* (2003) in another study concluded that of the 20 most exported herbals 12 are from arid zone. Likewise-additional 11 species of highest domestic trade value could also be cultivated. Species prioritized are *Aloe vera, Commiphora wightii, Mucuna pruriens, Cyperus rotundus, Ocimum sanctum, Tinospora cordifolia, Tephrosia purpurea, Cassia angustifolia, Adhatoda vasica, Lepidium sativum, Citrullus colocynthis, Gymnema sylvestre.*

Trade and economic potential

The trade volume of these plants and associated fiscal quantum is poorly known because of unorganized nature of this business. Some information is however available in respect of medicinal plants which are consumed in traditional health care, home remedies and Ayurvedic system. There are registered cottage industries as well as 9500 registered herbal industries in India. Through some 6000 plants are of known medicinal value, their quantity used has been only a guestimate. One such assessment by National Medicinal Plant Board in 2006-2007 (Ved and Goraya 2007) revealed 960 medicinal plants species used to manufacture 1289 botanical raw drugs. Besides, herbal products are also used in making herbal teas, functional foods, neutraceuticals, phytochemicals, flavours and fragrances, aroma therapy, culinary herbs, spices, herbal cosmetics, herbal dyes, colorants, coolants as well as in magico-religious purposes. Thus, many of these underutilized plants species have current use value for most vulnerable population across the world as well as potential value which is largely unknown at present.

Marichamy *et al.* (2014) estimated that the domestic trade of the AYUSH industry is of the order of US $ 80 to 90 billion (1US$ = Rs.50). As per information given by the Minister of State in the Ministry of Commerce and Industry Dr. E.M. Sudarsana Natchiappan in Lok Sabha on 9th Dec, 2013, export of herbals and products from India was worth US $298.59 in 2010-11 and reached 395.58 $ in 2012-13 i.e. a growth rate of 15-16% per year (Table 1). This underlines the value of these species in providing much needed cash for life support.

Table1. India's Export of Ayush, Medicinial herbs and their value added products (values in USD million)

Sl.No.	Commodity	2010-11	2011-12	2012-13
1.	Ayush	156.96	182.18	163.44
2	Medicinal herbs and their value added products	141.63	176.12	232.14
	Total	298.59	358.30	395.58

Evidently, a viable commercial potential therefore, exists in growing and processing of herbals as well as their collection from wasteland, as weeds from croplands and forest areas following good collection practices and good agricultural practices.

Trade of medicinal plants from Indian arid zone

A study on trading of ethno-medicinal plants in the Indian arid zone revealed that there are 131 ethno-medicinal plants out of 682 species in the entire Indian arid zone. A survey of purchase and sale price of different herbals was made by visiting traders. Knowing sale price was always possible but purchase price was never revealed. However, discussion with traders revealed that the difference between procurement and sale price of herbals could range from

15-25%. A comparison of selling rates of these herbals in Delhi and Indian arid zone revealed tremendous difference. The percent increase in sale price of raw herbals in arid zone was up to 100% in respect of 12 species, 200% in respect of 3 species and 300-500% in respect of 5 species (Kumar *et al.* 2005). The marketing of herbals in arid zone is highly opportunistic, exploitative of both plants and people, informal, monetarily ineffective and hence not widely practiced by farmers in this area. Moreover, the whole trade suffers from unethical practices by middleman who exploit these pharmacies and other users by creating artificial scarcities. They also exploit herbal collectors and growers by offering a meager price and by creating artificial situation of glut in the market. Besides, adulteration and substitutions are common maladies of this trade.

Value addition for enhanced income and livelihood

One of the ways to overcome this problem is to process them for crude or clarified extract for marketing. This adds value to the product and gets a better price to grower. Some medicinal plants used as crude extracts / tinctures include (1) Belladonna from *Atropa belladonna*, (2) Ipecac from *Cephaelis ipecacuanha,* (3) Opium from *Papaver somniferum,* (4) Henbane from *Hyoscyamus niger,* (5) Stramonium from *Datura stramonium,* (6) Liquorice from *Glycyrrhiza glabra,* (7) Rhubarb from *Rheum officinale,* (8) Podophyllum from *Podophyllum emodi* and (9) Capsicum oleoresin from *Capsicum annuum.* Next step in value addition is isolating active principle that fetches much more profit beside being pure. Some medicinal plants of which active principles have become commodity of trade in health care industry are listed in below (Table 2).

Evidently, for all these segments of drug, perfumery, flavouring and cosmetic industries, mostly the higher plants are one of the raw materials and come from either the natural resources i.e., forests through forest contractors employing local or tribal people or from the drug farms maintained by the growers of aromatic and medicinal plants.

Table 2. Active principles and their pharmalogical activity of some selected medicinal plants

Sl. No.	Plant	Active principle	Pharmacological activity
1	*Rauvolfia serpentina*	Reserpine	Hypotensive
2	*Catharanthus roseus*	Ajmalicine Vinblastine Vincristine	Vasodilator Anticancer Anticancer
3	*Cassia angustifolia, Cassia acutifolia*	Sennosides	Laxative
4	*Plantago ovata*	Psyllium mucilage	laxative
5	*Glycyrrhiza glabra*	Glycyrrhizic acid	Antiinflammatory
6	*Berberis sp.*	Berberin	Antidiarrhoeal
7	*Digitalis lanata*	Digitoxin Lanatoside	Cardiotonic
8	*Taxus baccata*	Taxol	Anticancer
9	*Podophyllum emodi*	Podophyllotoxin Etoposide Tenoposide	Anticancer
10	*Artemisia annua*	Artemisinin	Antimalarial
11	*Valeriana wallichii*	Valepotriates	Sedative, tranquilizer
12	*Silybum marianum*	Silymarin	Antihepatoxic
13	*Commiphora wightii*	Oleo gum resin	Hypolipidaemic

Quality control and its enhancement through value chains

One of the most important grey area in the entire herbal trade is the quality of herbal medicines. Understanding when, where and how the quality is compromised is possible only if we break down external and internal linkages within production, processing and trade network. Besides, one also has to minimize the ecological threats because of over collection of medicinal plant species from natural habitats. This is so because collection and marketing of medicinal plants from wild is an important source of livelihood for marginal, landless and poor farmers in all developing countries. Bringing many of these plants into cultivation with adequate technical and research backstopping even does not some time result in desired output. In order to be successful in developing a value chain of medicinal plants via cultivation mode would require integration of primary producers in the entire process. This results in removing market access barriers for better commercialization and is also attractive to companies as they can have greater control over supply, quality and pricing.

Aloe vera value chain - a case study

A value chain integrating cultivators to sellers exists in respect of *Aloe vera* in Nakhtarana taluka of Bhuj district in Gujarat. Some traditional farmers pooled their monetary shares to purchase the processing plant for making *Aloe vera* juice and its other products like soap, shampoo and moisturizer. They also started growing *Aloe vera* on a small area of their total land holding. This is cut and brought to processing unit where on the basis of per kg biomass of *Aloe vera* they are given fixed amount. The processed products are made by employing experts to ensure matching quality standards of the industry. Then, these are hooked up with post offices for over the counter sale to general public. Their products are being exported because of better quality. Thus, the growers are also processers and sellers of the finished products to domestic consumer via post offices and to foreign consumers via exports. The profit every year is shared amongst the farmers who initially pooled the money to setup the processing plant. The cost has been fully recovered and now the grower association is earning profit.

There is another model of value chain of *Aloe vera* in Moondra taluka, Gujarat wherein a single farmer has organically grown *Aloe vera* and extracted juice to finally develop its different flavours like lemon, orange, mango and so on. A distribution counter for its sale has been opened in Mumbai and it is now making roaring business.

In Rajasthan such a model has yet to come up. In Jodhpur and Nagaur districts there are large numbers of *Aloe vera* growers who periodically cut its leaves and sell it to a middle man who in turn supplies to processing units. The cost paid to farmers is Rs. 2-3 per kg of fresh leaves, whereas middleman charges around Rs. 7-10 per kg from processing units. The cost of processing, packaging, transportation is included by the processers who ultimately sell it at approximately Rs. 250 per liter. Hence from Rs. 4 to Rs. 250/- is the cost increase along the value chain.

But in this model, farmers are paid very low. They have to cut *Aloe vera*, upload in truck, which is not only labour intensive but also more in volume. In a variant of this model, value addition was introduced at farmer's level. They now peel its leaves, collect the pulp in plastic drums reducing the volume to half and bring this to processors. Farmers are paid Rs.14 per kg of this pulp by the processor. This small value addition has, therefore, enhanced their income. Having realized the potential of these medicinal plants either raw or processed and marketed through value chain, it is important to assess their role in conservation of natural resources and climate change risk mitigation so as to maintain and enhance livelihood opportunity.

Natural resource conservation in agroforestry system integrating medicinal plants *Cape verde case study (Tavares, 2008)*

In addition to have pure cropping of medicinal plants, attempts have also been made to use them in soil and water conservation specially as vegetative barriers.

It is a technique which uses the structure of a cross-slope barrier of *Aloe vera* to combat soil erosion by decreasing surface wash and increasing infiltration. *Aloe vera* is a durable herbaceous plant which was planted in the form of living barriers to recover degraded slopes on the Cape Verde Islands. The plants were closely planted along the contour to build an efficient barrier for retention of eroded sediments and superficial runoff. The living hedges of *Aloe vera* stabilized the soil, increased soil humidity by improving infiltration and soil structure. Groundwater was recharged indirectly. Soil cover was improved, and thus evaporation and erosion reduced. Implementation was relatively simple. The contour lines were demarcated using a water level. Seedlings were planted along one line at a distance of 30-50 cm between plants; spacing between the rows varied between 3-5 m according to the slope. The technology is applied in sub humid and semi-arid areas, on steep slopes with shallow soils, a poor vegetation cover and high soil erosion rates. These areas are generally used by poor subsistence farmers for rainfed agriculture with crops such as maize and beans, which are considered inappropriate for such slope angles. On slopes steeper than 30% the living barriers are often combined with stone walls (width 40-50 cm height 80-90 cm). The plants stabilized the stone risers, making this combined technology one of the most efficient measures for soil erosion control on the Cape Verde Islands. It can be grown with any crop, is available for any farmer; establishment and transport are simple; its green leaves are not palatable for livestock, the plant is extremely resistant to water stress and grows on any bioclimatic zone on the island. Furthermore, Aloe is known for its multiple uses in traditional medicine. Ecologically, it provides a host of ecological benefits as listed below:

i. Reduction of slope length
ii. Improvement of ground cover
iii. Improvement of topsoil structure (compaction)
iv. Stabilization of soil (e.g. by tree roots against landslides)
v. Increase of groundwater level, recharge of groundwater (expected)
vi. Sediment retention / trapping, sediment harvesting
vii. Increase of biomass (quantity)
viii. Control of raindrop splash
ix. Reduction of slope angle
x. Increase of surface roughness

Livelihood Enhancement

Its importance in providing off-farm income was also recognised. If rainfed then it enhances income over 50%, but in irrigated situations income enhancement was up to 30-40%. Besides, it enabled providing access to service and infrastructure as it helped accumulate on slope of more than 60%, the soil behind the barrier up to the depth of 55 cm. It helped reduce risk of production failure, increased crop yield and prevented loss of land (about 8% if the production area is 1 ha). Other benefits are listed below:

Socio-cultural benefits Socio-cultural disadvantages

i. Improved cultural opportunities (additional crops possible)

ii. Improved conservation / erosion knowledge

iii. Improved food security / self sufficiency

Ecological benefits

i. Improved harvesting / collection of water

ii. Reduced surface runoff and suspended sediment

iii. Reduced soil loss

iv. Improved soil cover

v. Increased biomass / above ground C

vi. Increased water quantity

vii. Increased soil moisture

viii. Increased water availability

Recurring expenditure

This is maintenance/recurrent neutral and is balanced very positive. The structure does not require costly maintenance; it is simply controlling the spacing of the barrier (vegetative control) and punctual replanting.

Acceptance/adoption

95% of land user families (380 families; 9% of area) have implemented the technology with external material support. 5% of land user families (20 families; 1% of area) have implemented the technology voluntary. There is little trend towards (growing) spontaneous adoption of the technology.

In conclusion, this case study proved that using this medicinal plant helped recover the degraded land and also enabled in increasing the production area.

The vegetation of the area between the barriers will make the recovery and protection of the soil layer is stronger. It also stabilized the soil making it more resistant to the impact of the rain water. Sedimentation behind the barriers is favored along the time due to the continued growth of the plant *Aloe vera.* It Improves the thickness of the soil leaving it stronger and more productive. The mineral and organic matter retained behind the lines of *Aloe vera* will promote an increase in the thickness of the soil, improving also the volume of water retained in the soil, resulting in better root development. Therefore, the process of soil formation is best done. Reduction of the production area, which is occupied by bands of *Aloe vera* can be managed by cutting leaves of *Aloe vera* plants leaving remaining plant growing and the cut leaves provide additional income.

Indian arid zone-case study

Experiments were carried out in Jodhpur and Bikaner. In CAZRI, CRF, an area of 0.6 ha having slope of 3% on both north and south facing slope was selected and divided equally into five plots on each side to have each plot of 10m × 40m. These were given land management treatments of two medicinal plants: *Cassia angustifolia* and *Aloe vera* as follows:

T-1 : Control (Barren without any plantation)

T-2 : *Cassia angustifolia* at 0.5 x .5 m without any vegetative barrier

T-3 : *Cassia angustifolia* sown at 0.5x.5 m and provided with one vegetative barrier of *Aloe vera* (in three rows at 50 cm spacing planted in a staggered way)

T-4 : *Cassia angustifolia* sown at 0.5 x 0.5 m and provided with two vegetative barriers of *Aloe vera* as in 2 above

T-5 : *Aloe vera* at 1x1 m without any vegetative barrier.

Total rainfall during 2005 was 283 mm and seasonal rainfall, 250.1 mm. Of the total 30 rainfall events (0.2 mm – 72.7 mm per event) only five rainfall events could generate runoff. Results revealed that runoff (%) and sediment load ($g\ l^{-1}$) were highest in untreated plots followed by that in Senna alone plot, Senna with single row of *Aloe vera*, Senna with two rows of *Aloe vera* and minimum in Aloe vera alone plot. Ca, Mg, Na, K, CO_3, HCO_3 and Cl showed mixed trends in different treatments. These were maximum in runoff from Senna alone and Aloe alone plots followed by Senna with single row of Aloe and Senna with two rows of Aloe.

Earlier experiments on vegetative barriers have also revealed that perennial grasses like *Lasiurus sindicus, Cenchrus ciliaris, C. setigerus, P. antidotale, P. turgidum. Saccharum bengalense* and *Vetiveria zizanioides* were more effective in conserving soil from erosion by water as run off volume and peak discharge were reduced significantly (up to 96-97% from 22-28%) (Sharma *et al.* 1999). In another study on sandy plains of Kalyanpur in Barmer, due to *Cassia angustifolia* barrier, 36.5% more moisture was stored (Gupta and Rathore 2002).

Similarly, these two plants *Aloe vera* and *C. angustifolia* were grown on sandy landscape in Bikaner to assess their impacts on wind borne soil loss. Bare plots had maximum soil loss (248.3 t ha^{-1}) in April – July compared to the soil loss under *C. angustifolia* (8.8 t ha^{-1}). The vegetation cover of Aloe resulted in net deposition (179.2 t ha^{-1}).

It can be said that like in Cape verde, the studies in Indian arid conditions also proved the ecological, economic and livelihood benefits of growing such medicinal plants.

Conclusion

Changing climate including aberrant weather does pose a risk to survival due to declining livelihood opportunities. The extant and severity of such risks are more expected in arid and semiarid zones. But these very areas also have the solutions to mitigate such risks in terms of large number of adapted species having economic and ecological value. These not only help withstand climatic vagaries but also enhance livelihood opportunities. All that we need is to priorities the species as per agroclimatic conditions and inculcate in the existing cropping system to diversify it and make them multilayered agroforestry systems so as to derive multiple benefits on one hand and attain resilience to impending global climate change.

This diversification has to go hand in hand with market intelligence. Selection of the new crop should be based on requirement of the market and availability of a standard value chain. Arrangements like buybacks or collective cultivation and sale legally binds the purchasers to buy entire produce at a predefined price. If unit price is related to unit cost, then a competitive and fair price to growers also ensures quality raw material without adulteration. The other approach is to organize collective cultivation through self-help groups (SHGs) or any other group. Value addition by way of cleaning, grading and packaging can enhance earnings. Collective bargaining directly with end-users, eliminating the middlemen can get better remuneration to growers and collectors thus profiting them and enhancing their livelihood.

References

Gupta JP, Rathore SS (2002). Biomass Production and Rehabilitation of Degraded Lands in Arid Zone, CAZRI, Jodhpur, pp. 20.

Kumar Suresh, Parveen F, Mertia R (2003). Taxonomic analysis of medicinal plants in the Indian Thar: A prioritization strategy. J Econ Taxon Bot 27(1):160-169.

Kumar Suresh, Mathur Manish, Kushwaha Neelam, Goyal Sangeeta, Chauhan Aruna, Parveen F (2014). New traditional herbals from Indian arid zone for curing rheumatism. Asian Agri History 18(2):133.

Kumar Suresh, Parveen F (2004). Traditional Ethnomedicinal Plants in the Indian Thar: Their Status in Natural and Possibilities of their Cultivation and Trading. CAZRI, Jodhpur. pp. 216.

Kumar Suresh, Parveen F, Goyal S, Chauhan A (2005). Trading of ethnomedicinal plants in the Indian arid zone. Indian Forester 131(3): 371-378.

Marichamy K, Yasoth Kumar N, Ganesan A (2014). Sustainable Development in Exports of Herbals and Ayurveda, Siddha, Unani And Homoeopathy (Ayush) in India. Science Park Research Journal 1 (27).

National Medicinal Plant Board (2006). Cultivation of selected medicinal plants. NMPB, New Delhi, pp. 214.

Sharma KD, Joshi NL, Singh HP, Bohra DN, Kalla AK, Joshi PK (1999). Study on the performance of contain vegetative barrier in an arid region using numerical models. Agricultural Water Management 41:41-56.

Tavares Jacques (2008). Field Trip Guide, INIDA - Cabo Verde, 2008 / Técnicas de Conservação de Solos e Água em Cabo Verde, MPAR &CILSS, 1994 / Consercação de Solos e Água (Teoria e Prática), Sabino, António Adnino, 1991 Instituto Nacional de Investigação e Desenvolvimento Agrário (INIDA), Cape Verde.

Ved DK, Goraya GS (2007). Demands and supply of medicinal plants in India. Executive summary. National Medicinal Plant Board, New Delhi & FRLHT, Bangalore, pp.8.

5

Conservation of Plant Genetic Resources through Agroforestry System

J.P. Singh[1], *V.S. Rathore*[2], *Venkatesan, K.*[1]

[1]*ICAR-Central Arid Zone Research Institute, Regional Research Station Jaisalmer, India*

[2]*ICAR-Central Arid Zone Research Institute, Regional Research Station Bikaner India*

Introduction

Low and erratic rainfall, extreme temperatures, long sunshine duration, high wind velocity, high evapotranspiration, low fertility of soil and paucity of water resources characterize hot arid region of India. Despite these hostile conditions, the region supports a large number of human and livestock population. However, the ever increasing human and livestock population and developmental activities exerts enormous pressure on natural resources and posing serious threat to sustainability of the region. In these hostile environmental conditions, several indigenous agroforestry systems are being practiced by farmers of this region (Shankarnarayanan *et al.* 1987). The agroforestry, which is a dynamic, ecologically based natural resource management practices that, through the integration of trees and other tall woody plants on farms and in the agricultural landscape, diversify production for increased social, economic and environmental benefits (ICRAF 2000). A number of studies have recognized the multiple benefits of agroforestry that include carbon sequestration, reducing forest degradation, biomass production, food security, income diversification, improvement and maintenance of soil biodiversity, provision of wildlife corridors, and a host of other ecological services and social benefits. In the last three decades, agroforestry has been widely promoted in the tropics as a natural resource management strategy that attempts to balance the goals of agricultural development with the conservation of soils, water, local and regional climate, and, more recently, biodiversity (Izac and Sanchez 2001). The present paper highlights on the integration of woody perennials (shrubs/trees) in various production systems through agroforestry approach for their conservation and sustainable utilization of plant genetic resources in hot arid environment.

Woody perennials in various production systems

The combined protective-productive systems like, integration of trees into farming systems rooted in the principles of ecology, productivity, economics and sustainability, are now generally referred as agroforestry (Harsh 2006). When the crops fail during severe drought period, most of the trees are able to provide fuel, fodder, fruit and other products being drought resistant species. Perennial woody plant species, whether native or introduced, generally provide a range of products and services in addition to wood (Ben Salem and Palmberg 1985), and are integral parts of the smallholder farming systems, as to some extent, they could satisfy the technical, economic, social and cultural needs of the smallholder farmers (Nitis 1992). Since ancient times, woody perennials have been the integral part of the traditional agro forestry system (TAFS) in hot arid region, which were developed by farmers over years of experience to suit ecological and socio-economic conditions with a view to attain stability in production. Malhotra (1984) classified agroforestry systems in eight zones within arid Rajasthan (Table 1). This classification is pre-dominantly based on climatic consideration, particularly that of rainfall zones. These agroforestry systems have involved independently in diverse socio-economic and cultural set ups of different zones.

Table 1. Agroforestry system in Arid Zone of Rajasthan

S. No.	Zone	Shrubs/trees with crops
1	Jaisalmer-Pugal-Girob (below 150 mm isohyets)	In extremely low rainfall zone, *Ziziphus nummularia* (Bordi) is most predominant with agricultural crops. *Prosopis cineraria* (Khejri) trees are very few in number.
2	Barmer-Osian- Bikaner Belt	Slightly higher rainfall zone; supports Khejri along with shrub Bordi.
3	Jodhput-Nagaur-Shekhawati Belt	Higher rainfall than zone 2 and therefore Khejri is most predominant while Bordi is grown with crops to a very limited extent.
4	Nawalgarh-Sikar	This zone has a further higher rainfall, Khejri continues to be predominant in agricultural fields but *Ailanthus excelsa* (Ardu) is also found.
5	Pali	Along with good average annual precipitation, water table is high and

(*Contd.*)

		irrigation facilities are available. Therefore, *Acacia nilotica* subsp. *cupressiformis* (syn. *Acacia cupressiformis*) locally known as Khajoor Babool and other *Acacia nilotica* species (Babool) are most abundant.
6	Arid and Semi arid areas bordering eastern Ajmer	Khejri continues to grow in this region but towards Ajmer region, *Acacia leucophloea* (Reonj) predominates due to suitability in pediment soil types.
7	Jalore District	Along with higher water table, brackish to saline water are in abundance, which suits to *Salvadora* species locally known as Jal.
8	Sirohi and Sojat tracts of Pali	Water table is moderate and irrigation facilities available. Babool and Khejri form the predominant tree cover in the crop fields.

Source: Malhotra 1984

The recent classification of traditional extensive agroforestry systems of Thar desert proposed by Saxena (1994) *viz*., *Calligonum – Haloxylon - Leptadenia* zone, *Ziziphus - Capparis* zone, *Calotropis - Calligonum – Clerodendrum* zone, *Prosopis- Ziziphus- Capparis* zone, *Salvadora- Prosopis- Capparis* zone, *Prosopis-Tecomella* zone, *Prosopis* zone and *Prosopis-Acacia* zone. The woody plant species adapted in various above systems are compatible with annual crops and grasses in rainfed as well as irrigated conditions. The important woody perennial species suitable for various agroforestry systems for hot arid region have been presented in Table 2.

Table 2. Woody perennial species of hot arid region to be utilized in various production systems

S.N.	Plant Species (Family, and local name)	Habit	Production system	Economic aspect	Remarks
1.	*Acacia jacquemontii* Benth. [Mimosaceae] (Bawli, Bui bawli)	Large shrub	Live hedge	Fodder, gum, fuel wood	In sandy areas in Bikaner, Jodhpur and Jaisalmer districts
2.	*Acacia leucophloea* (Roxb.) Willd. [Mimosaceae](Reonj, Urajio)	Tree	Agri-silvi	Fodder, fuelwood	Pali district
3.	*Acacia nilotica* (L.) Del. subsp. *indica* (Benth.) Brenan [Mimosaceae]	Tree	Agri-silvi, Khadins bunds	Fodder, gum, timber, medicine	Ganganagar, (Desi Babul, Kikar) Hanumangarh, Jalore, Pali, Barmer, Jodhpur and Jaisalmer districts
4.	*Acacia nilotica* (L.) Del. subsp. *cupressiformis* (J.L. Stewart) Ali and Faruqi [Mimosaceae] (Khajoor Babool)	Tree	Agri-silvi, boundary plantation	Fodder, timber	Important in Pali district
5.	*Acacia senegal* (L.) Willd. [Mimosaceae] (Kumat, Kumata)	Tree	Shelterbelt, Boundary plantation	Food, fodder, gum	Gravelly rocky, sandy areas in Jaisalmer, Jodhpur and Barmer districts
6.	*Balanites aegyptiaca* (L.) Delile [Balanitaceae] (Hingoto, Hingot)	Shrub to small tree	Live hedge	Browse, medicine	Dry parts of Jodhpur, Jaisalmer, and Bikaner districts
7.	*Calligonum polygonoides* L. [Polygonaceae] (Phog)	Shrub to small tree	Live hedge, Agri-silvi, Shelterbelt	Food, fodder, fuel wood, medicine	Sandy areas in Bikaner, Jaisalmer, Barmer and Jodhpur districts
8.	*Capparis decidua* (Forsk.) Edgew. [Capparidaceae](Ker, Kair) shrub to small tree	Agri-horti	Food, browse, medicine	Gravel, sandy plains in Jaisalmer, Bikaner, Barmer and Jodhpur districts	

9.	*Clerodendrum phlomidis* L.f. [Verbenaceae] (Arni, Yerna)	Large shrub	Live hedge	Medicine, browse	Sandy areas in Bikaner, Barmer and Jodhpur districts
10.	*Commiphora wightii* (Arnott) Bhandari [Burseraceae] (Guggul)	Large shrub	Boundary plantation	Medicine, browse	Gravelly sites in Jaisalmer and Jodhpur districts
11.	*Cordia dichotoma Forst. f.* [Boraginaceae] (Lasora)	Tree	Boundary plantation	Food, fodder, medicine	Barmer, Bikaner Pail, Jalore, Sirohi, Jodhpur and Jaisalmer districts with different irrigated and rainfed crops.
12.	*Cordia gharaf* (Forsk.) Ehrenb. and Aschers. [Boraginaceae] (Goondi)	Shrub to small tree	Agri-silvi, Agri-horti	Food, fodder, gum, medicine	Sandy, gravelly, areas in Jaisalmer, Jodhpur and Bikaner districts
13.	*Dichrostachys cinerea* (L.) Wt. and Arn. [Mimosaceae](Nutan)	Shrub	Live hedge/farm boundary, Silvi-pasture	Browse, fuelwood	Jodhpur and Bikaner districts
14.	*Euphorbia caducifolia* Haines [Euphorbiaceae] (Thor, Dandathor)	Dendroid shrub	Live hedge	Food, Browse, fuelwood	Rocky/gravelly terrain of Jodhpur and Jaisalmer districts
15.	*Grewia tenax* (Forsk.) Fiori [Tiliaceae] (Gangeran, Chabeni)	Shrub	Live hedge	Food, fodder, medicine	Gravelly areas in Jaisalmer, Barmer, Jodhpur and Pali districts
16.	*Lycium barbarum* L.[Solanaceae] (Mural)	Spinous shrub	Live hedge	Browse, medicine	Bikaner, Jodhpur, Barmer and Jaisalmer districts
17.	Maytenus emarginata (Wild.) Ding Hou [Celastraceae]	Shrub to small tree	Live hedge	Browse, medicine [Celastraceae]	Gravel, rocky, sandy small tree places in Jodhpur, Barmer, Jaisalmer, Bikaner, Churu, Jhunjhunu and Sikar districts

18.	*Mimosa hamata* Willd.[Mimosaceae] (Jinjani)	Shrub	Live hedge	Browse, medicine	Sandy areas Bikaner, Jodhpur, Jaisalmer and Nagaur districts
19.	*Moringa concanensis* Nimmo [Moringaceae] (Sarguro)	Large tree	Agri-horti	Food, medicine	Gravelly areas of Jodhpur districts
20.	*Opuntia elatior* Mill. [Cactaceae] (Hathhathor, Nagphani)	Jointed shrub	Live hedge/ field bunds	Food, fodder	Bikaner, Jodhpur and Jaisalmer districts
21.	*Prosopis cineraria* (L.) Druce [Mimosaceae] (Khejri, Jand, Sami)	Tree	Agri-silvi/ Agri-horti	Food, fodder, medicine, timber	Sandy areas in Bikaner, Barmer, Churu, Ganganagar, Jaisalmer, Jodhpur, Hanumangarh, Jhunjhunu, Nagore, Jalore and Sikar districts
22.	*Salvadora oleoides* Decne. [Salvadoraceae] (Meethajal, Barapilu)	Shrub to small tree	Agri-silvi/Agri-horti	Food, browse, medicine	Gravelly areas, in Jalore, Barmer, Jodhpur, Jaisalmer and Pali districts
23.	*Salvadora persica* L. (Kharajal, Chhotapilu)	Shrub to small tree	Agri-silvi,Shelterbelt/ windbreak	Food, browse, medicine	Saline areas in Barmer, Jalore and Pali districts
24.	*Tamarix ericoides* Rottl.[Tamaricaceae] (Javuro)	Shrub	Live hedge/ Windbreak	Fuelwood, basket making	Saline and Khadin areas in Bikaner, Barmer and Jaisalmer districts
25.	*Tecomella undulata* (Sm.) Seem [Bignoniaceae](Rohiro, Rohida)	Tree	Agri-silvi, boundaries of agricultural fields	Timber, fodder, medicine	Barmer, Jaisalmer, Bikaner and Jodhpur districts
26.	*Ziziphus mauritiana* (Burm.f.) [Rhamnaceae] (Ber, Bor)	Small tree	Agri-horti	Food, fodder	Sandy areas Jodhpur, Barmer and Bikaner districts
27.	*Z. nummularia* (Burmf.) Wt. and Arn. [Rhamnaceae] (Borti, Bordi)	Shrub	Agri-silvi/Agri-horti	Food, browse/fodder, fuelwood	Open sandy places in Barmer, Bikaner, Jaisalmer, Jodhpur, Churu, Jhunjhunu and Sikar districts

Khadin (runoff farming system): a specific system in hot arid region

In extreme arid region of Jaisalmer, a unique method of farming known as Khadin cultivation has been practiced traditionally since ages. This system is based on rainwater harvesting and its conservation to grow agricultural crops. It involves water harvesting from the shallow rocky surface into low lying farm land during monsoon period and subsequently growing crops when water recedes in winter (Kolarkar *et al.* 1983). The Paliwals of Jaisalmer first started it in the 15th century (Sehgal 1973). Khadins are very vital and unique traditional farming system in hot arid region particularly in Jaisalmer district. These are the important reservoir for harvesting rainwater. *Acacia nilotica, Prosopis cineraria, Salvadora oleoides*, *Tamarix* species, *Ziziphus nummularia* etc. are the important woody perennials traditionally grown on bunds in Khadin areas. There is need to conserve the available diversity of woody perennial in the system. Emphasis has to also be given on integration of fruits species like grafted ber (*Ziziphus mauritiana*), Gunda (*Cordia dichotoma*) and Khazoor (*Phoenix dactylifera*) in upper reach of Khadins for enhancing productivity in *Khadins*. The catchment zones of Khadins are largely barren and the same can be planted with suitable woody perennials for conservation and sustainable utilization.

Live fence/ hedge

Live fences are narrow lines of trees or shrub species planted on farm boundaries or between pastures, fields, or animal enclosures whose primary purpose is to control the movement of animals and/or people. Live fences usually are composed of a single row of trees or shrubs that are closely planted at uniform distances and may support barbed wire, although sometimes they arise from natural regeneration underneath fence lines. Live fences are the integral component of traditional farming system in hot arid region and make the sub-system of agroforestry. Apart from their protective function, live hedges play a vital role in livelihood of farmers by providing variety of economic products e.g. food, fodder, fibre, fuel wood, medicine, gum etc. As live fence system moderates the extreme temperatures and effects of high wind velocity, thereby, often resulting in different vegetation composition in crop field. Many important plant species and predators of various insect pests find the shelter in live hedges. Particularly, several key climbing species need the support of hedges for their survival. So, their vital role in protection, changing the microclimate of crop field and by supporting number of faunal and floral species, help in conserving and improving the overall agro biodiversity of the arid ecosystem.

The woody perennial species viz., *Acacia jacquemontii, A. senegal, Calligonum polygonoides, Calotropis procera, Capparis decidua, Clerodendrum phlomidis*, *Euphorbia caducifolia, Grewia tenax,*

Leptadenia pyrotechnica, Lycium barbarum, Maytenus emarginata, Mimosa hamata, Opuntia elatior, Salvadora oleoides, S. persica, Tamarix ericoides, Ziziphus nummalaria, etc. are traditionally utilized as live fences as per landform in western Rajasthan. Recently, species like *Lycium barbarum* attracted attention in arid zone for its medicinal importance. Considerable diversity in *L. barbarum* exists in its plant type in rangelands in western Rajasthan and also used as live hedge. *Clerodendrum phlomidis* is another important arid species reported as a potential source of natural antioxidant (Gokani *et al.* 2011). Raja *et al.* (2010) assessed the pharmaceutical potential of this species. They reported extensive use of this species in number of ethno veterinary practices in various parts of India. Presently this is one of the traditional woody medicinal plants in arid region which promoted as live fence, due to its demand of raw materials for pharmaceuticals.

Farmers also have their opinion in regard to positive and negative effect of these live fence species on crops. Therefore, maintaining live fence system in crop fields has prime and yet unrealized role in agro-biodiversity conservation in future climate change scenario. Moreover, it is utmost important to document, collect and evaluate the diverse germplasm of adapted live hedge woody perennials for selection of elite genotypes.

Windbreak /shelterbelts

Erection of windbreaks/shelterbelts along the boundaries of crop fields helps to reduce wind velocity and also reduces injury to the tender seedlings due to sand blasting and hot desiccating wind. Elite material of suitable species for wind break/ shelter belt plantation (Table 1) can be utilized for conservation and overall environmental services.

Role of agroforestry in conserving plant genetic resources

As much as 90 percent of the biodiversity resources in the tropics are located in human-dominated or working landscapes. Agroforestry improves diversity of genetic resources in working landscapes in at least three ways (Garrity 2004):

(i) Intensification of agroforestry systems can reduce exploitation of nearby or even distant protected areas,

(ii) Expansion of agroforestry systems can increase biodiversity in working landscapes, and

(iii) Agroforestry increases the species and within-species diversity of trees in farming systems.

The agroforestry helps to avoid overexploitation of plant resources of natural habitats and hence helps to conserve the plant genetic resources. The agroforestry in deforested regions helps to conserve species diversity by serving as habitats, corridors, or stepping stones for plant and animal species, adding structural and floristic complexity to the agricultural landscape and enhancing landscape connectivity.

The integration of woody perennials in various production systems enhances the structural diversity of the landscape, interrupting the monotony of pastures and crop fields and adding vertical and horizontal complexity. The woody perennials particularly fruiting tree species attracts birds, primates and other frugivores. Many of the birds that visit trees regurgitate or defecate seeds while perched in the trees, thereby dispersing seeds from forest patches into agricultural areas and enhancing both the abundance and species richness of seed input.

The modified microclimatic conditions (reduced solar irradiation and reduced temperature and humidity fluctuations) below woody perennials canopy may be more favorable, and the soils have better physical (structure and water infiltration), chemical and biological properties compared to that of open field, resulting in higher seed germination and plant establishment. The favorable micro-site conditions under woody perennials conducive for establishment and survival of plant species, and thus they enhance plant diversity. Woody perennials enhance the floristic diversity retained in the landscape by harboring diverse epiphyte communities, particularly if the woody perennials are relicts of the original vegetation of the region.

For example, the Khejri (*Prosopis cineraria*) based traditional Agroforestry Systems in western Rajasthan (Table1) shows the spread of Khejri on several of agricultural lands and, thus, of its importance in tree based agriculture (Malhotra 1984). Satyanarayan (1964) and Saxena (1977) reported that Khejri forms climatic climax of western Rajasthan and dominate the alluvial areas while Bordi (*Ziziphus nummularia*) is one of the main co-dominant on the plain areas of arid and semi arid zones. Gupta (1992) reported 30-40 trees/ha in areas around Jodhpur and as high number as 104 trees/ha in Sikar region without any adverse effect on crop production. Mann and Saxena (1980) reported the effect of different densities of Khejri on the major crops grown in different habitats. Even as high a tree density as 80 tree/ha showed an improvement in crop production in sandy plains. The increase in yield beside other factors could be due to build up of soil fertility (Aggarwal and Lahiri 1977). Experiments conducted by Shankarnarayan *et al.* (1987) on agri-silviculture show that *P. cineraria* has no adverse effect on grain production of mung bean (*Vigna radiata*) and guar (*Cyamopsis tetragonoloba*).

Further, in Rajasthan Orans (sacred woodlands) have been a source of natural wealth and regarded as a symbol of prosperity for the community that owned it. Moreover, Orans also played an important role in promoting a flourishing livestock based economy and growth of livestock rearing communities in Rajasthan. Ecologically valuable species of Orans perform key functions in the ecosystem and thereby contributing to support / enhance biodiversity. Generally, these species are selected and valued by the local communities for cultural or religious reasons.

Conclusion

The woody perennials viz., shrubs and trees is under severe stress due to change in land use pattern and other anthropogenic factors. The region has several adapted multipurpose woody shrubs and trees which are important source of genes targeting climate change. Therefore, there is need to collect, identify the elite genotypes and integrate these in various production systems. It will not conserve the genetic resources but will supply the products of economic importance.

Thus integration of these species has tremendous scope for increasing income and employment to rural masses. Many shrub species provide raw material to rural and cottage industries, which is lifeline of rural economy. Raw materials for pharmaceuticals, fibres, gums and resins, tannins can be exploited from these species. Some of the woody climbing species of medicinal importance like Giloy (*Tinospra cordifolia*), may be grown in Agri-Silvi system. There is need to identify suitable species and their integration in agroforestry system. Live fence component is one of the important sub-systems of agroforestry for conservation of genetic resources. For example *Calligonum polygonoides* locally called as Phog has been the important species of the Thar desert, is now facing threat of its survival. Conservation of available diversity of such pioneer hot arid species in traditional agroforestry systems is very much needed. It is utmost important to create greater awareness amongst the farmers regarding the multipurpose use of woody perennials for their ecological and socio-economic role in conservation and sustainable utilization of plant genetic resources.

References

Aggarwal RK, Lahiri AN (1977). Influence of vegetation on the status of the organic matter and nitrogen of the desert soils. Science and culture, 43:535.

Ben Salem B, Palmberg C (1985). Place and role of trees and shrubs in dry areas. In: Wickens GE, Goodin JR, Field, DV (eds), Plants for Arid Lands, George Allen and Unwin (Publishers) Ltd. London, pp. 93-102.

Garrity DP (2004). Agroforestry and the achievement of the Millennium Development Goals. Agroforestry Systems 61:5–17.

Gokani RH, Rachchh MA, Patel TP, Lahri SK, Santani DD, Shah MB (2011). Evaluation of anti-oxidant activity (*in vitro*) of *Clerodendrum phlomidis* Linn.f. suppl. root. Journal of Herbal Medicine and Toxicology 5(1):47-53.

Gupta JP (1992). Role or agroforestry for sustainable production in arid areas. Lecture delivered at 2nd annual group meeting cum symposia on agroforestry held at College of Agriculture, Nagpur, March 4-6.

Harsh LN (2006). Interactions of various components of agro forestry in arid regions.In: Compendium of Winter School on Sustainable Farming Systems for Arid and Semi-arid Ecosystems. CAZRI, Regional Research Station, Bikaner

ICRAF (2000). Paths to prosperity through agroforestry. ICRAF's corporate strategy, 2001–2010. Nairobi: International Centre for Research in Agroforestry.

Izac AMN, Sanchez PA (2001). Towards a natural resource management paradigm for international agriculture: the example of agroforestry research. Agricultural Systems 69: 5–25.

Kolarkar AS, Murthy KNK, Singh N (1983). Khadin: A method for harvesting water. Journal of Arid Environment 6:59-66.

Malhotra SP (1984). Traditional agroforestry practices in arid zone of Rajasthan. In: Shankarnayaran, KA (ed) Agroforestry in Arid and Semi-Arid Zones of India CAZRI, Jodhpur, pp. 263-266.

Mann HS, Saxena SK (1980). Role of *khejri* in Agroforestry. In: Mann HS, Saxena SK (eds) Khejri (*Prosopis cineraria*) in the Indian Desert. Its role in Agroforestry, CAZRI, Jodhpur, pp. 64-67.

Nitis IM (1992). Fodder shrubs and trees in smallholder farming systems in the tropics. In: Jong de R, Nolon T, Bruchem J van (eds) Natural Resource Development and Utilization-Future Research and Technology Management in Soil-Plant-Animal-Human Systems, Proc Commission of European Communities Coordination workshop held at the International Agriculture Centre, Wageningen. pp. 39-51.

Raja Muthu, Kumaradoss MohonMaraga, Mishra SR (2010). Comprehensive review of *Clerodendrum phlomidis*: a traditionally used bitter. Journal of Chinese Integrative Medicine 8:510-524.

Satyanarayana Y (1964). Habitat and plant communities of Indian desert. In: Proc. Symp. Problems of Arid Zone. Ministry of Education, Govt. of India, New Delhi, pp. 59-68.

Saxena SK (1977). Vegetation and its succession in the Indian desert. In: Desertification and its Control, ICAR, New Delhi, pp. 176-192.

Saxena SK(1994). Traditional agroforestry systems in agro-ecological zones of western Rajasthan. Annals of Arid Zone 33:279-285.

Sehgal KK (1973). Rajasthan district Gazetteers- Jaisalmer,8:82-121, Jaipur, India. Director, District Gazetteers, Govt. of Rajasthan, pp. 236-266.

Shankarnarayan KA, Harsh LN, Kathju S (1987). Agroforestry in arid zones of India. Agroforestry Systems 5:69-88.

6

Mitigating Climate Change Through Efficient Agroforestry Methods

Deepak Kumar

Assistant Professor, Amity Institute of Geoinformatics & Remote Sensing (AIGIRS)
Amity University, Noida – 201303, Uttar Pradesh, India

Introduction

Agroforestry, the word coined in the early seventies, has made its place in all the developed and the developing countries of the world and is now recognized as an important approach to ensuring food security and rebuilding resilient rural environments (Luedeling *et al.* 2014). Agroforestry applies to private agricultural and forest lands and communities that also include highly erodible, flood-prone, economically marginal and environmentally sensitive lands (Continents 1986). The typical situation is agricultural, where trees are added to create desired benefits. Agroforestry allows for the diversification of farm activities and makes better use of environmental resources (Taylor *et al.* 2007). Agroforestry is defined as a land use system which integrates trees and shrubs on farmlands and rural landscapes to enhance productivity, profitability, diversity and ecosystem sustainability. It is a dynamic, ecologically based, natural resource management system that, through the integration of woody perennials on farms and in the agricultural landscape, diversifies and sustains production and builds social institutions. India has been an all-time leader in agroforestry (Chavan *et al.* 2015).

Agroforestry is typically a land use management system in which trees or shrubs are grown around or among crops or pastureland.It defines an intensive land management system that optimizes the benefits from the biological interactions created when trees and/or shrubs are deliberately combined with crops and/or livestock. Within each agroforestry practice, it provides a continuum of options available to landowners depending o-n thcir own goals (e.g., whether to maximize the production of interpolated crops, animal forage, or trees) (Carsan *et al.* 2014). Agroforestry systems include both traditional and modern land-use systems where trees are managed together with crops and/or animal production systems in agricultural settings. They are dynamic, ecologically based, natural resource management systems that diversify and sustain production in order to increase social, economic and environmental benefits for land users at all scales.There is a growing body of scientific literature that demonstrates the gains accruing from agroforestry adoption. These advantages and believes agroforestry can contribute to improving the environment and the lives of people.

As world population increases, the need for more productive and sustainable use of the land becomes more urgent (Azapagic 2004). According to the United Nations, more than 7 billion people populated the Earth in 2011 and this number is expected to go up to 9.3 billion by the mid-century (Jónsdóttir 2011). To meet the demand for food by 2050, production will have to increase by over 60%. These figures, coupled with current problems borne out of past and existing non-sustainable land use practices, provide the case for changing the way we manage lands and our production of agricultural and tree goods (Luedeling *et al.* 2014).

Agroforestry typically requires adequate planning and know-how than simpler land uses because the system takes consideration of the diverse contradictory needs like grazing needs of cattle versus a tree's need to have its roots undisturbed. Best examples of the practice agroforestry include alley cropping, silvopasture, windbreaks, riparian buffer strips and forest farming. Agroforestry leads to the intentional integration of trees and shrubs into crop and animal farming systems to create environmental, economic, and social benefits. It works on the principle of the four *"i"s* encompassing *Intentional, Intensive, Integrated,* and *Interactive.* Its multifunctional propertiesprovide asolution to address the issues relating to environmental, economic or social (Mbow, Smith *et al.* 2014). Broadly any agroforestry practices deliver objective towards:

1. To inspire and magnify tree plantation in complementarity with crops and livestock to advance productivity, employment, income and livelihoods of rural households, especially the smallholder farmers.

2. To safeguard and soothe ecosystems to promote resilient cropping systems for minimizing the risk during extreme climatic events.

3. To suffice the requirement of raw material by industries to save foreign exchange.

4. To complement the availability of agroforestry products such as the fuel-wood, fodder, non-timber forest produce for the rural and tribal populations to reduce the pressure on existing forests.
5. To promote ecological stability in the vulnerable regions.

Mitigation for climate change issues is primary concerns for international agencies for improving the living standards of human's effecting consequently to the ecosystems. The less effective global mitigation measure is to reduce anthropogenic greenhouse gas (GHG) emissions and to increase the GHG sinks. Enhancing the adaptive capacity and/or reducing vulnerability to climate change impacts area taking advantage of the positive opportunities resulting from agroforestry sector. Both aiming to reduce the negative human and ecosystem impacts of climate change, the measures are different in their specific objectives, scope, time dimension, and level of collaboration required for it. Agroforestry supports to conserve and protect natural resources from mitigating non-point source pollution, controlling soil erosion, and creating wildlife habitat. The benefits of agroforestry add up to a significant enhancement of the economic and resource sustainability of agriculture. The agroforestry practices create both economic and environmental profitability in several ways like:

1. The total output per unit area of tree/ crop/livestock combinations is greater than any single component alone.
2. Crops and livestock protected from the damaging effects of wind are more productive
3. New products add to the financial diversity and flexibility of the farming enterprise.

The agroforestry is more useful for the farmer to generate constant revenue, which is not possible with traditional methods. It allows the diversification of farmhouse activity to make better use of environmental resources. The agroforestry practices have the following the environmental sensitivities:

1. *Advancement of natural resources:* the total production from an agroforestry plan is greater than the discrete production obtained by a separate cropping pattern on the same area of land. The weeds, which are naturally present in young forestry plantations are replaced by harvested crops or pasture; maintenance is less costly and environmental resources are better used.
2. *Improved control over cultivated areas:* the agroforestry contributes to retreating the cultivated area of land. The intensification of environmental resource use by agroforestry systems is not resulting in

more crop products.

3. *Creation of creativelandscapes:* Agroforestryprovides a truly innovative landscaping potential to improve the public image of farmers to society.
4. *Stabilize the greenhouse effect:* constitution of an effective system for carbon sequestration, by combining the maintenance of the stock of organic material in the soil, and the superimposition of a net fixing wooden layer.
5. *Protection of soil and water:* Improvement of biodiversity, especially by the abundance of *"edge effects"* to permit the integrated protection of crops by their association with trees, chosen to stimulate the hyperparasite (parasites of parasites) population of crops.

Owing to an increase in the population of human and cattle, there is increasing demand for food as well as fodder, particularly in developing countries like India (Dutta *et al.* 2015). Almost all forms of agroforestry systems exist across Indian Ecozones ranging from humid tropical lowlands to high-altitude and temperate biomes, and per-humid rainforest zones to parched drylands (Dwivedi *et al.* 2007). The country ranks foremost among the community of nations not only in terms of this enormous diversity and long tradition of the practice of agroforestry but also in fostering scientific developments in the subject (Nair and Nair 2014). India has been classified into eight broad agro-ecological regions. Since the neighboring countries have almost the same agro-climate as in the adjoining parts of India, an extrapolation of these agroecological zones to the countries of the Indian subcontinent gives a practical understanding of the entire subcontinent (Basu, 2014). These agro-ecological regions are shown in Figure 1, and their major features are described below. Agroforestry has a long tradition in the Indian subcontinent. The socio-religious fabric of the people of the subcontinent is interwoven to a very great extent with raising, caring for and respecting trees (Tiwari, Sharma, Narain, Raj 2007). Trees are integrated extensively in the crop- and livestock production systems of the region according to the agro-climatic and other local conditions (Project, 1963).

To cope with some of the challenges and tackle the problems of food and environmental security, the capabilities of agroforestry is required to be fully exploited. Presently some initiatives are being started by the governments, farmers, non-government organizations, and industries in the subcontinent to advance suitable agroforestry systems. The additional constant effort is essential to develop agroforestry systems suitable for each agro-ecological region.

Methodology/Conceptual Framework

Land-use opportunities increase resilience to reduce thevulnerability of contemporary societies to livelihood improvement and adaptation to environmental change. Agroforestry has the wider potential for livelihood improvement through the wider potential production of food, fodder, and firewood as well as mitigation of the impact of climate change. Theagroforestry systems have a sole contribution towards (i) biodiversity conservation; (ii) yield of goods and services to society; (iii) augmentation of the carbon storage in agro-ecosystems; (iv) enhancing the fertility of the soils, and (v) providing social and economic well-being to people. Agroforestry systems contribute variously to ecological, social and economic functions to natural ecosystems. To promote the well-being of the society, an organization of multifunctional agroforestry requirements to be strengthened by innovations for useful species and ethnic forestry systems. The forthcoming investigation is required to reduce many of the uncertainties that remain to carefully test the main functions attributed for agroforestry against alternative land-use options in order to know unequivocally. There is acuriosity to combine adaptation and mitigation measures to provide win–win solutions to climate change. Agroforestry systems offer convincing collaborations for the same. The experiential evidence from several studies supportagroforestry systems to enhance capacity through more efficient water utilization, improved microclimate, enhanced soil productivity and nutrient cycling, control of pests and diseases, improved farm productivity, and diversified and increased farm income while at the same time sequestering carbon. Although these appears precise hopeful, transactions may arise both at the farm and landscape scales. Some of the prospects of the efficient agroforestry schemes for mitigating climate change can be summarized as:

Climate-Smart Landscapes

The global challenges of climate change, food security, and poverty alleviation requires enhancing the adaptive capacity and mitigation potential of agricultural landscapes. However, adaptation and mitigation accomplishments tend to be advanced distinctly due to a variety of technical, political, financial, and socioeconomic constraints. Many agricultural systems can provide paybacks if they are designed and managed appropriately in context to their landscape. Many of the activities needed for agricultural landscapes are in line with the sustainable agriculture, but it explores a different dimension for achieving it. Mitigation actions in agricultural landscapes deal with significant benefits beyond the scope of climate change towards food security, biodiversity conservation, and poverty alleviation. Achieving these goals require transformative deviations in current policies, institutional arrangements, and funding mechanisms to foster broad-scale adoption of climate-smart approaches in agricultural landscapes. Despite the growing appreciation to track mitigation in agricultural systems and the current climate change perspective is a most significant factor to be resolved. Pursuing these activities, however, limits the potential to take advantage to minimize trade-offs across actions designed for mitigation benefits. It also leads to potential inadequacies in the use of funding for an integrated management approach to agricultural landscapes to address climate issues for ensuring the provision of food, water, and other ecosystem services.

Bioenergy from Agroforestry

Ingenious bioenergy systems can contribute to mitigating climate change through increased energy access along with rural poverty relief. Sufficient technical assistance and land management, farm yields and income can be increased to strengthen food security along with improved carbon sequestration. Agroforestry systems and practices address most of these risks to play a significant role in the ecological production of various bioenergy including efficient solid biomass, biogas, liquid biofuels, etc. The potential of such integrated approaches to providing multiple benefits, including the coproduction of food, animal feed, and organic fertilizers, while respecting economic, social, and environmental sustainability indicators can be efficiently achieved. The perspective of bioenergy from agroforestry systems progresses livelihoods and food security. The trees also play avital role for energy tosuggest ways of making bioenergy as a mainstream energy source. It mentions agricultural systems to describe various actions undertaken to answer questions concerning to constraints and motivations of farming systems.

Social Adaptation To Climate Variability

Environments provide important services that can help people adjust to climate variability and change. The role of ecosystem-based adaptation (EBA) deals with forests and trees to highlight the cases in which forests and trees support adaptation: (1) forests and trees providing goods to local communities facing climatic threats; (2) trees in agricultural fields regulating water, soil, and microclimate for more resilient production; (3) forested watersheds regulating water and protecting soils for reduced climate impacts; (4) forests protecting coastal areas from climate-related threats; and (5) urban forests and trees regulating temperature and water for resilient cities.

Climate change will affect human well-being in many parts of the world and effective adaptation is needed even under the most stringent mitigation scenarios. The role of ecosystem goods and services in societal adaptation to climate variability and change has received renewed recognition. Ecosystem-based adaptation (EBA) is an anthropocentric approach, in which ecosystem services are conserved or restored to reduce the vulnerability of people facing climate change threats (Tadesse *et al.* 2014). Ecosystem services are the benefits people obtain from ecosystems and can be classified as provisioning services (e.g., timber and firewood), regulating services (e.g., water regulation), and cultural services (e.g., recreation). Pilot projects under operation could assist as learning sites and existing information could be systematized and revisited with a climate change adaptation lens (Coppin *et al.* 2004). Examples of EBA include the restoration of mangrove shelterbelts for the protection of coastal settlements against storms and waves and the conservation of forested watersheds for the reduction of flood risk. Some of the studies can be exhibited as:

a) ***Case 1:*** *Forests And Trees Providing Goods To Local Communities Facing Climatic Threats:* Forest and tree products, such as timber and non-timber forest products (NTFPs; for example charcoal, firewood, wild fruits, mushrooms, roots, and fodder) constitute important safety nets and are part of income diversification strategies for many communities in developing countries facing increased climate variability and climate hazard risks. Rural communities use forest products as part of their coping strategies (i.e., in reaction to stresses) when crops fail due to drought (Wagner, Chhetri and Sturm 2014).

b) ***Case 2:*** *Trees in Agricultural Fields Regulating Water, Soil, and Microclimate for More:* Smallholder farmers and agriculture are threatened by rainfall and temperature variability. Trees in agricultural fields can help maintain production under a variable climate and also protect crops against climate extremes. Agroforestry (combining trees and shrubs with crops and/or livestock) is increasingly recognized as an effective approach for minimizing production risks under climate variability and change (Dutta *et al.* 2015). With their deep root systems, trees are able to explore larger soil depths for water and nutrients, which will be beneficial to crops in times of drought. Their contribution to increased soil porosity, reduced runoff, and increased soil cover leads to increased water infiltration and retention, and reduction of moisturestress during low rainfall (Bauer, Loffelholz and Wilson 2008). On the other hand, excess water is pumped out of the soil more rapidly in agroforestry plots due to their higher evapotranspiration rates.Nitrogen-fixing species are used extensively by communities in the drought-prone regions of Rajasthan, India, to help secure grain production under inadequate rainfall (Weng, Rajasekar, and Hu 2011). The benefits of agroforestry systems can be significantly enhanced through thetraditional soil and water conservation techniques where farmers practicing such techniques (e.g., stone dikes and soil pits filled with organic matter) were found more likely to regenerate and protect trees in their fields (Carsan *et al.* 2014). The trees protect irrigated plots from wind and water erosion, they regulate microclimate by decreasing insolation and evapotranspiration, and the agroforestry system needs less water than rice. Trees can be beneficial for cash crops such as coffee and cacao as well. Coffee is sensitive to microclimate fluctuations—e.g. the optimal temperature range for Arabica coffee is 18–21°C. Shade trees control temperature and humidity fluctuations and can also provide protection from wind and storm events that defoliate coffee trees (Van Noordwijk *et al.* 2014).

Interactions between Carbon Sequestration and Shade Tree Diversity

Agroforestry systems have ample potential to conserve native biodiversity to provide ecosystem services. Specifically, agroforestry systems have the potential to preserveinherent tree diversity and sequester carbon for climate change mitigation. However, little research has been conducted on the temporal firmness of species diversity and aboveground carbon stocks in these systems or the relation between species diversity and aboveground carbon sequestration. Carbon sequestration has apositive correlation with initial species richness of shade trees. Species diversity of shade trees did not change significantly but carbon stocks increased due to tree growth illustrate the opportunity for synergies between biodiversity conservation and climate change mitigation.

Agroforestry as alternative land-use production systems

The knowledge of agroforestry has advanced considerably to breedseveral agroforestry systems. An in-depthstudyof interactive processes of some agri-silvicultural systems has been undertaken and quantified. It has been found that the presence of woody species can enhance nutrient cycling, and can improve soil productivity, soil conservation and soil biotic and faunal activities (Williamson *et al.* 2014). Most agroforestry systems institute ecologically and biophysically sustainable land use systems, which are highly sustainable and economically viable, in highly complex and specialized types. Agroforestry systems have perspective of stabilizing the sloping lands and buffer zones around forest reserves, for recovering degraded lands, to improve the productivity of the bush fallow system. Rapid progress in biophysical research, field application of the science of agroforestry is still minimal. Therefore, the possibilities of exploring the multiple contributions of trees to food security, rural income generation, diversity of products and ecosystem conservation within sustainable agroforestry contextsare desired. Research efforts in the new millennium need to focus more on practical research on socio-economic to enhance the beneficial application of the science in the near future to small holder farmers.

Meta-Analysis and discussions

Global climate change is increasingly recognized as the greatest global threat facing humanity. For the majority of the world's population, the persistent problems of food insecurity, rural poverty, and the struggle to develop and sustain new sources of economic growth must now be considered against a backdrop of uncertainty and change in historical climatic patterns. Separately and together, governments and international organizations need to respond to the immediate concerns of extreme poverty, environmental degradation, and climate change. Under mounting time pressures, there is an urgent need to evolve win-win solutions that address both these immediate local and long-term global threats (Gao *et al.* 2014).

Carbon sequestration in agricultural soils and woody perennials, and the transfer of carbon credits through market structures represent one such win-win opportunity. Reforestation and agroforestry systems propose the greatest perspective to eradicate large amounts of carbon from the atmosphere. Indeed, such offsets are the only way to confiscate existing atmospheric carbon dioxide as all former mitigation measures can only lessen future emissions. Trees develop in all but the most extreme conditions (e.g., deserts and arctic). Also, many tree species also yields surplus high-value products like edible fruits and leaves, fodder for livestock, gums and oil-bearing nuts for human and industrial uses, including feedstock used in the manufacture of biofuels which offers an impeccable prospect for generating a win-win situation through removing carbon from the atmosphere and providing new sources of income. The methodology focuses on the co-development of market value-chains for sequestered carbon and secondary products. Most of the carbon market trade involves emission reduction credits but there is also growing interested in the use of trees and forests for absorbing carbon dioxide from the atmosphere (Anderson *et al.* 2008). There is substantial data that forests and agroforestry (the planting of trees on farms) in developing countries provide substantial benefits to rural dwellers, national economies, and the environment. Trees deliver a range of products for home use such as food, timber, firewood, medicines, and fodder as well as products for sale, boosting farm incomes, rural economies, and national exports. Trees on farms and in forests also offer a range of environmental services, such as conserving biodiversity, reduced soil erosion and sedimentation in rivers and lakes, and increased soil fertility (Sheoran, Sheoran and Poonia 2010). Moreover, what is often unrecognized is that while theforested area is declining in developing countries, tree cover on farms is rapidly increasing, as farmers substitute for the tree products they formerly accessed from forests and seize market opportunities for selling tree products (Tadesse *et al.* 2014). The agricultural land now accounts for over double the area of forested land in

Africa, giving justification to the slogan that, "the future of trees is on farms (Mbow, Van Noordwijk *et al.* 2014). As agroforestry binds the benefits of trees for agriculture, provides a good example of a strategy for mitigation and adaptation to sequester carbon to increase the resilience of agricultural systems by providing both income and production security.

Tree cover significantly subsidizes more to carbon pool on agricultural lands than estimated.In a global analysis of tree cover on agricultural lands using remote sensing data, it was found that the total biomass of carbon on agricultural lands to be four times higher than Inter governmental Panel on Climate Change (IPCC) default values, with trees contributing 75% of the biomass carbon (Luedeling *et al.* 2014). These results show that existing tree cover - thus far ignored in most global and regional calculations - makes a major contribution to the carbon pool on agricultural lands. Consequently, a substantial and significant correction on current estimates based on IPCC default values is warranted.

Agroforestry may offer both climate change mitigation and adaptation benefits to smallholder farmers. In addition to income from mitigation and from the production and sale of agroforestry products, trees on farms are a critical component of climate-smart agriculture in many systems. Trees regulate moisture – moderating drought or heavy precipitation – and soil temperature. Trees contribute to soil fertility by adding nutrients – in the case of nitrogen-fixing species, for example – and they contribute to increasing soil organic matter. Increasing biomass carbon on agricultural lands through agroforestry may also improve biodiversity, water quality, and in some cases, hydrological cycles (Nutini *et al.* 2014). Short term indirect benefits of agroforestry system comprises microclimate moderation and natural resources conservation (because villagers will be cutting fewer trees in the forest area) and consequential long-term benefits will be carbon sequestration (storage of carbon as a biomass) because various treespecies of agroforestry- acts as a sequester for carbon as primary forest trees and far greater than grass and crops. Agroforestry systems enable agriculture land can survive extreme weather conditions, soil erosion. It prevents deforestation, promotes soil and water conservation, nutrition recycling supplies raw material to wood based industries effects less import of wood impacts to current account deficit reduced.

Table 1. Different physiographic zones along with tree cover and geographical area

Physiographic zones	Geographical area (sq.km.)	Tree cover - area (sq.km.)	Tree cover as percentage of geographical area (%)
Western Himalayas	329255	8091	2.46
Eastern Himalayas	74618	324	0.43
North East	133990	2243	1.67
Northern Plains	295780	9473	3.2
Eastern Plains	223339	5444	2.44
Western Plains	319098	7497	2.35
Central Highlands	373675	9150	2.45
North Deccan	355988	7559	2.12
East Deccan	336289	11157	3.32
South Deccan	292416	8002	2.74
Western Ghats	72381	3847	5.32
Eastern Ghats	191698	4051	2.11
West Coast	121242	9427	7.78
East Coast	167494	6504	3.88
Total	3287263	92769	2.82

Table 1 illustrates the different physiographic zones along with their geographical extends and tree cover area. It narrates the fact that only 2.82 per cent is covered with tree cover which provides the immense possibility of adopting agroforestry practices. In partial support of the earlier arguments, the above table and statistics argue the fact that the potential agroforestry systems can contribute to the maximum potential for forthcoming demands. Trees are affected by climate variability and change, and in turn can influence regional climate by varying atmospheric processes, including water budgets. Trees on farms have substantial effects on livelihoods, both by refining ecosystem services or functions and by increasing or diversifying farm income and food and nutritional security. Agroforestry can, therefore, be considered as *"climate-smart"* because it combines improved livelihoods with mitigation for adaptation to climate change. It investigates the effects of trees on reducing farmers' vulnerability to climate variability to change their contribution to greenhouse gas mitigation. These involve biophysical, socio-economic and policy-related research to assess greenhouse gas fluxes of different agroforestry systems at the landscape scale; estimating costs and benefits as well as synergies and trade-offs to enable successful scaling up in different contexts. It also identifies the effects of trees on climate through modeling of land cover and land use change to answer queries related to climate finance, embeds agriculture and forestry for policy frameworks to support tree-based sustainable bioenergy solutions.

Conclusion

The agricultural practices in developing countries are rapidly shifting towards mitigation of global climate change. Agricultural practices in the developing countries require the modest selection of the best technology. The agroforestry is in principle economically and environmentally viable investment for a developing country. In inference, it can be underlined that these schemes are in line with the present-day requirements to achieve all requirements. A *"comprehensive"* idea represent all main characteristics all technical features, data inconsistencies, and purposes, to explore the future of the global and regional climate mitigation measures for the human and natural environment. The presence of trees in an agricultural landscape can also have a positive influence on the growing environment. Via facilitation processes, the shading of crops by tree canopies can actually be beneficial during periods of drought. Also, there has been some documentation of water availability being increased in the soil in the presence of trees. The agroforestry acts as acarbon sink to playing amajor role in greenhouse gasses sink cause of climate change. Agroforestry sequesters carbon outside forestry system, reduces pressure on forests and help conserve biodiversity and also generates employment in rural areas, reducing the dependence on the local community on forests. Agroforestry also boxes the affluent energy in the atmosphere by converting the energy into biomass, which acts as astore of environmental carbon hence more forests means less volatile carbon in the environment.

Agroforestry, not only facilitate for carbon sink, rather provide various goods and services to the peasant farmers, while facilitating in terms of fuelwood (due to incomplete combustion) and fodder, it may act as a source also. For instance, the trees in agroforests offer shade for both companion crops and the farmer against the rising temperatures to shelter the crops alongside the destructive effect of raging storms. The presence of plants on the farms ensures income divergence through the endowment of surplus resources like fruits, nuts, timber, vegetables, fodder, etc. These help the farmer cushion the effect of climate change impact like crop failure. Agroforestry plays a crucial role in climate change mitigation especially due to its tree component, as trees accumulate CO_2 (which is the most predominant GHG) in their biomass. Therefore, it not only supports in climate change mitigation but also climate change adaptation.

References

Anderson, MC *et al.* (2008). "A Thermal-Based Remote Sensing Technique for Routine Mapping of Land-Surface Carbon, Water and Energy Fluxes from Field to Regional Scales." Remote Sensing of Environment 112(12): 4227–41.

Azapagic, Adisa (2004). "Developing a Framework for Sustainable Development Indicators for the Mining and Minerals Industry." Journal of Cleaner Production 12(6): 639–62.

Basu, Jyotish Prakash (2014). "Agroforestry, Climate Change Mitigation and Livelihood Security in India." New Zealand Journal of Forestry Science 44(Suppl 1): 1-10.

Bauer, Marvin E, BcLoeffelholz, and Bruce Wilson (2008). "Estimating and Mapping Impervious Surface Area by Regression Analysis of Landsat Imagery." Remote Sensing of Impervious Surfaces: 612–25. http://books.google.com/books?hl=en&lr=&id=W jbUuuliObQ C&oi=fnd&pg=PT31&dq=Estimating+and+Mapping+Imperv ious+ Surface+Area+ by+Regression+Analysis+of+Landsat+Imagery&o ts=V4lvqT0 pJ4&sig= fIW LeSjiAZG63_mNyoHnwHhLIMs.

Carson, Sammy *et al.* (2014). "Can Agroforestry Option Values Improve the Functioning of Drivers of Agricultural Intensification in Africa?" Current Opinion in Environmental Sustainability 6(1): 35–40. http://dx.doi.org/10.1016/j.cosust.2013.10.007.

Chavan, SB *et al.* (2015). "National Agroforestry Policy in India: A Low Hanging Fruit." Current Science 108(10): 1826–34.

Continents, Land B-Y (1986). "Agroforestry for Sustainable Agriculture." (Table 3): 4–11.

Coppin, P *et al.* (2004) "Review ArticleDigital Change Detection Methods in Ecosystem Monitoring: A Review." International Journal of Remote Sensing 25(9): 1565–96.

Dutta, Dipanwita *et al.* (2015). "Assessment of Agricultural Drought in Rajasthan (India) Using Remote Sensing Derived Vegetation Condition Index (VCI) and Standardized Precipitation Index (SPI)." The Egyptian Journal of Remote Sensing and Space Science 18(1): 53–63. http://www.sciencedirect.com/science/article/pii/S1110982315000095.

Dwivedi, RP *et al.* (2007). "Socio-Economic Analysis of Agroforestry Systems in Western Uttar Pradesh." 7(September): 18–22.

Gao, Li *et al.* (2014). "Spatiotemporal Variability of Carbon Flux from Different Land Use and Land Cover Changes: A Case Study in Hubei Province, China." Energies 7(4): 2298–2316. http://www.mdpi.com/1996-1073/7/4/2298/ (April 15, 2014).

Jónsdóttir, Eydís Mary (2011). Methods "Methods for Sustainability Assessment/ : Sustainability Indicators." The university of Iceland.

Luedeling E, Roeland Kindt, Neil I. Huth and Konstantin Koenig (2014). "Agroforestry Systems in a Changing Climate-Challenges in Projecting Future Performance." Current Opinion in Environmental Sustainability 6(1): 1–7. http://dx.doi.org/10.1016/j.cosust.2013.07.013.

Mbow C, Pete Smith, *et al.* (2014). "Achieving Mitigation and Adaptation to Climate Change through Sustainable Agroforestry Practices in Africa." Current Opinion in Environmental Sustainability 6(1): 8–14. http://dx.doi.org/10.1016/j.cosust.2013.09.002.

Mbow C, Meine Van Noordwijk, *et al.* (2014) ."Agroforestry Solutions to Address Food Security and Climate Change Challenges in Africa." Current Opinion in Environmental Sustainability 6(1): 61–67. http://dx.doi.org/10.1016/j.cosust.2013.10.014.

Nair PKR and NairVimala D (2014). "Solid-Fluid-Gas: The State of Knowledge on Carbon-Sequestration Potential of Agroforestry Systems in Africa." Current Opinion in Environmental Sustainability 6(1): 22–27. http://dx.doi.org/10.1016/j.cosust.2013.07.014.

Nutini, Francesco, Mirco Boschetti, Gabriele Candiani, and Stefano Bocchi (2014). "Evaporative Fraction as an Indicator of Moisture Condition and Water Stress Status in Semi-Arid Rangeland Ecosystems." Remote Sensing 6: 6300–6323.

The project, Agro-forestry (1963). "Farm Forestry." Nature 200(4913): 1267–1267.

Sheeran V, SheeranAS, and Poonia P (2010). "Soil Reclamation of Abandoned Mine Land by Revegetation/ : A Review." International Journal of Soil, Sediment and Water 3(2): 1–21.

Tadesse, Getachew, Erika Zavaleta, Carol Shennan, and Margaret Fitz Simmons (2014). "Prospects for Forest-Based Ecosystem Services in Forest-Coffee Mosaics as Forest Loss Continues in Southwestern Ethiopia." Applied Geography 50: 144–51. http://www.sciencedirect.com/science/article/pii/S014362281400054X (April 3, 2014).

Taylor, Publisher *et al.* (2007). "International Journal of Remote Sensing The Effect of Changing Environmental Conditions on Microwave Signatures of Forest Ecosystems/ : Preliminary Results of the March 1988 Alaskan Aircraft SAR Experiment." IEEE Transactions on Geoscience and Remote Sensing (December 2014): 1119–44.

The Tate OF (2005). "Realizing the Economic Benefits of Agroforestry/ : Experiences,Lessons, and Challenges." Agroforestry Systems: 88–97.

Tiwari JC, Sharma AK, Narain P, Singh R (2007). "Restorative Forestry and Agroforestry in Hot Arid Region of India:A Review." Journal of Tropical Forestry 23(1&II): 1–16.

Van Noordwijk, Meine *et al.* (2014). "Pricing Rainbow, Green, Blue and Grey Water: Tree Cover and Geopolitics of Climatic Teleconnections." Current Opinion in Environmental Sustainability 6(1): 41–47. http://dx.doi.org/10.1016/j.cosust.2013.10.008.

Wagner, Melissa,NetraChhetri and Melanie Sturm (2014). "Adaptive Capacity in Light of Hurricane Sandy: The Need for Policy Engagement." Applied Geography 50: 15–23. http://www.sciencedirect.com/science/article/pii/S0143622814000101 (March 19, 2014).

Weng, Qihao, UmamaheshwaranRajasekar and Xuefei Hu (2011). "Modeling Urban Heat Islands and Their Relationship With Impervious Surface and Vegetation Abundance by Using ASTER Images." IEEE Transactions on Geoscience and Remote Sensing 49(10): 4080–89. http://ieeexplore.ieee.org/lpdocs/epic03/wrapper.htm?arnumber=5764519.

Williamson, David *et al.* (2014). "A Potential Feedback between Landuse and Climate in the Rungwe Tropical Highland Stresses a Critical Environmental Research Challenge." Current Opinion in Environmental Sustainability 6(1): 116–22.

7

Carbon Sequestration in Agroforestry Systems in Arid Regions of Rajasthan

M. L. Soni, V. Subbulakshmi, N. D. Yadava and K. R. Sheetal

ICAR-Central Arid Zone Research Institute Regional Research Station Bikaner-334004, India

Introduction

Agroforestry is the science of designing and developing integrated self-sustainable land management system, which involves introduction or retention of woody components such as trees, shrubs along with agricultural crops including pasture/animals simultaneously or sequentially, on the same unit of land and at the same time, to meet the ecological as well as socio-economic needs of local people. A typical agroforestry system will allow economic and ecological interactions between woody and non-woody components and also increase, sustain and diversify the total land output. Agroforestry systems are very important for the area currently under agriculture, the number of people who depend on land for their livelihoods, and the need for integrating food production with environmental services (Garrity, 2004; Makundi and Sathaye 2004). Agroforestry provides a unique opportunity to combine the twin objectives of climate change adaptation and mitigation. Although agroforestry systems are not primarily designed for carbon sequestration, there are many recent studies that substantiate the evidence that agroforestry systems can play a majorrole in storing carbon in above ground biomass (Sathaye *et al.* 2001; Montagnini and Nair 2004; Mutuo *et al.* 2005) and in soil and in below ground biomass (Nair *et al.* 2009). The amount of carbon stored in biomass and soils can be increased further by avoiding carbon releasing practices, such as deforestation, or by adopting land management practices that increase the amount of carbon stored in plant and soil. The increasing carbon storage in agroforestry systems are expected to increase carbon accumulation in the biomass of planted trees, and provide inputs of lignin rich litter that decomposes slowly to stabilize soil organic carbon (Montagnini and Nair, 2004). Some of the earliest assessments of national and global terrestrial carbon dioxide sinks reveal two beneficial attributes of agroforestry systems (a) direct near term storage (decadesto centuries)in trees and soils and (b) the potential to offset immediate GHG emissions associated with deforestation and subsequent shifting cultivation (Dixon, 1995).

The atmospheric concentration of CO_2 has increased from 310 ppm in 1950, to above 400 ppm in 2014, through changes in the land use and other developmental activities (IPCC, 2014). Tropical deforestation and forest degradation are considered to be an important source of GHG contributing to 17.4% of the global emissions (IPCC, 2007). Land use practices such as afforestration, reforestation, natural regeneration of forests and agroforestry help reduce CO_2 concentration (Canadell and Raupach, 2008). The role of forestry and agroforestry in reducing atmospheric CO_2 concentration and lowering the emissions rate of greenhouse gases (GHG) has led this system more functional (Mutuo *et al.* 2005). Globally, climate negotiations have highlighted the importance of land use sectors in mitigating the climate change. Agriculture alone accounts for 10-12% of the total global anthropogenic emissions of GHGs with an estimated non CO_2 GHG emission of 5120-6116 Mt CO_2 eq. yr^{-1} in 2005 (IPCC, 2007). Since agricultural lands are often intensively managed, they offer many opportunities to improve agronomic practices, nutrient and water management, land use practices to fit the land managers' objectives of carbon sequestration. The total carbon sequestration potential of global croplands is about 0.75-1Pg yr^{-1} or about 50% of the 1.6-1.8 Pg yr^{-1} lost due to deforestation and other agricultural activities (Lal and Bruce, 1999). Significant increases in carbon storage can be achieved by moving from lower biomass land uses to tree based systems such as forests, plantation forests and agroforestry (Roshetko *et al.* 2007).

Carbon sequestration potential of Agroforestry systems – world

Agroforestry has played a significant role in enhancing land productivity and improving livelihoods in both developed and developing countries. Although carbon sequestration through afforestation and reforestation of degraded natural forests has long been considered useful in climate change mitigation, agroforestry offers some distinct advantages. The planting of trees along with crops improves soil fertility, controls and prevents soil erosion, controls water logging, checks acidification and eutrophication of streams and rivers, increases local biodiversity, decreases pressure on natural forests for fuel and provides fodder for livestock (Makundi and Sathaye, 2004). It also has the ability to enhance the resilience of the system for coping with the adverse impacts of climate change. Agroforestry for carbon sequestration is attractive because: (i) it sequesters carbon in vegetation and in soils depending on the pre-conversion soil C, (ii) the more intensive use of the land for agricultural production reduces the need for slash-and-burn or shifting cultivation, (iii) the wood products produced under agroforestry serve as substitute for similar products unsustainably harvested from the natural forest, (iv) to the extent that agroforestry increases the income of farmers, it reduces the incentive for further extraction from the natural forest for income augmentation, and finally, (v) agroforestry practices may have dual mitigation benefits as fodder species with high nutritive value can help to intensify diets of methane-producing ruminants while they can also sequester carbon (Thornton and Herrero 2010). Agroforestry systems can have indirect effects on carbon sequestration as it helps decrease pressure on natural forests that are the largest sinks of terrestrial carbon, they also conserve soils and thus enhance carbon storage in trees and soils. Effects of agroforestry practices on the soil carbon pool indicated a rate of increase by 2-3 Mg Cha^{-1} yr^{-1} (Garg, 1998).

According to the IPCC (2007) agroforestry systems offer important opportunities of creating synergies between both adaptation and mitigation actions with a technical mitigation potential of 1.1-2.2 Pg C in terrestrial ecosystems over the next 50 years. The biomass produced in the system mitigate while the moderation of meteorological parameters help in adaptation (Monteith *et al.* 1991), the temperature (air and soil) remains less under plantations than the projected 2–4.5°C with a best estimate of 3°C by IPCC (2007). It has been reported that 630 million hectares area would be available for agroforestry, which has the potential to sequester 586 Mt C per year by 2040 (Watson *et al.* 2000). Annual carbon mitigation of about 0.072 Gt has been projected by Singh and Lal (2000) with conservative productivity of 5.5 t per hectare per year through short rotation plantations on 40 million hectares. The carbon in the above ground and below ground biomass in an agroforestry system is generally much higher than the equivalent land use without trees (i.e. crop land without any trees). The estimates of potential for carbon storage in different kinds of agroforestry systems in the worldare provided in Table 1. In Southeast Asia, agrisilvicultural systems have the capacity to store 12-228 Mg C ha^{-1} in humid tropical lands and 68-81 Mg C ha^{-1} in dry lowlands. Highest potential for carbon storage can be observed for North American silvi pastoral systems with a range of 90-198 Mg C ha^{-1} (Murthy *et al.* 2013). The potential to sequester carbon in above ground components in agroforestry systems is estimated to be 2.1×109 Mg C yr^{-1} in tropical and 1.9×109 Mg C yr^{-1} in temperate biomes (Oelbermann *et al.* 2004).

Table 1. Carbon storage potential of agroforestry systems in different eco-regions of the world

Continent	Eco region	System	Potential (Mg C ha^{-1})
Africa	Humid tropical high	Agrosilvicultural	29-53
South America	Humid tropical low	Agrosilvicultural	39-102
	Dry lowlands	Agrosilvicultural	39-195
Southeast Asia	Humid tropical	Agrosilvicultural	12-228
	Dry lowlands	Agrosilvicultural	68-81
Australia	Humid tropical low	Silvipastoral	28-51
North America	Humid tropical high	Silvipastoral	133-154
	humid tropical low	Silvipastoral	104-198
	Dry lowlands	Silvipastoral	90-175
Northern Asia	Humid tropical low	Silvipastoral	15-18

Carbon sequestration potential of agroforestry systems - India

The carbon storage capacity in agroforestry varies across species and from region to region (Newaj and Dhyani 2008; Murthy *et al.* 2013)). Carbon sequestration in different agroforestry systems occurs both below ground, in the form of enhancement of soil carbon plus root biomass and above ground as carbon stored in standing biomass. Substantial regional variability is associated with biomass production. Estimates of the total carbon storage under different agroforestry systems in India are presented in Table 3. The information about area, tree resources, their volume, farmers' preferences of species under agroforestry is hardly available. Forest Survey of India has been carrying out inventory of trees outside the forest since 2002. Table 2 shows the state/UTs wise tree green cover and carbon stock estimates under agroforestry systems in India.

Table 2. Regional estimates of total carbon storage under different agroforestry systems in India

Region	Agroforestry system	Components	Total carbon storage (t C ha^{-1})
Semi-arid	Silvipasture (5 yr old)	*Acacia nilotica* + natural pasture	9.5-17.0
		A. nilotica + established pasture	19.7
		Dalbergia sissoo + natural pasture	12.4
		D. sissoo + established pasture	17.2
		Hardwickia binata + natural pasture	16.2
		H. binata + established pasture	17.0
North-western India	Silvipasture (6 yr old)	*Acacia/Dalbergia/ Prosopis* + *Desmostacya*	6.8-18.5
		Acacia/Dalbergia/Prosopis + *Sporobolus*	1.5-12.3
Central India	Block plantation (6 yr old)	*Gemelina arborea*	24.1-31.1

(*Contd.*)

Arid (Rajasthan)	Agrisilviculture (8 yr old)	*Emblica officinalis* + *Vigna radiata*	12.7-13.0
		Colophospermum mopane + *Vigna radiata*	4.7-5.3
		Hardwickia binata + *Vigna radiata*	8.6-8.8
Semi-arid	Agrisilviculture (11 yr old)	*Dalbergia sissoo* + crop	26.0
North-western Himalayas	Silvipasture		2.17
	Agrihortipasture		1.15
	Hortipasture		1.08

Source: Basu 2014

Some of the earliest studies of potential carbon storage in agroforestry systems and alternative land use systems for India had estimated a sequestration potential of 68-228 Mg C ha^{-1} (Dixon *et al.* 1994), 25 t C ha^{-1} over 96 M ha of land (Sathaye and Ravindranath 1998). But this value varies in different regions depending on the biomass production (Pandey 2007). Studies done by Jha *et al.* (2001) showed that agroforestry could store nearly 83.6 t C ha^{-1} up to 30 cm soil depth, 26% more carbon compared to cultivation in Haryana plains. However, the magnitude of carbon sequestration from forestry activities would depend on the scale of operation and the final use of wood. The carbon sequestration potential of agroforestry systems has been established theoretically; however field measurements to validate these concepts are limited. The inherent variability in the estimates of potential carbon storage in agroforestry systems and the lack of uniform methodologies has made comparisons difficult (Jose 2009).

There is also clear evidence to suggest that the type of agroforestry system greatly influences the source or sink role of the trees (Montagnini and Nair 2004). For example, agri-silvi-cultural systems where trees and crops are grown together are net sinks while agro silvi-pastoral systems are possibly sources of GHGs. Practices like tillage, controlled burning, manuring, application of chemical fertilizers and frequent soil disturbance for raising crops in agroforestry systems can lead to significant emissions of GHGs (Montagnini and Nair 2004). The Greening India mission under the National Climate Change Action Plan targets 1.5 M ha of degraded agricultural lands and fallows to be brought under agroforestry; about 0.8 M ha under improved agroforestry practices on existing lands and 0.7 M ha of additional lands under agroforestry (Puri and Nair 2004). Much of the opportunity to store carbon through afforestation in India will occur through agroforestry on agricultural lands due to the fact that majority of arable land in India is being cultivated (Ravindranath, 2007). Agroforestry provides the best example of promoting mitigation and adaptation synergy in addressing climate change (Garrity *et al.* 2006). The carbon in the above ground and below ground biomass in an agroforestry system is generally much higher than the equivalent land use without trees (i.e. crop land without any trees).

Table 3. State / UTs wise tree green cover and carbon stock estimates under agroforestry systems

States/UTs	Tree green cover in agroforestry (km²)	Carbon stock (m tons)
Andhra Pradesh	8224	21.58
Arunachal Pradesh	2610	3.73
Assam	3922	6.51
Bihar	4570	9.75
Chhattisgarh	4535	19.41
Delhi	23	0.06
Goa	280	0.71
Gujarat	11591	26.41
Haryana	1333	3.45
Himachal Pradesh	2303	5.09
Jammu and Kashmir	2728	6.03
Jharkhand	3358	14.37
Karnataka	6090	18.33
Kerala	3803	9.97
Madhya Pradesh	6745	17.83
Maharashtra	11806	29.78
Manipur	606	0.78
Meghalaya	1876	2.42
Mizoram	464	0.59
Nagaland	1037	1.34
Odisha	5136	17.88
Punjab	1635	4.03
Rajasthan	8373	15.18
Sikkim	128	0.21
Tamil Nadu	4590	11.76
Tripura	576	0.74
Uttar Pradesh	7082	18.48
Uttarakhand	1966	4.21
West Bengal	4018	8.78
Andaman and Nicobar Islands	49	0.14
Chandigarh	0	0.00
Dadra and Nagar Haveli	56	0.17
Daman and Diu	9	0.02
Lakshadweep	16	0.01
Puducherry	16	0.05
Total	111554	279.83

Source: ISFR 2013

Factors influencing carbon sequestration in agroforestry

The amount of carbon sequestered largely depends on the agroforestry system being practiced. Other factors influencing carbon storage in agroforestry systems include tree species, system management, environment and socio-economic aspects. Further, the amount of carbon in any agroforestry system depends on the structure and function of different components within the systems put into practice.The fact that agroforestry systems can function as both source and sink of carbon (Montagnini and Nair 2004), because the integration of trees results in greater CO_2 sequestration from theatmosphere through the process of photosynthesis andlead to an enhancement of carbon storage in tree components for long term storage (Nair 2012; Arora *et al.* 2013) and soil including below ground biomass (Sharma *et al.* 2014).There is also clear evidence to suggest that the type of agroforestry system greatly influences the source or sink role of the trees. For example, agrisilvicultural systems where trees and crops are grown together are net sinks while agro silvipastoral systems are possibly sources of GHGs (Kandji *et al.* 2006). Practices like tillage, controlled burning, manuring, application of chemical fertilizers and frequent soil disturbance can lead to significant emissions of GHGs.

Potentials of the arid zones for carbon sequestration

The arid lands are prone to desertification and trees and shrubs can be grown on these lands to prevent further soil degradation and to produce biofuel. These trees also enhance soil organic carbon content and thereby increase soil carbon storage. About 104 million ha land area in the world is affected by strong and extreme soil erosion. Assuming a low rate of increase of soil organic carbon content at 40-60 kg C ha^{-1} yr^{-1} the potential of C sequestration in soil is estimated at 0.004-0.006 Pg C yr^{-1}. There are an additional 930 million ha of salt affected soils in the arid and semi arid regions of the world. Adoption of reclamation measures on these soils could potentially increase above and below ground biomass production and increase soil organic carbon (SOC). Assuming that the rate of SOC increase through adoption of reclamation measures is 200-400 kg C ha^{-1} yr^{-1}, the potential for C sequestration is 0.186-0.372 Pg C yr^{-1} (Izaurralde *et al.* 2000). Thus rehabilitation of arid lands can contribute to a great extent to solving the green house problem of the world.

The carbon sequestration potential of the arid region is high, due to their large area and their current low carbon content. The carbon sequestration in soil is strongly affected by the root production (Matamala *et al.* 2003). The trees in the arid zones put forth a large volume of below ground biomass in the form of roots. The tree roots play an important role in adding organic matter to the soil. About 25% of the total living biomass of trees is in roots which continuously add organic matter to soil by death and decay of roots. It is more important in arid zone where climax tree species like *Prosopis cineraria* that have very deep root system which can reach up to 70 m depth. Thus in arid zones large volume of carbon gets sequestered in lower layers, that have high resilience. Narain (2008) reported planting trees and grasses in the degraded lands of arid zone can help to increase the soil carbon stock from 24.3 Pg to 34.9 Pg. Land degradation is also a source of greenhouse gases.

Prevalent agroforestry systems in arid regions of Rajasthan

In arid zones, vegetation is typically sparse, and is comprised of perennial and annual grasses, other herbaceous plants, shrubs and small trees. The native plant species have adaptations that enable them to reproduce, grow and survive in the most inhospitable edapho-climatic conditions. Some plants have evolved special root systems, while others have unique leaf characteristics that allow them to withstand prolonged periods of drought.The number of tree species is very limited in arid zones, and in general, they are very slow growing due to limitations of environmental conditions, but nowhere in the world are they are so intricately associated with the life of human beings. To evade or minimize the adverse affects of frequent droughts, the native peoples in arid zones have often developed production systems in which woody perennials have a very important role, both from a productivity as well as a resource conservation point of view (Tewari 2000).The naturally growing species considered and protected in farm lands of arid regions of Rajasthan *are Prosopis cineraria, Tecomella undulata, Acacia senegal, Zizyphus mauritiana, Z. nummularia, Salvadora oleoides, S. persica, Capparis decidua, Calligonum polygonoides, Crotalaria sp., Leptadenia pyrotechnica, Calotropis procera,* etc. Other multipurpose tree species of agroforestry are *Acacia tortilis, A. luecoploea, A. nilotica, Ailanthus excelsa, Azadirachta indica* etc. The common grasses forlivestock and wildlife are *Lasiurus sindicus, Cenchrus ciliaris, C. setigerus, C. biflorus, Dichanthium annulatum, Panicum turgidum, Sporobolus sp*. etc. The western Rajasthan is divided into different rainfall zone depending on the vegetation types and the associated agricultural crops and grasses (Table 4).

Agroforestry plays an important role in the economy of arid regions due to high risk involved with arable farming, which is affected by low and highly variable rainfall, low soil fertility and high velocity wind. The people of western Rajasthan, since ages has developed a variety of site specific agroforestry system. The farmers allowed growing scattered trees and shrubs in their agriculture fields or grazing fields to sustain their life. In fact, such integration of arable crops with trees in the farming systems is a unique, combined, protective-productive system that works on the principles of ecology, productivity, economics, and sustainability. They consider these trees as boon in the region particularly during drought when rainfed crops fail. The trees provide fodder, fruit, vegetable, fuel wood, timber and fiber for sustaining rural livelihood. Agroforestry provides 62% of the fodder, fuelwood and timber requirement of the rural people. Thus, trees have a very important place in the life of people in the arid zone of India, as they are directly related to the livelihood of inhabitants, and also provide the important service of climate moderation in an inhospitable environment (Sharma and Gupta 1996).

Table 4. Tree/shrubs and associated crop/grass species in extensive agroforestry systems in the hot Indian arid zone of Rajasthan

Habitat	Annualrainfall (mm)	Tree/shrub speciesassociation	Associated crop/grass
Sand dunes, Inter dunes	110-150	*Calligonum -Haloxylon - Leptadenia*	Pearl millet, Cluster bean, *Lasiurus sindicus*
Rocky, gravellypadiments	150-200	*Zizyphus- Capparis*	Pearl millet, Green gram, Mothbean, Cluster bean / *Cymbopogonjwarncussa, Aristida* spp.,*Cenchrus ciliaris*
Sand, gravellypediments	200-250	*Calotropis-Calligonum –Clerodendrum*	Pearl millet, Cluster bean, Greengram, Moth bean, Sesame, *Cenchrus ciliaris*
Alluvial plains,soils, often-with "Kankar pans" at 80-150 cm soil depth	250-300	*Prosopis – Zizyphus – Capparis*	Pearl millet, Cluster bean, Greengram, Moth bean, Sesame, *Cenchrus ciliaris with C. setigerus*
Alluvial plains but soils are moderately saline	250-300	*Salvadora –Prosopis– Capparis*	Cluster bean, Pearl millet, Sesame and Wheat (irrigated areas) with *Cenchrus setigerus, Sprobolus* spp.
Sandy plains (rainfed)	275-325	*Prosopis – Tecomella*	Pearl millet, Cluster bean, Greengram, Moth bean/ *Cenchrus ciliaris* and *C. setigerus*
Alluvial plains (rainfed)	300-350	*Prosopis*	Pearl millet, Cluster bean, Greengram, Moth bean, *Cenchrus ciliaris*
Alluvial plains (irrigated)	300-350	*Prosopis – Acacia*	Sorghum, Cumin, Pearl millet, Mustard, Wheat

Source: Harsh *et al.* 1992

Most dominant agroforestry system in western Rajasthan is *Prosopis cineraria* based which cover about 47% of the total area (Table 5). *Ziziphus nummularia* based agroforestry occupy about 28% of the total area of western Rajasthan. Other systems involving *Acacia nilotica, Tecomella undulata* and *Acacia tortilis* occupy about 25% in combine.

Table 5.Tree component involved in the various agroforestry systems in arid region ofRajasthan

System	Surface Area, (square km)	%	District
Prosopis cineraria	158,647	46.3	Barmer, Bikaner, Churu Ganganagar, Jaisalmer, Jalore, Jodhpur, Junjhunu Nagaur, Sikar
Ziziphus nummularia	94,440	27.6	Barmer, Bikaner Jaisalmer, Jodhpur
Acacia nilotica sub sp*indica*	49,594	14.5	Ganganagar, Jodhpur Nagaur, Pali
Tecomella undulata	22, 937	6.7	Barmer
Acacia tortilis	16,672	4.9	Ganganagar
Total	342,290	100.0	

Source: Tewari *et al.* 2007

In the light of increasing pressure on land resources, Central Arid Zone Research Institute (CAZRI), Jodhpur initiated systematic studies on agroforestry systems in late 1970. Since then a number of improved agroforestry practices in order to enhance overall productivity and economic returns of the farming communities have been developed and standardized. Following improved practices have been found promising and remunerative, and easy to fit in existing traditional agroforestry systems.

Agrisilvicultural system

Prosopis cineraria: *Prosopis cineraria* is socially and culturally accepted tree in the farm lands in western Rajasthan. Its density varies widely in the farm land ranging from 1 tree per ha.in Jaisalmer to greater than 100 tree per ha. in Jhunjhunu district of Rajasthan. Both legume and non-legume crops are grown with this species. Pearl millet, moong bean, moth bean, cowpea, cluster bean or guar are grown in areas where meanannual rainfall is 400 mm or less. *P. cineraria* has synergetic effect on crops and improves soil fertility as compare to sole agriculture crop. As per the present rate of leaf forage and firewood, a farmer may reckon an annual return of Rs 3000-3500 per ha, in addition to the crop harvest. A well grown un-lopped *Prosopis cineraria* tree produces 2.5-2.8 kg fruit. *P. cineraria* provides utilizable biomass of 19.96 tons ha^{-1} including leaf fodder of 0.85 tons ha^{-1} at 12 year age (208 tree/ha) as the additional output.

Tecomella undulate: *Tecomella undulata* is another important tree of farm land and known as Marwar teak because of its timber value. A field experiment was conducted in AFRI, Jodhpur to evaluate the tree-crop interaction as affected by various tree densities (D1 - 417, D2 - 278 and D3 - 208 t ha^{-1}) associated with mung bean and pearl millet. It was found that the density of 278 stems ha^{-1} was the most favourable for crop production at the age of 6-7 years and 208 stem/ha. later at 10-11 year. Loss in SOC is less (3.2 to 35%) in agroforestry plots compared to that in control plot (56%).

***Acacia tortilis*:** This tree is a very fast growing and grows in almost all types of habitats. Harsh *et al.* (1992) reported increased production of mung bean, cluster bean and forage sorghum in association with established trees of *Acacia tortilis* but with pruning of roots by digging trenches to avoid competition. Sharma *et al.* (1992) however, reported decrease in yield of pearl millet and cluster bean grown in association with four year old *Acacia tortilis* planted in 5 m x 5 m spacing. The reduction could possible be due to root competition however, does not seem practically feasible. Acacia tortilis, however, is an excellent tree for rehabilitation of wastelands due to its fast grazing nature.

***Acacia albida*:** This tree has been introduced in arid regions from East Africa. It is a tall and straight growing tree and sheds its leaves in monsoon season and therefore does not seem to compete with crops. Harsh *et al.* (1992) reported108.2 cm mean annual height increment of this tree grown in association with mung bean and cluster bean at 5m x 5m, 10m x 5m and 10m x 10m spacings. Gupta (1992) reported maximum dry matter production of pearl millet grown in association with this tree at 10m x 10m spacing. Spacing less than this have been found to adversely affect crop besides obstructing the movement of tractors for land preparation etc.

***Acacia nilotica var. Cupressiformis*:** This is a tall tree with branches growing upward and it does not cast its shade on the crops growing underneath. Studies conducted at Pali have shown this tree to be highly compatible with crops. Moon beam, cluster bean and sorghum were successfully grown without reduction in yield in association with this tree planted at 5m x 5m spacing.

***Holoptelia integrifolia*:** This is an important fodder tree. Eight years old plantsplanted in 5m x 5m spacing at Jodhpur markedly reduced the grain and dry matter yield of mung bean and cluster bean (Paroda and Muthana 1979). Lopping, however improved the yields.

B. Silvipasture

Silvipastoral system with suitable species of trees and grasses help in increasing the land productivity and also maintain environmental potentialities. Planting the trees either on the field boundary or in rows in association with grasses provides valuable leaf fodder during scarcity or lean period (Gill 2003). Moreover, deep root system of trees binds the soil, reduces erosion and extracts moisture from deeper strata of the soil (Ahuja 1984). Perennial grasses besides providing fodder to the livestock also prevent soil erosion and ameliorate the soil health. Further, the integration of legume and application of nitrogen also improves the productivity and quality of fodder. Arid region of Rajasthan is reported to have 89 species of grasses. The species like *Lasiurus sindicus* is efficient builder of biomass with energy use efficiency of 1.4-2.0% (Harsh *et at.* 1992). In an improved silvipasture, *Hardwickia binata* was taken as tree component at 3m x 3 m spacing with *Cenchrus ciliaris* grass. Results of 9 years study revealed that average carrying capacity of the practice was 4.1 sheep ha^{-1} yr^{-1} against 3.7 for sole *C. ciliaris* pasture and 1.6 for sole *H. binata* plantation. In this type of improved silvipasture practice, in addition to grass + top feed production to the tune of 3.06 t ha^{-1} yr^{-1}, 0.26 t ha^{-1} yr^{-1} of fuel wood is also obtained. Muthana *et al.* (1980) reported 7.4, 3.2 and 1.0 q ha^{-1} dry forage yield of grass with coppiced plants of *Acacia tortilis* planted at 6 x 6, 4.5 x 4.5 and 3 x 3m spacings respectively. Higher grass production under Khejri could be due to reduced competition for moisture and increased organic matter, available nitrogen, phosphorus and micro nutrients content in soil (Aggrawal *et al.* 1976)

C. Agri-silvi-pastoral system

The wastelands and rangelands of this region are presently not capable of producing sufficient quantity of fodder for animals even for body maintenance. It is due to continuous over-exploitation of grazing lands, frequent droughts, low and erratic rainfall distribution, people negligence etc. It is fact that the improvement in animal husbandry is directly correlated with the improvement of native pastures and in this context, adoption of agri-silvi-pastoral system specially in cultivable wastelands may play a vital role in solving the problem of fodder shortage for animals in addition to fulfil the other daily needs of ever increased human and animal population of the region (Hazra 2014). Sharma (2015) observed that for getting higher and remunerative productivity in cultivable wasteland of arid tropics of north-western India, grain crop of cluster bean or moth bean and sewan grass, could be grown in association of multipurpose tree species (*Prosopis cineraria, Ailanthus excelsa* and *Tecomella undulata*).

D. Agri-horti-silvicultural systems

The best combination of horti and silvicultural tree species have the potential in improving soil, sequestering carbon and enhancing food production. Combination of silvicultural species *Prosopis cineraria, Ailanthus excelsa* and *Colophospermum mopane* were planted at on a farmers' land near Bilara, Jodhpur district of Rajasthan with horticultural species *Ziziphus mauritiana, Cordia myxa,* and *Emblica officinalis* and intercropped with wheat (*Triticum aestivum*). *Z. mauritiana* + *P. cineraria* were the best combination because of less competition of these tree species with wheat crop. In addition, this system provided fruit, fodder and fuel wood and helped farmers to get greater benefits as well as to control land degradation. The results emphasized on selecting suitable combination of horti and silvicultural species based on scientific knowledge of its potential in soil fertility improvement and structural root architecture and the adoption of management strategies to increase overall production on sustained basis (Singh *et al.* 2013).

E. Agrihorticulture

In this system fruit trees are grown with agricultural crops. Leguminous crop sown under ber (*Zizyphus mouritiona cv. Seb*) plantation produced 0.2 t ha^{-1} of grain and 0.8 t ha^{-1} quality ber fruits from same land unit even when seasonal rainfall is 200 mm, thus rendering a drought proofing mechanism to the system. The density of ber plants were kept 400 individuals/ha (Gupta 1997). The economics of this improved system indicated that in case of sole leguminous crop (mung bean) farming, the net profit per hectare was Rs. 4800 ha^{-1}, however, in case of ber intercropping, the profit was to a tune of Rs. 8000 ha^{-1}. Pomegranate another fruit tree has been found compatible with pearl millet, mung bean, isabgol, sorghum and cumin in the irrigated areas of western Rajasthan

F. Hortipasture

Integration of fruit trees with pasture (grasses and or legumes) in the same unit of land is called as Horti-pastoral system. This system acts asone of the best and economic alternative system for class IV and V type of land (Singh, 1996; Sharma 2004). It can supply the protective food (fruit) for humanbeing and fodder for animal and thus help in bridging the wide gap between the supply and demand of fruit and fodder. Ber based horti-pasture have proved highly remunerative in Thar desert on farmers' field. Ber trees were planted at a spacing of 6x6 m and grass *Cenchrus ciliaris* was introduced between tree rows after third year. On an average, the dry grass production was 1.55 t ha^{-1} $year^{-1}$ (Tewari *et al.* 1999). The fruit, leaf fodder and fuel wood production from ber was 2.77, 1.87, 2.64 t $ha^{-1}yr^{-1}$, respectively.

Carbon stocks in agroforestry systems in arid Rajasthan

A. Carbon sequestration in tree biomass

A study was conducted to compare carbon accumulation in both tree biomass and soil (0-30 cm in depth) in a six year old agri-silvi-horti system grown on a farmer field in arid region of Rajasthan. Silvicultural species were *Prosopis cineraria*, *Ailanthus excelsa* and *Colophospermum mopane* along with

Zizyphus mauritiana, Cordia myxa and *Emblica officinalis* horticultural species planted alternate to each other. These were intercropped with wheat (*Triticum aestivum*). The highest carbon content was 45.84% recorded in *C. mopane* and lowest was 43.61% in *A. excelsa* trees. Average carbon stock was highest in *P. cineraria* based agroforestry than other two silviculture species. It was more in agroforestry than in sole horti- and silvi-species as well as agriculture plots (Singh and Singh 2015).

Under waterlogged saline soils in IGNP area in Lakhuwali (Hanumangarh, Rajasthan), Soni *et al.* (2012) reported that among the three dominant species grown in water logged saline areas, the above ground biomass was highest in *Acacia nilotica* (132.1 Mg ha^{-1}) followed by *Eucalyptus spp.*(77.6 Mg ha^{-1}) and *Acacia tortilis* (40.6 Mg ha^{-1}). Maximum carbon storage was observed by *A. nilotica* (66.5 Mg ha^{-1}) followed by *Eucalyptus* spp. (38.8 Mg ha^{-1}).

B. Soil organic carbon enhancement

A high level of organic matter accumulation was reported under *Prosopis cineraria*, over a period of about 14 years (Aggarwal and Lahiri 1977). Traditional *Prosopis cineraria* based systems lead to a 50% increase in SOC largely due to leaf litter (Venkateswaralu 2010). In a study conducted at Jodhpur reveals that silvipasture and silviculture are the better option for increasing SOC sequestration. Agroforestry and pearl millet-legume sequence are the other options in minimizing SOC depletion and CO^2 emission (Singh *et al.* 2007). Bhati and Joshi (2007) reported that in an agroforestry system at CAZRI Jodhpur SOC (%) was more under leguminous trees namely *Acacia albida* (0.179) and *Prosopis cineraria* (0.165) compared to non-leguminous trees namely *Tecomella undulata* (0.138) and *Ziziphus mauritiana* (0.138). Considering the carbon input in soil, components of plant both above the ground namely leaf and stem and below the ground namely root are critical. In low rainfall zone large volume of carbon gets sequestered through root specially the tree roots. Among different tree species tried viz. *P. juliflora, A. tortilis, P. cineraria, A. senegal* and *Capparis decidua*, organic carbon increased from 0.03 to 0.47 per cent under P. cineraria based windbreak areas in western Rajasthan, the minimum was observed under *P. juliflora* cover (Sharma and Gupta 1989). Based on the study conducted at Jodhpur it was found that the growth of *Prosopis cineraria* and *Tecomella undulata* has increased the soil fertility status with respect to organic carbon (Aggarwal *et al.* 1976). Highest soil organic carbon in autumn (October) and lowest in summer has been observed under three year old agroforestry systems comprising *Emblica officinalis, Hardwickia binata* and *Colophospermum mopane* tree species with *Vigna radiata* as intercrop at Jodhpur, Rajasthan (G. Singh and Gupta 2000). The decrease in organic carbon during cropping and late spring to summer resulted probably due to decomposition of organic matter owing to enhanced tillage activity, moisture availability and increased microbial population (Venkateswarulu and Aggrawal, 1980) during monsoon. Decrease in SOC during summer (June) could be due to increased soil temperature (Raich, 1983). Considering the plots, SOC was greatest under *C. mopane* (0.211%) followed by *H. Binata* (0.204%), whereas, it was significantly ($P<0.01$) low in control (0.156%). Significantly higher SOC ($P<0.01$) in planted area coincides with the periods of litter production from the tree species and soon after the crop harvest. Research results during the past two decades show that three main tree-mediated processes determine the extent and rate of soil improvement in agroforestry systems. These are (i) increased nitrogen input by N_2–fixing trees, (ii) enhanced availability of nutrients resulting from production and decomposition of tree biomass, and (iii) greater uptake and utilization of nutrients from deeper layers of soils by deep-rooted trees. Silvopastoral systems improve the soil organic carbon and other nutrients of degraded soil (Soni *et al.* 2008, Shamsudheen *et al.* 2009).

Agroforestry system management for carbon sequestration

Growth rate differences among tree species and the native vs. exotic species controversies are among widely debated but not yet resolved biological issues related to C sequestration by trees in agroforestry systems (Nair *et al.* 2009). However, planting of nativespecies, because of their supposedly better adaptability to local conditions, would be superior to exotic ones for use in such plantations. Another aspect of uncertainty is the differences in wood quality of species in relation to their C-accumulation rates. In astudy of 32 neotropical species in the Amazon, Elias and Potvin (2003) found that the pioneer species (e.g., *Ochroma pyramidale*) presented some of the highest and non pioneers exhibited some of the lowest C values. However, the wood of slower growing species is usually of higher specific gravity than that of faster-growing species, such that the slow growing species may accumulate more C in the long-term (Bunker *et al.* 2005; Redondo-Brenes and Montagini 2006). The more valuable, high specific gravity wood also constitutes a longer term sink for fixed C (e.g., construction timber, furniture, wood crafts) than low specific gravity wood used for short lived purposes such as packaging cases and poles. Mixed plantings of N_2 fixing tropical species and commercial timber trees have been reported to produce more above ground biomass or volume production compared to their monoculture stands (Kumar *et al.* 1998).

Other silvicultural aspects such as stand density and rotation length may also influence biomass production and the perceived carbon sequestration potential of species. Overall, high density stands sequester larger amounts of C than lower density stands. Although these findings per se do not imply that mixed species planting is not important, they suggest that choice of species and it's management are critical to promoting C sequestration. This may, however, create conflicts with plantation management objectives such as timber, highlighting the need for stand density regulation approaches that are in sync with land management objectives (Kumar *et al.* 1995). Design of planting schemes to make trade-offs between generating ecological services (e.g., C sequestration) and goods (e.g., timber) is indeed a major silvicultural challenge.

Carbon-sequestration programs and rurallivelihood security

The CDM under the Kyoto Protocol allows industrialized countries with a GHG reduction commitment to invest in mitigation projects in developing countries as an alternative to what is generally more costly in their own countries. This offers an economic opportunity for subsistence farmers in developing countries, the major practitioners of agroforestry, for selling the C sequestered through agroforestry activities to industrialized countries; it will be an environmental benefit to the global community at large as well. Projects under the CDMs have the dualmandate of reducing GHG emissions and contributing to sustainable development. Carbon trading is also rapidly expanding, now that the World Bank and other institutions have established funds to facilitate the establishment of CDM projects (World Bank 2004). Industrialized countries consider CDM as apotential source for low-cost emission credits, while developing countries hope it may attract new and additional investment for sustainable development. Potentially there are twoways in which farmers could benefit from entering into contracts to sequester C: (1) farmers would be compensated for the C they sequester, based on the quantity of C sequestered and the market price of C; (2) farmers would benefit fromany gains in productivity associated with the adoption of C sequestering practices (Nair *et al.* 2009).

The success in the implementation of the agroforestry project for GHG mitigation will depend on the farmers' willingness to participate in the project. Several reasons have been recognized in support of introducing C-sequestration benefits into smallholder agroforestry practices in developing countries: (1) the sequestration service does not need to be physically transported, thus, it can benefit people in remote areas, most of whom are very poor. (2) There are no quality differences: a molecule of carbon is the same wherever it is located; so the problem often faced by smallholders in not being able to achieve the quality required by international markets in agricultural commodities does not apply here. Furthermore, even small amounts of additional income would make a great difference for these subsistence farmers who have very limited alternate employment opportunities to make such additional cash income (Takimoto *et al.* 2008).

Conclusion

Addressing land degradation through improved farming practices in arid region therefore also offers climate change mitigation benefits. Increasing the tree component in the farming system will sequester atmospheric carbon in the plant biomass and in the soil, thereby mitigating climate change. Reclamation of salt affected soils, sand dune stabilization, and rehabilitation of rangeland would further lock the atmospheric carbon in biomass and provide stability to the ecosystem. Globally, C trading is rapidly expanding, and the CDM of the Kyoto Protocol offers an attractive economic opportunity for subsistence farmers in developing countries, the major practitionersof agroforestry, for selling the C sequestered through agroforestry activities to industrialized countries. It will be an environmental benefit to the global community at large as well. The political environment is also favorable for enhancing small holder involvement in GHG-mitigation projects. The successin the implementation of such projects will depend onthe farmers' willingness to participate in the project. Implementation of National Agroforestry Policy (2014) in India, which regards tree-crop interface an important option to support climate smart agriculture to scale up the practical realization of carbon benefits with appropriate technological and market support, would open up the new avenues for the farmers.

References

Aggarwal RK, Gupta JP, Saxena SK, Muthana KD (1976). Studies on soil physico-chemical and ecological changes under twelve years old five desert tree species of western Rajasthan. Indian Forester 102:863–872.

Aggarwal RK, Lahiri AN (1977). Influence of vegetation on the status of organic carbon and nitrogenin desert soils. Science and Culture 43:533–535.

Ahuja LD (1984). Range management in agroforestry system in arid region of India. In: Shankarnarayan KA (ed) Agroforestry in Arid and Semi-Arid Zone, Central Arid Zone Research Institute, Jodhpur,pp. 161-166.

Arora G, Chaturvedi S, Kaushal R, Nain A, Tewari S, Alam NM, Chaturvedi OP (2013). Growth, biomass, C stocks and sequestration in age series *Populus deltoides* plantations in Tarai region of central Himalaya. Turkish Journal of Agriculture and Forestry 38: 1–11.

Bhati TK, Joshi NL (2007). Farming systems for sustainable agriculture in Indian arid zone. In: Vittal KPR *et al.* (eds) Dryland Ecosystem- Indian Perspective, pp.35-52.

Bilas Singh, Bishnoi Mahipal, Baloch Mana Ram, Singh G (2014) Tree biomass, resource use and crop productivity in agri-horti-silvicultural systems in the dry region of Rajasthan, India. Archives of Agronomy and Soil Science 60(8):1031–1049.

Bilas Singh, Singh G (2015). Biomass production and carbon stock in a silvi-horti based agroforestry system in arid region of Rajasthan. Indian Forester 141(12).

Bunker DE, De Clerk F, Bradford JC, Colwell RK, Perfecto Y, Phillips OL, Sankaran M, Naeem S (2005). Species loss and above-ground carbon storage in a tropical forest. Science 310: 1029–1031.

Canadell JG, Raupach MR (2008). Managing forests for climate change mitigation. Science 320: 1456-1457.

Dixon RK (1995). Agroforestry systems: sources or sinks of greenhouse gases? Agroforestry Systems 31:99-116.

Dixon RK, Brown S, Houghton RA, Solomon AM, Trexler MC, *et al.* (1994) Carbon pools and fluxes of global forest ecosystems. Science 263:185-190.

Elias M, Potvin C (2003). Assessing inter- and intra-specific variation in trunk carbon concentration for 32 neotropical tree species. Canadian Journal of Forest Research 33:1039–1045.

Garg VK (1998). Interaction of tree crops with a sodic soil environment: Potential for rehabilitation of degraded environments. Land Degradation and Development 9: 81-93.

Garrity D, Okono A, Grayson M, Parrott S (2006) World Agroforestry into the Future Nairobi: World Agroforestry Centre.

Garrity DP (2004). Agroforestry and the achievement of the millennium development goals. Agroforestry System 61:5-17.8.

Gupta JP (1992). Role or agroforestry for sustainable production in arid areas. Lecture delivered at 2nd annual group meeting cum symposia on agroforestry held at College of Agriculture, Nagpur, March 4-6.

Gupta JP, Rao GGSN, Ramakrishna YS, Ramana R (1984). Role of shelterbelts in arid zone. Indian Farming 34(7):29-30.

Harsh LN, Tewari JC, Burman U, Sharma SK (1992). Agroforestry in arid regions. Indian Farming 45:32-37.

Hazra CR (2014). Feed and forage resources for sustainable livestock development. Range Management and Agroforestry 35:1-14.

Indu K Murthy, Gupta Mohini, Tomar Sonam, Munsi Madhushree, Tiwari Rakesh, Hegde GT, Ravindranath NH (2013). Carbon sequestration potential of agroforestry systems in India. Journal Earth Science and Climate Change 4(1):1-7.

IPCC (2007). Summary for policy makers. In: Solomon S *et al.* (eds) Climate Change 2007: The Physical Science Basis Contribution of Working Group I to the Fourth Assessment Report of the Intergovernmental Panel on Climate Change Cambridge University Press, Cambridge, UK and New York, USA, pp. 996.

IPCC (2014). Summary for policymakers. In: Field *et al.* (eds)Climate Change 2014: Impacts, Adaptation, and Vulnerability. Part A: Global and Sectoral Aspects. Contribution of Working Group II to the Fifth Assessment Report of the Intergovernmental Panel on Climate Change,Cambridge University Press, Cambridge, UK and New York, USA, pp. 1-32.

ISFR (2013). India State of Forest Report.Forest Survey of India (Ministry of Environment and Forests) Kaulagarh Road, P.O –IPE, Dehradun, India.

Izaurralde RC, Rosenberg NJ, Lal R (2000). Mitigation of climate change by soil carbon sequestration: issues of science, monitoring and degraded lands. Advances in Agronomy 70:1-75.

Jha MN, Gupta MK, Raina AK (2001). Carbon sequestration: Forest soil and land use management. Annals of Forestry 9:249-256.

Jose S (2009). Agroforestry for ecosystem services and environmental benefits:an overview. Agroforestry Systems 76:1-10.

Jyotish Prakash Basu (2014). Agroforestry, climate change mitigation and livelihood security in India. New Zealand Journal of Forestry Science 44 (Suppl 1): S11.

Kandji ST, Verchot LV, Mackensen J, Boye A, Van NM, *et al.* (2006). Opportunities for linking climate change adaptation and mitigation through agroforestry systems. In: Garrity DP, Okono A, Grayson M, Parrott S (eds) World Agroforestry into the Future, World Agroforestry Centre (ICRAF), Nairobi, Kenya. pp. 113-121.

Kumar BM, Kumar SS, Fisher RF (1998). Intercropping teak with Leucaena increases tree growth and modifies soil characteristics. Agroforestry Systems 42:81–89.

Kumar BM, Long JN, Kumar P (1995). A density management diagram for teak plantations of Kerala in peninsular India. Forest Ecology, Manage 74:125–132.

Lal R, Bruce J (1999). Thepotential of world cropland to sequester c and mitigate the greenhouse effect. Environment Science and Policy 2:177-185.

Makundi WR, Sathaye JA (2004). GHG mitigation potential and cost in tropical forestry-relative role for agroforestry. Environment, Development and Sustainability 6:235-260.

Matamala R, Gozaler-Meler MA, Jastrow JD, Norby RJ, Sclesinger WH (2003). Impacts of fine roots turnover on forest NPP and soil carbon sequestration potential. Science 302:1385-1387.

Montagnini F, Nair PKR (2004). Carbon sequestration: An under exploited environmental benefit of agroforestry systems. Agroforestry Systems 61:281-295.

Monteith JL, Ong CK, Corlett JE (1991). Microclimatic interaction in agroforestry systems. Forest Ecology and Management 45:31–44.

Murthy IK, Gupta M, Tomar S, Munsi M, Tiwari R (2013). Carbon sequestration potential of agroforestry systems in India. Journal of Earth Science and Climate Change 131:1–7.

Muthana KD, Sharma SK, Raina Ashok, Meena GL (1980). Silvi-pastoral studies. Annual Progress Report, CAZRl, Jodhpur.

Mutuo PK, Cadisch G, Albrecht Palm CA, Verchot L (2005). Potential of agroforestry for carbon sequestration and mitigation of greenhouse gas emissions from soils in the tropics. Nutrient Cycling in Agro ecosystems 71:43-54.

Nair PKR (2012). Carbon sequestration studies in agroforestry systems: a reality check. Agroforestry systems 86:243–253.

Nair PKR, Mohan Kumar B, Vimala D Nair (2009). Agroforestry as a strategy for carbon sequestration Journal of Plant Nutrition and Soil Science 172:10–23.

Nair PKR, Nair VD, Kumar BM, Haile SG (2009). Soil carbon sequestration in tropical Agroforestry systems: a feasibility appraisal. Environmental Science and Policy 12:1099-1111.

Narain P (2008). Dryland management in arid ecosystem. Journal of the Indian Society of Soil Science 56:337-347.

National Agroforestry Policy (2014). Government of India, Department of Agriculture & Cooperation, Ministry of Agriculture, New Delhi.

Newaj R, Dhyani SK (2008). Agroforestry for carbon sequestration: Scope and present status. Indian Journal of Agroforestry 10:1-9.

Oelbermann M, Voroney RP, Gordon AM (2004). Carbon sequestration in tropical and temperate agroforestry systems: a review with examples from Costa Rica and southern Canada Agriculture. Ecosystems and Environment104: 359-377.

Pandey DN (2007). Multifunctional agroforestry systems in India. Current Science 92: 455-463.

Paroda RS, Muthana KD (1979). Agroforestry practices in arid desert. Paper presented at Seminar on "Agro-forestry" held at Imphal (Manipur).

Redondo-Brenes A, Montagnini F (2006). Growth, productivity, aboveground biomass, and carbon sequestration of pure and mixed native tree plantations in the Caribbean lowlands of CostaRica. Forest Ecology and Management 232:168–178.

Sathaye JA, Makundi WR, Andrasko K, Boer R, Ravindranath NH (2001). Carbon mitigation potential and costs of forestry options in Brazil, China, India, Indonesia, Mexico, Philippines and Tanzania. Mitigation and Adaptation Strategies for Global Change 6:185-211.

Sathaye JA, Ravindranath NH (1998). Climate change mitigation in the energy and forestry sectors of developing countries. Annual Review of Energy and Environment 23:387-437.

Shamsudheen M, Devi Dayal, Meena SL, Bhagirath Ram (2009). Improvement of soil properties under silvipastoral systems in the Kachchh region of arid Gujarat. In: 4th World Congress on Conservation Agriculture: Innovations for Improving Efficiency, Equity and Environment, (February 4-7), National Academy of Agricultural Sciences, New Delhi, India. pp.253-254.

Sharma H, Gupta MK, Chauhan PS (2014). Carbon sequestration:status of sequestered soil organic carbon under different land use in Jhalwar district of Rajasthan. Indian Forester 140:780–785.

Sharma AK, Gupta JP (1996). Agroforestry systems for the hot arid regions of India. In: proceedings of IUFRO-DNAES Conference on Resource Inventory Techniques to support agroforestry and environment activities. October 1-3, Chandigarh. pp. 259-262.

Sharma BD, Gupta IC (1989). Effect of trees cover on soil fertility in western Rajasthan. Indian Forester 115(5):348-354.

Sharma BM, Rathore SS, Gupta JP (1992). Compatibility studies on *Acacia tortilis* and *Zizyphus rotundifolia* with some arid zone crops. Indian Forester

Sharma KC (2015). Performance of different grain legumes and pasture grasses under agri-silvi-pastoral system in arid tropics of India. Range Management and Agroforestry 36 (1):41-46.

Sharma SK (2004). Hortipastoral based land use systems for enhancing productivity of degraded lands under rain fed and partially irrigated conditions. Uganda Journal of Agricultural Sciences 9:320-325.

Singh G, Gupta GN, Kuppusamy V (2000). Seasonal variations in organic carbon and nutrient availability in arid zone agroforestry systems. Tropical Ecology 41(1): 17-23.

Singh R, Lal M (2000). Sustainable forestry in India for carbon mitigation. Current Science 78(5):563–567.

Singh RP (1996). Alternate land use system for sustainable development. Range Management and Agroforestry 17 (2):155-177.

Singh Surendra Kumar, Kumar Mahesh, Sharma Brij Kishore, Tarafdar JC (2007). Depletion of organic carbon, phosphorus, and potassium stock under a pearl millet based cropping system in the arid region of India. Arid Land Research and Management 21:119–131.

Soni ML, Beniwal RK, Garg BK, Tanwar SPS, Burman U, Yadav P, Yadava ND (2012). Above ground biomass production and carbon storage during restoration of water logged saline soils of IGNP area. Crop Improvement (Special Issue): 555-556.

Soni ML, Beniwal RK, Yadava ND, Talwar HS (2008). Spatial distribution of soil organic carbon under agro forestry and traditional cropping system in hyper arid zone of Rajasthan. Annals of Arid Zone 103-106.

Takimoto A, Nair PKR, Alavalapati JRR (2008). Socio economic potential of carbon sequestration through agroforestry in the West African Sahel. Mitigation Adaptation Strategy for Global Change, 13, pp.745–761.

Tewari JC, Bohra MD, Harsh LN (1999). Structure and production function of traditional extensive agroforestry system and scope of intensive agroforestry in Thar Desert. Indian Journal of Agroforestry 1(1): 81-94.

Tewari JC, Sharma AK, Naraian P, Singh Raj (2007). Restorative forestry and agroforestry in hot region of India: A review. Journal of Tropical Forestry 23:1-16.

Thornton PK, Herrero M (2010). Potential for reduced methane and carbon dioxide emissions from livestock and pasture management in the tropics. In: Proceedings of the National Academy of Sciences of the United States of America, 107(46): 19667-19672.

Venkateswaralu J (2010). Rainfed agriculture in India: Research and development scenario.Indian council of Agricultural Research, New Delhi pp. 508.

Venkateswarulu B, Aggrawal RK (1980). Influence of different management practices on micro-organism of desert soil. Indian Journal of Microbiology 20: 149-151.

Watson R, Noble IP, Bolin B, Ravindernath NH, Verado DJ, Dokken DJ (2000). Land use change and forestry. Cambridge University Press, Cambridge, pp. 375.

World Bank (2004). Carbon finance at the World Bank. http://carbonfinance.org.

8

Geospatial Technology: An Effective Tool for Mapping, Monitoring and Decision Support in Agroforestry for Sustainability

Mahesh Kumar Gaur, R.K. Goyal and J.S. Chouhan

ICAR-Central Arid Zone Research Institute, Jodhpur, India

Introduction

India is the seventh-largest country in the world, with a total area of 3,287,263 square kilometres, roughly 2.5 percent of global landmass but is home to 17.9 percent of global human population and 1/5 of the livestock. Most of this increase in human population has occurred in second half of 20th century. Similarly, India has the highest number of cattle (210.2 m), buffaloes (111.3 m) and goats (150.4 m) in the world. It ranks second in sheep (74 m), fifth in chickens (866 m) and sixth in ducks (26 m). The increased population and associated increased demand of resources has led to significant impact on land use-land cover and has consequence in negatively impacting various ecosystem services i.e., hydrology, soil conservation, wildlife support, biodiversity and environmental protection, these agroforestry systems used to provide. Agroforestry has traditionally been a way of life and livelihood in India for centuries. Now it is a modern science inviting deliberate management of trees on farms and surrounding landscape (Bargali *et al.* 2009; Parihaar *et al.* 2015). In India, agroforestry practices are prevalent in different agro-ecological zones and occupy sizeable areas. However, the type and composition and extent vary extensively from place to place and region to region because of wide-ranging topography, biophysical characteristics, land uses and socio-economics (Singh *et al.* 2012). Actually, agroforestry is an ancient forest land use management system in which agriculture is integrated with tree system to increase overall productivity in the short, medium and long term (in comparison with sole forest land), biodiversity (in comparison with sole agricultural land) and sustainability of land (multi-production system). As such, agroforestry combines agriculture and forestry technologies to create more integrated, diverse, productive, profitable, healthy and sustainable land use systems.

Depending on the dominant and prevailing land use systems, these agroforestry systems canlargely be categorised into agro-silvicultural, agro-silvi-horticultural, silvo-pastoral and agro-silvo-pastoral. Examples prevailing in hot arid zone from each land use systems are given in the subsequent paragraphs (Gaur *et al.* 2017).

Satellite Remote Sensing and GIS for Natural Resource Management

Remote sensing and Geographic Information System (GIS) has emerged as powerful tool for planning and decision support in the area of agricultural research and management. Satellite-derived data products are attractive for monitoring because of the synoptic view of the landscape they provide. A host of products are available for the contiguous areas that address tree cover, but each has limited utility with regard to narrow linear plantings, or sparse cover, such as pasture or rangeland with trees. Agroforestry is characterized by high diversity and complex interactions that require management for multiple objectives, alternatives, and social interests over varied landscapes (Ellis, Bentrup AND Schoeneberger, 2004).

A chief use of remotely sensed data is to produce a classification map of the identifiable or meaningful features or classes of the land cover types in a scene. In the field of remote sensing, image classification is a process in which pixels or the basic element unit of an image are assigned to classes. Bycomparing pixels to one another and to those known identity, it is possible to assemble groups of similar pixels into classes and produce a thematic map. Image classification is defined as the process of creating thematic maps from satellite imagery (DeFries *et al.* 1999). A thematic map is an informational representation of an image which conveys information regarding the spatial distribution of particular theme say vegetation (Campbell, 1996). The objective of image classification is to classify each pixel of an image into land cover categories. In the case of hard classification, each pixel is assigned to only one class. However in fuzzy or soft classification, a pixel is associated with many land cover classes.

The resolution of imagery selected for monitoring should be appropriate to the features to be observed. O'Neill *et al.* (1996) recommend the grain size for map elements be one-fifth to one-half the size of the features of interest. There are many examples of high-resolution imagery used in natural resource monitoring applications of small targets. Laliberte *et al.* (2004) used QuickBird imagery (61-cm panchromatic and 2.4-m multispectral) to assess shrub encroachment.

Geographic Information Systems (GIS) are tools for acquiring, managing, analysing and visualization of spatially explicit information. GIS transform varied datasets into easy-to-read and easy-to-access maps and generate required information. In addition, the advantages are numerous. GIS architectures have traditionally focussed on a static environment in which users perform spatial analysis. Further, GIS offers a sound technical approach for documenting, monitoring, and evaluating the impact of agricultural farming activities. "Successful design of agroforestry practices hinges on the ability to pull together very diverse and sometimes large sets of information (i.e., biophysical, economic, and social factors), and then implementing the synthesis of this information across several spatial scales from site to landscape" (Ellis, Bentrup & Schoeneberger 2004 p. 401). Campagna (2006) draws attention to the fact that economic, social and environmental processes are inherently spatial due to their occurrence in space and their temporal changes. She argues that GIS is an essential tool that can be used to offer effective support for spatial planning and decision making while addressing these issues. According to Tomlinson (2007), there is an abundance of reliable technologies and spatial data that significantly widens the scope of potential GIS applications. "Geographic information systems integrate seemingly disparate information quickly and visually, which facilitates communication, collaboration, and decision making" (Tomlinson 2007). Ellis, Bentrup and Schoeneberger (2004) discusses the development of decision support tools (DSTs) including GIS in agroforestry (Table 1).

GIS has been applied to the management of agroforestry practices in various parts of the world in a wide range of agro-ecological regions from tropical to temperate. The use of information systems in agroforestry began with development of databases to aid in guiding plant selection. With the changing needs, improved technologies, and data proliferation, their application has evolved to monitoring, precision farming, and modelling different systems and predicting outcomes of different scenarios (Ellis, Bentrup & Schoeneberger, 2004).

The perpetual advancements in technology and availability of more information about agroforestry systems have seen not only the rise of internet-based GIS, but also better functionalities to meet the needs of diverse audiences. These applications range from simple demonstrations and references to GIS use, to complex geoprocessing tools used to solve spatial problems and support decision making. The internet is affecting GIS in three major areas of GIS data access, spatial information dissemination, and GIS processing and analysis. It enhances

Table 1. GIS-based decision support tools used in agroforestry

Decision support tool	Description	References
Agroforestry System Suitability in Africa	Spatial analysis using climate, soil land use and other spatial data alongside plant species data to determine species and agroforestry suitability	Booth *et al.* 1989; Booth *et al.* 1990; Unruh and Lefebvre, 1995
Agroforestry System Suitability in Ecuador	GIS Spatial analysis to determine suitable areas of *Annona cherimola* agroforestry systems in Southern Ecuador.	Bydekerke *et al.* 1998
Agroforestry System Assessment in Nebraska	Spatial suitability assessment for willow and forest farming agroforestry systems in a Nebraska watershed	Bentrup and Leininger, 2002
Agroforestry Parklands in Burkina Faso	Spatial analysis of dynamics of agroforestry parklands and species distribution due to human impacts	Bernard and Depommier, 1997
Agroforestry Planning Tool in China	Hybrid DST integrating GIS data, regression models plus expert knowledge to assess biophysical, social and economic suitability of *Paulownia* intercropping agroforestry systems	Liu *et al.* 1999
PLANTGRO (Plantation and Agroforestry Species Selection Tool) Hybrid	Plantation and agroforestry species selection tool integrates GIS and expert system on plant growth	Booth, 1996; Hackett and Vanclay, 2003
SEADSS (South eastern Agroforestry Decision Support System)	Landscape and site-scale agroforestry planning and species selection DST for landowners and extension agents of Southeast US that integrates GIS, tree and shrub database and expert knowledge	Ellis *et al.* 2003
Conservation Buffer Planning Tools for Western Cornbelt Region, USA	Suite of GIS, economic models and visualization tool for landowners and resource managers to evaluate agroforestry strategies in Midwest Cornbelt region of the USA	Bentrup *et al.* 2003

Note: Adapted from Computer-based tools for decision support in agroforestry: Current state and future needs (p.8), by Ellis EA, Bentrup G, & Schoeneberger MM (2004)

accessibility and reusability of GIS analysis tools, and enables users to work on GIS data using web browsers without installing GIS software on their local machines (Peng & Tsou 2003).

Normalized Differential Vegetation Index (NDVI)

With mid-resolution optical satellite imagery conifer versus broad-leaf and deciduous versus evergreen deciduous in the canopy may be distinguished by their reflectance (over seasons). In cases of a limited number of forest tree species, forest and TOFs (trees out of forests) with a species reference even without field check (e.g. beech TOF and forest) can be labelled for differentiation. But it will not be possible when there are more than 3-5 canopy species. However, TOFs in agro-silvo-pastoral systems are often a limited selection from the forest trees, or planted exotics (e.g. Eucalyptus, Pines) may allow for identification on imagery.

High resolution optical satellite imagery or aerial photography are useful for the identification of canopy and TOF species by crown shape. Again this only functions in case of few (dominant) tree species and rather locally as crown shape of the same species tend to differ between climatic zones and type of use.

By the way, agro-forestry refers to use, not cover. The most appropriate approach maybe to refer to LULC categories, but often this also causes considerable confusion in statistical practice. Further, a pixel-based Spectral Angle Mapper (SAM), nearest neighbour and membership function classifiers of the object oriented classification has also been applied for the discrimination of canopy. It is based on the collection of end-members through the regions of interest of the land cover classes and the assignment of each pixel in the image according to their similarity to the class statistical signature.

Hyperion dataset improves classification results for both the overall and the individual forest type accuracy, in particular for the selected optimum Hyperion band combination (Table 2). Hyperspectral remote sensors collect data of surface in hundreds of narrow, adjacent spectral bands. This imagery is an effective tool to detect material quality characteristics of objects (eg different plant species) and could provide relatively more information that multispectral imaging (Smith 2006).With its 242 potential bands, 10-nm band widths and a spatial resolution of 30 m, the sensor provides data that have a high spectral resolution and a superb spectral discriminating ability. This allows the identification of small differences in similar spectral responses between forest species.

Table 2. Main characteristics of Hyperion Sensor

Feature	Hyperion
Satellite	
Orbit altitude (km)	705
Orbit inclination °	98.3 sun-synchronous
Speed, km/s	7.1
Equator crossing time	10:30–10:15 a.m.
Orbit time	98.9 min
Revisit time	16 days
Swath width	7.5 km
Digitization	16 bits
Spatial resolution	VNIR30 m; SWIR 30 m
Bands	VNIR 70m and SWIR172 m

Certain reflectance values in the electro-magnetic radiation are useful to create vegetation indices, which correlate with the changes in biomass (Silleos *et al.* 2006). Plants reflect the visible (VIS) band but in the near infrared (NIR) the reflectance increases depend on the chlorophyll content of leaves. Using the reflectance from the RED (630-690 nm) and the NIR bands (760-900 nm), the green mass may be defined by the Normalized Differential Vegetation Index (NDVI):

$$NDVI = (R_{NIR} - R_{RED}) / (R_{NIR} + R_{RED})$$

Values of NDVI varies between -1 to +1 and dense vegetation can be easily separated from non-vegetation areas. Naturally, it is also possible to determine the sparse vegetation or moderate vegetation.

The demand for object oriented analysis has increased with the increase in hardware capability and availability of high spatial resolution images. Object-oriented analysis classifies objects instead of single pixels (Zhou *et al.* 2008). The idea of classifying objects stems from the fact that most image data exhibit characteristic texture that is neglected in conventional classifications (Jobin *et al.* 2008; Mallinis *et al.* 2008). Object oriented classification method has its appropriate techniques. The meaningful primitive objects, obtained by segmentation, can be classified through two methods: sample based classification by nearest-neighbour classifier and rule based classification through the membership function technique. In the nearest-neighbour method, the primitive objects are classified through their similarity to training units or segments for each class. The rest of the objects in the image are assigned to their nearest sample in each class (Definiens Imagine, 2004). In the membership function method, segments are classified by membership functions, which are based on fuzzy sets of object features. Also, the interpreter can define thresholds for the assignment of objects to each class by suitable attributes through fuzzy sets (Shataee *et al.* 2004; Matinfar *et al.* 2007).

The airborne LiDAR survey

Laser scanning is an active surveying technology for obtaining detailed -elevation, structural -information about the land surface. The field survey generates apoint cloud consisting of millions of points, with evaluation value of each point. This point cloud is useful to get spatial 3D information about the objects, surfaces. High laser point density provides to create high resolution digital elevation model (DEM) or digital surface model (DSM) (Wagner 2007). This kind of remote sensing technology is suitable for *inter alia* measure the high and diameter of trees, forecast estimate the biomass production, and create the runoff conditions in a large extension of area.

Role of Sub-Pixel Remote Sensing for Assessing Tree Cover

With its synoptic perspective and consistent observing capabilities, remote sensing is a viable tool for systematic mapping of tree cover. Indeed, individual trees and clusters of trees can be readily identified on very high spatial resolution (0.5 m–4 m) images provided by commercial satellite sensors. However, the cost and data volume associated with very high spatial resolution images make their use problematic for covering larger geographic areas. On the other hand, medium, and coarse resolution data, while easily and freely available, do not provide enough spatial detail to resolve trees, leaving us with the challenge of inferring tree cover using sub-pixel analysis methods, which extract information on one or more components of mixed pixels (Chuvieco and Huete 2010). Several of these methods have been used in research to estimate tree and other green vegetation cover fractions at regional or global scales. They can be broadly divided into two categories: (1) linear mixture models, which rely on pure endmembers for model calibration, and (2) regression tree algorithms, which make use of the entire continuum of tree cover as training data to produce a set of linear models (Fig. 1).

De Fries *et al.* (2000) used linear mixture modelling to generate a 1-km global percent tree cover dataset from Advanced Very High Resolution Radiometer (AVHRR) data. Hansen *et al.* (2003) produced a 500 m global percent tree canopy cover map from Moderate Resolution Imaging Spectrometer (MODIS) data using a regression tree algorithm and training data derived from Landsat. Rokhmatuloh *et al.* (2005) mapped tree cover for Africa at 1-km resolution, also using a regression tree algorithm. A number of finer resolution studies were conducted at regional scales with multispectral Landsat databased on linear mixture (Souza and Barret 2000; Lu *et al.* 2003; Wang and Qi 2005) as well as regression tree (Huang *et al.* 2001) modelling approaches. The vast majority of the regional studies emphasize on forested ecosystems, and global scale tree cover products also seemto perform better in regions of denser tree

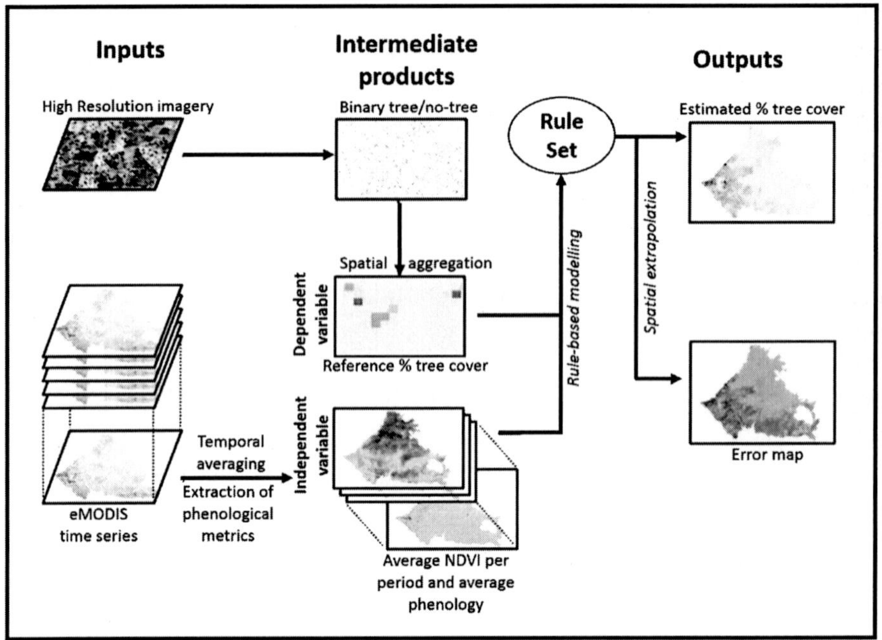

Fig.1. Schematic diagram of the data processing stream leading to the estimated percent tree cover map (adapted & modified after Herrmann *et al.* 2013)

cover than in sparsely vegetated dry lands. For their area of interest, the semiarid Sahel zone of West Africa with its rather sparse tree cover, neither the 500 mMODIS vegetation continuous fields (VCF) product (De Fries *et al.* 2000) nor the most current 250 m VCF product could accurately portray the spatial pattern of tree cover. Its limitations for describing patterns of tree cover in savannas were also acknowledged by Favier *et al.* (2012).

High Temporal Resolution Data: MODIS

The Moderate Resolution Imaging Spectrometer (MODIS) is characterized by a high temporal frequency of acquisitions—the MODIS sensors on the Terra and Aqua satellite platforms provide daily morning and afternoon overpasses respectively—as well as significantly improved spatial resolutions, geo-locational accuracy, and atmospheric corrections compared to the legacy Advanced Very High Resolution Radiometer (AVHRR) sensor (Townshend and Justice 2002). Since its launch in 2000, MODIS data products have been used by the land remote sensing community for a number of research applications including vegetation monitoring, crop yield estimation, and burn scar identification (Garcia-Moran *et al.* 2012). One of the most widely used data products from this sensor is the Normalized Difference Vegetation Index (NDVI), a normalized ratio of reflectances in the red and near infrared portions of the electromagnetic spectrum ((NIR – red)/(NIR + red)), which is sensitive to chlorophyll content and can be used as an indicator of the amount of actively photosynthesizing, green vegetation (Tucker 1979; Tucker and Nicholson 1999). The eMODIS (expedited MODIS) NDVI is a pre-processed dataset which is provided by the US Geological Survey Earth Resources Observation Systems (EROS) Data Centre in response to needs expressed by the vegetation monitoring community for a more user-friendly regional scale satellite data set. In historical processing mode, the eMODIS NDVI is a temporally smoothed (Swets *et al.* 1999) time series composited in 10-day intervals every five days at 250 m spatial resolution. Its shorter compositing intervals compared to the standard MODIS NDVI (16 days) make this product particularly useful for capturing rapid changes in vegetation phenology and subtle differences in the phenologies of woody and herbaceous cover.

The estimation of percent tree cover from eMODIS NDVI data depends on slight differences in overall greenness and seasonality between trees and non-tree land/soil cover types. The considerable inter-annual variability in timing and amounts of rainfall, however, leads to an equally variable vegetation response from year to year, which can exceed the often subtle differences between land/soil cover types. *Timesat* software (Eklundh and Joensson 2010) is excellent for extracting a number of phenological metrics from the eMODIS NDVI timeseries. *Timesat* fits into local polynomial functions to time series of NDVI data, from which the metrics could be extracted based on a definition for the start and end of the growing season.

The Spatial Decision Supporting System (SDSS)

Two types of criteria are used for the Spatial Decision Supporting System: constraints and factors. Constraints are those logical criteria that limit analysis, so 1 or 0 Boolean logical value is added to each investigated decision factors. Normally, this logical values is ideal for distinguishing those land use areas, which could be suitable or unsuitable for forestation under any condition. Factors are criteria that define some degree of suitability for all geographic regions. Both open-source and professional software are useful to create the site selection model to determine the potential areas of forestation.

In generally in the most geographic regions, important factors are the elevation of the surface, hydrological conditions and soil type. In this case of conceptual model building, the constraint layers may be: forest, built-up area, farm, watercourses, channels, and highway roads. These land uses cannot be directly forested but in the immediate vicinity of them the forestation can be done –if there are no other limitation constraints. Thus, by creating a uniform buffer zone around such areas and then these are merged into one "Constraints" layer, and erased out from the study area.

Dense vegetation most probably also contains tree hedges, isolated trees, shrubs etc. which cannot be classified into agro-forest land use. Similarly the merged constrains dense vegetation should be erased out from the area. Thereafter, dense vegetation of pastures and arable land is calculated separately in order to determine the extension of the potential forestation sites there.

Agroforestry Studies in India

Since early seventies, Landsat of USA spurred the use of remote sensing for natural resource inventory, monitoring and management planning. The Indian RS programme comprising of Indian Remote Sensing Satellite series started with launch of IRS-1A in 1988. Later, a virtual institutional framework for using RS data for national development was set up in the form of National Natural Resource Management System (NNRMS). In less than two decades, the IRS series grew into specialized ResourceSat, OceanSat and CartoSat series of satellites with sensors designed for land and water resource inventory and mapping, oceanography and cartography and large-scale mapping applications respectively. This capability covers panchromatic data at less than 1-m resolution, panchromatic stereo data at 2.5m resolution and multispectral data at 5.8, 22.5 and 56m resolutions with different repeat cycles. Over the years, India has developed a large vibrant and nationally relevant RS application programme.

Land use/land cover (LULC) are a dynamic phenomenon. It is influenced by many parameters, particularly monsoon and anthropogenic activities. It is also integral component for understanding the interactions between human activities and the environment. With increasingly intensifying social and economic development, the local ecological environment has changed dramatically. Therefore, land use/land cover change detection is very important for better understanding of land use dynamics in relation to agroforestry systems/ practices for sustainable management.

National Land Use-Land Cover Database (LULC database, 2005-06; 2011-12; 2015-16) is based on 23.5 m resolution IRS LISS-III data and does not explicitly map agroforestry land use. It includes forest categories only but it does not explicitly define any other agroforestry practices. National Remote Sensing Centre (NRSC) produces the geospatial Cropland Data Layer (CDL) primarily using Indian Remote Sensing Resourcesat-1 (IRS-P6) Advanced Wide Field Sensor (AWiFS) satellite data. Narrow tree plantings may appear as pixels with very low tree cover because *tree (*agricultural and forest*) plantation* is derived from 500-m Moderate-resolution Imaging Spectro-radiometer (MODIS) satellite data. The trade-off of broad-scale coverage provided by MODIS *tree* *plantation* or LULC database, is the pixel size or minimum mapping unit is frequently too large to effectively capture narrow plantings of trees.

The district-wise land use land cover (LULC) analysis of all the 15 agro-climatic zones, using the 22 fold LULC classification system was completed using satellite data for the period 1988 – 89, 2005-06; 2011-12 for the Planning Commission of India. The IRS - LISS-I data of kharif (July-October) and rabi (November–March) are used to generate details on crop land in kharif (July-October) and rabi seasons, the area under double crop, fallow lands, different types of forests, degradation status, wasteland, water-bodies, etc., using hybrid methodology i.e., visual as well as digital analysis approach. Out of 442 districts in the country, 274 districts were analysed using visual techniques and remaining 168 districts by digital techniques from 1988-98 data (Sudhakar and Kameshwara Rao 2010) (Fig. 2, Table 3).

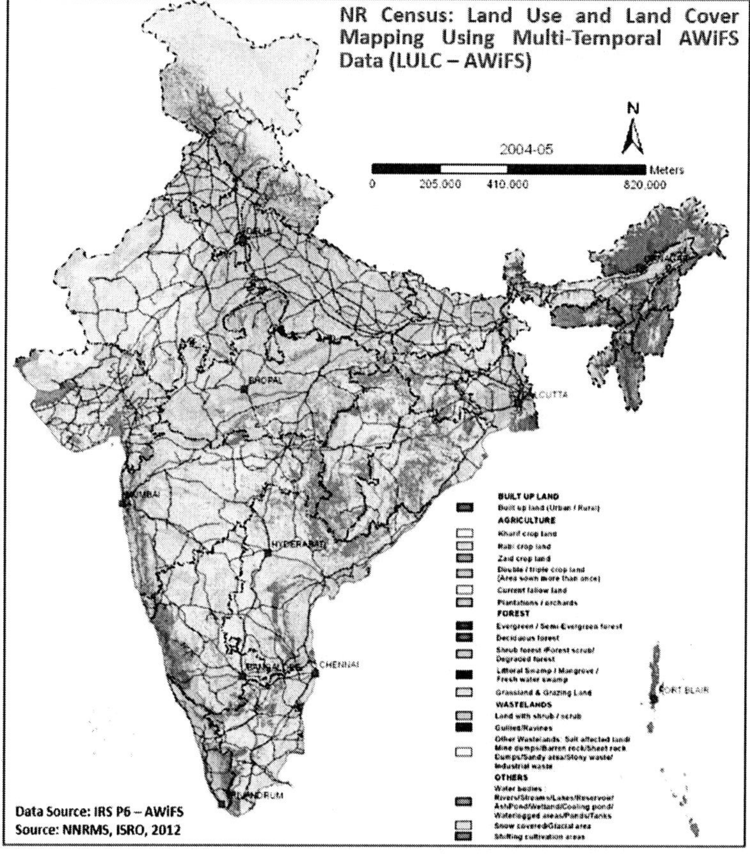

(*Contd.*)

Fig. 2. NR Census: Land Use and Land Cover Mapping Using Multi-temporal AWiFS Data

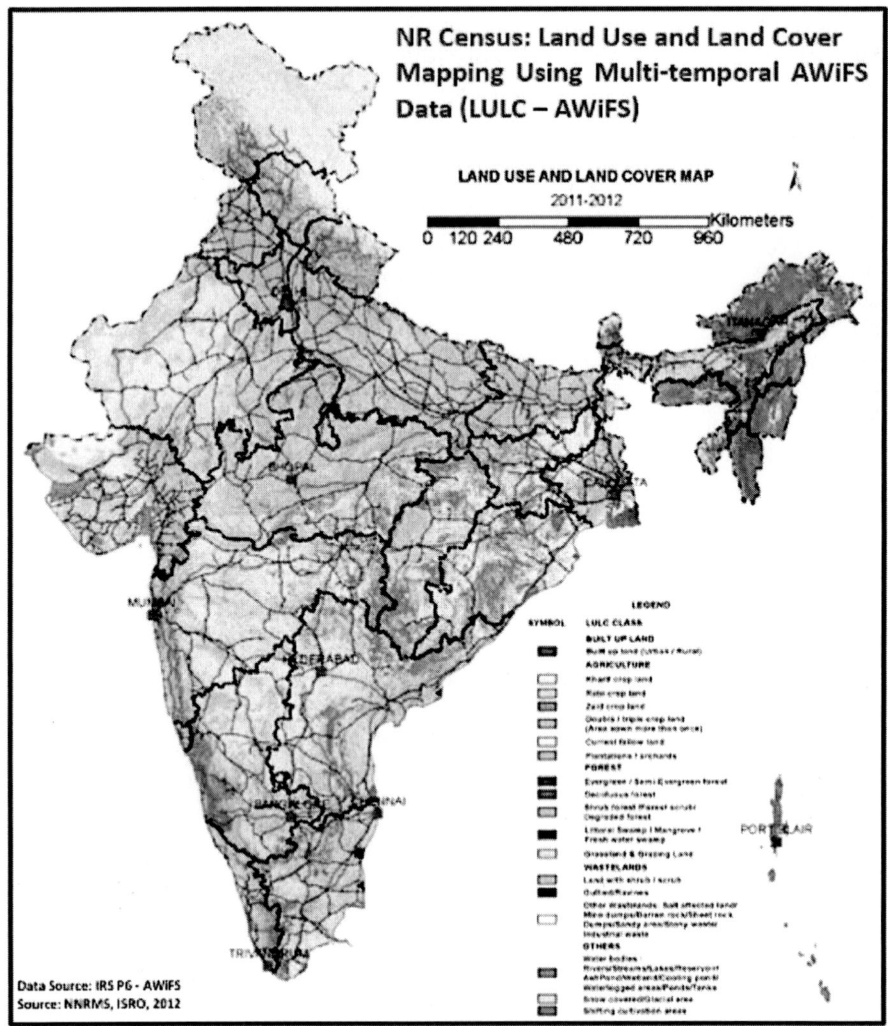

In India, agroforestry area through remote sensing and preliminary estimates were also attempted by Rizvi *et al.* (2014, 2013). Ahmad *et al.* (2007) developed a model for prediction of area under agroforestry in Yamunanagar district of Haryana State. Statistical evaluation and identification of villages that have potential for agroforestry was done by Ahmad *et al* (2010) using GIS. Remote Sensing has become an effective tool for mapping and monitoring of agriculture, forestry and other earth features. A strategy of forest/ non-forest cover mapping based on remotely sensed data and GIS was demonstrated by Maurya *et al.* (2013). Remote sensing technique has been used for crop forecasting (Chaudhary 1999), crop inventory (Salman and Saha 1998), estimation of acreage and crop production (Maurya 2011), calculating carbon sequestration (Tripathi *et al.* 2010). With the advent of remote sensing, future planning and management of natural resources has considerably broadened.

Table 3. Land Utilization in India (2004 – 05 and 2011–12)

LULC Classes (2004-05)	Area (lakh ha.)	LULC Class (2011-12)	Area (lakh ha.)	% Change
Kharif	607.16	Kharif	495.32	-3.40
Rabi	299.89	Rabi	214.7	-2.59
Zaid	26.95	Zaid	11.69	-0.46
Double	427.85	Double	724.17	9.01
Fallow land	426.41	Fallow land	298.95	-3.88
Plantation	62.01	Plantation	82.32	0.62
Evergreen/Semi-evergreen	161.82	Evergreen/Semi-evergreen	173.05	0.34
Deciduous	342.44	Deciduous	352.89	0.32
Wastelands				
Shrub/degraded/scrub	158.81	Shrub/degraded/scrub	145.19	-0.41
Littoral swamp/ Mangrove/ Freshwater swamp	5.36	Littoral swamp/ Mangrove/ Freshwater swamp	5.05	-0.01
Gullied /Ravines	12.20	Gullied /Ravines	10.27	-0.06
Rann area	DNA	*Rann* area	19.56	DNA
Snow cover/glacial/ cloud	47.41	Snow cover/glacial/ cloud	66.46	0.58
Scrub land	181.05	Scrub land	203.17	0.67
Other wasteland	341.40	Other wasteland	292.3	-1.49
Grassland & Grazing land	83.76	Grassland & Grazing land	76.5	-0.22
Shifting cultivation area	2.45	Shifting cultivation area	2.19	-0.01
Built up (rural &urban)	19.0	Built up (rural &urban)	27.59	0.26
Water bodies	81.07	Water bodies	85.85	0.15
Total Geographical Area	3287.23	Total Geographical Area	3287.26	

Source: http://bhuvan-noeda.nrsc.gov.in/

Ahmad *et al.* (2012a, b) developed an Objective Spatial Analytic Hierarchy Process (OSAHP) for identification of potential agroforestry areas using GIS. Ahmad *et al.* (2012a, b) has discussed management of agricultural resources using Remote Sensing and GIS technologies.

Various studies have been conducted in India on LULC change. But, these are scattered particularly in regions like Western and Eastern Ghats, Himalayas and North-eastern states. In Western Ghats over the past centuries the changes in landscape is mainly due to plantations (tea, coffee, rubber, teak etc.) and some due to anthropogenic pressure. Rate of deforestation in the Western Ghats was to be 0.57% annually during the period 1920-1990 (Menon and Bawa 1998). There was an annual decline of 0.90% in natural forest cover in Kerala for the period 1961-1988 (Prasad *et al.* 1998). The data from Agastya Malai region, Western Ghats indicating a five-fold increase in forest loss from the periods 1920-1960 to 1960-1990, also suggest that the rates may be increasing (Ramesh *et al* 1997). Jha *et al.* (2000) have assessed rate of deforestation in the southern Western Ghats, which has lost a quarter of its forest cover between 1973 and 1995.The land use pattern besides having economic implications has also important ecological dimensions in relation to agroforestry. The available land is broadly classified into two on the basis of its use, viz.

(i) agricultural land, and

(ii) non-agricultural land.

(i) Agricultural land (also cropped area): It denotes the land suitable for agricultural activities/ production, both crops and livestock and as well as agroforestry. It includes net sown area, current fallows and land under miscellaneous trees crops and groves. Agricultural land in India totals a little over 50 percent of the total geographical area in the country. This is the highest among the large and medium-sized countries of the world. Cropped area, particularly kharif season is again highly vulnerable as it is highly dependent on the vagaries of monsoon. It is clear from the Table 2 that crop area of kharif season had declined by 3 percent in 2011-12 over 2004-05 due to variability in the precipitation amount and time across the country. Decline in precipitation also influenced crop area during rabi and zaid seasons. Thus, these again showed a negative growth. It is interesting to note that area under double/triple cropping showed a positive growth in areas (i) where irrigation facilities were available, and (ii) good quality groundwater potential existed for cash crops and cereals. Land abandoned as a result of shifting cultivation is excluded.

The growth performance of agriculture in Madhya Pradesh (7.6%), Chhatisgarh (7.6%), Rajasthan (7.4%), Jharkhand (6.0%) and Karnataka (5.6%), was much higher than that of Punjab (1.6%), Maharashtra (2.0%), Tamil Nadu (2.2%), West Bengal (2.8%), Uttar Pradesh (3.3%) and Haryana (3.3%). High coefficient to variation (>2) was observed in the case of Himachal Pradesh and Maharashtra. Increasing demand for industrialization, urbanization, housing and infrastructure is forcing conversion of agricultural land to non–agricultural uses; the scope for expansion of the area available for cultivation is limited.

This indicates:

a. The influence of favourable physical factors (like size, extent of plains and plateaus, etc.), and

b. The extension of cultivation to a large proportion of the cultivable land.

The ultimate irrigation potential in the country is estimated at about 140 million hectares. Of this, about 58.5 million hectare is from major and medium irrigation sources, and 81.5 million hectare is from minor irrigation sources (about 64.1 million hectare from groundwater irrigation and 17.4 million hectare from surface water). Groundwater provides about 70 percent of irrigation and 80 per cent of the drinking water supplies. There are lot of inefficient water usage in irrigation which is also leading to environmental degradation via water logging and induced salinity.

But, because of the large population of the country, the per capita arable land (i.e. land suitable for agriculture) is low: 0.16 hectares against the world average of 0.24 hectares. About 15 per cent of the sown area is multi-cropped. While, most of the multi-cropped area is irrigated, only one-fourth of the gross cropped area is irrigated. The security provided by the irrigation facilities is a major factor in the intensive application of labour and other inputs to obtain high yields. Given the fact that land-holding size is shrinking, tree farming combined with agriculture is perhaps only way forward to optimize farm productivity and thus enhance livelihood opportunities of small, landless and women. Agroforestry interventions can be a potent instrument to help to achieve 4 per cent sustained growth in agriculture (Anonymous 2014).

(ii) Non-agricultural land: This includes

a. land under forests and permanent pastures,

b. land under other non-agricultural uses or built-up area (like towns, villages, roads, railways, etc.) and

c. land classified as cultivable waste as well as barren and uncultivated land of mountain and desert areas.

Area under forests includes all lands classed as forest under any legal enactment dealing with forests or administered as forest, whether state-owned or private, and whether wooded or maintained as potential forest land. The area of crops rose in the forest and grazing lands or areas open for grassing within the forests are also included under the forest area. There has been appreciable increase in the forest area up to the year 2011-12. It increased from 40.48 million hectares in 1950-51 to 68.48 million hectares in 2011-12.As per the assessment of the Forest Survey of India (2013), the total forest and tree cover of the country is 78.92 million hectare, which is 24.01 percent of the geographical area of the country. As compared to the assessment of 2011, there is an increase of 5871 km^2 in the forest cover of the country. The majority of the increase in the forest cover has been observed in open forest category mainly outside forest areas. The maximum increase in forest cover has been observed in West Bengal (3810 km^2) followed by Odisha (1444 km^2) and Kerala (622 km^2).

Madhya Pradesh has the largest forest cover of 77,522 km^2 in terms of area in the country followed by Arunachal Pradesh with forest cover of 67,321 km^2. In terms of percentage of forest cover with respect to total geographical area, Mizoram with 90.38 percent had the highest forest cover in terms of percentage of forest cover to geographical area followed by Lakshadweep with 84.56 percent. The present assessment also reveals that 15 States/UTs had above 33 percent of the geographical area under forest cover. Out of these States and UTs, eight states namely Mizoram, Lakshadweep, A&N Island, Arunachal Pradesh, Nagaland, Meghalaya, Manipur and Tripura had more than 75 percent forest cover while 7 States namely Goa, Sikkim, Kerala, Uttarakhand, Dadra & Nagar Haveli, Chhattisgarh and Assam had forest cover between 33 to 75 percent.

Wasteland is degraded and unutilised class of land which is deteriorating on account of natural causes or due to lack of appropriate water and soil management. It includes: barren, degraded forest, gullied land, ravenous land, salt affected land, salt encrustation, shifting cultivation, snow covered area, steep sloping area, waterlogged area, upland without scrub, and land with scrub. Wasteland can result from inherent/imposed constraints such as location, environment, chemical and physical properties of the soil or financial or management constraints. There has been a slight reduction of 3.88 percent of the area due to reclamation efforts. Spatial information on wastelands with respect to their nature, magnitude of degradation, extent, spatial distribution and temporal behaviour is a pre-requisite for development and implementation of plans for their rehabilitation. Jammu and Kashmir has more than 50 percent of its area under wastelands. There are two states viz., Himachal Pradesh and Sikkim that have wastelands ranging between 40-50 percent. Five states viz., Uttarakhand, Rajasthan, Nagaland, Manipur and Mizoram have wastelands ranging between 20 to 40 percent, while in Meghalaya wastelands occupy only 15 to 20 percent. There are 9 states that accounted for the extent of wastelands ranging between 10 to 15 percent. Eight other states have wastelands ranging between 5 to 10 percent. Only 3 States and Union Territories have less than 5 percent area under wastelands.

Agroforestry in arid zone of India

Arid zone of Rajasthan constitute Barmer, Bikaner, Churu, Ganganagar, Hanumangarh, Jaisalmer, Jalore, Jhunjhunun, Jodhpur, Nagaur, Pali and Sikar districts.Though agroforestry is not a well-defined part of land use systems in revenue records and surveys of Thar Desert region, still the land use category of permanent pastures/grazing lands is true example of extensive silvi-pastoral systems and entire agriculture is predominantly practiced in association with woody species (agri-silviculture) which vary from location to location (Table 4). Thus, Thar desert region of western Rajasthan supports the farming systems, which are in fact traditionally agroforestry based ones.

Table 4. Agro-Ecological Regions

Macro region	Micro region	Districts included	Area (ha)
Arid	Kharif mono cropping	Barmer, Jodhpur, Churu, Jaisalmer, Bikaner	13,340,545 (64.06%)*
	Irrigated kharif and rabi cropping	Ganganagar Hanumangarh	2,063,654 (9.91%)
Transitional,	Kharif monocropping	Jhunjhunun, Sikar, Nagaur	3,129,600 (15.03%)
(between arid and semi-arid)	Irrigated kharif and rabi cropping	Pali and Jalore	2,289,680 (11.00%)

* Values in parenthesis indicate the percentage of the total area of the districts.
Source: Agricultural Statistics of Rajasthan (2010-11), Directorate of Economics and statistics, Govt. of Rajasthan, Jaipur

Tewari (1997) described large tracts of land having widely scattered trees of 'khejri' (*Prosopis cineraria*), 'Rohida' (*Tecomella undulata*), Jal/Pilu (*Salvadora* spp.) and shrubs like *Zizyphus* spp. with crops of food grain or pasture species in Thar desert region of western Rajasthan as best examples of extensive agroforestry systems. Narayan and Tewari (2005) described that "multipurpose tree species" such as *P. cineraria, A. nilotica, Zizyphus* spp. etc. providing large number of end-use products besides contributing food security in drought conditions. These are important source of income in traditional agroforestry system.

Gravelly rocky plains exposed and buried pediments and rocky plateau of Jaisalmer and adjoining parts of Jodhpur district constitute this zone. Gravelly plains of Jodhpur and Nagaur districts also have similar vegetation. The entire zone has good stands of *Z. nummularia* and *C. decidua* with varying density ranging from 250 to 350 individual per ha (Shankar 1984). In most of the villages, common grass like *L. sindicus* is found to be growing in association with woody *Z. nummularia* and *C. decidua,* and serve as extensive silvi-pastoral system. Whereas few pockets, with some soil accumulation, are cultivated for Kharif crops like pearl millet, cluster bean and moth bean in association with these two woody species (agri-silvicultural system). In this zone, generally *Zizyphus* shrubs are not harvested and natural leaf fall from December to April provide fodder for grazing animals.

Rainfall appears to be governing factor for evolution of traditional agroforestry system and in crop production in this region. Over 400 mm rainfall zone predominant system is *Prosopis cineraria – Acacia nilotica* based; between 300 and 400 mm rainfall zone *P. cineraria* based; between 200-300 mm rainfall zone *Zizyphus* spp. – *P. cineraria* based; and in less than 200 mm rainfall zone *Zizyphus* spp. – *P. cineraria* – *Salvadora* spp. based. The tree/shrub density of system forming species is also clearly governed by rainfall regime. Data indicates that with decrease in rainfall, the density of woody component of the system substantially declines, and as well as the productivity of arable crops and grasses. In the arid western Rajasthan encompassing Pali, Jalore, parts of Jodhpur and Barmer districts, rainfall gradient of >500 to 200 mm is from south-eastern margin to western part.

Aggarwal and Kumar (1990) observed that in the agroforestry practice, the utility is due to tree roots which move deep exploiting large soil volume often inaccessible by herbaceous plant roots. A part of nutrients is deposited back on the surface through litter fall. An increase in the availability of N, P and K and crop growth (Aggarwal *et al* 1993) beneath the *P. cineraria* as compared to the adjacent area has been observed. Singh and Lal (1969) reported that the forage species produce higher biomass under *P. cineraria* tree canopy due to high fertility status. Ramakrishna and Shastri (1996) under arid conditions of Jodhpur, reported that the rainfall interception losses from 13 years old *Acacia tortilis* and 7 years old *Holoptelia integrifolia* were 14 to 33 per cent and 3 to 8 percent, respectively. A study of soil moisture regimes below five desertic trees showed maximum moisture under *Tecomella undulata* followed by *Prosopis cineraria, Acacia senegal, Albizzia lebbek and Prosopis juliflora* (Gupta and Saxena 1978). As a result, the dry forage yield under *P. cinerraria* was 2.3 t ha^{-1} as against 1.66, 1.32, 0.85 and 0.78 t ha^{-1} obtained under *Tecomella undulata, Albizzia lebbek, Prosopis juliflora* and *Acacia senegal*, respectively.

Gupta *et al* (1984) established three shelterbelts of 300m length with rows of *A. tortilis, C. siamea* and *P. juliflora* in each belt at 150 m apart to study the efficacy of the tree species in arresting the wind speed at 2H, 5H and 10H in the leeward side of the shelterbelt. Reduction in wind speed was more in the monsoon period (July-September) than in summer period the reduction being maximum at 2H distance from the tree. Shankarnarayan *et al* (1987) observed that some of the tree species suitable for environment improvement through shelterbelts are *A. tortilis, P. juliflora, C. siamea, A. indica, Albizzia lebbek, Acacia nilotica* var. *cupressiformis, Tamarix articulata,* etc.

Remote Sensing and GIS: Multispectral remote sensing images of Resourcesat-2/LISS IV (spatial resolution 5.6 m) were analysed for land use and land cover patterns. Geo-referenced standard PAN+LISS IV Mx merged scenes for the period 2011–12 were procured from the National Remote Sensing Centre, Hyderabad. Pre-processing of these scenes includes layer stacking, sub-setting with district boundary and mosaicing. Shapefile of district boundaries was provided by NRSC, Hyderabad.

Maximum likelihood method of supervised classification was applied for assessment of LULC in selected districts. It has been quite accurate for classifying the study area for the extraction of agroforestry fields. On the basis of comprehensive ground truthing, training sites are required to be generated in the supervised classification technique. The equation for the maximum likelihood/ Bayesian classifier is as follows:

$$D = \ln \ln (a_c) - [0.5 \ln (|\mathrm{cov}_c|)] - [0.5(X - M_c)^T (\mathrm{cov}_c^{-1})(X - M_c)]$$

Where,

D = weighted distance (likelihood)

C = a particular class

M_c = the mean vector of the sample of class c

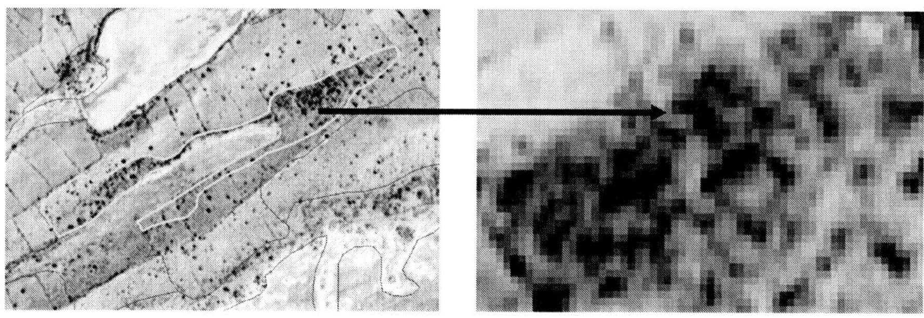

(a) Barmer (71°7’24.019"E 24°41’6.484"N)

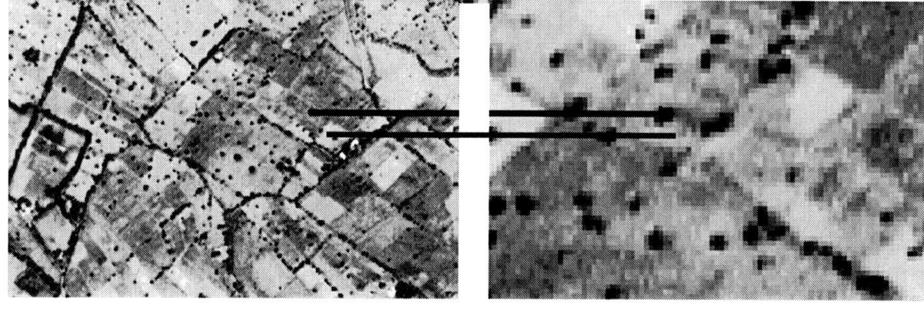

(a) Udaipur (73°21’38.684"E 25°9’23.224"N)

Fig. 4. (a) examples of classified images from Resourcesat-2

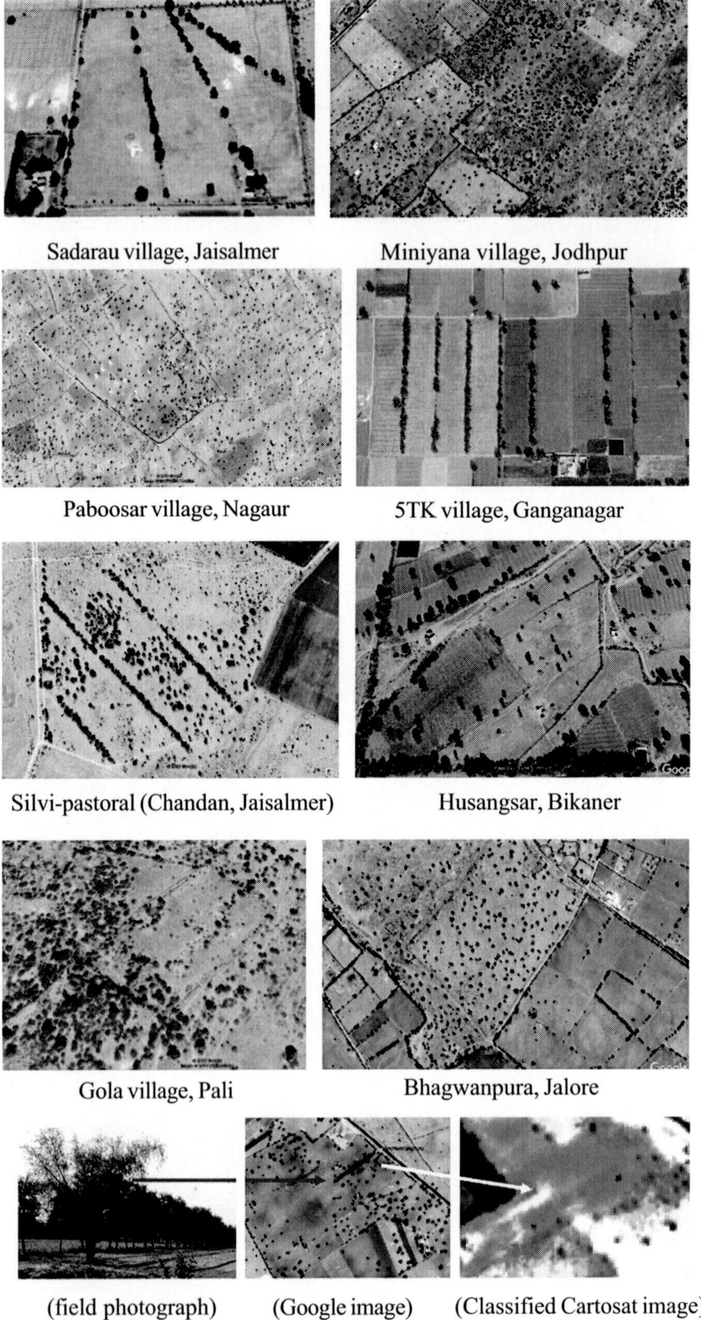

Fig. 4. (b) Images captured from Google Earth and field

X = measurement vector of the candidate pixel

a_c = percent probability that any candidate pixel is a member of class c

$|Cov_c^{(-1)}|$ = Determinant of Cov_c

cov_c^{-1} = inverse of Cov_c

n_l= natural logarithm function

T = Transposition function

These scenes were classified into 9 classes, viz. cropland, grassland, wasteland, plantation, agroforestry, forest, scrub land, built-ups, water bodies andsandy area (Figure 6a). In this classification, agri-silviculture/agri-horticulture systems and block plantations was accounted for in agroforestry class. Such pixel based methods account for single major feature occurring in a pixel, even if more than one features/land covers are present. Therefore, sub-pixel method of classification was applied on agricultural land because agroforestry exist on agricultural land only. Agricultural land including cropland and fallow land was masked from false colour composite (FCC) of the districts. Resultant image consisted of pixels of five categories (i) pixels covering trees plus cropland, (ii) pixels covering fallow land plus trees, (iii) pixels covering trees only, (iv) pixels covering cropland only and (v) pixels covering fallow land only. Pixels of first three categories represented agroforestry in real sense and their total area gave an estimate of area under agroforestry. Other agroforestry systems like boundary plantations or scattered trees on farm lands are also accounted because tree canopy cover within pixel is less than 50%. To overcome this constraint, Imagine Subpixel Classifier was applied.

Prospective Benefits of Agroforestry

Attempts have been made under agroforestry are to optimize the use of land for agricultural production on a sustainable basis at the same time meeting other needs from forestry (Fagbemi 2002). Nitrogen-fixing and non-nitrogen-fixing trees thrive adequately in agroforestry with annual crops, presents a farming system in which arable crop yields can be enhanced. The tree rooting system brings about stability that can lead to soil conservation. What is needed would be mutual interaction and proper management techniques that would reduce the adverse effects that may result when trees are integrated into agro-ecosystem (Connor 1983).

In India, large area is available in the form of farm boundaries, bunds, waste lands where agroforestry system can be adopted. This system permits the growing of suitable tree species in the field where most annual crops are growing well. Hence, by growing trees and crops on agricultural land, over all resources are utilized efficiently. Agroforestry system has vast potential to generate employment opportunities. It provides raw material for the cottage and small scale industries. It further helps in maintaining ecological balance through soil and water conservation, soil improvement (Tewari *et al* 2014). It also helps in meeting various needs of growing population.

Also, Alao and Shuaibu (2011) from their studies have demonstrated, the inherent advantages in agroforestry accruable to farmers. It has been estimated that agroforestry alone are likely to contribute about 25 million tonnes per annum of food produce by 2025 (Table 5).

Table 5. Contribution of agroforestry produce to national demand by 2025

Product Category	Total demand	Contribution from agroforestry			Contribution from agroforestry (%)
		Traditional	Improved	Total	
Timber (million cu m)	116	52	48	100	86
Fuel wood (million t)	330	60	13	73	22
Fodder (million t dry)	1040	100	12	112	11
Fruits (million t)	43	1.5	4.4	5.9	14
TBOs (million of bio-diesel)	10	0.5	5.4	5.9	59
Food (million t)	308	7	15	22	7
Bioenergy (MW)	16000	4000	1000	5000	3
Forest cover (mha)	100	10	15	25	25

Source: NRCAF (2007)

Studies conducted on agroforestry systems in arid regions of India have been described by Shankarnarayan *et al.* (1987). The introduction of *Acacia tortilis*, a comparatively fast-growing tree, in the arid zone has helped in the stabilization of sand dunes and early establishment of shelter belts. *A. tortilis* with the grass *Cenchrus ciliaris* in a silvo-pastoral system earned maximum returns among the different combinations tried. In another experiment, the growth rates of *P. cineraria* and *Acacia albida* and their influence on the associated crops of mung bean and guar were studied. *A. albida*was found to grow much faster with a height increment of 110 cm in three years compared to only 15 cm in the case of *P. cineraria* (Table 6).

It has many contributions like rehabilitation of degraded land, increased farm productivity and capability of conserving natural resource and it is an option to increase the forest cover to 33% in the country (Anonymous 2014). Besides meeting the subsistence need of food, fruits, fibre and medicines, this farming practice meets almost half of the demand of the fuel wood, two-thirds of the small timber, 70–80% wood for plywood, 60% raw material for paper pulp and

9–11% of green fodder requirement of livestock (Dhyani 2013). Also, agroforestry practices have enhanced overall biomass productivity from 2 to 10 t $ha^{-1}y^{-1}$ in rainfed areas in general and the arid and semiarid regions in particular. Agroforestry is also providing livelihood opportunities through lac, apiculture/ sericulture, and natural gums and resins production (Dhyani 2012).

Table 6. Grain yield of mung bean (*Vigna mungo*) and guar (*Cyamopsis tetragonoloba*) under tree species (1983-1985)

Treatments	Mung (10^2 kg ha^{-1})			*Guar* (10^2 kg ha^{-1})		
	1983	1984	1985	1983	1984	1985
Prosopis cineraria at 5 x 5 m	7.85	8.50	0.68	3.60	7.15	1.16
P. cineraria at 5 x 10 m	7.44	6.08	0.75	4.45	5.14	1.28
P. cineraria at 10 x 10 m	8.68	7.90	1.11	3.70	6.53	1.34
Acacia albida at 5 x 5 m	8.08	9.30	0.81	3.32	6.11	0.65
A. albida at 5 x 10 m	7.08	6.70	0.61	5.08	5.37	0.76
A. albida at10 x 10 m	7.92	7351	0.74	0.34	6.34	1.28
Control (crops without trees)	6.40	7.30	1.06	3.20	6.41	0.95

Source: NRCAF (2007)

Conclusion

The satellite-derived data products provide a means for monitoring trees over large areas. However, estimates of forest land are inconsistent across various datasets. Trees in agricultural settings are undoubtedly under-represented in satellite-derived datasets because sensor resolutions are too coarse to consistently capture narrow linear plantings of trees. For the mapping of very narrow tree plantings, such as single- or double-row windbreaks, remote sensing approaches that utilize existing imagery sources with resolutions finer than 5 m are needed. There is an incremental step toward monitoring trees in agricultural landscapes. Although NRSC, Hyderabad is a nationwide source of all kinds of satellite imagery, the lack of consistency between resolution and availability of cloud-free data is an operational challenge to mapping tree cover at a high-resolution on a broad scale. Further, using Random Forests approach could be a viable operational tool for mapping tree cover from existing imagery sources.

So, the goal of application of geospatial technologies in agroforestry is to optimize productivity and conservation benefits within a set of integrated land use practices. It has emerged as a robust land use, which advocates crop diversification, soil and soil-water conservation, cycling of organic matter and sequestration of CO_2 in plant and soil. It evolves a synergy between agricultural production and forestry that is beneficial for increased food production, sustainable wood production and improvement of the quality of the soil. Improved agroforestry systems (AFS) such as improved fallows and other AFS provide benefits that contribute to rural livelihoods, improved socioeconomic status and ecosystem functioning of land use systems. Recently, there is an increasing recognition of the contribution of agroforestry to improve ecosystem services and livelihoods especially in rural areas. In brief, usages of satellite images for agroforestry mapping, monitoring and management are crucial in obtaining information. Satellite imagery data offer considerable advantages in agroforestry studies and are useful to monitor agroforestry ecosystems. Options to use either optical high resolution satellite images or hyperspectral data depend on the needs. The output of a satellite imagery analysis can provide accurate information to facilitate conservation planning and policy-making as the technology is currently in an advanced stage. This is in fact, a win-win situation.

References

Aggarwal RK, Kumar P (1990). Nitrogen response to pearl millet (*Pennisetum typhoides* S & H) grown underneath *Prosopis cineraria* and adjacent open site in an arid environment. Annals of Arid Zone, 29:289-293.

Aggarwal RK, Kumar P, Raina P (1993). Nutrient availability from sandy soils underneath *Prosopis cineraria* (Linn. Macbride) compared to adjacent open site in an arid environment. Indian Forester, 19: 321-325.

Ahmad T, Rai A, Singh R (2012a). Objective spatial analytic hierarchy process for identification of potential agroforestry areas using GIS. Mod Assist Stat Appl 7(1):65–73

Ahmad T, Sahoo PM, Rai A (2012b). Managing agricultural resources. Geosp Today 11(6): 22–25.

Ahmad T, Rai A, Singh R (2010). Statistic al evaluation of development of villages' potential for agroforestry using GIS. Adv. Appl. Res, 2(2):157–163.

Ahmad T, Singh R, Rai A, Kant A (2007). Model for prediction of area under agroforestry in Yamunanagar district of Haryana. Ind.J.Agric. Sci, 77(1):43–45.

Alao JS, Shuaibu RB (2011). Agroforestry practices and preferential agroforestry trees among farmers in Lafia Local Government Area, Nataraja State, Nigeria. Waste Management and Bioresource Technology 1(2)12-20. www.waoj.com

Anonymous (2014). National Agroforestry Policy. Government of India.

Bargali SS, Bargali K, Singh L, Ghosh L, Lakhera ML (2009). *Acacia nilotica* based traditional agroforestry system: effect on paddy crop and management. Curr. Sci., 96(4), 581–587.

Campagna M (2006). GIS for sustainable development. Boca Raton: Taylor & Francis Group.

Chowdhury QI (1999). Bangladesh Overview. In. Chowdhury QI (Ed). Bangladesh state of Environment report 1998. Forum of Environment Journalists of Bangladesh (FEJB), Dhaka. 3-14.

Chuvieco E, Huete A (2010). Fundamentals of satellite remote sensing; FLCRC Press: Boca Raton, FL, USA.

Connor BJ (1983). Plant stress factors and their influence on production of agroforestry plant association. In: Plant Research and Agroforestry (PA Huxley: ed) ICRAF, Nairobi, pp. 401-426.

DeFries RS, Hansen MC, Townshend JRG, Janetos AC, Loveland TR (2000). A new global 1-km dataset of percentage tree cover derived from remote sensing. Glob. Change Biol., 6, 247–254.

Dhyani SK, Handa AK, Uma (2013). Area under agroforestry in India: an assessment for present status and future perspective. Indian Journal of Agroforestry,15(1): 1–11.

Dhyani SK (2012). Agroforestry interventions in India: focus on environmental crises and livelihood security. Indian Journal of Agroforestry, 13(2): 1–9.

Ellis EA, Bentrup G, Schoeneberger MM (2004). Computer-based tools for decision support in agroforestry: Current state and future needs. Agroforestry Systems, 401-421.

Fagbemi T (1997). Agroforestry for sustaining agricultural production in the tropics. In: Strategies and tactics of sustainable agriculture in the tropics (MA Badejo and AO Togon, eds) STASAT 1:45-68.

Favier C, Aleman J, Bremond L, Dubois MA, Freycon V, Yangakola JM (2012). Abrupt shifts in African savanna tree cover along climatic gradient. Glob. Ecol. Biogeogr., 21, 787–797.

Gaur Mahesh K, Tewari JC, Chouhan JS (2017). Agroforestry in India with special emphasis on land use pattern in different agro-ecological zones (in press). In: Ecological Interactions in Agroforestry, published by Indian Ecological Society, Ludhiana.

Gupta JP, Saxena SK (1978). Studies on the monitoring of dynamics of moisture in the soil and the performance of ground flora under desertic communities of tree. Indian Journal of Ecology, 5: 30-36.

Gupta JP, Rao GGSN, Ramakrishna YS, Ramana Rao BV (1984). Role of shelterbelts in arid zone. Indian Farming, 24(7): 29-31.

Hansen MC, DeFries RS, Townshend JRG, Carroll M, Dimiceli C, Sohlberg RA (2003). Global percent tree cover at a spatial resolution of 500 meters: First results of the MODIS vegetation continuous fields algorithm. Earth Interact., 7:1–15.

Herrmann Stefanie M, Wickhorst Andrew J, Marsh Stuart E (2013). Estimation of tree cover in an agricultural parkland of Senegal using rule-based regression tree modelling. Remote Sens., 5, 4900-4918.

Huang C, Yang L, Wylie B, Homer C (2001). A strategy for estimating tree canopy density using Landsat 7 ETM+ and high resolution images over large areas. In: Proceedings of Third International Conference on Geospatial Information in Agriculture and Forestry, Denver, CO, USA, 5–7 November 2001.

Jha CS, Dutt CBS, Bawa KS (2000). Deforestation and land use changes in Western Ghats, India. Curr. Sci., 79: 231-238.

Labiberte AS, Fredrickson EL, Rango A (2007). Combining decision trees with hierarchical object-oriented image analysis for mapping arid rangelands. Photogrammetric Engineering and Remote Sensing 73(2):197-207.

Lu DS, Moran E, Batistella M (2003). Linear mixture model applied to Amazonian vegetation classification. Remote Sens. Environ., 84: 456–469.

Maurya AK, Tripathi SK, Soni S (2013). A strategy of forest/non-forest cover mapping of Achanakmar-Amarkantak biosphere reserve, Central India: based on remotely sensed imagery and GIS data. International J. Remote Sensing & Geoscience, 4(2):50-54.

Maurya AK, Tripathi SK, Soni S, Soni PK (2011). Estimation of acreage & crop production through Remote Sensing & GIS technique. Proceeding of International Seminar "Geosapatial World Forum 2011" held on 18-21 Jan 2011 at HICC Hyderabad. http://www.geospatialworldforum.org/2011/proceeding/pdf/AbhishekFullPaper.pdf.

Menon S, Bawa KS (1998). Deforestation in the tropics: Reconciling disparities in estimates for India. Ambio, 27: 567-577.

Narain P, Tewari JC (2005). Trees on agricultural fields: a unique basis of life support in Thar Desert.In : Tewari VP and Srivastava RL (eds) *Multipurpose Trees in the Tropics:* Management and Improvement Strategies. Arid forest Research Institute, Jodhpur, pp. 516-523.

NRCAF (2007). Perspective Plan: Vision 2025.NRCAF, Jhansi.

O'Neill RV, Hunsaker CT, Timmins SP, Jackson BL, Jones KB, Riitters KH, Wickham JD (1996) Scale problems in reporting landscape pattern at the regional scale. Landscape Ecology 11:169-180.

Parihaar RS, Bargali K, Bargali SS (2015). Status of an indigenous agroforestry system: a case study in Kumaun Himalaya. Indian *J. Agric.* Sci., 85(3), 442–447.

Peng ZR, Tsou MH (2003). Internet GIS. New Jersey: John Wiley & Sons, Inc.

Prasad SN, VijayanL, Balachandran S, Ramachandran VS,Verghese CPA (1998) Conservation planning for the Western Ghats of Kerala: I. A GIS approach for location of biodiversity hotspots. Curr. Sci., 75: 211-219.

Ramakrishna YS, Shastri ASRAS (1996). Mixed cropping of annuals and woody perennials: an analytical approach to productivity and management. In: Tree-Crop Interactions- A Physiological Approach (Eds Chin K Ong and Peter Huxley). CAB International, U.K. and ICRAF, Nairobi, Kenya, pp. 25-49.

Ramesh BR, Menon S, Bawa KS (1997). A vegetated based approach to biodiversity gap analysis in the Agastya Malai region, Western Ghats, India. Ambio, 26: 529-536.

Rizvi RH, Dhyani SK, Newaj R, Karmakar PS, Saxena A (2014) Mapping agroforestry area in India through remote sensing and preliminary estimates. Indian Farming, 63(11), 62–64.

Rizvi RH, Dhyani SK, Newaj R, Saxena A, Karmakar PS (2013). Mapping extent of agroforestry area through remote sensing: Issues, estimates and methodology. Indian J. Agroforestry, 15(2): 26-30.

Rokhmatuloh DN, Al Bilbisi H, Tateishi R (2005). Percent tree cover estimation using regression tree method: A case study of Africa with very high resolution Quick Bird images as training data. IEEE Trans. Geosci. Remote, 3, 2157–2160.

Salman MA, Saha SK (1998). Crop inventory using remote sensing and geographical information system (GIS) techniques. Journal of Remote Sensing and Environment, 2: 19-34.

Shankar V, Dadhich NK, Saxena SK (1984). Agroforestry in arid and semi-arid zone.CAZRI, Jodhpur, 295 p.

Shankarnarayan KA, Singh KC (1987). Agroforestry in arid zones of India. Agroforestry Systems, 5: 69-88.

Shankarnarayan KA, Harsh LN, Kathju S (1987). Agroforestry in the arid zones of India. Agroforestry Systems, 5: 69-88.

Silleous NG, Alexandridi TK, Gitas IZ, Perakis K (2006). Vegetation indices: Advances made in biomass estimation and vegetation monitoring in the last 30 years. Geocar to International, 21 (4):21-28.

Singh AK, Kumar P, Singh R, Rathore N (2012). Dynamics of tree-crop inter face in relation to their influence on microclimatic changes - A review. Hort. Flora. Res. Spectrum, 1(3): 193-198.

Singh KS, Lal P (1969). Agroforestry for sustainable development in arid zones of Rajasthan. International Tree Crop Journal, 9: 203-212.

Souza C, Barreto P (2000). An alternative approach for detecting and monitoring selectively logged forests in the Amazon. Int. J. Remote Sens., 21, 173–179.

Sudhakar S, Kameshwara Rao SVC (2010). Land use and land cover analysis, in PS Roy, RS Dwivedi and D Vijayan (eds.) Remote Sensing Applications. NRSC/ISRO Publication.

Tewari JC, Moola Ram, Roy MM, Dagar JC (2014). Livelihood improvements and climate change adaptations through agroforestry in hot arid environments. In JC Dagar *et al* (eds), Agroforestry systems in India: Livelihood security and ecosystem services, Advances in Agroforestry Series 10. Springer India. 155-83.

Tewari JC, Tripathi D, Narain P (1997). Farm women and their participation in extensive arid silvi-pastoral system: an assessment. In: Silvipastoral systems in arid and semi-arid ecosystems (Eds MS Yadav, Manjit Singh, SK Sharma, JC Tewari and U Burman), UNESCO: ICAR, CAZRI, Jodhpur, pp. 449-457.

Tomlinson R (2007). Thinking about GIS: Geographic Information System planning for managers (3rd Edition ed.). Redlands: ESRI Press.

Tripathi NK, Bauchkar PR (2005). Open GIS based wireless spatial data logger for flood mitigation. (www.isprs.org/proceedings/XXXV/congress.yf/papers/954.pdf)

Wagner W, Roncat A, Melzer T, Ullrich A (2007) Waveform analysis techniques in airborne laser scanning. ISPRS Vol. XXXVI (Part 3/ W52), 413-417 pp.

Wang C, Qi J (2005). Cochrane, M. Assessment of tropical forest degradation with canopy fractional cover from Landsat ETM+ and IKONOS imagery. Earth Interact., *9*, 1–18.

Woodcock CE, Strahler AH (1987). The factor of scale in remote sensing. Remote Sensing of Environment 21:311-332.

Young A(1986). Effects of trees on soils. In Amelioration of soil by Trees. Commonwealth Science Council, London. Technical Publication 190:28-41.

9

Alternate Land use Systems in Western Rajasthan

N. D. Yadava and V. S. Rathore

ICAR-CAZRI, Regional Research Station, Bikaner-334004, Rajasthan, India

Introduction

The green revolution in mid-sixties stirred by research based new technological developments involving new materials, models and ways of organizing farm inputs (water, fertilizer, chemical etc) and government's policies transformed the agriculture dramatically. The outcome chewed a many fold increase in production and productivity viz. food grain production of 211 million tonnes (which was only 74.23 million tonnes in 1966-67) and food grain productivity of 1697 kg ha^{-1} (which was only 644 kg ha^{-1} in 1966-67). The country thus became self-sufficient in food production despite tremendous pressure to sustain 16% of world human population and 10% of cattle population with just 2.4% of total land.

The per capita availability of agricultural land in India was 0.46 hectares in 1951, which decreased to 0.15 hectares in 2000 as against the global average of 0.6 ha. Number of persons per hectare of net cropped area was 3 in 1951, 6.5 in 2000 and is estimated at 8 persons in 2025. This situation of rapidly declining land to man ratio is likely to worsen further owing to competitive demand for food, fibre, fuel, fodder, timber and developmental activities such as urbanization and industrialization, special economic zones, mining, road construction and reservoirs etc. The rainfed lands suffer from a number of biophysical and socio-economic constraints, which affect productivity of crops and livestock. These include low and erratic rainfall, land degradation and poor productivity (Katyal 1994), low level of input use and technology adoption, inadequate fodder availability with low productive livestock (Singh 1997), and resource poor farmers and inadequate credit availability.

Farming system is a natural resource management unit operated by a farm household, and includes the entire range of economic activities of the family members (on-farm, off-farm agricultural as well as off-farm non-agricultural activities) to ensure their physical survival as well as their social and economic well-being. This broad definition is of importance, as the farm family takes decisions considering not only the farming possibilities, but also the off-farm employment opportunities. Thus, there is need to develop suitable integrated farming systems for such farmers since single crop production enterprises are subject to a high degree of risk and uncertainty because of seasonal, irregular and uncertain income and employment to the farmers.

The Alternate land use system approach introduces a change in the farming techniques for maximum production in the cropping pattern and takes care of optimal utilization of resources. The farm wastes are better recycled for productive purposes in the integrated system. A judicious mix of agricultural enterprises like dairy, poultry, piggery, fishery, sericulture etc. suited to the given agro-climatic conditions and socio-economic status of the farmers would bring prosperity in the farming. An integrated farming system allows us to use some of the advantages of nature, and ecology, as opposed to relying on chemistry to solve all our production issues.

Advantage

- In an integrated system, livestock and crops are produced within a coordinated framework. The waste products of one component serve as a resource for the other. For example, manure is used to enhance crop production; crop residues and by-products used to feed the animals, supplementing often inadequate feed supplies, thus contributing to improved animal nutrition and productivity.
- The result of this cyclical combination is the mixed farming system, which exists in many forms and represents the largest category of livestock systems in the world in terms of animal numbers, productivity and the number of people it serves.
- Integration of allied activities will result in the availability of nutritious food enriched with protein, carbohydrate, fat, minerals and vitamins.
- Integrated farming will help in environmental protection through effective recycling of waste from animal activities like piggery, poultry and pigeon rearing.
- Animals play key and multiple roles in the functioning of the farm, and not only because they provide livestock products (meat, milk, eggs, wool, and hides) or can be converted into prompt cash in times of need. Animals

transform plant energy into useful work: animal power is used for ploughing, transport and in activities such as milling, logging, road construction, marketing, and water lifting for irrigation.

Modern high-input agriculture

- Overuse of natural resources, causing depletion of groundwater, and loss of forests, wild habitats, and of their capacity to absorb water, causing waterlogging and increased salinity.
- Contamination of the atmosphere by ammonia, nitrous oxide, methane and the products of burning, which play a role in ozone depletion, global warming and atmospheric pollution.
- Contamination of food and fodder by residues of pesticides, nitrates and antibiotics.
- Contamination of water by pesticides, nitrates, soil and livestock water, causing harm to wildlife, disruption of ecosystems and possible health problems in drinking water.

Sustainability of agricultural systems

Sustainable Agriculture refers to an agricultural production and distribution system that should:

(i) Achieve the integration of natural biological cycles and controls.

(ii) Protects and renews soil fertility and the natural resource base.

(iii) Reduces the use of non-renewable resources and purchased (external or off-farm) production inputs.

(iv) Optimizes the management and use of on- farm inputs.

(v) Provides an adequate and dependable farm income.

(vi) Promotes opportunity in farming family and farm communities,

(vii) Minimizes adverse impacts on health, safety, wildlife, water quality and the environment.

To summarize, the sustainability of any system of food or fibre production systematically pursues the following goals:

(i) Productive and Profitable

(ii) Conserves Resources and protects the environment

(iii) Enhances health and safety

(iv) Low input methods and skilled management

Agriculture situation of arid region

Seventy per cent area is mostly affected by medium to severe wind erosion and over exploitation of groundwater (nearly 50 % of ground water resources has already been utilized) and it has been estimated that up to 2020 the major part of arid zone will be devoid of economically viable groundwater. In semiarid regions also the situation is not very sound. The per capita land availability (0.08 ha to 0.15ha) will not be economical under traditional cropping system. The present gap between fodder availability and requirement of 60 and 137 million tons in terms of dry and green fodder will increase up to 69 and 145 million tons by 2020. The resources are going to be shortened day by day, which needs to be monitored, and change in the cropping system is must for remunerative agriculture.

As a conclusion in the sustainable farming systems, one may identify at least five conceptualizations of sustainable agriculture and farming systems listed as under:

1. A sustainable farming system is a system in which natural resources are managed so that crop yields do not decline over time.
2. A sustainable farming system is a system in which natural resources are managed in such a way so that the stock of natural resources does not decline over time.
3. A sustainable farming system is one that satisfies minimum conditions of ecosystem stability and resilience over time.
4. A concept related to sustainable farming systems is HNV farming systems, which are likely to be of importance from a nature-conservation point of view.
5. Sustainable agriculture is organized so that the necessary support services (credit, and alternate land use systems.

Need based alternate land use system

Among the several needs of a farmer Food always remains the first priority item, although fodder requirement is more as compared to food. Some of the need - based alternate land use systems matching the land capability classes are discussed below.

Alternate land use systems

Sl.No.	Food (arable land) II and III	Fodder (non - arable land) IV and V	Fuel/timber / fiber (marginal degraded land) VI and VII
1.	Alley cropping Agro horticulture	Horti - pastoral silvi pastoral	Tree farming timber- cum-fibre (TIMFIB)
2.	Agro - horticulture	Silvi - pastoral	
3.	Intercropping with NFTs	Ley farming Pasture management	

Rainfed agri-horti system

The system is most suitable for the rainfed region of western Rajasthan where surface runoff is harvested into the constructed Tanka in the field. These Tanka is designed on the basis of the rainfall pattern and average receipt of runoff during the rainy season. The number of fruit trees should be calculated on the basis of water requirement of the trees and water availability in the Tanka during the season. Generally low water requiring crops and fruit trees should be grown for better success of the system. Three-year-old plantation of budded ber (200 trees ha^{-1}) was found to perform well in association with moong bean even during the low rainfall situation (210 mm yr^{-1} rainfall).

Rainfed agri-pasture system

When the climatic situation is the very unfavorable, crops are adversely affected by climatic vagaries like drought or the fruit trees attained their full growth and crops cannot be taken in between the rows of fruit trees, the option remains to grow the most efficient and popular grasses like *Lasuirus sindicus* (deep rooted) or *Cenchrus ciliaris* (shallow rooted) which are sown once and continue to yield up to the 5-6 years or more. The shadow effect of the trees the grass vegetative growth become more and more tender forage at the green stage could be harvested under horti-pasture system.

Mixed sowing of various dryland crops is a common practice by the farmers of arid zone which hampers efficient management of the crops right from the sowing to harvesting and threshing. This problem can be overcome by sowing the crops/perennial grasses in strips of 2:1 or 3:1 or 4:1 ratio. The perennial grass components, besides imparting stability to crop *Cenchrus ciliaris/C. setigerus* and *Dichanthium anulatum* were compatible (Daulay and Bhati, 1989). These crops, besides giving full yield of grass component gave the bonus yield of grain (130-280 kg ha^{-1}) and crop straw fodder (350- 600 kg ha^{-1}). However, the inter cropping of grasses and drylands crops is sometimes not feasible in farmers micro farming situation and under these conditions strip copping of grasses and *Kharif* legumes in 1:2 ratio is recommended with a strip width of 5 meter (Bhati, 1997). The strip cropping of grasses and *Kharif* legume may help in increasing the quantity and quality of fodder.

Grass based strip-cropping system

In strip cropping, the grasses and crops are grown in different width of strips as per the soil characteristics and rainfall situation of the region. The protective strips of perennial grasses are grown in the strips at right angle s to the general directions of the prevailing winds. The strip ratio of 1:4 (3m: 12m) was found the most protective (checking the soil erosion) cum productive system and economical in terms of yield and monetary benefits under Bikaner climatic situation. The cluster bean crop grown in between the left strip produced 1.04 q ha^{-1} more yield than their sole cropping. The grass (*Cenchrus ciliaris*) yield of 34.04 q ha^{-1} was also recorded in normal rainfall (275 mm yr^{-1}) situation. In grass + pearl millet strip cropping pearl millet produced 3.0 q ha^{-1} of grain yield as compared to its sole cropping. The alternate strips of *Sewan* (*Lasiurus sindicus*) + cluster bean in 1:3 ratio (3m: 12 m) strips gave the highest dry matter of 70 q ha^{-1} of grass along with the cluster bean yield of 7.09 q ha^{-1} over 1:1 and 1:4 ratios strips.

Choice of crops in agri-pasture systems: Based on 100 years of climatic analysis and changing food habits of people of arid region in last 50 years, it has been advocated to reduce the area in pearl millet to 40% and put rest 60% of the total holding under kharif legumes and oilseeds (Bhati, 1997). Chauhan and Faroda (1979) found alternate row spacing of *Cenchrus ciliaris*/*Cenchrus setigerus* and *Lablab purpureus* better than other ratio. Inclusion of grass legume in pasture not only produced higher forage yield of high nutritive value, but also generated higher income. Among pasture legumes *Clitoria ternatea* and *Lablab purpureus* showed good compatibility with *Lasiurus sindicus* and Cenchrus ciliaris (Bhati *et al.*, 1986).

Growth performance of crops under agri-pasture system: Soni *et al.* (2007) reported that the plant height of cluster bean and moth bean in strip cropping with grass (5:15 meter grass crop ratio) did not vary significantly but number of pods of both the crops was significantly higher in agri-pasture system over their sole cropping (Fig. 1).

Productivity of crops under agri-pasture system: Grain and straw yield of Cluster bean and Moth bean was observed more in strip cropping system as compared to sole cropping. There was an increase of 47.2 % grain yield in cluster bean and 32.0 % in Moth bean under strip cropping system over sole cropping (Soni *et al.*, 2007). The main advantage of the strip cropping is that erosion resistant strip reduces the speed of prevailing wind, traps the saltating sand particle and reduced wind erosion. The strip of Sewan (*Lasiurus sindicus*) and castor (*Ricinus communis*) established at right angle to prevailing wind direction controlled the wind erosion effectively (Mishra 1971).

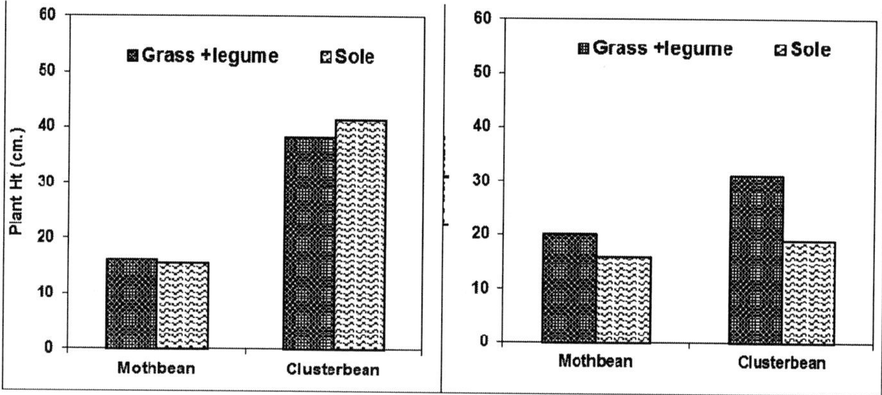

Fig. 1. Growth of arid legumes inter cropped with grass and sole cropping

Irrigated agri-horti system: The commencement of IGNP and tube wells with good quality water in the region now it becomes possible to go for irrigated horticulture. The precious and costly water could be used judiciously for higher water use efficiency and more profit as well. The arid situations are not very conducive for the irrigation having very high water application losses due to sandy soils with high infiltration/Percolation rate and high evaporation due to high temperature. Therefore for better use of available water the pressurized irrigation system like drip or sprinkler system can be used very successfully with better out put under these circumstances. These systems can be used in two ways (i) fruit crops under drip irrigation and arable crops as rainfed and (ii) fruit trees and arable crops under irrigated condition. In both the situation the competition for the water is always become more due to the intensification of the system.

Productivity of agri-horti-silvi system under sprinkler irrigation system: The height of all the 12 years old trees (citrus, mopane and shisham) in Agri-horti-silvi system was highest with intercropping of Aloe vera – Aloe vera rotation which was at par with intercropping of green gram-mustard crop rotation whereas tree stem diameter and canopy cover was highest in intercropping of cluster bean-barley under sprinkler irrigation .The system productivity in terms of cluster bean equivalent yield (CEY) was highest (6576.11 kg ha^{-1})in cluster bean-barley intercropping which was significantly higher over rest of the rotations and their sole cropping. The productivity of intercrop cluster bean (1692.5 kg ha^{-1}), green gram (886.3 kg ha^{-1}) and Aloe Vera (10.36 ton ha^{-1}) was recorded in intercropping with citrus planted at 5x6m spacing in agri-horti-silvi system. The yield of intercropping of green gram and cluster bean with mopane and shisham planted at 5x12m spacing were 159.78 and 43.72% higher over 5x6m spacing, respectively (Table 1).

Table 1. Growth and yield of trees and intercrops under agri-horti silvi system in micro-irrigation system.

Trees	Tree growth in agri-silvi-horti system			Yield of inter crops in agri-silvi-horti system			
	Plant height (cm)	Collar dia (cm)	Canopy cover (m2)	Moth bean /green gram (kg ha-1)	Cluster bean (kg ha-1)	Aloe vera (t ha-1) (Kg ha-1)	Lasiurus sindicus
Drip irrigation system (tree)							
Citrus	327.9	13.33	14.36	367.5	522.5	75.62	2667.5
Bael	486.3	15.89	19.25	239	440	56.73	1520
Gonda	337.7	18.23	14.99	207.5	452.5	47.66	1610.5
Sole	-	-	-	273.8	423.8	-	-
C.D.5%	41.43	.87	2.45	49.5	35.02	13.47	649.5
Sprinkler irrigation system							
Citrus	334.4	15.56	14.42	886.3	1692.5	10.36	-
Mopane	437.5	22.63	32.13	370	495	9.03	-
Shisham	551.3	17.30	18.53	763.5	1485	7.12	-
Sole	-	-	-	602.5	1135.0	-	
C.D.5%	35.33	1.74	2.73	33.4		226.35	-

Productivity of agri-horti system under drip irrigation system: All the fruit trees (citrus, bael and gonda) under drip irrigation system the plant height, stem diameter and canopy was highest with intercropping of moth bean and significantly higher over their sole planting. Highest yield of moth bean (367.5 kg ha^{-1}), cluster bean (522.5 kg ha^{-1}) and Aloe vera (75.6 t ha^{-1}) was recorded in intercropping with citrus (Table 1).

Summary

In arid zone as per the weather and climatic situation it has been proved that the single cropping of annual crops under rainfed condition is very risky, generally non profitable for the farmers. The inclusion of perennials like trees and grasses under different alternate land use systems provide a basis as an insurance against complete crop failure. The main precaution should be taken into consideration for the selection of crops and species for their combinations during cultivation providing there should very less competition between the crops/trees/ grasses in together.

References

Bhati TK (1997). Integrated farming systems for sustainable agriculture on drylands. In: Sustainable Dryland Agriculture, CAZRI, Jodhpur, pp. 102-105.

Bhati TK, Bhati GN, Shankarnarayan KA (1986). Legume-shrub grasslands in western Rajasthan. Annual Progress Report, CAZRI, Jodhpur.

Chauhan DS, Faroda AS (1979). Studies on establishment of mixed pasture of Cenchrus and Dolichos lablab. Forage Research 5:1-4.

Katyal JC, Das SK (1994). Rainwater Conservation for Sustainable Agriculture. Indian Farming, pp. 65-70.

Mishra DK (1971). Agronomic investigations in arid zone. In: Proceedings of symposium on Problems of Arid Zone, CAZRI, Jodhpur, pp. 165-169.

Singh RP, Das SK (1984). Timeliness and Precision Key Factors in Dryland Agriculture. Project Bulletin 9, CRIDA, Hyderabad, pp. 1-29.

Soni ML, Yadava ND, Beniwal RK, Singh JP, Sunil Kumar (2007). Production potential of arid legumes under grass based strip cropping system in arid rainfed condition of western Rajasthan. Journal of Arid Legumes 4 (1): 9-11.1

10

Improving Livelihood Through Agroforestry in Hot Arid Region

J.C. Tewari, Kamlesh Pareek and Shiran K.

Central Arid Zone Research Institute, Jodhpur, Rajasthan, India

Introduction

Tree growing in hot arid region is basically concerned with the management of trees for conservation and for limited production objectives like wood for fuel, poles and fencing material; leaves for livestock fodder; and pod/seeds for many types use in human diet. The role of trees to conserve the fragile ecosystems of hot arid regions has been well recognized (Mann and Muthana 1984). The trees also provided so many services to mankind, which make them an intricate part of man-livestock-agriculture continuum, the lifeline of hot arid regions (Saxena 1997).

The hot Indian arid zone is spread in 31.7 million hectare area of which major part is in northwestern India (28.57 m ha) and some in south India (3.13 m ha). Arid western Rajasthan covers 61 percent of total hot arid regions of the country and thus, forms the principal hot arid region of the India. This principal hot arid region is better known as Thar desert.

The image of desert one has of the vegetative less, dense unbroken mantle of sand and inaccessible terrain condition is, however, shattered upon entering the Thar desert in most of the parts barring extreme western fringes. Sparsely distributed trees and other woody taxa with underneath growth of arable crops (especially during 'kharif' season, as agriculture is pre-dominantly rain fed) and /or grasses in long stretches interspersed with distantly distributed village settlement and "Dhanis" (a unique settlement pattern characteristic of Thar desert, meant for living of agriculturists families during active cropping period away from the villages but, nearby their field) are actual features of the region

(Tewari et al. 1999). Thus, rural folk in this principal hot arid region of the country have been practising arable cropping in association with scattered trees on crop fields since time in memorial. As most trees are drought resistant, these are still able to provide fuel, fodder and other products when arable crops fail due to drought which is a common feature in the region.

Traditional agroforestry systems entail excellent 'multipurpose' species such as *Prosopis cineraria*, *Acacia nilotica*, *Zizyphus* species, etc., providing large number of end-use products. While contributing to food security in difficult time, trees on agricultural fields are important source of income. Boffa (1999) opined that multipurpose trees of traditional agroforestry systems are important for rural food security and income generation, and also for ensuring social and cultural stability to some extent. In addition, these systems actively maintain a rich pool of the forest genetic diversity on farmer's land. In addition to traditional agroforestry systems, the improved agroforestry systems developed in last 35 years have tremendous potential to improve livelihood of stakeholders.

Traditional agroforestry systems

Historians are generally of the view that organized settlements in the region began in the 4th century BC, when number of tribes migrated from fertile Indo-Gangetic plains into this environmentally hostile, but otherwise secure tract in response to waves of the invasions from the west. The settlements increased and expanded, and by the 6th or 7th century AD much of the desert region of western Rajasthan was not only settled but also politically organized (Tod 1832). Since then land use pattern of the region has witnessed several changes. The mixed crop-livestock farming, mixed livestock-crop farming and livestock farming form the spectrum of economic activities in arid regions of India ranging from settled agriculture (in true sense, settled agroforestry) to nomadism (particularly, pastoralism). Tewari (1997) described large tracts of lands in Thar desert region of Rajasthan having widely scattered trees/shrubs of various species in association with crops of food grain and fodder as the best example of traditional agroforestry.

The people of the region have evolved agroforestry based drought protective mechanism through their ingenuity and centuries old experience, which has descended from one generation to other. Depending upon climatic, edaphic, socio-economic and cultural situation, drought hardy woody perennials, which are multipurpose in nature, are integrated in farming systems to develop productive systems in form of traditional agroforestry. A detailed study of two transects by Tewari *et al.* (2014) clearly indicated that rain fall primarily governed the structure of traditional agroforestry systems, though local edaphic and socio-economic-cultural factor also play some roles. In upper transect, which ran from western border of Jaipur district to extreme western part of Bikaner-Ganganagar districts the major traditional agroforestry systems were: *Prosopis cineraria-Acacia nilotica* based (over 400 mm rainfall zone), *P. cineraria* based (between 300 and 400 mm rainfall zone), *Zizyphus* spp. *P. cineraria* based (between 200-300 mm rainfall zone) and *Zizyphus* spp. *P. cineraria – Salvadora* spp. based (less than 200 mm rainfall zone) (Narain and Tewari 2005).

The lower transect in arid western Rajasthan encompassed Pali, Jalore, parts of Jodhpur and Barmer district in rainfall gradient of >500 to 200 mm from south-eastern margin to western part (Anonymous 2008). Four distinct agroforestry systems have been identified in the study region. In south-eastern margin, where total annual rainfall was >500 mm, the pre-dominant agroforestry system was *Azadirachta Indica* and *Acacia nilotica* var. *cupressiformis* based. This system extends from extreme eastern part of Pali district to western part of Jalore district. In the rainfall zone >400-500 mm, *Prosopis cineraria – Azadirachta indica* based system was prevalent. In rainfall zone of 300-<400 mm, the major traditional agroforestry system was *P. cineraria - Tecomella undaulata - Salvadora* spp. based. In western part of Barmer district (rainfall zone >200-300mm), the most prevalent system was *P. cineraria – Salvadora oleoides- T. undulata* based.

Some improved agroforestry systems

Improved agroforestry practices

In the light of increasing pressure on land resources, Central Arid Zone Research Institute (CAZRI), Jodhpur initiated systematic studies on agroforestry systems in late 1970s. Since then a number of improved agroforestry practices in order to enhance overall productivity and economic returns of the farming communities have been developed and standardized. Following improved practices have been found promising and remunerative, and easy to fit in existing traditional agroforestry systems.

(a) Agro-horticulture

Leguminous crop sown under ber *(Zizyphus mauritiana* cv. Seb) plantation produced 0.2 t ha^{-1} of grain and 0.8 t ha^{-1} quality *ber* fruits from same land unit even when seasonal rainfall is 200 mm (Table 1), thus rendering a drought proofing mechanism to the system. The density of ber plants were kept 400 individuals ha^{-1} (Gupta 1997). The economics of this improved system indicated that in case of sole leguminous crop (mung bean) farming, the net profit per hectare was Rs. 4800/-, however, in case of ber intercropping, the profit was to a tune of Rs. 8000/- per ha.

Table 1. Improved agroforestry practices: agri-horticultural *(Ber*+ Mung bean)

Treatment	Annual rainfall (mm)	Fruit yield (kg ha^{-1})	Grain yield (kg ha^{-1})	Net profit (Rs ha^{-1})
Sole crop	200	-	520	4800
Intercropped with Ber	200	800	200	8000

(b) Horti-pastoral

Ber based horti-pasture have proved highly remunerative in *Thar* desert on farmers' field. *Ber* trees were planted at spacing of 6x6 m and grass *Cenchrus ciliaris* was introduced between tree rows after third year. On an average, the dry grass production was 1.55t/ha/year (Tewari *et al.* 1999) (Table 2). The fruit, leaf fodder and fuel wood production from ber was 2.77, 1.87, 2.64 t ha^{-1} yr^{-1}, respectively.

Table 2. Improved hortipastoral (Ber+ *Cenchrus ciliaris*) on farmers' field

Year	Dry grass (t ha^{-1})	Tree products		
		Fuelwood (t ha^{-1})	Leaf fodder (t ha^{-1})	Fruit yield (t ha^{-1})
1	1.50	3.10	2.13	3.10
2	1.67	2.74	1.80	2.74
3	1.42	2.46	1.71	2.49
4	1.66	2.49	1.63	2.85
5	1.48	2.43	2.01	2.80
6	1.57	2.63	1.94	2.67
Average	1.55	2.64	1.87	2.77

(c) Silvo-pastoral

Arid region of Rajasthan is reported to have 89 species of grasses. The species like *Lasiurus sindicus* is efficient builder of biomass with energy use efficiency of 1.4-2.0% (Harsh *et al.* 1992). In an improved silvo-pasture, *Hardwickia binata* was taken as tree component at 3m x 3 m spacing with *Cenchrus ciliaris* grass. Results of 9 years study (Table 3) revealed that average carrying capacity of the practice was 4.1 sheep ha^{-1} yr^{-1} against 3.7 for sole C. *ciliaris* pasture and 1.6 for sole *H. binata* plantation. In this type of improved silvo-pasture practice, in addition to grass + top feed production to the tune of 3.06 t ha^{-1} yr^{-1}, 0.26 t ha^{-1} yr^{-1} of fuel wood is also obtained.

Table 3. Production potential of *Harwickia binata* + *Cenchrus ciliaris* based improved silvopasture

Year after planting	Grass yield ($t\ ha^{-1}$)	Leaf fodder yield ($t\ ha^{-1}$)	Total fodder yield ($t\ ha^{-1}$)	Carrying capacity ($sheep\ ha^{-1}\ yr^{-1}$)
1	0.82	-	0.82	1.9
2	1.23	-	1.23	2.9
3	1.64	-	1.64	3.9
4	2.05	-	2.05	4.9
5	1.64	-	1.64	3.9
6	1.64	-	1.64	3.9
7	1.24	2.31	3.55	8.5
8	0.84	0.66	1.50	3.6
9	0.84	0.66	1.50	3.6
Average	1.33	0.40	1.73	4.1

Agroforestry for enhancing livelihood options in arid regions of India

Agroforestry contributes to livelihood improvement in arid regions of India, where people have a long history of accumulated local knowledge. Arid regions are particularly notable for ethano-forestry practices and indigenous knowledge system on growing trees on farm lands. The economics of previously discussed traditional agroforestry systems of arid western Rajasthan indicated that net B:C ratio of such systems is generally on positive sides (Table 4).

Table 4. Economics of traditional agroforestry systems in upper transect of arid western Rajasthan (Base Year 2002)

AF system	Expenditure ($Rs\ ha^{-1}$)	Returns ($Rs\ ha^{-1}$)			Gross returns ($Rs\ ha^{-1}$)	Net Returns ($Rs\ ha^{-1}$)	Net B:C ratio
		Crops	Fuel wood	Leaf fodder			
P. cineraria-*A. nilotica* based	1850	4103	1230	870	6203	4353	2.3
P. cineraria based	1550	3670	600	420	4690	3140	2.0
Z. spp. - *P. cineraria* based	1550	1506	620	600	2726	1176	0.7
Z. spp. - *P. cineraria* – *Salvadora* spp. based	1500	1400	500	500	2400	900	0.6

In fact, multipurpose trees in agroforestry systems provide a rational formula for drylands. In hot arid regions trees on farmlands are the organisms with an ecological investment strategy (Oldeman 1983). Matter and energy are immobilised in long lived trees and this majority of accumulated production is harvested at longer interval than in case of arable crops.

Jujube (Ber) is a very useful tree for agroforestry of arid regions. The leaves make excellent fodder, crude protein 13-17% and fiber 15%. Leaves are also food for silkworms. The branches and twigs are excellent fuel wood having calorific value 4878 K cal/kg. Root bark and leaves have 7% and 2% tannin, respectively and are sometimes mixed with other sources for tanning leather. The fruits provided by all the improved cultivars are delicious and rich source of vitamin A, C and B complex. Thus cultivation of jujube in arid regions offers solution to many problems. Tewari *et al.* (2001) demonstrated that agroforestry, particularly improved jujube based horti-pasture system can be major source of livelihood in many parts of arid regions (Table 5).

Table 5. Economic returns (in terms of gross income) from a Jujube based horti-pastoral system

Particular	Production (t ha^{-1})	Returns (Rs.)
Dry grass	1.55	3,100
Fuel wood	2.64	2,640
Leaf fodder	1.87	7,480
Fruit	2.77	33,240
Total	-	46,640

This system provides production in terms of human food, livestock feed and cooking energy needs to communities. Such multifunctional agroforestry systems provide social wellbeing to the people in addition to goods and services.

The most spectacular evidence in context of livelihood improvement potential of agroforestry in arid region came through *Acacia senegal* based localized traditional agroforestry system in parts of Barmer and Jodhpur districts (15 villages). *A. senegal* is source of gum Arabic which has very high commercial value. In addition to harvesting crop grain for food and crop straw for fodder, the farmers harvested substantial quantity of gum Arabic through small intervention by CAZRI.

From 2008-09 to 2013-14, farmers in more than 45 target villages of arid western Rajasthan earned revenue of more than Rs. 3.74 crores from the sale of gum Arabic produced using CAZRI technology. In this way since 2008-09 per year an additional revenue flow to the tune of Rs. 1.60 lakh $village^{-1}$ yr^{-1} maintained regularly till date in said villages. Moreover, during 2008-09 to 2013-14, CAZRI generated revenue of more than Rs. 14.73 lakh by way of sale of CAZRI gum inducer (Table 6). This self sustaining technology is now expanding in other parts involving other gum/resin yielding tree species in addition to *Acacia senegal.*

Table 6. Gum Arabic production and economic returns in 45 target villages of Barmer, Jodhpur and Nagaur district of arid western Rajasthan

Particular	Year						Total
	2008-09	2009-10	2010-11	2011-12	2012-13	2013-14	
Number of *A. senegal* trees treated (in thousands)	12.1	20.95	22.61	27.5	30.00	34.17	147.33
Production of gum Arabic by farmers (t)	5.45	10.48	7.67	11.00	12.00	13.67	60.27
Total income earned by farmers (Rs. Lakh)	27.23	52.38	38.33	77.00	84.00	95.69	374.63
Revenue generated by CAZRI (Rs. Lakh)	1.21	2.10	2.25	2.75	3.00	3.42	14.73

These examples demonstrate the potential of agroforestry in arid regions for breaking the poverty and food insecurity circle. In addition to providing food and feed, there are numerous other non-timber tree products (excluding gums and resins) collected from trees on farm lands. *Prosopis cineraria* trees in addition to providing fuel and fodder give nutritious vegetable in form of pod, locally known as '*Sangri*'. Likewise seeds of *Acacia senegal* are also used in preparing vegetable. During drought years their value is increased tremendously.

Success story: transforming subsistence level, rural economy through ber (*Ziziphus mauritiana*) based diversified agroforestry system

In environmentally inhospitable hot arid tract, Shri N. K. Jaisalmeria of village – Manaklao, located 25 km north of Jodhpur by adopting CAZRI technology package transformed the nature of undulating sandy unproductive landscape of 4.5 ha into a highly productive diversified agroforestry system is perhaps the best success story of CAZRI. Arable farming is by far the dominant pursuit of this densely populated principal arid zone of India. Though experience over the generations, the farmers have evolved a mixed livestock-crop-tree farming system. However, land productivity is low and permits no more than a subsistence living in most years and severe shortage of food and fodder in subnormal rainfall years such as occur in two out of five years on an average. Whereas for example in a good rainfall year, a hectare of land produces 5 q of grains and 12 q of fodder, a poor year gives hardly any grain and just 2-3 q of fodder. Assets and cash reserves of the rural population are negligible. Serous indebtedness follows and despite governmental support, under-nourishment of human population and livestock mortality are a common feature. Moreover, arable farming is somewhat over exploitative of land resources. With these considerations, Mr. Jaisalmeria started cultivation of ber fruits on his land with the technical assistance of CAZRI, Jodhpur.

He first established fairly large Ber orchards with 750 trees under rainfed condition and secondly a ber nursery was established by him for supply of reliable budded ber plants. The periphery of entire 4.5 ha area was covered by MPTS like *A. tortilis, P. juliflora* and other woody species, which formed a very good shelter belt within six years. Superior biomass like *P. cineraria, Tecomella undulata* and *Azadirachta indica* growing within the orchards are protected. *Cenchrus ciliaris* grass is planted between ber fruit trees. One hectare area is left to grow pearl millet which provided grain for human consumption and straw for livestock consumption. Eight goats are reared within the system without purchasing any kind of fodder. Farm waste is used to prepare compost within the system. The component and functioning of this diversified agroforestry system of Mr. Jaisalmeria is given in Figure 1.

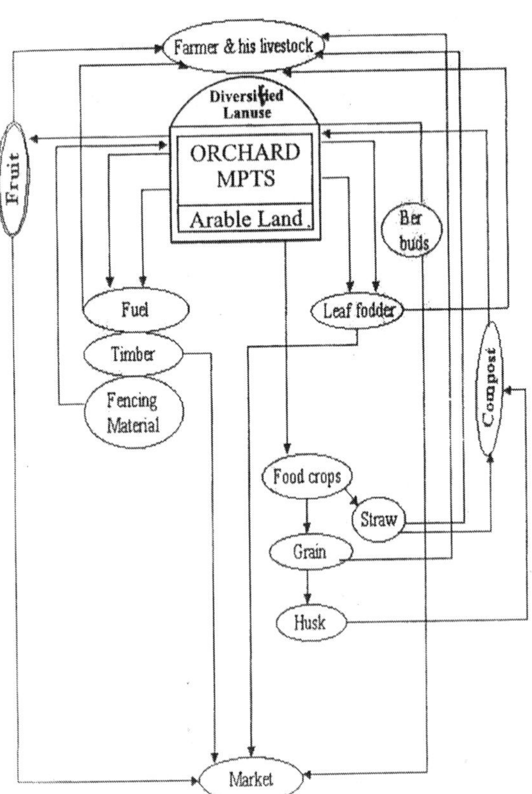

Fig. 1. Diversified agroforestry system: A schematic representation

The system is producing ber fruits to the tune 10.7 t ha^{-1}, fuel wood (dry) 4 t ha^{-1} and fodder about 2.4 t ha^{-1} in each year. This system is capable of providing employment to the tune of 331 man days per ha per year. The system is self-contained and the efforts of Mr. Jaisalmeria have been recognized in national and international forums. He was invited to South-East Asia UNCCD head quarter at Bangkok, Thailand to present his success story of dry land development during 2001. In the year 2003, he was awarded Krishi Shiromani Award of Ministry of Agriculture, Govt. of India and N. G. Ranga Award of ICAR for diversified agriculture during the year 2005.

The success of Mr. Jaisalmeria's efforts can be better judged on the basis of following important issues:

- *Land use issue:* The programme was implemented on a sandy undulating land of hot arid zone having negligible productivity of rainfed crops. His efforts demonstrates how such unproductive lands in the desert areas can be transformed into an efficient production system, which in fact, is a sustainable land use in hot arid zone. This integrate land use having ber fruit tree plantation protected by shelterbelts of MPTS like *A. tortilis* and *P. juliflora* coupled with utilization of interspace by planting other drought hardy plant species of economic value not only provides sustainable production but also leads to soil conservation and improvement in the fertility level.
- *Social and economic issues:* The land which was once not providing an income of Indian Rs. 500/- (US$ 10) per ha per year is returning (net income) an average Indian Rs. 20,000 (US$ 400) per year. The cost : benefit ratio is 1:5.5, which is economically quite sound. Though his effort, the technological package to utilize such unproductive arid lands in an integrated manner of perennial crop based system has been extended to 90,000 ha area from extreme north-west part of the country to southern states. At his farm the employment generation is in a tune of 1200 man days per ha per year. Considering the same rate of man days requirement, it is estimated that these plantations are providing employment generation in order of 1.08 million man days per year in the country. The technological package is very simple and acceptable to the farmers with various levels of family economies i.e., poor to rich ones. In country like India, especially in rural sector of arid areas, where huge labour force facing acute un-employment problems, the employment generation through ber based diversified agroforestry system could play vital role in improving the rural economy.

- *Social capital issue:* A very high order of adoption of said technological package in hot arid areas of the country has been changed the economies of many poor farmers. Due to adoption of integrated ber based diversified land use, the lands which have previously very poor or of no value, is now being considered worth Lakhs of Indian Rupees. This has direct impact on social status of the farmers in traditionally complex socio-economic-cultural web of rural life of India, who have successfully adopted the technological package. Moreover, a high rate of employment generation in environmental inhospitable arid tracts, especially for women folk, as they are considered most efficient in harvesting and grading the ber fruits is playing significant role in the women empowerment.
- *Policy related issues:* Once the success of his enterprise attracted the attention of farmers, subject matter specialists and policy makers, Horticulture departments of various state governments, especially those ones having area under hot arid zone, started giving serious thoughts to ber crop and in due course of time ber has been recognized as one of the important fruit crops. To promote the ber plantations, provision of subsidies and loans have been made by various state governments. Now many banks like National Bank for Agriculture and Rural Development, Gramin (village) banks, etc. and National Horticulture Mission have also made the provisions for short-term and long-term loans for raising ber based diversified production system.

Epilogue

The hot arid regions of India are economically and environmentally disadvantaged part of the country with unique problems. In fact, ecosystems in this part of the country are highly fragile and large liabilities causing severe impediments in development programs. However, despite of hostile climatic and edaphic conditions Indian hot arid zone is highly vegetated and tree species diversity is perhaps much higher in comparison of hot arid zones of other parts of world. Moreover, the tree species which have been introduced successfully in hot arid regions of the country from iso-climatic region of the world or from other drier parts were thoroughly evaluated on the basis of information available on climatic homologues, potential inherent plasticity and genetic variability before recommending them for plantation and agroforestry programs. In this part of the country rural folk need tree produce daily, however, in the economic balance of tree use, the cost of tree products is closely related to their availability.

The majority of tree species which are found in hot Indian arid zones, are multipurpose and they have the ability to satisfy the expectations and aspirations of rural folk regarding the productions of their basic needs i.e., fuel, fodder, timber, food (fruits pod^{-1} seed^{-1}) and other products like exuded gum, products of medicinal value, etc. Agroforestry with suitable tree species in different arid land forms thus assumes much significance for desertification control and ecosystem services.

Though combined productive-protective systems or agroforestry systems are not a well defined part of land use system, however, in true sense entire hot arid zone represents the vegetation complex in the form of agroforestry. The tree components play a vital role in both productivity and sustainability of these agroforests. The characteristics of trees which are generally considered to be environmentally beneficial are: ability to utilize incoming solar radiation throughout the year, which otherwise will be lost as arable crops are sown only during monsoon period; the capacity to enrich micro sites by depositing litter in the topsoil, which can then be exploited by more shallowly rooted species; and a capacity to modify the microclimate, which can bring about favorable effect on the soil and associated plant species.

References

Anonymous (2008). Annual Progress Report.CAZRI (ICAR), Jodhpur, India, pp.166.

Boffa JM (1999). Agroforestry parklands in sub-Saharan Africa. Food and Agriculture Organization of the United Nations, Rome, pp. 230.

Gupta JP (1997). Some alternative production systems and their management for sustainability. In: Gupta JP, Sharma BM (eds) Agroforestry for Sustained Productivity in Arid Regions, Scientific publishers, Jodhpur, India, pp. 31-39.

Harsh LN, Tewari JC, Burman U, Sharma SK (1992). Agroforestry in arid region.Indian Farming (Special issue on environmentally sound biotechnology) 45:32-37.

Mann HS, Muthana KD (1984). Arid Zone Forestry, CAZRI Monograph No. 23, CAZRI, Jodhpur, pp. 48.

Narain P,Tewari JC (2005) Trees on agricultural fields: a unique basis of life support in Thar Desert. In: Tewari VP, Srivastava RL (eds) Multipurpose Trees in the Tropics: Management and Improvement Strategies. Arid Forest Research Institute, Jodhpur, pp. 516-523.

OldemanRAA (1983). The design of ecologically sound agroforests. In: Huxley PA (ed) Plant Research and Agroforestry. International Council for Research in Agroforestry (ICRAF), Nairobi, Kenya, pp. 173-207.

Saxena SK (1997). Traditional agroforestry systems of western Rajasthan. In: Gupta JP, Sharma BM (eds)Agroforestry for Sustained Production in Arid Regions, Scientific Publishers, India, pp. 21-30.

Tewari JC (1997). Important fodder trees of arid zone and their management. In: Gupta JP, Sharma BM (eds)Agroforestry for Sustained Production in Arid Regions, Scientific Publishers, India,pp. 147-153.

Tewari JC, Bohra MD, Harsh LN (1999). Structure and production function of traditional extensive agroforestry systems and scope of agroforestry in Thar desert. Indian Journal of Agroforestry 1(1):81-94.

Tewari JC, Ram Moola, Roy MM, Dagar JC (2014). Livelihood improvements and climate change adaptations through agroforestry in hot arid environments. In: Dagar JC, Singh AK, Arunanchalam J (eds) Agroforestry systems in India: Livelihood Security and Ecosystem Services, Springer,pp. 155-183.

Tewari JC, Tripathi D, Narain Pratap (2001). Jujube: A multipurpose tree crop for arid land farming systems. The Botanica 51:121-126.

Tod Col J (1832). Annals of Antiquites of Rajasthan (2 volumes). Routledge and Kegan Paul, London.

UNEP (1992). World Atlas of Desertification. UNEP, Edward Arnold (A Division of Hodder & Stoughton), London-New York - Melbourne-Auckland, pp. 69.

11

Conservation Tillage and Crop Residue Management in Relation to Dynamics of Soil C and N Under Climate Change Scenario - I

N. S. Pasricha

Potash Research Institute of India, Gurgaon, Haryana, India

Introduction

The rate of increase in anthropogenic CO2 concentration of the atmosphere is increasing at an alarming rate from 1.7 µmol mol-1 yr-1 in 2009 (Tans 2009) to 2.2 µmol mol-1 yr-1 in 2013 (WMO 2013). At this rate, its level in the atmosphere will reach around 450 µmol mol-1 by the turn of century from its current concentration of 400 µmol mol-1, leading to increased global warming and concomitant climate change. There is, thus, a very strong scientific interest in finding ways to slow or reverse this trend. Mitigation strategy of diverting CO2 from atmosphere to soil by adopting appropriate agricultural practices such as conservation agriculture (CA) is a recognized such method (Lal 2008 a, b). Broadly, CA has three major components- reduced tillage, crop residue management, and intensive cropping. Conservation tillage practices which include reduced tillage (RT) and no-tillage (NT) have been amply demonstrated as the agricultural practices that are very helpful in this direction (West and Post 2002; Franzluebbers 2010). Crop residue management, along with crop intensification with diverse cropping are the other important components of CA which significantly help in conserving more C in soil in organic combination as soil organic carbon (SOC). Storing or sequestering C in natural and agricultural ecosystems, thus, has the potential to offset a significant portion of the future atmospheric increase in atmospheric CO2concentration. Annual mitigation potential of reduced or no-tillage and residue management strategies, as estimated by Smith *et al.* (2007 a, b) are placed at 0.17 to 0.86 Mg CO2 equivalent $ha^{-1}yr^{-1}$ in temperate dry climate and 0.53 to 1.12 Mg CO2 equivalent ha-1yr-1 in temperate moist climate. Adequate soil moisture content in soil can greatly facilitate such sequestration process. Crop residue management and wise tillage practices, by affecting soil immobilization and stabilization processes, can also greatly facilitate efficient utilization of N from fertilizer, crop residue and soil organic matter, thereby, preventing its possible leakage to the environment.

Due to prevailing high temperatures, such mitigation potential levels are likely to be relatively low in tropics and subtropics. This suggests a slower rate of soil organic carbon (SOC) accrual in tropical and subtropical agro-ecosystem especially in dry and hot climate conditions. Moreover, pre-sowing soil preparation by conventional tillage (CT) contributes towards greater carbon dioxide (CO2) release to the atmosphere from accelerated decomposition of soil organic matter (SOM) and loss of soil organic carbon (SOC). Loss of SOC through soil OM oxidation is generally more pronounced in the tropical and subtropical regions due to the thermic soil temperature regime and extended duration of high temperatures. So the accumulated SOC is generally low in these soils as compared to temperate regions. However, Wood *et al.* (1991) concluded that, even in the OM-depleted tropical and subtropical soils, increased soil C can be achieved in these thermic-udic regions with proper combination of conservation tillage, residue management and cropping practices. For example, the adoption of NT management regimes in Brazil has greatly increased during the last few decades. In the 2008 cropping season, Brazilian growers harvested around 100 million Mg of grain under NT regimes from more than 25 million ha representing >50% of Brazilian agricultural production. In the northwestern part of India, where intensive rice-wheat rotation is commonly followed under irrigated conditions, there is great scope in stocking C in the soil by tillage and residue management system. Cereal yields in this region are quite high and generate larger amounts of crop residues that have higher lignin contents and probably decompose more slowly. In-field burning of the crop residue in this cropping system is a common practice in the region which leads, not only to large scale emissions of CO_2 but also the practice deprives the soils from important source that can help maintain/ increase the much needed SOC in these light textured soils. Thus, one of the key ecosystem services that can be provided by CA practices in agricultural soils is C sequestration and concomitant improvement in soil health and moderation of climate both in temperate and subtropical moist agricultural regimes (Lal, 2008a).

Future food production targets can only be met when potential benefits of improved germplasm are combined with improved resource management technologies. This includes enhancement of soil quality and land use value through increase in soil organic matter (SOM) reserves and conservation of water in the root-zone. In high input cropping systems like rice-wheat system, improved soil quality through CA practices is reflected in improved efficiency of external inputs (fertilizer, water, pesticides, labour, and improved seed) resulting in lower production costs. To sustain high input cropping systems, it is necessary to improve the soil health by increasing its organic carbon content and protect it from degradation because of excessive mining of nutrients and depletion and pollution of ground water. Conservation agricultural practices involving tillage options and residue management improve and maintain the ability of soil to maintain high productivity over a period of time. It improves the quality of environment by containing the leakage of nitrates to the ground- and surface water resources. This means that soil health is protected from degradation and its physical, chemical, and biological properties are improved. Such management practices allow optimum productivity, stability and sustainability of agriculture with higher level of biological diversity supporting internal nutrient cycling system and resilience to wide climatic disturbances.

Impact of long-term conventional/traditional tillage methods of management

Soil fertility decline, soil structural degradation, and large scale soil erosion loss are some of the consequences of exploitation of soils in a most unsustainable conventional manner. Over a period under conventional agriculture systems, the soils in tropical and subtropical regimes have come to possess very low levels of stored SOC, which results in their low water-holding capacity, poor microbial activity and diversity, and low nutrient supplying capacity, conditions that limit productivity and in-put use efficiency. This situation can be practically overcome with CA management to increase the SOC and N stocks of soils. Better conservation of soil moisture under reduced tillage or no-tillage (NT) can lead to greater grain yield and N use, especially in dry land agriculture where rainwater can be more effectively conserved in the soil profile through NT and crop residue retention practices which helps in preventing fallowing. Conservation systems that maintain crop residues on the soil surface improve infiltration rate many fold, and reduce evaporation and conserve water for crop growth, making more efficient use of limited water supplies under dry land conditions (Todd *et al.*, 1998).

In the thickly populated South Asian countries of India, Pakistan, Bangladesh and Nepal, there are least chances for arresting soil degradation and restoration of fertility by resorting to practices such as planted forests or perennial pastures. Due to increasing population, the land available per capita is rather decreasing in these countries. The only option for these countries is to strictly follow the practices of conservation agriculture if they want to sustain their agriculture system. Even if alternative energy and green house gas mitigation strategies become available in future, we should still encourage CA soil management systems that sequester C in soil, because increased SOC in the soil improves soil quality which is reflected in increased crop productivity and more efficient use of water and other inputs. Continuous NT treatment is an environmentally appealing alternative to conventional tillage (CT) with a view to obtaining good crop yields in future when the rains become more erratic and relatively higher atmospheric temperature due to global warming and climate change.

Necessity for conservation agriculture practices under climate change situation

Atmospheric CO2 has increased by almost a quarter to its present level of 400 μmol mol-1 in the last 50 years (WMO 2013). However, recent data indicate that atmospheric levels of CO2 have risen by 35% faster since 2000 than scientists have anticipated, due, in part to decreased ability of the global green cover to absorb emitted C. The concentration of atmospheric CO2, at nearly 350 μmol mol-1 in 1990, was projected to increase at a rate of 1.5 to 1.7 μmol mol-1 yr-1 (Watson *et al.* 1990), however, latest rate of increase at 2.2 μmol mol-1 yr-1 (WMO 2013) is showing that GHG emissions to the atmosphere have rather accelerated in the past few decades despite the best efforts to contain them. At this rate, its level in the atmosphere will reach between 600-700 μmol mol-1 by 2100, much higher than the earlier estimate of 450 μmol mol-1. The elevated [CO2] level in the atmosphere, however, represents an upsurge of an essential resource for plant species of economic importance. There is wide spread evidence that elevated CO2 increases plant growth due to more efficient photosynthesis process (Kimball 1983; Newton 1991). However, increase in CO2 concentration is accompanied by increase in atmospheric temperature and change in rainfall pattern. Karl *et al.* (2009) predict a warming trend of 1.5 to 2.0oC and slight increase in precipitation over the next century. The practice of conservation agriculture (CA) may become more relevant under the impending conditions of climate change, when erratic rainfall pattern in future may result in greater frequency of high intensity rains i.e., say rainfall >50 mm received in < 48 h. It is a cause of concern because of inability of the soil to maintain infiltration rates high enough to absorb high-intensity rainfall events. This may lead to severe soil erosion and nutrient losses due to increased run-off losses of water. To control this problem, CA (crop residue retention, minimum or no- tillage, intensive cropping, and crop rotation) would become more relevant in future to contain runoff, prevent soil erosion and improve water infiltration rates and to maintain a higher soil profile water content. Increased stored water with continuous CA practice will also prove helpful in preventing short spells of mid-season droughts (when rainfall is delayed or there is non-availability of timely irrigation). Thus, response of agricultural crops to future climatic changes will be strictly dependent on wise management practices.

Even the modest projected increase in temperature to the extent of 0.80 °C during the next 50 years will increase soil water evaporation and crop transpiration to a considerable extent. This may lead to increasingly more soil water deficits in future. Water availability will, thus, become more prominent in crop production in future. This can adversely impact the crop production unless mitigation measures such as CA practices of crop residue retention and minimum/ no-tillage are used. Interaction of water with CO_2 and temperature will have to be more precisely understood for better adaptation of cropping systems to climate change. There is general agreement that CO_2 level will increase to near 450 μmolmol-1, and temperature by 0.8 to 1.2° C by year 2050. Increased atmospheric temperature may have more deleterious effect than the positive effect of increased [CO_2] especially under tropical and sub-tropical conditions. Any increase in temperature at a time when wheat crop is at grain filling stage in the months of February and March, will shorten crop's life cycle, it will shorten reproductive phase duration and will result in reduced yield. Potential impact of climate change on the wheat production in India, if temperature increases by 0.8°C over the next 50 years, has been assessed by Ortiz *et al.* (2008). According to them, more than 50% of the area under wheat in India which is currently classified as high potential, irrigated, low-rainfall mega-environment would be reclassified to heat-stressed, irrigated, short-season production mega-environment. This particular area currently accounts for about 15% of world's wheat production and would undergo significant reduction in yield unless measure practices of conservation agriculture in tandem with improved germplasm are developed and adapted to the projected climate regime. Without such an adaptation, the impact on the production potential would drastically alter the ability of the country to produce a sufficient food supply for its population.

It is projected that there will be an increase in the number of days when temperature will be higher than the climatic-normals by 5°C. Similarly, the nights would be warmer in the sense that the minimum temperatures would also rise. This may decrease the duration of the growing season of crops especially in tropical and subtropical countries. Such an accompanied increase in temperature can more than offset the increased plant growth due to increased [CO_2] concentration in the atmosphere. Increasing temperature from 28/21 to 37/30° C can decrease the rice yield from 10.4 to 1.0 Mg ha-1 even under conditions of 660 μmolmol-1 CO_2 concentration through spike-let sterility due to higher temperature at grain-forming stage (Baker and Allen, 1993). Increased

CO2 in the atmosphere can cause decreased transpiration by manipulating the opening of stomata.

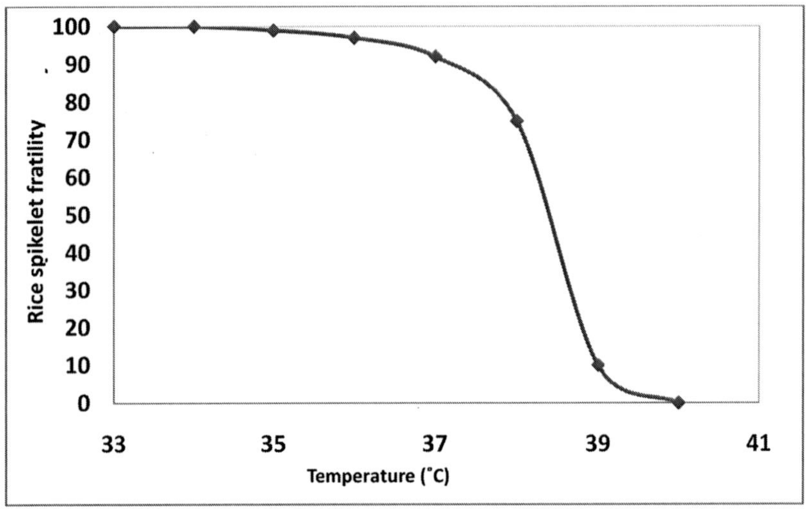

Source: Horie *et al.* 1997

Fig.1. Effect of atmospheric temperature on rice spike-let fertility

Under conditions of reduced transpiration, plant temperatures will be higher and gross thresholds for damage more often. Rice spike-lets show a sigmoid response to air temperature from 100% fertility at 330C to 0% at 400C (Horie *et al.* 1997). Since 330C air temperature does commonly occur in the rice season in India, rice crop is on the threshold of declining fertility with temperature (Fig.1) in future. Crop residues retained on the soil surface in CA practices' can help overcome this problem to some extent by moderating soil temperature due to mulching effect of retained crop residue.

Tillage effects on organic C and total N stocking in soil

Stocking more organic C in soil with the adoption of CA practices of no-tillage (NT) and residue retention has been proposed as a potential way to slow down the rising atmospheric CO_2 concentration resulting from accelerated use of fossil fuel as source of energy and food production by conventional soil cultivation methods. One goal of soil C sequestration is to increase the mass of C stored in agricultural soils so as to check its increasing concentration in the atmosphere and prevent global warming and climate change effects of this green house gas (GHG). Lack of soil aeration because of lack of plowing in NT causes specific changes in soil biological and physical conditions that results not only in the increase of soil C stocks but also influences soil N dynamics as C and N in the soil organic matter are biologically linked together and complement each other.

Using global data base of 67 long-term field experiments, it has been estimated that change from conventional tillage (CT) to NT could sequester an average 40 to 68 g C m-2yr^{1} (West and Post 2002). Similarly, NT practice with residue retention on soil surface has resulted in a significant increase in SOC pool in the top 0 to 30 cm layer after 43 yr of continuous corn crop (Ussiri and Lal, 2009). After a 50 yr of stubble-mulch tillage, Rasmussen *et al.* (1989) found that the soil of eastern Oregon had 33% more SOM in the top 7.5 cm with NT system than those soils that were conventionally plowed. Similarly, in an earlier observation, organic C and N in the top 7.5 cm soil were 26 and 32% higher, respectively, in two stubble-mulch systems than in conventionally plowed (Rasmussen and Rhode 1988). Vanden Bygaart *et al.* (2008) estimated annual C sequestration rates when agricultural soils are converted to NT in Canada to vary from 60 to 160 kg C ha-1yr^{-1}.

Addition of SOC is both due to retention/addition of crop residues and decreased decomposition of SOM in NT as compared with CT. Similarly, soil organic C sequestration has been reported as 0.45±0.04 Mg ha-1yr^{1} more with conservation tillage compared with CT crop land for a long-term investigation for 10±1 yr in Southwestern U.S.A (Franzluebbers, 2010). Establishment of perennial pastures, where soil disturbance is virtually zero, is more efficient in this respect and sequestered almost double the amount at 0.84±0.11 Mg C ha-1yr^{-1}. In a review of literature on the effect of CA on the SOC accumulation in the soil for several field experiments run for 10 to 12 tears in south-western US, SOC increased from 25.6±1.4 in CT to 30.2±1.4 Mg ha^{-1} in 21±1 cm soil depth.

In subtropical and tropical conditions, CA practice is rather more important as SOC contents actually decrease more rapidly with CT due to prevailing higher temperature. Conservation agricultural practices of NT and residue retention impacts on SOC build up is sensitive to climate and influence the changes in SOC in the order: tropical moist > tropical dry > temperate moist > temperate dry (Olge *et al.*, 2005). NT helps to protect the OC by stabilizing it in micro aggregates with in macro aggregates (Six *et al.*, 2004). Such occluded particulate OC in soil micro aggregates contributes to long-term soil C-sequestration in soils. However, in many such studies, there is not adequate base line data, including corrections for unequal soil mass due to changes in soil bulk density with conservation agriculture treatments. In a long-term tillage and residue management study, the change in the SOC pool (0-20 cm depth) after 16 yr showed a net loss of -6.3 Mg C ha^{-1} from the CT (moldboard plow) treatment and net gains of 5.1 and 10.8 Mg C ha^{-1} in NT and RT treatments, respectively (Halpern and Madarmootoo, 2010). These authors calculated changes in SOC pools on an equivalent soil mass basis, so the values reflect an actual gain or loss from the site in the 16 yr period, independent of changes in soil density resulting from tillage operations. Conservation tillage (rotary tillage and no-tillage) treatments may increase SOC and TN contents in the upper soil layers, but this may not be the case in lower soil layers where NT did not increase SOC and TN storage over CT (moldboard plow with residue) when calculated on the soil profile basis. In NT, most of the SOM accumulation is mainly near the soil surface. Soils in Northwestern India are often light-textured, have inherently low SOC content, often <1%. Therefore, even small changes in its concentration with adoption of conservation agricultural practices can significantly affect soil properties and agricultural productivity in this, otherwise, highly intensively cultivated irrigated agro-eco zone. Yaduvanshi and Sharma (2008) reported average SOC of 3.17 g kg-1 in 0-15 cm soil layer at the end of their 3 yr study in the NT plots as compared to 2.84 g kg-1 in CT treatment in sandy loam soil in Karnal, Haryana. They observed slight increase in grain yield of wheat with lesser amount of irrigation water in NT and reported a substantial saving of 7.22 cm of irrigation water.

Effect of fertilizer-N application on carbon sequestration

Carbon sequestration in the soil can be positively increased through CA practices such as reduced tillage or no tillage with crop residue management. Such sequestration significantly increases further with N fertilization to crops. Organic matter storage in soil is directly related to the amount of C input through residue retention/incorporation, below-ground biomass and rhizo-depositions. Nitrogen-fertilization, especially, by increasing crop productivity and corresponding amount of crop residue and root mass returned to the soil, affects favourably the SOC stocking. Also the availability of N in soil greatly influences the dynamics of soil decomposers, and thus, can affect C recycling in soil. In one of the studies, it was found that decomposition of corn residue in soil was greatly hindered when N content decreased below 30 mg N kg^{-1} soil but at N level >60 mg N kg^{-1} soil, there was no such hindrances (Recous *et al.*, 1995).

There can be a differential effect of applied N on decomposition of crop residue and mineralization of native SOC. A long mean residence time of corn residue-derived C was observed under NT and high N fertilizer than when residue were incorporated by plowing (Clapp *et al.*, 2000). Applied N only fastens the crop residue decomposition without affecting the mineralization rate of SOC (Moran *et al.* 2005). Such acceleration in the decomposition rate of native SOC is more confined to its lighter fractions, however SOC present as heavy fractions are in fact further stabilized with addition of fertilizer N (Neff *et al.* 2002). Similarly, long-term NT had been found to have higher SOC stock than minimum tillage (MT) and CT, and long-term medium to high N-fertilization tended to increase or maintain SOC contents (Powlson *et al.*, 2010). Increased crop production including crop residue and increased C inputs with N fertilization are expected to increase SOC stocks, but this may not always happen as opined Halvorson *et al.* (2002). Though crop yields of corn and soybean were almost 10% higher in moldboard plowing (MP) than under NT, which led to higher estimates of residue-C inputs in tilled soil, nevertheless, there was no significant difference in SOC stocks between the two tillage treatments (Poirier *et al.* 2009). When considering whole profile, SOC storage may not be affected by N-fertilization despite greater corn yields and larger estimates of total residue C inputs with increased N rates in MP. Similar significant interaction between tillage and N fertilization were also reported by McCarty and Meisinger (1997) and Allamras *et al.* (2004). These studies all reported that N fertilization increased the SOC storage and concentration between 0 and 20 cm depth in NT soils but not in conventionally tilled soils. The decomposition of crop residues can be limited by mineral N when they are incorporated into the soil by plowing. The mixing of mineral N with plowing in CT or MP may result in further decrease in native SOC content. Piorrier *et al.* (2009) observed that the plots under MP and those fertilized with 160 kg N ha^{-1} received the highest estimated amount of C inputs from crop residues but showed lowest amounts of SOC in the 0-20 cm layer. This shows that increased mineralization induced by mineral N application in CT systems can more than offset the increase in C input that is induced through increased crop yields. Limited or even negative effect of long-term (40-50 yr) mineral N fertilization on SOC in plowed soils were also noted by Khan *et al.* (2007) despite large increases in residue C incorporation induced by N fertilization.

The response of SOC stocks to N fertilization as related to C inputs through crop residues may be different under semiarid conditions due to reduced availability of water. Under semiarid conditions, C inputs may be mainly from the above ground crop residues and roots may not respond much to N-fertilization. However, Morell *et al.* (2011a) observed greater response to C inputs through N -fertilization in NT than in MT or CT even under arid soil conditions. Soil organic C content on equivalent basis was found to be very significantly influenced by N-fertilization under different tillage treatments (Table 1). Increased amount of organic C sequestered on equivalent basis as a result of increased N levels of fertilizers can be assigned to increased biomass due to ample availability of N in soil.

Table 1. Soil organic C (SOC) (g kg^{-1}) in 0-5 cm depth and soil organic C content in equivalent soil mass (SOCesm, $Mgha^{-1}$) in 0-10 cm depth under different tillage systems and N fertilization levels

Tillage treatment	SOC (g kg^{-1})				SOCesm ($Mgha^{-1}$)		
	ZN	MN	HN	Mean	ZN	MN	HN
NT	15.0	17.2	17.8	16.7	19.4	22.0	22.6
MT	11.5	12.9	12.6	12.3	18.0	19.8	18.8
CT	7.4	7.7	8.4	7.8	12.3	12.8	13.7

ZN = zero N application; MN = medium level N application (60 kg N ha^{-1}); HN = high level N application (120 kg N ha^{-1}).
Source: Morell *et al.* 2011a

Thirteen years after establishment of the experiment, the SOC stock under long-term NT was 3.9 Mg C ha-1 greater than under CT, and 4.3 Mg C ha-1 greater than under MT. Carbon dioxide fluxes decreased and SOC stocks in soil were 4 Mg C ha-1 greater with N fertilizer additions than the stocks on unfertilized plots. Increase in SOC and TN with N fertilization was more pronounced under NT than under MT or CT. Above ground crop residue addition may be essential to increases SOC content with adequate N-fertilization. N-fertilization increases C inputs from the above ground portions with N-fertilization was more pronounced under NT than under Mt and least under CT.

In temperate regions, N fertilization for 40 to 50 years of grain cropping decreased soil total N, although fertilizer rates always exceeded crop N removal in most cropping systems (Mulvaney *et al.* 2009). Dalal *et al.* (1995) earlier also observed similar results of decreased soil total N in N fertilized continuous cereal cropping systems over a 22 yr period. Higher soil total N as observed by Dalal *et al.* (2011) in their later studies due to retention of crop residue is attributable to the higher organic N inputs and greater N immobilization associated with residue retention than where the residue was burned, as well as a part of N lost as NH3 and N-oxides from residue burning. Residue retention without NT and NT without residue retention had no effect on soil total N content. Findings of Dalal *et al.* (2011) support the contention of Glover *et al.* (2010) to the extent that conversion of grasslands to annual cropping systems by opting CT system can decrease soil total N content, decrease root-biomass, and increase NO3-N accumulation/ leaching losses.

Thus, factors that influence plant growth like N-fertilization which increases crop productivity and amounts of residue returned to the soil, results in increased SOC stocks in soil. N-fertilization, can also affect C cycling in soil by influencing dynamics of soil decomposers. Moran *et al.* (2005) reported that the N additions at adequate rates (Recous *et al.* 1995) can increase the mineralization of crop residue C without increasing the mineralization of native SOC. Follett *et al.* (2013) observed that N fertilization slowed the losses of residue C and added to soil by corn crop in a long-term study of 8 yr. NT proved superior to CT in maintaining SOC. Jantalia and Halvorson (2011) and Halvorson and Jantalia (2011) reported that application of fertilizer N resulted in larger amounts of SOC with NT than with CT in irrigated corn in top 30 cm. Thus N-fertilizer applications required to get optimum crop yields also help in greater accumulation of SOC in soil.

Effect of intensive and diverse cropping on carbon sequestration

Cropping intensification combined with reduced tillage systems and optimum fertilizer management targeted to production level of the system can increase SOM content and improve soil quality. Intensively cultivated row crops that return more residue-mass and have wider C: N ratio can be considered as the key component to rebuilding SOC content (Hunt *et al.*, 1996). Intensification and diversification of cropping systems influence nutrient demand, cycling, and distribution with in the soil profile and increase yield potential by influencing nutrient availability. Conservation agriculture which involves reduced or no-tillage and crop residue retention on soil surface, can minimize run-off and soil erosion and is recommended as the best management for sustainable crop production. Conservation agricultural practices help in conserving more water in the soil profile by increasing infiltration and decreasing runoff and evaporation losses of water (Mupangwa *et al.*, 2007). With improved infiltration rate with NT system, more water can be stored in soil profile, which not only protect the crops from mid-season short-term droughts under conditions of delayed rainfall especially under dry-land conditions but can also help in improving cropping intensity. There has been a dramatic increase in the frequency of cropping in the dry land systems in many parts of the world under CA practices. This has been made possible by the adoption of minimum or reduced tillage systems. Thus NT and minimum tillage systems may allow producers in the semi arid areas to intensify cropping compared with traditional crop-fallow system. This can help to attain more efficient use of limited water supplies. Farehani *et al.* (1998) and Halvorson *et al.* (1999a) also reported that increasing the frequency of cropping to 2 out of 3 yr, 3 out of 4 yr, or even every year has been successful when minimum or NT systems are used. Thus crop intensification with NT and crop residue retention is both cause and effect.

No-till practice generally reduces the duration of irrigation; however, this decrease in duration of irrigation is limited to first irrigation only, which is associated with irrigation water flowing faster in untilled fields. This is especially beneficial for wheat crop because during December month, in tilled fields, excessive water applied in irrigation causes yellowing of the crop. Slightly higher grain yields in combination with limited water savings results in significantly higher water productivity for NT compared to CT wheat as observed form large number of field experiments in Haryana, India (Table 2). Small increase in yield combined with relatively modest irrigation savings in NT; imply that increased water productivity in these areas. A separate water use survey conducted within the context of Haryana showed more significant water savings attributable to NT technology effects (Erenstein *et al.* 2007a). This survey confirmed that NT for wheat saves irrigation time (6.4 h/growing season), saves irrigation water (340 m3/ha/growing season) and increases wheat grain yield (260 kg/ha). According to the same survey, total tube-well water volume applied to NT amounted to 2200 m3 compared to 2500 m3 for CT, statistically significant water saving of 13.4 % which was primarily achieved in the first irrigation. This implied significantly higher water productivity for NT wheat (1.5 kg/gross irrigation, m3) compared to CT wheat (1.3 kg/gross irrigation, m^3).

Table 2. Water productivity in wheat as affected by tillage practices in Haryana, and Punjab, India

Observation	Haryana		Punjab	
	NT (n=138)	CT (n=99)	NT (n=87)	CT (n=67)
Grain yield (Mg/ha)	4.38	4.21	3.24	3.36
Irrigation water productivity (Mg/ha)	1.43	1.35	1.07	1.07
Irrigation water productivity (kg/irrigation m^3)	3.11	2.65	1.67	1.47
Gross water productivity (kg/m^3) (rain + irrigation)	1.76	1.59	1.02	0.97

Source: Erenstein *et al.* 2007a

Residue mulching not only results in more stored water down the soil profile (0-1.2m) as compared to CT practice but also the soil cover provided by anchored crop straw plays an important role in regulating the soil surface temperature. It insulates the soil against temperature and saves soil moisture from evaporation losses. Due to this effect, the soil surface temperature in NT with residue retention can be significantly lower, often to the extent of 2 to 8 0C during day time in the summer season as compared to CT (Oliviera *et al.* 2001). During night time, the insulation effect of retained residue may lead to slightly higher temperature resulting in lower amplitude of soil temperature variation in NT. This effect on temperature is highest on the surface layer and decreases with depth. Dahiya *et al.* (2007) observed that mulching reduced average soil temperatures by 0.74, 0.66, 0.58 OC at 5, 15, and 30 cm depth, respectively. Thus crop residue retention greatly influences the total amount of water stored in the soil profile both by increased infiltration rate and decreased evaporation losses of water from soil surface through its effect on soil temperature. Pasricha (2013) observed almost 30% more stored water in the soil profile (0-1.2 m) after harvest of wheat with NT and crop residue retention in rice-wheat rotation (Table 3). This is despite the fact that NT plots received lower amount of total irrigation water (1758 k L ha^{-1}) as compared to CT (1874 k L ha^{-1}) from post seeding to harvest of wheat. Thus CA practices, due to increased water availability, can help in increasing the intensity of cropping, especially in dry land agriculture and riding the soil from the necessity of fallowing which has adverse effect on SOC. Conservation agriculture by storing more water in the soil profile, decreases the frequency and intensity of short midseason droughts and significantly improve the crop yields during the years of poor rainfall distribution.

Table 3. Profile moisture content by depth after wheat as a function of tillage treatment and residue retention

Soil depth(m)	Profile moisture content			
	CT		NT	
	(%)	(mm)*	(%)	(mm)*
0-0.15	1.74	3.92	2.25	5.06
0.15-0.30	3.80	8.55	4.52	10.17
0.30-0.60	5.03	22.64	6.42	28.89
0.60-1.2	7.34	66.06	9.84	88.56
0-1.2	-	101.17	-	132.68

*Assuming soil bulk density of 1.4g cm-3

Source: Pasricha 2013

Increased soil profile stored water by facilitating increased cropping intensity helps to overcome the problem from keeping the soil fallow, a management system which is responsible for rapid loss of SOC and N and soil erosion. There has been a dramatic decline in the SOC and soil TN in the semi-arid regions representing Great Plains of the USA during the last 50 yr of conventional cultivation (Peterson *et al.*, 1998). The estimated losses are in the range of 30-50% of the original SOC level. This region has been supporting conventional tilled dry land wheat cropping with alternate summer fallow. Less intensively cropping is in fact more prone to rapid SOM depletion in CT system and especially so if the cropping system includes summer fallow. Negative effect of fallowing is so much marked that the continued use of crop-fallow farming system may result in loss of SOM even with NT practice. Greater moisture and temperature during summer fallow can result in accelerated rate of SOM oxidation (Haas *et al.*, 1974). Annually cropped soils have greater C and N than soils that are summer fallowed (Campbell *et al.*, 2000). In one of the studies, an estimated

233 kg C ha^{-1} was sequestered each year in the annual cropping system with NT, compared with only 25 kg C ha^{-1} with mold board plow (MT) and a net loss of 141 kg C ha^{-1} with CT (Halvorson *et al.*, 2002). Thus conversion from crop-fallow to more intensive cropping systems utilizing NT will be needed to have a positive impact on reducing decomposition of soil OM and stocking more SOC and soil TN. In fact fallow period represents a time of high microbial activity and decomposition of SOM with no input of crop residue. Sherrod *et al.* (2003) reported that over the last 20 yr have shown that losses of SOC and N are reduced considerably by implementing measurement systems which include reduced tillage (RT) or NT, retention of crop residues on the soil surface and lesser or little frequency of fallowing. Thus, in arable cropping systems, SOM storage in soil is usually directly related to C inputs, root biomass and rhizo-depositions. Intensive cropping results in increased estimated residue C inputs, root biomass and rhizo-depositions. Effect of no-till systems of wheat-fallow (WF), wheat-corn-fallow (WCF), wheat-corn-millet-fallow (WCMF), and continuous cropping (CC) on soil organic carbon and total nitrogen levels after 12 years at three locations in Colorado, USA is shown in Fig. 2 and 3.

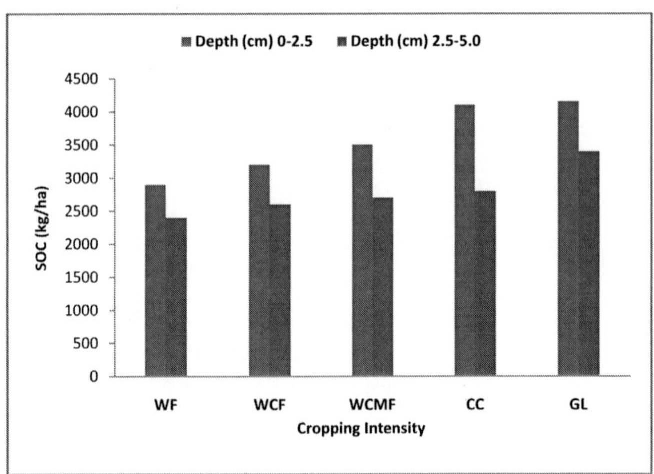

Fig. 2. SOC in 0-2.5 and 2.5-5.0 depths after 12 years under no-till management as affected by cropping intensity with grass as reference point (Adaped from Sherrod et. al, 2003).WF=Wheat-fallow; WCF=Wheat-corn=fallow; WCMF=Wheat-corn-millet-fallow, CC=Continuous corn; GL=Grass land.

Continuous cropping had 35% and 17% more SOC and N, respectively, than the WF system. Cropping intensity significantly influenced SOC and TN when summed to 10 cm depth, however, greatest impact was found in 0-2.5 cm depth and it decreased with depth. Annulized stover mass explained 80% of the variation in SOC and TN in 0-10 cm soil profile. Thus cropping systems that eliminated summer fallowing are responsible for maximizing the amount of SOC and TN sequestered.

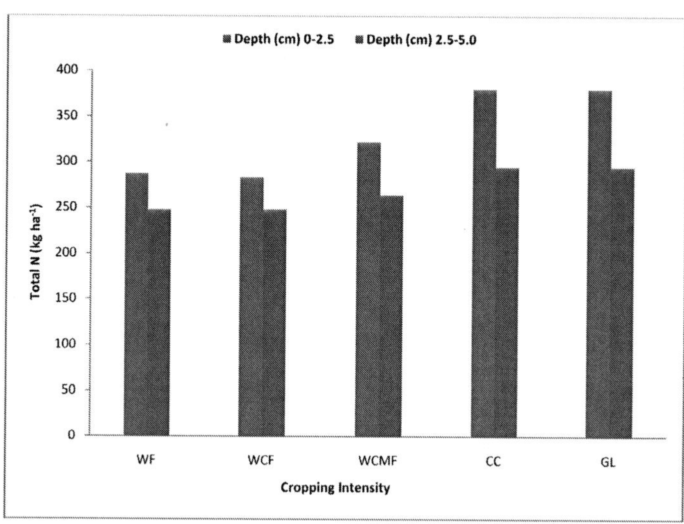

Fig. 3. Total N in 0-2.5 and 2.5-5.0 depths after 12 years under no-till management as affected by cropping intensity with grass as reference point (Adapted from Sherrod *et al*, 2003). WF=Wheat-fallow; WCF=Wheat-corn=fallow; WCMF=Wheat-corn-millet-fallow, CC=Continuous corn; GL=Grass land.

Cropping intensity increases SOC and TN because intensification results in greater amounts of biomass returned to the soil. Effect of soil intensification is very pronounced in building SOC and TN as apparent from the data in Fig. 2 and 3 that it almost approached the G reference (Grass) in 0-2.5 cm depth. Soil total N results followed similarly with the SOC results. This is expected because SOC and soil TN are biologically linked. A highly correlated relationship (r=0.99) between soil organic matter and total soil N with various tillage and cropping systems and soil depths has been observed by Ungar (1968). Generally stover production increases as system intensity increases, with CC producing approximately 70% more stover than WF cropping system. Stover inputs averaged over all cropping systems and other treatments like slope, position etc. had a strong relationship with SOC and TN in 0-10 cm depth (r2=0.80). The slope value for SOC was approximately 10 times the slope value found for TN. About 80% of the variability in SOC and TN in the 0-10 cm is accounted for by the annualized Stover production.

For clearer picture on the effect of tillage, cropping intensities and cropping systems on SOC and TN, it is necessary to calculate these values on equivalent basis as soil physical properties like soil bulk density is affected by the various CA treatments. Changes in the SOC and TN pool as shown in table 4 were calculated on an equivalent soil mass basis, so the values reflect an actual gain or loss from the site in the past 16 years, independent of changes in soil density resulting from operations.

Table 4. Soil organic C and TN (0-20cm) as affected by 16 yr of tillage and residue inputs in plots under continuous corn production

Treatment	SOC (Mg C/ha)		TN (Mg N/ha)
	1991	2007	2007
NT-HI	67.4±4.8	79.4±12.0	7.6±i.0
RT-HI	62.3±1.4	80.8±1.9	7.7±0.1
CT-HI	69.0±5.1	59.1±5.2	5.4±0.4
NT-LI	66.2±2.7	64.5±4.2	6.4±0.6
RT-LI	62.1±3.7	65.2±11.2	6.1±0.9
CT-LI	59.8±3.4	57.0±3.4	5.2±0,3
Contrast (significance probability)*			
CT vs. NT	-	P=0.085	P=0.024
CT vs. RT	-	P=0.066	P=0.028
CT vs. NT & RT	-	P=0.044	P=0.013
RT vs. NT	-	NS	NS
HI vs. LI	-	P=0.097	0.085

Significant at (P=, 0.1) and non-significant (NS) treatment effects. HI= high residue input; LI= low residue input; NT=no-till; RT=reduced tillage; CT= conventional tillage.

Source: Adapted from Halpern and Madaemootoo 2010

Retention of crop residue is expected to contribute to the SOC pool. A 10-20% of corn residue was reported to get converted SOC, however, Blanco-Canqyi and Lal (2007) showed that approximately a third of all C from wheat residues returned to a silty loam soil during a 10- yr experimental period was sequestered in the SOC pool. Such an effect is also determined by the texture of the soil, sandy soils have shown a limited capacity to accumulate SOC compared with clay rich soils.

Practices of CA, have added advantage when followed in dry land conditions. Reducing the amount of tillage conserves surface residues and improves infiltration of water into the soil and by protecting from surface evaporation losses, conserves more water in the soil profile. This in turn allows for more intensive cropping and reduced frequency of summer fallow in the semi-arid regions. This makes it possible to have more stable and higher yields than the traditional practices in the semiarid regions and is often termed as eco-farming system. Papendick and Miller (1977) reported that wheat yield had the potential to increase up to 20% with conservation of an additional 2 cm of water in a 25-cm precipitation zone. Conservation agricultural practices are thus synergistic with intensive cropping, as lack of soil disturbance helps in stocking greater quantity of SOC and N and also optimizing water use efficiency (WUE). WUE, defined as kg of grain per hectare produced per mm of available growing-season water (0-120 cm) increases with increase in the available N in soil. Higher SOC and total soil N under NT is attributable to increasing soil water storage, which in turn increases the amount of plant biomass returned to the soil. Continuous cropping minimizes the opportunity for accelerated rates of SOM oxidation and most closely simulates perennial system in which balance between immobilization and mineralization processes results in minimum loss of C and N and maximum accumulation of SOM.

Frequency of conservation tillage systems with residue retention for a number of different cropping systems has increased in recent years because of a number of factors including their potential to reduce losses and / or sequester SOC. Compared with CT; SOC under NT was 36, 60, and 62% greater for continuous wheat, sorghum-wheat-sorghum, and wheat-sorghum, respectively. In the cultivated silt loam, SOC stock was higher under NT (20.7 Mg ha^{-1}) followed by MT (17.3 Mg ha^{-1}) and plow tillage (PT) (16.8 Mg ha-1) for 0 - 10 cm depth. Long-term use of NT practices are highly sustainable and result in high SOC and water stable aggregates (WSA), low bulk density, and greater available water capacity than MT or PT. Consistent differences are not observed in deeper soil depths between CT and NT. In a long-term experiment (1989-2004) on SOC as affected by tillage and cropping systems, Varvel and Wilhelm (2010) reported that no-till SOC reserves ranged from 4.8 to 11.6 Mg ha-1 greater than SOC reserves in other tillage systems (chisel, disk, ridge-till, sub-till) after 15 yr of experimentation in the 0 to 30 cm depth in continuous corn cropping. Continuous corn cropping as compared to corn-soybean resulted in greater SOC concentrations and reserves in all the tillage systems. A significant interaction between tillage treatment and cropping system on SOC reserves in surface 0-7.5 cm depth is demonstrated in Fig. 4. It seems likely that a significant interaction between tillage treatment and cropping system is observed between

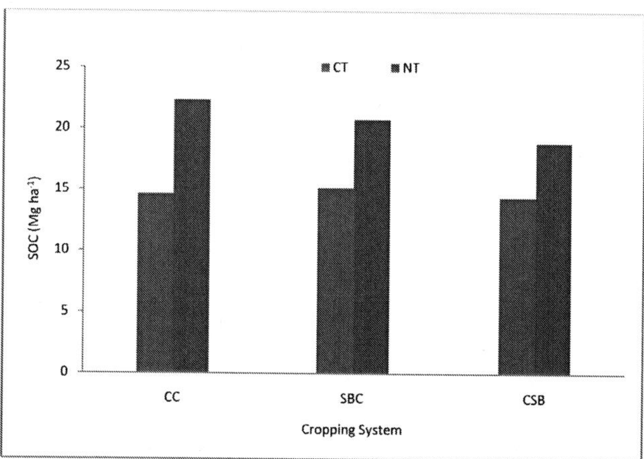

Fig. 4. Soil organic carbon reserves in the surface 0-7.5 cm depth for three cropping systems as influenced by no-till treatment. CC=continuous corn, SBC= soybean corn, CSB= corn soybean. *Source*: Vavel and Wilhelm 2010

SOC reserves for three cropping systems ranked CC>SB-C>C-SB in no-till treatment, SOC reserves were greater in CC and least in C-SB with intermediate value for SB-C in the no-till treatment.

Greatest increase in SOC levels were observed in systems which contributed highest levels of residue with least amount of soil disturbance, which strongly supports the adoption and use of no-till technology for soil sustainability.

However, NT plays prominent role in sequestering SOC. In conventionally tilled long-term cropping system experiments, even with increased cropping intensity and diverse cropping, SOC levels either decreased or at the best maintained after long-term cropping. Thus, some form of conservation tillage is necessary even if there is intensive and diverse cropping system. SOC sequestration tends to increase as soil disturbance decreased and also as cropping system diversity increased. General conclusion is that changes in cropping system diversity did not increase SOC sequestration as much as switching to RT or NT systems.

Effect of crop residue retention/incorporation on SOC and N stocking in soil

Crop residues retained on the soil surface in CA, in general, serve a number of beneficial functions, including soil surface protection from erosion, enhancing infiltration and cutting run-off rate, decreasing surface evaporation losses of water, moderating soil temperature and providing substrate for the activity of soil micro-organisms, and a source of soil organic carbon (SOC).

Even to-date a major portion of crop residues are either burned or removed for other purposes. Rice-wheat is an important cropping system followed on more than 10 million ha in the Indo-gangetic plains of the country. Crop residue burning in rice-wheat production system is although a quick, labour-saving practice to get rid of residue that is viewed as a nuisance by farmers. Residue-burning, however, has several adverse environmental and ecological impacts. The burning of dead plant material adds a considerable amount of CO2 and particulate matter to the atmosphere and can reduce the return of much needed C and other nutrients to soil. Lack of soil surface cover due to burning or removal of the crop residues increases the loss of mineral and organic matter-rich surface layer in run-off. Crop residues returned to the soil, on the other hand, help increase SOM levels, which facilitate greater infiltration and help storing greater amount of water in the soil profile. Crop residues provide substrate to soil organisms which help in recycling of the plant nutrients. Managing crop residues by burning also removes the evaporation barrier that residues provide due to their mulching effect, causing soil to dry up quickly. Burning also removes the insulating effect of residues over the soil surface causing temperatures to fluctuate more widely. Thus burning of crop residues is increasingly being looked upon as an unacceptable management practice because of the additional CO_2 and particulate matter that is expelled into the atmosphere.

In 1995, the Intergovernmental Panel on Climate Change (IPCC) determined that agriculture was directly responsible for approximately 20% of the annual anthropogenic emissions of the greenhouse gases. Rice-based cropping systems are very common under intensive irrigated agro-eco zones. Introduction of rice straw incorporation or retention often confounds methods of cultivation. While in-field burning is the traditional disposal method of straw and stubbles in the tropical and subtropical countries of South and Southeast Asia, however, because of its pollution effect, there is great pressure on evolving methods other than burning of the straw. Some important rice growing states in the country like Punjab have promulgated legislation against straw burning, but this practice is continuing unabated because of poor implementation of the law. Agricultural soils, can be however, actually serve as a significant C sink, rather than a C source at least until the maximum capacity to store C in soil is achieved, if improved residue management and reduced tillage systems are adopted. Therefore, alternative residue management options must be developed and adopted to minimize or eliminate the traditional residue burning practices.

In low-land rice, soil surface is kept submerged with water for most part of the rice growth period. This results in anoxic conditions. The O_2 is used by facultative anaerobic bacteria causing reduction conditions in the soil. In the absence of aeration, the crop residue is allowed to decompose under reduction conditions and mineralization of residue N does not proceed beyond NH4-N. So a significant proportion of total mineral N present in the submerged rice soils is in the form of NH4-N. This cationic form saves N loss through leaching and also gaseous losses in the form of N2O through denitrification since there is no NO3-N formation (Pasricha 2010). Nitrogen is the most important in rice system, accordingly, 67% of the total fertilizer application goes to rice crop and responsible for more than 45% increase in grain yield compared with genetic (29%) and other factors (24%). Effect of such changes in soil N supply on N use efficiency and seasonal uptake in flooded rice crop has shown increased plant N following straw incorporation indicating increase in plant available N (Table5).

Table 5. Soil and fertilizer N recovery (kg ha-1) in rice plants as affected by rice straw management

Straw treatment	Total plant (grain + straw)			
	1997			1998
	Total N	Soil N	Fertilizer N	Total N
Burn	169	97	72	141
Incorporation	195	121	74	174

Source: Eagle *et al.* 2001a

Additional benefits following the incorporation of organic material (straw and stubbles) such as mineralization of other nutrients and improved soil quality, may lead to an increase in total N accumulation in the crop by supplying other limiting nutrients and increasing microbial activity. Thus, the combination of non-burning of crop residue followed by NT planting is feasible set of alternative rice residue management practices that will likely result in the building of soil OM and increased C sequestration, both of which will improve the long-term sustainability of rice-wheat rotation in the northwestern India.

Alternative uses of crop residues

For small and marginal farmers, the crop residues are of much value and they cannot afford to leave them on soil surface to promote soil protection, thus potentially limiting CA benefits and adoption. Continuous corn has been reported as best cropping system so far SOC stocking in soil with NT is concerned. In the sub-mountainous regions of the country, where corn is important crop, but its residue, after harvest is a valuable fodder for live-stock in this region. So the farmers in this region will not leave the residue on the soil surface as part of CA. Although, the soils in this region are highly prone to erosion losses through instant run-off of water received mostly as high intensity rains. Under such situations, following a mid-way path will be more convenient for the adoption of CA practices. In an investigation to evaluate the benefits of NT with less than 10% crop residues cover during the rainy season, it has been found that even, this much small amounts of crop residues retained on the soil surface, NT resulted in 26% higher SOC than under CT (Mchunu *et al.*, 2011). They found that NT reduced the soil erosion losses by 68% and SOC losses by 52%. Less erosion in NT compared to CT is attributable to greater occurrence of indurate crusts under NT, which is less prone to soil losses. These results show the potential of NT even under low crop residue cover (<10%) to significantly reduce soil and SOC losses by run-off water under small and marginal scale agriculture and at the same time help building SOC stocks.

Tillage effects on labile fraction of SOC

Soil organic matter is of prime importance for sustainable soil productivity and agricultural management practices impact significantly not only its amount but also its quality by influencing the chemical properties of SOM. Active fraction of SOC which is also called labile-organic C plays an important role in several soil chemical and biological processes such as nutrient cycling, detoxification of soil applied pesticide chemicals and energy supply to soil micro organisms. Liability of SOM is defined as the relative ease and rate with which it is decomposed by microbes, which depends on both chemical recalcitrance and physical protection from accessibility to the action of micro-organisms. These fractions are reported to be more sensitive to changes in soil management practices for crop production. Labile organic C fractions in soil constitute soil microbial biomass C (SMBC), mineralizable OM, dissolved OM, permanganate oxidizable OM, and particulate OM. These fractions have a greater turnover rate, in other words, they have shorter mean residence time in the soil. Due to this reason, the labile pool has a smaller size, ranging not more than 1 to 20% of the total SOC.

Labile C compounds are more transient and therefore more measurably affected by tillage and residue management in contrast to total SOC pool. Tendency for greater C max value in RT and NT treatments relative to the CT treatment, suggested that in situ, C max values ranged from 949 to 2765 mg C kg^{-1} soil, representing 4 to 11% of total SOC in labile pool and is readily mineralized (Halpern *et al.*, 2010). Labile C and N fractions are physically protected in aggregates under NT system. Labile C and N fractions (MBMC, and potentially mineralizable C and N) are significantly correlated with total SOC pool, MBMN and N max in 5-20 cm depth. These correlations indicate that measurements of labile fractions can be best indicators of management-induced changes in the total SOC pool. It has been observed that the MBMC and MBMN are concentrated in soil surface layer of 0-5cm depth were greater in the NT treatment than the CT treatment with intermediate values for RT (Table 6).

Table 6. Microbial biomass C (MBMC, mg C/kg) and N (MBMN, mg N/kg) concentrations in 0-5 and 5-10 cm depth of plots under continuous corn production after 16 yr of tillage and residue input treatment

Treatment	0-5 cm		5-10 cm	
	MBMC	MBMN	MBMC	MBMN
NT-HI	417±36	95±39	263±36	31±3.0
RT-HI	361±30	65±1.5	332±32	53±3.9
CT-HI	211±33	21±5.7	245±18	25±1.6
NT-LI	286±33	32±5.5	268±32	21±2.9
RT-LI	219±72	21±5.0	276±33	25±1.5
CT-LI	190±13	20±2.4	277±54	2±1.03
Treatment effects				
Tillage	P=0.008*	P=0.036*	P=0.857 NS	P=0.022*
Residue	P=0.010*	P=0.060*	P=0.527 NS	P=0.013*
Tillage x residue	P=0.251 NS	P=0.122 NS	P=0.536 NS	P=0.031*

*Treatment effect significant at p=0.05, NS= treatment effects not significant. HI= high residue input; LI= low residue input; NT=no-till; RT=reduced tillage; CT= conventional tillage.

Source: Halpern *et al.*, 2010

No-till practice significantly increases the size of SOC and labile SOC, compared with CT, especially in 0-5 cm. Greatest differences in mineralized C between

CT and NT are observed in surface 0-5 cm depth, where NT has been reported to increase mineralization by as much as 35% compared to CT. Mineralized N is highly correlated with mineralized C irrespective of the cropping systems, but is more closely related to SMBN in 0-15 cm, indicating that biomass may also serve as a significant source of labile N (Fig. 5). Intensified cropping also increases labile C pools, which generally decease with depth. Labile pools are highly correlated with each other and SOC, but their slopes are significantly different, being lowest for dissolved OC and highest for hydrolysable C.

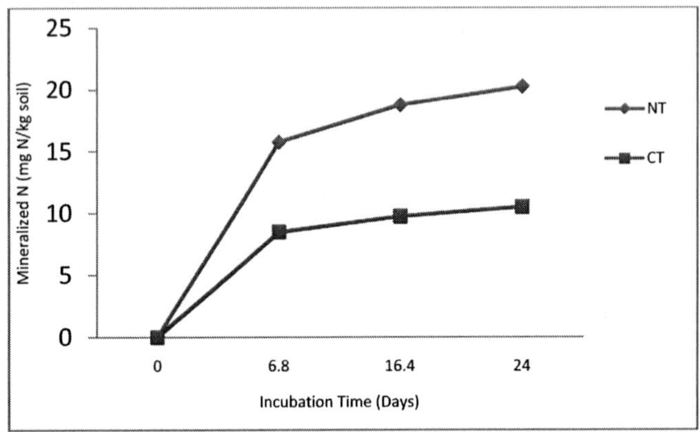

Fig. 5. Cumulative N mineralized during 24-d incubation of soil samples from 0-15 cm depth as affected by tillage (CT=conventional tillage; NT=no-till) (Adapted from Dou *et al.*, 2008).

Sensitivity of labile C to changes in management practices

To investigate the sensitivity of labile or active SOC, such as soil microbial biomass C (SMBC), dissolved organic C, and hydrolysable C, particulate C, to change with management practices, Dou *et al.* (2008) sampled soil in a 20 yr experiment with tillage, CT, and NT, and N-fertilization treatment. They observed that NT significantly increased the size of SOC and all labile SOC pools compared with CT, especially in the surface 0-5cm layer. Intensified cropping and crop rotation also increase SOC and all labile pools, which generally decrease with depth. The C concentration in microbial biomass, floatable organic C (light fractions) or macro-aggregate-associated organic matter (OM) is considered to be a more sensitive indicator of change in SOC as a result of conservation agricultural (CA) practices. Like Dou *et al* (2008), Jastrow *et al* (2007) also reported that no-till significantly increased the SMBC in the 0 - 30 cm depth, especially in the surface 0-5 cm layer. On an average for different crop rotations, SMBC in 0-5 cm was 27% greater than CT, but was approximately 14% lower than under CT at 5-15 cm depth. This difference in SMBC can be explained on the basis of the effects of litter availability. Residue retained on soil surface in NT provides easily available substrate for soil micro-organisms, which accounts for higher SMBC in the surface soil. Moreover, NT provides wetter soil conditions which encourage greater microbial activity. Increased cropping intensity which affords to provide greater residue-C input causes increased SMBC (33%) due to slow decomposition under NT as compared with CT. Microbial biomass and potentially mineralizable C and N were also correlated with SOC pool, which supports the use of these labile fractions as indicators of management induced changes in the SOC pool.

No-till management increases microbial biomass in agricultural soils compared with CT, CT systems on the other hand are reported to decrease potentially mineralizable C and N (Woods and Schuman, 1988) and the soil's ability to immobilize and thus conserve mineral N. Doran (1980) reported that greater quantity of potentially mineralizable N under NT compared with CT in the upper surface layer of 0-7.5 cm, is associated with larger microbial biomass. Nitrogen supplying potential of the soil increases after NT practice and can be successfully utilized for a few growing seasons, as does the closely related content of active N fractions. The soil microbial biomass is first to decline following tillage compared to NT treatment. Nelson *et al.* (2006) stated that a combination of the residue management and NT cultivation favours a nutrient-rich environment for the soil micro-organisms to survive and flourish. MBMC is an overall indicator of soil microbiological activity of the soil environment and may provide information related to soil nutrient processing and crop nutrient acquisition.

Over a period of time, straw incorporation results in sustained greater microbial biomass C and N. An increase in SMBM can affect C and N sequestration rates of fertilizer and crop residue through greater immobilization of and conversion to stable SOM as well as through greater mineralization of stabilized SOM-C and N. Long-term straw incorporation/retention and soil submergence during the growth period of rice in summer can alter mummification process, thereby, affecting N sequestration rates in SOM fractions and its subsequent turn over. Thus long-term straw management in the flooded rice can affect the size and stability of the soil N supply. Residual N remaining in soil (20-40%) after cropping has limited availability to subsequent crops (<15%) primarily because of stabilization into resistant fractions of SOM (Kelley and Stevenson 1995). There are studies that examine tillage method effects on soil microbial biomass-N (SMBM-N) especially over the growing season. An increased biomass-N during early growing season of wheat in 0-5 cm layer under NT than CT, and then decline to similar level at the end of the growing season is common observation. Considerably higher concentration (>50%) of SMB existed in the surface layers under NT, compared to CT, which generally persists during growing season, suggesting that crop N availability should have been affected, likely through immobilization into microbial tissue.

Humic acid (HA) from CT is less aliphatic and more aromatic than HA from NT. Humic acid has been found to be more reactive in the top soil (0-5 cm) under NT than under CT. Both SOC and its light fractions were higher in the 0-5 cm soil of NT than CT treatment. These observations emphasize that long-term tillage management can significantly change the characteristics of both physical and chemical fractions of SOM. After 4 yr of straw management, Bird *et al.* (2002) observed that soil incorporation of straw increased mobile humic acid and light fraction C, and N compared with burned straw. N-fertilizer sequestration in mobile humic acid and light fraction-C compounds was greater with straw incorporation compared with burned. Additions of fertilizer N have been reported (Neff *et al.* 2002) to accelerate decomposition of light fractions of SOC but tend to further stabilize SOC present in heavier fractions. A nitrogen addition especially influences the decomposition of high-quality residue with low lignin content but inhibits decomposition of low quality residues with high lignin content.

Tillage effects on soil CO_2 fluxes

Soil CO_2 flux and quantification of C inputs provide insights to the processes that regulate soil C sequestration. Most of the CO_2 produced in soil is released to atmosphere as a flux of CO_2. Pre-sowing soil preparation by conventional tillage (CT) contributes towards greater carbon dioxide (CO_2) release to the atmosphere from accelerated decomposition of soil organic matter (SOM) and loss of soil organic carbon (SOC). This CO_2 from soil comes collectively from mineralization of SOM and plant residue, and autotrophic respiration of roots. The amount of CO_2 gas, thus released is almost 10 times the total CO_2 released from burning of fossil fuel in the whole world (Morell *et al.* 2011a). Chemical and biological processes which account for substantial amount of total global CO_2 emissions are estimated at 77 to 98±12 Pg C yr^1 (Bond-Lamberty and Thomson 2010). Soil CO_2 fluxes both from forest ecosystem and agro-ecosystems are significantly determined by seasonality and depend on soil temperature and moisture. In agro-ecosystem, C budgets in soil are studied in strategies to build SOM and for mitigating climate warming by containing soil CO2 fluxes. Since, soil organic matter decomposing is an oxidation process, any disturbance of the soil through tillage activities will facilitate decomposition resulting in greater soil CO_2 fluxes. For example, average across two years (2007 and 2008) growing season, CO_2 emissions were greater for CT-continuous corn system (2259 kg CO_2-C ha-1) than NT-continuous corn system (2046 kg CO2-C ha^{-1}). These results of Halvorson *et al.* (2010) are similar to those of Mosier *et al.* (2006) and Alluvione *et al.* (2009).

Effect of N-fertilization on tillage-induced CO2 fluxes

The general observation is that CO_2 flux increases with N-fertilization. Brye *et al.* (2006) observed a 37.6% higher soil surface CO_2 flux from CT than from NT in wheat-soybean production system. They further indicated the importance of N in retaining C in soil and reported a 6.1 % higher CO_2 flux from a low than the high N rate treatment. On the other hand, adequate management on croplands, such as conservation tillage systems and adequate nutrient supply may increase SOC stock. Morell *et al.* (2011) in a study on the effect of tillage and N fertilization on soil CO_2 fluxes and SOC concentration has observed that there was greater soil CO_2 flux under minimum tillage (MT) and NT in dry years, and greater under CT and minimum in MT and NT during the wet growing seasons. Such an effect is mainly through the availability of N as influenced by available moisture content in the soil. Surprisingly, Morell *et al.* (2011) did not find any reverse relationship between soil CO_2 fluxes and SOC stocking in soil. In fact, soil organic matter decomposition is not the only source of soil CO_2 fluxes; a major portion of it must be coming from the root respiration, a process unrelated to SOC build up in soil. This could also be attributed to differential effect of N-fertilization on below ground root growth and above ground vegetative growth under semi arid conditions which can explain to some extent the non-relationship between soil CO_2 flux and SOC stocks in soil. However, different contributions of both microbial-derived CO_2 and root-derived CO_2 to the total CO_2 flux could be the main reason responsible for the lack of relationship between soil CO_2 flux and SOC socks.

Effect of temperature and moisture content on CO_2 flux

Both soil temperature and soil moisture content have generally been shown to greatly control soil surface CO_2 flux in a variety of climate regimes. Retention of crop residue on the soil surface affects both soil surface temperature and moisture content in the soil profile. Soil temperature has been observed to control soil surface CO_2 flux more than soil moisture parameters (Hendrix *et al.*, 1998). Under normal moisture content in the soil profile, additional increase in moisture supply is expected not to affect the soil surface CO_2 flux. The product between soil water content and soil temperature explained 75 and 94% of the seasonal variability of soil CO_2 flux, respectively (Morell *et al.*, 2011). Specifically, conversion from CT to NT has the potential to reduce soil surface CO_2 flux because of favourable effect of this treatment on soil temperature and moisture content.

Effect of heterotrophic respiration on CO_2 flux

Converting from conventional to conservation tillage or NT process can greatly reduce heterotrophic respiration through its influence on decomposition process of SOM. Oorts *et al.* (2007) indicated that enhancement in heterotrophic respiration and subsequently CO2 emissions by CT over NT was in the range of 1.5 to 2.45 g C m-2 d-1 throughout the whole growing season for the corn crop. Thus, NT can be regarded as a sure management practice that enhances C sequestration. Chang *et al.* (2012) through simulation models showed that SOC sequestration in the slow pool i.e., +10.7 and -11.5 g C m-2yr-1 for NT and CT respectively. Thus, in agro-ecosystems, size of the soil CO_2 flux can be considerably reduced by adopting CA practices. Removal or in-field burning of the above ground crop residue and tillage leads to large scale CO_2 evolution and SOC depletion. This is evident from increased CO_2 fluxes from soil as observed by a number of researchers in CT practices. Thus continuous CT may ultimately lead to soil quality degradation.

Conclusions

Adoption of conservation agricultural practices of reduced/no-tillage and crop residue management has been immensely reported to help in mitigating global warming effect of GHG by storing more C in soil. This practice has several other benefits including reduced energy input, greater water storage in the soil profile, reduced run-off and soil erosion losses, reduced soil surface evaporation losses, moderating soil temperature, and reducing turning point in timely seeding of crops in rotation.

Soil C and N stocks are increased in NT compared to CT; such C gains are a sink for atmospheric CO_2, which contributes to decreased net GHG from agro-ecosystems. Long-term field experiments revealed more accrual of SOC and N with NT in temperate than in tropical and subtropical climate. Such accumulations are amply affected by available moisture and nitrogen content in soil.

No-till management practices has the potential of conserving enough water in the soil profiles of medium to fine textured soils as to successfully facilitate reducing fallow periods and increasing cropping intensity especially under limited water regions of dry land and rain fed agriculture. Increased cropping intensity has been reported to enhance SOC and N in the NT agro-ecosystem. After 12 years of continuous cropping systems (without any fallow season), SOC and TN were 35 and 17% greater, respectively, than amounts found in the wheat-fallow cropping system. Fallowing is reported to enhance microbial activity resulting in greater CO2 evolution and increased rate of soil organic matter decomposition. Continuous cropping, on the other hand, more closely simulates natural perennial systems in which balance between N immobilization and mineralization processes results in minimum N loss and maximum accumulation of SOM.

Like intensive cropping, different cropping systems also affect the SOC reserves in each tillage system differently. Continuous corn, continuous soybean, and soybean-corn have been found to result in more accumulation of SOC and TN as compared with other cropping systems when conservation tillage with residue retention on the soil surface is used.

Tillage systems strongly affect the distribution of SOC and soil TN in soil profile. There is more SOC and TN stratification with NT and RT treatments which results in their accumulation mainly in the surface soil layers. Such stratification strongly influences the infiltration rate, closer the accumulation to the surface greater the rate of infiltration.

The labile fractions of SOC and N in the microbial biomass, mineralizable OM, dissolved OM, permanganate oxidizable OM, and particulate OM, are reported to be more sensitive to management practices. Their concentrations are greater at soil surface as a result of CA practices of reduced/ no-tillage and crop residue retention. Both NT and increased cropping intensity has significant effect on total SOC, TN, and labile C pools in soil under long-term management implementation. Soil microbial biomass C, mineralizable C, and N, particulate organic matter, acid hydrolysable C, and dissolved organic C significantly increase with NT, especially in the surface soil layer.

Soil water conservation under NT and RT treatments during seasons allow for higher C inputs and better conditions for microbial decomposition compared to CT. Long-term N fertilization and tillage affects CO_2 emissions. Rapid SOM decomposition under CT is attributable to greaterr aeration caused by constant soil disturbance by conventional tillage. The product between soil water content and soil temperature can explain between 75-94% of seasonal variability of soil CO_2 evolution. Despite long-term NT and N fertilization which lead to significant SOC sequestration, soil CO_2 flux and SOC stocks may not be related. Different contributions of both microbial-derived CO_2 and root-derived CO_2 to the total CO_2 flux among treatments could be the ain reasons responsible for lack of relationship between CO_2 flux and SOC stocks.

References

Allamaras RR, Linden DR, Clapp CE (2004). Corn-residue transformation into root and soil carbon as related to nitrogen, tillage, and stover management. Soil Sci Soc Am J 68:1366-1375.

Alluvione F, Halvorson AD, del Grosso SJ (2009). Nitrogen, tillage and crop rotation effects on carbon dioxide and methane fluxes from irrigated cropping systems. J Environ Qual 37:1337-1344.

Blanco-Canqui H, Lal R (2007). Soil structure and organic carbon relationships following 10 years of wheat straw management in no-till. Soil Tillage Res 95:240-254.

Bond-Lamberty B, Thomson A (2010). Temperature-associated increases in the global soil respiration record. Nature 464: 579-582.

Braye KR, Longer DE, Gbur EE (2006). Impact of tillage and residue burning on carbon dioxide flux in a wheat-soybean production system. Soil Sci Soc Am J 70:1145-1154.

Campbell CA, Zentner RP, Liang BC, Roloff G, Grgorich EC, Blomert B (2000). Organic C accumulation in soil over 30 yr in semiarid southwestern Saskatchewan- Effect of crop rotations and fertilizers. Can J Soil Sci 80:179-192.

Chang KH, Bartlett P, Wagner-riddle C (2012). Using Day CENT to simulate carbon dynamics in conventional and no-till agriculture. Soil Sci Soc Am J 77:941-950.

Clapp CE, Allmaras RR, Layese MF, Linden DR, Dowly RH (2000). Soil organic carbon and 13C abundance as related to tillage, crop residue, and nitrogen fertilization under continuous corn management in Minnesota. Soil Tillage Res 55:127-142.

Dahiya R, Ingwersen J, Streck T (2007). The effect of mulching and tillage on the water and temperature regimes of a loes soil: Experimental findings and modeling. Soil Tillage Res 96:52-63.

Dalal RC, Strong WM, Weston EJ, Cooper JE, Lahane KJ, King AJ, Chicken CJ (1995). Sustaining productivity of vertisol at Warra, Queensland, with fertilizer, no-tillage, or legumes I. Organic matter status. Aust J Exp Agric 35:903-913.

Dalal RC, Wang W, Allen DE, Reeves S, Menzies NW (2011). Soil nitrogen and nitrogen-use efficiency under long-term no-till practice. Soil Sci Soc Am J 75:2251-2261.

Doran JW (1980) Soil microbial and biochemical changes associated with reduced tillage. Soil Sci Soc Am J 44:765-771.

Dou F, Wright AL, Hons FM (2008). Sensitivity of labile organic carbon to tillage in wheat-based cropping systems. Soil Sci Soc Am J 72:1445-1453.

Eagle AJ, Bird JA, Hill JE, Horwath WR, van Kessel C (2001). Nitrogen dynamics and fertilizer N use efficiency in rice following straw incorporation and winter flooding. Agron J 93:1346-1354.

Erenstein O, Malik RK, Singh S (2007). Adoption and impacts of zero tillage in irrigated rice-wheat systems of Haryana, India. New Delhi, India: CIMMYT India and RWC.

Farehani HI, Peterson GA, Westfall DG (1998). Dry land cropping intensification; a fundamental solution to efficient use of precipitation. Adv Agron 64:197-223.

Franzluebbers AJ (2010). Achieving soil organic carbon sequestration with conservation agricultural systems in the Southeastern United States. Soil Sci Soc Am J 74:347-357.

Glover JD, Culman SW, DuPont SP *et al.* (2010). Harvested perennial grasslands provide ecological benchmarks in agricultural sustainability. Agric Ecosyst Environ 13:3-12.

Haas HJ, Willis WO, Bond JJ (1974) Summer fallow in the western US.In: USDA Cons. Res. Rep. No. 17. US Gov. Print. Office, Washington, DC,pp. 2-35.

Halvorson AD, Del-Grosso SJ, Alluvione F (2010). Tillage and organic nitrogen source effects on nitrous oxide emissions from irrigated cropping systems. Soil Sci Soc Am J 74:436-445.

Halvorson DH, Wienhold BJ, Black AL (2001). Tillage and nitrogen fertilization influence grain and soil nitrogen in an annual cropping system. Agron J 93:836-841.

Halvorson DH, Wienhold BJ, Black AL (2002). Tillage, nitrogen and cropping system effects on soil carbon sequestration. Soil sci. Soc. respiration in conventional and no-till agroecosystems under different winter cover crop rotation. Soil Tillage Res 12:135-148.

Halvoson AD, Black AL, Krupinsky JM, Merrill SD (1999). Dryland winter wheat response to tillage and nitrogen within an annual cropping system. Agron J 91:702-707.

Helpern MT, Madramootoo GA (2010). Long-term tillage and residue management influences soil C and N dynamics. Soil Sci Soc Am J 74:1211-1217.

Hendrix PF, Chun-Ru H, Groffman P (1998). Long-term tillage and residue management influences soil carbon and nitrogen dynamic. Soil Sci Soc Am J 74:1211-1217.

Horie T, Ohnishi M, Angus JF *et al.* (1997). Physiological characteristics of high yielding rice inferred from cross location experiments. Field Crops Research 52: 55-57.

Hunt PG, Karlen DL, Matheny TA, Quisenberry VL (1996). Changes in carbon content of a Norfolk loamy sand after 14 years of conservation or conventional tillage. J Soil Water Conserv 51: 255-258.

Jastrow JD, Amonette JF, Bailey VI (2007). Mechanisms controlling soil carbon turnover and their potential application for enhancing carbon sequestration. Clim Change 80:5-23.

Karl TR, Melilo JM, Peterson TC(ed.) (2009). Global climate change impacts in the United States. Cambridge Univ. Press, New York.

Kelley KR, Stevenson FJ (1995). Forms and nature of organic N in soil. Fert Res 42:1-11.

Khan SA, Mulvaney RL, Ellsworth TL, Boast CW (2007). The myth of nitrogen fertilization for soil carbon sequestration. J Environ Qual 36:1821-1832.

Kimball BA (1983). Carbon dioxide and agricultural yield. An assemblage of 430 prior observations. Agron J75:779-788.

Lal R (2008a). Carbon sequestration. Philos. Trans R Soc London Ser B 363:439-453.

Lal R (2008b). Promise and limitations of soils to minimum climate change. J Soil Water Conserv 63:113A-118A.

McCarty GW, Meisinger JJ (1997). Effects of nitrogen fertilizer treatments on biologically active N pools in soils under plow and no-tillage. Boil Fertl Soils 24:406-412.

Mchunu CN, Lorentz M, Jewitt G, Manson A, Chaplot V (2011). No-till impact on soil and organic carbon erosion under crop residue scarcity in Africa. Soil Sci Soc Am J 75:1503-1512.

Moran KK, Six J, Horwath WR, van Kessel C (2005). Role of mineral nitrogen in residue decomposition and stable soil organic matter formation. Soil Sci Soc Am J 69:1730-1736.

Morell FJ, Cantero-Martinez C, Lampurlanes J, Plaza-Bonilla D, Alvaro-Fuentes J (2011). Carbon dioxide flux and organic carbon content: Effect of tillage and nitrogen fertilization. Soil Sci Soc Am J 75:1874-1884.

Mosier AR, Halvorson AD, Reule CA, Liu XJ (2006). Net global warming potential and greenhouse gas intensity in irrigated cropping systems in northern Colorado. J Environ Qual 35: 1584-1598.

Mulvaney RL, Khan SK, Ellsworth TR (2009). Synthetic nitrogen fertilizers deplete soil nitrogen: A global dilemma for sustainable cereal production. J Environ. Qual 38: 2295-2314.

Mupangwa W, Twomlow S, Walker S, Hove L (2007). Effect of minimum tillage and mulching on maize yield and water content of clayey and sandy soils. Phys Chem Earth 32: 1127-1134.

Neff J, Townsend AR, Gleixner G, Lehmans SJ, Bowman W (2002). Variable effects of nitrogen on the stability and turnover of soil carbon. Nature 419: 915-917.

Nelson MA, Griffith SM, Steiner JJ (2006). Tillage effects on nitrogen dynamics and grass seed crop production in western Oregon, USA. Soil Sci Soc Am J 70: 825-831.

Newton PCD(1991). Direct effect of increasing carbon dioxide on pasture plants and communities. NZ J Agric Res 34:1-24.

Ogle SM, Breidt EJ, Pausitian K (2005). Agricultural management impacts on soil organic carbon storage under moist and dry climatic conditions of temperate and tropical regions. Biogeochemistry 72: 87-121.

Oliveira JCM, Timm LC, Tominaga TT (2001). Soil temperature in a sugar-cane crop as a function of management system. Plant Soil 230: 61-66.

Oorts K, Merckx R, Grchan E, Labrcuche J, Nicolardot B (2007). Determinants of annual fluxes of CO2 and N2O in long-term no-tillage and conventional tillage systems in Northern France. Soil Tillage Res 95: 133-148.

Ortiz R, Sayre KD, Govaerts B, Gupta R, Subbarao GV, Ba T (2008). Climate change: Can wheat beat the heat. Agric Ecosyst Environ 126:46-58.

Papendick RI, Miller DE (1977). Conservation tillage in the Pacific Northwest. J Soil Water Conserv 32: 49-56.

Pasricha NS (2010). Nutrient management under water-logged soil conditions. p. 132-189. In: Gurbachan Singh *et al.* (eds) Strategic Nutrient Management in Water-Logged Soils, CSSRI, Karnal, India.

Pasricha NS (2013). Success story of resource conservation technology in Indo-gangetic plains. In: Ghosh SK *et al.* (eds) Resource Conservation Technology in Pulses, Scientific Publishers (India), pp. 32-41.

Peterson GA, Halvorson AD, Havlin JL, Jones OR, Lyon DJ, Tanaka DL (1998). Reduced tillage and increasing cropping intensity in the Great Plains conserve soil C. Soil Tillage Res 7:207-218.

Poirrier V, Angers DA, Rochette P *et al.* (2009). Interactive effects of tillage and mineral fertilization on soil carbon profile. Soil Sci Soc Am J 73: 255-261.

Powlson DS, Jenkinson DS, Johnston AE *et al.* (2010). Comments on Synthetic nitrogen fertilizers deplete soil nitrogen: A global dilemma for sustainable cereal production. J Environ Qual 39:749-752.

Rasmussen PE, Rohde CR (1988). Long-term tillage and nitrogen fertilization effects on organic nitrogen and carbon in a semiarid soil. Soil Sci Soc Am J 52:1114-1117.

Rasmussen PE, Collins HP, Smiley RW(1989). Long-term management effects on soil productivity and crop yield in semiarid region of eastern Oregon. Columbia Basin Agric. Res., Oregon, Agric. Exp. Stn. Spec. Bull. 675. Oregon State Univ., Pendleton, OR.

Recous S, Robin D, Darwis D, Mary B (1995). Soil organic nitrogen availability: Effect on maize residue decomposition. Soil Biol Biochem 27:1529-1538.

Sherrod LA, Peterson GA, Westfal DG, Ahuja LR (2003). Cropping intensity enhances soil organic carbon and nitrogen in a no-till agroecosystem. Soil Sci Soc Am J 67:1533-1543.

Six J, Bossuyt H, Degryze S, Denef K (2004). A history of research on the link between (micro)aggregates, soil biota, and soil organic matter dynamics. Soil Tillage Res 79: 7-31.

Smith PD, Martino Z, Cai Z *et al.* (2007a). Agriculture. In: Mertz P*et al.* (eds.) Mitigation. Contribution of Working Group III to the fourth assessment report of the Intergovernmental Panel on Climate Change. Cambridge Univ. Press, Cambridge, UK.

Smith PD, Martino Z, Gwary CD, Janzen H, Kumar P *et al.* (2007b). Agriculture. In: Mertz P*et al.* (eds.) Climate change 2007; Mitigation. Cambridge Univ. Press, Cambridge, U.K.pp. 497-540.

TansP (2009). Trends in atmospheric carbon dioxide: Recent Mauna Lao CO2. Available at www.esrl.noaa.gov/gmd/ccgg/trends/ Earth Syst. Res. Lab., Boulder, CO.

Todd RW, Klocke NL, Hergert GW, Parkhurst AM (1998) Evaporation from soil influenced by crop shading, crop residue, and wetting regime. Trans ASAE 34: 461-466.

Unger PW (1968). Soil organic matter and nitrogen changes during 24 years of dry land wheat tillage and cropping practices. Soil Sci Soc Am Proc 32:427-476.

Ussiri DAN, Lal R (2009). Long-term tillage effects on soil carbon storage and carbon dioxide emissions in continuous corn cropping system from an Alfisol in Ohio. Soil Tillage Res 104:39-47.

Vanden Bygaart AJ, McConkey BG, Angers DA *et al.* (2008). Soil carbon change factors for the Canadian agriculture national greenhouse gas inventory. Can J Soil Sci 88: 221-236.

Varvel GE, Wilhelm WW (2010). Long-term soil organic carbon as affected by tillage and cropping systems. Soil Sci Soc Am J 74:915-921.

Watson RT, Rodhe H, Oesheger H, Siegenthaler U (1990). Greenhouse gases and aerosols. In: HoughtonJT *et al.* (eds) Climate Change: The IPCC scientific assessment. Cambridge Univ. Press, Cambridge.pp. 1-40.

West TO, Post WM (2002). Soil carbon sequestration rates by tillage and crop rotation: A global data analysis. Soil Sci Soc Am J 66:1930-1946.

WMO (2013). Green house gases in the atmosphere reach new record. Annual GHG Bulletin.

Wood CW, Edwards JH, Cummins CJ (1991). Tillage and crop rotation effects on soil organic matter in a Typic Hapludult of northern Alabama. J Sustain Agric 2:31-41.

Yaduvanshi NPS, Sharma DR(2008). Tillage and residual organic manure/chemical amendment effects on soil organic matter and yield of wheat under sodic water irrigation. Soil Tillage Res 98:11-16.

12

Conservation Tillage and Crop Residue Management in Relation to Dynamics of Soil C and N Under Climate Change Scenario–II

N.S. Pasricha

Potash Research Institute of India, Gurgaon, Haryana, India

Effect on SOC and N losses through runoff and soil erosion

With increase in population, agriculture is being extended to even marginal soils which were hitherto uncultivated. Such soils are more seriously prone to erosion losses (FAO 2008), mostly because of their topographical position. Overall, soil organic matter and nutrient rich surface soil erosion could become more serious threat in the wake of changing climate in future; therefore, it is necessary to find suitable remedy measures to sustainable food production while protecting the soil from erosion.

Reduced-tillage or NT with crop residues or surface cover crops has been recognized as the best management systems for reducing the runoff and checking soil losses through erosion. Soil erosion from the agricultural field is 1 to 2 times greater with CT when compared with soil erosion under native cover (Montgomery, 2007). Long-term erosion under these conditions can exceed the rate of natural replenishment through the processes of soil formation. With the help of radioactive ^{137}Cs, it has been calculated that with conversion to NT and residue retention, there occurred an 87% reduction in soil erosion as compared to CT (Schuller *et al.,* 2007). Even on slightly sloping lands, NT is very effective in controlling soil erosion. No-till systems of crop production are progressing world-wide, both under large-scale and small–scale agriculture. In NT, there is minimum mechanical disturbance of the soil and also such a system provides a kind of permanent cover consisting of either cover crops or crop residues.

Minimum tillage or NT systems of crop production are considered to be credible agricultural systems that protect the organic C and N rich surface soil from erosion, and also help increasing or maintaining SOC stocks. In NT, soil surface cover with retained crop residues prevent soil surface sealing, prevent splash erosion, enhance soil porosity through improved biological activity and diversity and decrease run-off by increasing soil infiltration.

Mchunu *et al.* (2011) evaluated run-off, soil losses and SOC and N losses under CT and NT in the subtropical sub humid area in Natal, Africa; they found that the soil surface under NT was dominated by structural crusts, which plays important role in preventing runoff, increasing infiltration and preventing soil erosion. Mean SOC content in 0-20 cm layer was 11.1 compared with 14.8 g C kg^{-1} soil under NT. Soil surface coverage under NT is generally dominated by structural crusts, whereas the coverage under CT exhibits a higher proportion of sedimentary and run-off depositional crusts. Average run-off during three rainfall events with cumulative rainfall of 39 mm and maximum rainfall intensity of 24 mm h^{-1} was 25% in NT and 36% in CT. Mean soluble C in run-off was 0.9 g L^{-1} for NT and 2.2 g L^{-1} for CT. Thus soil organic matter remains protected to run-off losses in NT systems. Soil losses were 3.1 times greater under CT (301.5 g m^{-2}) than under NT (96.8 g m^{-2}). Remarkable reduction in soil erosion, to the extent of between 30 and 99% has also been reported by Zheng *et al.* (2004). In about two-third of the review in their study, NT has reduced soil erosion by >60%. Valentin *et al.* (2008), in a small experiment of 0.6 ha catchment in a sloping land, showed that sediment yield with NT decreased from 5.7 to 0.7 t $ha^{-1}yr^{-1}$. Thus there is as much as 87.7% reduction in the sediment load of the runoff water passing over the sloping land managed under conservation agriculture of crop residue retention and NT. Overall reduction in soil erosion under NT is commonly attributed to the retained crop residues on the soil surface. These residues decrease the impact of rain drops and thus favour infiltration. In a rainfall simulator tests on vertisols, it has been found that time-to-pond, final infiltration rate and total infiltration were significantly higher in conservation agricultural practices of NT and crop residue retention in these soils. The abundance of apparently continuous soil pores from the soil surface to depth is, perhaps, responsible for higher infiltration in NT. On the other hand, high-surface density crusts in CT greatly hinder the infiltration process. Soil cover plays crucial role in rainfall infiltration. Soil with 100% soil cover resulted in complete infiltration of a 60 mm rainfall event (Roth *et al.*, 1988), but only 20% of the rain infiltrated when soil was kept bare. Due to similar reasons, fallowing drastically cuts the total amount of rain fall infiltration. Zhang *et al.* (2007a) in a 24 yr old experiment found that in NT the infiltration rate with residue retention was 3.7 times that of CT with residue burned in Australia. Higher infiltration with residue cover and NT drastically decreases the runoff and concomitant reduction in loss of SOC and nutrients due to decreased surface soil erosion loses.

The NT treatment with crop residue retention imparts the soil surface with peculiar structural crusts which have been shown to be very effective in checking soil erosion. In the sloping lands of Laos and under clayey soil conditions, it has been observed that water erosion decreased with increasing proportion of such structural crusts (from 24.33 to 21 t $ha^{-1}yr^{-1}$ for soil and from 1.46 to 0.31 t $ha^{-1}yr^{-1}$ for SOC), not only due to greater hardness of these crusts but also because of greater occurrence of algae on soil surface which affords physical protection through binding and gluing. Moreover, under NT, a greater proportion of SOC and total N are present as microbial biomass-C and microbial biomass-N, which also make it resistant against erosion losses.

In case where crop residues are removed or only a small fraction is allowed to remain on the soil surface, the differences in run-off between CT and NT may not be very significant and that predominance of sedimentary and depositional crusts under CT and structural crusts under NT may not vary in the two tillage treatments. Thus another important aspect of the CA practices is that despite no additional soil surface coverage by the mulch, NT can reduce soil losses by as much as 68% and SOC losses by around 50% which is a significant contribution of no-till practice alone. The decrease in overall soil and SOC erosion with NT, where there is scarcity of soil surface cover, can be explained due to development of structural crusts under NT which have been shown to have greater resistance to detachment. This shows that under such conditions, structural crusts developed due to NT treatment are more effective in checking soil loss and SOC erosion than the surface cover, which facilitates infiltration and checks run-off velocity. Thus NT with scarce residue appears a credible alternative to CT. Thus, residue scarcity due to it alternate use as animal feed, should not be viewed as a major limiting factor for the implementation of NT since results of the study in Africa indicate significant benefits for soil protection and potential mitigation of green house gas emissions even under sparse or no soil cover with crop residues.

Effect on soil microbial activity, soil aggregation, and soil physical properties

Soil microbial population is the primary catalytic agents involved in decomposition of soil organic compounds. They are considered as key indicators of soil health and quality as they play a vital role in C and N cycling in soil. Therefore, maintaining more vibrant soil microbial activity and diversity is fundamental for sustainable agriculture management. The quantity and quality of SOC is the primary factor determining the size and diversity of microbial activity in soil.

Tillage has been reported to reduce microbial diversity, resulting in different soil microbial communities under conservation and conventional tillage systems. Thus, cropping systems that include NT, crop residue retention, and crop rotation, can increase the overall size of microbial biomass and its activity and diversity compared with conventional practices that include traditional tillage, crop residue removal or in-field burning crop residues.

The transformation and storage of plant residue C and N in soil is a function of biological, chemical, and physical factors and these factors are heavily influences by tillage practices. With similar C inputs, the higher total C in the NT surface layers may be a function of not only more substrate concentrated on the surface, but also differential C processing by fungi (Table 1). As total C and total N are highly correlated, similar accumulations in total N were seen here in these and other such studies.

Table 1. Soil organic C and soil total N dynamics as influenced by tillage system

Soil tillage	Depth (cm)	Total C	Total N
		g kg^{-1} soil	
NT	0-5	18.98	1.74
CT	0-5	12.04	1.09
NT	5-15	14.20	1.31
CT	5-15	12.04	1.09

Source: White and Rice 2009

Dry aggregate size distribution is markedly influenced by no-till practice in the presence of retained crop residue when compared with CT tillage treatment. No-till practice also improves the mean weight diameter (MWD) of the soil aggregates which has more positive effect on their water stability. There is a close relationship between SOC and MWD of soil aggregates. Thus increase in soil organic matter content with conservation agricultural practices increase resistance and resilience to deformation of macro aggregates in soil and help in maintaining the macro-porosity intact in soil.

In NT with residue retention, soil is less prone to structural deterioration. In CT, on the other hand, the soil structural components are weaker and cannot resist the slaking effect of water. Plant roots and fungal hyphae play crucial role in binding the soil and organic carbon into aggregates. They are constantly disrupted in CT treatment, resulting in the destruction of soil aggregates. No-till treatment, in comparison with CT, induces soil conditions that result in increased macro-fauna population increasing the potential on soil aggregation. Soil aggregation plays an important role in increased C sequestration in soil under CA practices.

Soil microbial activity is an integral component of C sequestration and soil aggregation as the abundance of arbuscular micorrhizal (AM) and saprophytic fungi are often correlated with mass of macro aggregates. In fact, AM fungal hyphae forms a kind of web to encase macro aggregates and glycoprotein produced by AM fungi may act as an adhesive to bind soil aggregates. Metabolic products from saprophytic fungi, such as melanin, chitosan, and quinones are more resistant to further decomposition and become chemically bound to clay minerals with in macro aggregates. Kumar *et al.* (2012) reported that percentage of total water stable aggregates (WSA) >2000 um in soil under NT (47%) was significantly higher than under MT (38%) or PT (34%). Thus soil management strategies that increase or maintain soil fungi, such as NT, could result in greater C sequestration. White and Rice (2009) in their investigations on tillage effects on microbial and soil C dynamics during plant residue decomposition revealed that reduced soil disturbance, e.g., NT management facilitated relatively more fungal growth and resulted in higher C sequestration rates. They observed that the microbial community structure was significantly affected by the addition of crop residues and that the effect was different for NT and CT. Total Gram positive and Gram negative bacterial and fungal phospholipid fatty acids were found to be higher under NT 0-5 cm during the most period of residue mineralization compared with CT 0-5 cm or 5-15 cm depths. These data clearly indicate higher biological activity associated with NT soils than under CT, and increased retention of plant C and N in macro aggregates (>1000 um) (Table 2). The authors observed significant tillage x aggregate size interaction for aggregate N (Table 2).

Table 2. Retention of plant residue C and N in soil aggregates under no-till and conventional tillage

Tillage treatment	Aggregate size (mm)	Mean aggregate C (% remaining)	Mean aggregate N (% remaining)
CT	>1000	10.80	2.28
	250-1000	7.02	2.06
	53-250	4.60	0.53
	20-50	3.88	0.29
NT	>1000	22.50	14.10
	250-1000	8.20	2.29
	53-250	3.10	0.43
	20-50	3.84	0.43

Source: White and Rice 2009

A higher amount of actively metabolizing fungi could result in the buildup of recalcitrant metabolites, such as melanins, chitosan, and quinones, and this could lead to a greater amount of C sequestration through protection by fungi by formation of recalcitrant by products and /or increased aggregation. Increased SOC in NT soils is related to increased soil aggregation, as soil aggregates provide chemical and physical protection against plant residue decomposition and alter C mineralization kinetics.

It is known that the NT continuous sorghum soil had greater SOC (McVay *et al.*, 2006), higher abundance of fungal hyphae responsible for greater aggregation than the CT system. White and Rice (2009), as they summarized their observations on tillage effects on microbial and C dynamics show that decomposition was accompanied by a pulse in microbial phospholipid fatty acids (PLFA) with increase in Gm+ and Gm- bacterial and fungal PLFA in NT0-5 cm, R+, CT0-5 cm, R+CT5-15 cm. Significant greater amounts of residue ^{13}C and ^{15}N were found within macro-aggregates and could represent particulate OM or microbial-derived OM. Overall , the greater amount of microbial biomass present in 0-5 cm NT soils was presumably due to high total carbon content related to surface retained crop residues. With a higher microbial biomass, greater amount of soil fungi are present, which could lead to increased soil C sequestration through production of recalcitrant by-products and through increased soil aggregate formation.

Sharifi *et al.* (2013) observed that the long-term (14 yr) NT practices resulted in greater quantity of SOC in the surface (0-5 cm) soil layer compared with CT practices and also a greater quality of SOC as indicated by increased respiration of CO_2 per unit SOC. They did not notice any evidence of a change in the soil catabolic activity associated with tillage system, therefore changes in microbial activity due to tillage system appears to be related only to changes in the quantity and quality of SOC and not to change in microbial function. Zhang *et al.* (2007), however, in an earlier long-term study showed that while increases in MBMC and MBMN mineralization occurred under NT systems in Brazillian Oxisols, the microbial community under CT was more catabolically active than under NT systems. Incorporation of crop residue in CT did, however, increase the MBMC and total recovery of ^{14}C as $MBM^{14}C$ plus respiration of $^{14}CO_2$, suggesting that incorporation of crop residues in CT systems increased the amount and efficiency of utilization of added crop residue. The proportion of SMBC did not change with depth, ranging from 5-8%. However, compared with SOC or SMBC, soil microbial biomass N (SMBN) declined faster with depth. Spedding *et al.* (2004) also reported greater decline in SMBN with increasing depth. Microbial biomass quality often remains unaffected in the tillage treatment as revealed by no change in MBMC/N ratio in tillage treatments. The MBMC/N ratio is controlled by the ratio of fungi/bacterial biomass which also remains unaffected by tillage treatment. Different crops may affect soil macro-aggregate formation in soil differently depending on their root system. Wheat, because of its more horizontal growing root systems, results in significantly more large macro-aggregates in soil than maize crop. Wheat crop has relatively higher plant population which gives relatively denser root net work as compared to maize.

Table 9. Distribution of water stable aggregates (WSA), mean weight diameter (MWD), mean tensile strength of aggregates, soil penetration strength, gravimetric moisture content, available water capacity (AWC) in 0-10 cm and 10-20 cm soil depth maintained under long-term tillage and crop management for two sites in Ohio, USA

Soil property	Soil depth							
	0-10 cm				10-20 cm			
	PT	MT	NT	WL	PT	MT	NT	WL
Wooster (well-drained silt loam soil)								
WSA(>2000mm) (%)	34	38	47	43	23	27	38	32
MWD (mm)	0.82	1.50	2.30	2.53	0.53	0.87	1.25	1.77
TS (kPa)	306	255	183	123	335	310	268	157
SPR (MPa)	3.58	3.30	2.82	0.92	4.46	4.41	3.98	1.84
GWC (kg/kg)	0.15	0.15	0.17	0.22	0.13	0.14	0.15	0.16
AWC (m3m-3)	8.30	10.20	13.40	14.20	6.4	8.7	10.1	13.2
Hoytville (poorly-drained clay loam soil)								
WSA(>2000mm) (%)	47	48	68	67	38	45	56	54
MWD (mm)	1.53	2.46	2.74	2.84	1.58	1.90	2.56	2.35
TS (kPa)	279	224	185	122	378	391	320	150
SPR (MPa)	3.55	3.74	2.50	1.02	5.08	4.03	3.88	1,36
GWC (kg/kg)	0.15	0.16	0.19	0.20	0.13	0.15	0.18	0.18
AWC (m3m-3)	8.0	11.2	15.6	15.8	7.70	8.70	11.4	13.5

PT, plow tillage; MT, minimum tillage; NT, no-tillage; WL, woodlot areas with undisturbed soil
Source: Kumar *et al.*, 2012

Influence of tillage options on soil properties in relation to soil type, crop, climate and duration has been reported by Kumar *et al.* (2012). After a long-term cropping with NT, MT, and PT, they observed that in cultivated silt loam soil, stock of SOC was higher under NT (20.7 Mg ha^{-1}) followed by MT (17.3 Mg ha^{-1}), and PT (16.8 Mg ha^{-1}) for 0-10 cm depth. Soil bulk density for this depth was lower under NT by 8 and 3% than PT and MT respectively. The percentage of total soluble aggregates (WSA) > 2000 mm in soil under NT (47%) was significantly higher than under MT (38%) or PT (34%). A similar trend observed for the clay loam soil. Thus long-term use of NT practices are highly sustainable and result in higher SOC and WSA, low bulk density, and greater available water content (AWC) than MT and PT (Table 9). The results in the table on different soil properties show that long-term and continuous use of NT practices is highly sustainable and influenced the soil properties favourably. The NT system resulted in higher SOC stocks that improved or maintained soil tilth and enhanced the aggregate stability and available water capacity of the soil. In contrast, aggregates of soils under conventional tillage were less stable because of significantly lower SOC concentrations.

Tillage effects on SOC and N stratification in soil

Conservation agricultural practices of soil surface retention of crop residue combined with reduced or no-tillage has the potential of conserving C and N in the agro-ecosystems as SOC and N. However, such sequestration of SOC and N may not be uniformly distributed in the soil profile. Long-term CA systems typically develop a highly stratified vertical distribution of SOC with time. It has been reported that significant increase in SOC with CA systems are usually confined to the surface 30 cm and even more typically to the surface 15 cm soil. Data in Table10 show a long term effect of tillage treatments on the depth wise distribution of SOC and TN. It is clear from these data that while no-till treatment shows a stratified accumulation of both SOC and TN over a period of time, but no definite trend is observed with moldboard plow.

Table 10. Soil organic C (SOC) and total N concentrations at the 0-5, 5-15, and 15-30 cm depth intervals following 30 yr of tillage practices

Tillage treatment	SOC (g/ kg)			TN (g/kg)		
	0-5 cm	5-15 cm	15-30 cm	0-5 cm	5-15 cm	15-30 cm
NT	35.7	26.6	22.9	2.8	2.1	1.8
MP	23.9	24.1	24.5	1.9	1.9	1.9

NT=no-till; MP= moldboard plow

Source: Gal *et al.*, 2007

Chang *et al.* (2012) through simulation modeling showed that NT systems can be regarded as a sure management practices that enhance C sequestration, such sequestration dynamics are, however, associated with vertical variability of soil texture, bulk density etc. NT generally, results in an increase in SOC in the 0-10 cm sampling layer, however, the effect of C sequestration may not be significantly different for the lower sample layers between CT and NT. With continuous CA practices, the accumulation becomes more glaring in the surface layer as is evident from the increasing ratio of SOC in 0-15/15-30 cm. On the basis of large number of reported information, it can be concluded that the ratio of SOC at 0-15 cm to that at 15-30 cm ranged from 2.4 at initiation to 3.1 at the end of 5 yr and to 3.6 at the end of 12 yr. The stratification ratio of SOC was 0.9 and 3.8 under CT and NT respectively, at the end of 1 yr, 1.1 and 3.8 at the end of 2 yr, and 1.2 and 3.9 at the end of 3 yr in a tillage x cropping system experiment (Franzluebbers and Stuedmann 2008). Similar values for stratification ratio of SOC (0-15/15-30cm) were reported for four measurements during 30 yr of experimentation with moldboard plowing (1.7±0.2), NT management (3.4 ± o0.1) and grass sod (4.1±0.2) (Diaz-Zorita and Grove 2002). In a survey of 5 Southeastern states of United States, the stratification ratio of SOC (0-15/ 12.5-20 cm) averaged 1.4 with CT and reached a plateau of 2.8 with in 10 yr of conservation tillage cropland, and a plateau of 4.2 with perennial pasture. The ratio is time dependent and increases with years of CA practices. In three different soil types, the stratification ratio of SOC (0-2.5/7.5-15 cm) was observed to be linearly related to the number of years of continuous NT, initially 1.5 following CT and increasing to 3.6 with 14 yr of NT. Majority of the C stored under CA management in the Ultisols of Southeastern United States occurred within surface 5 cm.

No-till, as usual was found to promote SOM stratification in top layers compared to CT, however, when 0 to 0.3 m or deeper soil layers are considered, soil under both NT and CT practices were usually found to contain similar amounts of SOC. This appears to occur especially in subtropical and tropical regions, although only limited information is available on soil changes under long term CT and NT practices in these regions. Other studies also indicated that SOC and TN stocks under NT were more stratified with greater accumulation near the soil surface (Gal *et al.*, 2007; Yang *et al.*, 2008). On the basis of published literature of 23 investigations with 47 site comparisons, Angers and Eriksem-Hamel (2008) showed that in most of the cases, NT had significantly greater SOC storage than full-inversion tillage (FIT) in the surface layer, whereas at 21- to 25 cm and 26- to 35 cm layers, the average SOC stocks were significantly greater under FIT than NT. Thus continuous NT practices may lead to profile stratification of SOC, and TN pools, degree of stratification with depth expressed as ratio, could indicate soil ecosystem function.

Both tillage and fertilizer management influences SOC storage profile (Poirier *et al.*, 2009). In a long-term study on the interactive effect of tillage and mineral fertilizer management on soil C profile, it has been observed that NT enhanced the SOC contents in the soil surface layer, but moldboard plow resulted in greater SOC content near the bottom of the plow layer. Nitrogen and P fertilization with moldboard plow increased the estimated crop C inputs to the soil. Highest SOC stocks of 0-20 cm soil layer were observed in the NT treatment with highest N rates.

Such organic C enriched surface soil fosters productivity, regulates terrestrial water flow, cycles plant nutrients through diverse biological activity, and creates biologically active and diverse warehouse of soil micro-organisms. This reconfigures soil matrix into relatively stable structure with permanent bio-pores (channels), a process that is important in achieving high water infiltration. Stratification of SOC through CA systems proves a boon for infiltration of water in to the soil. For example, with doubling of SOC uniformly throughout a 12 cm depth increased the infiltration by 27% in Typic Kanhaludult, but increased more than 200% when SOC was concentrated in the surface 3 cm of the soil (Franzluebbers 2010). Thus, SOC stratification is apparently integrally linked to abatement of runoff and soil erosion. With decrease in the run-off velocity, soil erosion loss decreases and was found to be inversely related to the calculated SOC stratification ratio in the field studies with small-plot rainfall simulations (4.6–5.5 m^2). Number of water-catchment and other field-plot investigations have also documented positive influence of conservation tillage and pasture management on soil stabilization and avoidance of water and nutrient run-off.

Effect on N-immobilization/mineralization

Oxidation of organic matter N into mineral forms from the soil organic amendments and crop residues through the process of mineralization plays a significant role in meeting a part of the total N needs a crop. This amount can range from 20 to 80 % in a growing season. The amount of N mineralized, however, depends on soil texture besides soil management practices. In CA, where crop residues are retained on the soil surface and seeding is done with minimum or no-tillage, this mineralization process is drastically hampered, rather, under such conditions, immobilization process can set in resulting in decrease in the availability of N for the crop growth and crop may suffer because of shortage in N supply. This is important as it leads to reduced N supply to the crop especially during the initial years of conservation agricultural practice. Thus, optimum fertilizer-N management in crop production by taking into account N immobilization will have agro-economic performance implications. Lower availability due to immobilization in CA can reduce crop yield and lower crop quality especially in cereals by decreasing protein content in the grain. This is expected to happen during the initial years of converting from CT to NT. However, long-term practice of conservation agriculture can result in important changes in active and stable soil organic C and –N fractions that can profoundly influence soil N dynamics. Long-term use of practice of retaining crop residues on soil surface and use of reduced tillage or NT, overtime, can lead to a substantial increase in mineralization of soil N potential. The Increased mineralizable soil N potential with continuous practice of CA can lead to higher N present as microbial biomass and also in the mineral form. Thus, management–induced changes in the size and quality of mineralizable N should be considered in developing best N programs for crop production to achieve highest nitrogen use efficiency (NUE).

Use of conservation tillage has been reported to increase short-term N immobilization due to slower plant decomposition process when tillage is limited. Thus, retention of crop residue on the surface with no-till or reduced tillage tends to decrease grain yield of the cereals, especially during initial years of conservation agriculture. This is due to N immobilization and delayed N release caused by an increase in the C/N ratio. This is the reason, in conservation tillage, fertilizer N rates are often higher by as much as 25% to prevent yield limitations (Randall and Bandel 1991) and grain quality from short-term N immobilization. Andraski and Bundy (2008) observed that increasing residue levels lowered early season soil NO_3-N production, and reduced corn yields without additional application of N, and suggested that applying about 30 kg N ha^{-1} of extra N will provide yield benefits in NT corn residue systems. It is possible that lower yields were in part due to decreased N mineralization associated with conservation tillage. Lamb *et al.* (1985) found that soils of stubble-mulch tillage system accumulated only about 70% as much NO_3-N as conventionally tilled soils. In NT system in which crop residue is allowed on the soil surface only results in its slow decomposition. Under such situation, besides decreased release of residue N and other nutrients, an immobilization of soil N can cause marked reduction in availability of N from crop use. This may result in application of slightly higher N rates to attain equal grain yields for NT systems compared with CT production system at least for the first few years after converting to NT. Residual NO_3-N level has a tendency to be higher in CT systems than in the NT system, providing a higher level of available N at the same N fertilizer rate in CT system than in NT system, thus contributing towards a slightly higher yields in CT than in NT systems. Average net N mineralization as observed by Nelson *et al.* (2006) at two sites were 56.7 and 91 kg N $ha^{-1}yr^{-1}$ for NT and 155 kg N $ha^{-1}yr^{-1}$ for CT, the total annual net nitrification was significantly (P=0.05) greater in tilled soils.

Wheat N content can be significantly reduced in conservation tillage treatment, suggesting reduced N mineralization. Reduced mineralization is caused by increased N immobilization associated with higher residue systems as discussed earlier in this section. However, there are observations of absence of tillage x N interaction which suggests that greater N immobilization may not be the sole factor in grain yield or quality reduction with conservation tillage. Pasricha (2013), observed no significant difference in the grain yield of wheat in two tillage treatments. No-till wheat had lower N concentration (0.352% in straw; 1.973% in grain) than corresponding CT wheat (0.521% in straw; 2.44% in grain) indicating lower N availability with NT (Table 11). Total N uptakes in wheat at harvest were lower in NT (144.6 kg ha^{-1}) as compared to CT plots (184.63 kg ha^{-1}), percent N derived from fertilizer at 44.62 was, however, significantly higher than CT treatment, which were only 36.25. The rate of N applied was 150 kg/ha, which seems not to be adequate to meet N requirement of NT crop. The crop has, perhaps, less access to soil derived N, therefore, requires higher fertilizer N than CT crop. A lower uptake of N in NT plots suggests a lower net mineralization in NT plots. This is also apparent from greater % ^{15}NEXC in NT treatment. In the absence of adequate soil-derived N, crop plants derived greater N from fertilizer source in NT plots. This could be ascribed to immobilization of the N, as residue decomposition is very slow in NT plots due to lack of mixing.

Table 11. Nitrogen content and uptake in wheat as influenced by tillage treatment and residue retention

Tillagetreatment	N content (%)		% ^{15}N-EXC		N uptake (kg ha^{-1})	N derived from fertilizer	
	Straw	Grain	Straw	Grain		(%)	(kg ha^{-1})
CT	0.522	2.551	3.64	3.60	184.63	36.25	66.94
NT	0.352	1.973	4.20	4.52	144.65	44.62	64.54
CD(Pd" 0.05)	0.078	0.235	0.31	0.205	21.25	4.36	NS

Source: Pasricha 2013

Wheat is one of the most important crops in the Indo-gangetic Plains; about 15% of the world wheat production is attributed to this region. Any conversion from CT to NT on large scale without additional fertilizer N application can result in large scale malnutrition due decreased grain protein content in CA managed systems at least in the initial years of adoption. Therefore, an additional N application at appropriate growth stage of the crop may become necessary to overcome this problem. Similarly in Southern humid Pampa region of Argentina, in order to increase or maintain productivity of wheat in these regions, an efficient use of fertilizer N in tandem with conservation tillage practices is necessary. Wuest and Cassman (1992 a, b), Sarandon *et al.* (1997), and Falotico *et al.* (1999) emphasized the need for fertilization timing, tillage system and environmental conditions that may modify grain N yield, N accumulation pattern, and N remobilization in wheat.

Soil analysis data on ^{15}N and total N content of soil (0- to 0.30 m) under two tillage treatments (Table 10) show that soil total N content at 1622 kg ha^{-1} in NT plots was greater by about 112 kg ha^{-1} than CT plots at 1510 kg ha^{-1}. This is due to greater mineralization of N in tilled plots and its utilization by wheat. This shows that in tilled plots, because of more mineralization of organic N, there is higher availability and consequent higher uptake in the crop, the amount left in soil is lower than no-till plots. In CT plots, greater levels of soil N have been mineralized due to unrestricted aeration in the plots. The SOM is lost more easily in tilled plots. In NT plots, on the other hand, such SOM remains intact for greater period of time and thus is helpful in improving the soil health. Moreover, most of the remaining mineralized N present in the tilled soil after the harvest of wheat is invariably in NO_3form which is liable to be lost through leaching and denitrification as soon as these soils are submerged during the land preparation for transplanting of following crop of rice after wheat in rice-wheat rotation. Masses of soil total N in the upper 0.3 m layer under NT were higher by 7% and fertilizer N remaining unutilized by crop was higher by 33% than in CT (Table 12). This fertilizer N is probably

Table 12. Soil N content in 0-30 cm layer after harvest of wheat crop as influenced by tillage treatment

Tillage treatment	Depth	Soil total N content		Fertilizer N remaining in soil	
	(m)	(%)	(kg ha^{-1})	(% of soil N)	(kg ha^{-1})
Conventional-till	0-0.15	0.0444	889	2.08	17.84
	0.15-0.30	0.0311	622	0.427	2,65
	0-0.30	-	1511	-	20.49
No-till	0-0.15	0.0489	978	2.494	24.39
	0.15-0.30	0.0322	644	0.441	2.89
	0-0.30	-	1622	-	27.23

Source: Pasricha 2013

present as soil microbial biomass and, on the other hand, even at lower level of fertilizer N remaining in soil in CT plots, a part may be present in mineralized form of NO_3-N, which may be lost through leaching when soils are flooded for field preparation for following rice crop (Nelson *et al.* 2006) and another part may be lost as N_2O gas due to low oxidation potential of the submerged soils (Pasricha 2010). However, because of potential increase in infiltration with NT, a better understanding of tillage effects of NO_3-N leaching is required. Weed and Kanwar (1996) found that although NT produced more percolation water, the flow-weighted NO_3-N concentration and NO_3-N masses in percolation from CT were higher or equivalent to that from NT. Thus, there is likelihood of apparent retention of more C along with N per unit C input in NT treatment. Zhu and Toth (2003) recommended NT as the best management practice for reducing erosion in agricultural productivity of soil. Therefore, NT or reduced tillage practice is a good option for sustaining rice-wheat cropping system and can safeguard the productivity of Indo-gangetic soils.

In a survey on the effect of tillage practices on the wheat yield on farmer's field in Haryana and Punjab, Erenstein *et al.* (2007), did not observe significant differences in grain yield in NT and CT treatments. On the basis of average grain yield in Haryana, NT gave higher yield of 4.38 Mg ha^{-1} (average of 138 locations) than 4.21 Mg ha^{-1} for CT (average of 99 locations). However, general lack in yield increase in grain yield in Punjab largely reflects that NT-induced time savings in the land preparation did not translate into timelier crop establishment. Such an effect may not be very glaring if date of seeding in both the treatments is before mid November. In Haryana, date of planting was 10th Nov. for NT and 12th Nov. for CT, while in Punjab; the dates of planting were rather irrational to the CA practice, Nov. 27 for NT and Nov. 24 for CT. The problem with large scale adoption of the practice is that farmers expect higher yields from any new technology. Although in NT, there is savings on land preparation, however, yield differences are not attractive to the farmers. They overlook the long-term benefits of improved soil health with the CA practice. The lack of a significant yield effect has undermined widespread CA acceptance and is a major factor explaining poor adoption in these states which are considered food bowl of the country.

The story of N immobilization is different in Vertisols which occupy a significant area in the country. These soils are generally difficult to manage because of their poor physical characteristics, including high shrink-swell potential, high water holding capacity, high plasticity, exceptionally high strength when dry, and very limited range of soil water content in which soil tillage operations can be performed. There low capacity for infiltration when wet leads to high run off rates. These soils are readily eroded because these soils have a tendency to stake into fine aggregates. There was no indication that short-term N immobilization reduced corn yield in NT compared with chisel tillage systems in vertisols of Blackland Prairie (Torbert *et al.* 2001b). It is likely that in vertisols, if a short-term N-immobilization occurs in these systems, it is not sufficient to either decrease grain yield or reduce N uptake by end of growing season. These observations are from well established NT systems that had been in place for 8 yr. This shows that the short-term N imbalance in NT systems, compared with CT systems does not prevail long-term. It indicates that for well established NT systems, there is no need to increase fertilizer N application rate to compensate for N immobilization in vertisol soils.

Effect on NO_3 accumulation/leaching losses

Recent spurt in concerns over environmental quality, energy conservation, and economics have increased the need to maximize crop utilization of applied fertilizer N and to reduce excess application that may contribute to stream or groundwater contamination. In the wake of greater concern about the environment quality, an understanding of N soil fertility in relation to crop production will lead to important fertilizer management and preserve water quality. Tillage enhances annual total N mineralization at sites which are relatively better drained. This results in more potential soil $NO_3$to be loaded when the crop demand is low with the high precipitation especially if the crop demand for N is low or for fallow years when the actively growing crop is lacking. No-till system, on the other hand, discourages soil nitrification process of SOM due to lack of access to free air. Bundy *et al.* (1995) showed that pre-seeding soil profile (0-90 cm) NO_3-N contents under continuous NT were 55 to 110 kg NO_3-N ha^{-1} less than under moldboard plow tillage (CT) in continuous corn. This indicates that less soil N is made available in NT treatment and thus there are fewer losses of N through NO_3 leaching.

Net N mineralization is, however, little affected by tillage in the more poorly drained soils. In the northwestern region of the country, rice crop grows from mid June to mid September or early October, when >90% of the annual Monsoon precipitation is received. This precipitation regime may facilitate soil NO_3 flushes to shallow water table and surface waters especially during the initial stages of plant growth when crop demand for N is low. Most of the soils of the region are light textured and well drained. During the winter months, when wheat crop is in the field, there is greater NO_3 flushing, and the residual NO_3 in the soil after harvest of wheat is subject to leaching to shallow ground waters when these soils are flooded for the preparation of rice planting in rice-wheat rotation. Thus with regard to increasing fertilizer N use efficiency (NUE), and reducing N losses and minimizing potential effects of off-site NO_3 movement on water quality, a better understanding of mineralization process in northwestern region is imperative as specific information for this region is lacking. It is especially important to understand how tillage and residue management, which can greatly influence soil NO_3 levels, affects soil N mineralization (N source) and N use by the crop (N sink).

Due to lack of free air, mineralization of SOM under submerged soil conditions does not go beyond NH_4^+ formation for low-land paddy cultivation. Ammonium ions accumulate in soil and not oxidized or nitrified to NO_3. This soil NH_4^+ obtained directly from the applied urea fertilizer or soil N mineralization, readily binds soil colloidal particles, making these ions relatively immobile. However, a part of this ammonium ions are dispersed as NH_4^+ -laden soil particles and make their way to water ways through the process of soil erosion. This fraction of NH_4^+ ions can be nitrified to NO_3 and can cause some water quality problems. Nitrate ions thus formed when not absorbed by microbes or plants, are relatively mobile and can easily move to surface water bodies and adversely affect aquatic wild life.

Many soil and water quality problems are associated with CT, along with other problems that affect water resources. Tillage systems have a significant effect on N dynamics by affecting N pools in the soil system. Soil disturbance during CT process and incorporation of surface residue increases soil aeration, which can cause increased rate of residue decomposition. This process impacts soil organic-N mineralization whereby readily available N for plants is increased. Under such situations, a considerable part of soil total N is present as NO_3-N.

Deep accumulation of NO_3-N in the soil profile represents a potential for NO_3-N leaching into shallow water tables. Halvorson *et al.* (2001) reported that CT and reduced tillage (RT) systems accumulated more soil NO_3-N down to 150 cm compared with NT system in wheat-fallow cropping system. They observed that CT and RT systems mineralized more N at soil surface due to soil disturbance than NT system. In another study, soil NO_3 in the 15-60 cm depth has been reported to be higher in CT than in NT system (Grant and Lafond 1992). This is again attributed to greater N mineralization in ploughed well-drained soils than in poorly drained soil compared with corresponding NT treatment. Tillage by mixing and incorporating crop residues in the soil tends to promote faster release of residue N than surface-placed (NT) residue. Conventional-tilled soil has been observed to contain 28 kg ha^{-1} more NO_3-N to 100 cm depth than NT soil (Soon *et al.* 2001). Moldboard plowing reduced the tile flow by an average of 2 cm of water depth compared with NT (Weed and Kanwar 1996). In the same study, the 3 yr average NO_3-N concentration of the tile water of NT was lower (21.9 mg L^{-1}) than the moldboard plow (36.9 mg L^{-1}), and the average NO_3-N loss from the NT system was 74 kg ha^{-1} less than the moldboard plow system.

Impact of tillage-induced enhanced infiltration rate on nitrate leaching

Although, increased infiltration in CA practices will help storing maximum water in soil profile, however, concerns have been raised regarding environmental soundness of this practice. Tillage and N management systems can have significant effect on N use by crops and NO_3-N movement through the soil profile. Potential water quality and NO_3-N loss problems associated with CT and applied N has prompted several investigations. In one of the investigations, strip-tillage and no-tillage resulted in lower residual NO_3-N build up than chisel plow in the 0-1.2 m soil profile after two yr of tillage implementation. Some researchers feel that increased infiltration rate in NT and residue coverage of soil surface which leads to increased infiltration may be accompanied by an increase in NO_3 leaching to ground water. There are reports on more leachate volume and NO_3-N masses in leachate collected below 106 cm soil depth in NT than CT (Tyler and Thomas 1977). In contrast Kanwar *et al.* (1985) noted that more NO_3^- was leached below 1.5m in CT than in NT using rainfall simulation on a loam soil. Similar results have been reported by Angle *et al.* (1993) and Ritter *et al.* (1993). There are other reports which show that although NT produced more percolation water, the flow-weighted NO_3-N concentration and NO_3-N masses in percolation water from CT were higher than equivalent to that from NT. Zhu *et al.* (2003) also observed greater NO_3-N in CT than NT both under no-fertilizer N and 100 kg N ha^{-1} treatments (Table 13). They used wick lysimeters, which have shown to collect approximately 100% of soil percolation water as compared to pan lysimeters which collected 40% during the growing season. Six yr annual leachate volume was greater in the CT than NT, for studying NO_3-N leaching under different tillage systems.

Table 13. Nitrate –N concentration in the leachate as affected by tillage and fertilizer N treatment (0 and 100 kg N ha^{-1})

Tillage treatment	NO_3-N conc. (mg N L^{-1})	
	No fertilizer N	100 kg N ha^{-1}
CT	237	312.5 (Av. Of 3 plots)
NT	190.5 (Av. Of 2plots)	250

Source: Zhu *et al.* 2003

Some areas in the Northwestern part of the country are intensively cultivated for rice-wheat under irrigated conditions. The soil texture mostly ranges from loamy sand to sandy loam. These alluvial soils are well-drained and thus are vulnerable to NO_3 leaching losses, especially under present situation of faulty irrigation methods due free electricity power to farmers for irrigation. This has led to the tendency for excessive application of irrigation resulting into the deep leaching losses of NO_3-N and contamination of ground water. Nitrate-N concentration in drinking water in some of these areas has been found to exceed maximum limit as per USEPA drinking water quality standards (10 mg $L^{-1}NO_3$-N). Increased infiltration with reduced tillage or no-till management systems can further accelerate this problem of NO_3-N leaching to environment in these well drained soils, especially in areas where rice-wheat rotation is followed. Residual N in soil can readily undergo oxidation to NO_3 when these soils are prepared for wheat cultivation after the harvest of rice. Subsequent losses of NO_3-N to ground water and surface water bodies through run-off can cause hypoxia in the lakes, and adverse health effects such as methemoglobinemi. Such a process of NO_3formation and subsequent pollution of ground- and surface-water bodies may not occur in areas where rice-rice rotation instead of rice-wheat rotation is followed.

Effect of N-fertilization on NO_3 accumulation/leaching

Influence of crop residues on plant available N in the soil depends on how these residues affect the net mineralization of the soil N sources. Agriculture is recognized as a source of nonpoint pollution and one major concern is the contamination of surface and ground waters with nitrates from chemically fertilized crop production systems. Nitrate pollution of ground and surface waters is common in area of intensive agriculture and where fertilizer N is often used in excessive amounts to ensure optimum crop productivity. It is thus considered that a high level of NO_3 in the water is due to agricultural production. However, there is scope for reducing these losses of NO_3 to environment by modifying the agricultural practices to minimize NO_3pollution problem by opting for CA. Tillage, crop residue management and crop rotations are management practices of CA that can profoundly influence the N dynamics of soil-plant system. When released in synchrony with crop N demand, crop residue N is a particularly desirable source of N because losses to environment are minimized. Long-term N-fertility studies have shown that residual NO_3-N levels increase when N-fertilization rates exceed that needed for optimum yields. Increasing levels of residual soil NO_3-N in the lower part of the root zone increases the potential of leaching of NO_3-N below the root zone and into the shallow water tables, creating environmental concerns. Residual NO_3-N in the 150 cm soil profile tended to be higher with CT and MT than with NT (Table 14) and further increased with fertilizer-N applications. In years of higher precipitation, NO_3-N movement below root zone was more pronounced. Results of these investigations indicate that NT with annual cropping may reduce the relative quantity of residual soil NO_3-N available for leaching compared with MT and CT systems even under situations of higher fertilizer N applications.

Over-fertilization is commonly employed to ensure maximum crop productivity both in developed and developing countries. This is particularly so in the irrigated areas where rice-wheat rotation is intensively followed as in the Indo-Gangetic plains of India, Pakistan, Bangladesh, and Nepal. Also in areas of assured and adequate rainfall during the growing season, over-use of N fertilizers is common. This excessive N is usually present in soil in the NO_3-N form and is freely leachable to ground waters. Besides excessive N fertilizer addition, tillage system also influences the level of NO_3. Summing up the results of the whole time span of 18 yr, the NO_3 concentrations were higher, on average, under CT compared to NT. Most of the NO_3 in the soil profile were found to be concentrated in 30- to 60 cm and 60- to 90 cm depths. Lower NO_3 under NT treatment is due to lower net N mineralization due to slower decomposition and more N-immobilization and nitrification differences. However, because of potential increase in infiltration with NT, better understandings of tillage effects on NO_3-N leaching is needed.

Table 14. Post-harvest NO_3-N (mg N L^{-1}) in 0-150 cm soil profile as a function of tillage and N treatments, averaged over 3 crop sequences (spring wheat, winter wheat, and sunflower) each year for 12 yrs (1985-96)

Tillage treatment	N-rates applied (kg/ha)			
	36	67	101	mean
CT	72.92	170.92	192.12	145.34
MT	86.83	103.89	174.33	121.66
NT	43.25	64.17	179.50	81.22

Source: Halvorson *et al.* 2001

Effect on denitrification / N_2O_2 emissions

Among the molecular and ionic species that exist in more than one oxidation states, nitrate-N is the first to be affected due to excessive soil wetness or poor aeration. In the absence of adequate free oxygen, facultative organisms in the soil start using nitrates as source of oxygen for their respiration, in the process, NO_3 is reduced to nitrogenous gaseous forms such as N_2O, NO and N_2 depending upon the intensity of reduction conditions. Thermodynamically, nitrous oxide is most likely form in which nitrogen from soil can be lost (Pasricha 2010). This is perhaps the reason for relatively low N-use efficiency in flood-irrigated rice crop as compared to wheat and other arable crops. NO_3-N is the precursor of N_2O gas. Although most of the soil N in submerged soils is in the NH_4^+ form, still such gaseous losses of N cannot be altogether prevented as even under excessively wet soil conditions; some micro-zones in the rhizosphere may exist as oxidized layers. Rice roots are known to exude oxygen in the root zone resulting in the formation of a thin oxidized area in their immediate vicinity. Any NH_4^+ which happens to come in this zone as result of diffusion during plant uptake will be oxidized to NO_3^-, a part of which may be absorbed by rice roots and other part may move to reduced zone again thorough diffusion and mass flow process and get lost after its reduction to N_2O gases.

$$2NO_3^- + 10\ H^+ + 8\ e^- = N_2O\uparrow + 5\ H_2O$$

N_2O is present in atmosphere in minute quantities in the air as a product of denitrification. Thermodynamics of NO^-_3 –N_2Osystem show that nitrate will not decompose to nitrous oxide in aqueous media in equilibrium with O_2 /air but will do so readily at potential (Eh=0.33 V or p^E = 5.63) at which oxygen just disappears from the soil submerged in water (Pasricha 2010). If this system is in equilibrium with O_2–H_2O system of the atmosphere, then the partial pressure of N_2O in equilibrium with 10^{-2} mole L^{-1} of nitrate at pH 7.0 is infinitesimally small at $10^{-32.16}$. But as the O_2 is depleted, the p^E falls to 5.63, the partial pressure of N_2O that will be in equilibrium with 10^{-2} moles of nitrate at pH 7.0 shoots to as high as $10^{31.84}$. Thus nitrate is highly stable with respect to air, but very unstable in the absence of O_2 thus thermodynamically N_2O is the highly probable product of denitrification of nitrate.

Release of N_2O gas as a result of agricultural activities has far greater impact on global warming and climate change than CO_2. Its concentration in the atmosphere is far less than that of CO_2; however, the global warming potential of N_2O is more than 300 times greater than CO_2. Therefore, even small changes in soil N_2O emissions can have a significant impact on the GHG balance of farms. Total emissions of N_2O from agriculture accounts for about 58% of the total anthropogenic emissions, with emissions from soils alone constituting about 38% of the total N_2O emissions due to agricultural activities as per estimates in the year 2005 (USEPA 2010). US agriculture alone contributes approximately 78% of the total emissions in that country (Halvorson *et al.* 2010). Global mean fertilizer-induced N_2O emissions amounted to approximately 0.9% of N-fertilizer applied (Bouwmans *et al.* 2002). Furthermore, global agricultural N_2O emissions are projected to increase by 35 to 60% by the year 2030 due to increase in intensity of cropping, increased N fertilizer use, and increase in animal manure production (FAO, 2003). Currently, it accounts for about 5% only of the total green-house effect (Smith *et al.* 2007b), but in view of its tremendous impact on global warming, agricultural practices leading to formation of N_2O and subsequently emission to the atmosphere, need to be more aggressively investigated.

In Canada, it is claimed that agricultural soils have lost almost 20- to 30% of their native organic C following the onset of their cultivation in traditional manner with intensive tillage systems (Jenzen *et al.*, 1998). It has been suggested that these soils, however, have stopped losing C and beginning to gain C with increasingly more use of more efficient reduced or NT and better fertilization practices, and better crop rotations. However, in fine textured clay rich soils which are prone to remain relatively more wet for long periods, the practice of NT that help reduce CO_2 emissions and sequester C into soils may lead to increased N_2O emissions, a gas that has a radiative forcing potential 310 times greater than CO_2 (IPCC 2001) as observed in earlier section. Thus, Long-term tillage and rotation practices can result not only in changes in soil properties and C stocks, but also in changes in N dynamics. The most prevalent hypothesis in the relevant literature to date is that, NT is expected to result in higher denitrification and N_2O emission rates because of higher moisture content, greater amount of SOC and relatively more microbial population near the surface. Omonode *et al.* (2007) reported significantly greater CO_2 emissions from chisel plow relative to moldboard plow and NT. Surprisingly; they observed a strong co-relation between N_2O and CO_2 emission. It seems, there is ample amount of NO_3 formation during aerobic decomposition of SOM under chisel plow system. However, possible soil saturation subsequently can cause reduction of NO_3 to N_2O. The authors did not try to determine NO_3 level in soil. In fact, a strong relationship between CO_2 and NO_3 level makes more sense.

During the growing season, N_2O emissions are related to soil temperature, volumetric soil water content, precipitation, and air temperature. Collectively, these factors can account for up to 31% of total variability associated with N_2O emission. NT management in relatively warm and wet periods or regions may result in N_2O emission rates similar to or slightly less than those in chisel plow and moldboard plow. Contrary to Omonode *et al.* (2011) contention, that SOC decomposition under enhanced aeration in chisel plow and moldboard plow would encourage N_2O emissions, in fact, N_2O formation would require anaerobic soil conditions when there is scarcity of free air or oxygen in the soil. One cannot expect poor aeration under chisel or moldboard plow systems under arable soil conditions. In NT systems, although surface soil is generally rich in SOC, but lack of adequate aeration does not allow the nitrification process to take place, oxidation process stops at NH_4^+ and does not proceed to the formation of NO_3. In the absence of NO_3 in NT system, there are virtually no chances for the production of N_2O, a process that can take place only through the denitrification reaction (Table 15).

Table 15. Cumulative N_2O emissions for the year 2004, 2005 and 2006 cropping seasons as influenced by tillage practices

Tillage treatment	CumulativeN_2O emissions (kg N//ha)			
	2004	2005	2006	3 yr average
NT	0.66	2.32	6.90	3.37
CP	0.60	4.41	18.01	7.25
MP	0.51	2.26	13.32	5.61

NT= No-till; CP= Chisel till; MP= Moldboard plow

Source: Omonode *et al.* 2011

Episodes of high N_2O emissions coincided with increases in NO_3 while lower emissions were associated with lowest NO_3concentrations suggested that denitrification was the main source of N_2O in the fine textured clayey/clay soils as observed by Rochette *et al.* (2008). High N_2O emissions can occur after harvest of wheat in rice-wheat rotation when soils are plowed in standing water incorporating residue and stubbles for planting of rice. Most of residual N present in the soil after harvest of wheat is in the NO_3 form. This NO_3 is subject to denitrification to N_2O when there is sufficient decomposing OM in soil and there is limited aeration because submerged soil conditions. High OM in the soil under wet conditions intensifies the reduction process (Pasricha 2010).

Rochette *et al.* (2008) stated that anticipated benefits of increased SOC stocks on net soil-surface GHG emissions after adoption of soil conservation practices can be more than offset by increases in soil N_2O emissions. They reported that annual emissions of N_2O were exceptionally high in the heavy soils, varying from 12 to 45 kg N_2O-N ha^{-1} during the 3 yr study. Such high emissions in these fine textured soils are associated with the dentrification process which is sustained by the decomposition of large SOM stocks (192 Mg C ha^{-1} in the top 0.5 m). On an average, NT practice more than doubled N_2O emissions as compared with moldboard plow. In the light textured loam soils, on the other hand, average emissions during 3 yr were relatively much low and similar in the NT and MP plots (Fig. 6). The influence of plowing in MP on N_2O flux in heavy clay soil is probably the result of increased soil porosity that maintained soil aeration and water content at levels restricting denitrification and N_2O production in the top 0.2 m. In the loam soil, average emissions during the 3 yr were similar in the NT and MP plots. The results of this study indicate that potential of NT for decreasing net GHG emissions may be limited in fine textured soils rich in OM that are prone to high water content and reduced aeration. Denitrification chances are usually increased in SOC- rich heavy and relatively wetter soils after several years of NT practices. This may result in increased rates of emissions of N_2O in NT soils resulting in more than offsetting the beneficial effect of stocking C in soil. Depending upon the soil texture, OM content and susceptibility to excessive wetness, the adoption of NT can show different results with respect

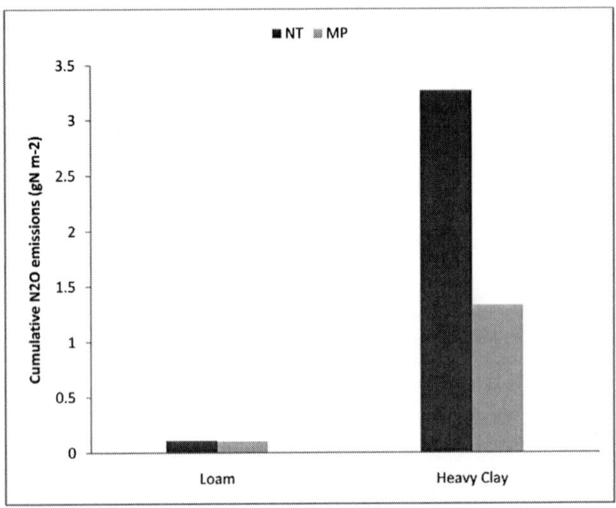

Fig. 6. Cumulative N_2O emission from heavy clay and loam soils under no-till (NT) and mold board plow (MP) (mean values for 2001, 2002, and 2003)
Source: Rochette *et al.*, 2008

to N_2O emissions.

In Canada, adoption of NT resulted in higher soil N_2O emissions in the eastern region, but lower emissions in the relatively coarse textured prairie region. Thus, there is positive interaction between tillage and soil texture on N_2O production and emission at soil surface. Similar reports of higher emissions under NT than under CT in fine-textured soils have been earlier reported by Chaudhary *et al.* (2002). Due to high radiative forcing of N_2O, the mean NT- induced N_2O emissions of 2 g N_2O-N $m^{-2}yr^{-1}$ corresponds to approximately 265 g CO_2-C equivalents $m^{-2}yr^{-1}$, a value which is more than 130 times greater than the expected annual soil C gain under NT.

Effect of fertilizer-N application on nitrous oxide emissions

Although, tillage practices have shown to have profound effect on N_2O emissions, but as much as about 59% of the total N_2O emissions occur shortly after fertilizer-N applications regardless of tillage or rotation practices. No-till practices over time build SOC stocks as has been discussed in other sections. Also soils under NT retain more moisture and remain relatively wetter than under CT, a condition that favours reduction process for NO_3. On the basis of reduction-oxidation equilibria in soils, most likely product of this reduction process is N_2O.The opinion on the use of CA practices that reduce CO_2 emissions from soil and increase the SOC level may differ as a green technology, as these increased SOC stocks in soil may stimulate soil N_2O emissions. This is particular critical in fine textured clayey soils and usually with imperfect drainage under humid climate. This may limit the adoption of conservation tillage or no-till agricultural practices designed to stock more C in soil and mitigate green house gases (GHG). Increasing fertilizer N application rates can significantly accentuate N_2O emissions. Almaraz *et al.* (2009) reported that higher N_2O fluxes occurred during the spring which was invariably associated with high rainfall event in heavy soils causing relatively more wetness. Application of N at high rates (180 kg N ha^{-1}) resulted in large peaks of N_2O irrespective of tillage practice during wet part of the season in the fine textured soils. There was not much difference in conventional tillage (CT) and NT so far CO_2 emissions are concerned but NT had higher cumulative N_2O emissions than CT. Thus changing from CT to NT under heavy soil conditions may increase GNG, mainly as a result of increase in N_2O emissions under relatively wetter soil conditions and adequate fertilizer N applications. This may greatly curtail the use efficiency of fertilizer N by NT in such soils. This negative effect of NT could be reduced by avoiding applications of excessive rates of fertilizer N especially when the soil conditions are wet or when precipitation is more intense.

Although fertilizer-N is necessary for crop yield and quality, its application also enhances the rates of N loss from the soil as N_2O. Nitrous oxide fluxes increase within days following the application of urea and reach peak level during the first 30 days following N fertilization with urea, then decline to near background level in about 40 days (Halvoson *et al.* 2010). Flux from controlled-release polythene coated urea was, however, much low but tended to extend over a greater period beyond 40 d. The potential for reduction of N_2O emissions with the use of controlled-release and stabilized N-fertilizer sources in NT system is substantial. The fertilizer-induced component of N_2O emission could be reduced substantially with NT, but the degree of reduction would also depend on cropping system, and, of course, site specific conditions. On the basis of average of two growing seasons, N_2O-N emissions from CT-continuous corn and NT-continuous corn systems as percentage of the N fertilizer applied varied significantly with N-source. In continuous-corn system, N loss as N_2O-N per unit of fertilizer N applied was 0.85% for CT and 0.33% for NT with urea as source of N-fertilizer.

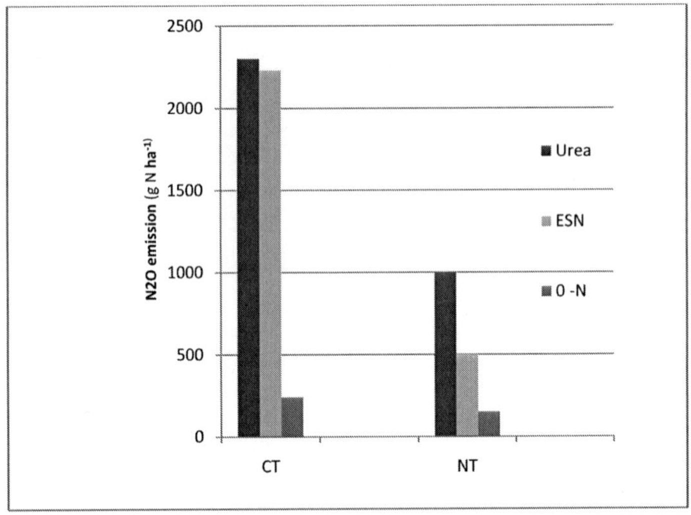

Fig. 7. Average cumulative N_2O emission of 2007-2008 growing seasons for the CT and NT with continuous corn rotation with no N, urea, or polymer coated urea (ESN) applied
Source: Halvorson *et al.* 2010

When source of N was polythene coated urea, these losses were 0.83% for CT and as low as 0.14% for NT (Fig. 7).

There was 83.5 % reduction in the N loss as N_2O in NT as compared to CT, when source of fertilizer N was polythene coated urea in the continuous corn system. Nitrous oxide emissions resulting from applications of commercially available enhanced-efficiency N fertilizers with emissions from conventional dry granular urea in irrigated cropping systems under CT and NT systems need more intensive investigations. Cumulative growing season N_2O emissions from urea and polymer-coated urea applications were not different under CT-continuous corn, but polymer-coated urea reduced N_2O emissions were 49% compared with urea under NT-continuous corn system. Compared with urea, stabilized urea source (Super U) reduced N_2O emissions by 27% in dry bean and 54% in corn in the NT-corn-dry bean rotation and by 19% in barley and 51% in corn in the NT-corn-barley rotation. Venterea *et al.* (2005) also reported earlier, greater N_2O emissions from anhydrous NH_3 and lower in urea-NH_4NO_3 (UAN).

Effect of soil moisture content, temperature, and soil reaction on nitrous oxide emissions

Mineralization of soil organic matter (SOM) results in the production of CO_2 and process of mineralization and denitrification results in the production of N_2O. These biological processes in soil are mediated by microbial activity which is closely controlled by soil climatic conditions like temperature, moisture content/ aeration etc. In clayey soils, microbial activity is at its peak at around 25-30^{0}C and plant residue decomposition is optimal when there is adequate soil aeration, i.e., when there is 55-60% of water-filled pore space (WFPS). Under these conditions, there is production of NO_3 through nitrification process. However, under wetter soil conditions, when WFPS is >60%, process of denitrification sets in with the production of N_2O that emanates to the atmosphere. Such situations lead to not only decreased fertilizer–N use efficiency but also contributes towards global warming by production of highly radiative N_2O gas. Higher amounts of plant-residue in heavy soils at saturated WFPS induce acute reduction conditions in soil which cause denitrification of the $NO_3$to N_2O. This is exactly what happens in rice-wheat rotation. After harvest of wheat, soil contains unused N which is wholly present in NO_3 form. Moment the soil, after harvest of wheat is submerged with the anchored wheat straw, a part of NO_3-N is lost through leaching and other part is denitrified to N_2O gas when the fields are submerged before the transplanting of rice nursery. Gaseous losses of N can also be in the form N_2 as reported by Rochester (2003) for alkaline vertisols, and at times this form can be substantial. A N_2O/N_2 ratio of 0.024 has been reported for these soils. Thermodynamically, N_2O is most likely being the gaseous form of NO_3 reduction, however, what are the circumstances in alkaline vertisols, where, instead, it is the gaseous N_2 which is the dominant form of N gaseous losses. In heavy clay vertisol soils, water movement through the profile is generally very slow, therefore, N leaching is normally considered to be non consequential. However, differential water flow through soil cracks may greatly contribute to water recharge of the soil profile, and therefore, could contribute to some N losses from soil through leaching. According to Torbert *et al.* (2001a), the largest N losses in such soils were most likely due to denitrification, since all of the soil conditions required for denitrification are present in these soils. Soil conditions in these soils are commonly saturated. In addition, these soils are often calcareous and denitrification process is promoted by elevated pH levels. Also, in these tillage systems, the residue of the previous year's crop causes more intense reduction conditions in the soil when residue decomposition takes place under anaerobic soil conditions (Pasricha and Ponnamperuma 1978)

Effect on fertilizer N-use efficiency

Soil organic N may represent a significant source of total N assimilated by crop plants. In rice crop, 50 to 80% of the total N assimilated by the crop may be coming from sources other than fertilizer N. Therefore, in order to improve, nitrogen use efficiency (NUE), residual N in soil from previous crop must be considered in the cropping system due to increases of this nutrient in soil, and risk to the environment. The carry-over effect of N could be substantial. The traditional rates of N used by farmers could be far higher than those recommended by government and university advisory services, with greater chances of its loss to environment through leaching and denitrification especially in excessively wet heavy textured organic matter-rich rice soils.

Burning of the straw is a common practice in the rice-wheat belt of Indo-gangetic plains for early vacation of the fields so that wheat sowing can be completed well in time. A transition from conventional method to conservation agricultural practices has prompted to know how the straw left in the field from open-field burning will influence N-immobilization-mineralization dynamics and their effect on long-term soil fertility. N-fertilizer use efficiency of rice is much less as compared with upland crops like wheat (40-60% one season recovery of applied N). Earlier reports on long-term conservation agricultural practices in cropping systems involving rice in tropical climate indicates increased plant-available soil N after 5 to 10 yr of straw incorporation. Mobile humic fraction ($Na_4P_2O_7$/NaOH extractable) is influenced by agronomic practices of tillage intensity and residue retention/incorporation and is more dynamic than humus associated with metal-oxides and mineral complexes. A clearer picture of sequestration rate and stability of N in humic and fuvic fractions may indicate the extent to which changes in the crop residue management practices alter the size and availability of soil N in rice ecosystem. A better understanding in the N dynamics is critical to develop straw-management practices that lead to an optimization of the long-term N supply in flooded rice soils. Smaller and more reactive fractions will account for the greater plant-available soil-N supply after 3-4 seasons of straw retention/incorporation compared with burning. Despite positive effect of straw incorporation on N economy in rice based cropping systems, in rice-wheat cropping system, where alternate submergence and draining of the soil has been shown to increase N losses compared with losses incurred in continuously submerged soils under rice-rice system.

There is dire need to use effective N management practices to increase its efficiency and prevent its leakage to environment; continuous increase in the cost of energy has accelerated the cost of N-fertilizers. This has made it even more critical to increase use efficiency of N fertilizers in terms of more numinous returns to the farmers. Improper management may result in NO_3-N accumulation leading to its subsequent loss to ground water through leaching. Such a process intensified in CT fields with faulty irrigation especially when the soils are light-textured in the range of loamy sand to sandy loam as is usually the case with soils in the state of Punjab. Excessive irrigation to these well drained soils, results in excessive leaching of NO_3-N present in soil as a result of rapid mineralization in tilled plots in wheat – rice cropping system. Free electricity/power supply to the farmers is the root cause of the excessive and faulty irrigation methods in the state. Earlier, farmers used to make small seed beds as to facilitate rapid irrigation. Now, in the wake of free water and free power, they make very big beds/plots, and it takes a lot more time for water to reach the other end of the field. As the soils are light textured and well drained, a lot of water seeps through the soil causing severe leaching losses of NO_3-N, and other soluble nutrients. Drinking water from ground water sources contain excessive level of NO_3-N beyond the permissible limit is playing havoc with lives of people in many of these areas. This is the root cause of low N-use efficiency in rice-wheat cropping system.

Lower NH_3 volatilization losses

With Zero Till Seed-cum-Fertilizer Drill, N–fertilizer along with other fertilizer nutrients is applied by the side of the seed. The usual form of N is urea, which can release NH_3 gas under slightly alkaline soil conditions. However, N losses as NH_3 gas is not expected from urea hydrolysis in CA practices, as any ammonia gas emanating from the subsurface placed urea is readily trapped in the upper moist soil layers. Several studies have shown decreased crop yield and apparent decrease in N use efficiency with surface broadcast urea, especially when the soils are alkaline in reaction. Urea applications below the soil surface in no-till system has been proposed to minimize losses through NH_3 gas volatilization.

Conservation agricultural practices, NT or RT, crop residue retention, intensive cropping and N fertilization frequently results in increased soil total N along with SOC. Keeney (1982) hypothesized that application efficiency of fertilizer N can be enhanced by synchronizing fertilizer application with plant uptake needs. Reeves *et al.* (1993) also reported that synchronization of residue N mineralization, fertilizer-N application time, and subsequent crop demand for N can improve N use efficiency of crops planted in conservation tillage system. Dalal *et al.* (2011) observed that significant effects of treatments on soil total N were confined to top 0- to 10 cm depth. They reported that although fertilizer N recoveries were only 46 to 59% during the period from 1969 to 2008, still total N exponentially declined in all the treatments of tillage and residue management. Mineral N in the soil profile (0- to 1.2 m) ranged from 68-406 kg N ha^{-1}. Crop residue retention and low rate of N application (30 kg N ha^{-1}) had greater N use efficiency (NUE) (35-44%) than residue burned and high rates of N application. Apparent fertilizer N recoveries in the soil plant system , as observed by Dalal *et al.* (1995) in cereal cropping systems over a long-term period of 22 yr, were between 34 and 64% under different tillage and residue management practices, and NUE in grain N uptake ranged from 6 to 17 %. Lower values as compared to global average of 34% have been attributed to significant NO_3-N (20-38%) in the soil profile (table 16) subject to deep leaching from the soil. NUE in grain uptake ranged from 16% in CT treatment with residue burned and N-application at the rate of 90 kg N ha^{-1}, similar were the results for NT with residue burned and N applied at the rate of 90 kg N ha^{-1}. But in NT, when residue was retained, this value increased to 49%. This higher NUE was, however, at the reduced rate of N application of 30 kg N ha^{-1}. So it seems increased value of NUE more due to lower level of applied N rather than the tillage treatment or residue management. Strong *et al.* (1996) also reported that NUE in wheat was similar in NT and HT over 4 yr periods using ^{15}N-labeled fertilizer on a vertisol. Vertisols have inherent shrinking/swelling character, vertical pedoturbation and cracking characteristics, which perhaps disarranged the often observed nutrient and SOC stratification with NT. Moreover, substantial amount of NO_3 may have lost through deep leaching. Dalal *et al.* (2011) also anticipated N loss through N_2O through denitrification as also reported by Wang *et al.* (2011) for this site with vertisol.

Table 16. Amounts of NO_3-N (kg N ha^{-1}) in soil as influenced by tillage and residue management at 90 kg N ha^{-1} application in wheat crop

Tillage treatment	Residue management	
	Residue retained	Residue burned
NT	128	244
CT	144	150

Source: Dalal *et al.*, 2011

An adequate strategy for enhanced NUE is to ensure N availability to crop plants when crop needs it or when water is available to enhance N uptake. This means timing of fertilizer N application is important. Application of fertilizer at growing stage increases wheat grain yield while late application increase grain protein concentration. Both these timings of application improve total N uptake by crop plants. Thus, N uptake during post-anthesis period can contribute to total grain N content. Melaj *et al.* (2003) observed that late fertilization at the time of tillering increased nitrogen derived from fertilizer (Ndff) recovery in whole plant and in grain, and this effect was more pronounced in NT than in CT (Table16).

Favourable external factors such as soil water and on-soil and air temperature also influence N uptake during grain filling stage. Both these factors are significantly affected by residue retention and tillage option in the conservation system. It is generally accepted that soil under NT improves water storage due to enhanced infiltration rate and lower surface evaporation losses due to the mulching effect of retained residue in conservation agricultural practices. Conservation agriculture in which crop residue is retained on the soil surface has a moderating effect on soil temperature. Andraski and Bundy (2008) observed a temperature decrease in soil surface from 1.7 to 4.0^{O} C with retention of residue on the soil surface. Greater the thickness of retained crop residue, larger is the decrease in temperature. Halvorson *et al.* (2010) reported that soil temperature tended to be 1 to 2°C lower in NT soils than CT soils in high temperature months of March-April, and with this trend reversing during the winter months of mid December to mid February.

Eagle *et al.* (2001 a, b) observed that FUE-15N was greater when straw was burned rather than incorporated in rice over the growing season and at final harvest. Incorporating straw compared with burning increased the soil N availability through increase in net N mineralization and corresponding dilution of fertilizer ^{15}N. Four seasons of residue incorporation increased C and N contents of the active, light and mobile humic fractions (Bird *et al.* 2002). Clearly, the incorporation of straw for a prolonged period of time changed the overall N dynamics and cycling of the soil and caused a net increase in the N supply power of the soil which is reflected in higher yields ant total N uptake of the unfertilized rice crop. Cassman *et al.* (1993) had earlier emphasized that the adjustment of fertilizer N application to better reflect soil N supply should be considered to increase FNUE in rice systems. Recovery of N in plant derived from fertilizer (Ndff) at maturity ranged from 21.9 to 70.4 kg ha^{-1} in the whole wheat plant (Melaj *et al.* 2003). These values implied a percentage recovery in the crop of 18.3 to 58.7% of the 120 kg ha^{-1} applied (Table 18). The lower value corresponded to N applied at sowing (120S) under NT, probably explained by high microbial N immobilization. Labeled N fertilizer recovery as reported by other scientists (Wuest and Cassman 1992a; Zepata and Hera 1995; Lopez *et al.* 2002) ranged from 18 to 68%. Approximated two third of the fertilizer N recovered in the whole crop is in grain (Table 17).

Table 17. Nitrogen derived from fertilizer (Ndff) at physiological maturity under no-till (NT) and CT, as affected by the time of 120 kg N ha^{-1} application. Data is average of two yr (1998 and 1999)

Tillage treatment	Ndff (kg N ha^{-1})			
	Whole plant		Grain	
	120S	120T	120S	120T
NT	37.76	58.46	27.11	43.32
CT	45.02	61.46	29.53	43.44

120 kg N/ha applied at sowing (120S) or at tillering (120T)
Source: Melaj *et al.*, 2003

From the data in the Table 17 it is clear that under NT, recovery of Ndff added at tillering was greater than that added at sowing. Under CT, less or equal Ndff was recovered by crop from tillering application compared with sowing application. Greater N immobilization is generally observed under NT during early crop stages. This explains the benefit in N recovery of fertilizer N when application was delayed under this tillage system. Effectively, the recovery of fertilizer N in the whole plant and in the grain was greater when N was applied at the grand growth stage of tillering.

Often, there are wide variations in the observed values of recovery of fertilizer N whether based on the ^{15}N isotope dilution or the N difference method. Such large differences in the recovery of fertilizer – N indicates the strong presence of an effect of added N interaction. Apparent effect of added N interaction is caused by mineralized unlabeled N which replaces fertilizer ^{15}N ions in solution. This process is biological and microbial driven, with concomitant N immobilization of added fertilizer N and mineralization of native soil N. These results suggest that microbial stabilization of fertilizer N leads to the enhanced apparent added N immobilization with incorporation of residue into the soil rather than its burning. Due to increased soil N availability as a result of continuous incorporation of straw over long-term period of time, FNUE declines when straw was incorporated compared with burning. Over a period of time, the incorporation practice in rice systems can allow to reduce fertilizer N rates. A reduction in fertilizer N input without reduction in yield would improve FNUE.

Bottlenecks in large scale adoption of conservation agricultural practices

Residue management and conservation tillage systems, although help containing soil erosion, sequester organic C, conserve more water in soil, moderate soil temperature but many farmers are still wary of adopting such systems. Our farmers always expect an increase in the yield with every new technology that they are offered to adopt. In the adoption of CA practice, initially such an increase in the yield may not be there, rather sometimes a decrease can take place because of such factors as transitional decrease in soil available N due to the process of immobilization particularly during the initial years of adoption of this technology. They ignore the long-term benefits of improved soil productivity and sustainability of their natural resources of soil, water and environment. Besides these basic difficulties, there are some inherent problems with no-till seeder-cum-fertilizer drill which stand in its large scale adoption. Large quantities of residue on the soil surface have traditionally been viewed as a nuisance, and have been reported to be associated with mechanical planting difficulties, poor seed germination and poor crop stand establishments, decreased effectiveness of the herbicides, release of growth-inhibiting allelopathic compounds, harbouring of insect-pests, and ultimately yield reductions. Greater quantities of surface residue in the stubble mulch treatment contributes to reduced crop germination and stand establishment because of poor seed-soil contact and less uniform seedbed conditions compared with usual plow tillage. Under semi-arid conditions, insufficient seed zone moisture can be a major limitation in the establishments of the crop, small decrease in the seed zone moisture can decrease the yield substantially. Inadequate crop establishment is because of non-perfect planting equipment. There has been a constant improvement in the equipment which has increased the probability of getting comparable yields through proper crop establishment. However, adoption of this technology is quite slow in this region and rest of the country, because no increase or sometimes decrease in the yields in NT as compared to CT, which is often attributed to poor germination and crop stand. The delay in seedling emergence in NT with large quantity of loosely lying residue left on the soil surface may cause excessive wet soil conditions along with cool surface conditions because of decrease in temperature due to mulching effect of surface-retained crop residue in the NT soils. The NT seeders need to be improved upon so as to be able to better cut through the surface residue and loosen the soil ahead of opener. Strip tillage has been shown to overcome some of these difficulties while still retaining all the benefits of NT. Al-Kaisi and Licht (2004) proposed strip tillage system which results in minimum disturbed narrow zone of 15-20 cm wide and 15-20 cm deep in the previous crop row, where as the inner-row area is left undisturbed.

Poor weed control in NT is one of the major reasons resulting in lower yields. It has complicated large scale adoption of the conservation agricultural practices; consequently, many farmers are reluctant to adopt this technology, despite possible long-term benefits of improved soil quality. Importance of weed control to the success of conservation tillage systems has been well documented. It follows that further increased efficiency of weed control and reduction in herbicide costs would go a long way toward making reduced tillage systems more practical. These and other specific problems with conservation tillage and residue management must be overcome before widespread adoption of such systems will occur in the country.

Conclusions

Another important aspect of the CA practices is that despite no additional soil surface coverage by the mulch, NT can reduce soil losses by as much as 68% and SOC losses by around 50% which is a significant contribution of no-till practice alone. The decrease in overall soil and SOC erosion with NT, where there is scarcity of soil surface cover, can be explained due to development of structural crusts under NT which have been shown to have greater resistance to detachment. This shows that under such conditions, structural crusts developed due to NT treatment are more effective in checking soil loss and SOC erosion than the surface cover, which facilitates infiltration and checks run-off velocity. Thus NT with scarce residue appears a credible alternative to CT. Residue scarcity should not be viewed as a major limiting factor for the implementation of NT.

NT practices of crop production can, however, also result in incremental N_2O emissions due to the denitrification process that can more than offset the effect of the soil CO_2 sink especially in fine textured soils which are prone to remain excessively wet during cropping seasons. Differences in N_2O emissions between NT and CT are greatest when differences in water-filled pore space between the two tillage treatments are high, suggesting that increased N_2O emissions in NT are the result of denitrification of NO_3-N. In most arable cropping systems, unused soil N is mostly present in NO_3 form which is subject to leaching and denitrifacation losses. In well-drained light textured soils, leaching losses of NO_3 are more pronounced while in fine texture soils which are prone to excessive wetness such losses are more in the form of gaseous N as N_2O. No-till practice may encourage NO_3 leaching losses by facilitating more infiltration in light textured soils while in fine textured soils, excessive wetness causes reduction conditions and results in more pronounced dnitrification losses of NO_3 to N_2O emissions. Due to these reasons, such losses are more from wheat-rice than from continuous rice-rice cropping system. Consequently, the potential of NT for decreasing net GNG emissions be limited in fine textured soils that are prone to reduced aeration under high water content situations.

Nitrous oxide emissions are significantly related to soil temperature, volumetric soil water content, precipitation, and air temperature. Collectively, these factors accounted for up to 31% of total variability associated with N_2O emissions. Thus N_2O emissions are expected to increase in future due increase in atmospheric temperature and total precipitation as predicted for climate change scenario. This situation as related to NT may become alarming because of the immensely high GHG effect of N_2O. The radiating effect of N_2O is 285 times that of CO_2.

Nitrous oxide production is immensely enhanced with increased use of common N-fertilizers like granular urea. However use of slow or controlled release N-fertilizers can significantly help in reducing N_2O emissions under NT crop production systems. Controlled-release and stabilized N-fertilizers have been shown to have the potential to reduce N_2O emissions from irrigated NT cropping systems when compared with commonly used granular urea. Nitrous oxide fluxes resulting from urea and super-U application peaked within days after application, whereas N_2O flux peaks from polythene coated urea granules (ESN) occurred 4 to 6 weeks after application and with flux peaks of much lower magnitude than with urea.

Soil profile NO_3-N levels are generally greater with CT than NT. Residual NO_3-N in the root zone tended to be low in the no-till and strip till than the chisel plow, which reduces the quantity of NO_3-N available for leaching in the NT systems. Most of the N in NT treatment is supposed to be present as part of SOM.

Due to lack of adequate aeration, mineralization process is slow in NT, rather immobilization of soil N occurs because of increased activity of microbes due to ample availability of crop residue retained on the soil surface under CA practices. This may temporarily cause low availability of N for crops under NT systems. Additional fertilizer-N has to be applied in the initial years of NT to overcome decreased yields and grain quality. Nitrogen removal in grain is a function of grain yield, which is affected by fertilizer-N rates, time of its application relevant to plant growth and tillage system. Fertilizer N application at tillering increases grain yield and N recovery when SOM and mineralization rate ensure adequate N availability for initial crop development. With retention/ incorporation of straw over years of CA, there is every likely hood of decreasing the fertilizer N rates to crop without decreasing the yields; a reduction in fertilizer N rates without reduction in yield would improve FNUE. Thus, residue retention and low rates of fertilizer N application at appropriate growth stage, e.g. tillering stage in wheat, increases N-use efficiency.

There is great potential to restore soil fertility and mitigate greenhouse gas emissions with the adoption of and improvements in conservation agricultural systems, e.g. continuous no-till, high crop residue retention without any fallowing. Thus the conservation agricultural practices have been recorded as one of the most effective agricultural strategies for sequestering atmospheric C in soil, conventional tillage, on the other hand, accelerates OC oxidation to $CO_{2.}$ Any introduction of C trading, in future, may well provide substantial financial incentives to farmers to adopt this green technology. As a result this practice over a period of time, land value as a provider of non-marketable ecosystem services will ultimately increase. Income generation may become progressively more and more assured and ultimately investment in agriculture may become more attractive. Land value thus upgraded, may favourably impact the rural economy and country's food security.

References

Al-Kaisi M, Licht MA (2004). Effect of strip tillage on corn nitrogen uptake and residual soil nitrate accumulation compared with no-tillage and chisel plow. Agron J 96: 1164-1171.

Andraski TW, Bundy LG (2008). Corn residue and nitrogen source effect on nitrogen availability in no-till corn. Agron J 100:1274-1279.

Angers DA, Eriksen-Hamel NS (2008). Full-inversion tillage and organic carbon distribution in soil profiles: A meta-analysis. Soil Sci Soc Am J 72:1370-1374.

Angle JS, Gross CM, Hill RL, McIntosh MS (1993). Soil nitrate concentration under corn as affected by tillage, manure and fertilizer application. J Environ Qual 22:141-147.

Bird JA, van Kessel C, Horwath WR (2002). Nitrogen dynamics in humic fractions under alternative straw management in temperate rice. Soil Sci Soc Am J 66:478-488.

Bouwmans AF, Bouwmans IJM, Batjes NH (2002). Modeling global annual N_2O and NO emissions from fertilized fields. Global Biogeochem Cycles 16: 6-1–6-13.

Bundy LG, Andraski TW, Oplinger ES (1995). Tillage and rotation effects on nitrogen availability. In:KellingKA (ed) Proc. 1995 WIs. Fert., Aglime and Pest Manage. Conf., Univ. of Wisconsin Extension, Madison.pp.87-99.

Cassman KG, Kropff MJ, Grant J, Peng S (1993). Nitrogen use efficiency of rice reconsidered: What are the key constraints? Plant Soil 156:359-362.

Chang KH, Bartlett P, Wagner-riddle C (2012). Using Day CENT to simulate carbon dynamics in conventional and no-till agriculture. Soil Sci Soc Am J 77: 941-950.

Choudhary MA, Akramkhanov A, Saggar S (2002). Nitrous oxide emissions from a New Zealand cropped soil: Tillage effects, spatial and seasonal variability. Agric Ecosyst Environ 93: 33-43.

Dalal RC, Wang W, Allen DE, Reeves S, Menzies NW (2011). Soil nitrogen and nitrogen-use efficiency under long-term no-till practice. Soil Sci Soc Am J 75:2251-2261.

Diaz-zorita M, Grove JH (2002). Duration of tillage management affects carbon and phosphorus stratification in phosphatic Paleudalfs. Soil Tillage Res 66:165-174.

Eagle AJ, Bird JA, Hill JE, Horwath WR, van Kessel C (2001a). Nitrogen dynamics and fertilizer N use efficiency in rice following straw incorporation and winter flooding. Agron J 93:1346-1354.

Eagle AJ, Bird JA, Horwath WR, Linquist BA, *et al.* (2001b). Rice yield and nitrogen utilization efficiency under alternative straw management practices. Agron J 92:1096-1103.

Erenstein O, Malik RK, Singh S (2007). Adoption and impacts of zero tillage in irrigated rice-wheat systems of Haryana, India. New Delhi, India: CIMMYT India and RWC.

Falotico JL, Studdert GA, Echeverria HE (1999). Nutricionnitrogenadadeltrigobajosiembradirecta y labranzaconvencional. (In Spanish, with English abstract) Cienc. Suelo 17:9-20.

FAO (2003). World agriculture: Towards 2015/2030. An FAO perspective. FAO, Rome.

FAO (2008). The state of food and agriculture 2008. Biofuels: prospects, risks and opportunities. FAO, Rome.

Franzluebbers AJ (2010). Achieving soil organic carbon sequestration with conservation agricultural systems in the Southeastern United States. Soil Sci Soc Am J 74:347-357.

Franzluebbers AJ, Stuedemann JA (2008). Early response of soil organic fractions to tillage and integrated crop-livestock production. Soil Sci Soc Am J 72:613-625.

Gal A, Vyn TJ, Micheli E, McFree WW (2007). Soil carbon and nitrogen accumulation with long-term no-till verses moldboard plowing overestimated with tilled-zone sampling depths. Soil Tillage Res 96:42-51.

Grant CA, Lafond GP (1994). The effects of tillage systems and crop rotations on soil chemical properties of a Black Chernozemic soil. Can J Soil Sci 74:301-306.

Halvorson AD, Del-Grosso SJ, Alluvione F (2010). Tillage and organic nitrogen source effects on nitrous oxide emissions from irrigated cropping systems. Soil Sci Soc Am J 74: 436-445.

Halvorson DH, Wienhold BJ, BlackAL (2001). Tillage and nitrogen fertilization influence grain and soil nitrogen in an annual cropping system. Agron J 93:836-841.

IPCC (2001) Houghton JT *et al.* (ed) Climate Change (2001). The scientific basis. Cambridge Univ. Press, UK.

Janzen HH, Campbell CA, Izaurrade RC *et al.* (1998). Management effects on soil C storage on Canadian prairies. Soil Tillage Res 47:181-195.

Kanwar RS, Baker JL, Laflen JM (1985). Nitrate movement through soil profile in relation to tillage system and fertilization application method. Trans ASAE 28:1802-1807.

Kumar S, Kadono A, Lal R, Dick W (2012). Long-term no-till impacts on organic carbon and properties of two contrasting soils and corn yield in Ohio. Soil Sci Soc Am J 78:1798-1809.

Lamb JA, Peterson GA, Fenster CR (1985). Wheat-fallow tillage system's effect on a newly cultivated grassland soils; nitrogen budget. Soil Sci Soc Am J 49:352-356.

Mchunu CN, Lorentz M, Jewitt G, Manson A, Chaplot V (2011). No-till impact on soil and organic carbon erosion under crop residue scarcity in Africa. Soil Sci Soc Am J 75:1503-1512.

McVay KA, Budde JA, Fabrizzi K, Mikha MM, *et al.* (2006). Management effects on soil physical properties in a long-term tillage studies in Kansas. Soil Sci Soc. Am. J. 70:434-438.

Melaj MA, Echeverria HE, Lopez SC *et al.* (2003). Timing of nitrogen fertilization in wheat under conventional and no-tillage system. Agron J 95:1525-1531.

Montgomery DR (2007). Soil erosion and agricultural sustainability. Proc Natl AcadSci USA 104: 13268-13272.

Nelson MA, Griffith SM, Steiner JJ (2006). Tillage effects on nitrogen dynamics and grass seed crop production in western Oregon, USA. Soil Sci Soc Am J 70:825-831.

Omonode RA, Smith DR, Gal A, Vyn TJ (2011). Soil nitrous oxide emissions in corn following three decades of tillage and rotation treatments. Soil Sci Soc Am J 75:152-163.

Omonode RA, Vyn TJ, Smith DR, Hegymegi P, Gal A (2007). Soil carbon dioxide and methane fluxes from long-term tillage systems in continuous corn and corn-soybean rotations. Soil Tillage Res 95:182-195.

Pasricha NS (2010). Nutrient management under water-logged soil conditions. In: Gurbachan Singh *et al.* (eds) Strategic Nutrient Management in Water-Logged Soils, CSSRI, Karnal, India, pp. 132-189.

Pasricha NS (2013). Success story of resource conservation technology in Indo-gangetic plains. In: Ghosh SK *et al.* (eds) Resource Conservation Technology in Pulses, Scientific Publishers (India), pp. 32-41.

Pasricha NS, Ponnamperuma FN (1978). Chemistry of submerged saline, alkali soils. I. Influence of salinity on the chemical and electro-chemical kinetics and growth of rice.IL RISO.

Poirrier V, Angers DA, Rochette P *et al.* (2009). Interactive effects of tillage and mineral fertilization on soil carbon profile. Soil Sci Soc Am J 73:255-261.

Randall GW, Bandel VA (1991). Overview of nitrogen management for conservation tillage systems: An overview. In: Logan TJ *et al.* (eds) Effect of Conservation Tillage on Groundwater Quality, Nitrogen and Pesticides. Lews Publ. Chelsea, MI., U.S.A. pp. 39-63.

Reeves DW, Wood CW, Touchton JT (1993). Timing nitrogen application for corn in a winter legume conservation-tillage system. Agron J 85:98-106.

Ritter WF, Scarborough RW, Chirnside AEM (1993). Nitrate leaching under irrigated corn. J Irrig Drain Eng 119:544-553.

Rochester IJ (2003). Estimating nitrous oxide emissions from flood-irrigated alkaline grey clays. Aust J Soil Res41:197-206.

Rochette P, Angers DA, Chantigny MH, Bertrand N (2008). Nitrous oxide emissions respond differently to no-till in a loam and heavy clay soil. Soil Sci Soc Am J 72:1363-1369.

Roth CH, Meyer B, Frede HG, Derpsch R (1988). Effect of mulch rates and tillage systems on infiltrability and other soil physical-properties of an oxisol in Parana, Brazil. Soil Tillage Res 11:81-91.

Sarandon SJ, Golik S, Chidichimo HO (1997). Accumulation y particiondelnitrogenoen dos cultivares de trigo pan ante la fertilizcionnitrogenadaensiembradirecta y labranzaconvencional. Rev FacAgron, Univ Nac La Plata 102:175-186.

Schuller P, Walling DE, Sepulveda A, Castillo A, Pino I (2007). Changes in soil erosion associated with the shift from conventional tillage to a no-tillage system, documented using (Cs)-137 measurements. Soil tillage Res 94:183-192.

Sharifi M, Zebarth BJ, Burton DL, Drury CF, Grant CA (2013). Mineralization of carbon-14-labeled plant residues in conventional tillage and no-till systems. Soil Sci Soc Am J 77:123-132.

Smith PD, Martino Z, Gwary CD, Janzen H, Kumar P *et al.* (2007b) Agriculture. In: Mertz B *et al.* (ed) Climate change 2007; Mitigation. Cambridge Univ. Press, Cambridge, U.K.pp. 497-540.

Soon YK, Clayton GW, Rice WA (2001). Tillage and previous crop effect on dynamics of nitrogen in a wheat-sorghum system. Agron J 93: 842-849.

Spedding TA, Hamel C, Mchuys GR, Madramootoo CA (2004). Soil microbial dynamics in maize-growing soil under different tillage and residue management systems. Soil Biol Biochem 36: 499-512.

Strong WM, Dalal RC, Weston EJ, Cooper JE, Lehane KJ, King AJ (1996). Nitrogen fertilizer residues for wheat cropping in subtropical Australia. Aust J Agric Res 47:695-703.

Torbert HA, Potter KN, Morrison JE Jr (2001a). Tillage intensity and fertility level effects on nitrogen and carbon cycling in a vertisols. Soil Sci Plant Anal 28: 699-710.

Torbert HA, Potter KN, Morrison JE Jr (2001b). Tillage system, fertilizer nitrogen rate, and timing effect on corn yields in the Texas Blackland Prairie. Agron J 93:1119-1124.

Tyler DD, Thomas GW(1977). Lysimeter mrasurements of nitrate and chloride losses from soil under conventional and no-tillage corn. J Environ Qual67:63-66.

Valentin C, Agus F, Alamban R *et al.* (2008). Runoff and sediment losses from 27 upland catchments in Southern Asia; Impact of rapid land use changes and conservation practices. Agric Ecosyst Environ 128:225-238.

Venterea RT, Burger M, Spokas KA (2005). Nitrogen oxide and methane emissions under varying tillage and fertilizer management. J Environ Qual 34:1467-1477.

Wang WJ, Dalal RC, Reeves SH, Balderbach-Bahl, Kiese R (2011). Greenhouse gas fluxes from the Australian subtropical crop land under long-term contrasting regimes. Global Change Biol.

Weed DAJ, Kanwar RS (1996). Nitrate and water present in and flowing from root-zone soil. J Environ Qual 25:709-719.

White PM, Rice CW (2009). Tillage effects on microbial and carbon dynamics during plant residue decomposition. Soil Sci Soc Am J 73:138-145.

Wuest SB, Cassman KG (1992a). Fertilizer nitrogen use efficiency of irrigated wheat: I. Uptake efficiency of pre-plant versus late-season application. Agron J 84: 682-688.

Wuest SB, Cassman KG (1992b). Fertilizer nitrogen use efficiency of irrigated wheat: II. Partitioning efficiency of pre-plant versus late-season application. Agron J 84:689-694.

Yang XM, Drury CF, Reynolds WD, Tan CS (2008). Impact of long-term and recently imposed tillage practices on the vertical distribution of soil organic carbon. Soil Tillage Res 100:120-124.

Zhang GS, Chan KY, Oates A, Heenan DP, Huang GB (2007). Relationship between soil structure and runoff/soil loss after 24 years of conservation tillage. Soil Tillage Res 92: 122-128.

Zheng FI, Merrill SD, Huang CH (2004). Runoff, soil erosion, and erodibility of conservation reserve program land under crop and hay production. Soil Sci Soc Am J 68:1332-1341.

Zhu Y, Fox RH, Toth JD (2003). Tillage effects on nitrate leaching measured by pan and wick lysimeter. Soil Sci Soc Am J 67:1517-1523.

13

Agroforestry: A Sustainable, Multifunctional and Diversified Production System for Hot Arid Zone of India

Archana Verma[1]*, Shiran K.*[1]*, J.C. Tewari*[1]*, Rajwant Kaur Kalia*[1] *Saresh N.V. and Shrawan Kumar*[1]

[1]*Central Arid Zone Research Institute, Jodhpur – 342 003, Rajasthan, India*

Introduction

Arid environs are delineated by extreme and harsh weather conditions due to excessive heat and inadequate, variable precipitation; however, contrasts in climate occur. These climatic disparities are the result of temperature variation, distribution and season of rainfall and the degree of aridity. Moreover, high evapo-transpiration, periodic droughts, low organic matter levels, different associations of vegetative cover and soil enhances the problems of these regions. In general, hot arid climates have excessive heat and strong prevailing winds, unhampered by obstacles on the ground and as result aeolian erosion is common with frequent seasonal drought occurrence (Sharma and Tewari 2005). These climatic limitations make very difficult place for the inhabitants to attain their livelihood. People of these areas have spent their life within these constraints for centuries. They have faced serious challenges due to lack of sufficient resources and very low productivity which has added to the vulnerability of the region.

People living in the arid regions practice the mixed farming combining rain-fed cropping, livestock rearing and other income generating activities to sustain their livelihood. However, due climatic uncertainties like long dry spells hinders the growth of crops and have negative effect on their yield. Climate change will further challenge the livelihoods of those living in these sensitive ecosystems and may result in higher levels of resource scarcity. Therefore, to sustain the livelihood of people agroforestry in arid ecosystems has played a vital role since ages.

Agroforestry is not a new practice in arid India; farmers are practicing cultivation of crops with trees and shrubs on their fields from years. Agroforestry systems combining tree/shrub, crop, grass and livestock have great scope and role in optimizing land productivity and environmental protection in the fragile ecosystems. Agroforestry that is traditionally followed in these regions is a complete, ecologically sustainable livelihood system (Roy and Tewari 2014). In modelling future designs for agroforestry the demands of the provincial people needs to be addressed. The components like trees, crops and livestock are to be integrated in such a way that their interaction complements each other.

Distribution and classification of world arid regions

There are many criteria to define a desert but perhaps the most important one is aridity- the lack of water as the main factor limiting biological processes. One of the most common approaches to measure aridity is through an estimator called the Aridity Index, which is simply the ratio between mean annual precipitation (*P*) and mean annual potential evapotranspiration (PET, the amount of water that would be lost from water-saturated soil by plant transpiration and direct evaporation from the ground; Thornthwaite 1948). Arid and hyper-arid regions have a P/PET ratio of less than 0.10; that is, rainfall supplies less than 10 per cent of the amount of water needed to support optimum plant growth (UNEP 1997, FAO 2004).

Table 1. Dryland classification of the world

Classification	Aridity index (P/PET)	Area ($km^2 \times 10^6$)	Area (%) of world total
Hyper-arid	<0.05	10.0	7.5
Arid	0.05-0.20	16.2	12.1

Source: UNEP 1997

The hot arid zone of India

Extent of distribution

The arid zone of India covers about 11.8% of the geographical area including 31,900 million km^2 of hot desert located in parts of Rajasthan (61%), Gujarat (20%), Punjab and Haryana (9%), and Andhra Pradesh and Karnataka (10%). The hot arid region of India lies between 24° and 29°N latitude, and 70° and 76 °E latitude (Roy *et al.* 2011). The Great Indian Desert or Thar Desert lies in western Rajasthan (19,084 million ha) which is the prime arid zone of India. Based on the climatic, edaphic and terrain characteristics, the region has been further subdivided into three sub-zones: (a) Arid western plain (zone I, 12,416 million ha), (b) Transitional plain of inland drainage (zone II, 3,699million ha), and (c) Transitional plain of Luni basin (Zone III, 2,969 million ha).

Climate/Atmospheric conditions

The climate in the arid regions has erratic rainfall, frequent droughts, high evaporation, intense heat and high velocity winds. The temperature is often 45°C to 50°C in peak of summers with cool and dry winters with temperature between 10°C to 14°C. The low erratic rainfall (100-300 mm~90% during July-September), high wind velocity(>30 km h^{-1} during sandstorms in summer) and high evapotranspiration adds to the agony of harsh weather of this region. Drought is and will remain a major determinant of agriculture practices in the region. These climatic factors become unfavorable for the growth and development of vegetation in these areas.

Landform and soils

Soils are dominantly sandy, with sand dunes of average height 10 m, interspersed with interdune plains of different sizes covering more than 60% area. The desert soils consist of aeolian sand (90 to 95%) and clay (5 to 10 %). Soils are generally very coarse in texture, low in water holding capacity and of low nutrient status (Mann and Muthana 1984). These are generally low in organic carbon (0.04-0.12 %), low to medium in available phosphorus (0.05-0.10 %) and nitrogen is originally low (0.07-0.20 %) but its deficiency is made up to some extent by the availability of nitrogen in the form of nitrates. Thus, the presence of phosphates and nitrates make them fertile soils wherever moisture is available.

Vegetation and its ecology

Vegetation of arid regions is very sparse with scattered thorny trees, shrubs and grasses. The Thar Desert of India exhibits 682 plant species (Bhandari 1978; Khan 1997). 588 species of dicots and 186 species of monocots and one gymnosperm have been recorded (Shetty and Singh, 1993). Most dominant family is Poaceae followed by Cyperaceae with 125 and 35 species, respectively. Three types of plant forms, ephemeral annuals, succulent perennials and non-succulent perennials are found in dryland environments (Folliott *et al.* 1995). The vegetative pattern changes with rainfall gradient from grassland with isolated trees in north western parts of both the provinces to open scrubland in the middle, and open-to-dense in pockets towards the region's south east parts of

Rajasthan. The vegetation falls under the category 'thorn forest type' or 'scrub forest type' (Champion and Seth 1968).

Land use

Agriculture is the main source of livelihood other than the livestock rearing for the people of arid regions. The crops that are commonly grown in the Thar Desert are pearl millet, cluster bean, green gram, moth bean and sesame. However, where irrigation sources are available like (khadins, tube-wells etc.) rabi crops like wheat, mustard, cumin and isabgol are also cultivated by farmers. The net sown area has been intensified up to 11.7 per cent up to 1990's as compared to the early 1980's. The forest area of the Rajasthan is only 8.03 per cent recording only 0.25 per cent increase from 2003-2004. This may be the result of planting of trees by professional foresters as well as local people for sustainable land use management under agroforestry which is the traditional practice of the region. Animal husbandry is also major economic activity of the rural peoples, especially in the arid and semi-arid regions of the Rajasthan. Development of livestock sector has a significant beneficial impact in generating employment and reducing poverty in rural areas. But, there was decline in the pasture and grazing lands because of the cultivation of marginal lands. The area under permanent pastures and other grazing lands has decreased up to 0.05 per cent since 2003-04.

Table 2. Different types of land use in Rajasthan (2012-13)

Land use	Area (thousand hectares)	Percentage
Total geographical area	34224	100
Fallow land	3894	11.37
Net area sown	17479	51.07
Total cropped area	23954	69.99
Forest	2750	8.03
Permanent Pasture& other grazing lands	1694	4.94
Culturable wasteland	4152	12.13

Source: Directorate of Economics and Statistics, DARE MOA and FW, India.

Traditional agroforestry systems: structure and attributes

Agroforestry is traditionally practiced by people of arid areas as failure of crops due to drought is unavoidable; they have to depend on the alternate source of livelihood like trees and livestock. Although farmers have not any planned geometry and pattern of planting but still they manage trees on their fields and grow crops along with them. In fact, such integration of arable crops with trees in the farming systems is a unique, combined, protective-productive system that works on the principles of ecology, productivity, economics, and sustainability.

Because most trees are drought resistant, they are still able to provide fuel, fodder, fruit and other products, when the crops fail, as frequent droughts are a common phenomenon (Sharma 2003). On the basis of rainfall four type of traditional agroforestry has systems have been observed in arid regions (Table 3).

Table 3. Traditional agroforestry system in arid region of western Rajasthan

Rainfall (mm)	Agroforestry system	Tree/shrubs (No ha^{-1})	% density of prominent species
>400	*P.cineraria - A. nilotica*	31.4	80.5
300-400	*P.cineraria* based	14.2	80.0
200-300	*Zizyphus spp. - P.cineraria* based	91.7	100.0 (91.7 % *Zizyphus* spp.)
<200	*Zizyphus spp. - P.cineraria - Salvadora spp.* based	17.2	87.2 (65 % *Zizyphus spp.)*

Source: Roy *et al.* 2011

Agri-silviculture system

Agri-silviculture systems are those in which both crops and trees/shrubs are grown on same unit of land and products are obtained from both components. This is the most popular system in the areas receiving rainfall between 200-400 mm/yr. People raise the trees like *Prosopis cineraria*, *Zizyphus nummular*, *Tecomella undulata*, *Salvadora oleoides* etc., in the crop fields as well as bunds. Farmers grow crops like pearl millet, moth bean, mung bean, cluster bean etc. along with these tree species. Interestingly, many researchers has reported that yield of pearl millet which is the main food grain of the region has not affected by the canopy of woody components, although in some cases the yield has increased in cropping along with trees. In *Prosopis cineraria* based agroforestry system the yield and yield attributes of crops like pearl millet, *Brassica* spp., have been reported to be substantially increased (Kaushik and Kumar 2003; Bishnoi and Singh 2009; Roy *et al.* 2011).

Khadin cropping system

Khadin is traditional method of harvesting water for practising agriculture developed hundred years ago in Jaisalmer. This system is similar to some irrigation methods used in the Middle East and in Negev desert. The earthen embankment (100-300 m) is raised opposite to the hilly slopes to collect the runoff water in the lower valley areas. Spillways are made to drain out the runoff water. Rainfall of about 100 mm is enough to take one crop for farmers in the season. After the first rains the pearl millet crop is sown in Khadins and in case of failure of rains further the crop is grown on conserved moisture and if grain is not formed at least fodder can be harvested which is a way to safeguard in case of drought. If good rains occur water level reaches above ground; and when this water receds in early November, leaving the surface soil just moist and fit for sowing of rabi crops, like wheat, chickpea (gram), mustard etc. A study conducted at Central Arid Zone Research Institute, Jodhpur (India) has shown that even without the use of chemical fertilizers, the average crop yield in khadin ranges from 2.5-3.0 t ha^{-1} of wheat and 1.5-2.5 t ha^{-1} of chickpeas (Kolarkar 1990). Even during severe drought years, khadins may be used for getting a successful crop on stored soil profile moisture.

Silvipastoral system

Livestock rearing is the next important occupation of the people of arid areas after agriculture. Silvipasture in common is the practice of raising grasses along with fodder trees; but in Rajasthan these systems are developed traditionally by people. In very low rainfall areas growing crops is not feasible so people raise livestock to earn their livelihood. The piece of land in the village called as 'orans' or 'gochar' is left on the name of god and no cutting is allowed in that place. In these areas only animals are allowed to graze the grasses and browse the leaves of trees. Farmers in arid region in India encourage growing *Prosopis cineraria* (khejri), ber and babool trees in the crop fields. There is a common belief that crop productivity increases in association with khejri trees and in addition it provides a good feed for small and large animal production system (Patil and Pathak, 2013). Thus, silvipasture is an ideal system of arid and semi arid regions of Rajasthan as it assures the availability of forage for areas with big population of livestock and is suited to drought-tolerant regions; and also restores the natural rangelands with diversification of species and improving productivity.

Innovative agroforestry systems developed

In most agroforestry systems, the trees grown do not have the usual silivicultural recommendations in terms of spacing (Owonubi 2002). Given the reality of awareness among the farmers of multiple land use management, the need to improve on the existing agroforestry practices becomes necessary in the face of increasing population and limited nature of land (Alao and Shuaibu 2013). In the light of increasing pressure on land resources, CAZRI, Jodhpur initiated systematic studies on agroforestry systems in late 1970s. Since then a number of improved agroforestry practices in order to enhance overall productivity and economic returns of the farming communities have been developed and standardized. Following improved practices have been found promising and remunerative, and easy to fit in existing traditional agroforestry systems (Tewari *et al.* 2014).

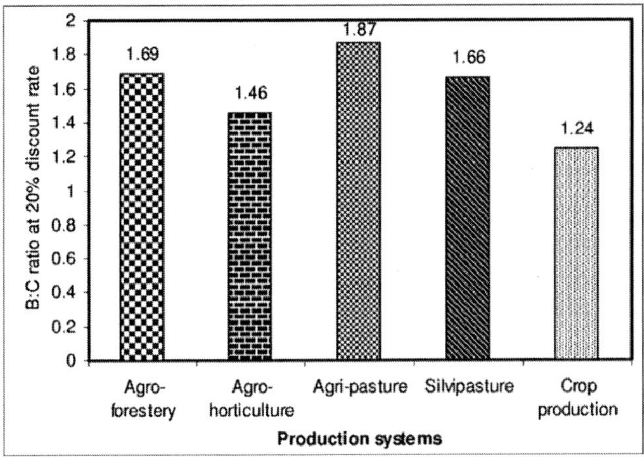

Fig. 1. Benefit: Cost ratio of different land use systems in arid region. (Bhati *et al.*, 2009)

Agrihorticulture

In traditional agri-horti system farmers raise crops like pearl millet, moth bean, mung bean, cluster bean along with *Zizyphus* spp. (Ber), *Cordia myxa* (gonda), *Capparis decidua* (Ker) trees (Table 4). The yield of Zizyphus based cropping system has been reported higher than the monoculture of either trees or crops by many workers (Table). The B: C ratio of Ber+ Mung was 2.02 and Ber + cluster bean was 2.15 which was comparatively higher than the sole ber (2.1), sole mung bean (1.42) and sole cluster bean (1.93) (Meghwal and Henry 2006). Intercropping in ber orchard produced higher grain yield of intercrops by 5-20% over their sole cropping and intercropping is promising particularly during juvenile period of fruit plantation (Gupta *et al.* 2000; Saroj *et al.* 2003; Singh *et al.* 2003; Bhandari *et al.* 2014). Similarly, in aonla-based cropping system, it has been demonstrated that model consisting of aonla + ber along with moth bean or fenugreek can be adopted as a sustainable model for nutritional and income security of the inhabitants (Awasthi *et al.* 2007). The possibilities of utilizing the inter-row spaces of fruit orchards of aonla (*Embelica officinalis*), ber (*Ziziphus mauritiana* Lam.) and pomegranate (*Punica granatum* L.) for growing legumes such as cluster bean, cowpea, and moth bean to serve as an additional source of income was evaluated in Bhuj (Gujarat). Among the 3 legumes evaluated, cluster bean performed well with orchards and highest benefit: cost ratio of 1.83 was obtained with ber + cluster bean system followed by ber + moth bean (1.65) (Dayal *et al.* 2015). In addition to ber, aonla, gonda fruit trees like guava, koronda and pomegranate are emerging as promising components of agrihorticulture system for enhancing the productivity of arid lands.

Table 4. Recommend agri-horti crops component for Indian arid regions

Growing conditions	Horticultural component			Crop component
	Highstorey	Medium storey	Ground storey	
Rainfed (150–300mm)	Bordi and Indian mesquite	Jharber	Cucurbits and guar	Guar, moth bean, pearl millet and sesame
Rainfed (300–500mm)	Indian cherry, Indian jujube and Indian mesquite	Jharber	Cowpea, cucurbits, guar and Indian bean	Cowpea, guar, green gram, moth bean, pearl millet and sesame
Irrigated	Bengal quince,	Guava, kinnow,	Brinjal, chilli, cole	Chickpea, green
	Indian gooseberry	karonda, lime,	crops, cucurbits gar-	gram, groundnut
	Indian jujube and	Pomegranate and	lic, okra, onion peas	Mustard and seed
	Indian mesquite	sweetorange	root/leafy vegetables and tomato	spices

Source: Bhandari *et al.* 2014

Table 5. Performance of Zizyphus in intercropping with different crops in arid regions of India

Tree	Crop	Yield		Region Jodhpur	Reference
		Sole	Intercropping		
Zizyphus mauritiana	Green gram		0.8 t ha^{-1}(fruit)		Gupta 1997
Zizyphus mauritiana	Cluster bean, mung bean, sesame	5.2 kg tree^{-1} (fruit)	148 kg tree^{-1} (fruit)	Pali, Rajasthan	Singh 1997
Zizyphus mauritiana	Indian aloe	13.55 q ha^{-1}	8.09 q ha^{-1}	Bikaner (Rajasthan)	Saroj*et al.* 2003
Zizyphus mauritiana	Pearl millet, green gram, cluster bean, sorghum	85.4 q ha^{-1}	91.4 q ha^{-1}	Dantiwada, (Gujarat)	Patel *et al.* 2003
Zizyphus mauritiana	Cluster bean, moth bean, mustard and brinjal	56.32 q ha^{-1} (fruit)	84.60 to 86.52 q ha^{-1} (fruit)	Bikaner (Rajasthan)	Arya *et al.* 2011
Zizyphus mauritiana	Cluster bean, moth bean, cow pea	41.25 kg tree^{-1}	42.35 kg tree^{-1}	Bhuj (Gujarat)	Dayal 2009

Hortipasture

In parallel to agrihorticulture system intercropping of fodder grasses along with horticulture trees intensifies the productiveness of the system. The introduction of *Cenchrus ciliaris* with Ber (6m×6m) was very profitable where grass production was the dry grass production was 1.55 t ha^{-1} $year^{-1}$ and the fruit, leaf fodder and fuel wood production from ber was 2.77, 1.87, 2.64 t ha^{-1} $year^{-1}$, respectively (Tewari *et al.* 1999). The silvi-pastoral system of *Ziziphus rotund folia* and *Cenchrus ciliaris* could sustain 554 tharpakar cattle days/ha with 60% pasture utilization (Tanwar *et al.* 2014). The growth and fruit production of aonla was not affected when raised with pearl millet + cowpea and produced fruit yield of 13.65 t ha^{-1}. Pearl millet (multicut) + cowpea in association with tree produced 14.4 and 30.4 t green fodder. There were higher net profit returns in intercropping as compared to sole crops and sole trees (Kumar *et al.* 2015). The yield of leaf fodder (3.93 kg $plant^{-1}$), fruit (36.84 kg $plant^{-1}$) and fuel wood (23.31 kg $plant^{-1}$) in ber (*Ziziphus mauritiana*) were higher in association *of Cenchrus setigerus* than *Cenchrus ciliaris*. The maximum gross return of (Rs.104 429 ha^{-1}), net returns of (Rs. 72 029 ha^{-1}) and benefit: cost ratio (2.21) was recorded in combination of *Cenchrus setigerus* and ber plants in hortipasture system (Meena 2015).

Agrisilviculture system

Tree based land used systems is the mainstay of arid regions because of harsh and inconsistent climate conditions. In the severe drought conditions when crop fails the trees act as source of survival for man as well as livestock. In a study cluster bean, mung bean, pearl millet was raised with *Prosopis cineraria* where the growth of trees was not affected by intercrops and *P. cineraria* did not adversely affect the yields of intercrops during the first two years of tree growth if tree density was maintained at 833 stems ha^{-1} (Gupta *et al.* 1998). In another study it was found that when crops are raised with trees the total productivity per unit land as well as the biological activities raises when compared to sole arable farming (Tanwar *et al.* 2014). The grain yield of pearl millet, green gram and cluster bean was higher in association with *Prosopis cineraria* as compare to arable farming (Table 6) (CAZRI 2014). The wider spacing (10 m × 10 m) of *Ailanthus excelsa* trees was not only favourable for the growth of associated crops like moth bean and pearl millet but the overall growth of trees was also enhanced. In a recent study the crops cluster bean and moth bean were cultivated with *P. cineraria*, *Ailanthus excelsa* and *Tecomella undulata*. Results indicated that tree species had no adverse effect on growth and grain yields of crops and vice versa and also the status of soil fertility gets improved. Slightly higher values of net returns Rs. 8450 and Rs. 14949 and B: C ratio 1.56 and 2.08 were observed with *khejri* plantation in two years as compared to other tree species (Sharma 2015).

Table 6. Grain and fodder yield (kg ha^{-1}) of kharif crops under various farming systems

Cropping System	Agroforestry system with *P. cineraria*		Arable farming	
	Grain	Stover	Grain	Stover
Pearl millet (HHB 67)	1730	5400	1460	3268
Green gram (K851)	386	870	340	916
Dew gram (RMO 40)	606	2000	650	1950
Cluster bean (RGC 936)	666	1830	600	1730

Silvipastoral system

The livelihood of rural people of arid region in India is largely dependent on livestock; therefore silvipastoral systems are the inherent part of land use systems in these regions. Thus, it is of utmost importance to enhance and sustain the productivity of various components of this system. In a study carried out in Kachchh region of Gujarat grasses namely *C. ciliaris* and *C. setigerus* were raised with trees neem, Acacia and subabul. Both the grass yield as well as tree growth was found to superior when grasses were grown with Neem tree. *Cenchrus ciliaris* grass was more productive in fodder yield when compare with *C. Setigerus*. The fodder yield of grasses did not did not differ significantly due to association of trees with grasses in a silvipasture system (Dayal *et al.* 2008). With 20 years of cultivation under silvipastoral systems, the soil organic carbon under grasses improved from 0.47 to 0.58 % (Shamsudheen *et al.* 2009). In a study on Integrated framing systems carried out in CAZRI, Jodhpur the highest dry matter yield was recorded in *Colophospermum mopane+ Cenchrus ciliaris* association (27.00 t ha^{-1}) followed by *Hardwickia binata + Cenchrus ciliaris* (26.39 t ha^{-1}) whereas, the yield of *Cenchrus ciliaris* when raised in sole pasture was 22.59 t ha^{-1} (Table 7). For obtaining higher productivity of quality fodder, inter-cropping of grasses with legumes in association with *H. binata* appears to be a highly suitable proposition for a silvipastoral system in an arid environment (Patidar and Mathur 2016).

Table 7. Fodder productivity under various silvipastoral systems

Silvipastoral systems	Tree density (trees ha^{-1})	Dry matter Yield (q ha^{-1})		
		grass	Leaf	Total
Hardwickia binnata + C. ciliaris	120	22.19	4.2	26.39
Colophospermum mopane+ C .ciliaris	198	14.3	9.41	23.71
C.mopane+ C .ciliaris	78	20.8	6.2	27.0
Ailanthus excelsa + C .ciliaris	90	16.97	2.34	19.31
Sole pasture (*C .ciliaris)*	-	22.59		22.59

Diversified agroforestry system

A farmer Shri N K Jaisalmeria of village Manaklao near Jodhpur by adopting technology of CAZRI turned 4.5 ha of undulating and barren land into high yielding diversified agroforestry system with the technical assistance of CAZRI. Though farmers in arid zone of India practice mixed farming but due to low rainfall and occurrence drought year twice in five years the production as well as productivity is very low. Therefore, Mr Jaisalmeria started the farming of ber trees on his land. He planted 750 trees of ber under rainfed condition and the orchard was surrounded by shelterbelt of trees like *Acacia tortilis*, *P. juliflora* and within orchard trees like *P. cineraria*, *Tecomella undulata* and *Azadirachta indica* were raised. He raised *Cenchrus ciliaris* grass between the ber trees and pearl millet in one hectare area to provide grain for household consumption and fodder for livestock. Eight goats were reared within the system and farm waste was used to prepare the compost. The system is producing ber fruits to the tune of 10.7 t ha^{-1}, fuel wood (dry) 4t ha^{-1}, and fodder about 2.4 t ha^{-1} year^{-1}. This system is capable of providing employment to the tune of 331 man days per ha per year. Shri N K Jaisalmeria has been recognized at many national and international platforms to present his success story (Tewari *et al.* 2014). The constituents and operation of the system is depicted in Fig 2.

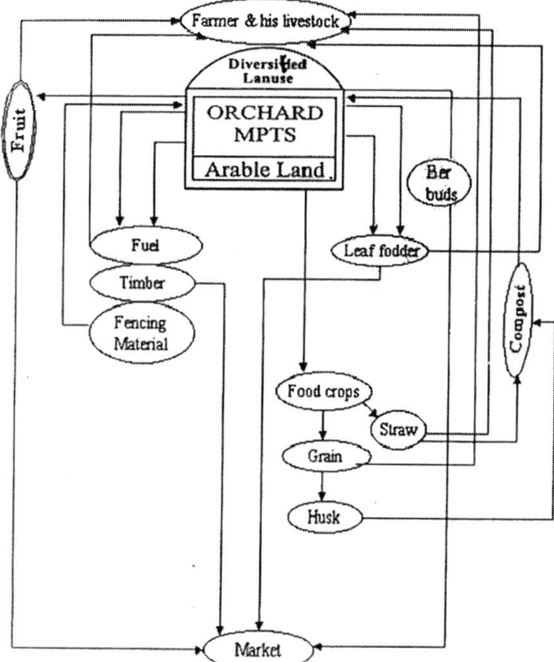

Fig. 2. Diversified agroforestry system developed by Mr. N. K. Jaisalmeria at village, Manaklao near Jodhpur

Tree crop interaction in agroforestry system

Agroforestry in arid regions can mitigate the impact and consequences of harsh climatic conditions on crop production. In agroforestry systems, trees and crops are grown in same land simultaneously which leads to various site specific positive and negative interactions between tree and crops. The various interaction between tree and crop in agroforestry system includes, reduced light intensity for crops due to shading of trees, root level competition for water and nutrients, nitrogen supply to crops from the nitrogen fixation and long term effect on erosion, soil organic matter content, soil compaction and leaching of nutrients (van Noordwijk and Lusiana 2000).

In the most of the agroforestry systems strong competition has been observed for the site resources. Especially in the arid regions, where the water is the limiting factor of crop growth, below ground interaction annual crops and perennial tree crops leads to yield reduction (Ong *et al.* 1991). Therefore, the central hypothesis of agroforestry is that "the trees must acquire resources that the crop would not otherwise get" (Cannell *et al.* 1996). To overcome negative aspects of tree-crop combinations, annual and perennial plants should have root systems that capitalize on different soil zones, the tree having deep roots and the crop, shallow roots. Agroforestry systems have a more closed nutrient cycling than monocultural systems and enrich the soil with nutrients and organic matter. This would require the interaction of the root systems of associated crops.

The advantage of agroforestry practices in the dry lands to be the exploitation of the complementary interactions between crops and mature trees grown for their marketable products. Such trees can be fitted into the landscape by exploiting the under-utilised areas within and around the farms, such as footpaths and home compounds and by integrating crops with boundary plantings, scattered trees.

Interaction of *Prosopis cineraria* with various crops in the arid regions of India is one of the keen example for the positive effects of agroforestry system. *Prosopis cineraria* based agroforestry system is naturally seen in the western Rajasthan and people knows the benefits of having tree in their field. The scientific study also proves the positive interaction of *Prosopis cineraria* with crops. The major crops in arid regions, pearl millet, cluster bean, cowpea and mung bean has shown higher yields with association of *Prosopis cineraria* (Kumar *et al.* 1992; Singh and Bishnoi 2013). The association improves vegetative growth and productivity of crops than in open area. The higher productivity under tree shows insignificant competition between tree and crop for site resources. In the case of *Prosopis cineraria*, it is having deep root system and thin layered canopy do not impose competition for the moisture and light for crop. But at the same time moisture retention seen for long time near to the tree and have improved soil fertility due to shade by the tree canopy in the hot arid condition. Deep rooted trees absorb nutrients from deeper layer of the soil and releases it in the upper layer through litter fall. This also improves growing conditions near tree (Toky and Bisht 1992). The presence of *Prosopis cineraria* in Kharif crops (Jaimini and Tikka 1998) shown less dehydration and improved soil fertility by nitrogen fixation and addition of organic matter. *Prosopis cineraria* shows good compatibility with crop due to early litter decomposition, nitrogen fixing ability, insignificant competition with crops for moisture and light.

Agroforestry and livelihood security in hot arid zone

Arid regions primarily characterized by low and erratic rainfall with high variability, shallow soils with low inherent fertility status, unabated land degradation and poor economic status of farmers. In this region successful arable cropping are limited & risky. The growing food insecurity and deteriorating livelihood situations in arid regions of India call for concerted and consorted actions to take advantage of the high potential of agro forestry, among other systems, for promoting best land use practices, which increase the productivity (yield) of land, combine the production of crops, including the tree crops and forest plant and animals simultaneously or sequentially on the same unit of land, meeting the ecological and socioeconomic need of rural people. Agroforestry increase livelihood security through simultaneous production of food, fodder, and firewood and an increase in total productivity per unit area of land. It also ensures regular income of rural people and improve the socioeconomic condition of the farmers of these region.

The tree based system has multiple benefits in comparison to sole cropping under arid situation. Trees provide a range of products such as fruits, fodder, fuel wood and ecosystem services such as reducing soil and water erosion, carbon sequestration. A number of agroforestry systems were developed and some of them have been adopted by farmers in arid regions of the country. Tree based systems are reported to generate high net returns compared to annual crop systems. It provides reasonable returns during the years of low rainfall. Management practices, diversified components like animals and organic production practices, high value intercrops, high density planting, canopy management improve returns.

Agroforestry enhance livelihood in arid regions of India, where people have accumulated local knowledge. Arid regions are particularly notable for agroforestry practices and indigenous knowledge system on growing trees on farm lands. The economics of traditional agroforestry systems of arid western

Rajasthan indicated that net Benefit Cost Ratio of such systems is on positive sides. An important agroforestry system which is having high potential of livelihood improvement in arid region is the *Acacia senegal* based agroforestry system. It gives high returns to the farmers with lesser effort in management. Tree crop in the system i.e., A. senegal produces gum Arabic, highly demanded product in the households due to its medicinal properties. The system fulfils both human and animal needs for food by grains and straw from the crops cultivated along with *A. senegal*. The additional income is generated to farmers from the gum collection in the non-cultivating season also. Central Arid Zone Research Institute has done various research to improve the livelihood of the farmers in the region, also developed a gum inducer for the *A. senegal* by which farmers were able to collect substantial quantity of gum Arabic. This additional income generation from the underutilised area helped farmers for a sustainable livelihood in the arid conditions.

Prosopis cineraria, tree which occupies all over the arid regions of Rajasthan have immense significance in the livelihood improvement of farmers. The tree supports rural economy with its all parts, like no other wild vegetation does. The fruit of the tree is major and high valued vegetable, eaten as sangria, cooked as a delicious vegetable. The leaf litter of the tree improve the soil fertility and enhance crop growth under the tree compared to open areas. This tree also supplies fodder to the cattle as it shows increased the milk yield. *Prosopis cineraria* based agroforestry system improves the livelihood of the farmers by meeting their various requirement near to their households. The farmers are well aware of the uses of this tree and it is one of the species farmers always willing to keep in their fields.

Another major agroforestry system in the arid region is Ber based system. The farmers grow Ber, mainly for its fruit, which are very delicious and rich source of vitamin A, C and B complex. Along with economic returns from fruits it also supplies excellent fodder. The branches and twigs are used as fuel wood have high calorific value. Also tannin is produced from root bark and leaves used for the tanning leather. These additional benefits from tree component improves livelihood of the farmers in the arid region. Improved ber based horti-pasture system practised in arid regions is a major source of livelihood for the farmers in the area (Tewari *et al.* 2001). This system provides various needs of farmers such as human and animal food and cooking energy needs to the village communities. These multifunctional agroforestry systems provide goods and services directly to farmers and improves the social wellbeing of the people. During drought years, these trees were serving the rural community for their survival. Village community's needs were satisfied with the available tree sources. In the arid regions of Rajasthan farmers are keeping trees in their farm area are multipurpose in nature and they can satisfy their various basic needs up to a level. This shows importance of agroforestry in the arid regions, it's a life support system in the hot arid regions.

Adaptation role of agroforestry to climate change in hot arid region

Agroforestry is a spreading land use adaptation in the arid regions, which supports livelihood improvement through production of materials for the basic human needs like food, fodder and firewood along with mitigation and adaptation to climate change. Climate change is mitigated by agroforestry practices because of tree component, which results in permanent tree cover and varied ecological niches. So agroforestry systems in arid regions contribute to ecological and social and economic functions.

Climate change is major threatening issue identified in this century due to high greenhouse gas production happening in all over the world. Arid regions are experiencing highly erratic conditions which will become more severe in climate change. The permanent tree cover in the agroforestry system can minimise the harsh effect of climate change. It also enhances carbon storage through biomass accumulation in wood and soil and minimizes the negative impact of emission of greenhouse gases.

Agroforestry enhance the uptake of greenhouse gases or reduce its emission. It is having the potential to remove CO_2 from the atmosphere and carbon is locked in wood. Agroforestry in arid region sequesters carbon in vegetation and in soils. Agroforestry systems have the potential to sequester large amounts carbon in above and belowground soil compared to tree-less farming systems

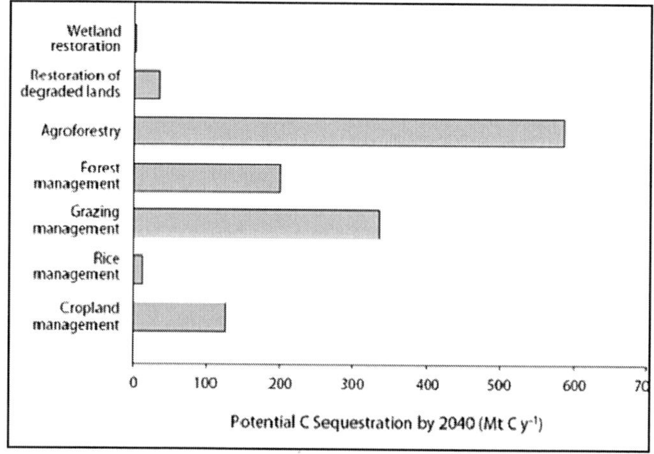

Fig. 3. Carbon sequestration of different land management options
Source: IPCC 2000)

practised in farmer's field. Agroforestry in the farmer field reduces the pressure and extraction of wood from the natural forests by providing them alternate wood supply from their own field. The improved income generation also in a way reduces farmer exploitation on forest. In India, average carbon sequestration potential in agroforestry system has been estimated to be 25 t c ha^{-1} (Sathaye and Ravindranath 1998).

Sustainability of agroforestry system in the arid regions is the strong asset for climate change adaptation. Climate change mitigation is done by trees through carbon sequestration. Carbon sequestration takes place as above ground and below ground biomass in the agroforestry system. When comparing with mono cropping system, agroforestry system helps in greenhouse gas–mitigation strategy through carbon sequestration through its greater efficiency in using resource (nutrients, light, and water) in capture and utilization (Nair *et al.* 2009).

Agroforestry has a particular role to play in mitigation of atmospheric accumulation of GHGs. of all the land uses analyzed in the Land-Use, Land-Use Change and Forestry report of the IPCC, agroforestry offered the highest potential for carbon sequestration (IPCC 2000). Agroforestry has such a high potential, not because it is the land use practice with the highest carbon density, but because there is such a large area that is susceptible for the land use change.

In the agroforestry systems, perennial component trees play protective role. It can enhance soil productivity through efficient nutrient cycling and nitrogen fixation. It also captures nutrients and water from deeper layers of the soil and improves the surface layer. Trees maintain woody biomass for long term and provide food, fibre and vegetative cover for soil. Trees in agroforestry system enhance the crop yields of coarse grains through a combination of mulching and water conservation. In the arid region of India, productivity of *Hordeum vulgare* (barley) was found to be positive in presence of *Azadirachta indica, Acacia albida, Prosopis cineraria* and *Tecomella undulata*. The grain yield was enhanced by 86.0% in combination with *P. cineraria, T. undulata* by 48.8%, *A. albida* by 57.9% *and A. indica* by 16.8% over the control. Biological yield was also high under trees in the system than that in the open area. Rich organic matter, moisture content and nutrient status was seen under tree canopy (Kumar *et al.* 1998).

Loss of biodiversity is a biggest threat happening in the world wide. This decrease ecosystem functioning and services. As compared to traditional mono-cropping agroforestry conserves the biodiversity. It provides habitat for species and can with stand certain level of disturbance. Also better livelihood to the people indirectly reduces the human pressure on the natural resources. Trees in the agroforestry system conserve biodiversity by providing erosion control and water recharge and thus it prevents degradation of habitat. High biodiversity and resulting ecosystem services in agroforestry also contribute to synergy between climate change adaptation and mitigation. Trees in agro-ecosystems in arid regions of Rajasthan have been found to support threatened cavity- nesting birds and offer forage and habitat to many other bird species. Biodiversity conservation may not be a primary goal of agroforestry systems. Nevertheless, in some cases traditional agroforestry systems in arid region found to support very good species diversity and also act as a buffer to parks and protected areas.

Adaptation to climate change is now unavoidable. Agroforestry qualifies as adaptation to climate change in arid regions. It plays a critical role in controlling soil erosion. Shelter belt on agricultural fields is an effective practise in controlling soil erosion in the arid regions. (Gupta *et al.* 1984). Agroforestry have capacity to enhance the water use efficiency. The perennial tree canopies in agroforestry systems intercept the rain water and reduce its velocity which in turn reduce soil and water runoff (Khan *et al.* 1995). In arid region where crop experience water scarcity, the enhanced efficiency in the use of rain water, improves the crop productivity. Agroforestry systems also provides an economically viable and environment friendly way to improve soil fertility. Tree crops in the system contribute to efficient nutrient cycling in the area. It also moderate microclimate in the cropping area of arid regions (Monteith *et al.* 1991). It gives crop favourable changes in microclimate of cropping area and have significant impact on photosynthesis, transpiration and soil water use.

Future needs

Agroforestry research should have as a high priority for the identification of appropriate tree species for their assigned role in the system in a particular environment. There are many local and exotic species available. It is the insight of the scientists to identify the specific tree for specific role in that ecosystem and to develop the agroforestry systems based on the available components so that sustainable production can be achieved without harnessing the resources. The systems thus composed should be passed through several farm trials on farmer's fields before extending on large scale. A number of systems failed to become popular with farmers because of the absence of the necessary input supply and marketing infrastructure. This may be crucial for agroforestry because overproduction of wood, fruit or fodder without any market can affect the farmers adversely (Singh 1987). Whether it is the problem of apple boxes in the hills, or fodder and fuel in the semi-arid/arid areas, or shelterbelts in deserts, or ground cover in high-rainfall areas, or cash crops in high input areas, trees can play a role and can be suitably integrated with the existing agricultural production system.

There are many tree species which are already introduced in the field of farmers and farmers are interested to adopt the other options. There is huge scarcity of fuel and fodder in villages; therefore farmers are ready to plant the species which provide them multipurpose benefits. Therefore determined attempt is required to develop the technology pertaining to particular site and policy makers to develop a suitable infrastructure for the disposal of diversified products from such a system.

Another challenge is to popularize the tree planting among the rural people to supply them fuel, fodder, food and other components. They have to be motivated to conserve their traditional agroforestry practices and enhancing them by selecting the tree species more useful for livelihood improvement. A constant cycle of production and harvest of wood products has to be maintained for sustainable livelihood of farmers. Region-specific agroforestry models for small, marginal and large farmers need to be developed. The value added tree products (NTFPs) and their processing are not widespread which needs attention here is a need to focus on methodologies for quantification and payment of ecosystem services (PES) in agroforestry system, which will link consumers of environmental services with suppliers.

Conclusion

Arid ecosystems are very fragile systems which are further restrained by abrupt and harsh weather conditions. Agroforestry was always an integral part of the farming system of these regions as alone arable farming is not dependable due to occurrence of periodic droughts. The traditional agroforestry practised by farmers is not so fruitful because of low land productivity which leads to the malnutrition of human and mortality of livestock as there is poor availability of resources. There are number of innovative systems developed by CAZRI as well as other institutes working in the region to increase the per unit land productivity by incorporating the components like fruit trees, arable grasses, multipurpose tree species, high yielding varieties of crops etc. Farmers have also adopted the various technologies to diversify their traditional low yielding farming systems to elevate their per unit productivity of land as well as farm income. But still there are long ways to go and there are challenges to sustain different agroforestry systems with high productivity vis-a-vis preventing the deterioration of environment. There is need to identify specific tree for specific role in particular ecosystem with several farm trials before disseminating on large scale. A constant cycle of production and harvest of wood products has to be maintained for sustainable livelihood of farmers and environmental services provided by them needs to be paid. There is need to develop the region specific agroforestry models for farmers with small land holdings. Therefore, these improved practises along with traditional agroforestry practices will play a pivotal role in sustaining the livelihood of people living in arid zone.

References

Alao JS, Shuaibu RB (2013). Agroforestry practices and concepts in sustainable land use systems in Nigeria. Journal of Horticulture and Forestry 5 (10) :156-159.

Arya R, Awasthi OP, Jitendra Singh, Singh IS, Manmohan JR (2011). Performance of component crops in tree-crop farming system under arid region. Indian J Hort 68(1): 6-11.

Awasthi OP, Saroj PL, Singh IS, More TA (2007). Fruit Based Diversified Cropping System for Arid Regions, CIAH Tech Bull, No 25, CIAH, Bikaner, pp. 18.

Bhandari DC, Meghwal PR, Lodha S (2014). Horticulture based production systems in Indian arid region. In: Nandwani D (ed) Sustainable Horticultural Systems, Sustainable Development and Biodiversity, Cham: Springer International Publishing, pp. 19-49

Bhandari MM (1978). Flora of Indian Desert. Scientific publishers, Jodhpur, India, pp. 471.

Bhati TK, Rathore VS, Singh JP, Beniwal RK, Nathawat NS (2009). Alternative Farming Systems for Hot Arid Regions. In: Kar Amal, Garg BK, Singh MP, Kathju S (eds) Trends in Arid Zone Research in India, Central Arid Zone Research Institute, Jodhpur, pp. 354-382.

Bishnoi M, Singh J (2009). Vegetative growth of pearl millet in *Prosopis cineraria* based agroforestry system in arid zones. Green Farming 2 (9): 642-644.

Cannell MGR, van Nordwijk M, Ong CK (1996). The central agroforestry hypothesis: the trees must acquire resources that the crop would not otherwise get. Agrofor Syst 34: 27-31.

Central Arid Zone Research Institute (2014). Annual progress Report. CAZRI, Jodhpur, Rajasthan, India, pp. 53-54.

Champion HG, Seth SK (1968). A Revised Survey of the Forest Types of India. Government of India, New Delhi.

Dayal D, Mangalassery S, Meena SL, Ram B (2015). Productivity and profitability of legumes as influenced by integrated nutrient management with fruit crops under hot arid ecology. Indian Journal of Agronomy 60(2): 297-300.

Dayal Devi, Shamsudheen M, Bhagirath Ram (2009). Alternative Farming Systems Suitable for Kachchh Region of Gujarat. Regional Research Station, Central Arid Zone Research Institute, Kukma-Bhuj, Gujarat, pp. 22.

Dayal Devi, Swami ML, Bhagirath Ram, Meena SL, Shamsudheen M (2008). Production potential of grasses under silvipastoral system in Kachchh region of arid Gujarat. In: National symposium on Agroforestry Knowledge for Sustainability, Climate Moderation and Challenges Ahead, abstracts, NRC-Agroforestry, Jhansi (UP), pp. 195.

FAO (2004). Carbon Sequestration in Dryland Soils. World Soils Resources Reports, No. 10, Food and Agriculture Organization, United Nations, Rome.

Ffolliott PF, Brooks KN, Gregersen HM, Lundgren AL (1995). Dryland Forestry: Planning and Management. John Wiley and Sons, Inc, New York, USA.

Gupta GN, Singh G, Kachwaha GR (1998). Performance of *Prosopis cineraria* and associated crops under varying spacing regimes in the arid zone of India. Agrofore Syst 40: 149–157.

Gupta JP (1997). Some alternative production systems and their management for sustainability. In: Gupta JP, Sharma BM (eds) Agroforestry for Sustained Productivity in Arid Regions. Scientific Publishers, Jodhpur, pp. 31–39.

Gupta JP, Joshi DC, Singh GB (2000). Management of arid agro-ecosystem. In: Yadav JSP, Singh GB (eds) Natural Resource Management for Agricultural Production in India.

Indian Society of Soil Science, New Delhi, pp. 553–668.

IPCC (2000). International Panel on Climate Change Special Report on Land Use, Land Use Change and Forestry. Summary for Policy Makers, Geneva, Switzerland.

Jaimini SN, Tikka SBS (1998). Khejri based agro forestry system for dry land areas of north and north western Gujarat. Indian J For 21(4): 331-332.

Kaushik N, Kumar V (2003). Khejri based agroforestry system for arid Haryana, India. J Arid Environ 55: 433-440.

Khan TI (1997). Conservation of biodiversity in western India. Environmentalist 17: 283-287.

Kolarkar AS (990). Khadin - a sound traditional method of runoff farming in Indian desert. In: workshop, Water: a scarce resource- a cultural symbol, Max Mueller Bhawan Goethe Institute, Bombay.

Kumar V, Yadav HD, Sharma HC (1992). Agroforestry, the suitable farming system for arid and semi-arid region. Haryana Farm 21(2):15-16.

Kumar Sunil, Shukla AK, Singh HV (2015). Efficient utilization of interspaces of aonla (*Emblica officinalis* G.) orchard through intercropping under rainfed condition. Range Manage Agrofor 36(2):188-193.

Kumar A, Hooda MS, Bahadur R (1998). Impact of multipurpose trees on productivity of barley in arid ecosystem. Annals of Arid Zone 37: 153-157.

Mann HS, Muthana K D (1984). Arid Zone Forestry. Central Aid Zone Research Institute, Jodhpur, India, pp.3.

Meena LR (2015). Productivity and economics of ber (*Ziziphus mauritiana*) based hortipasture system as influenced by integrated nutrient management under rainfed condition of Rajasthan. Forage Research 41(3):176-182.

Meghwal PR, Henry A (2006). Economic evaluation of different agrohorticulture systems under rainfed condition of arid zone. In: Kumar D, Henry A, Vittal KPR (eds) Legumes in Dry Areas, Indian Arid Legume Society and Scientific Publishers, Jodhpur, pp. 299-303.

Monteith JL, Ong CK, Corlett JE (1991). Microclimatic interactions in agroforestry systems. For Ecol Manage 45:31–44.

Nair PKR, Kumar BM, Nair VD (2009). Agroforestry as a strategy for carbon sequestration. J Plant Nutr Soil Sci 172:10–23.

Ong CK, Corlett JE, Singh RP, Black CR (1991). Above and belowground interactions in agroforestry systems. For Ecol Manage 45: 45-58.

Owonubi JJ (2002). Disappearing forests: a review of the challenges for conservation of genetic resources and environmental management. J Forest Res Manage 1:11-20.

Patel BM, Patel S, Patel SK, Patel SB (2003). Intercropping studies in ber *(Zizyphus mauritiana* LAMK.). Agric Sci Digest 23 (2): 113 – 115.

Patidar M, Mathur BK (2016). Enhancing forage production through a silvi-pastoral system in an arid environment. *Agrofor Syst* (in press).

Patil NV, Pathak KML (2013). Approaches for improvement of livestock production system as affected by climate change in dry lands. Annals of Arid Zone 52 (3 and 4): 275-286.

Roy MM, Tewari JC, Moola Ram (2011). Agroforestry for climate change adaptations and livelihood improvements in Indian hot arid regions. International Journal of Agriculture and Crop Sciences 3(2): 43-54.

Roy MM, Tewari JC (2014). Agroforestry in arid regions for sustainable livelihoods. Indian Farming 63 (11): 9-18.

Saroj PL, Dhandhar G, Sharma BD, Bhargava R, Purohit CK (2003). Ber (*Ziziphus Mauritiana* L.) based agri-horti system: a Sustainable land use for arid ecosystem. Indian J Agrofor 5: 30-35.

Sathaye JA, Ravindranath NH (1998) Climate change mitigation in the energy and forestry sectors of developing countries. Annual Review of Energy and Environment 23: 387–437.

Shamsudheen M, Devi Dayal, Meena SL, Bhagirath Ram (2009). Improvement of soil properties under silvipastoral systems in the Kachchh region of arid Gujarat. In: 4th World Congress on Conservation Agriculture: Innovations for Improving Efficiency, Equity and Environment, February 4-7, National Academy of Agricultural Sciences, New Delhi, India, pp. 253-254.

Sharma AK, Tewari JC (2005). Arid zone forestry with special reference to Indian hot arid zone, in forests and forest Plants, from Encyclopedia of life support systems (EOLSS). Developed under the auspices of the UNESCO, Eolss Publishers, Oxford, UK, (http://www.eolss.net).

Sharma AK (2003). Arid Zone Agroforestry: Dimensions and Directions for Sustainable Livelihoods IUFRO-archive.boku.ac.at/iufro/taskforce/tfscipol/chennai-papers/faksharma.pdf.

Sharma KC (2015). Production potential of fodder crops sequences in association with ber (*Zizyphus mauritiana* Lamk.) under agri-horticulture system in hot arid ecosystem of western India. Range Mgmt Agrofor 35 (2): 188-192.

Shetty BV, Singh V (1993). Flora of Rajasthan Vol III. Kolkata: BSI. 861–1246.

Singh GB (1987). Agroforestry in the Indian subcontinent: past, present and future. In: Steppler HA, Nair PKR (eds) Agrofrestry a Decade for Development. ICRAF, Nairobi, pp. 117-140.

Singh J, Bishnoi M (2013). Performance of bean crops under Khejri (*Prosopis cineraria*) canopy in agroforestry systems of arid zone. International J Agri Sc Res 2(12):318-321.

Singh RS (1997). Note on effect of intercropping on growth and yield of ber (*Z. mauritiana* Lamk.) in semi-arid region. Current Agriculture 21(1/2): 117-118.

Singh RS, Gupta JP, Rao AS, Sharma AK (2003) Micro-climatic quantification and drought impacts on productivity of green gram under different cropping systems of arid zones. In: Naraian P, Kathju S, Kar A, Singh MP, Kumar P (eds) Human Impact on Desert Environment, Scientific Publishers, Jodhpur, pp.74-80.

Tanwar SPS, Kumar S, Roy MM (2014). Integrated farming systems. In: Roy MM, Kumar S, Tripathi RS, Saha D., Das T (eds) ENVIS, Desert Environment, CAZRI, Jodhpur. Newsletter 16 (2).

Tewari JC, Bohra MD, Harsh LN (1999). Structure and production function of traditional extensive agroforestry systems and scope of agoroforestry in Thar desert. Indian J Agrofor 1(1):81–94.

Tewari JC, Moola Ram, Roy MM, Dagar JC (2014). Livelihood improvements and climate change adaptations through agroforestry in hot arid environments. Dagar JC *et al.* (eds.) Agroforestry Systems in India: Livelihood Security and Ecosystem Services, Advances in Agroforestry pp. 155-183.

Tewari JC, Tripathi D, Narain P (2001). Jujube : a multipurpose tree crop for arid land farming systems. The Botanica 51:121-126.

Thornthwaite CW (1948). An approach toward a rational classification of climate. Geographical Review 38: 55–94.

Toky OP, Bisht RP (1992). Observations on rooting patterns of important agroforestry trees growing in arid climate of north-western India. Agrofor Syst 18: 245-263.

UNEP (1997) World atlas of desertification (2nd edition). United Nations Environmental Programme, Nairobi, Kenya.

Van Noordwijk M, Lusiana B (2000). Wa Nu LCAS version 2.0, Background on a model of water nutrient and light capture in agroforestry systems. International Centre for Research in Agroforestry (ICRAF), Bogor, Indonesia.

Young A (1987). The environmental basis of agroforestry. In: Proceedings of an International Workshop on The application of Meteorology to Aagroforestry Systems Planning and Management, pp. 36-55.

14

Role of Plant –Microbe Interaction in Agroforestry

Anjly Pancholy and S.K. Singh

ICAR-Central Arid Zone Research Institute, Jodhpur-342003, Rajasthan, India

Introduction

Agroforestry system (AFS) is a form of sustainable land use that combines trees, shrubs, crops and livestock so as to increase and diversify farm and forest production along with conserving natural resources. Microorganisms play a crucial role in improving nutrient availability to plants, thereby decreasing the dependence on chemical fertilizers and achieve sustainable agriculture through agroforestry (Araujo *et al.* 2012). Arbuscular mycorrhizal (AM) fungi, plant growth-promoting rhizobacteria (PGPR), and the association of rhizobia with leguminous plants are symbiotic associations of high economic importance for increasing agricultural production. Yadav *et al.* (2010) studied soil biological properties under different tree based traditional agroforestry systems in a semi-arid region of Rajasthan, India and concluded that biological properties can be optimized in the soil under AFS. Many authors have reported that soil microbial biomass and microbial diversity are greater in the AFS due to the ameliorative effects of trees and organic matter inputs and the differences in litter quality and quantity and root exudates (Mungai *et al.* 2005; Sørensen and Sessitsch, 2007). Also, the presence of a large and diverse soil microbial community is crucial to the productivity of any agroecosystem. Moreover, more than one plant species in AFS have been reported to have shown a larger diversity and/or abundance of mycorrhizal fungi than monocultures (Cardoso and Kuyper 2006) and more efficiency in biological fixation of the nitrogen, especially in tropical soils (Freitas *et al.* 2010). Also, soil microbial biomass has other important functions in the soil such as nutrient cycling and the degradation of pollutants (Watanabe and Hamamura 2003; Araújo and Monteiro 2006).

Agroforestry Systems and soil microbial biomass

AFS are being increasingly recommended as a sustainable form of land use because they are believed to provide the optimum level of food production, a supply of firewood, and cash benefits, while maintaining soil fertility (Heuvelop *et al.* 1988). There is more accumulation of organic matter in soils under AFS through the inclusion of different crops and permanent vegetation cover, which in turn increases the soil microbial biomass. This is expected to improve the biological properties of degraded soils due the high quantity of organic residues. Additionally, AFS have been recognized as an alternative for the rehabilitation of degraded areas through the use of different tree species with crops. Many tropical tree species can fix atmospheric nitrogen and may therefore increase the soil nitrogen content. The large root system of trees is capable of accumulating nutrients from a large volume of soil, whereas fallen litter concentrates nutrients near the soil surface. Trees may also enhance the above- and belowground microclimate around plant roots and may alter the soil biological properties. Tangjang *et al.* (2009) observed seasonal variation in bacterial and fungal colony forming units (cfu) per gram dry soil at three different soil depths under traditional AFS in Northeast, India (Table1).

Table 1. Seasonal variations in bacterial cfu (x10^5 g^{-1} dry soil) and fungal cfu (x10^3 g^{-1} dry soil) at three soil depths

		Bacterial (cfu)				Fungal (cfu)			
Sites	Soil depth	W	S	R	A	W	S	R	A
Harmutty	0-10	107.33	169.67	171.41	123.67	30.67	84.32	31.67	193.00
	10-20	88.33	118.60	193.33	121.33	29.67	72.50	35.33	145.00
	20-30	38.66	87.72	76.67	41.67	16.00	60.00	21.20	46.00
Nirjuli	0-10	97.33	148.70	146.67	109.69	23.67	59.00	24.00	101.00
	10-20	95.66	121.00	156.67	97.33	21.33	53.84	34.50	87.00
	20-30	48.33	54.20	68.00	70.00	15.67	38.60	19.17	30.00
Doimukh	0-10	72.67	133.16	129.33	46.20	21.00	69.20	20.67	104.00
	10-20	41.33	101.52	148.70	2.67	21.00	35.00	34.00	88.33
	20-30	19.67	71.06	49.00	9.67	22.00	34.00	23.00	46.00

W: Winter; S: Summer; R: Spring; A: Autumn
Source : Tangjang *et al.* (2009)

Both the amount of microbial biomass carbon and the enzymatic activities were greater in soils under AFS than in conventional systems. The greater microbial biomass reflected the response of the increased input of organic matter to the soil under the AFS. Kaur *et al.* (2000) studied the effects of the monocropping of rice, forestry, and agroforestry on the soil microbial biomass in northern India and observed that the soil microbial biomass was increased by 42% (microbial carbon) and 13% (microbial nitrogen) in AFS compared to monocropping. They attributed the higher soil microbial biomass to the high quantity of carbon released by the AFS.

Table 2. Soil microbial biomass carbon (MBC) and microbial biomass nitrogen (MBN) averaged across the seasons and their relationships with total soil carbon and nitrogen in different management systems

Treatments	MBC (mg g^{-1} soil)	MBN (mg g^{-1} soil)	Microbial C:N	MBC Soil C (%)	MBN Soil N (%)
Control	76.13f	11.06e	6.88a	1.21c	1.51d
Rice–berseem	96.14e	18.61cd	5.16cd	2.29cd	4.04ab
Acacia	143.40b	31.45a	4.56d	2.31b	4.08ab
Acacia+rice–berseem	153.40a	32.57a	4.71d	2.26d	3.79bc
Eucalyptus	109.12d	15.78de	6.92a	2.32ab	3.36c
Eucalyptus+rice–berseem	133.80c	21.78bc	6.14ab	2.23b	3.69bc
Populus	131.10c	21.70c	6.04bc	2.57a	4.48a
Populus+rice–berseem	150.13a	25.62b	5.86bc	2.24b	3.28c
LSD(($p<0.05$)	4.86	3.38	0.731	0.194	0.525

For each column, values not marked with the same letter are significantly different ($p<0.05$).
Source: B. Kaur *et al.* (2000)

Agroforestry systems and biological nitrogen fixation

The biological nitrogen fixation (BNF) process is an economically attractive and ecologically sound method to reduce chemical fertilizer nitrogen input and improve the availability of organic nitrogen through biological fixation. BNF by associative diazotrophic bacteria is a spontaneous process where soil nitrogen is limited and adequate carbon sources are available. However, successful use of legumes is dependent upon formation of effective symbioses with root nodule bacteria subject to availability of environment capable of supporting the growth of these microorganisms. Biological systems that are capable of fixing nitrogen are classified as nonsymbiotic or symbiotic, depending on the involvement of one or more organisms, respectively (Hubbell and Kidder 2003). For leguminous species with potential use in agroforestry systems, efficient nitrogen-fixing rhizobia have been selected and are available for inoculant production. New species of rhizobia or bradyrhizobia are continuously isolated, and large collections of these isolates are being developed. These rhizobia are of economic importance in low-input sustainable agriculture, agroforestry, and land reclamation (Balachandar *et al.* 2007). Previously, it was generally accepted that legumes (and the non-legume genus *Parasponia*) are nodulated exclusively by members of the family Rhizobiaceae in the á-proteobacteria, which includes the genera *Azorhizobium*, *Bradyrhizobium*, *Mesorhizobium*, *Rhizobium*, and *Sinorhizobium* (Sprent 2001). Subsequently, however, several other species of á-proteobacteria have been shown to nodulate legumes, such as strains of *Methylobacterium* that nodulate *Crotalaria* and *Lotononis*, (Jaftha *et al.* 2002);

Table 3. The α- and β-Proteobacteria, nodulating legumes and their taxonomical position.

Non-rhizobial species	Host legume	Phylogenetic group	Taxonomical position		Reference
			Family	Order	
Methylobacterium nodulans	*Crotalaria glaucoides*	α -Proteobacteria	methy lobacteriaceae	Rhizobiales	Sy *et al.* (2001)
Blastobacter denitrificans	*Aschynomene indica*	α -Proteobacteria	Bradyrhizobiaceae	Rhizobiales	Van Berkum and Eardly (2004)
Devosia neptuniae	*Neptunia natans*	α -Proteobacteria	Hyphomicrobiaceae	Rhizobiales	Rivas *et al.* (2002)
Phyllobacterium trifolii	*Trifolium repans*	α-Proteobacteria	Phyllobacteriaceae	Rhizobiales	Valverde *et al.* (2005)
Ochrobactrum lupini	*Lupinus albus*	α -Proteobacteria	Brucellaceae	Rhizobiales	Trujilo *et al.* (2005)
Agrobacterium like strains	*Phaseolus vulgaris*	α -Proteobacteria	Rhizobiaceae	Rhizobiales	Mhamdi *et al.* (2005)
Burkholderia tuberum, B. phymatum	*Mimosa*	β -Proteobacteria	Burkholderiaceae	Burkholderiales	Vandamme *et al.* (2002)
Cupriavidus taiwanensis	*Mimosa*	β -Proteobacteria	Burkholderiaceae	Burkholderiales	Chen *et al.* (2001)
Herbaspirillum lusitanum	*Phaseolus vulgaris*	β -Proteobacteria	Burkholderiaceae	Burkholderiales	Valverde *et al.* (2003)

Source: Balachander (2007)

Blastobacter denitrificans, which nodulates *Aeschynomene indica* (van Berkum and Eardly 2002); and *Devosia* strains that nodulate *Neptunia natans* (Rivas *et al.* 2002). Several other non-rhizobial species, belonging to α- and β-subgroup of Proteobacteria such as *Phyllobacterium, Ochrobactrum, Agrobacterium, Cupriavidus, Herbaspirillum, Burkholderia* and some γ-Proteobacteria have been reported to form nodules and fix nitrogen in legume roots (Table3).

Agroforestry systems and plant growth-promoting rhizobacteria (PGPR)

PGPR in combination with efficient rhizobia, could improve growth and nitrogen fixation by inducing the occupancy of the introduced rhizobia in the nodules of the legume (Tilak *et al.* 2006). Silva *et al.* (2006) found out that some *Bacillus* strains with effective Rhizobia resulted in enhanced nodulation in cowpeas (*Vigna unguiculata* L.). Araújo and Hungria (1999) observed similar results in soybeans (*Glycine max* L.), and Figueiredo *et al.* (2008) in beans (*Phaseolus vulgaris* L.) (Table 4). According to Saravana-Kumar and Samiyappan (2007), *Bradyrhizobium* enhanced nodulation and growth of legumes in combination with active 1-aminocyclopropane- 1-carboxylate (ACC) deaminase-expressing PGPR.

Table 4. Effect of co-inoculation with *Rhizobium tropici* CIAT 899 and different PGPR (G1) on nodulation (nodule numbers (NN), nodule dry matter (NDM), individual nodule (IN) and specific nodulation (SN) in the common bean (*Phaseolus vulgaris* [L.])

Treatments (G1)	Nodulation			
	NN (nodule pot^{-1})	NDMa (mg pot^{-1})	IN (μg nodule^{-1})	SN (nodule g RDM^{-1})
CIAT 899 alone	40.3 a	0.370	9.40 a	32.55 b
CIAT 899 + DMS 13796	33.0 a	0.283	9.25 a	30.32 b
CIAT 899 + 65 (E)180	52.0 a	0.310	6.41 a	52.23 ab
CIAT 899 + Loutit (L)	54.7 a	0.322	5.46 a	71.41 a
CIAT 899 + DMS 27	43.0 a	0.353	8.87 a	80.59 a
CIAT 899 + DMS 13411	61.0 a	0.287	4.77 a	56.25 ab
CIAT 899 + DMS 704	64.3 a	0.339	5.53 a	63.85 ab
CIAT 899 + DMS 24	55.3 a	0.310	5.08 a	68.56 ab
CIAT 899 + DMS36	75.4 a	0.350	10.17a	57.82 ab
% CV	11.85	17.99	30.06	16.04

Footnote: CIAT 899 alone (*Rhizobium tropici*); DSM 13796 (*Bacillus endophyticus*); DSM 27 (*B. pumilus*); DSM 704 (*B. subtilis*); DSM 13411 (*Paenibacillus lautus*); DSM 24 (*P. macerans*); DSM 36 (*P. polymyxa*) ; Loutit (L) (*Paenibacillus polymyxa*); 65E180 (*Bacillus* sp.); and one control (without inoculation).

Source: Figueiredo *et al.* (2008)

Agroforestry systems and arbuscular mycorrhiza fungi

Symbiotic arbuscular mycorrhiza (AM) fungi and rhizobia are two of the most important plant symbionts as they play a key role in natural ecosystems and influence plant productivity, plant nutrition, and plant disease resistance (Demir and Akkopru, 2007). Mycorrhizae benefit the host through the mobilization of phosphorus from non-labile sources, whereas Rhizobium fixes nitrogen (Scheublin and Van der Heijden, 2006). AM fungi have widespread presence in nodulated legumes and play an important role in improving nodulation and rhizobial activity within the nodules (Barea *et al.* 2005). The interaction between AM fungi and rhizobia appears mutually beneficial for the development of both the partners (De Boer *et al.* 2005). Suitable combinations of AM fungi and rhizobia may increase plant growth and resistance to pathogens (Aysan and Demir 2009; Artursson 2006). The bacteria involved in the establishment of mycorrhiza or their function were designated as mycorrhiza helper bacteria (MHB) by Garbaye (1994) and are currently the most investigated group of bacteria that interact with mycorrhizas. The MHB strains that have been identified to date belong to many bacterial groups and genera, such as Gram-negative proteobacteria (*Agrobacterium*, *Azospirillum*, *Azotobacter*, *Burkholderia*, *Bradyrhizobium*, *Enterobacter*, *Pseudomonas*, *Klebsiella*, and *Rhizobium*), Gram-positive firmicutes (*Bacillus*, *Brevibacillus*, and *Paenibacillus*) and Grampositive actinomycetes (*Rhodococcus*, *Streptomyces*, and *Arthrobacter*) (Frey-Klett *et al.* 2007) (Table 5).

Table 5. Examples of synergistic interactions between bacteria and AM fungi, potentially leading to enhanced plant growth

Bacterial species	AMF species	Effect	Reference
Gram +, low G+C			
Bacillus pabuli	Glomus clarum	↑ f.g., ↑ s.g., ↑ r.c.	Xavier and Germida (2003)
Bacillus subtilis	G. intraradices	↑p.s., ↑ r.c.	Toro *et al.* (1997)
Paenibacillus validus	G. intraradices	↑f.g.	Hildebrandt *et al.* (2002)
Paenibacillus sp.	G. mosseae	↑f.g., ↑s.g.,↑r.c. + i.p.p.f.	Budi *et al.* (1999)
Gram +, high G+C			
Corynebacterium sp.	G. versiforme	↑s.g.	Mayo *et al.* (1986)
Streptomyces orientalis	Gigaspora margarita	↑s.g.	Carpenter-Boggs *et al.* (1995)
γ-Proteobacteria			
Enterobacter sp.	G. intraradices	↑p.s., ↑r.c.	Toro *et al.* (1997)
Pseudomonas sp.	G. versiforme	↑s.g.	Mayo *et al.* (1986)
Pseudomonas sp.	Endogone sp.	↑f.g., ↑r.c.	Mosse (1962)

(Contd.)

Pseudomonas sp.	G. mosseae	↑f.g. + i.p.p.f.	Barea *et al.* (1998)
Pseudomonas aeruginosa	*G. intraradices*	↑p.s. ↑p.s.	Villegas and Fortin (2001; 2002)
Pseudomonas putida	*G. intraradices*		Villegas and Fortin (2001; 2002)
Pseudomonas putida	Indigenous mix of AMF	↑r.c.	Meyer and Linderman (1986)
Rhizobium meliloti	G. mosseae	↑Nitrogen fixation rates	Toro *et al.* (1998)

Footnote: f.g., fungal growth; s.g., spore germination; r.c., AM fungal root colonization; p.s., phosphate solubilization; i.p.p.f., inhibition of plant pathogenic fungi; ?, enhanced.
Source: Artursson 2006.

Conclusion

Agroforestry Systems promote the permanent input of litter to increase the organic matter content of the soil and positively influence the soil microbial community by providing a large source of carbon and energy. The improvements in the soil microorganism status are important because microorganisms provide many functions in a soil ecosystem, including organic matter decomposition, nitrogen fixation, uptake of phosphorus by mycorrhiza, and the promotion of plant growth. They are involved in fundamental activities that ensure the stability and productivity of both agricultural systems and natural ecosystems. Arbuscular mycorrhizal (AM) fungi and bacteria can interact synergistically to stimulate plant growth through a range of mechanisms that include improved nutrient acquisition and inhibition of fungal plant pathogens. These interactions may be of crucial importance within sustainable, low-input agricultural cropping systems that rely on biological processes rather than harmful agrochemicals to maintain soil fertility and plant health. Although there are many studies concerning interactions between AM fungi and bacteria, further studies are required to understand the underlying mechanisms behind these associations, so that optimized combinations of microorganisms can be applied as effective inoculants within sustainable crop production systems. Certain co-operative microbial activities can be exploited, as a low-input biotechnology, to help sustainable agriculture and agro forestry.

References

Araujo Ademir, Luiz Leite, Bruna De Iwata, Mario De Lira, Gustavo Xavier, *et al.* (2012). Microbiological process in agroforestry systems. A review. Agronomy for Sustainable Development, Springer Verlag/EDP Sciences/INRA 32(1):215-226.

Araújo ASF, Monteiro RTR (2006). Microbial biomass and activity in a Brazilian soil plus untreated and composted textile sludge. Chemosphere 64:1043-1046.

Araújo FB, Hungria M (1999). Nodulação e rendimento de soja coinfectada com Bacillus subtilis e Bradyrhizobium japonicum/Bradyrhizobium elkanii. Pesq Agropec Bras 34: 1633-1643.

Artursson Veronica, Roger D Finlay, Janet K Jansson (2006). Interactions between arbuscular mycorrhizal fungi and bacteria and their potential for stimulating plant growth. Environmental Microbiology 8(1):1-10.

Aysan E, Demir S (2009). Using arbuscular mycorrhizal fungi and Rhizobium leguminosarum, Biovar Phaseoli against Sclerotinia sclerotiorum (Lib.) de bary in the common bean Phaseolus vulgaris L. Plant Pathol J 8:74-78.

Balachandar D, Raja P, Kumar K, Sundaram SP (2007). Non-rhizobial nodulation in legumes. Biotech Mol Biol Rev 2:49-57.

Barea JM, Werner D, Azcon-Aguilar C, Azcon R (2005). Interactions of arbuscular mycorrhiza and nitrogen fixing simbiosis in sustainable agriculture. In: Werner D, Newton WE (eds) Agriculture, Forestry, Ecology and the Environment. Kluwer, The Netherlands.

Barea JM, Andrade G, Bianciotto VV, Dowling D, Lohrke S, Bonfante P, *et al.* (1998). Impact on arbuscular mycorrhiza formation of pseudomonas strains used as inoculants for biocontrol of soil-borne fungal plant pathogens. Appl Environ Microbiol 64:2304-2307.

Budi SW, Van Tuinen, Martinotti DG, Gianinazzi S (1999). Isolation from the *Sorghum bicolor* mycorrhizosphere of a bacterium compatible with arbuscular mycorrhiza development and antagonistic towards soil borne fungal pathogens. Appl Environ Microbiol 65: 5148-5150.

Cardoso MI, Kuyper TW (2006). Mycorrhizas and tropical soil fertility. Agric Ecosys Environ 116:72–84.

Carpenter-Boggs L, Loynachan TE, Stahl PD (1995). Spore germination of *Gigaspora margarita* stimulated by volatiles of soil-isolated actinomycetes. Soil Biol Biochem 27:1445-1451.

Chen W-M, Laevens S, Lee TM, Coenye T, de Vos P, Mergeay M, Vandamme P (2001). Ralstonia taiwanensis sp. nov., isolated from root nodules of Mimosa species and sputum of a cystic fibrosis patient. Int J Syst Evol Microbiol 51:1729-1735.

De Boer W, Folman LB, Summerbell RC, Boddy L (2005). Living in a fungal world: impact of fungi on soil bacterial niche development. FEMS Microbiol Rev 29:795-811.

Demir S, Akkopru A (2007) Using of arbuscular mycorrhizal fungi (AMF) for biocontrol of soil-borne fungal plant pathogens. In: Chincholkar SB, Mukerji KG (eds) Biological Control of Plant Diseases. Haworth Press, USA, pp.17-37.

Figueiredo MVB, Martinez CR, Burity HA, Chanway CP (2008). Plant growth-promoting rhizobacteria for improving nodulation and nitrogen fixation in the common bean (*Phaseolus vulgaris* L.). World J Microbiol Biotechnol 24:1187-1193.

Freitas ADS, Sampaio EVSB, Santos CERS, Fernandes AR (2010). Biological nitrogen fixation in tree legumes of the Brazilian semi-arid caatinga. J Arid Environ 74:344-349.

Frey-Klett P, Garbaye J, Tarkka M (2007). The mycorrhiza helper bacteria revisited. New Phytol 176:22-36.

Garbaye J (1994). Helper bacteria: a new dimension to the mycorrhizal symbiosis. New Phytol 128:197-210.

Heuvelop J, Fassbender HW, Alpizar L, Enriquez G, Falster H (1988). Modelling agroforestry systems of cacao (*Theobroma cacao*) in Costa Rica. II. Cacao and wood production, litter production and decomposition. Agrofor Syst 6:37-48.

Hildebrandt U, Janetta K, Bothe H (2002). Towards growth of arbuscular mycorrhizal fungi independent of a plant host. Appl Environ Microbiol 68:1919-1924.

Hubbell DH, Kidder G (2003). Biological nitrogen fixation. SL-16, soil and water science department, Florida cooperative extension service, institute of food and agricultural sciences. University of Florida, USA, pp 1-4.

Jaftha JB, Strijdom BW, Steyn PL (2002). Characterization of pigmented methylotrophic bacteria which nodulate Lotononis bainesii. Syst Appl Microbiol 25:440-449.

Kaur B, Gupta SR, Singh G (2000). Soil carbon, microbial activity and nitrogen availability in agroforestry systems on moderately alkaline soils in northern India. App Soil Ecol 15:283-294.

Mayo K, Davis RE, Motta J (1986). Stimulation of germination of spores of *Glomus versiforme* by spore-associated bacteria. Mycologia 78:426-431.

Meyer JR, Linderman RG (1986). Response of subterranean clover to dual inoculation with vesicular-arbuscular fungi and a plant growth-promoting bacterium, *Pseudomonas putida*. Soil Biol Biochem 18:185-190.

Mosse B (1962). The establishment of vesicular-arbuscular mycorrhiza under aseptic conditions. J Gen Microbiol 27: 509-520.

Mungai NW, Motavalli PP, Kremer RJ, Nelson KA (2005). Spatial diversity in temperate alley cropping systems. Biol Fertil Soils 42:129-136.

Rivas R, Velázquez E, Willems A, Vizcaíno N, Subba-Rao NS, Mateos PF, Gillis M, Dazzo FB, Martínez-Molina E (2002). A new species of Devosia that forms a unique nitrogen-fixing root nodule symbiosis with the aquatic legume Neptunia natans. Appl Environ Microbiol 68:5217-5222.

Saravana-Kumar D, Samiyappan R (2007). ACC deaminase from Pseudomonas fluorescens mediated saline resistance in groundnut (*Arachis hypogea*) plants. J Appl Microbial 102:1283–1292.

Scheublin TR, Van der Heijden MGA (2006). Arbuscular mycorrhizal fungi colonize nonfixing root nodules of several legume species. New Phytol 172:732-738.

Silva VN, Silva LESF, Figueiredo MVB (2006). Atuação de rizóbios com rizobactérias promotora de crescimento em plantas na cultura do caupi (Vigna unguiculata L. Walp). Acta Sci Agron 28:407-412.

Sørensen J, Sessitsch A (2007). Plant-associated bacterial-lifestyle and molecular interactions. In: Van Elsas JD, Jansson JK, Trevors JT, Sprent JI (eds) Nodulation in Legumes. Royal Botanic Gardens, Kew, London.

Sprent JI (2001). Nodulation in legumes. Royal Botanic Gardens, Kew, London.

Sy A, Giraud E, Jourand P, Garcia N, Willems A, de Lajudie P, Prin Y, Neyra M, Gillis M, Boivin-Masson C, Dreyfus B (2001). Methylotrophic Methylobacterium bacteria nodulate and fix nitrogen in symbiosis with legumes. J Bacteriol 183:214-220.

Tangjang S, Arunachalam K, Arunachalam A, Shukla AK (2009). Microbial population dynamics of soil under traditional agroforestry systems in Northeast India. Res J Soil Biol 1:1-7.

Tilak KVBR, Rauganayaki N, Manoharachari C (2006). Synergistic effects of plant-growth promoting rhizobacteria and Rhizobium on nodulation and nitrogen fixation by pigeonpea (Cajanus cajan). Europ J Soil Sci 57:67-71.

Toro M, Azcón R, Barea JM (1997). Improvement of arbuscular mycorrhiza development by inoculation of soil with phosphate-solubilizing rhizobacteria to improve rock phosphate bioavailability (^{32}P) and nutrient cycling. Appl Environ Microbiol 63: 4408-4412.

Toro M, Azcón R, Barea JM (1998). The use of isotopic dilution techniques to evaluate the interactive effects of *Rhizobium* genotype, mycorrhizal fungi, phosphate-solubilizing rhizobacteria and rock phosphate on nitrogen and phosphorus acquisition by *Medicago sativa*. New Phytol 138: 265-273.

Van Berkum P, Eardly BD (2002). The aquatic budding bacterium Blastobacter denitrificans is a nitrogen-fixing symbiont of Aeschynomene indica. In: Finan TM, O'Brian MR, Layzell DB, Vessey JK, Newton WE (eds) Nitrogen fixation: global perspectives. CAB International, New York, pp. 520.

Vandamme P, Goris J, Chen WM, Vos P, Willems A (2002). Burkholderia tuberum sp. nov. and Burkholderia phymatum sp. nov., nodulate the roots of tropical legumes. Syst App Microbiol 25:507-512.

Villegas J, Fortin JA (2001). Phosphorus solubilization and pH changes as a result of the interactions between soil bacteria and arbuscular mycorrhizal fungi on a medium containing NH_4^+ as nitrogen source. Can J Bot 79: 865-870.

Villegas J, Fortin JA (2002). Phosphorus solubilization and pH changes as a result of the interactions between soil bacteria and arbuscular mycorrhizal fungi on a medium containing NO_3^- as nitrogen source. Can J Bot 80:571-576.

Watanabe K, Hamamura N (2003). Molecular and physiological approaches to understanding the ecology of pollutant degradation. App Soil Ecol 31:120-135.

Xavier LJC, Germida JJ (2003). Bacteria associated with *Glomus clarum* spores influence mycorrhizal activity. Soil Biol Biochem 35: 471-478.

Yadav RS, Yadav BL, Chhipa BR, Dhyani SK, Ram M (2010). Soil biological properties under different tree based traditional agroforestry systems in a semi-arid region of Rajasthan, India. Agroforest Syst 81:195-202.

15

Rainwater Management for Climate Resilience in Arid Region

R.K. Goyal and Mahesh Kumar Gaur

ICAR-Central Arid Zone Research Institute, Jodhpur-342003, Rajasthan, India

Introduction

Rainfall is the principal source of water in arid areas, which augments soil moisture, groundwater and surface flows. Agriculture and several of the other economic activities in arid areas depend on rain. This region is devoid of any well defined perennial river system. Under such circumstances every drop of water becomes very precious. Of the total water use about 85% of water is used for irrigation and remaining 15% is used for drinking, industrial and other purposes. About 65% of irrigation water and 30-40% of drinking water is subjected to serious losses. Hence, increasing water use efficiency coupled with increasing availability of water through rainwater harvesting and management is of prime importance for sustainable development of arid areas. Rainwater harvesting, its conservation and efficient utilization can solve problem of water scarcity to the greater extend. Rainwater harvesting in small ponds (nadis), underground tanks (tankas), Khadins (Low lying areas) etc. is an age-old tradition in arid zone of Rajasthan. These traditional rainwater harvesting structures vary in design, shape and size. These traditional methods have undergone several changes for being more economical and efficient. Central Arid Zone Research Institute (CAZRI), Jodhpur since its inception in year 1959 is continuously working for development and refinement of water efficient techniques for crop production, individuals and for the communities for development in hot arid zone of India.

Rainwater harvesting for arid lands

Since primary source of water in arid zone is rainwater, so for any improvement in availability of water, catchment conditions become crucial. Catchment is the base for harvesting rainwater in form of runoff. Runoff is highly dependent on catchment's shape, size, slope, and type etc. beside rainfall characteristics. Some general modifications can greatly help in enhancing the runoff percentage from catchments.

Techniques for enhancing runoff from catchments

1. Simple earth smoothing and compaction helps increasing runoff from catchment areas. Success is generally greater on loam or clay loam soils. Care must be taken to reduce the slope and/or the length of slope to lessen runoff velocity and thereby reducing runoff.
2. Removal of stones and boulders and unproductive vegetation from catchment helps in uninterrupted flow, enhances runoff to collection site.
3. Land shaping into roads and collection of water in channels.
4. Sandy soils have low water holding capacity. Spreading of clay blanket to the soil surface reduces the infiltration and consequently accelerates runoff.
5. Chemical treatments like wax, asphalt, bitumen and bentonite prevent downward movement of water, which augments runoff.

Khadin system of rainwater harvesting for crop production

Khadin is a unique practice of water harvesting, moisture conservation and utilization in hyper arid region of Rajasthan. This system was designed and developed by the Paliwal Brahmins of Jaisalmer (Rajasthan) in the 15th century. This system has great similarity with the irrigation methods of the people of Iraq around 4500 BC and later of the Nabateans in the Middle East. A similar system is also reported to have been practiced 4,000 years ago in the Negev desert, and in southwestern Colorado 500 years ago. The main feature of khadin is a very long (100-300 m) earthen embankment built across the lower hill slopes lying below gravelly uplands. Sluices and spillways allow excess water to drain off. The khadin system is based on the principle of harvesting rainwater on farmland and subsequent use of this water-saturated land for crop production. The ratio of farmland and catchment areas is regulated to be about 1:10 so that a suitable moisture supply is uniformly maintained. It is suitable for deep soil surrounded by some natural rock outcrops constituting catchment area. CAZRI has developed khadin of 20 ha areas in Baorali-Bambore watershed with surplus arrangements (Fig.1). Before construction of khadin, uncontrolled runoff from upper catchment used to wash away seeds, fertilizers, and standing crops besides loss of valuable water. After construction of khadin, farmer could take excellent kharif and rabi crops (Narain and Goyal 2005). Collecting water in a khadin aids the continuous recharge of groundwater aquifers. Studies of groundwater recharge through khadins in different morphological settings suggest that 11 to 48 % of the stored water contributed to groundwater in a single season. This replenishment of aquifers means that sub-surface water can be extracted through bore wells dug downstream from the khadin. The average water-level rise in wells bored into sandstone and deep alluvium was 0.8 m and 2.2 m, respectively (Khan, 1996).

Fig.1. Khadin with wastewier at Baorali-Bambore watershed (Jodhpur)

Design package and guidelines for khadin construction

Central Arid Zone Research Institute, Jodhpur has prepared the design package and guidelines for construction of khadin by users agencies;

1. Khadin may be defined as a water harvesting system used for runoff farming on stored soil profile moisture.

2. The catchment may be classified on the basis of infiltration rate. In the areas where infiltration rate is less than 5 cm hr^{-1} may be considered as good catchment. The delineation of catchment should be done on the cadastral/village map or G.T. sheet through reconnaissance survey.

3. The average rainfall of over 30 years available at the nearest rain gauging stations should be considered for working out the catchment yield. Log Pearson III method or strange table should be used.

4. For calculation of flood discharge upto 480 ha area Rational Formula and above 480 ha Dicken's Formula may be used.

5. Khadin may be constructed in a area where soil is fine textured, medium to deep with high soil moisture retention capacity. Soil should be free from salinity.
6. In order to have economic design the ponding depth over sill level at the khadin bund may vary from 0.65 to 1.10 m with overall average of 0.60 m.
7. The flood lift may be adopted as 0.3 m.
8. During the ponding period from July-October there will be wave action therefore, a free board of 0.5 m may be considered.
9. The side slopes of the bunds may be generally kept 2.5:1 (D/S) and 2:1 (U/S). However, these would be governed by the type of soil, angle of repose, bund cross section and its safety factor.
10. The top width of khadin bund may be calculated by appropriate formula and not for constructing inspection road.
11. A murrum capping of 7.5 cm thick layer be provided over the bund section for protection against wind and rain erosion.
12. The head outlet sluice of appropriate size may be provided in the khadin bund for the release of the standing water if any before the rabi sowing.

Rainwater harvesting through tanka

Tankas (small tanks) are underground tanks, found traditionally in most part of western Rajasthan. They are generally built in the main house or in the courtyard. *Tanka* is a circular hole made in the ground, lined with fine polished lime for collection of rainwater primarily from rooftop of individual house. The water collected in small tankas is generally used for drinking purposes, however bigger tanka can be used for providing supplemental/life saving irrigation to horticultural plants. During subnormal rainfall when tanka does not get adequate rainwater, water is hauled in camel/bullock cart from nearby wells/nadis to fill the household tankas. For rainwater management, CAZRI has designed underground tanka of 10 m^3 to 600 m^3capacities for different rainfall and catchment conditions (Goyal and Issac 2009). These tankas were successfully constructed in Jhanwar, Sar, Baorali-Bambore (Jodhpur district) and Kalyanpur (Barmer district) villages. Harvested water of these tankas was used to provide life saving irrigation to plants. The Benefit cost ratio of tanka ranged from 1.25 to 1.40 under different uses (Goyal *et al.* 1995, 1997; Goyal and Sharma, 2000).

Tanka should be constructed at appropriate site. If rainwater is to be collected from rooftop, its location should be constructed near to place of intended use. If rainwater is to be collected from natural catchment then tanka should be constructed at one side of depression area for maximum runoff and safe disposal of excess water. In arid area of western Rajasthan, a murrum layer is reported in sub surface strata at many places. Special care is needed when tanka is to be constructed at these sites. Murrum has a tendency of swelling after getting some moisture and causes cracks especially in sidewalls. To avoid these cracks surrounding of whole tanka should have an envelope of 5 cm sand around sidewalls. In case of little leakage from sidewall, sand envelope of 5 cm thickness of sand will absorb the pressure exerted by the swelling of murrum around sidewalls and will prevent the cracks in sidewalls. Circular tankais more economical in comparison to rectangular tanka of same capacity in term of cost of materials. Further rectangular tanka has tendency for development of cracks in four corners due to uneven distribution of pressure whereas in circular tanka pressure distribution is even thus less chance of cracks in sidewalls. Cement concrete is preferred over masonry construction due to cost and life span especially for larger tanka of capacity over 100000 litters. However masonry construction is equally good for small capacity tanka and does not require trained workers for construction as in case of cement concrete. Depth of tanka should be equal to diameter of tanka. Capacity should be based on no. of members in family and livestock. A 21000 litres capacity tanka is sufficient to meet the drinking water requirement of family of 6 persons for round the year (Fig. 2). Approximately 300 m^2 area is needed as catchment under assured minimum rainfall of 180 mm with a runoff coefficient of 0.4 to fill it. 5-10 cm sand filling between masonry wall and earth is recommended for longer life span of tanka particularly under clay and heavy soils. To avoid entry of any pollutant from outside, tanka is covered at top with stone slabs and cement concrete flooring

Fig. 2. Improved tanka of 21 m^3 capacity at village Berania (Dungarpur) under TSP in 2014

with an opening for withdrawal of water. Inlet and outlet is provided with weld mesh to avoid entry of any foreign material and small ruminants. A uniform slope of 2-3 per cent towards tanka is provided for harvesting maximum possible runoff. The improved design of tankas have a lifespan of more than 30 years. Bigger size tankas are more economical from construction and additional income generation point of view. The improved tanka design developed at CAZRI has wide acceptability in the region, which has been widely replicated in large numbers under Rajeev Gandhi National Drinking Water Mission.

Quality of collected rainwater

The cleanest water is always that which falls freely from the sky. The natural water cycle is very efficient in screening out contaminants that are normally found in ground water and other sources. Rainwater does not come in contact with the soil, and so it does not contain contaminants such as harmful bacteria, dissolved salts, minerals or heavy metals. Rainwater is healthy and is soft water so, among other things, you will use less soap. Roof-collected rainwater can be made safe and potable by adopting some simple measures such as cleaning of rainwater storage structure and catchment, diversion of first flush and coarse rainwater filters. The quality of rainwater further improves with time after the rain, mainly due to sedimentation and bacteria die-off. It takes an average of 3.5 to 4 days to achieve a 90% reduction in *E.coli* numbers. It has been proved that people drinking tank rainwater are at lower risk of many diseases than those drinking public mains water.

Rainwater harvesting through nadi

Nadis are village ponds used to store runoff water from adjoining natural catchments during the rainy season. In arid Rajasthan nadi system of water harvesting is the oldest practice and still the principal source of water supplies for human and livestock consumption. Across Rajasthan, most nadis have a capacity of between 1,200 to 15,000 m^3. Water availability in nadi ranged from 2-12 months after the rains. Since nadis received runoff from sandy and eroded rocky basins, large amounts of sediments used to deposit regularly in them, resulting in quick siltation. High evaporation and seepage losses through porous sides and bottom, heavy sedimentation due to biotic interference in the catchment and contamination are major bottlenecks. Evaporation losses ranged from 55 to 80 per cent of the total losses in various environments. Seepage losses are greatest during the rainy season (July-September) when nadi is completely filled. To overcome these problems, CAZRI has developed design for improved nadis with LDPE lining on sides and bottom keeping surface to volume ratio 0.28 and provision of silt trap at inlet.

The site selection of nadi is based on availability of natural catchment and its runoff potential (Fig. 3). The location of the nadi had a strong bearing on its storage capacity due to catchment and runoff characteristics. Nadis are 1.5 to 4.0 m deep in dune areas and those in sandy plains vary from 3 to 12 m. In addition, planting suitable tree species around the nadi creates an oasis in the desert and improves the local environment.

Fig. 3. Improved nadi with inlet and outlet

Rainwater harvesting thorough micro-catchment for tree establishment

Micro-catchment technique is particularly suitable for establishment of trees. In this technique a circular catchment of 1 to 1.5 m radius is constructed around the tree (Fig. 4). The catchment is compacted by roller or any other heavy machine. A slope of 5-10 per cent is provided in catchment towards tree for directing flow of water. The catchment can also be lined with locally available materials such as polythene sheets, lime mortar, stone pieces, grasses, etc. for higher runoff generation. It is reported that the plants with micro-catchment

Fig. 4. Circular micro-catchment for tree establishment

have better chances of establishment in rainfed conditions as compared to conventional plantation technique (Ojasvi *et al.* 1999). In another study Sharma *et al.* (1986) suggested that conversion of canopy area into runoff catchment may be just sufficient for improving its soil moisture profile.

Contour vegetative barriers

Contour vegetative barriers are hedgerows of perennial grasses or shrubs planted at a regular interval on contours for conserving soil and water in sloping lands. Suitable grass species are grown along contours at suitable vertical interval to intercept part of runoff and to control erosion in agricultural fields having flat to slight undulating topography. The contour vegetative barrier moderates the velocity of overland flow and traps silt at low cost, and augment production of food, fuel and fodder or fibre from lands by growing suitable vegetation species. In recent years, contour vegetative barriers have found acceptability among the farmers as these are cheaper over mechanical measures and are protective while being productive. Contour vegetative barriers can be easily established across a wide spectrum of soil-climatic conditions. Selection of species depends upon purpose of barrier, site-specific conditions, particularly soil and climatic variables. The spacing between plant-to-plant and row-to-row is governed by vegetation species to be planted as barrier. In general the plant-to-plant spacing is kept at 20 to 30 cm. Predetermined or 0.5 to 1 m vertical interval between the barriers has been found effective for soil and water conservation. Generally, paired row of barrier planted in staggered form across the slope proves more effective.

Generally, dominant grass or shrub species of the region should be preferred for vegetative barrier. Among grasses *Cenchrus ciliaris, Cenchrus setigerus, Saccharum bengalense, Vetiveria zizanioides, Lasiurus sindicus, Panicum antidotale* and *Panicum turgidum* can be effectively used for soil and water conservation in arid areas. Shrubs like *Leptadenia pyrotechnica, Ipomoea carnea* and *Euphorbia antisyphylitica* can also provide good protection against water and wind erosion.Contour vegetative barriers of *C. jwarancusa, C. ciliaris* and *C. setigerus* transplanted at 0.30 m apart on contours at 0.6 to 1.0 m vertical interval in sandy loam soil of Jodhpur (Rajasthan) have performed well and formed effective barriers in reducing soil erosion and increasing soil moisture storage. In a study conducted during 1992-1994 at 19 farmers' fields near Jodhpur rooted slips of local eight species of perennial grasses (*C. ciliaris, C. setigerus, C. jwarancusa, L. sindicus, P. antidotale, P.turgidum, S. bengalense* and *V. zizanioides*) and seedling of six species of shrubs (*Agave americana, Aloe barbadensis, Barleriaprionitis, E. antisyphylitica, I. carnea* and *L. pyrotechnica*) were transplanted at 1 m vertical interval on contours across the slope. Result indicated that perennial grass species performed the best and formed effective barrier against soil erosion. Runoff volume and specific peak discharge were reduced by 28 to 97 per cent and 22 to 96 per cent, respectively (Sharma *et al.* 1999; Tiwari and Kurothe 2006). In another study conducted at Kalyanpur (Barmer district) during 1998, vegetative barrier of *L. sindicus, Saccharu mmunja* and *Cassia angustifolia* were established at horizontal interval of 30 m. The moisture data revealed 36.5, 72 and 54.2 per cent higher moisture storage as compared to control in *C. angustifolia, L. sindicus* and *S. munja* respectively (Gupta and Rathore, 2002).

References

Goyal RK, Sharma AK (2000). Farm Pond: A well of wealth for the dryland dwellers. Intensive Agriculture XXXVIII (5-6): 12-14.

Goyal RK, Issac VC (2009). Rainwater Harvesting through Tanka in Hot Arid Zone of India. Central Arid Zone Research Institute, Jodhpur, CAZRI Bulletin, pp. 33.

Goyal RK, Ojasvi PR, Bhati TK (1995). Economic evaluation of water harvesting pond under arid conditions. Indian Journal of Soil Conservation 23(1): 74-76.

Goyal RK, Bhati TK, Ojasvi PR (2007). Performance evaluation of some soil and water conservation measures in hot arid zone of India. Indian Journal of Soil Conservation 35(1): 58-63.

Gupta JP, Rathore SS (2002). Biomass Production and Rehabilitation of Degraded Lands in Arid Zone. Central Arid Zone Research Institute, Jodhpur, pp. 20.

Khan MA (1996). Inducement of groundwater recharge for sustainable development. In: Proceedings of the 28th Annual Convention, Indian Water Works Association, Jodhpur, India, pp. 147-150.

Narain P, Goyal RK (2005). Rainwater harvesting for increasing productivity in arid zones. In: National Symposium on Efficient Water Management for Eco-friendly Sustainable and Profitable Agriculture, Abstract, Indian Society of Water Management and Indian Agriculture Research Institute, and Water Technology Centre, New Delhi, pp. 141-142.

Ojasvi PR, Goyal RK, Gupta JP (1999). The micro-catchment water harvesting techniques for the plantation of Jujube (*Zizyphus mauritiana*) in an agroforestry system under arid conditions. Agricultural Water Management 41: 139-147.

Sharma KD, Joshi NL, Singh HP, Bohra DN, Kalla AK, Joshi PK (1999). Study on the performance of contour vegetative barriers in an arid region using numerical models. Agricultural Water Management 41: 41-56.

Sharma KD, Pareek OP, Singh HP (1986). Micro-catchment water harvesting for raising jujube orchards in arid climate. Transactions of American Society of Agriculture Engineering 29(1): 112-118.

Tiwari SP, Kurothe RS (2006). Effect of vegetative barriers on soil and nutrient losses at 2% slope on agricultural lands of reclaimed Mahi ravines. Indian Journal of Soil Conservation 34(1):37-41.

16

Horticulture Based Agroforestry in Arid Region of Rajasthan

A.K. Shukla, M.B. Noor mohamed, Keerthika A, Dipak Kumar Gupta B.L. Jangid and P.L. Regar

ICAR- Central Arid Zone Research Institute (CAZRI), Regional Research Station Pali, Marwar, Rajasthan – 306 401, India

Introduction

Arid regions of the world are diverse in terms of climate, soils, vegetation, animals and life style and activities of people. The binding feature of all arid regions in the world is aridity. Of the total area of arid zones of the world, Africa, accounts for 46.1 % followed by Asia (35.5 %). Majority of rest 19.4 % of arid zones are spread over in Australia, North America (Mexico and Southern part of USA), and South America. The Indian arid zone covers around 12 % of country's geographical area occupying 31.8 million ha of land of which major part is in northwestern India (28.57 Mha) and some in southern India (3.13 Mha). It covers parts of Andhra Pradesh, Gujarat, Haryana, Karnataka, Maharashtra, Punjab and Rajasthan states of India (Table 1 and Fig. 1). The arid regions of Rajasthan, Gujarat, Punjab, and Haryana together constitute Great Indian Desert, better known as Thar. As arid western Rajasthan accounts for 61% of hot arid region of country, therefore it is considered principle hot arid region. Major part of it occurs between Aravalli ranges on the east and southeast and Thal desert of Pakistan (Thal desert is simply the western extension of Thar, only name has been changed) which is spread up to Suleman Kithara ranges in extreme west.

The region is characterized by frequent drought and extremes of aridity due to low to erratic rainfall (100-420 mm yr^{-1}), extremes of temperature (-5 to 50°C), long sun shine duration (6.6-10 hours), low relative humidity (30-80%), high wind velocity (9-13 km h^{-1}) and high evapo-transpiration (1500-2000 mm). Nearly

70% of the region is occupied by light textured sandy to sandy loam soils. Poor fertility, low water holding capacity, high erosion and undulating topography pose problem to soil management in this region. On irrigated lands the ground water availability is meager and problematic particularly due to deep water table, high salinity and negligible recharge. Sand dunes are a dominant land formation of the region. More than 58% of the area is sandy and intensities of dunes vary place to place. In general, soils contain 1.8-4.5% clay, 0.4-1.3% silt, 63.7- 87.3% fine sand and 11.3-30.3% coarse sand. Vegetation constitutes primary source of life support where animal husbandry being major vocation of people that depends entirely on natural vegetation, besides being of direct economic relevance, provides stability to wind prone sandy friable surface covering nearly two-third of the region. Further, inhospitable climate, too deep or too shallow soils with low moisture and poor fertility, deep underground water, which is often brackish or saline, coupled with intense biotic pressure permits specialized plants, which are well adapted to these climatic, edaphic and biotic adversities and fluctuations. With the increasing biotic pressure, most of the arid and semi-arid regions are confronted with the challenges of producing more per unit land with uncertain and dwindling supplies of water.

Table 1. Distribution of arid regions in different states of India

State (s)	Area (million hectares)	Total Percentage
Rajasthan	19.61	61.00
Gujarat	06.22	19.60
Punjab and Haryana	02.73	09.00
Andhra Pradesh	02.15	07.00
Karnataka	00.86	03.00
Maharashtra	00.13	00.4
Total	31.70	-

Bhandari *et al.* (2014)

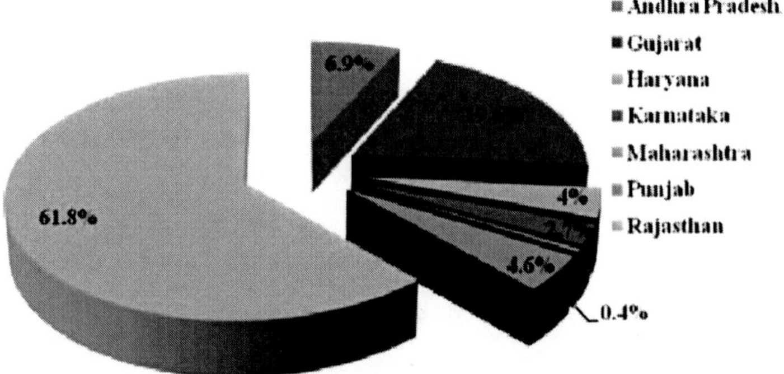

Fig. 1. Distribution of hot arid region in different states of India
Source: Bhandari *et al.*, 2014

The production and life support system in this part of hot Indian arid zone are constrained by climatic limitations. The human population density of this part of the country is quite high (127 person km^{-2}) as compared to global average of 6–8 persons km^{-2} for arid zones. Large tracts of lands in Thar desert region of Rajasthan having widely scattered trees/shrubs of various species in association with crops of food grain and fodder as the best example of traditional agroforestry. To overcome this adverse effect, agroforestry is way of life in arid western Rajasthan and this approach has been widely advocated as a means to harmonize use of scarce inputs so as to make production system sustainable and climate smart. There are many pathways through which agroforestry qualify as an adaptation to climate change, especially in arid regions of the country. In fragile ecosystems of hot arid regions of India, agriculture alone cannot be a dependable enterprise, hence the desert dwellers with their traditional wisdom are integrating forestry into farming since ages in order to confer stability and generate assured income (Narain and Tewari 2001).

Fruit based diversified cropping system for arid region

In the hot arid region of Rajasthan, agricultural development is a challenging task. Crop yield under arid conditions is low and uncertain. The climate is hostile, soils are infertile and unproductive, and biotic pressure is intense. Agriculture is subject to uncertainty due to erratic and low rainfall. About 95% area of the arid western Rajasthan depends upon monsoon rains for crop production. Raising trees and shrubs in the arid zone ensures against drought due to production of wood, leaves, pods or fruits even under such difficult conditions. In such circumstances, growing fruit trees with crops can meet the need of desert dwellers in terms of balanced diet, fuel-wood, fodder and other useful products. Fruit based cropping system is now considered to be the most ideal strategy to provide food, nutrition and income security to the people (Awasthi and Pareek, 2008). Integration of annual crops with fruit trees yields multiple outputs that ensure production and income generation (Randhawa 1990; Osman 2003) in a sustainable manner. Natural resources (land, solar radiation, water, soil) as well as socio-economic inputs (labour, credit, power, market infrastructure) are efficiently utilized. Crop productivity under tree canopy is enhanced due to improved soil fertility and ameliorative influence of shade by reducing understorey temperature and evapotranspiration (Bunderson *et al.* 1990). Incorporation of fruit trees into the cropping system is more remunerative. It increases the resilience of the system over time. It enables to

i) Maximize system productivity on annual basis,

ii) Utilize resources with high efficiency through due consideration of various interactions and direct, residual and cumulative effects occurring in soil-plant-atmosphere continuum,

iii) Intensify input use *vis-à-vis* quality of environment,

iv) Impart sustainability of farm resources and environment in long term perspective.

Some of the horticultural crops which is suitable in arid region of Rajasthan has given below.

1. Indian Jujube (*Ziziphus mauritiana* Lamk.)

The Indian jujube (Ber) of family Rhamnaceae is one of the most ancient cultivated fruit trees in north Indian plains. It grows even on marginal lands or inferior soils where most other fruit trees either fail to grow or give very poor performance. It is regarded as the king of arid zone fruits and also as poor man's apple. There are three main species found in the country. The *Z. mauritiana* is the main species of commercial importance with its several varieties. *Z. nummularia* is prized for leaves (rich in protein) which provide fodder (Pala) for livestock. The third one, Z. rotundifolia also bears edible fruits but of smaller size. It is used as rootstock for commercial Indian jujube. The seeds contain saponins, jujubogenin and obelin lactone. Jujube fruits contain fairly high amount of vitamin C, besides vitamin A, B, protein, calcium and phosphorus (Jawanda and Bal 1978).

It is a perennial hardy fruit tree which gives income from multiple products such as fruits, fodder and fuel wood even in severe drought conditions to the resource deficient farmers. It is the only fruit crop which can give good returns even under rainfed conditions and can be grown in a variety of soils and climatic conditions ranging from sub-tropical to tropical.

2. Indian Mesquite (*Prosopis cineraria* (L.) Druce)

Indian mesquite (khejri) of family Fabaceae is an important component of farming system and plays significant role in the economy of Indian desert. It is found growing in the arid and semi-arid parts of Rajasthan, Gujarat, Haryana, Punjab, Delhi and some parts of southern India. This tree grows well in all sorts of climatic constraints which is evidenced by the fact that new foliar growth, flowering and fruiting occur during extreme dry months (March–June) when most other trees of the desert remain leafless or dormant. Because of its multiple economic value and suitability in agro forestry systems, it is conserved in arable land where its population is regulated by the farmer. All arid land forms except hills and saline depression receiving an average rainfall of 150–500 mm are having good density of the tree. The immature pods are rich in crude protein, carbohydrates and minerals. Duhan *et al.* (1992) recorded 18 % crude protein, 56 % carbohydrates, 0.4 % each of phosphorus and calcium and 0.02% iron in immature pods, which are used as vegetable both fresh as well as after dehydration, while ripe dried pods having 9–14 % crude protein and 6–16 % sugar (Arya *et al.* 1991) can be powdered and used in the preparation of bakery items such as biscuits and cookies. The variability in pod quality traits is most important from the horticultural quality point of view. Diversity was observed in pod characteristic such as taste, tenderness, fiber content, color, length, thickness, seed number, seed size, protein content and mineral constituents (Pareek 2002).

3. Indian Gooseberry (*Emblica officinalis* Gaertn.)

The Indian gooseberry (Aonla) of family Euphorbiaceae is being cultivated in India since Vedic Era. As a result of intensive research and development, it has attained commercial status and also proved to be potential fruit crop for arid ecosystem. It is hardy, prolific bearer and highly remunerative even without much care and can be grown in variable agro-climatic and soil conditions. The fruits are recognized for their nutritive, medicinal and therapeutical values and are rich source of vitamin C (4–9 mg/g^{-1}), pectin, iron, calcium and phosphorus. The fruit is the main ingredient in Chayvanprash and triphala used in Ayurvedic medicine.

4. Indian Cherry (*Cordia myxa* L.)

Indian cherry of family Boraginaceae, locally known as lasoda is another important fruit plant suitable for arid and semi-arid regions of India. Its fruits and other parts have multiple uses in human health, nutrition and other uses. Green unripe fruits are important as fresh vegetable and pickles during April–May when availability of conventional vegetables is scarce. The species is also important ecologically in providing vegetative cover as tree component of arid farming system, preventing soil erosion and promoting biodiversity. The advantage with this species for agro forestry system is that it offers least competition with rainy season crops since its fruiting season is during summer season when main crops are already harvested. This plant also offers scope in using harvested rain water for fruit production since it requires irrigation only for 2–3 months period during summer season (April–June).

5. Pomegranate (*Punica granatum* L.)

Pomegranate (Anar) of family Lythraceae is an economically important commercial fruit crop of arid and semi-arid regions. Commercial plantations of pomegranate exist in Maharashtra, Gujarat, Rajasthan, Andhra Pradesh and Karnataka owing to its preference for arid climate. Its xerophytic characteristics and hardy nature makes it suitable crop for dry, rainfed, pasture and undulating land, where other fruit crops cannot grow successfully. Besides, being a favorite table fruit it is also used for preparation of juice and squash. Dried seeds give an important condiment coined as anardana. It also has medicinal value and rind is being used for dyeing cloths.

6. Kair (*Capparis decidua* (Forsk.) Edgew)

Kair is a multipurpose, perennial, woody shrub or small tree of family Capparaceae which grows widely without much care in the Thar Desert of western Rajasthan. It is much branched, leafless bushy and thrives well in the most adverse climatic conditions and in the soils of poor fertility. It is highly suitable for stabilizing sand dunes and controlling soil erosion by wind and water. Due to its xerophytic adaptive nature, the plant grows successfully under harsh climatic conditions. Its berry shaped unripe fruits are rich in carbohydrates, proteins and minerals used as fresh vegetables and in the preparation of pickles. Dehydrated fruits are used in the off season as vegetable either alone or in combination with other dried vegetables. In general, it is highly valued by inhabitants of hot arid areas (Meghwal and Vashishtha 1998).

7. Karonda (*Carissa carandas* L.)

Karonda is an evergreen spiny shrub or a small tree up to 3 m height and suitable for arid tropics and sub-tropics. It grows successfully on marginal and wastelands. The plant is also useful for making attractive thorny dense hedge around any fruit orchard. It yields a heavy crop of attractive berry like fruits which are edible and rich in vitamin C and minerals especially iron, calcium, magnesium and phosphorus. Mature fruit contains high amount of pectin and, therefore, besides being suitable for making pickle, it can be exploited for making jelly, jam, squash, syrup and chutney, which are of great demand in the international market. Its main flowering season is March–April with fruits maturing during August–September which enables the plants to make best use of monsoon rain. However, some varieties/plant types also flower during October–November.

8. Bengal Quince (*Aegle marmelos* (Linn.) Correa)

Bengal quince (Bael) of family Rutaceae is an indigenous hardy fruit crop and can be grown successfully in dry areas. It is well known for its nutritional and therapeutic properties. The ripe fruits are laxative and unripe ones are prescribed for diarrhea and dysentery and are in great demand for native system of medicine such as Ayurvedic. Various chemical constituents, viz. alkaloids, coumarins and steroids have been isolated and identified from different parts of bael tree such as leaves, wood, root and bark by various workers. The marmelosin content of fruit is known as the panacea of the stomach ailments.

9. Kinnow (*Citrus reticulata* Blanco)

Kinnow mandarin is a hybrid cultivar of citrus developed by crossing King (*Citrus nobilis*) with Willow leaf (*Citrus deliciosa*) and is extensively grown in Punjab and Rajasthan states of India. This easy peel citrus developed has assumed special economic importance and export demand due to its high juice content, special flavor, and as a rich source of vitamin C. Its beautiful golden orange color, abundant juice, excellent aroma and taste have contributed greatly to the success of this fruit.

10. Kachri (*Cucumuscallosus* (Rottl.) Cogn)

Kachri, a drought hardy cucurbit grows naturally and is also cultivated with rainfed crops. It is a short duration crop; flowering starts just after 30–35 days of sowing and produces 2–8 kg per vine small sized edible fruits of high nutritive value with little care. The fruits are rich in minerals specially calcium and used fresh in garnishing of vegetables. Most desert inhabitants store fruits round the year as dried slices, whole dried or in powder form for use. This is generally grown in all farming systems in combination with other rainy season crops without any competition for available soil moisture.

Major horti based agroforestry systems in Rajasthan

Based on the diversity of fruit, vegetable and other crops grown in arid region, the major cropping systems that emerge at the macro level are (Singh and Verma 2010):

1. Horti-Agri system
2. Horti-Pastoral system
3. Horti-Agri-Pastoral system

(i) Horti-Agri system

Growing of vegetable crops, pearl millet, moth bean, cluster bean, and gram between khejri, ber, lasoda, pilu and kair is a prevalent traditional Horti-Agri system in the arid region. Growing of fruit trees as an overstorey component and ground storey crops in association enforce the dual purpose of moisture and nutrient extraction in subsurface as well as above the soil surface (Table 2). Singh (1980) reported that in arid condition of western Rajasthan about 25-40 per cent water penetrates into murram sub-stratum during the rainfall or irrigation which isof no use for moderately rooted crops/grasses but inclusion of fruit trees under different alternateland use system due to their deep rooting characteristics, the water can be efficiently utilized by the fruit trees.

Table 2. Recommended Agri-horti crops component for Indian arid regions

Growing conditions	Horticultural component			Crop component
	High storey	Medium storey	Ground storey	
Rainfed (150–300 mm)	Bordi and Indian mesquite	Jharber	Cucurbits and guar	Guar, moth bean, pearl millet and sesame
Rainfed (300–500 mm)	Indian cherry, Indian jujube and Indian mesquite	Jharber	Cowpea, cucurbits, guar and Indian bean	Cowpea, guar,green gram, mothbean, pearl milletand sesame
Irrigated	Bengal quince, Indian gooseberry Indian jujube and Indian mesquite	Guava, kinnow, karonda, lime, pomegranate and sweet orange	Brinjal, chilli, colecrops, cucurbits, garlic,okra, onion, peas, root/leafy vegetables and tomato	Chickpea, greengram, groundnut mustard and seedspices

Source: Bhandari *et al.* (2014)

(1) *Ber* based system

Growing of vegetable crops, pearl millet, moth bean, cluster bean, and gram between the trees of *Ziziphus nummularia* (Jharber), *Ziziphus rotundifolia* (Bordi) and *Z. mauritiana* is a prevalent traditional practice in the arid region. Crops like pearl millet, cluster bean and moth bean grow in association with jharber even on soil sediments in rocky plateaus near Jaisalmer and Jodhpur. Cluster bean has been found to be a good intercrop at all the locations. In agrihorticulture study involving ber plants and mung bean during subnormal years when rainfall was 51.3% less than the long term average of 360 mm at Jodhpur, the yield of mung bean grown with ber was reduced by 44.2% whereas under sole crop mung bean yield was reduced by 51% (Faroda 1998). Gupta *et al.* 2000 reported that three-year-old plantation of ber (*Z. mauritiana*) at a density of 400 plants/ha in association with greengram performed very well with a seasonal rainfall of 210 mm. Intercropped green gram yielded only 160 Kg ha_{-1} as against 620 kg ha^{-1} from pure crop. The fruit yield from the intercropped system increased the net profit to 2,886. This system is however generally recommended for areas with rainfall more than 250-300 mm.

Leguminous crop sown under ber (*Ziziphus mauritiana* cv. Seb) plantation produced 0.2 t ha^{-1} of grain and 0.8 t ha^{-1} quality ber fruits from same land unit even when seasonal rainfall is 200 mm, thus rendering a drought proofing mechanism to the system. The density of ber plants were kept 400 individuals ha^{-1} (Gupta *et al.* 1997). The economics of this improved system indicated that in case of sole leguminous crop (mung bean) farming, the net profit per hectare was Rs. 4800, however, in case of ber intercropping, the profit was Rs. 8000 ha^{-1} (Table 3). (Roy *et al.* 2011).

Table 3. Improved Agroforestry practices in Horti-Agriculture system (Ber+Mung bean)

Treatment	Annual Rainfall (mm)	Fruit Yield (Kg/ha)	Grain yield (Kg/ha)	Net profit (Rs. /ha)
Sole crop	200	-	520	4800
Intercropped with Ber	200	800	200	8000

Roy *et al.* (2011)

(2) Aonla-based system

Aonla is one of the important fruit crops of arid region. However, despite its adaptability in arid climate, productivity and economic return of the crop are affected by low temperature (-2°C) and frost (CIAH 2006) which cause economic loss to the growers. In order to mitigate the risk of total crop failure, suitable multistory crop combinations were evaluated at Central Institute for Arid Horticulture during 2004-2008 in the inter space of aonla orchard which could generate extra income, improve productivity, ameliorate and improve ecological situations (Awasthi *et al.* 2009). Moth bean grown during *kharif* season was a common ground storey crop grown in rotation with *rabi* crops i.e., fenugreek, chick pea, mustard and cumin. Growth parameters in terms of plant height, stem girth, canopy spread and canopy volume of aonla was recorded to be significantly more with intercrops compared with its sole plantation. Higher grain and straw yield were recorded in moth bean-chickpea (497, 1250 kg ha^{-1}) and moth bean-fenugreek (465, 1161 kg ha^{-1}) crop sequence. Amongst the *rabi* crops, grain yield of fenugreek, chickpea, mustard and cumin were higher by 28.05, 38.11, 19.96 and 36.05%, respectively, when grown in association with aonla compared to its sole crops. The highest net profit (28260/ha) was obtained from moth bean-cumin cropping system, followed by moth bean-chickpea (25024/ha)cropping system. Moth bean-chickpea intercropping with aonla supplemented 22.01, 5.00 and 27.90 kg ha^{-1} nitrogen, phosphorus and potassium through crop residues, followed by moth bean, fenugreek crop sequence. In a similar study, Arya *et al.* (2010) recorded significant differences inyield levels of perennial as well as ground storey components in multi species cropping models as compared to sole cropping. The growth and yield in perennial components were more under the multispecies cropping system, i.e. aonla-ber-karonda-cluster bean-brinjal and aonla-ber-karonda-cluster bean-fallow. Minimum yield was recorded in sole perennial crops. Plant height, number of branches/plant, pods/silique/fruits/plants and yield was found to be superior in multispecies cropping systems of aonla-ber-karonda-cluster bean-brinjal and aonla-ber-karonda-mothbean-Indian mustard as compared to sole cropping except mustard where a reverse trend was observed.

(3) Khejri based system

Prosopis cineraria, commonly known as *khejri* is considered as a "Kalpvriksha" of the Indian desert. It is one the chief indigenous tree of the Indian north-western plains and gently undulating ravine lands. The trees have monolayer canopy and deep root system and provide nutritious edible pods and fuel wood to the human beings, nutritious forage to the animals and fertility to the poor soils. Farmers usually grow jowar, pearl millet, moth bean and cluster bean between the scattered khejri trees in the arid region. Puri *et al.* (1992, 1994) observed the yield of chickpea also improved and resulted in higher soil fertility and moisture conservation. Maximum net return (15,197 ha^{-1}) and benefit-cost ratio (3.73) were obtained when pearl millet in kharif was followed by toria in *rabi* between the khejri trees (Kaushik and Kumar 2003). Yield of fodder crops during both *kharif* and *rabi* seasons was higher in association with *khejri* trees as compared to sole cropping (Table 4).

Table 4. *Prosopis cineraria* based cropping system

Location	Average Rainfall (mm)	Crops
Bikaner	240	Jowar, Pearl millet, Moth bean, Cluster bean and Sesame
Jodhpur	290	
Barmer	350	
Nagaur	362	
Pali	490	

Source:Awasthi *et al.* (2007)

(4) Cordia based system

Lasora (*Cordia myxa*.) also known as cherry of the desert, owing to its higher productivity, suitability to adverse soil and climatic conditions and high processing value is now becoming popular as monoculture and horti-agri system. *Cordia myxa* has been observed to be a suitable tree in association with which several crops can be grown (Table 5).

Table 5. Cordia based agroforestry system in arid zone

Location	Average Rainfall (mm)	Crops	
		Rainfed	Irrigated
Jodhpur (Pipad)	290	Pearl millet, Cluster bean Wheat, Green gram	Rapeseed mustard,
Pali and adjoining areas	490	Vegetables, Pearl millet, cluster bean, taramira	Raya, Wheat Green gram
Jalore, Sirohi	434-544	Vegetables, cluster bean, Pearl millet	Rapeseed Mustard
Bikaner, Barmer	243-350	Pearl millet	Rapeseed Mustard

Source:Awasthi *et al.* (2007)

(ii) Horti-pastoral system

A combination of fruit trees and pasture species, commonly known as "Horti-pastoral" system is one of the several ways to satisfy human needs and alleviate cattle hunger. The system involves growing of fruit trees and grasses in combination during the initial and later years of trees growth as per the suitability of the species. Fruit trees usually form the first tier where as grasses are grown as ground storey crop. Such system sustains the requirement of rural masses for fodder through leaves, grasses and fruits, which is a much-needed component in human diet. In such system it is however, important to ensure that forage crops do not effect the growth and development of the main crop. Trees such as Jharber or bordi are commonly grown in the rural areas and is an important top feed species. Even in drought years when crop fail, the top feed of this species comes to the rescue of the farmers. The trees are heavily lopped during winter for itsleaf fodder known as "pala". One-hectare area of jharber may yield about 125 kg of pala. The perennial grasses such as sewan, anjan and dhaman besides karad (*D. annulatum*), *P. antidotale* and *Schima nervosum* are found in natural pasturelands, gochar, oran and range lands and give high forage yield under rainfed conditions.

Suitable ber based pastoral combinations for arid regions are ber-anjan; jharber-anjan; jharber-sewan and boradi-anjan. Results of growing different grass species between productive and unproductive ber trees with varying degree of reports have been reported. Ber based-pastoralstudies on sandy rangelands of Rajasthan revealed that *Cenchrus ciliaris- Ziziphus mauritiana* system produced 1.2 t ha^{-1} forage and did not affect the fruit yield (Sharma and Diwakar 1989). Ber trees were planted at spacement of 6X6 m and grass *Cenchrus ciliaris* was introduced between tree rows after third year. On average, the dry grass production was 1.55 t ha^{-1} yr^{-1} (Tewari *et al.* 1999). The fruit, leaf fodder and fuel wood production from ber was 2.77, 1.87, 2.64 t ha^{-1} yr^{-1}, respectively (Table 6).

Table 6. Horti-Pastoral system in Farmers field

Year	Dry grass (t/ha)	Tree products		
		Fuel wood (t ha^{-1})	Leaf fodder (t ha^{-1})	Fruit yield (t ha^{-1})
1	1.50	3.10	2.13	3.10
2	1.67	2.74	1.80	2.74
3	1.42	2.46	1.71	2.49
4	1.66	2.49	1.63	2.85
5	1.48	2.43	2.01	2.80
6	1.57	2.63	1.94	2.67
Average	1.55	2.64	1.87	2.77

Source: Tewari *et al.* (1999)

(iii) Horti-Agri-Pastoral system

Horti-Agri-Pastoral system refers to the combinations of fruit crops along with other agricultural or fodder crops in a manner that farmers gets some return throughout the year. This is however, variable and depends on the demand of a farmer and agro-climatic situations. For instance, if a farmers approach is livestock oriented, he can diversify his crops such as Ber-Kair-Grass or Ber-Pilu-Lana (*H. salichornicum*) or Ber-Drumstick-Grass, Through these systems the farmers can obtain fodder through ber, pilu, lana (a preferred feed for camels) and moringa leaves as top feed whereas fodder through grasses. Lana (*Haloxylon salichornicum*) is very common andpreferred feed by the camels. It is naturally found in saline patches of arid region. The cropmodels have to be critically adjusted seeing into the rooting behaviour, fruiting period and food-cum-fodder demand particularly during the lean period. The crop-pasture components selected in the models can be made to interact synergistically to increase the productivity and generate higher net returns. While growing agricultural and fodder crops in association with fruit trees due consideration should be given on tree morphology and other characteristic so that there is asymbiotic association between over storey and ground storey components.

Role of horti based agroforestry in arid region of Rajasthan

1. Production and Livelihood Improvement

Agroforestry systems are helpful in maintaining soil productivity at optimum levels over a long period of time, when compared to agricultural crops alone, because the leguminous trees used in agroforestry systems fix nitrogen. Leaf litter also generally aids micro-nutrients in the soil. Combining agricultural crops with trees helps in increasing the productivity of the land. Studies were initiated at Arid Forest Research Institute, Jodhpur (AFRI) in 1991 on several agroforestry models with varying tree densities and crop combinations. Intercropping and a low tree density at 400 trees ha^{-1} was beneficial for *P. cineraria* (khejri) compared to an uncropped and high tree density of 1600 trees ha^{-1}. In another experiment, it was observed that forage yield decreased with increasing density of P. cineraria trees (Ahuja 1980).

In view of the increasing fodder and fruit requirement by rural people for themselves and their livestock, studies with fodder and fruit yielding tree species in agroforestry system were started at AFRI in July 1994. Three tree species, viz. *Emblica officinalis*, *Colophospermum mopane* and *Hardwickiabinata*, were planted. The crop was harvested and the yield of mung bean was recorded. The highest yield was found under *Emblica officinalis* (330 kg ha^{-1}) followed by *C. mopane* (313 kg ha^{-1}) and *H. binata* (303 kg ha^{-1}), respectively (ICFRE 1995). The characteristics of under grown crops also determine the impact of canopy cover. Agroforestry gives more income to the farmer per unit area of land than pure agriculture or forestry. Agroforestry contributes to livelihood improvement in arid regions of India, where people have a long history of accumulated local knowledge. Arid regions are particularly notable for ethano-forestry practices and indigenous knowledge system on growing trees on farm lands. The economics of previously discussed traditional agroforestry systems of arid western Rajasthan indicated that net B:C ratio of such systems is generally on positive sides.

Table 7. Economics of traditional agroforestry systems of arid western Rajasthan

Agroforestry system	Expenditure (Rs. ha^{-1})	Returns (Rs. ha^{-1})			Gross returns (Rs. ha^{-1})	Net returns (Rs. ha^{-1})	Net B:C ratio
		Crops	Fuel wood	Leaf fodder			
*P. cineraria- A. nilotica*based	1850	4103	1230	870	6203	4353	2.3
P. cineraria	1550	3670	600	420	4690	3140	2.0
Z. spp. – P. cineraria based	1550	1506	620	600	2726	1176	0.7
Z. spp. – P. cineraria – Salvadora spp.	1500	1400	500	500	2400	900	0.6

2. Nutrient content of important fruit trees from arid zone of Rajasthan

Forests have provided food and shelter to man since ages. About 20% of the plants occurring in the forests are reported to have direct utility to mankind.

Around 600 plant species in Indian forests are enumerated to have food value. Arid zone vegetation comprises a wide range of edible fruit bearing and food producing species viz., *Capparis decidua* (Kair), *Cordia dichotoma*(lasora), *Ziziphus mauritiana* (ber), *Ziziphus nummularia* (Bordi), *Salvadora oleoides* (Jal), *Balanites aegyptiaca* (Hingota), *Prosopis cineraria* (Khejri) etc. which play an important role in the nutrition of children in rural and urban areas alike and are relished by them. Most of these fruits are rich sources of protein and energy. Kair is a rich source of fibre, vitamin A and vitamin C. Ber is richer than apple in protein, phosphorous, calcium, carotene and vitamin C. However, they are often undervalued and underutilized as more exotic fruits become accessible. Also most of these are not cultivated and there is only scant and dispersed knowledge about them. Their production and consumption provides a dietary supplement as well as commercial opportunity. Potential fruit species from arid region are reviewed in context with their nutrient contents (Rathore 2009).

3. Soil and environmental improvement

Agroforestry practices play an important role in improving the fertility of the soils. Increases in amount of nitrogen and phosphorus and other macro- and micronutrients are reported under the growth of 14 years old plantation of *P. cineraria,* as compared to a bare site and *Prosopisjuliflora* of the same age (Shankarnarayan, 1984). It is reported that the forage species produce higher biomass under *P. cineraria* tree canopy due to a high fertility status (Aggarwal *et al.*1976; Singh and Lal 1969). Soil profile characteristics have been studied in *P. cineraria* and *A. nilotica* plantations (Singh and Lal, 1969). The silt + clay increased up to a depth of 120 cm, and at subsequent depths the reverse trends were observed. However, in open field devoid of trees, the increase in silt + clay was found only up to a depth of 90 cm (Singh and Lal, 1969). Such improvements in soil physical conditions resulted in the maintenance of higher soil moisture beneath these trees (Gupta and Saxena, 1978).

Windbreaks and shelterbelts are known to have beneficial effects on agricultural production throughout the world (Frank *et al.* 1976). Experimental shelterbelts of *A. nilotica* and *Dalbergia sissoo* were established over 102 km in square blocks at the Central Mechanized Farm of Central Arid Zone Research Institute (CAZRI), Jodhpur at Suratgarh in Bikaner District, Rajasthan. Reduction in wind speed was greater in the monsoon period (July-September) than the summer period before the monsoon (April-June). There was a 5-14% reduction in evapotranspiration from April to July. In general, the use of shelterbelts brought about a 50% reduction in the magnitude of wind erosion. In the *Cassia siamea* shelterbelt, the soil loss was 184.3 kg ha^{-1}, while in bare soil (i.e. without any shelterbelt) it was 346.8 kg ha^{-1} (Roy *et al.* 2014).

4. Agroforestry systems play a critical role in controlling soil erosion

There are several ways by which climate change manifests soil degradation. Higher temperatures and drier conditions lead to lower organic matter accumulation in the soil resulting in poor soil structure, reduction in infiltration of rain water and increase in runoff and erosion (Rao *et al.* 1998) while the expected increase in the occurrence of extreme rainfall events will adversely impact on the severity, frequency, and extent of erosion (WMO 2005). The woody component in agroforestry systems helps in reducing soil erosion which is the most harmful abiotic stress in arid region (Table 8). It was observed if rows of trees are planted right angle to wind direction a tremendous amount of soil drop can be checked from agricultural fields. The shelterbelts on agricultural fields form a type of agroforestry practice (Gupta *et al.* 1984).

Table 8. Effect of different type of shelterbelts on soil erosion

Type of shelterbelt	Total amount of soil loss kg ha^{-1}		
	Year I	Year II	Mean
Prosopis juliflora	93.2	609.3	351.2
Cassia siamea	91.5	277.1	184.2
Acacia tortilis	106.0	494.1	300.0
Agricultural field without trees	262.7	831.0	546.8

Source: Roy *et al.* (2014)

5. Carbon sequestration through horti based agroforestry systems

Carbon sequestration in different agroforestry systems occurs both belowground, in the form of enhancement of soil carbon plus root biomass and aboveground as carbon stored in standing biomass (Fig. 2). Average sequestration potential in agroforestry has been estimated to be 25t C ha^{-1} over 96 million ha of land in India and 6-15 t C ha^{-1} over 75.9 Mha in China (Ravindranath *et al.* 2008). Watson *et al.* (2000) estimated carbon gain of 0.72 Mg C ha^{-1} yr^{-1} on 4000 million ha land under agroforestry, with potential for sequestering 26 Tg C yr^{-1} by 2010 and 45 Tg C yr^{-1} by 2040. Proper design and management of agroforestry practices can make them effective carbon sink. As in other land-use systems, the extent of C sequestered will depend on the amount of C in standing biomass, recalcitrant C remaining in the soil and C sequestered in wood products. Average carbon storage by agroforestry practices has been estimated as 9, 21, 50 and 63 Mg C ha^{-1} in semiarid, sub-humid, humid and temperate regions (Montagnini and Nair 2004).

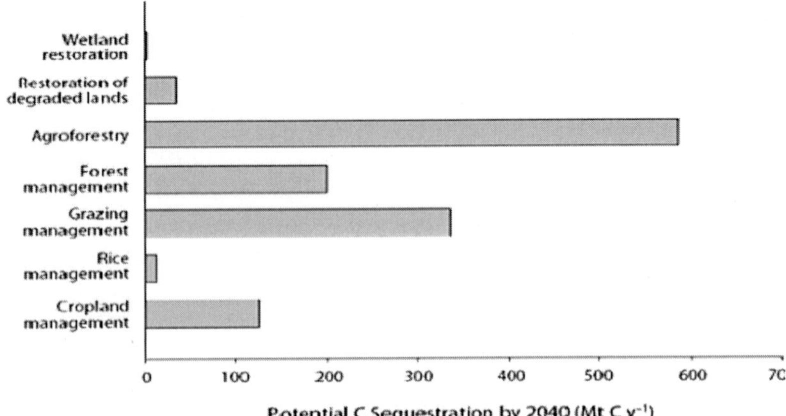

Fig. 2. Carbon sequestration of different land management options
Source: IPCC (2000)

References

Aggarwal RK, Gupta JP, Saxena SK, Muthana KD (1976). Studies on soil physicochemical and ecological changes under twelve-year-old five desert tree species of western Rajasthan. Indian Forester 1(2): 863-872.

Ahuja LD (1980). Grass production under khejri. Monograph No. 11, Central Arid Zone Research Institute, Jodhpur, Rajasthan, India.

Arya R, Awasthi OP, Singh J, Arya CK (2010). Comparison of fruit based multi-species cropping system under arid region of Rajasthan. Indian Journal of Agricultural Sciences 80 (5): 423-426.

Awasthi OP, Pareek OP (2008). Horticulture-based cropping system for arid regions-a review. Range Management and Agroforestry 29 (2): 67-74.

Awasthi OP, Saroj PL, Singh IS, More TA (2007). Fruit based diversified cropping system for arid regions. Technical bulletin: 25. CIAH, Beechwal, Bikaner, Rajasthan.

Awasthi OP, Singh IS, More TA (2009). Performance of intercrops during establishment phase of aonla (*Emblica officinalis*) orchard. Indian Journal of Agricultural Sciences 79 (8): 587-91.

Bhandari, Meghwal PR, Lodha S. (2014). Horticulture Based Production Systems in Indian Arid Regions. Nandwani D (ed) Sustainable Horticultural Systems, Sustainable Development and Biodiversity 2, Springer International Publishing, Switzerland.

Bunderson WT, Wakeel AE, Saad Z, Hashim I (1990). Agroforestry practices in Western Sudan. In: Budd W *et al.* (eds.) Planning for Agroforestry New York, Elsevier Science Publisher.

CIAH (2006) Annual Report (2005-2006). Plant genetic resource management in aonla. In: Research Achievements. CIAH, Bikaner, pp. 7-8.

Duhan A, Chauhan BM, Punia D (1992). Nutritional value of some nonconventional plant foods of India. Plant Foods Hum Nutr 42: 193-200.

Faroda AS (1998) Arid zone research: an overview. In: Faroda AS, Manjeet Singh (ed) Fifty Years of Arid Zone Research in India. CAZRI, Jodhpur, pp. 1-16.

Frank AB, Harris DG,Wills WO (1976). Influence of windbreaks on crop performance and snow management in North Dakota. In: Tinus RW (ed) Shelterbelts on the Great Plains, Proceedings of Symposium, GPAC Publication No. 78, Denver Co, USA, pp. 41-48.

Gupta JP, Rao GGSN, Ramakrishna YS, Ramana R (1984). Role of shelterbelts in arid zone. Indian Farming 34 (7): 29-30.

Gupta JP, Saxena SK (1978). Studies on monitoring of dynamics of moisture in the soil and the performance of ground flora under desertic communities of trees. Indian Journal of Ecology 5:30-36.

ICFRE (1995). Annual Report of Indian Council of Forestry Research and Education, Dehradun, India.

IPCC (2000) Special Report on Land use, Land use change and forestry. Summary for Policy, Geneva, Switzerland, pp. 20.

Jawanda JS, Bal JS (1978). The ber, highly paying and rich in food value. Indian Horticulture 3(3): 19-21.

Kaushik N, Kumar Virendra (2003). Khejri (*Prosopis cineraria*) based agroforestry system for arid Haryana, India. Journal of Arid Environment 55: 433-450.

Montagnini F, Nair PKR (2004). Carbon sequestration: An under exploited environmental benefits of agroforestry systems. Agroforestry Systems 61:281-295.

Narrain P, Tewari JC (2005). Trees on Agricultural fields: a unique basis of life supports in Thar desert. In: Tewari VP, Srivastava (eds) Multipurpose Trees in the Tropics: Management and Improvement Strategies. Arid Forest Research Institute, Jodhpur, pp.516-523.

Osman M (2003). Alternate landuse systems for sustainable production in rainfed areas. In: Pathak PS, Newaj Ram (eds) Agroforestry : Potential and Opportunities, pp. 177-181.

Puri S, Kumar A, Singh S (1994) Productivity of *Cicer arietinum* (chickpea) under a *Prosopis cineraria* agroforestry system in the arid regions of India. Journal of Arid Environments 27(1): 85-98.

Puri S, Kumar Anil, Kumar A (1992). Management and establishment of *Prosopis cineraria* in hot deserts of India. In: Integrated Land Use Management for Tropical Agriculture: Proceedings Second International Symposium Queensland, Module 2, 46: 1-46.

Randhawa KS (1990). Pulse based cropping system in juvenile orchards. In: International Symposium on Natural Resource Management for Sustainable Agriculture.

Rathore M (2009). Nutrient content of important fruit trees from arid zone of Rajasthan. Journal of Horticulture and Forestry 1(7):103-108.

Ravindranath NH, Chaturvedi RK, Murthy IK (2008). Forest conservation, a forestation and reforestation in India: implications for forest carbon stocks. Current Science 95 (2): 216-222.

Roy MM, Tewari JC, Moola Ram (2011). Agroforestry for climate change adaptations and livelihood improvement in Indian hot arid regions. International Journal of Agriculture and Crop sciences 3(2): 43-54.

Shankarnarayan KA (1984) Influence of *Prosopis cineraria* and *Acacia niloticaon* soil fertility and crop yield. In: Shankarnarayan KA (ed) Agroforestry in Arid and Semi-Arid Zones, Central Arid Zone Research Institute, Publication No. 24, Jodhpur,Rajasthan, India, pp. 59-66.

Sharma SK, Diwakar GD (1989). Economic evaluation of hortipastrol system on arid region of western Rajasthan. Indian Journal of Range Management 10 (2): 119-122.

Singh KS, Lal P (1969) Effect of khejri *(Prosopis cineraria)* and Babool *(Acacia arabica)* trees on soil fertility and profile characteristics. Annals of Arid Zone 8: 33-36.

TewariJC, Bohra MD, Harsh LN (1999). Structure and production function of traditional extensive agroforestry system and scope of intensive agroforestry in Thar desert. Indian Journal of Agroforestry 1(1): 81-94.

Watson RT, Noble IR, Bolin B, Ravindranath NH, Vetardo DJ, Dokken DJ (eds) (2000). Land use, Land use change, and Forestry. Intergovernmental Panel on Climate Change, Cambridge University Press, NY.

WMO (2005). Climate and Land Degradation. WMO No. 989, World Meteorological Organization.

17

Role of Small Ruminants and Agro Forestry System in Arid Zone

A.K. Patel

ICAR-Central Sheep and Wool Research Institute, Arid Region Campus, Bikaner, Rajasthan, India

Introduction

Livestock sector plays an important role in the rural economy at national level and this sector is emerging as a driving force in agriculture by contributing towards the income and employment of rural households. Livestock contributed 16% to the income of small farm households that indicates importance of livestock sector as a whole. In 2010-11, livestock generated outputs worth Rs 2075 billion (at 2004-05 prices). Agriculture sector grew at low rate of 4.7% during 2013-14 and contributed 13.9% to National GDP (Economic Survey 2013-14). The livestock sector contributed over 4.1 per cent of the total GDP in 2012-13, that nearly accounts for 29.5% of the agricultural GDP (Economic Survey 2013-14) indicating a significant contribution of the livestock. Sheep and goats are an important livestock species for a large number of rural poor, especially socially backward, marginal and landless labourers in India. Sheep are mostly reared for wool and meat purpose. India ranks third in sheep population and account for nearly 6% of world sheep population, currently India has 65.06 million sheep as per Livestock census 2012. The wool production is almost stagnant in the country for last one decade, however, a slight increase was achieved in wool production to about 48 m kg in 2013-14 from previous year 46.1 m kg (BAHS 2014). Despite of low wool production in country the handmade carpet exports have crossed Rs. 8559.01 crores during 2014-15 fiscal year. Wool industry provides employment to about 2.6 million people. Out of total production of raw wool in the country about 70% is carpet grade, 20% coarse grade and 10% apparel grade (Karim and Shakyawar, 2011). Two thirds of wool is used in manufacture of garments and about one third in carpets, upholstery and rugs (Naqvi, 2013). Besides this sheep also significantly contributes to mutton production which increased to 441.14 m kg. The goat population over the past few decades has increased 186.4% (from 47.2 m in 1951-52 to 135.2 m in 2012). According to Compound Annual Growth Rate (CAGR) of sheep and goat during (1992-2003) in the country shows that their population has increased at a fast rate in some of the districts (Kumar *et al.* 2010). This is a main factor that makes goat as most desired species of animal for meat production in the country. The goats and their products contribute Rs 17661.7 crore annually to the national economy.

The dry areas of arid and semi arid region of the country are more favourable for livestock and pasture production system. The harsh climatic conditions prevailing in these regions, e.g. low and erratic rainfall, high evapo-transpiration, high solar radiations, frequent droughts etc make them unsuitable for crop farming. A major part of land is covered under arid zone (31.8 lakh km^2) in the country. The 62% of hot arid region lies in the state Rajasthan and 1/3rd area of arid zone of Rajasthan are wastelands, of which 50% are grazing lands and 45% are sandy wastes (Balak Ram *et al.* 2005). The production from available grazing lands is hardly 300-400 kg/ha biomass. This low productivity is due to deterioration and over grazing. Further, the occurrence of droughts are common in arid zone which affects agricultural and forage production. This coupled with degraded grazing lands causes feed and fodder deficit for large population of livestock in arid zone. The fodder deficit of western Rajasthan is estimated to be as high as 60% of the demand and ranges from 55% in western districts to 69% in central districts and 72% in eastern districts (Pratap Narain and Kar 2005). This situation is further aggravated during drought years. The deficiency of dry and green fodder was estimated to be 35.9% and 79.9% during drought year (Patidar and Saxena,2007).

The dry areas of arid and semi arid nature require a planned effort to create sufficient feed base in terms of feed and fodder to sustain the productivity of livestock in the region and it is more essential to rejuvenate the natural rangelands by practicing the methods of reseeding, protection and grazing management.In order to combat the problems of overall nutritional deficiency there is need to develop and popularize agroforestry system in arid and semiarid lands for enhancing forage biomass even in low and erratic rainfall years.

Climate change and agriculture, forestry and livestock productivity

Climate change has become a serious environmental, social and economic threat particularly to natural ecosystems, biodiversity, and livelihood of agricultural and forest fringe communities. Climate change is becoming a key driver in agriculture and forest management, but its complexity and magnitude threaten the sustainability. Agriculture along with forestry and animal husbandry can play an important role in climate change adaptation, through diversified land-use practices, livelihoods and sources of income. Sustainable agroforestry has a tremendous potential to serve as a tool in adapting climate change, protecting ecosystems and livelihoods, and creating a foundation for economic and social development. Climate change is expected to alter temperature, precipitation, atmosphere carbon dioxide levels and water availability which will affect the productivity of crops and livestock systems (Hatfield *et al.* 2008).Climate change could increase thermal stress for animals affecting animal health, reproduction; production and overall productivity and reduce profitability. Climate changes, could impact the economic viability of livestock production systems worldwide. Heat stress negatively affects reproductive function. Normal oestrus activity and fertility are disrupted in livestock during summer months. The livelihood of the rural poor in developing countries depends critically on local natural resource-based activities such as crop and animal production. Due to negative weather impact on livestock rearing, the poor farmers whose principal livelihood security depends on these animals performance is directly at stake. Housing and other management technologies are available through which climate impacts on livestock can be minimized. Under the changing climatic scenario, the concept of multiple stresses emerges as a potential threat to livestock production and its survival. Hence, there is need to prioritize the research to tackle the multiple stress including nutrition stress and sustain the productivity of animals in change climate scenario.

Forage scenario of arid Rajasthan

The fodder production varies widely depending on the cropping pattern, socio-economic conditions, and type of livestock. The cattle and buffaloes are normally fed on the fodder available from cultivated areas, supplemented to a small extent by harvested grasses and top feeds. The three sources of fodder supply are crop residues, cultivated fodder and fodder from common property resources, viz. forests, permanent pastures and grazing lands. The deficiency in feed and fodder is identified as one of the major constraints in achieving desired level of livestock productivity. The requirement and availability of fodder during different situations is described in Table 1. The acute fodder shortage in central and eastern districts is due to extensive cropping, fragmentation of land holding, intensive cultivation and shrinking of pasture and grazing lands.

Table 1: Requirement and availability of fodder in arid zone of Rajasthan

Fodder	Requirement	Availability (mt)		Deficit (%)	
		During normal year	During drought year	During normal year	During drought year
Dry fodder	22	14.00	7.68	36	65
Green fodder	41	12.78	10.71	69	74
Concentrate	13	2.66	0.67	80	94

The gap between the demand and supply will continue to widen if appropriate strategies are not initiated to enhance biomass and forage production substantially.

The research outcome of last five decades of research institutions in the country for pasture establishment and utilization techniques for various sub-agro-climatic regions showed that pasture grasses like *Cenchrus ciliaris, Cenchrus setigerus, Lasiurus sindicus, Dichanthium annulatum* and *Panicum antidotale* are promising for higher forage production (Daulay and Bhati 1989).

Silvipasture systems in arid zone

Traditionally silvipasture is used for grass farming in forest areas, which accommodate millions of livestock for grazing on the herbaceous growth. Extending this principle, silvipasture is the conscious practice of cultivation of forage annuals along with fodder trees and shrubs somewhat similar to agroforestry. Farmers have encouraged and protected the Khejri (*P. cineraria*) and Bordi (*Z. nummularia*) in their crop fields to lop them in winter for top feed and fuel wood. Nair (1993) defined the silvipastoral system as the land use system in which trees and or shrubs are combined with livestock and pasture for forage and fuel wood production on the same unit of land.Silvipastoral system has special significance in arid and semi-arid regions, where there is lower vegetation and high rate of deforestation and land degradation due to low and erratic rainfall.

The productivity level of ruminants is by and large extremely low, because of poor quality and seasonal nature of forage supply in the arid region.The problem becomes more acute in dry seasons.In such situations, forage based silvipastoral system could be a promise. Besides, trees in silvipastoral system also provide shade to the grazing animals and serve as fences, hedge etc wild breaks.

Suitable silvipastoral models for different land forms and rainfall situations

The selection of suitable species of trees/shrubs, grasses and legume for establishment of silvipastoral system depend on agro climatic condition of the

(A) Suitable species for silivipastoral system under different landform			
Type of land	**Suitable species**		
	Grasses	**Legumes**	**Trees/shrubs**
Desert land and sand dunes	*Lasiurus sindicus,Cenchrus ciliaris, Cenchrus setigerus*	*Lablab purpureus, Clitoria ternatea, Atylosia scarabaeoides*	*Acacia tortilis, Acacia nilotica, Acacia senegal, Prosopis cineraria, Prosopis juliflora, Azadirachta indica, Ziziphus nummularia, Colophospermum mopane Dichrostachys nutan, Colligonum polygonoides*
Ravine lands	*Cenchrus ciliaris, Cenchrus setigerus, Dichanthium annulatum, Pannisetum pedicellatum, Saccharam spontaneum, Chrysopogon fuluus*	*Stylosanthes* spp., *Stizolobium decrigenum Macroptilium atropurpureum, Atylosia* spp.	*Acacia tortilis, Acacia nilotica, Acacia catechu, Albizia lebbek, Albiziaamara, Dalbergia sissoo, Zizyphus* spp., *Ficus* spp., *Emblica officinalis, Eugenia jambolana*
Cultivable waste lands	*Cenchrus ciliaris, Cenchrus setigerus, Pennisetum polystachyon, Panicum antidotale, Sehima neruosum, Chrysopogon fuluus, Dischanthium annulatum*	*Macroptilium atropurpureum, Glycine jaranica, Clitoria ternatea*	*Albizia spp., Hardwickia binnata, Leucaena leucocephala, Acacia* spp. *Sesbania* spp., *Dichrostachys nutans, Prosopis cineraria, Ziziphus mauritiana*
Salt affected area	*Cynodon dactylon, Paspalium notatum, Chloris gayana, Lasiurus sindicus, Brachiaria mutica, Sporolobus marginatus, Urochloa*	*Glycine jauanica, Macroptillum* spp. *Phaseolus junatea, Stylosanthes* spp.	*Acacia tortilis,* A. *nilotica, Prosopis juliflora, Salvadora* spp., *Ziziphus* spp. *Sesbania* spp., *Albizia amara, Atriplex* spp.
(B) Suitable species of grasses and trees on the basis of rainfall			
Rainfall Zone	**Grasses**	**Legumes**	**Trees/shrubs**
150 to 300	*Lasiurus sindicus, Cenchrus ciliaris*	*Lablab purpureus, Clitoria ternatea*	*Prosopis juliflora, Acacia tortilis, Acacia senegal, Ziziphus nummularia. Colligonum polygonoides, Dichrostachys nutans*
300 to 500 mm	*Lasiurus sindicus, Cenchrus cilaris, Cenchrus setigerus, Panicum antidotale*	*Lablab purpureus, Clitoria ternatea*	*Acacia. tortilis, A. nilotica, P. cineraria, Tecomella undulata, Colophospermum mopone, Ziziphus spp. Colligonum polygonoides, Azadirachta indica, H. binnata, Ailanthus exelsa, Albizia lebbek*
Above 500 mm	*C. ciliaris, Dichanthium annutatum, Chrysopogon fulvus, Sehima nervosum, Panicum antidotale, Heteropgon contortus*	*Stylosanthes*	*Acacia nilotica, Acacia catechu, Dalbergia sissoo, Leucaena leucocephala, Albizia spp., Alianthus excelsa, Hardwickia binata*

regions. The following trees/shrubs and pasture species for different land forms and rainfall zones are identified.

Forage production of various silvipastoral models

In traditional silvipastoral system, trees/shrubs are permitted to grow at random in fields on the boundary of the fields or community grazing land with grasses of low productivity and nutritive value. In arid zone of Rajasthan, Ahuja (1980) recorded highest grass yield (1.1 to 1.5 t ha^{-1}) under Khejri and lowest (0.6 to 0.7 t ha^{-1}) under Kumat plantation of 14 to 18 year old with substantial contribution of *Cenchrus* and *Eleusine* species as under storey perennial grasses. The productivity and carrying capacity of grazing land can be increased through improved silvipastoral system. At Jhansi, Pathak *et al.* (1996) reported that the degraded lands produced hardly up to 1 t ha^{-1} yr^{-1} biomass, which can be improved up to 10 t/ha through silvipastoral system. A study conducted at NRCAF, Jhansi revealed that an average total biomass yield of 12.62 t ha^{-1} yr^{-1} recorded under silvipastoral system consisting of *A. amara* and *L. leucocephala* as trees, *P. cineraria* as shrub and *C. fulvus* + *S. hamata* as pasture was about 4 time higher than yield (3.16 t ha^{-1} yr^{-1}) obtained from natural grassland (Rai *et al.* 1999). Silvipastoral studies undertaken at CAZRI, Pali involving four tree species, (*Acacia tortilis, Azadirachta indica, Albizia lebbek* and *Holoptelia integrifolia*) and four grasses (*C. ciliaris, C. setigerus. D. annlatum* and *Panicum antidotale*) revealed non significant difference in the dry fodder yield under different tree species, however, the mean dry forage yield was maximum in *Dichanthium annulatum* (2.8 t ha^{-1}) followed by 2.51 t ha^{-1} in *Cenchrus ciliaris* and 2.2 t ha^{-1} in *Panicum antidotale* (Muthana and Shankarnarayan 1978). Silvipastoral studies at Jodhpur indicated that sowing of *Cenchrus ciliaris* under the cover *H. integrifolia* did not interfere with normal growth with an additional dry forage yield of 0.7 to 1.2 t ha^{-1}. In another study Harsh *et al.* (1992) reported that a mixture of *C. fulvus* and *Sehima nervosum* grass produced 3.25 t ha^{-1} forage under canopy of five yea old *Hardwickia binata.* Similarly Patidar (2006) recorded maximum herbage yield with *L. sindicus* + cowpea (9.0 and 3.1 t ha^{-1}) green and dry forage yield) followed by sole 25 (8.9 and 3.08 t ha^{-1}) green and dry fodder yield and *C. ciliaris* + cowpea system (8.46 and 2.86 t ha^{-1}) green and dry forage yield grown in association with *C. mopane* and *H. binnata.* The forage yield was not different due to tree species however, application of 40 kg N t ha^{-1} enhanced fodder yield by 13-14% over control. Yadav and Poonia (2003) recorded maximum biomass yield of grasses and tree from silvipastoral system with *Dichrostachys nutans* followed by *Prosopis cineraria* at field in arid climatic condition of Rajasthan. In a horti-pastoral studies on sandy rain fed of Rajasthan, Sharma and Diwakar (1989) reported that *Cenchrus ciliaris* in association with *Ziziphus mauritiana* system produced 1.2 t/ha forage yield and did not affect the fruit yield.

Lots of work on silvi-pasture and agroforestry systems was carried out to get maximum forage yield in semi-arid zone at CSWRI, Avikanagar. The highest dry fodder production (59.06 q ha^{-1}) was recorded with mixture of *Cenchrus* and *Dolichos*. Whereas lowest yield (28.82 q ha^{-1}) from *Cenchrus setigerus* alone was recorded. From mixture of *Cenchrus* and *Clitoria*, 45.2 q ha^{-1} dry fodder was obtained. The intercropping of cowpea and moth found to increase 5 and 2 times higher dry fodder yield of *Cenchrus* (Sharma and Chand 2012). Maximum dry fodder (41.0 q ha^{-1}) from single cutting was obtained when carpet legume was sown at spacing of 30x30 cm. In low rainfall year (375 mm) the highest production was recorded from mixture of pearl millet and carpet legume in the ratio of 1:2, whereas, in high rainfall year (1000mm) the maximum fodder was obtained from the mixture of maize and carpet legume. Harvesting of *Clitoria* forage at 40 days interval produced maximum dry matter yield. Maximum grain yield was produced with pure pearl millet followed by pearl millet + *Clitoria* and maximum dry fodder yield was obtained when pearl millet was inter cropped with Cenchrus. Introduction of cowpea in rainfed cereal gave maximum crude protein yield followed *Dolichos* and cluster bean. Paired cropping was superior to normal cropping. Among *Cenchrus* spp., *Cenchrus ciliaris* 358 produced higher yield. *Cenchrus ciliaris* and *Clitoria ternate* produced maximum dry matter during December.

Leaf fodder production- an additional source of feed in dry areas

Studies on lopping intensity (Bhimya *et al.* 1964) revealed that heavy intensity of lopping adversely affected the growth of *Khejri.* Further recurrent lopping reduced that leaf fodder yield irrespective of lopping perhaps due to successive reduction in the overall surface of new shoots. Need for a period of rest between two loppings of *Khejri* is, therefore, prima facie obvious for sustained yield of loppings (leaf fodder of *Khejri).* It has been suggested that in arid area lopping of *Acacia* spp., particularly *A. nilotica* should be regulated on a cycle of 4 years with restriction on lopping of thicker (diameter of 1.9 cm and above) branches. In Madhya Pradesh and Tamil Nadu lopping of *Hardwickia* spp. were permissible only in times of scarcity. Sharma and Gupta (1981) studied the effect of seasonal lopping on the top feed production and growth of *P. cineraria.* They observed that winter was the ideal season for lopping as it improved plant height and bole diameter of the tree as compared to trees lopped in rest three seasons i.e. spring, summer and monsoon.

Natural, single, double and multi-tier systems were studied for fodder production in Avikanagar (Sharma and Chand 2012). Cenchrus grass yield from all the systems were higher over natural pasture. Growth of ardu (*Ailanthus exelsa*) was better in two-tier system. The production for 5 years in natural land, single tier, two tiers and three tier silvi-pastures was to the tune of 16.59, 19.20 and 23.74 q ha^{-1}, respectively. The three and two tier systems provided 9.5 and 6.25 q ha^{-1} tree fodder. The average net income over four years was highest in ardu pasture then neem, siris and babool pasture systems. The highest green fodder yield (154.69 q ha^{-1}) was in multi-tier system (Ardu + pearl millet + cowpea) followed by two-tier; however in multi-tier, maximum dry fodder yield was with sole pearl millet followed by Cenchrus + cowpea. The highest dry fodder yield from crop and grasses and their combinations was in single-tier system followed by two-tier system. Growth parameters and yield of Cenchrus were comparatively higher in association of ardu tree in comparison to other fodder trees.

Alternate feed resources for arid livestock

Arid regions have many feed resources in the form of agricultural produce, weeds, byproducts from agro-processing methods etc. Tumba (*Citrullus colocynthis*) seed cake a byproduct of the oil extraction industry is nutritionally rich as it contains 16-22 per cent crude protein. Inclusion of 25 % tumba seed cake (TSC) in concentrate feed lowers the cost of animal feeding by 18-35 per cent without any adverse effect on the production, general health and reproductive performance (Mathur *et al.* 1989; Mathur *et al.* 2000). Salty shrub Lani (*Salsola baryosma*) at vegetative stage can be a good source of fodder to animals. The acceptability and palatability of fresh cut shoots showed that Lani is palatable (Mathur *et al.* 2007). Lana (*Haloxylon salicornicum*) seeds having 18.60 per cent crude protein could replace about 25 per cent of the conventional sesamum cake in the concentrate for lactating cows (Anonymous 2007). The powder of *Prosopis juliflora* pods can be used up to 35 per cent in the concentrate of lactating goats with maintenance of milk production without having any detrimental effect on health (Mathur *et al.* 2003). Similarly *P. juliflora* pod husk can be used up to 50 percent level in the concentrate along with tumba (*Citrullus colocynthis*) seed cake as low cost ration in the sheep without any adverse effect on animal health (Mathur *et al.* 2002).

Fresh leaves of *Hardwickia binata* are palatable to goats and sole feeding supports the body weight growth of growing kids (Patil *et al.* 2009). Feeding fresh leaves of *Colophospermum mopane* is possible in goats but it can be fed at 35-40 per cent of whole ration (Patil *et al.* 2007). The palatability of dry *C. mopane* leaves was low and decreased with progress of feeding from 15 to 5 per cent from 1st to 5th week and the traditional local *P. cineraria* leaves were better source of supplementation even in the dry form supporting the milk yield of goats. (Mathur *et al.* 2006).

Thornless cactus *Opuntia ficus indica* introduced in Indian arid region has fodder value as maintenance feed and was observed to reduce the water requirement if fed along with the dry roughages in goats, sheep and growing cattle. In addition, its high mineral content may reduce the mineral requirement, as arid animals often suffer from mineral imbalance (Mathur *et al.* 2009; Meghwal *et al.* 2010).

The weeds in the rocky and sandy habitat of arid region have a distinct relative preference index for grazing sheep and goats. Patil *et al.* (2005) studied the preference of these animals to different weeds and found the same to be in the order: Kanti (*Tribulus terristris*), Kagio (*Tetrapogon tenellus*), Santo (*Trianthema protulacastrum*), Lolaru (*Digeria muricata*), Bekario (*Indigofera cordifolia*) and Gangan (*Grewia tanax*).

Animal performance on desert top feeds

By and large, the consumption of *Z. nummularia* leaves is more than that of *P. cineraria* leaves, in both sheep and goats, the latter consuming considerably higher than the sheep (Ghosh and Bohra 1984).The leaves of *A. excelsa, Z. nummularia* and *P. cineraria* are high in total digestible nutrients (64%, 50%, and 41.0% respectively), while the digestible crude protein (DCP) value of *A. excelsa* leaves is higher (16%) than the more palatable *P.cineraria* in sheep (3.1%) and goats (5.5%) (Bohra 1980). The DCP of Z. *nummularia* in sheep and goats were 4.1 and 3.6 per cent, respectively (Bohra *et al.* 1999). From the point of nutritive value, *A. excelsa* leaves can be considered the best feed among all desert tree leaves. However, this tree does not grow as extensively as *P. cineraria* or *Z. nummularia* in this region. On the basis of per unit of feed material in general, top feeds provide more energy than the grasses, as the digestibility of cellulose from desert grass *C. ciliaris* (Bohra and Ghosh 1977) is higher than that of the tree leaves presumably because of the high lignin content of the top feeds.

The nutritive value in terms of digestible crude protein and total digestible nutrients of different feeds differ with animal species. *Z. nummularia* has high nutritive value for camel followed by the sheep and goats, while *P. cineraria* has the highest value for goats, followed by camel and the sheep. The browsing rams and grazing bucks reared under a silvipastoral system having *C. ciliaris* as a dominant grass and supplemented with loppings of *Zizyphus rotundifolia* exhibited substantial body weight gains (Patil *et al.* 2009). In the arid zone a substantial area is salt affected, where nothing can come up, except a few halophytic plants like *Salicornia bigelovii,* which can be cultivated under flood irrigation of brackish well water (Bohra *et al.* 2009).

The desert top feeds, especially, *P. cineraria* and *Z. nummularia* leaves although have appreciable quantities of crude protein but their DCP value is very low because of tannins and lignin in these feeds. (Bohra *et al.* 1999), which not only reduces the palatability, but also hinders utilization of dietary proteins by the animals as indicated by *in situ* degradability study of *P. cineraria* leaves in sheep rumen (Mathur *et al.* 1998). Attempts have been made to improve the nutritive value *P. cineraria* leaves by treatment with ferric chloride. For this purpose, over night soaking of leaves with 0.5N aqueous sodium carbonate solution, followed by washing with water, proved to be the best method for detanning *P. cineraria* leaves removing about 94 per cent tannins of these leaves (Bohra and Goyal 1986).Other desert plants also yield appreciable quantity of palatable pods. *P. juliflora* leaves are not relished by the animals, but its pods are highly palatable and relished by almost all the species of the livestock. Feeding of pods after suitable processing like crushing and grinding is possible. *A. tortilis* pods contain 5.7 per cent DCP and 62 per cent TDN and the *P. juliflora* pods, 7 per cent DCP and 75 per cent TDN (Mathur and Bohra 1993). These pods contain appreciable quantity of micro-minerals, too.

Supplementary feeding in grazing animals

Supplementary feed blocks

As the livestock in arid regions are mostly range managed where dry grasses in the ranges and pastureland, and crop residues in the fallow lands are available, the animals suffer from deficiency of essential nutrients including fermentable energy, protein, minerals as well as carotene except during monsoon. There exist different means of supplementation of essential nutrients to the livestock. Appropriate formulations of multi-nutrient feed blocks(MNB) using locally available feed resources wheat bran, guar korma, *Bajra* husk and *Ardu* leaves, *Prosopis juliflora* ground pods, sugar cane molasses, urea, vitaminized mineral mixture, dolomite, common salt, deoiled soya bean meal have been developed (Bohra *et al.* 2012). And as a binder locally available organic product-guar gum powder is being used. The supplementary feed blocks meant for the grazing ruminants have been tested in cattle, sheep and goats and were found to be economical source of nutrient supplementation to support the production functions of milk yield, body weight gain and wool yield. During 16 weeks feeding trial the MNB supplemented sheep group recorded 3.6 per cent gain over the sheep maintained on roughage diet. The supplementary feed blocks were also found to improve the nutrient utilization in cattle. In a digestibility trial study on Rathi cows, the digestibility coefficient for DM and crude protein were found better (Mondal and Bohra 2001). The supplementary feed blocks in the arid areas were found to improve the production performance of lactating animals especially cows, buffaloes and goats. The farmers also reported improvement in the conditions of Pica in the animals suffering from mineral deficiency along with increase in feed and water intake. In case of cattle and buffaloes, a 2 kg block offered as a lick lasted for about 7 and 5 days, respectively (Patel *et al.* 2003). Supplementation of MNB enhanced 11.5% in daily milk yield of buffaloes under field in Nagaur district of Rajasthan (Patel *et al.* 2012, 2014).

Supplementary multi-nutrient mixture

For the small grazing ruminants that do not lick the above blocks, supplementary multi-nutrient mixture formulations were developed using the same ingredients. Feeding these mixtures to goats and sheep after grazing hours was found to improve body weights and milk yield appreciably (Rohilla *et al.* 2009; Patel *et al.* 2006)

Complete feeds and feed/fodder banks for arid livestock

A low cost balanced diet has been formulated for ruminants, which does not require these animals to be sent for grazing in the rangelands. The alternate feeding diet for ruminants is based on crop residues and locally available agro-industrial products like straws, fallen tree leaves, by-products from other crops like cotton seeds, maize, groundnut, subabul, forest products like dry grass, pods, which can be blended to enable balanced supply of nutrients to the animals. The complete feed rations have been formulated area wise as per availability of various agricultural residues.

To improve the livestock productivity, total mixed ration (TMR) or complete feed blocks (CFB) have been developed. In an evaluation trial the lactating Tharparkar cows fed TMR prepared out of local fodder *Cenchrus ciliaris* and concentrates available in the area were found to produce milk more economically compared to cows fed on local grass and supplemented with pelleted cattle feed (Mathur *et al.* 2006). The results of trials conducted at CSWRI, Avikanagar under intensive feeding on 50:50 roughage and concentrate ratios of complete feed indicated that the feed efficiency was higher than 20 per cent because of higher carcass fats. In this study, the crossbred weaner lambs maintained on these diets had higher ADG of 180 g compared to 127 g ADG in native lambs. In another study the lambs fed intensively on 50:50 RC based complete feed attained 33.5 kg finishing weight with 11% feed efficiency at 6 months of age (Shinde *et al.* 1995). Higher growth (20%) was also observed in Magra male lambs due feeding of complete fodder block under field condition (Sawal *et al.* 2015). In a study conducted on Marwari kids, Patil *et al.* (2006) observed that the kids weaned at 2 months age had comparatively better growth rate than the kids weaned at 3 and 4 months age when these were fed on complete feed diets having roughage to concentrate ratio 30:70 utilizing Khejri leaves and Masoor straw in equal proportion as a roughage source and local material in the concentrate mixture. The cost of feeding was economical for kids reared on CFB after weaning at 2 months age (Rs. 65.37 kg^{-1}) as compared to those weaned at 3 months (Rs. 70.40 kg^{-1}) and at 4 months (Rs. 71.27 kg^{-1}). Similarly in the intensive management system the Marwari kids fed on CFB along with goat paneer whey gained higher body weight than the kids fed on CFB only. The results of the above studies indicate that goat and sheep can be reared intensively without sending the animals for grazing and it can achieve the targeted body weights during minimum possible time. It also helps to achieve the expanded use of crop residues and locally available agro-industrial by products by formulating complete rations.

The technology of complete feeds is appropriate for the arid region livestock encountering frequent droughts. In such situations advanced planning could be of great help to have enough stores of complete feeds as feed banks, which can be mobilized to places of deficiency.

References

Balak Ram, Narain P, Sharma JR, Nagaraja R, Jayanti S (2005). Wasteland Management of Arid western Rajasthan. CAZRI Jodhpur and MRSC, Hyderabad.

Bhagmal, Pathak PS, Upadhyaya VS, Gupta JN, Suresh G (2006). Forage crops and grasses. In: Hand Book of Agriculture ICAR, New Delhi, pp. 1128-1161.

Bhimya CP, Kaul RN, Ganguli BN (1964). Studies on lopping intencities of *Prosopis cineraria.* Indian Forester 90 (1):19-23.

Bohra HC (1980). Nutrient utilization of *Prosopis cineraria* (Khejri) leaves by desert sheep and goats. Annals of Arid Zone 19: 73-81.

Bohra HC, Ghosh PK (1977). Effect of restricted water intake during summer on the digestibility of cell wall constituents, nitrogen retention and water excretion in Marwari sheep. Journal of Agricultural Science 89: 605-608.

Bohra HC, Ghosh PK, Goyal SP (1999). Nutrient availability from *Ziziphus nummularis* leaves in desert sheep ad goats. Journal of Eco-physiology 2:129-134.

Bohra HC, Goyal SP (1986). Chemical detaining of Prosopis cineraria leaves. In: 5th Animal Nutrition Workers Conference, Udaipur: Animal Nutrition Society of India.

Bohra HC, Patel AK, Kaushish SK (2009). Palatability, digestibility of various constitutes and nitrogen retention in Marwari sheep offered *Salicornia bigelovii* biomass and *Cenchrus ciliaris* straw mixed diet. Indian Journal of Animal Sciences 79 (10): 1050-1053.

Bohra HC, Patel AK, Rohilla PP, Mathur BK, Patil NV, Misra AK (2012). Feed production technologies for sustainable livestock production in arid areas. CAZRI, Jodhpur.

Daulay HS, Bhati TK (1989). Crops and pasture production potential of sandy soils. In: Review of Research on Sandy Soils, CAZRI, Jodhpur, pp. 245-262.

Deb Roy R, Pathak PS (1974). Silvipastoral management. Indian Farming 24:41-42.

Ghosh PK, Bohra HC (1984). Palatability, digestibility and nutritive value of some important top feeds of arid and semi-arid regions of India. In: Shankarnarayan KA (ed) Agroforestry in Arid and Semi-arid Zones. Central Arid Zone Research Institute, Jodhpur.

Harsh LN, Tewari JC, Burman U, Sharma SK (1992). Agroforestry in arid regions. Indian Farming 40: 32-37.

Hatfield J, Booke K, Fay P, Hahn LC, Izaurralde BA, Kimball T, Mader J, Morgan D, Ort W, Polley A, Thomson, Wolfe D (2008). Agriculture. In: The Effects of Climate Change on Agriculture, Land resources, Water resources and Biodiversity in the United States. A report by the US climate change science program and the subcommittee on global change research. Washington, DC, USA, pp. 362.

Karim SA, Shakyawar DB (2011). Proceedings of National Seminar cum Workshop on Recent R& D Initiatives and Development Schemes of Wool and Woollens, Mumbai.

Kumar S, Kareemulla K, Venkateswarlu B (2010). Small ruminant production and sustainable rural livelihood. In: Climate Change and Stress Management: Sheep and Goat Production. pp. 687-702.

Mathur BK (2003). Livestock: human need for sustainability in arid environment. In: Human Impact on Desert Environment. Pratap Narain, Kathju, S, Amal Kar, Singh MP, Praveen-Kumar (eds) Arid Zone Research Association of India and Scientific Publishers, Jodhpur, India, pp. 506-514.

Mathur BK, Bhati TK, Tewari JC (2006). Comparative palatability of *Prosopis cineraria* v/s *Colophospermum mopane* leaves in lactating Marwari goat in arid region. In: National Symposium on Livelihood Security and Diversified Farming Systems in Arid Region, Arid Zone Research Association of India, CAZRI, Jodhpur, pp.51-52.

Mathur BK, Bohra HC (1993). Nutritive value of *Prosopis juliflora* leaves and pods. In: Proceedings of the Workshop on Potential of *Prosopis* spp. for Arid and Semi Arid Regions of India, Central Arid Zone Research Institute, Jodhpur.

Mathur BK, Bohra HC, Patel AK, Kaushish SK (1998). In situ rumen degradability of khejri (*Prosopis cineraria*) leaves in Marwari sheep. In: Proceedings of Golden Jubilee Seminar on Sheep, Goat and Rabbit Production and Utilization for Maximizing Production Efficiency, Indian Society for Sheep and Goat Production and Utilization, Jaipur, CSWRI, pp. 27.

Mathur BK, Mittal JP, Prasad S (1989). Effect of tumba cake (*Citrullus colocynthis*) feeding on cattle production in arid region.Indian Journal of Animal Sciences 59: 1464-1465.

Mathur BK, Mittal JP, Shiv Prasad, Mathur AC (1991). Need of conserving Tharparkar breed of cattle in Indian desert. Indian Farming 41 (3): 4-6.

Mathur BK, Patel AK, Kaushish SK (2000). Utilization of non-conventional resource: tumba (*Citrullus colocynthis*) seed cake of desert for goat production. Indian Journal of Animal Sciences70 (4): 431-433.

Mathur BK, Patil NV, Mathur AC, Bohra HC, Bohra RC, Sharma KL (2009). Comparative mineral status of jodhpur district villagers' cattle v/s institute farm managed tharparkar cattle in hot arid zone. In: Proceedings of Animal Nutrition World Conference, New Delhi, India, pp.61.

Mathur BK, Patil NV, Mathur AC, Bohra HC, Patel AK, Bohra RC (2006). Effect of total mixed ration prepared from local feed resources on feed intake, milk production efficiency and blood profile of Tharparkar cattle in arid region. In: VI Biennial Conference of Animal Nutrition Association on Strengthening Animal Nutrition Research for Food Security, Environment Protection and Poverty Alleviation, Sher-e-Kashmir University of Agricultural Sciences and Technology, Jammu, pp.77.

Mathur BK, Shiv Prasad, Mittal JP (1989). Effect of tumba cake (*Citrullus colocynllus*) feeding on cattle production in arid region. Indian Journal of Animal Science 59:1464-65.

Mathur BK, Singh JP, Beniwal RK, Singh NP (2007). Utilization of salty shrub-Lani (*Salsola baryosma*) of arid region as drought feed for goats. In: International Tropical Animal Nutrition Conference, NDRI, Karnal, pp.33.

Mathur BK, Siyak SR, Bohra HC (2003). Feeding of vilayati babool (*Prosopis juliflora*) pods powder to Marwari goats in arid region. Current Agriculture 27(1-2):57-59.

Mathur BK, Siyak SR, Paharia Shailesh (2002). Incorporation of *Prosipis juliflora* Pod husk as ingredient of low cost ration for sheep in arid region. Current Agriculture 26:1-2, 95-98.

Meghwal PR, Patel AK, Patidar M, Roy MM (2010). Introduction, utilization and further potential of cactus pear in Indian arid zone. In: VIIth International Congress on Cactus Pear and Cochineal, Agardir., Morocco, pp. 175-176.

Mondal BC, Bohra HC (2001). Effect of multi-nutrient feed block supplementation on feed intake and nutrient utilization in Rathi heifers. In: Proceedings of X Animal Nutrition Conference on Emerging Nutritional Technologies for Sustainable Animal Production and Environment Protection, Abst. N.D.R.I., Karnal, pp. 95.

Muthan KD, Shankaranarayan KA (1978). Scope of silvipastral system in arid region. In: Silver Jublee, Souvenir on Arid Zone Research, CAZRI, Jodhpur, India.

Nair PKR (1993). An Introduction to Agroforestry. Kulwar. Academic Publisher, The Netherland.

Naqvi SMK (2013). Present status of carpet wool production in India viz a viz world. In: Interactive Meeting on Prospects in Improving Production, Marketing and Value Addition of Carpet Wool, Arid Regional Campus, CSWRI, Bikaner, pp. 1-7.

Narain Pratap, Kar Amal (2005). Drought in Western Rajasthan Impact, Coping Mechanism and Management Strategies, CAZRI, Jodhpur, PP. 45.

Patel AK, Bohra HC, Bhati TK (2003). On farm trial of multi nutrient block on buffaloes in arid region. In: Proceeding of the 4th Asian Buffalo Congress on Buffalo for Food Security and Rural Employment, New Delhi, Vol II, pp.167.

Patel AK, Bohra HC, Rohilla PP (2006). On-farm trial of nutrient mixture supplementation in lactating goats under semi-arid conditions. Indian Journal of Animal Production and Management 22 (3-4): 157-159.

Patel AK, Bohra HC, Patidar M, Roy MM (2012-14). Multi-nutrient feed block: a strategic alternative supplement for lactating buffaloes under field conditions in arid zone. Indian buffalo J. 10-12 (1 & 2): 40-43.

Pathak PS, Gupta SK, Singh P (1996). IGFRI Approaches: Rehabilitation of Degraded Lands. IGFRI, Jhansi.

Patidar M (2006). Strip cropping of grasses and legume in silvipastoral system. In: National Symposium on Livelihood Security and Diversified Farming System for Arid Region, Abstr, CAZRI, Jodhpur, pp. 33.

Patidar M, Sexena Anurag (2007). Paryavarnsarankshankeliyecharautpadhan. Kheti 60 (3): 18-22.

Patil NV, Mathur BK, Bohra HC, Patel AK (2005). Relative preference index for arid forages for goat and sheep. In: National Seminar on Conservation, Processing and Utilization of Monsoon Herbage for Augmenting Animal Production, Indian Society for Sheep and Goat Production and Utilization, RRS, CSWRI, Bikaner, pp.190.

Patil NV, Mathur BK, Patel AK, Bohra HC, Khan MS (2006). Growth performance of arid breed kids weaned at different ages and received on complete feed block. In: Proceedings of Animal Nutrition Workers Association Conference, SKUAT, Jammu, pp. 78.

Patil NV, PatelAK, Khan MS, Bhati TK (2009). Potential assessment of silivipasture system of arid Rajasthan based on its traditional use for rearing growing lambs and kids. In: Proceedings of World Conference on Animal Nutrition Preparedness to Combat Challenges, New Delhi, pp. 278.

Rai P, Solanki KR, Rao GR (1999). Silvipasture research in India. A Review 1:107-120.

Rohilla PP, Patil NV, Bohra HC, Patel AK (2009). Effect of supplemental feeding of urea molasses mineral block (UMMB) on growth performance of Marwari lambs. In: Proceeding of National Symposium on Organic Livestock Farming- Global Issues, Trends and Challenges, West Bengal University of Animal and Fishery Sciences, Kolkata, pp 87.

Sawal RK, Patel AK, Chopra A, Narula HK, Ayub M (2015). Growth performance of Magra lambs fed multi-nutrient blocks under field conditions of hot arid Rajasthan. The Indian Journal of Small Ruminants 21 (2) 336-337.

Sharma SC, Roop Chand (2012). Research Contributions 1962 to 2012 of Central Sheep and Wool Research Institute, Avikanagar. Shinde AK, Swarnkar CP, Prince LLL (eds.).

Sharma SK, Diwakar GD (1989). Economic evaluation of hortipastoral system in arid region of west Rajasthan. Indian Journal of Range Management 10:119-122.

Sharma SK, Gupta RK (1981). Effect of seasonal lopping on the top feed (long) production and growth of Prosopis cineraria (Linn.) Druce. Indian J. Forestry 4 (4):253-255.

Shinde AK, Karim SA, Singh NP, Patnayak BC (1995). Growth performance of weaner lambs and kids under intensive and semi intensive feeding management. Indian Journal of Animal Sciences 65: 630-633.

Venkatateswarlu J, Anantharam K, Purohit ML, Khan MS, Bohra HC, Singh KC, Mathur BK (1992). In: Forage 2000 AD, The Scenario for Arid Rajasthan. CAZRI, Jodhpur, pp. 32.

Yadav GL, Poonia TC (2003). Production potential of grasses between existing stocks of trees under silvipastoral system in semi arid watershed area. In: National Symposium on Sustainability, Advancement and Future Thrust Area of Research on Forage, CCS, HAU, Hisar, pp. 53.

18

Traditional Agroforestry System for Global Warming Adaptation in Arid Rajasthan

G. Singh

Division of Forest Ecology, Arid Forest Research Institute, New Pali Road Jodhpur-342005, Rajasthan, India

Introduction

Agriculture has always been vulnerable to fragile environmental conditions of dry areas, where droughts of varying amplitude and frequency influenced by insufficient and uncertain rainfall are common features. Recent anthropogenic activities have accelerated the rate of green house gas (GHGs) accumulation in atmosphere influencing atmospheric temperature and local patterns of temperature and precipitation, which ultimately affect agricultural production and food securities. About 35.5 percent of the total world's population resides in dry areas, which is facing the problems of land degradation and food securities. Expansions of global dry land will further increase the population affected by water scarcity and land degradations. Water scarcity will also increase with increase in the aridity affecting agricultural production and people livelihood in this region. Because of the role of agriculture in the social and economic progress vulnerability of agricultural systems to climate change has received considerable attention (Kumar and Gautam 2014; BE 2016). Carbon dioxide (CO_2) is the most important GHGs with highest growth-rate during the last 10 years, i.e. 1.9 ppm (part per million) per year. At present, the amount of CO_2 in the atmosphere is more than 400 ppm (https://www.co2.earth/monthly-co2). Other gases contributing to global warming are methane, nitrous oxide and chlorofluorocarbons. Burning of fossil fuels, land use and agriculture are the major causes of the increase in green house gases (GHGs) in past 250 years. Different agricultural processes like rice cultivation and enteric fermentation in cattle comprise 54% of methane emissions, 80% of nitrous oxide emissions and major percentage of carbon dioxide (Senapati *et al.* 2013).

The existing literature suggests that the overall impacts of climate change on agriculture especially in the tropics will be highly negative, although there may be minor increases in crop production in the short term in some areas (Gautam and Sharma, 2012). The processes directly influenced by climate change are soil water, carbon and nitrogen cycles, crop growth and development, and incidence of weeds, pests, and diseases. These effects are marked by an increase in heat stress, evapo-transpiration and rates of plant photosynthesis, shortening of seasons, and reduced water use because of reduced stomatal conductance with increased concentration of CO_2 in atmosphere (Wullschleger *et al.* 2002; Mahato 2014). A substantial increase in water-use efficiency in temperate and boreal forests over the past two decades have also been reported (Keenan *et al.* 2013).Because of vulnerability of agriculture to uncertain climatic conditions people of dry areas in general and arid zone in particular are being promoting and maintaining trees and shrubs on their farmlands to meet the requirements of fuelwood and fodder during crop failure. Likewise, a considerable knowledge has been generated on integrating trees on farmlands for food and wood production in tropical (Shankarnarayan *et al.* 1987; Mbow *et al.* 2014) and temperate regions (Palma *et al.* 2007; Smith 2010; Qaisar *et al.* 2014) both.

Tree on farmlands has shown potential to increase and sustain food production per unit area (Khan and Tewari 2009; Singh *et al.* 2007; Sileshi *et al.* 2008). It is considered as a promising approach in adapting to climate change and enhancing livelihood security (Glover *et al.* 2012; Jat *et al.* 2011), largely because the trees are associated with enhancing and sustaining soil health and therefore crop production (Singh 2009; Barrios *et al.* 2012). Though most of the trees are well accepted on farm lands in the arid regions of north-western India, but most of the species are decreasing in their population due to over exploitation, mechanization of agriculture, reduced rate of regeneration and extreme environmental conditions. Trees in arid zones provides shelter and food for wildlife and contributes to efficient resource use, control of soil erosion and sequester carbon by enriching soil organic carbon (SOC) and nutrient levels in addition to other environmental and ecological services in the region. Agri-silvi system consisting of trees as woody component with various dry land crops has been observed promising due to improved soil fertility, microclimate and moisture availability (Aggarwal *et al.* 1976; Harsh and Tewari 1993). Growing tree with crops increases the total productivity per unit of land as compared to sole arable cropping in arid region (Singh 2005).

Table 1. Different land uses in India and Rajasthan during 1990-91 and 2011-12

S No	Land Use Classification	India (m ha)	Rajasthan (m ha)	
		2011-12	1990-91	2011-12
i	Areas under forest	70.01	2.35	2.75
ii	Area under non agricultural uses	26.45	1.49	1.88
iii	Barren and un-culturable land	17.28	2.79	2.39
iv	Permanent pastures and other grazing land	10.24	1.91	1.69
v	Miscellaneous trees crops and groves not included in the net area sown	3.16	0.02	0.02
vi	Culturable waste land	12.58	5.57	4.17
vii	Fallow lands other than current fallows	11.00	1.93	1.85
viii	Current fallow	15.28	1.81	1.48
ix	Net area sown	139.93	16.38	18.03
	Total reported area for land utilization	305.94	34.25	34.27

Source: GoR (1991, 2012)

Climate change adaptations and mitigation

Agroforestry practices have been part of strategies to improve natural resource management and an adaptation to climatic conditions of dry regions (Khan and Tewari 2009; Ong and Kho 2015). The practices are more effective in providing regulating, supporting and cultural ecosystem services as compared to other land use practices (Pagella and Sinclair 2014). Some of them are microclimatic buffering, amelioration of soil structure and water infiltration, reduction of overland flow, regulation of the water cycle and sequestering carbon in both wood and soils (Singh 2010; Kuyah *et al.* 2013; Verchot *et al.* 2008; Rosenstock *et al.* 2014). This helps farmers adapt to climate change through the risk-mitigating effects of additional farm products derived from trees, positive microclimatic effects through shading and enhanced farm income through tighter nutrient and water cycles (Singh *et al.* 2000; Garrity *et al.* 2010).Agroforestry is also measures of adaptation and mitigation, the two major policy responses to climate change. While adaptation seeks to lower the risks posed by the consequences of climatic changes, mitigation addresses the root causes, by reducing GHGs emissions to the atmosphere. Both the approaches are necessary, because even if emissions are dramatically decreased in future, adaptation will still be required to deal with the global changes that have already been set in motion (Tubiello and van der Velde 2004). Adaptation activities work well to limit damage from low-to -medium warming, while mitigation actions work on longer timescales. Thus implementation of adaptation and mitigation actions jointly across a wide range of land and water resource management solutions can provide both adaptation benefits in the short term and long-lasting mitigation benefits in the longer term.

Introducing trees in agricultural farms is useful in increasing soil carbon status because the presence of trees affects carbon dynamics directly or indirectly. Trees improve soil productivity through ecological and physicochemical changes that depend upon the quantity and quality of litter reaching to the soil surface and the rate of litter decomposition and nutrient release (Prescott, 2010; Ngatia *et al.* 2014). Furthermore, tree provides an assurance to the farmers towards agricultural production in normal rainfall year, while in famines and drought years they provide top feed for livestock. In addition they conserve soil carbon, which otherwise goes to atmosphere in absence of agroforestry products. Various interacting factors through which a tree influences carbon stock in the soil under agroforestry are addition of litter, maintenance of higher soil moisture content, reduced surface soil temperature, proliferated root system, enhanced biological activities and decreased risk of soil erosion (Blevines and Frye, 1993; Singh and Rathod 2002). Further, various factors like competition for land due to intense population pressure, rapidly depleting fuel wood resources, time-scale for orchard operations, dispersed distribution of benefits from forestry and seasonal shortage of labour are putting enormous pressure on non- renewable land resource leading to shift in emphasis towards agroforestry based land use system (Rekha *et al.* 2007).

Climate change adaptation

Adaptation in general is the adjustments that society or ecosystems make to minimize or prevent the negative impacts of climate change that may be local or regional in nature (NRC 2010). It also includes purposefully modifying the developmental interventions to ameliorate environmental conditions and to ensure people livelihoods. Adaptation to short-term climate variability and extreme events serves as a starting point for reducing vulnerability to longer-term climate change (Spanger-Siegfried and Dougherty 2005). Adjusting or restructuring in anticipation of adverse effects of climate change and taking appropriate actions for minimizing the damage or taking advantage of opportunities that may arise out of climatic variations are also adaptation. Adoption of conservation agriculture, crop management practices and utilization of suitable crop varieties or tree species are the ways to intensify crop production and sustain the rural livelihoods and are important means of adaptation to climate change as well as combating desertification and securing livelihood (Smith *et al.* 2014). Other important adaptations include diversification of income generation activities, *in situ* moisture conservation, rainwater harvesting and recycling to enhance water availability and enhanced productivity (Gupta 1995; Singh 2012), efficient use of irrigation water, conservation agriculture, energy efficiency in agriculture and use of poor quality water for increasing vegetation cover and biomass

(Venkateswarlu and Shanker 2009). Different strategies to climate change adaptation include activities that provide additional income and ensure food security to rural communities, for instance, forestry management and agroforestry techniques, good agricultural practices that conserve soil and water resource; and properly scaled bio-energy projects for rural communities. Human societies have repeatedly exhibited a strong capacity for adapting to different climates and environmental changes throughout the ages.

Climate change mitigation

Climate change mitigation means using new technologies and renewable energies, making older equipment more energy efficient, or changing management practices or consumer behaviour (www.unep.org). Greenhouse gases accumulating in the atmosphere can be reduced by supply-side mitigation options (i.e., reducing GHG emissions per unit of land/ animal, or per unit of product), or by demand-side options (i.e., changing demand for food and fibre products, reducing waste *etc.*). Climate change and climate extremes influences people for land use conversion and corresponding soil carbon sequestration (Li and Wu 2010). An estimate indicates that removal of atmospheric CO_2 from the Land Use, Land Use change and Forestry (LULUCF) sector offset about 13% of total U.S. greenhouse gas emissions, where forests (including vegetation, soils, and harvested wood) accounted for about 88% of LULUCF CO_2 flux in 2013. In this sector, there exists opportunities both to reduce emissions and increase the potential to sequester carbon from the atmosphere by enhancing sinks adopting different management practices (Table 1). Though there are various strategies to mitigate climate change, but no strategies are complete or successful without reducing emissions from agriculture, forestry, and other land uses. However, climate change mitigation depends on awareness of the problem, capacity to change, and the willingness to do so (Burton 2007). Land-based carbon sequestration offers the possibility of large-scale removal of greenhouse gases from the atmosphere by way of plant photosynthesis. Major strategies for reducing and sequestering terrestrial greenhouse gas emissions are:

(i) Enriching soil carbon,

(ii) Farming or retaining woody perennials,

(iii) Climate friendly livestock production,

(iv) Maintaining and protecting natural forests and habitat,

(v) Restoring degraded watersheds and rangelands, and

(vi) Improved use of biomass and fuel.

Agroforestry as carbon sinks

Agroforestry is a promising land use practice to maintain or increase agricultural productivity while preserving or improving fertility. From the perspective of climate change and the global carbon cycle, agroforestry practices are attractive by two ways: they directly store carbon in tree components, and they potentially slow deforestation by reducing the need to clear forest land for agriculture. Carbon (C) constitutes almost 50% of the dry weight of branches and 30% of foliage, but the greater part of C sequestration (around 2/3) occurs belowground, involving living biomass such as roots and other belowground plant parts, soil organisms, and C stored in various soil horizons (Nair *et al.* 2010). A survey data on tree growth and wood production converted to estimates of carbon storage indicates that median carbon storage by agroforestry practices was 9 tons C ha^{-1} in semi-arid, 21 tons C ha^{-1} in sub-humid, 50 tons C ha^{-1} in humid and 63tons C ha^{-1} in temperate eco-regions (Schroeder 1993). World over, it has been estimated that the potential area suitable for agroforestry from land-use change could be up to 630 million ha, which would result in storing 586 mt C yr^{1} by 2040, mostly in developing countries (Watson *et al.* 2000).It is now becoming an attractive entrepreneur (Makundi and Sathaye 2004) because:

i. It sequesters carbon in vegetation as well as in soils depending on the pre-conversion soil carbon.

ii. The more intensive use of the land for agricultural production reduces the need for slash-and-burn or shifting cultivation as observed in relatively higher rainfall region.

iii. The wood products produced under agroforestry serve as substitute for similar products unsustainably harvested from the natural forest.

iv. The extent to which agroforestry increases the income of farmers, it reduces the incentive for further extraction from the natural forest for income augmentation.

v. Agroforestry practices have dual mitigation benefits as fodder species with high nutritive value that help intensifying diets of methane-producing ruminants while they can also sequester carbon (Thornton and Herrero 2010).

Carbon accumulation in biomass

Transformation of low productive crop lands into agroforestry can triple carbon stocks in about 25-year period (Watson *et al.* 2000). Agroforestry is an attractive option for climate change mitigation as it sequesters carbon in vegetation and soil, produces wood, serving as substitute for similar products that are unsustainably harvested from natural forests, and also contributes to farmers' income (Sudha *et al.* 2007). In India, average sequestration potential in agroforestry has been estimated to 25 tons C per ha (Sathaye and Ravindranath 1998) but there is substantial variation in different regions depending upon the biomass production. This system holds more carbon as compared to degraded lands. Above ground biomass accumulation in an agroforestry system in central Himalayan has been found to 3.9 tons ha^{-1} yr^{-1} compared with 1.1 tons $ha^{-1}yr^{-1}$ at the degraded forestland (Maikhuri *et al.* 2000). The strip plantations in Haryana sequestered 15.5 tons C ha^{-1} during the first rotation of 5 years and 4 months (Ram *et al.* 2011).

The total growing stock of India's forests and trees outside forests is estimated as 6047.15 million m^3 i.e., 4498.73 million m^3in the recorded forest area and 1548.42 million m^3 outside the forests accounting for about total carbon stock as 6663 million tons in India (FSI 2011). The role of trees outside forests in carbon balance has been considered only recently, and it stores about 934 Tg C or 4 tons C ha^{-1} (Kaul *et al.* 2011). The net annual carbon sequestration rates for fast growing short rotation agroforestry crops such as poplar and Eucalyptus have been reported at 8 tons C $ha^{-1}yr^{-1}$ and 6 tons C $ha^{-1}yr^{-1}$, respectively (Kaul *et al.* 2010). Studies from Punjab suggest that at a rotation of seven years, poplar timber carbon content appears 23.57 tons ha^{-1} and an equal amount may be contributed by roots, leaves and tree bark (Chauhan *et al.* 2010). In smallholder bamboo farming system in Barak Valley, Assam (Nath and Das 2011), a traditional homegarden system, carbon estimate in aboveground vegetation ranged from 6.51 (2004) to 8.95 (2007) tons ha^{-1} with 87%, 9% and 4% of the total carbon stored in culm, branch and leaf respectively. In tropical homegardens of Kerala, average above-ground standing stocks of carbon ranged from 16 to 36 tons ha^{-1}, where small homegardens often have higher carbon stocks on unit area basis compared to large- and medium-sized ones (Kumar 2011).

Allocation of biomass into stem has been observed varying between 38.3 and 56.6% in different girth categories, whereas in root component it varies from 23% to 48% of the total biomass (both above-ground and below-ground biomass) of *T. undulata* tree. Allocation to twig and leaves varies from 11% to 16% and 1% to 5% of the total biomass. On an average, contribution of different components in total biomass is about 47%, 14%, 4% and 35% for stem, twig, leaf and root. Below-ground (root) to above-ground biomass ratio varies from 0.30 to 0.78 and have observed decreasing with increase in tree size. There is an increase in stem, twig and leaf biomasses and decrease in root biomass with increase in total biomass of the trees of *T. undulata* as indicated by decrease in root to total biomass ratio in relation to total biomass. This ratio for other components indicates an increasing trend (Fig 1a). The increases in biomasses of different components viz. stem, twig, leaves and root in relation to increase in total biomass indicate relatively greater increase in stem biomass followed by root, twig and leaves (Fig 1b). The coefficients of linear lines are 0.513, 0.286, 0.154 and 0.045 for the respective component.

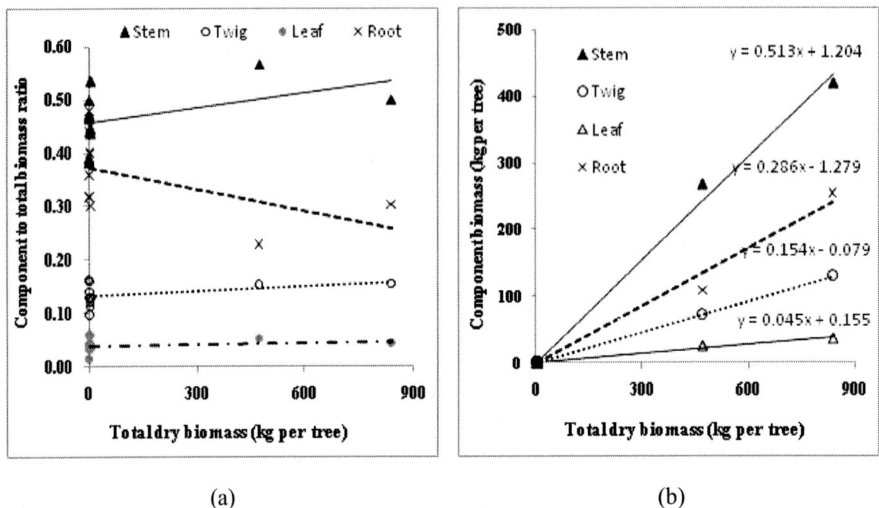

(a) (b)

Fig 1. Allocation of biomass in different components of *T. undulata* trees in western Rajasthan

During 12 years of experimentation, carbon accumulation per tree in *P. cineraria* was higher at tree density of 208 tree ha^{-1}as compared to the tree density of 278 and 416 trees ha^{-1} (Table 2). Carbon accumulation in above-ground biomass ranged from 5.29 tons C ha^{-1} to 9.76 tons C ha^{-1}, whereas in roots it ranged from 1.70 to 3.10 tons C ha^{-1} (Singh 2004). Root biomass was significantly high for the trees 208 tree ha^{-1} compared to 278 and 416 trees ha^{-1} plots (Singh *et al.* 2007). Thus stand of *P. cineraria* maintained at tree density of 208-tree ha^{-1} could not only maintain higher crop production but also accumulate higher carbon in the form of tree biomass.

Table 2. Carbon storage in 12-year-old tree of *P. cineraria* influenced by varying stem density in June 2002

Tree density	Carbon (kg tree^{-1})						ton C ha^{-1} (no. ha^{-1})		
	Stem	Branch	Leaf	Above	Root	Total	Above	Root	Total
417	8.83	3.50	0.35	12.68	4.08	16.76	5.29	1.70	6.99
278	21.67	6.66	0.71	29.04	8.50	37.54	8.07	2.36	10.43
208	34.08	11.62	1.23	46.93	14.91	61.84	9.76	3.10	12.86

Source: Calculated from Singh *et al.* (2007)

Enrichment of soil carbon

Soil organic carbon (SOC) is the carbon associated with soil organic matter, which is the organic fraction of the soil made up of decomposed plant and animal materials as well as microbial organisms. However, it does not include fresh and un-decomposed plant materials like straw and litter lying on the soil surface. Soil organic carbon is important for all three aspects of soil fertility like chemical, physical and biological fertility (Lal 2010; Olson *et al.* 2014). Soil carbon can also be present in inorganic forms, e.g. lime or carbonates in some soils in the drier areas including Rajasthan. According an estimate 30 - 60% of the atmospheric carbon dioxide absorbed by plants is deposited into the soil as organic matter particularly in the form of bud sheaths for protecting the delicate root tips and as a range of other root excretions, which is 5-33% of daily photo-assimilates (Bardgett *et al.* 2005). However, the quantity of exudates is likely greatest in fast-growing plant species, especially in those, which show highly branched fine root systems (Personeni and Loiseau 2004). Tree integration, no-tillage, organic manuring, soil and water conservation etc. provide high rates of SOC sequestration (Singh 2005; Swart *et al.* 2003; Wang *et al.* 2015). For example soil C pool rate of 2- 3 Mg C ha"[1] yr"[1] have been observed in sodic soils in northern India under agroforestry practices (Garg 1998). A SOC sequestration rate of 3.37-6.58 tons ha^{-1} y^{-1} (Average 4.93 tons ha^{-1} y^{-1}) has also been observed in 0-40 cm soil layer under afforestation and rainwater harvesting during 65 moths of restoration of degraded hills (Singh 2013). Nair *et al.*2009 a, b) estimated soil organic C stocks ranging from 6.9 to 302 Mg ha^{-1} and the variation was because of differences between systems, ecological regions, and soil types. However, the study revealed a general trend of increasing soil carbon sequestration in agroforestry when compared to other land-use practices, with the exception of forests. Most important management practice that influences soil C sequestration is the insertion of trees in agricultural systems. In general, greater C content are observed nearer to the trees (Singh 2009). Carbon sequestration potential of tropical agroforestry systems is estimated between 12 and 228 Mg ha^{-1} with a median value of 95 Mg ha^{-1}, whereas other estimates based on the global status of the area suitable for the agroforestry suggests that 1.1-2.2 Pg C (petagramme carbon) could be stored in the terrestrial ecosystems over the next 50 years (Albrecht and Kandji 2003). Another estimates of C stored in agroforestry systems range from 0.29 to 15.21 Mg $ha^{-1}yr^{-1}$ aboveground and 30 to 300 Mg C ha^{-1} up to 1-m dcpth in thc soil (Nair *et al.* 2010). Saha *et al.*2010) found 30% and 114% greater carbon in homegardens in India than in coconut plantations and rice paddies, respectively. In the Northeast of Spain, Howlett *et al.* (2011) studied the levels of soil C in silvopastoral systems composed of different species, and found that systems with birch (*Betula pendula*) showed greater levels of soil carbon than systems with pine (*Pinus radiata*).

Tillage is reported to cause significant loss of surface soil organic carbon and disruption of the process of macro-aggregate formation, increasing bio-available sources of carbon from smaller aggregates. However, Tonucci *et al.* (2011) observed that on the short term, carbon levels in pasture soils can be greater than under systems with trees due to the faster turnover of the grass root system, as well as the greater bulk density of soils in pasture systems, that can lead to higher carbon values than that in less dense soils. The lag in achieving positive results in terms of soil carbon accumulation under trees are similar to what has been observed in the transition to continuous no-till agriculture. Derpsch *et al.* (2008) recorded a clear increase in soil organic matter only after 5–10 years of adoption of this system of cropping. Climate conditions also affect carbon accumulation in soils, since temperature and humidity greatly affect the activity of microbial communities and the breakdown of organic matter (Bertin *et al.* 2010). In agro-ecosystems of Indo-Gangetic Plains about 69% of soil carbon in the soil profile is confined to the upper 40 cm soil layer where C stock ranges from 8.5 to 15.2 tons C ha^{-1}. Agricultural soils of Indo-Gangetic Plains contain 12.4 to 22.6 tons ha^{-1} of organic C in the top 1 m soil depth (Singh *et al.* 2011). In fact 5 to 10 kg C ha^{-1} can be sequestered in about 25 years in soils of extensive tree-intercropping systems of arid and semiarid lands to 100-250 kg C ha-^{1} in about 10 years in species-intensive multistrata shaded perennial systems and homegardens of humid tropics (Nair *et al.* 2009).

A decreasing trend in soil organic carbon (SOC) with soil depth in both sole agriculture (control) and the plots with *Emblica officinalis*, *Hardwickia binata* and *Colophospermum mopane* trees have been observed (Singh 2005). SOC was lowest in control plot compared to the tree integrated plots in the three soil layers. Soil organic carbon was highest in *E. officinalis* plot except in 1996 and 1998. In most of the observations, low SOC was recorded in *C. mopane* plot. The rotation-cropping plot had high SOC till the year 1999, whereas fixed cropping plot of *Vigna radiata* had high SOC after 1999. Temporal changes in soil organic carbon (0-75 cm soil layer) indicated a decreasing trend from 1994 to 2002 in 0-25 and 25-50 cm soil layers in the control plot. However, an increasing trend in SOC was observed from 1998 and 2000 in 0-25 cm and 25-50 cm soil layer, respectively in *E. officinalis* plot. In *H. binata* and *C. mopane* plots the increase in SOC was started from the year 2000. Initial soil carbon density in July 1994 was 3.02 kg m^{-2} (0-100 cm soil depth), which decreased to 1.34 kg m^{-2} in 2002 in sole agricultural crop plot, where a loss of 1.68 kg m^{-2} (i.e., 16.6 Mg SOC) was observed. However, the trend of decrease in soil carbon density was less under agroforestry systems (Table 3). The increase in SOC stock was noticed from the sixth in *E. officinalis* and in the seventh year in *H. binata* and *C. mopane* plots. Integration of tree species resulted in a sequestration of 12.7, 8.8, 4.7 Mg ha^{-1} in fixed crop plot and 13.0, 8.6 and 5.3 Mg ha^{-1} of SOC in rotational crop plot under *E. officinalis*, *H. binata* and *C. mopane* plots, respectively when compared with the carbon SOC in control plot in the ninth year. Thus, land management actions can also enhance the uptake of CO_2 or reduce its emissions and have the potential to remove a significant amount of CO_2 from the atmosphere if the trees are harvested, accompanied by regeneration of the area, and sequestered carbon is locked through non-destructive use of such wood.

Table 3. Temporal changes in carbon density (kg m^{-2}) in 0-100 cm soil layer under the influence of different tree species. CP: cropping pattern, FC: fixed crop of *Vigna radiata*; and RC: rotation crop of legume and non-legume

Treat	CP	Initial	1994	1995	1996	1997	1998	1999	2000	2001	2002
Control	-	3.02	2.01	1.86	1.87	1.87	1.85	1.70	1.49	1.35	1.34
E. officinalis	FC	3.02	3.09	3.00	2.91	2.78	2.63	2.63	2.84	2.88	2.87
	RC	3.02	2.96	2.60	2.72	2.87	2.66	2.67	2.73	2.76	2.97
H. binata	FC	3.02	2.81	2.73	2.82	2.81	2.52	2.24	2.27	2.31	2.40
	RC	3.02	3.09	2.96	2.96	3.09	2.73	2.36	2.22	2.27	2.31
C. mopane	FC	3.02	3.12	2.81	2.77	2.60	2.48	2.33	1.89	1.89	1.97
	RC	3.02	3.30	2.82	2.85	2.81	2.58	2.36	1.89	1.89	1.93

Source: Singh (2005)

Though difference in soil organic matter between the canopy zone of *P. cineraria* and *T. undulata* trees under different crop sequences want, but it was relatively greater under *P. cineraria* than under *T. undulata* indicates the impacts of leguminous tree though it might also be due to crops effects (Singh 2004). It is indicated by effects of crop sequences on soil organic matter, which in generally higher in top soil layer than in the deeper soil layers with soil depth (Table 4). Mung bean-mung bean plot (legume rotated by legume) showed greater soil organic matter, whereas it was relatively low in the plot with mixed cropping of mung bean and sesame (*Sesamum indicum*).

Table 4. Soil organic matter (%, w/w) after crop harvest affected by crop sequences and tree species

Crop sequences	*P. cineraria*				*T. undulata*			
	0-25	25-50	50-75	Mean	0-25	25-50	50-75	Mean
Tree only	0.35	0.22	0.12	0.23	0.34	0.22	0.11	0.22
Mung-mung	0.38	0.26	0.19	0.28	0.35	0.23	0.16	0.25
Sesame–Sesame	0.36	0.22	0.12	0.23	0.33	0.21	0.12	0.22
Sesame- mung	0.34	0.21	0.12	0.22	0.33	0.21	0.11	0.22
Senna – Senna	0.36	0.23	0.12	0.24	0.34	0.21	0.12	0.22
Guar-Guar	0.35	0.22	0.13	0.23	0.33	0.21	0.12	0.22
Mung + Sesame	0.32	0.22	0.12	0.22	0.31	0.21	0.11	0.21
Mean	0.35	0.23	0.13	0.24	0.33	0.21	0.12	0.22

Trees add organic matter to the soil system in various manners, whether in the form of roots or litterfall or as root exudates in the rhizosphere (Bertin *et al.* 2003). These additions are the main substrate for a vast range of organisms involved in soil biological activity and interactions, with important effects on soil organic carbon, nutrients and soil fertility. However, the contributions of leguminous trees are relatively greater as compared to non-leguminous trees. For example, per cent soil organic carbon in an agroforestry system has been observed greater under *Acacia albida* (0.179) and *P. cineraria* (0.165) than to non-leguminous *T. undulata*(0.138) and *Z. mauritiana* (0.138) (Bhati and Joshi 2007). Less biomass input and more soil disturbance result in buildup of less carbon in soil under the legume crop, i.e. mung bean (0.125). However, oxidation of soil organic carbon (i.e., 200-850 g per year) under the prevailing temperature and management influence SOC accumulation (Kar *et al.* 2007; Singh *et al.* 2007). A study on SOC depletion during 1975 to 2002 has been reported to the level of 10.4% in the central part of western Rajasthan, whereas a decline in SOC in one meter deep soil of western Rajasthan has been observed at 78 million tons, i.e. ~825 million tons in 1975 to ~ 747 million tons in 2002 (Kar *et al.* 2009).

Maintaining and farming woody perennials

Perennial crops, grasses, shrubs and trees constantly maintain and develop their root and woody biomass that helps in increasing the associated carbon in addition to providing vegetative cover to the soils (De Deyn *et al.* 2008; Singh *et al.* 2013a). In arid regions where soil moisture restricts plant growth, the traits that enable the survival and growth under the extremes of precipitation and temperature regimes through opportunistic or persistence strategies (Ogle and Reynolds 2004) and the traits that govern carbon distribution throughout the soil profile (Jobbagy and Jackson 2000; Schenk and Jackson 2002) directly influence carbon storage through primary productivity and carbon stabilization in soil. A tree transpire about 500 kg of water for each kg of carbon fixed on annual basis in arid areas, where about 60% of this carbon returns to the atmosphere by respiration and loss of water vapour through stomata is far more than 1000 times the net carbon gain making an unfortunate tradeoff in the region (Sabaté and Gracia 2011). There is significant potential to substitute annual tilled crops with perennials, particularly for animal feed and vegetable oils, as well as to incorporate woody perennials into annual cropping systems in the form of agroforestry systems. Managing and promoting trees or woody shrubs (*Zizyphus* spp., *Punica granatum* etc.) in agro-ecosystems, ethnoforests, and trees outside forests mitigate GHGs emissions and provides a better climate change mitigation option (Table 5). It helps to attain food security and secure land tenure in developing countries, increasing farm income, restoring and maintaining above-ground and below-ground biodiversity, corridors between protected forests, as CH_4 sinks, maintaining watershed hydrology, and soil conservation (Singh *et al.* 2007; Singh *et al.* 2013b). Promoting woodcarving industry facilitates long-term locking-up of carbon in carved wood and new sequestration through intensified tree growing.

Restoring woody vegetation on cleared land, restoring croplands by reconverting them to rangelands, restoring vegetation to bare soils, and restoring soil stability could all increase carbon sequestration as well as storage (Brooker *et al.* 2013). Providing incentives may encourage farmers and land users to maintain natural vegetation through product certification, payments for climate services, securing tenure rights, and community fire control (Asbjornsen *et al.* 2013).Several studies suggest that introducing or reintroducing broadleaf trees have favourable impact on carbon sequestration on marginal lands or degraded forestland seeing that they have large root systems that promote the growth of forbs and grasses beneath the canopy (De Deyn *et al.* 2008; Singh *et al.* 2011). Rate and magnitude of carbon sequestration however, depends upon soil type, species, nutrient management and climate, but it may not be always positive (Johnson and Curtis 2001). For example, litter produced by invasive exotic plants may differs from native plant litter in quality and quantity affecting litter decomposition and soil respiration in ways that depend on whether exotic and native plant litters decompose in mixtures. Exotic *Alternanthera* produces rapidly decomposing litter which accelerates the decomposition of native plant litter in litter mixtures and enhances soil respiration rates affecting total soil carbon storage negatively (Zhang *et al.* 2014). According to a prediction, mitigation potential of agriculture and forestry observed to be about 4-18 billion tons CO_2 equivalent by 2030 (Table 5). In this highest expectation is from forests and its conservation followed by agriculture.

Table 5. Mitigation potential in agriculture and forestry by 2030

Land use	2030 reductions billion tonsCO_2e
Global	15-25
Agriculture	1.5-5.0
Reduction of non CO2 gases	(0.3-1.5)
Agro forestry	(0.5-2)
Enhanced soil carbon sequestration	(0.5-1.5)
Forest	2.5-12
REDD+	(1-4)
SFM	(1-5)
FR including A/R	(0.5-3)
Bio-energy	0.1-1.0
Total	4-18

Source: FAO (2005)

Restoring rangelands

Extensive areas of the world have been denuded of vegetation through land clearing for crops or grazing and from overuse and poor management and the degradation of our environment have exceeded the rate of conservation (Cairns Jr. 1998). Degradation has not only generated a huge amount of GHGs emissions, but local people have lost a valuable livelihood asset as well as essential watershed functions. Extent of degradation of forest and grazing lands can be understood by a study carried out in Jodhpur district of Rajasthan, where average organic carbon stock (both live biomass both above- and below-ground, SOC in top 1 m soil layer) ranged from 16.16 tons ha^{-1} in pastureland to 26.83 tons ha^{-1} in sacred groves showing significant spatial variation ranging from 14.43 tons ha^{-1} in Bilara to 25.67 tons ha^{-1} in Balesar range (Singh 2015). Restoring vegetative cover on degraded lands can be a win-win-win strategy for addressing climate change, rural poverty, and water scarcity (Arnalds A. 2004; Singh and Rathod 2002; Singh *et al.* 2013a). Improved land management could offset a quarter of global emissions from fossil fuel use in a year and one need to pursue land use solutions in addition to efforts to improve energy efficiency and speed the transition to renewable energy for enhancing carbon storage. Likewise, use of native species selected in the vicinity of the working area, as well as the implementation of soil obtained from the nearby forest fragments, could allow better restoration and carbon sequestration (Araujo *et al.* 2014). For all of these, watershed approach is an important element, which help both in adaptation and mitigation to climate change as soil and water conservation, moderation of the runoff, enhancing ground water recharge and minimizing floods even during high intensity rainfall enhance vegetation recovery and carbon sequestration (Bhati 1997).Intensification and protection of rangelands from grazing also improves vegetation characteristics and soil properties. Rong *et al.* (2014) recorded improvement in vegetation biomass and soil properties by removal of sheep grazing, but a less impact was observed on the species richness and diversity in extremely dry region. Li *et al.* (2014) observed increase in vegetation cover, height and above- and belowground biomass, soil organic carbon and total nitrogen concentration and a decrease in soil pH, electrical conductivity and soil bulk density in the protected area.These authors also found recovery of vegetation in 6 years after fencing, soil pH in 8 years, soil organic carbon in 16 years, total nitrogen in 30 years and total phosphorus concentrations in 19 years after fencing.It was recommended to reduce stocking rate by 1/3 of the current carrying capacity, managing grazing regime as 1-year of grazing followed by a 2-year rest to sustain the current status and application of N- fertilizer to shorten the differed period depending on the rate of application (Li *et al.* 2014). Application of treatment like grazing plus ploughing resulted in the least SOC

(15.30 Mg C ha^{-1}), whereas protection from grazing and shrub removal led to 28.49 Mg C ha^{-1} (Daryanto *et al.* 2013).

Conclusion and recommendations

Predicted climate change is going to exacerbate the vulnerability of agriculture furthermore in dry areas. Agroforestry is a traditionally tested strategy among the many considered for adapting to and mitigating of climate change effects. It is an important carbon sink but the extent of carbon sequestration depends upon soil types, tree species, age of trees and shrubs, geographic location, local climatic factors, and management regimes. It has potential to sequester large amounts of above- and belowground carbon compared to tree-less farming systems in the form of both tree biomass and soil carbon. Management of agriculture lands or other community wastelands by integrating trees or afforestation is an attractive option for enhancing carbon stock and reducing CO_2 gas emission for better incentives. Moreover, enhancing soil carbon will increase soil fertility and improve crop production as an adaptation in addition to the benefits obtained from the various tree products.

References

Aggarwal RK, Gupta JP, Saxena SK, Muthana KD (1976). Studies on soil physico-chemical and ecological changes under twelve years old five desert tree species of western Rajasthan. Indian Forester 102(12): 863-872.

Albrecht A, Kandji ST (2003). Carbon sequestration in tropical agroforestry systems. Agriculture, Ecosystems and Environment 99:15-27.

Araujo ICL, Dziedzic M, Maranho LT (2014). Management of the environmental restoration of degraded areas. Braz Arch Biol Technol 57(2): Curitiba.

Arnalds A (2004). Carbon sequestration and the restoration of land health. Climate Change 65: 333-346.

Asbjornsen H, Hernandez-Santana V, Liebman M, Bayala J, Chen J, Helmers M, Ong CK, Schulte LA (2013). Targeting perennial vegetation in agricultural landscapes for enhancing ecosystem services. Renewable Agriculture and Food Systems.

Bardgett RD, Bowman WD, Kaufmann R, Schmidt SK (2005). A temporal approach to linking aboveground and belowground ecology. Trends Ecol Evol 20:634–641.

Barrios E, Sileshi GW, Shepherd K, Sinclair F (2012). Agroforestry and soil health: linking trees, soil biota and ecosystem services. In: Wall DH *et al.* (eds) Soil Ecology and Ecosystem Services, Oxford University Press, Oxford, UK, pp. 315-330.

BE (2016). Climate Change affects Indian Agriculture. Business Economics, February 2.

Bertin C, Yang X, Weston LA (2003). The role of root exudates and allele chemicals in the rhizosphere. Plant and Soil 256:67–83.

Bhati TK (1997). Sustainable Farming Systems for Natural Resource Conservation in Arid Natural Resource Conservation in Arid Watersheds and Index Catchments. CAZRI, Jodhpur.

Bhati TK, Joshi NL (2007). Farming systems for sustainable agriculture in Indian arid zone. In: Vittal KPR *et al.* (eds) Dryland Ecosystem-Indian Perspective, pp. 35-52.

Blevins RL, Frye WW (1993). Conservation tillage: an ecological approach to soil management. Adv Agron 51:33-78.

Brooker K, Huntsinger L, Bartolome JW, Sayre NF, Stewart W (2013). What can ecological science tell us about opportunities for carbon sequestration on arid rangelands in the United States? Global Climate Change 23:240-251.

Burton D (2007). Evaluating climate change mitigation strategies in south east Queensland. Urban Research Program Research Paper.

Cairns Jr (1998). Eco-societal restoration: rehabilitating human society's life support system. In: Rana BC (ed). Damaged Ecosystems and Restoration, World Scientific, Delhi.

Chauhan SK, Sharma SC, Chauhan R, Naveen G, Ritu (2010). Accounting poplar and wheat productivity for carbon sequestration in agri-silvicultural system. Indian Forester 136(9): 1174-1182.

Daryanto S, Eldridge DJ, Throop HL (2013). Managing semi-arid woodlands for carbon storage: Grazing and shrub effects on above- and belowground carbon. Agriculture, Ecosystems and Environment169:1-11.

De Deyn GB, Cornelissen JHC, Bardgett RD (2008). Plant functional traits and soil carbon sequestration in contrasting biomes. Ecology Letters.

Derpsch R (2008). No-Tillage and Conservation Agriculture: A Progress Report. In: Goddard T, Zoebisch M, GanY, Ellis W, Watson A, Sombatpanit S (eds)No-Till Farming Systems, World Association of Soil and Water Conservation. Special Publication No. 3:7-39.

FSI (2011). India State of Forest Report. Forest Survey of India, Dehradun.

Garg VK (1998). Interaction of tree crops with a sodic soil environment: Potential for rehabilitation of degraded environments. Land Degrad Dev 9:81-93.

Garrity DP, Akinnifesi FK, Ajayi OC, Weldesemayat SG, Mowo JG, Kalinganire A, Larwanou M, Bayala J (2010). Evergreen agriculture: a robust approach to sustainable food security in Africa. Food Secur 2(3):197-214.

Gautam HR, Sharma HL (2012). Environmental degradation, climate change and effect on agriculture. J Kurukshetra 60:3-5.

Glover JD, Reganold JP, Cox CM (2012). Agriculture: plant perennials to save Africa's soils. Nature 489(7416):359-361.

GoR (2012). Statistical Abstract. Directorate of Economics and Statistics, Rajasthan, Government of Rajasthan, Jaipur.

GuptaGN(1995). Rain-water management for tree planting in the Indian desert. J Arid Environ 31:219-235.

Harsh LN, Tewari JC (1993). Sand dune stabilization, shelterbelts and silvi-pastoral plantation in dry zones. In: Sen AK, Kar A (eds) Desertification and Its Control in the Thar, Sahara and Sahel Regions, Scientific Publishers, Jodhpur, pp. 269-279.

Howlett DS, Mosquera-Losada MR, Nair PKR, Nair VD, Rigueiro-Rodrigues A (2011). Soil carbon storage in silvopastoral systems and a treeless pasture in northwestern Spain. Journal of Environmental Quality 40:825–832.

Jat HS, Singh RK, Mann JS (2011). Ardu (*Ailanthus* spp.) in arid ecosystem: A compatible species for combating with drought and securing livelihood security of resource poor people. Indian J Traditional Knowledge 10(1).:102-113.

Jobbagy EG, Jackson RB (2000). The vertical distribution of soil organic carbon and its relation to climate and vegetation. Ecol Appl 10:423-436.

Johnson DW, Curtis PS (2001). Effects of forest management on soil C and N storage: meta analysis. Forest Ecology and Management 140:227-238.

Kar A, Moharana PC, Singh SK (2007). Desertification in arid western India. In: Vittal KPR *et al.* (eds) Dryland Ecosystem-Indian Perspective, pp. 1-22.

Kar A, Mohrana PC, Raina P (2009). Desertification and its control measures. In: Trends in Arid Zone Research in India. pp. 1-47.

Kaul, M, Mohren G, Dadhwal V (2010). Carbon storage and sequestration potential of selected tree species in India. Mitigation and Adaptation Strategies for Global Change 15(5): 489-510.

Kaul M, Mohren G, Dadhwal V (2011). Phytomass carbon pool of trees and forests in India.Climatic Change108(1):243-259.

Keenan TF, Hollinger DY, Bohrer G, Dragoni D, Munger JW, Schmid HP, Richardson AD (2013). Increase in forest water-use efficiency as atmospheric carbon dioxide concentrations rise. Nature 499:324-328.

Khan MA, Tewari JC (2009). Watershed management for fuelwood and fodder security in a traditional agroforestry system of arid western Rajasthan. J Tropical Forestry 25:1-10.

Kumar BM (2011). Species richness and aboveground carbon stocks in the homegardens of central Kerala, India. Agriculture, Ecosystems and Environment 140:430-440.

Kumar R, Gautam HR (2014). Climate change and its impact on agricultural productivity in India. J Climatol Weather Forecasting 2:109.

Kuyah S, Dietz J, Muthuri C, van Noordwijk M, Neufeldt H (2013). Allometry and partitioning of above-and below-ground biomass in farmed *Eucalyptus* species dominant in Western Kenyan agricultural landscapes. Biomass Bioenergy 55:276-284.

Lal R (2010). Enhancing eco-efficiency in agro-ecosystem through soil carbon sequestration. Crop Science 50:S120-131.

Li YQ, Han JJ, Wang SK *et al.* (2014). Soil organic carbon and total nitrogen storage under different land uses in the Naiman Banner, a semiarid degraded region of northern China. Canadian J Soil Science 94: 9–20.

Li M, Wu JJ (2010). Predicting China's land-use change and soil carbon sequestration under alternative climate change scenarios. http://ageconsearch. umn.edu/handle /61671.

Mahato A (2014). Climate Change and its Impact on Agriculture. International J Scientific and Research Publications 4(4):1-6.

Maikhuri RK, Semwal RL, Rao KS, Singh K, Saxena KG (2000). Growth and ecological impacts of traditional agroforestry tree species in Central Himalaya, India. Agroforestry Systems 48:257-271.

Makundi WR, Sathaye JA (2004). GHG mitigation potential and cost in tropical forestry – relative role for agroforestry. Environment, Development and Sustainability 6:235–260.

Mbow C, Smith P, Skole D, Duguma L, Bustamante M (2014). Achieving mitigation and adaptation to climate change through sustainable agroforestry practices in Africa. Curr Opin Environ Sustain 6:8-14.

Nair PKR, Kumar BM, Nair VD (2009a). Agroforestry as a strategy for carbon sequestration. J Plant Nutrition and Soil Science 172:10–23.

Nair PKR, Nair VD, Kumar BM, Showalter JM (2010). Carbon sequestration in agroforestry systems, In: Donald LS (ed) Advances in Agronomy 108:237-307.

Nath AJ, Das AK (2011). Carbon storage and sequestration in bamboo-based smallholder homegardens of Barak Valley, Assam. Current Science 100:229-233.

NgatiaLW, Reddy KR, Nair PKR, Pringle RM, Palmer TM, Turner BL (2014). Seasonal patterns in decomposition and nutrient release from East African savanna grasses grown under contrasting nutrient conditions. Agric Ecosystem and Environ 188:12-19.

NRC (2010). Adapting to Impacts of Climate Change. America's Climate Choices: Report of the Panel on Adapting to the Impacts of Climate Change. National Research Council. The National Academies Press, pp. 292.

Ogle K, Reynolds JF (2004). Plant responses to precipitation in desert ecosystems: integrating functional types, pulses, thresholds, and delays. Oecologia 141:282–294.

Olson KR, Al-Kaisi MM, Lal R, Lowery B (2014). Experimental consideration, treatments, and methods in determining soil organic carbon sequestration rates. Soil Sci Soc Am J 78:348-360.

Ong CK, Kho R (2015). A framework for quantifying the various effects of tree-crop interactions. In: Black C, Wilson J, Ong CK (eds), Tree–Crop Interactions: Agroforestry in a Changing Climate, CABI, pp.1–23.

PagellaTF, Sinclair FL (2014). Development and use of a typology of mapping tools to assess their fitness for supporting management of ecosystem service provision. Landsc Ecol 29(3): 383–399.

Palma J, Graves A, Burgess P, van der Werf W, Herzog F (2007). Integrating environmental and economic performance to assess modern silvo-arable agroforestry in Europe. Ecol Econ 63(4):759–767.

Personeni E, Loiseau P (2004). How does the nature of living and dead roots affect the residence time of carbon in the root litter continuum? Plant Soil 267:129-141.

Prescott CE (2010). Litter decomposition: what controls it and how can we alter it to sequester more carbon in forest soils? Biogeochemistry 101:133

Qaisar KN, Khan PA, Banyal R (2014).Temperate Agroforestry for Sustenance and Climate Moderation. ICAR, New Delhi.

Ram J, Dagar JC, Lal K, Singh G, Toky OP, Tanwar VS, Dar SR, Chauhan MK (2011). Biodrainage to combat waterlogging, increase farm productivity and sequester carbon in canal command areas of northwest India. Current Science 100:1673-1680.

Rekha KB, Rao PM, Mahavishnan K (2007). Agri-silvicultural system studies- a review. Agric Rev 28(2):142-148.

Rong Y, Yuan F, Ma L (2014). Effectiveness of Exclosures for Restoring Soils and Vegetation Degraded by Overgrazing in the Junggar Basin, China 60, pp. 118-124.

Rosenstock T, Tully K, Arias-Navarro C, Neufeldt H, Butterbach-Bahl K, Verchot L (2014). Agroforestry with N_2-fixing trees: sustainable development's friend or foe? Curr Opin Environ Sustain 6:15–21.

Saha SK, Nair PKR, Nair VD, Kumar BM (2009). Soil carbon stock in relation to plant diversity of homegardens in Kerala, India. Agroforestry Systems 76:53-65.

Saha SK, Nair PKR, Nair VD, Kumar BM (2010). Carbon storage in relation to soil size-fractions under tropical tree-based land-use systems. Plant and Soil 328:433–446.

Schenk HJ, Jackson RB (2002). Rooting depths, lateral spreads, and below-ground D above-ground allometries of plants in water-limited ecosystems. J Ecol 90:480-494.

Schroeder P(1993). Agroforestry systems: integrated land use to store and conserve carbon. Climate Research 3:53-60.

Senapati MR, Behera B, Mishra SR (2013). Impact of climate change on Indian agriculture & its mitigating priorities. Am J Environ Protection 1.4:109-111.

Shankarnarayan KA, Harsh LN, Kathju S (1987). Agroforestry in the arid zones of India. Agroforestry Systems 5:69-88.

Sileshi G, Akinnifesi FK, Ajayi OC, Place F (2008). Meta-analysis of maize yield response to woody and herbaceous legumes in sub-Saharan Africa. Plant Soil 307:1–19.

Singh AK (2010). Probable agricultural biodiversity heritage sites in India: VII. The arid western region. Asian Agri-History 14(4):337–359.

Singh G (2004). Agro forestry research for sustainable production in arid and semi arid regions of Rajasthan. Project Completion Report. Submitted to Indian Council of Forestry Research and Education (ICFRE)., Dehradun.

Singh G (2005). Carbon sequestration under an agri-silvicultural system in the arid region. Indian Forester 131:543-552.

Singh G (2009). Comparative productivity of *Prosopis cineraria* and *Tecomella undulata* based agroforestry systems in degraded lands of Indian Desert. J Forestry Research 20:144-150.

Singh G (2012). Enhancing growth and biomass production of plantation and associated vegetation through rainwater harvesting in degraded hills in southern Rajasthan. New Forests43: 349-364.

Singh G (2013). Production and functional aspects of agroforestry for enhancing livelihood supports in Indian dry regions. In: Pandey CB, Chaturvedi OP (eds) Agroforestry: Systems and Prospects, New India Publishing Agency, Pitampura, New Delhi, pp. 181-221.

Singh G, Rathod TS (2002). Plant growth, biomass production and soil water dynamics in a shifting dune of Indian desert. Forest Ecology and Management 171:309-320.

Singh G, Gupta GN, Kuppusamy V (2000). Seasonal variations in organic carbon and nutrient availability in arid zone agroforestry systems. Trop Ecol 41(1):17-23.

Singh G, Mishra D, Singh K, Parmar R (2013). Effects of rainwater harvesting on plant growth, soil water dynamics and herbaceous biomass during rehabilitation of degraded hills in Rajasthan, India. Forest Ecology and Management 310:612-622.

Singh G, Mutha S, Bala N (2007). Effect of tree density on productivity of a *Prosopis cineraria* agroforestry system in North Western India. J Arid Environments 70:152-163.

Singh H, Pathak P, Kumar M, Raghubanshi AS (2011). Carbon sequestration potential of indo-gangetic agroecosystem soils. Tropical Ecology 52:223-228.

Singh K (2015). Effect of land use types on floral diversity and carbon stock in Jodhpur district of Rajasthan. Theses submitted to FRI University, Dehradun for award of Ph.D. degree.

Singh S, Park J, Litten-Brown J (2011). The Economic Sustainability of Cropping Systems in Indian Punjab: A Farmers' Perspective. http://ageconsearch.umn.edu/bitstream/116007 / 2/Singh_Sukhwinder_557a.pdf.

Smith J (2010). The history of temperate agroforestry. Progressive farming trust limited. Trading as the organic research centre, elm farm. http://orgprints.org/18173/1/ History_of_agroforestry_v1.0.pdf.

Smith P, Bustamante M, Ahammad H, Clark H, Dong EA *et al.* (2014). Agriculture,forestry and other land use. In: Climate Change: Mitigation of Climate Change. Contribution of Working Group III to the Fifth Assessment Report of the Intergovernmental Panel on Climate Change. Cambridge University Press, Cambridge, UnitedKingdom and New York, NY, USA.

Spanger-Siegfried E, Dougherty B (2005). User's Guidebook. In: Lim B, Spanger-Siegfried E (eds) Adaptation Policy Framework for Climate Change: Developing Strategies, Policies and Measures, UNDP and GEF.

Sudha P, Ramprasad V, Nagendra M, Kulkarni H, Ravindranath NH (2007). Development of an agroforestry carbon sequestration project in Khammam district, India. Mitigation and Adaptation Strategies for Global Change 12:1131-1152.

Swart R, Robinson J, Cohen S (2003). Climate change and sustainable development: expanding the options. Climate Policy 3S1: S19–S40.

Thornton PK, Herrero M (2010). Potential for reduced methane and carbon dioxide emissions from livestock and pasture management in the tropics. In: Proceedings of the National Academy of Sciences of the United States of America, 107, pp. 19667-19672.

TonucciRG, Nair PKR, Nair VD, Garcia R, Bernardino FS (2011). Soil carbon storage in silvopasture and related land-use systems in the Brazilian Cerrado. J Environmental Quality 40:833–841.

Venkateswarlu B, Shanker AK (2009). Climate change and agriculture: adaptation and mitigation strategies. Indian J Agron54:226-230.

Verchot LV, Brienza S, de Oliveira VC, Mutegi JK, Cattânio JH, Davidson EA (2008). Fluxes of CH_4, CO_2, NO, and N_2O in an improved fallow agroforestry system in eastern Amazonia. Agric Ecosyst Environ 126:113–121.

Wang G, Welham C, Feng C, Chen L, Cao Fuliang (2015). Enhanced soil carbon storage under agroforestry and afforestation in subtropical China.Forests 6:2307-2323

Watson L (2010). Portugal gives green light to pasture carbon farming as a recognized offset. Australian Farm Journal44-47.

Watson RT, Noble IR, Bolin B, Ravindranath NH, Verardo DJ, Dokken DJ (Eds.). (2000). Land use, land-use change and Forestry. A Special Report of the IPCC. Cambridge University Press, New York.

Wullschleger SD, Gunderson CA, Hanson PJ, Wilson KB, Norby RJ (2002). Sensitivity of stomatal and canopy conductance to elevated CO2 concentration-interacting variables and perspectives of scale. New Phytologist 153:485-496.

Zhang L, Wang H, Zou J, Rogers WE, Siemann E (2014). Non-native plant litter enhances soil carbon dioxide emissions in an invaded annual grassland. PLOS ONE 9(3):e92301.

19

Water Budgeting and Management for Production in Tree Based System

R.K. Goyal

ICAR-Central Arid Zone Research Institute, Jodhpur-342003, Rajasthan

Introduction

Arid lands are among the world's most fragile ecosystems, made more so by periodic droughts and increasing over exploitation of meagre resources. Arid and semi-arid lands cover around one-third of the world's land area and are inhabited by about one billion people, a large proportion of whom are among the poorest in the world (Malagnoux *et al.* 2007). Arid environments are extremely diverse in terms of their land forms, soils, fauna, flora, water, and human activities. Aridity is usually expressed as ratio of mean annual precipitation (P) to the mean annual potential evapotranspiration (EPT) where potential evapotranspiration is calculated by method of Penman, taking into account atmospheric humidity, solar radiation, and wind speed. UNEP (1997) has recognized four main classes of aridity: hyper-arid ($P/EPT < 0.03$), arid ($0.03 < P/EPT < 0.20$), semi-arid ($0.20 < P/EPT < 0.50$), and dry sub-humid ($0.50 < P/EPT < 0.65$). Of the total land area of the world, the hyper-arid zone covers 4.2 %, the arid zone 14.6 %, and the semi-arid zone 12.2 % (FAO 1989). Therefore, almost one-third of the total area of the world is "arid land".

More than 26% of the land surface are covered by forests. Roughly 6 % of the world's forest area (about 230 million hectares) is located in arid lands (FAO 2001). Trees outside forests (scattered in the landscape, in arable lands, in grazing lands, in barren lands and in urban areas) have a vital role in arid lands, although it is difficult to assess their extent. As per 2010 estimates of Food and Agriculture Organisation of the United Nations, India has 68 million hectares area (22% of the country's area) under forest cover. However 2013 Forest Survey of India states its forest cover increased to 69.8 million hectares by 2012. This represents an increase of 5,871 km^2 of forest cover in 2 years.

Forestry in India is more than just about wood and fuel. India has a thriving non-wood forest products industry, which produces latex, gums, resins, essential oils, flavours, fragrances and aroma chemicals, incense sticks, handicrafts, thatching materials and medicinal plants. About 60% of non-wood forest products production is consumed locally. About 50% of the total revenue from the forestry industry in India is in non-wood forest products category.

Indian arid zone

The arid region of India is spread in 38.7 million hectare (Mha) area out of which 31.7 Mha is under hot arid zone and 7 Mha under cold arid zone. The hot arid region occupies major part of north-western India (28.57 Mha) between 22°30' and 32°05' N latitudes and from 68°05' to 75°45' E longitudes, covering western part of Rajasthan (19.6 Mha, 69%), north-western Gujarat (6.22 Mha, 21%) and 2.75 Mha in south-western part of Haryana and Punjab (Faroda *et al.* 1999). The climate of Indian hot arid zone is characterized by an abundance of solar energy from cloud-less sky, high diurnal and seasonal temperature variations and annual and inter-annual irregular rainfall with long dry seasons associated with strong winds. Annual rainfall is approximately in the range of 100-500 mm, with a coefficient of variation varying from 40 to 80% (Rao and Singh 1998). The rainfall results largely from convective cloud mechanisms and is characterized by a relatively high intensity, short duration and limited aerial extent. The distribution of rainfall in space is very much influenced by the local terrain. With low ground level and little topographic relief, it is common to see rain evaporating before reaching the ground. The incoming radiation ranges from 15.12 - 26.50 MJ m^{-2} day^{-1} with very little cloud for most of the year. The annual potential evapotranspiration ranges from 1400 – 2000 mm yr^{-1} leading to a permanent negative water balance (Rao 2009).

Forests, trees and grasses are essential constituents of arid zone ecosystems and contribute to maintaining suitable conditions for agriculture, rangeland and human livelihoods. In providing goods (especially fuelwood and non-wood products) and environmental services to the rural poor and in contributing to the diversification of their household sources of income, forests and trees in arid zones boost poverty alleviation strategies and reduce food insecurity. Availability of water – surface water, groundwater and air moisture – is usually the main factor limiting natural distribution of trees in arid lands, along with climate (rainfall, temperatures and wind) and soil quality.

In modern industrial societies forests are sources of wood for paper, pulp, building, furniture, and energy. In developing countries wood is primary used for cooking and building. In the last decades protection and social functions of forests become more important. Forest areas act as recreation areas and as sources of cold and fresh air for urban areas. In high mountains forests additionally protect the terrain against erosion, landslides and avalanches. Today these forests are threatened by human activities like increasing usage of wood, agricultural use and air pollution.

Water balance of forest areas

In hydrology, a water balance equation is used to describe the flow of water in and out of a system. A system can be one of several hydrological domains, such watershed or a drainage basin. This equation uses the principles of conservation of mass in a closed system, whereby any water entering a system (via precipitation), must be transferred into either evaporation, surface runoff (eventually reaching the channel and leaving in the form of river discharge), or stored in the ground (Fig.1). A water balance can be used to help manage water supply and predict proportion of water used by different component of a closed system.

Water balance equation in a simplest form can be written as

$$P-E-Q = \Delta M$$

Where P= Precipitation

E= Evapotranspiration (Evaporation + Transpiration)

Q= Runoff

Δ M= Change in storage

Precipitation

Precipitation is water released from clouds in the form of rain, freezing rain, sleet, snow, or hail. It is the primary connection in the water cycle that provides for the delivery of atmospheric water to the Earth. Most precipitation falls as rain.

Evaporation

The process whereby liquid water is converted to water vapour and removed from the evaporating surface. Water evaporate from variety of surfaces such as lakes, rivers, pavement, soil and wet vegetation.

Transpiration

Transpiration is the evaporation of water from the surface of leaf cells in actively growing plants. This water is replaced by additional absorption in soil leading to continuous rise of water column in the plant's xylem. The process of transpiration provides the plant with evaporative cooling, nutrients, carbon dioxide entry and water to provide plant structure. Transpiration depends on:

A. Climatic factors

1. Temperature
2. Humidity
3. Sunshine duration
4. Wind speed
5. Vapour pressure

B. Plant/tree factors

1. Stomata- Pores in the leaf
2. Boundary layer- Thin air layer
3. Cuticles- Thin wax layer

Runoff

Runoff (also known as overland flow) is the flow of water that occurs when excess storm water, melt water, or other sources flows over the Earth's surface. This might occur because soil is saturated to full capacity, because rain arrives more quickly than soil can absorb it, or because impervious areas (roofs and pavement) send their runoff to surrounding soil that cannot absorb all of it. Surface runoff is a major component of the water cycle. It is the primary agent in soil erosion by water. Factors affecting runoff are;

1. Catchment area
2. Slope of catchment
3. Catchment orientation
4. Shape of catchment
5. Altitude of catchment
6. Stream pattern
7. Rainfall Intensity and duration
8. Vegetation on the surface
9. Land use pattern
10. Sub-surface conditions
11. Presence of reservoir/lakes/swamps

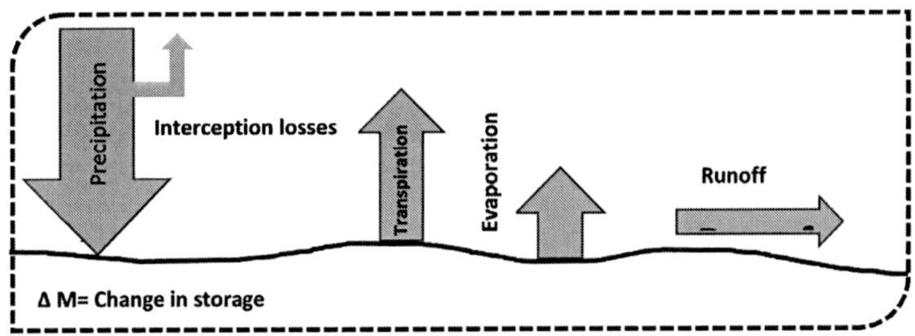

Fig.1. Water Balance

The water budget of forested catchments differs markedly from bare soils or grasslands by significantly higher interception losses due to higher leaf area per surface area.

Interception losses

Interception can be technically defined as portion of precipitation which, while falling, is intercepted by aerial portion of vegetation/ plant canopy and its subsequent return to the atmosphere through evaporation. The adhesive force between the water drops and the vegetation hold back the drop of water against gravity until they grow in size to over weigh and slip down. Vegetation can intercept as high as 50% of the rain that falls on its leaves. The leaves of deciduous trees commonly intercept anywhere from 20 to 30% of the falling rain. The interception losses depends

- Intensity and duration of the precipitation
- Density of tress
- Type of tree
- Season of the year
- Wind velocity at the time of precipitation

The general form of equation for interception loss can be written as

$$I_L = aP + b\left(1 - e^{(-p/b)}\right)$$

Where 'a' and 'b' are constants which depend on the factors of infiltration loss. 'a' varies between 0.01 and 0.2 and 'b' between 2.5 to 38% of rainfall. P is the precipitation depth in mm.

Some observations for forested catchment

- Availability of water is usually the main factor limiting natural distribution of trees in arid lands.

- Vegetation often modifies the intensity and distribution of precipitation falling on and through its leaves and woody structures.
- The water budget of forested catchments differs markedly from bare soils or grasslands by significantly higher interception losses due to higher leaf area per surface area.
- This filtering function of forests leads to better quality of usable water as well as to a more equally distributed runoff thus minimising erosion.
- The decreased runoff from forested area does not mean less supply of water in downstream because forested area delay the runoff and distribute them over longer periods. Conservation of large and dense forests is necessary to cope with global water problems in the future.
- In tropical and subtropical forests are responsible for the stability of soils because of their high evapotranspiration and water storage capacity. In these regions trees with their deep roots are often the only one vegetation which can spend food for humans and animals after long dry periods.

Water management for forested watersheds

The water resources of the arid region are scarce and because of low and erratic rainfall, replenishment is also very poor. The quantity and quality of water available from various sources such as surface water and ground water is not adequate even for drinking purposes. Apart from insufficient quantity, the ground water is moderate to highly saline over large area. Recurring droughts and consequent crop failures are regular events in this zone. Yet the demand for water is increasing steadily. However, with the adoption of certain dryland watershed technologies forested watershed can be managed to a greater extent for higher fuel and fodder production. Some of these dryland watershed technologies are as follows:

Contour furrowing

Contour furrowing is the most effective measure to reduce runoff and soil loss, increase in yield and commonly adopted in grasslands and forestlands. However, in very sandy soils or soils with heavy clay pan area, their benefit is limited. Contour furrows varying from 30-60 cm wide and 10-25 cm deep can be used. The shape varies from "V" to square, rectangular, or parabolic. The cross section and depth of furrows mainly depend on soil and equipment used for making them. Furrows spaced 8-10 m apart give a batter distribution of runoff water and higher yield of fodder. The effectiveness of contour furrows to hold water depends upon the degree of slope smoothness of the surface and accuracy in following contours and its life depends upon stability of soil and water storage capacity of the furrow, which can be estimated by following equation.

$$Q = \frac{WxD}{100HI}$$

Where, Q = Depth of runoff water stored in cm from unit area

W = width of furrow, cm

D = depth of furrow, cm

HI = horizontal spacing, m

Construction of contour furrow is always started from the ridge and progressively extended towards the valley. In a study conducted in arid part of Iran it is found that contour furrow and pitting has significantly helped in controlling soil erosion, increasing water penetration and soil moisture content and promoted propagation of *Hammada saliconica* species, a desirable plant species for both soil conservation and livestock grazing in the region (Jahantigh and Pessarakli 2009).

Contour trenches

A contour trench is a useful practice in forestry areas. This practice can be adopted in area which is unsuitable for cultivation but suitable for forestry. Normal standard size of a trench is 60 cm x 30 cm x 60 cm depth with an unexcavated portion 1.5 m after every 50-75 cm. Length spacing or vertical interval depends on the slope of land. Spacing may vary from 30-60 m. After the trenches are excavated to correct size, they are refill partially, and stocking the remaining excavated material as a small bund on the downstream side. The storage capacity of trench is be estimated by the following equation.

$$Q = \frac{WxD}{100HI\left(1 + \frac{X}{L}\right)}$$

Where, Q = depth of runoff from area in cm

W = width of trench in cm

d = depth of trench in cm

HI = horizontal interval in meters

X = gap between the trenches in m

L = length of trench in m

Mane *et al.* (2009) reported effectiveness of continuous contour trench for runoff control and recommended as best soil conservation practice on area having 7 to 8 % slope in Konkan region of India.

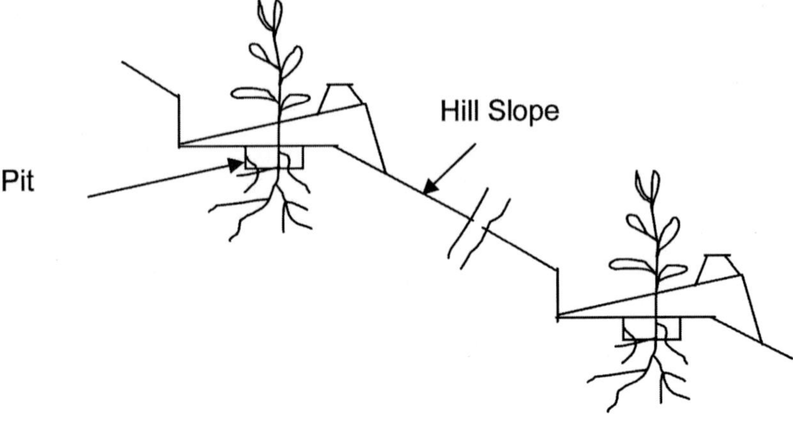

Fig 2. Cross section of a Gradoni

Gradonies

Gradonies are steeply inward-sloping narrow bench terraces constructed on contours. Usually, gradonies are suitable for aforestation in uniformly steep sloping lands (Fig-2). Based on the steepness of slope, vertical interval is kept from 1.0 to 1.5 m. The width of gradonies also varies from 1.0 to 1.5 m. The material dug from the inner side is heaped on the outer edge in order to make a berm of about 20 to 30 cm high with an inward slope of 7.5:1. In the middle of the gradonies, pits of 50 cm x 50 cm x 50 cm are made at spacing of about 3 m. On an average 1000 trees are planted within the gradonies and interspaces in one hectare area (Mahnot and Singh 1993).

Contour farming

Contour farming is beneficial on all slopes where line sown is adopted. All ridges and rows of plants are placed across the slope to form continual series of miniature barriers to water and offer maximum opportunity for infiltration. Contour operations reduce the power of the water to erode, suspend and carry away soil particles and increase the moisture storage. Increased uniform moisture storage can boost up the yield above 10%. Contour farming alone cannot control runoff volume from higher sloppy lands which may need bunding and grass waterways in natural drainage. For small fields with uniform slopes, one contour guideline is enough and where slopes are irregular two to three guidelines may be enough for farming operation. Contour farming is most effective on moderate slopes of

2 to 7%. Contour farming is reported to reduce soil loss up to 60 % in comparison to up and down farming on 1% slope (Smith and Wischmeier 1962).

Contour bunds

Contour bunds are narrow base trapezoidal earthen embankment on contour, 1.5 to 2 m wide, constructed across the slope to act as barriers to runoff, to form water storage area on their upslope side and to break up a long slope into short segments. Contour bunds are recommended up to 6% slope and rainfall of up to 600 mm. Contour bunds are not suitable for shallow soils having depth less than 7.5 cm. Spacing between contour bunds is usually expressed in terms of vertical interval (VI), which is the difference in elevation between two similar points on two consecutive bunds. The principle involved in fixing the spacing between two bunds is to keep the velocity of water below critical value in order to avoid scour. Vertical interval is generally expressed as a function of percent slope (s)

$$VI = 0.305\left(\frac{s}{a} + b\right)$$

Where V I = vertical interval between consecutive bunds in m

S = land slope in percent

'a' and 'b' are constant specific to particular region

For soils having good infiltration rates value of 'a' and 'b' are 3 and 2 respectively. For soils with low infiltration rate the value of 'a' and 'b' may be taken as 4 and 2 respectively. The horizontal interval (HI) is related to vertical interval (VI) by following equation

$$HI = \left(\frac{VI}{s} x100\right)$$

Depth of impounding in calculated on the basis of maximum rainfall of 24 hours for 10 years recurrence interval as follows

$$h = \sqrt{\frac{\mathrm{Re}\, x VI}{50}}$$

Where 'h' is height of impounding (m) Re is 24 hours maximum rainfall (cm) for 10 years recurrence interval. Total height of bund H can be find out by adding 20- 25% freeboard to height of impoundment. For light red loam and sandy loam soils, the side slope of both the sides is taken as 1.5: 1 whereas for sandy soils it is taken as 2:1. The line of seepage should not cross the bottom of the bund while deciding the other dimension of the bund. On average, contour bunds had 27 % higher soil moisture and 14 to 181 % higher fodder yield than flat surfaces on grasslands of western Rajasthan (Wasi-Ullah *et al.* 1972).

Graded bunds

Graded bunds are used to dispose off safely the excess water from the agricultural fields to avoid water stagnation. Graded bunds are suitable in areas where annual rainfall is 500 mm and if the soils are highly impermeable. Graded bunds usually have wide and shallow channels and earthen bund laid along a predetermined longitudinal slope. As graded bunds are essentially means for the safe disposal of excess water from cropped lands, suitable outlets are required to be constructed on graded bund. Draining of excess water from one plot to another through outlets provided in the bund require special attention since considerable amount of soil may be lost through these outlets. Provision should be made to arrest the silt and allow only clear water to flow away. The vegetated watercourse strengthens the system. In areas with high rainfall where the volume of water to be disposed from one plot to the next lower level plot is large and the vertical interval between the plots is reasonably high, a site specific outlet design like a pipe outlet, drop structure etc. must be provided for safe disposal of the excess water. Graded bund is reported to reduce the run-off from 20 to 4.8 per cent and soil loss from 24 to 4.12 t ha^{-1} yr^{-1}. Besides other benefits intercropping on contour resulted in 48 per cent higher grain yield (Singh *et al.* 1997).

Grass waterways

Grass waterways are developed for safe disposal of excess water from agricultural fields. These may be natural or man-made courses protected against erosion by suitable grass cover. Grass waterways are also used for channelizing and regulating runoff flows for water harvesting purposes. The best location for waterways is a natural depression or along valley line. These may also be constructed along field boundaries for safe disposal of excess rainfall from agricultural fields. Vegetative waterways may be located in all classes of lands except hard rocks, where construction may be difficult. The cross section of waterways may be trapezoidal, triangular or parabolic with shallow depth and flat side slope to facilitate easy movement of man, animal and machinery. The depth of waterways may be kept within 20 to 50 cm and side slope more than 4:1. The channel should have free board of 15 cm. The channel cross-section and bed slope should be such that the computed velocity is within permissible limit (Singh *et al.* 1990). Cost of construction of grass waterways depends upon type of soil, channel cross-section, length of channel and grass plantation technique.

Diversion drains

Diversion drains are constructed to divert the runoff and to allow it to flow at a non-erosive velocity to a suitably protected outlet. They are particularly adapted to areas already where trees or grasses are in dominance. Usually diversion drains are constructed to protect conservation structure lower down in the cultivated areas by regulation flows from upland non-agricultural areas. Diversion drains are provided across the slopes on slight gradients. Their capacity is worked out and designed for a 10 year recurrence interval storms.

Check dams

Check dams are masonry overflow barriers (weirs) constructed across seasonal streams. A check dam as such has a relatively limited storage capacity but a large volume of water can still be pumped from such storage as the stream continues to flow and the check dam serves the purpose of an ideal intake structure. A check dam, by storing the base flow, maintains a supply of water for recharge as well as for direct use beyond the monsoon period (Goyal and Narain 2006). It creates flooding of upstream area, which requires surplusing arrangements at suitable intervals to drain water. Check dam should be avoided in isolation. The number of check dams primarily depends upon the slope of the gully and the quantity of runoff. It may not be advisable to construct check dams on bigger streams with high gradient and where runoff is very high. The bigger streams should be treated with drainage line treatment like gabionic structures and boulder checks with masonry work to curtail the runoff.

Gabionic cheek dams

Gabionic check dams are useful in a locality where stones are readily available and their irregular shape makes them unsuitable for making loose stone cheek dams. If the expected water velocity is very high, gabion is recommended in place of loose rock dams. A gabion is rectangular shaped cage made of galvanized wire, which is filled with locally available boulders, rocks or stones. The gabion may be conveyed flat and are folded to shape at the construction site. Usually, gabions are 1 m wide and 0.75-12 m high with varying lengths ranging between 2-10 m. The gabionic check dams are constructed by connecting several gabions in horizontal and vertical direction. The gabionic check dams are very stable and semi-permanent in nature. These structures are flexible; they may even change shape automatically according to the streambed, even when the bed shape changes due to erosion, without losing stability.

Loose stone/dry stone masonry check dams (LSCD)

These structures are effective for checking runoff velocity in steep and broad gullies. These are suitable at upper reaches of the catchment. They have a relatively longer life and, usually require less maintenance. The bed of the gully is excavated to a uniform depth of about 0.3 m. Stones are then hand packed from the foundation level. Flat stones of size 20-30 cm are the best for construction and laid in such a way that all the stones are keyed together. Large size stones are placed at the centre of the dam and gaps between stones may be filled with small piece stones. The dam should go up to 0.3 to 0.6 meter into the stable portion of the sides of the gully to prevent end cutting. In the centre of the dam, sufficient spillway is provided to allow maximum runoff to discharge. LSCD constructed at 1 m V.I. in Jhanwar watershed area on 17 gullies proved to be very effective in controlling further extension of gullies (Goyal *et al.* 2007). Vangani *et al.* (1998) reported sediment deposition of 3.86 t ha^{-1} yr^{-1} against loose stone check dam in Osian-Bigmi watershed (Distt. Jodhpur).

Brushwood check dams

These check dams are constructed by using locally available brushwood and supported by wooden stakes and used in the small gully heads not deeper than 1 m. These check dam are of two types; single row post brush dam and double row post brush dam. Brushwood check dams are constructed in areas where wooden posts, brushwood etc., are available in plenty. These check dams can only be used in the small gully heads not deeper than 1 m. Single row post brush dam are made of single row of wood stakes to which long branches of trees are tied length wise along gully with their butt ends facing upstream while in double row post brush dam, the straw and brushwood are laid across the gully between two rows of wooden posts, the distance between the rows being not more than 0.9 m. The longest branches are laid at the bottom and the shorter length branches are laid above it till the required dam height (0.3 to 0.7 m) is obtained.

Earthen gully plugs

An Earthen gully plug is constructed out of local soil across the stream to check soil erosion and flow of water. It is suitable in the upper catchment areas having scope of water storage and where the soil for the embankment is close by. Depth of the gully should be less than 2 meter and gully bed slope is less than 10%. The site should have scope for side spillway. The bed of the gully is excavated to a uniform depth of about 0.3 m. Foundation trench is filled with compacted earth and earthen embankment is constructed by using local soil.

Conclusions

A successful application of any measures for improving production requires an integrated approach. Watershed management techniques of rainwater harvesting and conservation will not only solve the problem of water but will also save valuable money, which can be utilized for other welfare schemes. All the technologies discussed above are essentially site specific and different components need to be integrated as a holistic approach to conserve rainwater to provide water for different needs on sustainable basis. These technologies are time tested and are of proven soundness for extreme conditions such as of arid. The present day need is only to revive these technologies and apply them in the field on full scale.

References

FAO (1989) Arid Zone Forestry: A Guide for Field Technicians. FAO Conservation Guide - 20. Food and Agricultural Organization, Rome, Italy.

FAO (2001) Global Forest Resources Assessment 2000 - main report. FAO Forestry Paper No. 140. Rome.

Faroda AS, Joshi DC, Balak Ram (1999) Agro-ecological Zones of North-western Hot Arid Region of India. Central Arid Zone Research Institute, Jodhpur, pp.24.

Goyal RK, Pratap Narain (2006) Impact evaluation of watershed conservation measures in hot arid zone of India. In: conference on Natural Resources Management for Sustainable Development in western India (NRMSD-2006), Indian Association of Soil and Water Conservationist, Dehradun-Pune, pp.77-79.

Goyal RK, Bhati TK, Ojasvi PR (2007) Performance evaluation of some soil and water conservation measures in hot arid zone of India. Indian Journal of Soil Conservation 35 (1): 58-63.

Jahantigh Mansour, Mohammad Pessarakli (2009) Utilization of contour furrow and pitting techniques on desert rangelands: evaluation of runoff, sediment, soil water content and vegetation cover. Journal of Food, Agriculture and Environment 7:736-739.

Mahnot SC, Singh PK (1993) Soil and Water Conservation. Inter-cooperation Coordination Office, Jaipur, pp. 90.

Malagnoux M, Sene EH, Atzmon N (2007) Forests, trees and water in arid lands: a delicate balance. Unasylva 58: 24-29.

Mane MS, Mahadkar UV, Ayare BL, Thorat TN (2009) Performance of mechanical soil conservation measures in cashew plantation grown on steep slopes of Konkan. Indian Journal of Soil Conservation 37: 81-184.

Rao AS (2009) Climatic variability and crop production in arid western Rajasthan. In: Amal Kar, Garg BK, Singh MP, Kathju S (eds) Trends in Arid Zone Research in India. Central Arid Zone Research Institute, Jodhpur, pp. 41-61.

Rao AS, Singh RS (1998) Climatic features and crop production. In: Faroda AS, Singh Manjit (eds) Fifty Years of Arid Zone Research in India, Central Arid Zone Research Institute, Jodhpur, pp. 17-38.

Singh AK, Kumar AK, Katiyar VS, Singh KD, Singh US (1997) Soil and water conservation measures in semi-arid regions of south-eastern Rajasthan. Indian Journal of Soil Conservation 25 (3): 186-189.

Singh GC, Venkatramanan GS, Joshi BP (1990) Manual of Soil and Water Conservation Practices. Oxford and IBH Publishing Co. Pvt. Ltd, New Delhi, pp. 385.

Smith DD, Wischmeier WH (1962) Rainfall erosion. Advances in agronomy 14: 109-148.

UNEP (1997) United Nations Environment Program, World Atlas of Desertification 2ED. UNEP, London.

Vangani NS, Singh Surendra, Sharma RP (1998) Index catchment - a new concept for sustainable integrated development. Annals of Arid Zone 37 (2): 133-137.

Wasi-Ullah, Chakravarty AK, Mathur CP, Vangani N S (1972) Effect of contour furrows and contour bunds on water conservation in grasslands of western Rajasthan. Annals of Arid Zone 11(3/4):170-182.

20

Traditional Agroforestry in India: Problems and Prospects

S.K. Dhyani

ICAR-Natural Resource Management Division, Krishi Anusandhan Bhavan II PUSA, New Delhi-110012

Introduction

India has a long tradition of agroforestry which is widely practiced in all ecological and geographical regions of the country. Traditional agroforestry systems are broadly based on indigenous knowledge and the species are selected as a part of the cultural patterns of the community. The farmers and land owners in different parts of the country integrate a variety of woody perennials in the crop and livestock production systems depending upon the agro-climatic conditions and local requirements.

Agroforestry is an age old land use practice defined as land use systems in which woody perennials (trees, shrubs, palms, bamboo etc.) are grown on the same piece of land with herbaceous plants and/or animals, either in spatial arrangement or in time sequence and in which there are both ecological and economic interactions between the trees and non-tree components (Beets, 1989). Agroforestry systems (AFS) include both traditional and modern land-use systems where trees are managed together with crops and or/ animal production systems in agricultural settings. India is blessed with varied agro-climates, as a result there are huge variations in agroforestry systems in their structural complexity and species diversity, their productive and protective attributes and their socio-economic dimensions. Traditionally, agroforestry has been practiced by farmers to meet the household requirements of fruit, fodder, fuel and fibre (Dhyani *et al.* 2009), yet, over the years small-scale tree production has gained momentum due to development of marketing avenues, wood and food based processing and value addition and demand from industry.

Traditional agroforestry

Trees and forest are always considered as an integral part of the Indian culture. Shifting cultivation in India is prehistoric and partly a response to agro-ecological condition in the region. Most agroforestry systems have the intrinsic potential to provide food, fuel, fodder, green manure, plant derived medicines, and timber resources. The choice of species and planting techniques adopted in such systems also reflect the accrued wisdom and insights of the traditional people who interacted with the environment for long (Dhyani *et al.* 2005).

The agroforestry systems range from apparently simple forms of taungya and shifting cultivation to complex home-gardens: from systems involving sparse stands of trees on farmlands (e.g. *Prosopis cineraria* in arid regions of Rajasthan) to high-density complex multi-storied homesteads of Kerala: from systems in which trees play a predominantly 'service' role (e.g. shelter belts) to those in which they provide main saleable products (e.g. intercropping with plantation crops).Most of the systems are site-specific. A detailed account of various traditional agroforestry systems followed in different parts of the country is available in literature (Chauhan and Dhyani 1989; Dadhwal *et al.* 1989; Dhyani and Chauhan 1994; Pathak *et al.* 2000; Dhyani *et al.* 2005, 2009, 2014; Kumar *et al.* 2012). Some of the important traditional agroforestry systems/ practices in different region is presented in the following pages and summarized in Table 1.

Agroforestry in the Indian Himalayas

Agroforestry is the most natural way of life in this region, considering the general physiography, vegetation, existing agricultural practices and the attitude of the people. In the Himalayan region, agroforestry is practiced in arable lands and non-arable degraded lands, and even on bouldery riverbed lands. The age old practice of shifting cultivation which is still prevalent in the states of North-Eastern region except Sikkim and Arunachal Pradesh is yet another form of agroforestry. Farmers practicing shifting cultivation retain selected trees such as Khasi pine, Schima spp. etc. to grow during the fallow phase. In Sikkim, the indigenous tribes like Lepcha and Limbu used to collect large cardamom (*Ammomum subalatum*) from natural forests, which were later on domesticated and introduced in Alder (*Alnus nepalensis*) forests. This practice evolved more than two hundred years ago, and now it is being practiced in more than 48000 ha area in the north-eastern states (Dhyani 1998). Traditional agroforestry systems in north-eastern region indicated a spectacular increase in soil pH, organic-C, Ca, Mg, K, and P within 10-15 years of practice and the exchangeable Al, potential cause of infertility of these lands disappeared completely (Singh *et al.* 1994). Similar results were obtained when multipurpose tree species (MPTS) were evaluated in an extremely P-deficient acid Alfisol in Meghalaya (Dhyani *et al.* 1994).

In the Western Himalayas, 60 to 70% requirement of the firewood and fodder is met from the trees growing along the bunds of agricultural terraces or scattered trees on the pasture lands. *Grewia optiva*, *Morus alba*, *M. serrata*, *Toona cilata* and *Melia azedarach* are the common tree species on field bunds which meet the fuel-fodder requirement (Dadhwal *et al.* 1989).In the Himalayan region intercropping in fruit orchards is very common.

Indo-Gangetic Plains

The region is food bowl and contributes more than 50 per cent to the national food grain output. Farmers use trees according to their need as well as suitability of the species. Many common trees such as *Azadirachta indica*, *Acacia nilotica*, *Dalbergia sissoo*, *Prosopis cineraria*, and others are found grown very frequently on farm lands particularly along crop-field boundaries for meeting the fodder requirements. In sub-montane zone of Punjab, farmers plant *Leucaena* either in block or boundary plantation. In south-west Punjab, farmers prefer to grow *Melia* and *Eucalyptus* as boundary plantation.

A sizable area in the Indo-Gangetic plains is salt-affected. Farmers in their accrued wisdom and trial and error succeeded in selecting species which can withstand the problems of alkali soils such as *Prosopis juliflora*, *Acacia nilotica*, *Pithecellobium dulce*, *Emblica officinalis*, ber, Jamun (*Syzygium cumini*) and others. Similarly, farmers choose fodder trees such as *Syzygium cuminii*, *Terminalia arjuna*, *Dalbergia sissoo*, *Acacia nilotica* etc. utilize the seepage water from unlined canals causing waterlogging and salinization water for biomass production. In ravine lands, trees like *Acacia nilotica*, *A. catechu*, *Dalbergia sissoo*, *Azadirachta indica* and *Pongamia pinnata* are most effective in association with forage grasses.

Agroforestry in humid and sub-humid region

In humid and sub-humid regions, *Albizia* spp., *Gmelina arborea* and *Gliricidia* are the important fodder trees for different agroforestry systems. In most of the region, tree component is used along with livestock and poultry component, whereas in irrigated areas fish component is also incorporated. Agrisilviculture is common in Jharkhand and Chattisgarh. *Acacia nilotica*, *Terminalia arjuna*, *Butea monosperma*, *Albizia* spp. are grown in Chattisgarh while *Zizyphus mauritiana*, *B. monosperma*, *Aegle marmelos*, *Mangifera indica*, *Schleichera oleosa* (Kusum) in Jharkhand. Homestead Agroforestry is also being practiced using *Gmelina arborea*, *Artocarpus heterophyllus*, *Madhuca latifolia*, *Zizyphus mauritiana* etc. In Odisha, agri-silviculture (Cocos nucifera for boundary plantation, block plantation of *Casuarina equisetifolia*, *Anacardium occidentale*) and homesteads are commonly practiced. In this region, *Pongamia pinnata*, *Acacia nilotica* and *Dalbergia sissoo* are important component of silvipasture systems.

Agroforestry in arid and semi-arid regions

The people of the region have evolved agroforestry systems as a drought protective mechanism based on long term experience. Primarily, farmers in the region opted for tree component as part of the agriculture and grazing land management. At CAZRI, on the basis of rainfall, four types of major traditional agroforestry systems have been identified in upper transact of arid western Rajasthan extending from Danta - Ramgarh (transitional zone between arid and semi-arid region) to extreme western fringes of Bikaner/Ganganagar districts. Over 400 mm rainfall zone pre-dominant system is *Prosopis cineraria - Acacia nilotica* based; between 300 and 400 mm rainfall zone. P. cineraria based; between 200-300 mm rainfall zone *Zizyphus* spp. - *P. cineraria* based; and in less than 200 mm rainfall zone *Zizyphus* spp. - *P. cineraria - Salvadora* spp. based. The density decreased with decrease in rainfall. The density of *Prosopis cineraria* (found across all rainfall zones) tend to decrease with decreasing rainfall from east to west. It was very interesting that yield of pearl millet, main cereal crop of the region, below the canopy of woody components in any of system was not affected, rather in *Zizyphus* spp. - *P. cineraria* and *Zizyphus* spp. - *P. cineraria - Salvadora* spp. systems; it had better production.

The most important agroforestry practice from the region which developed about four hundred years ago is known from the Kangeyam tract of Tamil Nadu, where *Acacia leucophloea* + *Cenchrus setigerus* in silvipasture system was perfected. Similarly, in ravines of Yamuna and Chambal, trees, shrubs and bamboos with grasses were planted for rearing milk-producing Jamunapari breeds of goats and sheep. Scattered trees with khejri or mehndi in association with bajra, jowar and chillies were grown in the semi-arid area of Tamil Nadu. Among the many species selected by farmers for agroforestry in the region include *Acacia nilotica*, *A. tortilis*, *Albizia lebbek*, *A. procera*, *Dalbergia sissoo*, *Hardwickia binata*, *Azadirachta indica*, *Ailanthus excelsa*, *Prosopis cineraria*, *P. chilensis*, *Salvadora oleoides*, and fruit trees guava, ber, bael and aonla. Furthermore, *Ailanthus* and *Prosopis* based farming systems are important for livelihood security of the farming and rural community of semi-arid region of Gujarat and Rajasthan. These tree species not only meet the fodder requirement during lean period but also provide extra returns in cash due to industrial importance of Ailanthus as soft wood and Prosopis for higher calorific and nutritional value. Tree plantation continued as demarcation and control against wind erosion throughout the country. Plantation of khejri trees for various uses on-farm was a common practice in Rajasthan. Application of green manure to paddy fields was common in Madhya Pradesh and Uttar Pradesh.

Improvements in soil quality under agroforestry systems have been reported from research conducted at CAZRI where increase in amount of N and P and other macro and micro-nutrients are reported in 14 -yr old plantation of Khejri (*Prosopis cineraria*) as compared to a bare site of *Prosopis juliflora* of the same age. Forage species produce higher biomass under Khejri tree canopy due to a high fertility status.

Agroforestry systems for coastal and island regions

Low lying water logged marshy areas, flood plains, and ill-drained lands are the common features in the coastal areas swamps and river banks are occupied by the mangroves and associate halophytes. Plantation crops integrated with livestock and poultry and rice fields are main features of this region.

Farmers in this region particularly have been domesticating fruit trees and other agricultural crops around their dwellings for millennia, primarily to meet their subsistence needs. The best example of this is the tropical homegardens, which are essentially a complex integration of diverse trees with understorey crops. In Kerala, the homegardens typically have a high degree of biodiversity as the result of generations of selection by farmers and natures responses to those choices. Multistory homesteads or home-gardens were in existence in Kerela, Karnataka, Tamil Nadu, Tripura, Assam and other North Eastern states as an important agroforestry practice. This practice is still followed in these states.

In coastal area *Casuarina equisetifolia* and other trees were grown in association with crops on farm lands for cash and to generate small timber. Live hedges were common as an agroforestry practice in which Mehndi, *Agave sisalana* and *Euphorbia* species were common. In paddy-growing areas Pongamia glabra and *Sesbania grandiflora* were grown, lopped annually and their leaves applied to fields as green manure. In Western Ghats *Terminalia* leaves were harvested, spread on land, burnt and then paddy, ragi and millets were sown.

The above major agroforestry practices followed in different parts of India can be summarized as presented in Table 2. The practice of growing scattered trees on farmlands is quite old and has not changed much over centuries. Shifting cultivation, home gardens and plantation-based cropping systems are mostly practiced in humid tropical regions. Taungya, boundary plantations, live hedges, range land trees are agro-ecologically adapted to all regions. Boundary plantations

Table 1. Major types of agroforestry systems and nature of their benefits in different agro-ecological zones of India

Agroforestry system or Agro-ecological zone	Major benefits	Examples
I. Himalayan Region		
1. Trees in agricultural fields or on field bunds	Production of food, fruits, fodder, etc. and stabilization of bunds.	Grewia optiva and other trees in Jammu and Kashmir, Himachal Pradesh, Uttarakhand, Sikkim and other north-eastern hill region
2. Intercropping in fruit orchards.	Production of fruits, food etc.	Orange and other citrus, and guava inter-cropped with cereals, tuberous and rhizomatous crops in Sikkim, Meghalaya and other north-eastern states and with temperate fruits in western Himalaya.
3. Plantation crops under shade of trees	Food, spices, fuel-wood, timber etc.	Large cardamom, coffee under alder in Sikkim and Nagaland; tea under legume trees in Assam and West Bengal; betel vine and black pepper on arecanut and other trees in Meghalaya and other NE States.
4. Silvopastoral practices (fodder trees with pastures)	Fodder, fuelwood etc. Reclamation of degraded lands, production of fuel and fodder. Fuel, fodder, Timber	Tree leaf fodder from *Grewia*, *Celtis*, *Bauhinia*, *Albizia*, *Ficus* and other species in the Himalayas.
II. Indo-Gangetic Plains		
1. Trees for soil Reclamation (sodic and saline or degraded soils)	Fuel, fodder, timber etc.	*Prosopis chilensis, Acacia nilotica*, Parkinsonia aculeata and other species for problem soils.
2. Fodder trees in degraded grazing lands.	Fodder, soil conservation, fuel etc.	*Albizia lebbek, Bauhinia purpurea, Dalbergia sissoo* and other tree species on grazing lands.
3. Trees on boundaries of agricultural fields.	Fodder, fuelwood, fruits, cash, shade and minor products.	Populus, *Tamarindus indica*, *Bombax ceiba*, *Eucalyptus*, *Dalbergia sissoo*, etc.
4. Fodder banks and woodlots.	Reclamation of soil, fuel, fodder, minor products.	
III. Arid and Semi-arid Region		
1. Multipurpose trees In agricultural fields	Sand dune Stabilization	*Azadirachta indica*, *Melia azaderach*, *Albizia* spp., *Syzygium cuminii* etc. fodder trees. *Casuarina equisetifolia*, *Eucalyptus*, Bamboos, *Dalbergia sissoo*
2. Trees for reclamation of degraded soils.		
3. Windbreaks and shelter belts.		

		etc. woodlots.
4. Fodder banks 5. Trees on rangelands (silvopastoral systems)		Khejri (*Prosopis cineraria*), *Ziziphus* spp., *Acacia senegal*, *Ailanthus excelsa*, *Acacia nilotica*, *A. leucophloea etc.* tree species. *Albizia lebbek*, *Casuarina* sp., *Acacia tortilis*, *Azadirachta indica* and other tree species. *Acacia nilotica*, *A. senegal*, other *Acacias*, *Azadirachta indica*, *Cassia* spp., *Prosopis chilensis*, *Tamarix* spp. etc. *Prosopis cineraria*, *Ziziphus nummularia*, *Salvadora persica*, *S. oleoides*, *Acacia nilotica* etc. *Prosopis chilensis*, *Acacia senegal*, *A. tortilis*, *P. cineraria*, *Ziziphus* spp., *Azadirachta indica* etc. in pasture lands.
IV. Humid and Sub- Humid Region		
1. Home gardens/ Homesteads 2. Multi-tier system or plantation crop combination 3. Multipurpose trees in agricultural fields	Soil conservation, fodder, fuel, etc. Fodder, fuel shade, timber.	Homegardens in Kerala, Assam, W. Bengal have a mixture of trees, shrubs, herbs etc. coconut, arecanut, *Erythrina*, *Gliricidia*, with black pepper, cacao, coffee, cassava and other cash crops. Coconut, arecanut, *Erythrina* and other trees with coffee/banana, pineapple/ papaya, cacao/coffee and black pepper/ betel vine, large-cardamom. *Acacia auriculiformis*, *Aegle marmelos*, *Albizia* spp., *Anogeissus latifolia*, *Anthocephalus chinensis*, *Artocarpus* spp., *Casuarina equisetifolia*, *Diospyros melaxylon* etc.
V. Coastal and Island Region		
1. Plantation crop combination, multi- storeyed 2. Trees with aquaculture 3. Mangrove plantation as part of homesteads. 4. Shelter belts and wind breaks	Production of multiple outputs. Cash, and multiple outputs. Fodder, minor products Production of multiple outputs, cash	Coffee under *Erythrina lithosperma*, Cacao with coconut, black pepper on *Gliricidia*, *Grevillea robusta*, Cardamom under *Toona ciliata*, *Artocarpus* etc. Composite fish culture in ponds and multipurpose trees

5. Trees on boundaries of agricultural fields.	Fish, fuel, fodder, timber.	in homesteads.
	Shore protection, fuel, fodder, environmental protection.	*Acanthus ilicifolius*, *Avicennia officinalis*, *Carbera odollam*, *Rhizophora conjugata* etc.
	Shore or beach stabilization.	*Azadirachta indica*, *Casuarina equisetifolia*, *Prosopis chilensis*, *Acacia senegal* etc.
	Fodder, fuel, shade, minor products	*Aegle marmelos*, *Albizia* spp., *Azadirachta indica*, *Bamboos*, *Bombax malabaricum*, *Calliandra calothyrsus*, *Cassia* spp. etc.

are found in Uttar Pradesh, Gujarat, in part of South India, particularly the Nilgiri hills, Haryana, Himachal Pradesh, Bihar and Orissa. Woodlots are found in hilly areas whereas shelter belts are found in wind-prone regions like coastal and desert areas. In Andhra Pradesh, Tamil Nadu, Karnataka, Orissa, Punjab, Haryana, Gujarat and Assam woodlots are very common. Scattered trees on farmlands are found in all regions specially arid and semi-arid. Agro-silvo-pastoral practices are found in semi-arid regions of India. The homegardens are found in Kerala and Andaman and Nicobar Island.

Table 2. A summary of major agroforestry practices followed in different parts of India

Practice	Agro-ecological region or states
Fodder trees and Farm woodlots	Throughout the country
Intercropping or grasses with fruit trees	Sub-tropical and tropical; orchards in hilly regions
Seasonal forest grazing	Semi-arid and mountainous ecosystem
Boundary planting and live hedges	In all regions
Taungya and shifting cultivation	Eastern region and north-eastern states
Woodlots for soil conservation	In hilly areas, along sea coast and ravine lands
Tree planting for reclamation of saline soils and wastelands	Semi-arid and canal-irrigated regions, mostly in the northern and north-western regions
Scattered trees on farms, parklands	All regions, especially semi-arid and arid regions
Shelterbelts and windbreaks	In wind-prone areas, especially coastal, arid and alpine regions
Industrial plantations with crops	Intensively cropped areas in northern and north-western regions: Haryana, Himachal Pradesh, Punjab, Uttar Pradesh; Uttarakhand also in southern states (Andhra Pradesh, Karnataka, Kerala, and Tamil Nadu)
Shaded perennial systems with plantation crops	Mainly humid tropical region in the southern region; also in Assam, West Bengal and other NE states.
Homegardens or homestead	Mainly tropical west coast region, Kerala, southern Karnataka, Andaman and Nicobar Islands

A number of reports from different agro-climates have indicated improvements in soil quality under agroforestry systems which have a direct bearing on long-term sustainability and productivity of soil having a viable option for eco-restoration, maintenance of soil resources and obtaining ecosystem services in the form of good air and water quality in the area.

Agroforestry in recent years

Although agroforestry systems were predominant in many parts of the country traditionally, aimed at food production - either directly producing edible products or indirectly (facilitating enhanced and or sustained production). However, much has changed in respect of the attitudes of people towards nature and natural resource conservation. As a result many of these traditional systems have now culminated into simple subsistence agricultural systems in many situations, some of the systems are no more in vogue due to fragmentation of land holdings, natural resource degradation largely owing to growing demographic pressure and climate change effects. In present day conditions, the agroforestry techniques applied by the farmers appear to be poorly developed and exploitative. In most cases the trees are neither protected nor replanted or properly managed (Dadhwal *et al.* 1989). Besides, of late there have been serious disincentives to agroforestry adoption in terms of social, cultural, economic, and policy issues. Therefore, a need for understanding the science of agroforestry was felt. This resulted in initiative for organized research during the seventies of last century.

Agroforestry research in India

Organized research on agroforestry started in India with the establishment of the All India Coordinated Research Project on Agroforestry in 1983. The research initiatives gained further momentum with the commencement of forestry education programs in the State Agricultural Universities of India during 1985 or 1986 and the founding of the National Research Centre for Agroforestry (NRCAF) at Jhansi, UP, in 1988. The Centre is now upgraded as Central Agroforestry Research Institute (CAFRI) w.e.f. 01.12.2014. At present there are 37 centres of AICRP on Agroforestry representing all agro-climates of the country. In addition, Indian Council of Forestry Research & Education (ICFRE) also conducts agroforestry research through its research institutes and advanced research centres (ICFRE 2011). A number of business corporations, limited companies such as ITC, WIMCO, West Coast Paper Mills Ltd., Hindustan Paper Mills Ltd., and other institutions initiated agroforestry research with emphasis on production of improved planting material of the fast growing species (Dhyani *et al.* 2015) with an objective to meet the demand of wood based industries. The agroforestry research through the AICRP on Agroforestry was conceptualized with the six projects viz. Diagnostic survey and appraisal of existing farming system and agroforestry practices including farmers' preference; Collection and evaluation of promising MPTS for agroforestry interactions; Studies on management practices of agroforestry systems; Analyze economical relation of agroforestry systems; Explore the role of agroforestry in environment protection; and Studies on post-harvest technology, fishery, apiculture, lac, etc. in relation to agroforestry systems.

Diagnostic survey and appraisal under the project revealed that agroforestry practices abound in the country (Pathak *et al.* 2000), but there are considerable variability in the nature and arrangement of the components and the ecological

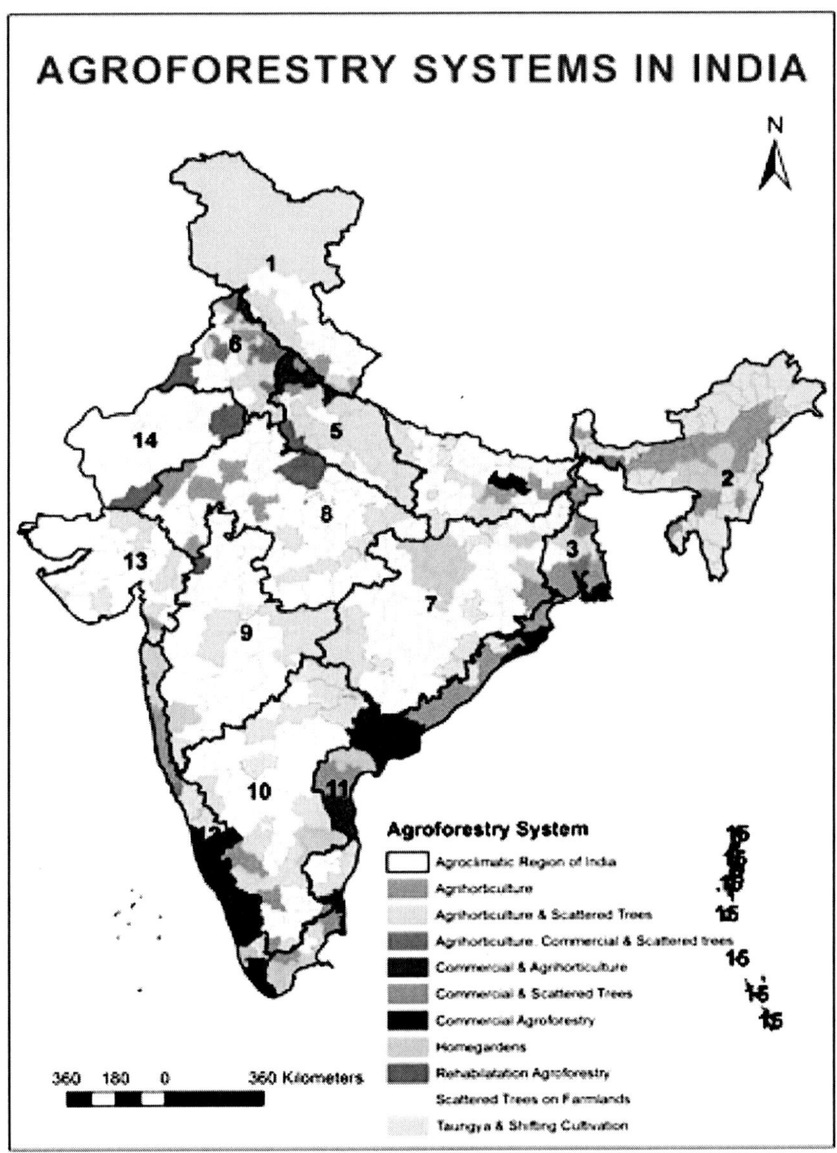

and socio-economic conditions under which such systems are practiced (Dhyani *et al.* 2009, 2014). Germplasm of 184 tree species has been collected and evaluated and improved accessions of poplars, *Eucalyptus*, *Dalbergia*, neem, *Acacia*, *Leucaena*, *Ailanthus*, *Pongamia*, *Casuarina* and *Mangium* hybrids have been identified and being supplied to farmers. Other research themes include development of volume tables and growth equations for estimating tree productivity, development of location-specific agroforestry practices for different agro-climates and wastelands and economic analysis of these systems, extent of agroforestry area and carbon sequestration potential of agroforestry systems etc. it was found that agrisilviculture followed by agri-horticulture and silvipasture are the prominent agroforestry systems from different agro-climates in the country (Fig. 1 in text box). A detailed account of agroforestry research and its achievements is reviewed recently by Handa *et al.* (2015).

Enhancing agroforestry adoption innovations and future challenges

The research results so far indicated that in general, agroforestry is capable of making available diversified foodstuffs, averting malnutrition, and providing organic food materials, for which there is an emerging market even in the developing countries. More recently, however, such systems are considered important from the perspective of augmenting economic returns to the growers. Agroforestry practices including the tree-based smallholder production systems offer great potential to create new jobs in the rural areas. As the great diversity of products from agroforests provides opportunities for development of small-scale rural industries and for creating off-farm employment and marketing opportunities. The spread of agroforestry such as poplar-based in north-western states, *Casuarina* and *Eucalyptus* based in southern states and *Ailanthus* based in western states, and the associated industrial development in these areas indicate the trend. Given this push for commercialization and the economic imperatives, agroforestry in many parts of the country is being both intensified (e.g., intensive tree and crop management practices) and simplified (e.g., fewer economically important species) (Kumar *et al.* 2012). In addition, when strategically applied on a large scale, with appropriate mix of species, agroforestry enables agricultural land to withstand extreme weather events, such as floods and droughts, and climate change (NRCAF 2013).

However, agroforestry growth and development is influenced by a number of factors. State regulations and permit systems for felling of trees and for timber transportation have been main interferences in marketing of timber grown on farmers' land by private investment and have been major discouragement to the expansion of agroforestry. Dhyani and Handa (2013a) strongly advocated for an urgent action to initiate National Agroforestry Policy and launching of a National Mission on Agroforestry for its promotion in the country. In addition, there has been shortage of long term studies on the successful traditional agroforestry systems and their functioning. A better understanding of insights into how and why farmers adapt and modify adopted systems, factors influencing the choice of tree species and other components is essential to understand system functioning.

Agroforestry coverage, contribution and potential

Unlike forest cover and area under crops, the exact area under agroforestry is not available. However, of late there have been a number of attempts to estimate the area under agroforestry. Forest Survey of India indicated 11.32 M ha (ISFR, 2013) under agroforestry, while Dhyani *et al.* (2013) estimated area under agroforestry as 25.32 Mha, or 8.2 per cent of the total geographical area of the country. Using geospatial technology, Central Agroforestry Research Institute (CAFRI), Jhansi gave preliminary estimates for extent of agroforestry in India. The estimates are 14.46 Mha when fallow land was not included and about 17.45 Mha with fallow land (Rizvi *et al.* 2014). Though, these are only preliminary estimates and the same will have to be verified once the methodology for agroforestry mapping using geospatial technologies is perfected. However, whatever the extent of area under agroforestry, it is already playing a vital role in the Indian economy by way of tangible and intangible benefits. It has helped in rehabilitation of degraded lands on one hand and has increased farm productivity on the other. At present agroforestry meets almost half of the demand of fuel wood, 2/3 of the small timber, 70-80 % wood for plywood, 60 % raw material for paper pulp and 9-11 % of the green fodder requirement of livestock, besides meeting the subsistence needs of households for food, fruit, fibre, medicine etc. In fact, agroforestry is also playing the greatest role in maintaining the resource base and increasing overall productivity in the rainfed areas in general and the arid and semi-arid regions in particular (Dhyani *et al.* 2013).

There is potential to increase agroforestry area in near future at least up to 53.32 Mha, representing about 17.57 % of the total reported geographical area (TRGA) of the country, thus making it a major land-use activity, after agriculture and forestry (Dhyani *et al.* 2013). This is more than the Greening India Task Force (GITF 2001) of the Planning Commission estimates of 43 million under agroforestry which includes 15 million ha of JFM, 10 million ha of irrigated agroforestry and another 18 million ha under rainfed agroforestry.

New Initiatives

Agroforestry, evergreen agriculture, and smallholder production systems have attracted considerable attention around the world, of late, and tree-based production systems are being promoted, the world-over (Kumar *et al.* 2012).In fact the agroforestry momentum is getting a good response in India. As per the latest FSI report (ISFR 2015), there is an increase of 110.34 M cum in total growing stock of the country as compared to last assessments (ISFR 2013). The noteworthy feature of this is the healthy contribution of 88.66 M cu m from the tree outside forests, which indicates agroforestry contribution. The potential of agroforestry to contribute to sustainable development has been recognized in many international policy declarations also. For example, the United Nations Framework Convention on Climate Change (UNFCCC) and the Intergovernmental Panel on Climate Change (IPCC) acknowledged it as a component of climate-smart agriculture and is frequently mentioned as having a strong potential for climate change adaptation and mitigation. The United Nations Convention to Combat Desertification (UNCCD) acknowledges agroforestry's potential to control desertification and rehabilitation. It is also seen as an important element in the ecosystem approach promoted by the Convention on Biological Diversity (CBD) for agrobiodiversity conservation.

National agroforestry policy

In India, agroforestry has been receiving greater attention by researchers, policy-makers and others for its perceived ability to contribute significantly to economic growth, poverty alleviation and environmental quality (Dhyani *et al.* 2014). Agroforestry is now recognized as an important part of the 'evergreen revolution' movement in the country. This all helped the country to launch National Agroforestry Policy (NAP 2014) (http://www.indiaenviro nmentportal.org.in/ content/389156/national-agroforestry-policy-2014/) on 10th February 2014 and became the first country in the world to have a National Agroforestry Policy (http://ccafs.cgiar.org/publications/ indias-new-national-agroforestry-policy). The policy is not only seen as crucial to India's ambitious goal of achieving 33 per cent tree cover but also to mitigate GHG emissions from agriculture sector.The NAP 2014 has the potential to substantially reduce poverty in rural India by creating employment in wood based industry, providing livelihood opportunities, help in conservation of natural resources and mitigation of climate change effects.

Implementation of NAP 2014

There are a number of schemes of Government of India, state governments and other organizations where agroforestry is recognized as a component and being promoted as such. Some of the schemes and programmes in which agroforestry is a component are Integrated Watershed Management Programme (IWMP)-now part of Pradhan Mantri Krishi Sinchai Yojana (PMKSY), Mahatma Gandhi NREGA (MoRD), National Horticulture Mission, National Bamboo Mission and National Mission on Medicinal Plants (DAC MoA) to name a few. Around 1.74 million ha forest land has been allotted to 4.66 million tribal and other forest dwellers under the aegis of FRA 2006. Agroforestry will play an important role to ensure regular income to the land allottee. Thus, roughly on average nearly Rs. 2000 crore (or more) is being incurred annually on tree planting/agroforestry through number of Central Sector Schemes and programmes (Dhyani *et al.* 2013). In the on-going Green India Mission (http://moef.nic.in), 3.0 million ha of degraded lands and fallows are to be brought under agro-/social forestry within ten years.

Since the NAP 2014 launched in 2014, considerable progress has been made in terms of putting it into practice. To implement the recommendations of the NAP 2014, an inter-ministerial committee has been set up. Department of Agriculture Cooperation and Farmers Welfare (DAC and FW) under the Ministry of Agriculture and Farmers Welfare (MOA and FW) is also playing a significant role in the promotion of agroforestry. It has taken a policy decision to include trees in all its programmes, and this will significantly increase tree-planting on farms, especially under schemes funded by the National Mission on Sustainable Agriculture (NMSA). This has so far approved the funding for 80,000 ha of new agroforestry projects (pers. comm.). Efforts are on to issue guidelines on the production and supply of high-quality planting material and accreditation of nurseries producing MPTS planting material.

Until recently, the felling, transit and processing of trees grown on farms required approvals and permits from government agencies, and this was a significant impediment to establishing agroforestry systems. In order to promote agroforestry, 20 multipurpose tree species (MPTS) which are commonly grown by the farmers were prioritized which can be exempted from the regulatory regime (Dhyani and Handa 2013). On 18th November, 2014, Ministry of Environment, Forest and Climate Change (MoEFCC) issued fresh guidelines to all States orUT governments for simplification of felling and transit regulation of tree species grown on non-forest/private lands. So far states such as Haryana, Punjab, Himachal Pradesh, Tamil Nadu, Gujarat and Madhya Pradesh have de-notified a number of tree species from felling and transit regulations, and this will make it much easier for landowners and farmers to practice agroforestry (pers. comm.). At the same there is a strong political support for agroforestry. The Prime Minister frequently uses the phrase Har Med par Ped, which means "trees on every field bund/boundary". Agroforestry is being implemented presently as a sub-Mission under the National Mission for Sustainable Agriculture (DAC&FW, Ministry of Agriculture, Govt. of India). However, a Sub-Mission on Agroforestry (SMAF) with an outlay of Rs. 990 crore has recently been approved in principle. The project is expected to assist all the states to scale up agroforestry in a targeted manner.

Carbon sequestration potential of agroforestry

Agroforestry has importance as a carbon sequestration strategy because of C storage potential in its multiple plant species and soil as well as its applicability in agricultural lands and in reforestation. Agroforestry can also have an indirect benefit on C sequestration when it helps to decrease pressure on natural forests, which are the largest sinks of terrestrial C. Another indirect avenue of C sequestration is through the use of agroforestry technologies for soil conservation, which could enhance C storage in trees and soils. For increasing the C sequestration potential of agroforestry systems practices such as- Conservation of biomass and soil carbon in existing sinks; improved lopping and harvesting practices; improved efficiency of wood processing; fire protection and more effective use of burning in both forest and agricultural systems; increased use of biofuels; increased conversion of wood biomass into durable wood products are advocated to be exploited to their maximum potential. Agroforestry thus contributes to the resilience of agriculture by adaptation and mitigation of climate change effects. In India, evidence is now emerging that agroforestry systems are promising land use system to increase and conserve aboveground and soil carbon stocks to mitigate climate change (Dhyani *et al.* 2009). Average sequestration potential in agroforestry in India has been estimated to be 25 t C ha^{-1} over 96 million ha (Sathaye and Ravindranath 1998).

Corporate social responsibility and agroforestry

The Corporate Social Responsibility (CSR) laws of India were modified on 1st April, 2014 and notified by the Ministry of Corporate Affairs through Section 135 and Schedule VII of the Companies Act, as well as the provisions of the Companies (Corporate Social Responsibility Policy) Rules, 2014. Accordingly, agroforestry became a legitimate part of the recognized CSR activities, number 6 (Ensuring environmental sustainability, ecological balance, protection of flora & fauna, animal welfare, agroforestry, conservation of natural resources and maintaining quality of soil, air & water) (http://finance.bih.nic.in/Documents/CSR-Policy.pdf.).

Post 2020 climate action plan

India intends to reduce the emissions intensity of its GDP by 33 to 35 % by 2030 from 2005 level and to create an additional carbon sink of 2.5 to 3 billion tones of CO_2 equivalent through additional forest and tree cover by 2030. Agroforestry will play a crucial role in meeting the INDC targets as there is no further scope to put more areas under forest land. The major share of the land to be brought under agroforestry will come from fallows, cultivable fallows, pastures, groves and through rehabilitation of problem soils. In addition bunds on agriculture lands are another potential area for agroforestry. Recently, a Green Highways (Plantation and Maintenance) Policy under National Green Highway Mission (NGHM) to develop 140,000 km long "tree-line" along both sides of national highways was formulated. This will further help in increasing green cover.

Carbon sequestration potential of existing agroforestry systems in India

CAFRI initiated estimation of carbon sequestration potential (CSP) of existing agroforestry systems (AFS) in the country. In first step it has surveyed in twenty six districts of ten selected states of India. Out of the surveyed districts, two districts of Rajasthan viz. Jhunjunun and Sikar are part of the study. The native *Prosopis cineraria* known as "khejri" was the most dominant tree species in both the districts. Other tree species such as *Tecomella undulata, Capparis decidua*, *Acacia tortilis*, *Ailanthus excelsa* and *Prosopis juliflora* were also common in the desert landscape. The observed average number of trees per hectare were 6.95 in Jhunjunun and 12.42 in Sikar. The observed number of trees on farmers' field in the two districts varied from 6.95 to 12.42 trees per hectare. The biomass in the tree component varied from 4.33 to 7.62 Mg DM ha^{-1}, whereas, the total biomass (tree and crop) ranged from 17.13 to 19.19 Mg DM ha^{-1}. The soil organic carbon ranged from 4.28 to 4.51 Mg C ha^{-1}. The average estimated carbon sequestration potential of the agroforestry systems (AFS) on farmer's field in the two district was 0.22 to 0.28 Mg C ha^{-1}yr^{-1} (Ajit *et al*. 2016).

At the country level, observed number of trees per hectare on farmers' field varied from 1.81 to 204 with an average value of 19.44 trees. The total biomass (tree and crop) ranged from 4.96 to 58.96 Mg DM ha^{-1} and the soil organic carbon ranged from 4.28 to 24.13 Mg C ha^{-1}. The average estimated carbon sequestration potential of the agroforestry systems (AFS) representing varying edapho-climatic conditions on farmer's field at country level was 0.21 Mg C ha^{-1} yr^{-1}. At national level, existing AFS are estimated to mitigate 109.34 million tons CO2 annually, which may offsets one-third (33%) of the total GHG emissions from agriculture sector (Ajit *et al*. 2016). However, the potential of agroforestry systems as carbon sink varies depending upon the species composition, age of trees, geographic location, local climatic factors and management regimes.

Table 3. Biomass accumulated in the tree or crop components and carbon sequestered under existing AFS in two districts of Rajasthan

Parameters		Observed no. of existing trees per ha in AFS at district level are given in parenthesis	
		Jhunjhunu(6.95)	Sikar (12.42)
Tree Biomass (above and below ground) Mg DM ha^{-1}	Baseline	4.33	7.62
	Simulated Biomass	10.04	18.74
Total Biomass (tree+ crop)			

(Contd.)

Mg DM ha^{-1}	Baseline	17.13	19.19
	Simulated	23.2	30.64
Soil carbon (Mg C ha^{-1})	Baseline	4.51	4.28
	Simulated	8.48	7.34
Biomass carbon (Mg C ha^{-1})	Baseline	7.58	8.64
	Simulated Carbon	10.48	14.11
Total carbon (biomass + soil) (Mg C ha^{-1})	Baseline	12.09	12.92
	Simulated	18.96	21.45
Net carbon sequestered in AFS over the simulated period of 30 years (Mg C ha^{-1})	Carbon sequestered	6.87	8.53
Estimated annual carbon sequestration potential of AFS in two districts of Rajasthan (Mg C ha^{-1}yr^{-1})	Carbon sequestered	0.22	0.28

Source: Ajit *et al.* (2016)

Agroforestry and ecosystems services

Agroforestry systems are believed to provide a number of ecosystem services (Dhyani and Handa 2013b). Trees with deep rooting systems in agroforestry systems can also improve ground water quality by taking up excess nutrients that have been leached below the rooting zone of agronomic crops. The other benefits include protecting crops, removing atmospheric carbon dioxide and producing oxygen, reducing wind velocity and thereby limiting wind erosion and particulate matter in the air, reducing noise pollution. However, long term studies on quantification of the ecosystems services from agroforestry are yet to be initiated. A study by CAFRI indicated scaling-up of integrated watershed management in drought prone rainfed areas with enabling policy and institutional support would promote equity and livelihood along with strengthening various ecosystem services while reducing poverty and building resilience in semi-arid tropics (Singh *et al.* 2014).

Way forward

The organized research efforts undertaken during the last more than three decades have clearly demonstrated the potential of agroforestry for resource conservation, improvement of environmental quality, rehabilitation of degraded lands and providing multiple outputs to meet the day to day demand of the rural population. However, the results obtained so far have to be validated on large scale and in farmer's fields. Also, the agroforestry models developed by the research institutions suitable for the diverse agro-climatic regions are to be scaled up and tested with sound database. The R and D on processing technologies for the agroforestry species is lacking. The status of organized agroforestry research and the way forward is presented in Table 4.

Table 4. Status of organized agroforestry research and way forward

Thrust areas	Status	Way forward
1. Diagnostic survey and appraisal of existing agroforestry practices and farmers' preferences.	Major agroforestry practices of systems identified in different agro-climatic zones valuable information collected on farmer's choice of species. Recurrent survey initiated at present by 9 centres.	In view of change in land use pattern, the recurrent survey needs to be conducted at an interval of 15-20 years.
2.Collection and evaluation of MPTS	The germplasm of 184 promising tree species was collected and evaluated Identification of priority tree species of agroforestry research for various agro-climates achieved. Germplasm has been registered at NBPGR by few Centres.	20 species identified at national level needs to be prioritized and further germplasm collection and evaluation to be strengthened All the promising germplasm to be registered at NBPGR
3. Tree selection and improvement	On 20 MPTS selection and improvement work done. Out of them, significant achievements made on Poplar, *Eucalyptus*, *leaucaena*, *Casuarina* and teak. The improved material has been utilized by the wood based industry for its raw material. The species such as *Acaia, Ailanthus, Albizia* spp., *Dalbergia, Azadirachta indica, Anogeissus, Pongamia,* bamboos, *Anthocephalus cadamba, Grewia, Hardwickia, Melia, Prosopis cineraria, Jatropha, Salix, Gmelina arborea* etc. have also gone to different stages of selction and improvement and some of them are under multi-location trials. Seed or clonal orchards for few species established.	DNA fingerprinting of the clonal material of a species to document the available variability. Standardization of clonal propagation techniques and mass multiplication of elite planting material Seed or clonal orchards for the identified species to be established. Multi-location trials to be conducted on all the promising species with identified promising germplasm. Harvest and post-harvest processing and value addition of natural gums, resins etc.
4. Development and management of agroforestry practices or systems	Traditional agroforestry practices have been improved with scientific interventions. AFS viz. agrisilviculture, agri-horticulture, agri-horti-silviculture, hortipastoral, silvipastoral, and specialized systems were developed for different agro-climatic regions. Most of the systems developed and recommended are location specific, besides in many situations these systems are not looked favourably by the farmers due to wide variations in the choice of species, crop preferences etc. also some of the systems developed have not completed a full rotation for the tree component, therefore biomass production and returns are only on the basis of extrapolation.	Full package of practices, choice of crops with trees in temporal sequence and information on insect pest disease and effect of aberrant weather conditions, which are now of common occurrence is to be made available to stakeholders. Agroforestry product research, new product development, new designs and quality standards would be evolved for downstream processing.
5. Economic analysis of agroforestry systems	Few efforts have been made for economic analysis of the AFS however, they are limited to B:C ratio and does not reflect true economics. Besides the analysis are based on extrapolation as most of the systems except for Eucalyptus and poplar based systems have not completed full rotation.	Systematic economic analysis of AFS needs to be done in different agro-climatic regions. Development and demonstration for adoption of AF models linked with market for enhancing productivity and profitability of small holding farmers.

6. Environment, wasteland and community land development through agroforestry interventions	Very limited efforts have been made by the network. However, watershed schemes being implemented in the country during the last six decades have generated valuable data but the contribution of agroforestry needs to be segregated.	Long term impact of identified AFS needs to be assessed in terms of employment generation, livelihood support, conservation of land and water, quality of produce, increase in biomass productivity and micro-climate improvement. Agroforestry based climate resilient agriculture promoted. Developing AF technologies for critical areas like arid and semi arid zones and other fragile ecosystems for higher productivity and natural resource management.
7. Integration of livestock, fishery, apiculture, lac etc. as component of agroforestry	Few efforts have been made. Sericulture, apiculture, lac, livestock and fishery components have been integrated. However, rigorous analysis of data is not done in most of the studies.	Integrated system approach for small and marginal farmers with all components for year round returns and livelihood security to be initiated.
8. Area mapping, extent of area under agroforestry	Few location specific attempts were done to map agroforestry areas but such efforts were lacking at national level. FSI in the latest report (2013) has given area estimates for agroforestry.	CAFRI has initiated efforts to map area under agroforestry with the use of geo-spatial technologies recently. Development of spectral signature library for agroforestry species or systems
9.Carbon sequestration potential, environmental benefits from AFS	Studies on carbon sequestration potential (CSP) of AFS have been initiated and published by few Centres, but country wide assessment is not available.	CAFRI has initiated studies on quantification of CSP of major AFS in cultivated areas in different agro-climatic regions. The efforts needs to be made to cover AFS in all land use systems. AFS for mitigation of climate change effects and management of stresses needs to be standardized. Developing mechanism for the benefit of CSP reaching to small holders.
10. Policy issues	Issues and concerns for agroforestry implementation have been identified and highlighted. The efforts culminated into launching of National Agroforestry Policy 2014.	Implementation of NAP2014 streamlined through appropriate guidelines. Trends in marketing of agroforestry products and services including market intelligence needed. Agroforestry role in poverty alleviation; women empowerment and livelihood support is to be identified and emphasized.

Source: Handa *et al.* (2015)

References

Ajit, Dhyani SK, Handa AK, Ram Newaj, Chavan SB, Alam B, Prasad R, Asha Ram, Rizvi RH, Jain AK, Uma, Tripathi D, Shakhela RR, Patel AG, Dalvi VV, Saxena AK, Parihar AKS, Backiyavathy MR, Sudhagar RJ, Bandeswaran C, Gunasekaran S (2016). Estimating carbon sequestration potential of existing agroforestry systems in India. Agroforestry Systems (in press).

Beets WC (1989). The Potential Role of Agroforestry in ACP Countries. CTA/EEC, Wageningen, Netherlands.

Chauhan DS, Dhyani SK (1989).Traditional agroforestry practices in N. E. Himalayan region. Indian J Dryland Agric Res and Dev 4(2):73-81.

Dadhwal KS, Narain P,Dhyani SK (1989). Agroforestry systems in Garhwal Himalayas of India. Agroforestry Systems 7(3): 213-225.

Dhyani SK, Chauhan DS (1994). Agroforestry practices of North-eastern hill region of India. In:Narain P*et al.* (eds) Agroforestry Traditions & Innovations CSWCRTI, Deharadun, pp. 19-23.

Dhyani SK, Singh BP, Chauhan DS, Prasad RN (1994). Evaluation of MPTS for agroforestry system to ameliorate infertility of degraded acid alfisols on sloppy lands. In: Singh Panjab *et al.* (eds) Agroforestry Systems for Degraded Lands, Oxford & IBH Publishing Co. Pvt. Ltd., New Delhi, Vol. I, pp.241-247.

Dhyani SK (1998). Alder-an essential component of agroforestry systems in the north-eastern hills of India. Agroforestry Today (ICRAF) 10(4):14-15.

Dhyani SK, Handa AK (2013). Agroforestry in India and its potential for ecosystem services. In:Dagar JC *et al.* (eds) Agroforestry Systems in India: Livelihood Security & Ecosystem Services, Advances in Agroforestry 10, Springer India, pp. 345-366.

Dhyani SK, Handa AK (2013a). India needs agroforestry policy urgently: issues and challenges. Indian J Agroforestry 15(2):1-9.

Dhyani SK, Handa AK (2013b). Agroforestry in India and its potential for ecosystem services. In: Dagar JC *et al.* (eds) Agroforestry Systems in India: Livelihood Security & Ecosystem Services, Advances in Agroforestry 10, Springer India, pp. 345-366.

Dhyani SK, Handa AK, Uma (2013).Area under agroforestry in India: An assessment for present status and future perspective. Indian J Agroforestry 15(1):1-11.

Dhyani SK, Handa AK, Uma (2015).Agroforestry research in India: present status and future perspective. In: DhyaniSK, NewajRam, Alam Badre,Dev Inder(eds) Organized Agroforestry Research in India: Present Status and Way Forward. Biotech Book Publisher, Delhi.

Dhyani SK, Ram Newaj, Sharma AR (2009). Agroforestry: its relation with agronomy, challenges and opportunities. Indian J Agronomy 54(3):249-266.

Dhyani SK, Kumar BM, Sikka AK (2014).Agroforestry research and development in India. Indian Farming 63(11):9-11.

Dhyani SK, Sharda VN, Samra JS (2005). Agroforestry for sustainable management for soil, water and environment quality: looking back to think ahead. Range Mgmt and Agroforestry 26(1):71-83.

GITF (2001). Report of the Task Force on Greening India for Livelihood Security and Sustainable Development, Planning Commission, Govt. of India. pp.254.

Handa AK, Dhyani SK, Uma (2015). Three decades of agroforestry research in India: retrospection for way forward.Agric Res J 52 (3):1-10.

ICFRE (2011). Agroforestry Research at ICFRE. Media & Publication Division, Directorate of Extension, ICFRE, New Forest, Dehradun. No. 133 ICFRE Book - 79/2011.

ISFR (2013). India State Forest Report. Forest Survey of India (Ministry of Environment and Forests and Climate Change), Dehradun, India.

ISFR (2015). India State Forest Report. Forest Survey of India (Ministry of Environment and Forests and Climate Change), Dehradun, India.

Kumar BM, Singh AK, Dhyani SK (2012). South asian agroforestry: traditions, transformations and prospects. In: The Future of Global Land Use, Advances in Agroforestry 9, Springer, pp. 359-389.

NRCAF (2013). Vision 2050. National Research Centre for Agroforestry, Jhansi, India, pp. 30.

Pathak PS, Pateria NM, Solanki KR (2000)Agroforestry Systems in India: A Diagnosis and Design approach. NRC for Agroforestry, ICAR, pp. 166.

Rizvi RH, Dhyani SK, Ram Newaj, Karmakar PS, Saxena A (2014) Methodology for mapping agroforestry area in India through remote sensing and preliminary estimates. Indian Farming (Special Issue): 63 (11):62-64.

Sathaye JA, Ravindranath NH (1998) Climate change mitigation in the energy and forestry sectors of developing countries. Annual Review of Energy and Environment 23:387-437.

Singh BP, Dhyani SK, Prasad RN (1994) Traditional agroforestry system and their soil productivity on degraded alfisols/ultisols in hilly terrain. In: Singh Panjab *et al.* (eds) Agroforestry Systems for Degraded Lands, Oxford & IBH Publishing Co. Pvt. Ltd., New Delhi, Vol. I, pp. 205-214.

Singh R, Garg KK, Wani SP, Tewari RK, Dhyani SK (2014) Impact of water management interventions on hydrology and ecosystem services in Garhkundar-Dabar watershed of Bundelkhand region, Central India. J Hydrology 509:132-149.

21

Complementarity in Growth Resources Sharing Makes Homegarden Agroforestry Sustainable

C.B. Pandey

Present address: Central Arid Zone Research Institute, Jodhpur Rajasthan -342 003, India

Introduction

Two types of interactions, i.e. competitive and complementary, are found in ecosystems and agroecosystems (Callaway 1998), which shape community structure, plant diversity and system's functions (Callaway 1995; Pugnaire *et al.* 1996; Bruno *et al.* 2003). Competitive interaction in ecosystems (Wedin and Tilman 1993) and agro-ecosystems (Pandey *et al.* 1999) is well documented. Mechanism of complementary interaction, however, is not much understood so far. Some studies (Bertness and Callaway 1994; Callaway 1998) argue that complementary interaction is facilitative, either facultative or obligatory, and occur generally in harsh physical environments. This suggests that habitat amelioration by neighbours is a common denominator of positive interactions (Bertness and Callaway 1994; Callaway and Walker 1997). Other studies report that complementary interaction is mutualistic (Boucher *et al.* 1982). Irrespective of whether facultative or mutualistic, it is now well established that complementary interaction is evolutionary (Bertness and Callaway 1994). But, it is still inconclusive how two species exist together in ecosystems and agroecosystems (Bertness and Callaway 1994).

Homegardens are a prominent land use system covering 63% of the arable land in the Andaman and Nicobar Islands of India (Basic Statistics 2001) where plant species of different life forms and heights, i.e. coconut and arecanut palms, clove and nutmeg tree spices, mango, banana, guava fruit trees, and other agroforestry trees grow together and form a multistorey structure similar to that found in Southeast Asia (Millate-E-Mustafa *et al.* 1996; Pandey *et al.* 2007). From the ground layer to upper stories, the gradient of light and humidity determines different niches that species exploit according to their own requirements (Fernandes and Nair 1986). Niche separation in the homegarden likely occurs due to complementary relations among the plant species for their growth resource sharing, which results in sustainable yield (Jensen 1993; Pandey *et al.* 2007; Soemarwoto 1987). Sustainability in the yields may also be attributed to the compatibility in the homegarden trees for belowground growth resource utilization (Jensen 1993; Pandey *et al.* 2000; Rhoades 1997). Overstorey trees in the homegarden generally extend their roots quite close to the trunk of their understorey plants, but understorey plants restrict their roots to a limited distance (Pandey and Venkatesh 2007). It is still not known how they separate their niches below the ground for utilization of growth resources, particularly nutrients.

We hypothesize that an overstorey tree in a homegarden agroecosystem intercepts light and, thereby, provide partial shade to its understorey tree crops (a facilitation effect), while the understorey tree crops, in return, exploit nutrients only from a limited space below the ground and, simultaneously, tolerate the presence of the overstorey tree. The overstorey tree, however, mines nutrients from its niche as well as from the niches of its understorey crops (intercrops). These facilitation, exploitation and tolerance mechanisms together allow the tree crops to coexist and exploit growth resources complementarily in the homegarden.

Homegardens include typically many species, life forms and involve several organic matter / nutrient input and output processes. Therefore, it is difficult to separate the species-specific interactions. Hence, to simplify species-specific interactions in the complex homegarden agroecosystem, the present study was carried out in a coconut (*Cocos nucifera* L.)–clove (*Eugenia cariophyllata* Thunb) and a coconut-nutmeg (*Myristica fragrans* Houtt. Nees) plantation, having similar tree crop geometry found in the homegardens. The objectives of the study were to examine: (1) aboveground growth resource (sunlight) sharing, and (2) patterns of fine root biomass distribution (belowground niche separation), and uptake of N, P and K by the coconut tree (main crop) and its intercrops (i.e. clove and nutmeg trees) at different distances between two tree crops (coconut-clove and coconut-nutmeg) in the plantation. The ultimate objective of the study, however, was to understand the mechanism of complementary interactions among the tree crops in the homegardens of the Andaman Islands of India.

Study area and experimental details

This study was conducted in a coconut-clove and a coconut-nutmeg plantation, both twenty-year old, at a research farm of Central Agricultural Research Institute (CARI) at Sipighat, South Andaman Island, India located at 10° 30'-13° 42' N and 92° 14'-94° 14' E and 315 m asl. Soils of the site are dystric fluvisols and the parent material is sandstone. The soils are well drained, sandy-loam, slightly acidic in reaction and moderate to poor in nutrients (Pandey *et al.* 2009). The Islands experience a true maritime climate throughout the year with a little variation in temperature from 23 to 30 °C; humidity varies from 71 to 85 %. December is the coolest month and May is the warmest month. Ten-year data (1997-2006) reveal that average annual precipitation is 3000 mm distributed over 8-9 months from May to January with 86 % occurring between May to November (e" 400 mm month^{-1}) (wet season). Three months, from February to April, are relatively dry (< 100 mm month^{-1}) (dry season) (Pandey *et al.* 2007). Native vegetation is tropical rainforest dominated by *Dipterocarpus* species.

The coconut-clove and coconut-nutmeg plantation, each about 7 ha, was established in 1983 at the research farm of CARI. Coconut trees (main crop) were planted at a spacing of 7.5 x 7.5 m and spice trees like clove and nutmeg were planted in a quincunx manner (one spice tree in the centre of four coconut trees planted at the corners of a square) as intercrops in the plantation. Fertilizer (N:P:K - 2:1.3:1) 200 g tree^{-1} and farmyard manure (FYM) 2 kg tree^{-1} were applied to the trees in an area of 50 cm diameter around each tree for four years in the beginning until the trees started bearing. Thereafter, leaves fallen on the ground floor were put around the respective tree every month to simulate soil fertility management that homegardeners (farmers) do in their homegardens. Leaves of coconut were cut into pieces before they were placed. Weeding was performed in the plantation once a year in October.

An association of each coconut-clove and coconut-nutmeg (quincunx in planting), similar in shape and size, replicated three times, was selected randomly in the plantation for sampling. In a preliminary study, a maximum of 90 % fine roots were found to be located in the soil to a depth of 20 cm soil depth (Pandey and Venkatesh 2007). In each association a line transect was laid randomly from a coconut to its intercrop at monthly interval. Total 12 transects (3 transects from a coconut to its intercrop x 4 coconut trees in the quincunx planting) were laid in an association over a year. A monolith (15 x 15 x 20 cm) was excavated along a transect at 0.75, 1.50, 2.65, 3.80 and 4.55 m distance from the coconut trees towards the intercrops. Thus, total 360 monoliths (12 transects x 5 distances x 2 coconut-intercrop associations x 3 replicates) were excavated from the plantation. The monoliths were washed with a fine jet of water using successive

2-mm and 0.5-mm mesh screens. Fine root (< 2 mm diameter) biomass (FRB) and coarse root (2 mm – 1 cm diameter) biomass were separated for each species. The fine and coarse roots of the species studied were different in colour and texture. FRB was further separated into live and dead root biomass. Separated live and dead fine root biomass, and coarse root biomass were dried separately at 60 ^{0}C in an oven to a constant weight. Fine root biomass production (FRBP) at each distance was calculated as the sum of increments in live root biomass between two successive months and the amount of dead root biomass, which exceeded the amount of decline in the live root biomass during the same period (Singh and Singh 1991). Oven-dried live and dead fine root biomass samples were pooled across the months, and ground to powder. The powdered samples were analyzed for nitrogen content by microkjeldahal method using Kel Plus auto N analyzer, P by colorimetric method and K content by flame photometer. Nitrogen, P and K uptakes at different distances by the studied tree crops in both associations were calculated as FRBP x concentrations of N, P and K in the FRB (Singh and Singh 1991). Intensity of light over the trees was measured by a Lux meter using a ladder. Light intensity in open was treated as the amount of light available over the coconut trees (Pandey *et al.* 2011). Height of the trees was measured using a long stick and a measuring tape.

The data were subjected to two-way ANOVA using SPSS statistical package to test for the significance of variation in root biomass (live, dead and total fine root biomass), fine root biomass production and N, P, K concentrations in fine roots, and the uptake of these nutrients by the studied trees. Treatment for the root biomass was: species 3, month 12 and distance 3. Treatment for the remaining parameters was: species 3 and distance 3 (2.65, 3.80 and 4.55 m where root biomass of both main-crop and intercrop was present). One-way ANOVA was performed to test for the significance of variation in root biomass and other studied parameters of coconut due to three distances (0.75, 1.50 and 2.65 m); at these distances root biomass and other parameters of the intercrops were not present. Number of replicates was three (replicates of coconut-intercrops associations) for both two-way and one-way ANOVA. Means of the parameters were compared using least significant difference ($P<0.05$).

Dimension of trees, structure, and physical environment of homegardens

Height of the studied intercrop tree/shrub species is 33 to 42 % lower than the main crop (Table 1). Coconut intercepts 32 and 30 % light above the nutmeg and clove canopies, respectively. Variation in the intercepted light occurs likely due to movement of leaves, which regulate sun rays to travel through spaces among leaflets and leaves to the intercrops. Pandey and Singh (2009) have reported that coconut-clove and coconut-nutmeg tree crop associations survive for a long time and provide sustainable, though low yields in the homegardens. This indicates that clove and nutmeg intercrops are compatible with coconut trees and co-exist as intercrops. Reduction in the intensity of light over the intercrops indicate that coconut intercepts light and, thereby, facilitate a conducive environment for the intercrops. Clove and nutmeg trees are known to prefer partial shade (Thankamani *et al.* 1994). This facilitative mechanism of aboveground growth resource (partial light) use in the coconut-spice tree associations has been reported among species in ecosystems in deserts (Arriaga *et al.* 1993), savanna and woodland (Guevara *et al.* 1992), mediterranean shrub-land (Fuentes *et al.* 1984), salt marshes (Bertness and Hacker 1994), grasslands (Greenlee and Callaway 1996), successional forests (Bertness and Callaway 1994), *Paulownia*- (Zhaohua *et al.* 1986) and *Grevillea*- (Huxley *et al.* 1994) based agroforestry systems. In successional forests young and small understorey plants survive under overstorey crops complementarily (Bertness and Callaway 1994). Plants have been found to have positive effects on each other by accumulation of nutrients, provision of shade, and amelioration of disturbances or protection from herbivores by some species, which enhance the performance of neighbouring species (Bertness and Callaway 1994).

Table 1. Dimensions of coconut and its intercrops, and light intensity availability to the tree crops in a coconut-clove and a coconut- nutmeg plantation at the South Andaman Islands of India

Tree	Height(m)	CBH(cm)	Canopy cover (m^2)	*Light intensity(Lux)
Coconut	12.0±0.03	44.9±1.41	98.7±1.09	4865±408
Nutmeg	6.9±0.55	39.6±0.92	48.3±0.72	3294±187
Clove	8.1±0.12	16.14±0.20	42.3±1.91	3387±195

*Intensity of light over coconut is the value recorded in open. Coconut is the top storey, hence light intensity present in open is assumed to be present over the coconut. CBH = circumference at breast height. Coconut (N = 24), Nutmeg (N = 3), Clove (N = 3). Height, CBH and canopy cover of all trees are presented on per tree basis

Distribution and seasonality in fine root biomass of homegarden trees

Total fine root biomass (live and dead) varies significantly ($P < 0.0001$) among the studied species, distance from the tree trunk and months. The fine root biomass, averaged across the months and distances, is the highest (127.9 g m^{-2}) in coconut and the lowest (40.7 gm^{-2}) in clove (Fig.1a, b). In both coconuts and intercrops, fine root biomass is higher closer to their trunks and declines with distance. The fine root biomass of both intercrops, i.e. nutmeg and clove is found up to middle distance (2.65 m) towards coconut in the coconut-nutmeg and coconut-clove associations. But, the fine root biomass of coconut extends quite close to the trunk (4.55 m) of its intercrops i.e. nutmeg and clove in the associations (Fig. 1a, b). The fine root biomass is 17-20 % of the total root biomass (coarse + fine).

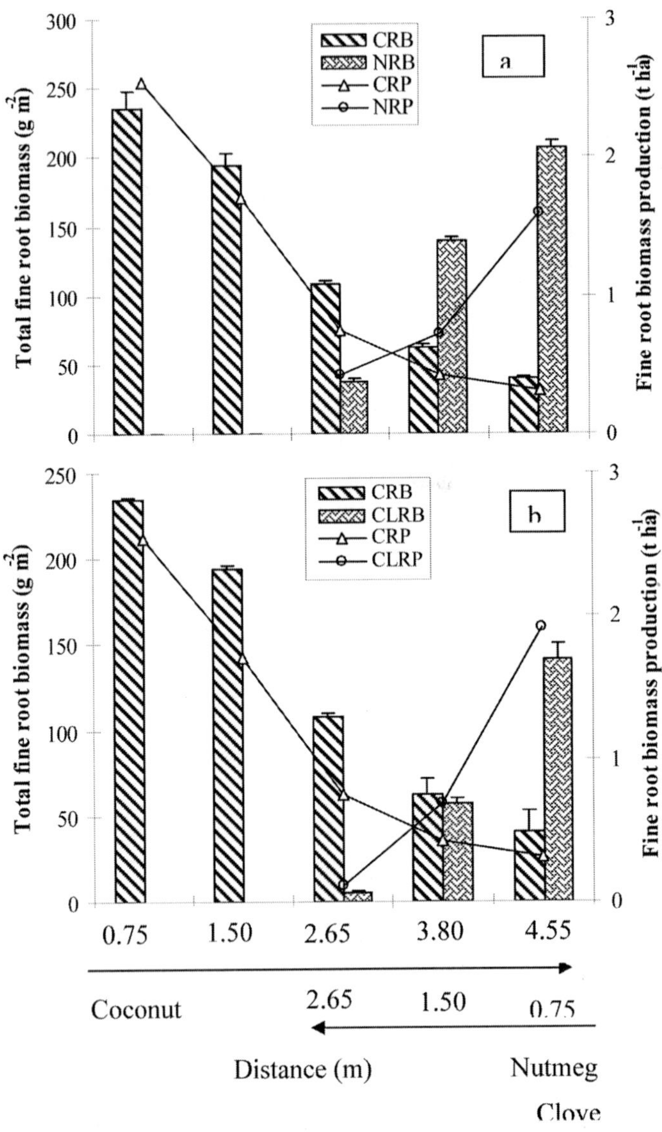

Fig.1. Fine root biomass production (RP) and total fine root (live + dead) biomass (RB) of (a) coconut (CRP and CRB) and nutmeg (NRP and NRB), and (b) coconut (CRP and CRB) and clove (CLRP and CLRB) at different distances between coconut-nutmeg and coconut-clove trees in a coconut-clove and a coconut- nutmeg plantation at the South Andaman Islands of India

Live and dead fine root biomass varies significantly ($P < 0.001$) with species studied and months. Live fine root biomass of coconut increases quickly in May, but declines sharply in June. It increases further from July and peaks in December (Fig. 2a, b, c). A decline in live fine root biomass is observed from January to April. On the other hand, the dead fine root biomass declines from May to December. Pattern of seasonal variation in live and dead fine root biomass of nutmeg is similar to that of the coconut tree, but live fine root biomass of clove increases sharply from February to April and, thereafter, declines gradually with little fluctuations until February. Dead fine root biomass varies across the months with small troughs and peaks. Pattern of seasonal variation in live and dead root biomass in all species is similar at all distances.

Occurrence of fine root biomass of coconut under the intercrops suggests that in the coconut-clove and coconut-nutmeg associations coconut trees extends their fine roots horizontally quite close to the trunk of their intercrops. Wiersum (1982) has reported that roots of coconut trees is located in upper depth of soils and are distributed extensively on the floor of homegardens in Indonesia. On the contrary, the fine roots of the intercrops are brittle and weak and are confined to a certain distance from their trunks as also observed by Wiersum (1982). This horizontal separation of belowground niches of the tree species seems evolutionary, which occurs probably to meet their nutrients and other growth resource requirements below the ground. Homegardens in Indian subcontinent have been developed since time immemorial (Kumar and Nair 2004). Farmers generally grow coconut trees in the premises of their houses in a wider spacing and spice trees as intercrops under the coconut trees. This probably force the trees to carve out their niches in a complementary manner over the time for growth resource acquisition. But, the mechanism of interaction below the ground in our study is found different from that of the above ground. We found that the intercrops derive benefit (partial shade) from coconut trees above the ground, but tolerate its presence, in return, in their niches below- the ground. This suggests that complementarity in nutrient sharing below the ground is tolerated unlike the facilitative mechanism above the ground. Connell and Slatyer (1977) are of the view that facilitative and tolerance mechanisms of complementary interaction may occur within the same plant community. Nelliat *et al.* (1974) have suggested that horizontally separated root systems can be the basis for complementarities in coconut-cocoa-pineapple multistorey agroforestry systems in Indonesia. Contrary to horizontal niche separation below the ground for nutrients acquisition in our homegardens, vertical niche separation has been reported to occur in *Paulownia*-based agroforestry system in China. *Paulownia* tree has the majority of its fine roots located at 40-100 cm depth, whereas its intercrops, namely wheat, maize and groundnut, above the tree root zone

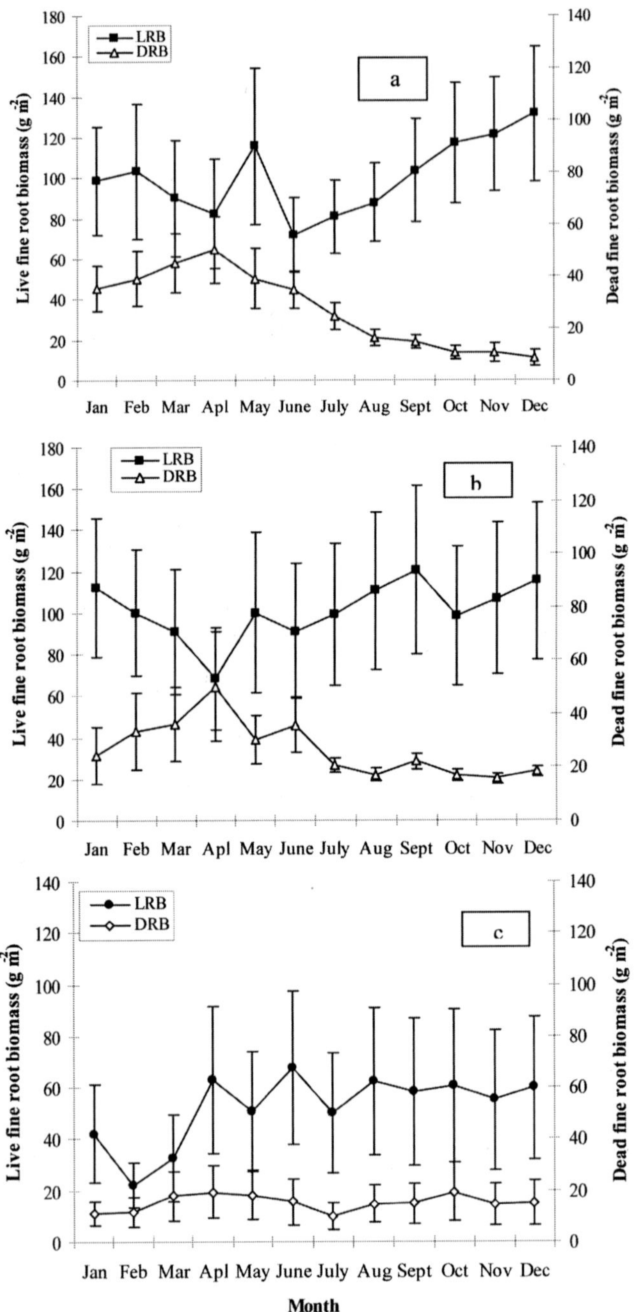

Fig. 2. Seasonal variation in live (LRB) and dead (DRB) fine root biomass of (a) coconut (b) nutmeg and (c) clove in a coconut-clove and a coconut- nutmeg plantation at the South Andaman Islands of India

(Zhaohua *et al.* 1986). This vertical niche separation probably makes the *Paulownia*-based agroforestry system complementary. Also, in *Grevillea*-based agroforestry system of Kenya, vertical niche separation occurs below the ground (Huxley *et al.*, 1994).

Fine root biomass production, and uptake of nutrients

Fine root biomass production varies significantly ($P < 0.001$) due to the species studied and distances. Fine root biomass production, at all distances is the highest in coconut and the lowest in clove. In both coconut-nutmeg and coconut-clove associations, fine root biomass production in all species is higher closer to the trunk and declines with distance (Fig. 1a, b). Concentration of N, P, and K in fine roots varies significantly ($P < 0.001$) with distance in all species. Generally it is higher closer to the trunk and declines with the distance in all species (Fig. 3 a, b). Concentrations of N, P and K in the fine root biomass of nutmeg and clove at all distances within their niches are greater than in the coconut roots

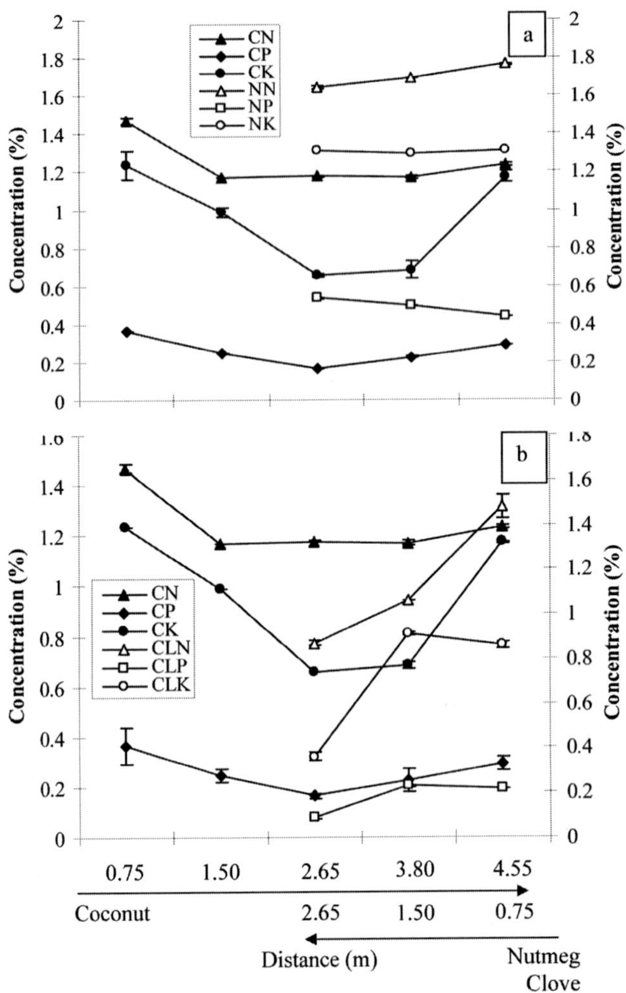

Fig. 3. Concentrations of nitrogen (N), phosphorus (P) and potassium (K) in the root biomass of (a) coconut (CN, CP and CK) and nutmeg (NN, NP and NK), and (b) coconut (CN, CP and CK) and clove (CLN, CLP and CLK) at different distances between coconut-nutmeg and coconut-clove trees in a coconut-clove and a coconut- nutmeg plantation at the South Andaman Islands of India

found there. Uptake of N, P and K by all species is greater closer to the trunk and declines with the distance. Uptake of N, P and K by the intercrops is up to the middle point (2.65 m) towards their main tree crop, but the main tree crop mines the nutrients quite close (4.55 m) to the trunk of its intercrops (Fig. 4a, b). Uptake of N, P and K by the coconut from the niches of the intercrops, however, is lower than that of the intercrops.

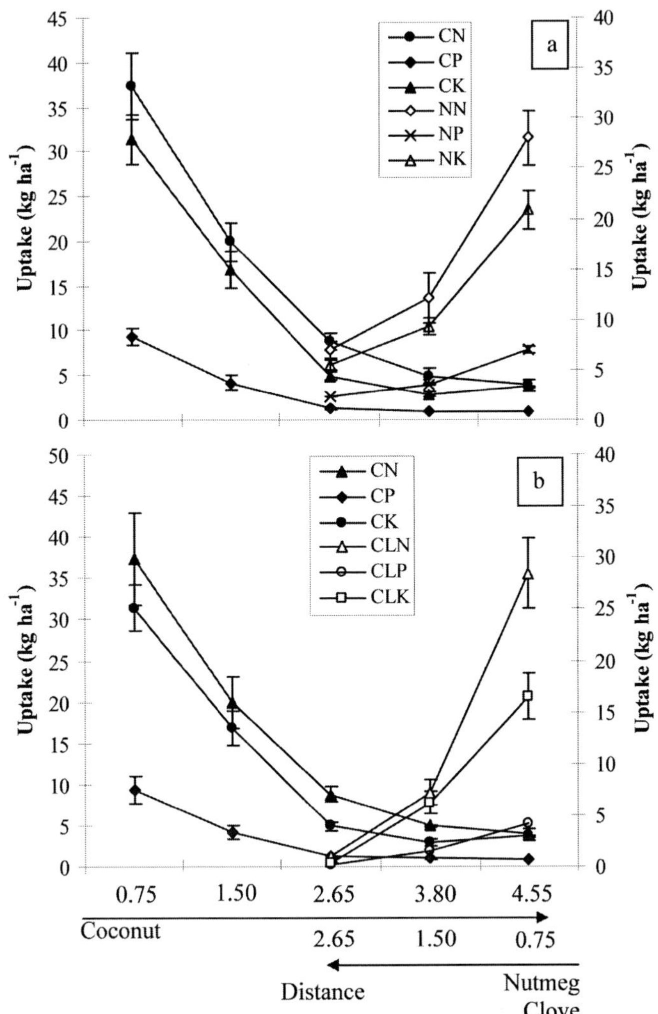

Fig. 4. Uptake of nitrogen (N), phosphorus (P) and potassium (K) by the (a) coconut (CN, CP and CK) and nutmeg (NN, NP and NK), and (b) coconut (CN, CP and CK) and clove (CLN, CLP and CLK) at different distances between coconut-nutmeg and coconut-clove trees in a coconut-clove and a coconut- nutmeg plantation at the South Andaman Islands of India

Uptake of N, P and K by the coconut from beneath the clove and nutmeg trees suggest that shelter / co-existence of the intercrops under the main crop (coconut) in the homegardens is not rent-free. Complementarity in the homegarden seems to have developed in such a way that the intercrops will have to share their belowground growth resources with the main crop. In our study, coconut, being evergreen, flowers and produces fruits every month, and probably possess a big-size sink for nutrients above the ground (leaves). Like the coconut, nutmeg is also evergreen and flowers and produces fruits every month, but not as vigorously as the coconut does. Clove, being semi-evergreen, flowers and produces fruits seasonally once a year. Comparatively, bigger sink size of coconut is also evident from the greater total uptake of N, P and K across the distances than the intercrops. This clearly suggests that the nutrient uptake potential and size of sink for nutrients in leaves can be the evolutionary forces, which lead to the niche separation among the studied tree species in such a way that understorey plants not only tolerate the presence of roots of the main crop but also pay lunch bills, though partial, of the over-storey tree. van Noordwijk *et al.* (1996) are of the view that uptake of water and nutrients are often directly proportional to the aboveground demand, i.e. the size of the leaf and the strength of the aboveground sink for nutrients (van Noordwijk *et al.* 1996). Wiersum (1982) speculate that a shallow but extensive root system of coconut in the upper depth of soils in a mixed stand probably help it to exploit belowground growth resources to the greatest extent.

A similar pattern of fine root distribution of the studied-species in both associations, and uptake of N, P and K by the coconut from the niches of its intercrops suggest that complementarity in the homegarden is not species-specific. Field observation suggest that in addition to the clove and nutmeg, several other trees like fruit trees, agroforestry trees, etc. co-existed with coconut and perform well in the homegardens (Pandey *et al.* 2007). This further corroborate that complementary interaction in the homegarden is not species-specific. Studies related to species-specificity in complementary interactions in homegardens are lacking. However, Callaway and D'Antonio (1991), and Callaway (1992) have found that survival of *Quercus agrifolia* seedlings is much higher under some shrub species than others although it has not been found to be the case for other species. In different ecosystems, distribution and abundance of many species have been found favourably altered by the presence of others (Callaway 1997). Callaway (1998), Kellman and Kading (1992) have found that larger plants are more likely to benefit from facilitative associations than smaller plants. Some other studies, however, report opposite results where younger or smaller plants depend on facilitative partnerships more than older or larger plants (Archer *et al.* 1988; Callaway and Davis 1993; Callaway *et al.* 1996).

Conclusion

This study concludes that coconut tree intercepts sunlight and provides partial shade to its shade-loving intercrops, i.e. clove and nutmeg in coconut-clove and coconut-nutmeg association, respectively, above the ground. Below the ground, the coconut tree extends its roots quite close to the trunk of its intercrops, but the intercrops extend their roots only up to a certain distance within the radius of their canopies. Thus, the main crop and its intercrops separate their niches horizontally. Clove and nutmeg exploit nutrients from their own niches, while coconut exploits nutrients from the niches of their intercrops in addition to its own niche. These observations suggest the occurrence of facilitative mechanism by the main crop to its intercrops above the ground, but exploitative mechanism below the ground. Further, the tolerance by the intercrops (presence of roots of the main crop in the niches of intercrops) make the coconut-clove and coconut-nutmeg associations complementary in the plantation and, therefore, in the homegardens.

References

Archer S, Scifres C, Bassham CR, Maggio R (1988) Autogenic succession in a subtropical savanna: conversion of grassland to woodland. Ecological Monograph 58: 111-127.

Arriaga L, Maya Y, Diaz S, Cancino J (1993) Association between Cacti and nurse perennials in a heterogeneous dry forest in northeastern Mexico. Journal of Vegetation Science 4: 349-356.

Basic Statistics (2001) Andaman and Nicobar Islands: Basic Statistics. Directorate of Economics and Statistics, Andaman and Nicobar Administration, Port Blair, India.

Bertness MD, Callaway R (1994) Positive interactions in communities. TREE 9(5):191-193.

Bertness MD, Hacker SD (1994) Physical stress and positive associations among marsh plants. American Naturalist 144: 363-372.

Boucher DH, James S, Keeler KH (1982) The ecology of mutualism. Annual Review of Ecology and Systematics 13:315-347.

Bruno JF, Stachowicz JJ, Burtness MD (2003) Inclusion of facilitation into ecological theory. Trends in Ecology and Evolution 18:119-125.

Callaway RM (1992) Effect of shrub on recruitment of Quercus douglasii and Quercus lobata in California. Ecology 73: 2118-2128.

Callaway RM (1995) Positive interactions among plants. Botanical Review 61: 306-349.

Callaway RM (1997) Positive interactions in plant communities and the individualistic continuum concept. Oecologia 112:143-149.

Callaway RM (1998) Competition and facilitation on elevation gradients in sub-alpine forests of the northern Rocky Mountains, USA. OIKOS 82:561-573.

Callaway RM, D'Antonio CM (1991) Shrub facilitation of coast live oak establishment in central California. Madrono 38: 158-169.

Callaway RM, DeLucia EH, Moore D, Nowak R, Schlesinger WH (1996) Competition and facilitation: contrasting effects of Artemisia tridentata on desert vs. montane pines. Ecology 77(7): 2130-2141.

Callaway RM, Davis FW (1993) Vegetation dynamics, fire and physical environment in coastal central California. Ecology 74: 1567-1578.

Callaway RM, Walker LR (1997) Competition and facilitation: a synthetic approach to interactions in plant communities. Annual Review of Ecology and Systematics 13: 315-347.

Connell JH, Slatyer RO (1977) Mechanisms of succession in natural communities and their role in community stability and organization. American Naturalist 111: 1114-1119.

Fernandes ECM, Nair PKR (1986) An evaluation of the structure and function of tropical home gardens. Agroforestry Systems 21:279-310.

Fuentes ER, Otaiza RD, Alliende MC, Hoffman A, Poiani A (1984) Shrub clumps of the Chilean matorral vegetation: structure and possible maintenance mechanisms. Oecologia 62: 405-411.

Greenlee JT, Callaway RM (1996) Abiotic stress and the importance of interference and facilitation in montane bunchgrass communities in western Montana. American Naturalist 148: 386-396.

Guevara S, Meave J, Moreno-Casasola P, Laborde J (1992) Floristic composition and structure of vegetation under isolated trees in neotropical pasture. Journal of Vegetation Science 3: 655-664.

Huxley PA, Pinney A, Akunda E, Muraya P (1994) A tree / crop interface orientation experiment with a Grevillea robusta hedgerow and maize. Agrofestry Systems 26: 23-45.

Jensen M (1993) Productivity and nutrient cycling of a Javanese home gardens. Agroforestry Systems 24: 187-201.

Kellman M, Kading M (1992) Facilitation of tree seedling establishment in a sand dune succession. Journal of Vegetation Science 3: 679-688.

Kumar BM, Nair PKR (2004) The enigma of tropical homegarden. Agroforestry Systems 61:135-152.

Millate -E- Mustafa MD, Hall JB, Teklehaimanot Z (1996) Structure and floristics of Bangladesh homegardens. Agrforestry Systems 33: 263-280.

Nelliat EV, Bavappa KV, Nair PKR (1974) Multistoreyed, a new dimension in multiple cropping for coconut plantations. World Crops 26:262-266.

Pandey CB, Singh AK, Sharma DK (2000) Soil properties under Acacia nilotica trees in a traditional agroforestry system in Central India. Agroforestry Systems 49:53-61.

Pandey CB, Venkatesh A (2007) Agroforestry for Sustainable Biomass Production in the Andaman and Nicobar Islands. Technical Report, Central Agricultural Research Institute, Port Blair, India.

Pandey CB, Pandya KS, Pandey D, Sharma RB (1999) Growth and productivity of rice (Oryza sativa) as affected by Acacia nilotica in a traditional agroforestry system. Tropical Ecology 40: 109-117.

Pandey CB, Singh L (2009) Soil fertility under homegarden trees and native moist evergreen forest in South Andaman, India. Journal of Sustanable Agriculture 33:303-318.

Pandey CB, Rai RB, Singh L, Singh AK (2007) Homegardens of Andaman and Nicobar, India. Agricultural Systems 92: 1-22.

Pandey CB, Srivastava RC, Singh RK (2009) Soil nitrogen mineralization and microbial biomass relation, and nitrogen conservation in humid-tropics. Soil Science Society of America Journal 73: 1142-1149.

Pandey CB, Verma SK, Dagar JC, Srivastava RC (2011) Forage production and nitrogen nutrition in three grasses under coconut tree shades in the humid-tropics. Agroforestry Systems 83: 1-12.

Pugnaire JI, Haase P, Puigdefabregas J (1996) Facilitation between higher plant species in a semiarid environment. Ecology 77:1420-1426.

Rhoades CC (1997) Single-tree influence on soil properties in agroforestry: lessons from natural forest and savanna ecosystems. Agroforestry Systems 35:71-94.

Soemarwoto O (1987) Home gardens: a traditional agroforestry system with a promising future. In: Steppler HA, Nair PKR (eds) Agorofrestry: A Decade of Development, ICRAF, Nairobi, Kenya, pp. 157-172.

Singh L, Singh JS (1991) Storage and flux of nutrients in a dry tropical forest of India. Annals of Botany 68: 275-284.

Thankamani CK, Sivaraman K, Kandiannan K, Peter KV (1994) Agronomy of tree spices (clove, nutmeg, cinnamon and all spice)- a review. Journal of Spice & Aromatic Crop 3(2):105-123.

van Noordwijk M, Lawson G, Soumare A, Groot JJR, Hairiah K (1996) Root distribution of trees and crops: competition and/or complementarity. In: Ong CK, Huxley P (eds) Tree-crop interactions: a physiological approach, CAB International, Wallingford, UK, pp. 319-365.

Wedin D, Tilman D (1993) Competition among grasses along a nitrogen gradient: initial conditions and mechanisms of competition. Ecological Monograph 63: 199-229.

Wiersum LK (1982) Coconut-palm root research in Indonesia. FAO Consultancy Report. INS/72/007.

Zhaohua Zhu, Chao Ching-Ju, Lu Xin-Yu, Xiong Yao Gao (1986) Paulownia in China: cultivation and utilization. Asian Network for Biological Science and IDRC, Singapore and Ottawa.

22

Livelihood and Climate Change Mitigation and Adaptation through Agroforestry

Sivakumaran Sivaramanan

Environmental Impact Assessment unit, Environmental Management and Assessment Division, Central Environmental Authority, Parisara Piyasa, 104 Denzil Kobbekaduwa Mawatha, Battaramulla, Sri Lanka

Introduction

Human civilization used to undergo several form of disasters throughout its life history. Most of such events are unpredictable, sudden and short in duration, and such events can easily classifiable as either natural (e.g. cyclones, earth quake, natural forest fires and tsunami) or manmade such as spread of toxic chemicals (e.g. Bhopal disaster 1984), nuclear explosions, civil war, diseases (Minamata, itaiitai, etc.). However, climate change is a real time prolonged disaster which is caused by both natural (caused by variation in solar irradiance, variations in orbital parameters of earth, volcanic activities, natural forest fire, methane release from thawing of permafrost from ocean floor, natural wetlands and cattle) and anthropogenic means (by greenhouse gases resulted from human activities-all kinds of combustion, industrial production of greenhouse gases and agricultural water lodging activities such as paddy cultivation and artificial wet lands, and deforestation). Climate change is not new to the earth and biosphere but it is virgin for man. Since the last ice age (7000 years ago) last three decades have shown rapid increase in global greenhouse gases mainly the CO_2 levels resulted in subsequent rise in average global temperature. In general climate change has both pros and cons on life hood depend on the geographic location on earth, type of life form, nature of locomotion and behavioral patterns. Generally mitigation can be drawn by taking necessary action to prevent or reduce the greenhouse gas emission, preventing deforestation, implementing afforestation, reforestation protecting marine plankton sinks and through all kind of manmade activities towards greening. However, the most relevant part to be considered here is obtaining most sustainable mitigation methods for the issue. Agroforestry is a widely improving area which combines forestry and agriculture in a broad spectrum to achieve multiple goals such as productivity, profitability and environmental stewardship leading towards environmental sustainability; this includes reduction of unnecessary land use for the hiking human population.

Impacts of climate change in livelihood

In tropics, livelihood assets showed negative impact. Mainly Islands and coastal population has vastly effected by rising sea levels. Increasing sea level causes flooding in low lying lands and evacuation e.g. five of Salomon Islands have already submerged due to 10 mm sea level rise in last two decades and six other Islands in the archipelago now facing the threat, situation in Maldivian Islands also the same. Sea water intrusion to groundwater in coastal regions is a threat in coastal areas in US and coastal plains of Hersek, Tas-köprü, and Altinova in Yalova region- Turkey, Egypt, Lebanon, Syria coastal aquifers in India,Sri Lanka, coastal aquifers in South Korea, Gulf of Mexico and Gulf of States. Main course for the sea water intrusion is overconsumption of ground water resources. Fresh water consumption has been increased all over the tropical region due to the rise in temperature and subsequent drought as described below:

Increase in desertification

In history main cause of disappearance of Mesopotamian civilization is desertification. Today, anthropogenic activities such as deforestation, shifting cultivation, overgrazing, frequent burning, prolonged mono cropping, use of chemical fertilizers, dumping of hazardous chemicals, war, overconsumption of groundwater resource, over exploitation, denudation (removal of cover plants for fuel wood, timber harvest or burning vegetation) cultivation in marginal land, soil compaction due to tillage activities cause degradation of soil and ultimately causing desertification. During the process of desertification soil undergoes following changes such as progressive loss of mature stabilizing plant cover, loss of agricultural crops and loss or removal of unconsolidated (loose) top soil layer. In nature, elevated population of wild grazing animals, running water and wind erosion cause soil erosion and consequently leading to desertification (Osman 2014). In Mediterranean climatic regions such as Southern Europe, South Africa and Western Australia precipitation gets reduced, soil moisture levels decline and ultimately productivity goes down. Loss of agricultural lands in United Arab Emirates, Egypt, Kuwait and Iraq, where 12-15 % of their fertile delta lands are lost due to drought (Sadek 2010). In addition to crop failure, pest outbreak also affects the economy.

Abrupt and extreme weather

Increased frequency of catastrophic events such as storms, typhoons, droughts, forest fires, land slides and flooding events could cause several adverse effects to man and the environment. This affect the agriculture and result in food shortages cause displacement of people from affected regions and spreading of diseases.

Flood and landslides: Both flood and landslides cause large death and injury in human population such events are increasing with the global climatic change in countries such as Bangladesh, Khartoum, Netherlands, Sri Lanka, Egypt and Sudan.

Hurricanes and Tornadoes: Ocean temperatures increasing due to global warming which subsequently increases the wind speed when maximum wind speed exceeds 74 miles per hour called hurricanes in Atlantic and typhoons in pacific e.g. Hugo hurricane in Oman. Tornadoes are more frequent in USA, causes mass destruction to lives, properties and crops (Union of Concerned Scientists, 2006).

Droughts: There are four categories of droughts such as meteorological (low precipitation), agricultural (lack of moisture for crop development), hydrological (abnormal surface & ground water supply) and socioeconomic (effect in the economy due to water scarcity) such events are common in Sahel region, more precisely East African countries such as Ethiopia and Sudan.

Forest fires: Earlier drying of forests leads to increased forest fires. Forest fires are more common in Australia and Indonesia during El-Niño events. Forest fires can naturally ignited by lightening, volcanic eruptions, spark from rock falls and spontaneous combustion. Anthropogenic slash and burn agriculture and exotic or invasive oily plants such as eucalyptus and pine trees naturally causes fires. It has been estimated between 1850 and 1980, 90-120 billion metric tons of CO_2 was released by forest fires ('Earth observatory,' n.d.); (McMichael 2003).

Heat waves: Heat waves killed more than 2500 people in India (by June 2015). Most affected regions are Andhra Pradesh, Telangana, Punjab, Uttar Pradesh, Odisha and Bihar. It also severely affected cattle and crop production.

Increase in epidemics: diarrhoea, cholera, dengue and malaria: There are direct physiological effect by heat and cold. High heat affects several in Indian states during the early 2015, sun stroke killed several people, continuous exposure can causes skin damage, eye disease, adverse effect on immune system and skin cancer, temperature increases blood pressure, viscosity and pulse, this increases the death related to cardio vascular disease and increased stress and malnutrition also adversely affect the health. Epidemics of water born (Diarrhoea, cholera and dysentery) and vector borne (falciparum malaria, vivax malaria, dengue, elephantiasis, yellow fever and West Nile fever, rodent borne diseases plaque, Lyme disease and tick born encephalitis and hanta virus pulmonary syndrome) diseases occur as flooding increases breeding places of mosquito vectors and also breakage/ blockage in water pipes, septic tanks, sewers, drainage and storm-water get leak and contamination in portable water sources. Developing countries such as Bangladesh often hit by such epidemics frequently after flooding event (McMichael 2003).

Impacts on biodiversity: Sustenance of plants and animals get affected and may leads to their extinction. Increased temperatures of land and ocean moved the habitat range of many species pole ward or upward from their current location such movements also accelerated by drought events and desertification. Species with restricted habitat requirement or sedentary (coral reefs) or limited climatic or geographical range (mountain top or Island habitats) are more vulnerable to climate change. This also may increase the net primary productivity as atmospheric CO_2 levels increases and opportunists (weeds) win the competition. Organisms of temperature dependent sex determination such as sea turtles, crocodiles, amphibians with permeable skin and eggs are more vulnerable. Species that are already at risk face extinction, many habitats such as wetlands, beaches, grass lands and seagrass beds disappear. Climatic change associated reduction in Arctic and Antarctic ice alter seasonal distribution, migratory pattern, nutritional and reproductive status of marine mammals, it also affects the plankton distribution, affects the marine food chain and causes loss of a keystone species which makes the entire food chain get collapsed. Long living species such as perennial trees slowly exhibit evidence of climate change and they slowly get recover. Changes in phenotype, breeding seasons, behaviour and patterns of migration (in birds and animals) are already observed (Adapted from Secretariat of the conservation on biological diversity, 2003).

Effects on coral reefs: Increasing temperature causes coral bleaching in various parts of the world and acidification of oceans affect the corals regard to their formation of skeleton, acidified waters cause difficulties in absorbing calcium from the water which is essential for shell formation and it also dissolves the reefs ('Tech Ocean Science', n. d.).

Shortage of water in already water scarce areas: This may increases additional energy expenditure for cooling and excavation of ground water or bringing river water or lack of fresh water also opens seawater desalination projects in draught affected areas. When U.S. has been strugglingby recent four year drought states such as California decided to open desalination plants in order to use sea water to mitigate the effects.

Permanent loss of glaciers and ice sheets: Ice sheaths in Greenland and Antarctica has shown decline in their mass. Greenland lost 150-250 cubic km of ice per year in the period between 2002 and 2006 and Antarctica lost about 152 km of ice in the period of 2002 to 2005. According to the 'NASA-GCC-Land ice' (2015) the loss of ice mass in Antarctica is at the rate of 147 billion metric tons of ice per year since 2003, this is 258billion metric tons per year in Greenland. Melting of permafrost leads to destruction of structures, landslides and avalanches.

Declining arctic sea ice: Snow plays a vital role to the environment by reflecting the sunlight back; this helps to reduce the warming. In addition, melting seasonal snow provides fresh water for the life and accrued soil moisture helps the growth of vegetation. However, increase melting of ice by global warming leads to spring time floods. According to the satellite data amount of spring snow cover in the northern hemisphere has declined over the last five decades. Arctic sea ice is declining at the rate of 13.3% per decade. Data further reveals, the lowest arctic ice extent was recorded in 2012 Antarctic ice shelves accounted for a mass loss of 1.089 trillion kilogram ice per year in the period between 2003 and 2008. Warm ocean waters melt the ice sheet from underneath (basal shelf melt) accounted for 55% of the ice shelf melts, it also changes the ocean currents (Shaftel 2015).

Reduced precipitation increases air pollution and high aero Allergens: allergen levels in dry areas increase in the air due to lack of precipitation in dry season. It also increases the human diseases of immune response such as allergy and respiratory diseases such as asthma and bronchitis rates, earlier blooming plants make this even worse.

Cultural heritage sites get destroyed rapidly by increased extremes of weather pattern: Increased air pollution elevates occurrence of acid rain which affects the calcareous and metal structures. Acid rain also affects the plants by causing chlorophyll loss which leads to loss of photosynthesis, organisms having calcium carbonate exo-skeleton and egg shells are also affected.

Acidification of oceans: Ocean acidification has lowered the pH of the ocean waters by about 0.11 units (SCOR 2009 as cited in 'Tech Ocean Science', n. d.). This is due to anthropogenic CO_2 emission and amount of CO_2 on upper layer of the ocean has been increasing by 2billion tons per year. Oceans have absorbed 1/3 of the CO_2 produced by human activities since 1800 and fossil fuel burning alone account for half of the CO_2 (Sabine *et al.* 2004 as cited in 'Tech Ocean Science', n. d.). If CO_2 emission levels continue unchanged, the future CO_2 levels will be high enough to lower the pH of ocean to 7.8 by the year 2100 (Royal Society 2005, cited in 'Tech Ocean Science', n. d.).

Increase in unemployment: Unemployment is common to both developing and developed countries. However, in underdeveloped and developing countries it may causes famine, decrease in wages, agricultural land loss, unprofitable fishing and halting of tourism, further, starvation, malnutrition and increased deaths in the areas of food shortage such as East African countries e.g. Ethiopia, Somalia, Sudan, etc. This also increases theft such as piracy off the coast of Somalia threatens international shipping navigation and business since 2005.

Emigration increases from poor or low lying countries to rich and wealthier nations: Migration to Europe, UK, USA, Canada, Australia and New Zealand from developing and underdeveloped word on a hike this includes illegal migration attempts. Every year hundreds of illegal migrants get caught and are returned to their home countries on their way to Australia by vessels, this is considered as illegal migration and they are imprisoned in their home countries.

Economical imbalance: Natural disasters caused by sea level rise, floods, cyclones, forest fires, landslides and drought cause destruction and severe damages to properties such as houses, roads, railways, airports, transportation, power lines, irrigation, drinking water purification systems, sea walls, bridges, dams, farm lands, etc. It hinders the food production and cause crop loss which makes intense burden in the agricultural countries. Increase epidemics also cause additional medicinal and disaster relief expenses and it also increases climate refugees, to handle increasing security threats and coping cost as to mitigate and bring down the level of greenhouse gases in the atmosphere also have an impact on countries economy and creates dependency on international disaster relief funding organization and developed countries. When hurricane 'Sandy' hit the East cost of U.S. and the Caribbean in October 2012 it affected millions of people and billion dollars worth assets (includes1.8 million structures and homes), the total economic loss exceeds $65 billion, tourism sector losses more than $ 1 billion and 10,000 jobs. Katrina in August 2005 caused destruction of assets worth an estimate of $ 40 billion and took a long time to recover; even eight years later New Orleans's population was only 72% of its pre-storm level reported by U.S. Census. Thai floods of 2011 affected the manufacturing of automobiles and electronics and caused estimated loss of $45 billion which heavily effected Thai's economic growth, companies such as Sony and Honda are among the most affected industries in the nation. Intergovernmental Panel Climate Change (IPCC) recently stated that even though it is difficult to measure the overall economic loss by the effects of climate change, roughly the aggregate loss of global economy have a more than 50% chance of being greater than 2% of Global GDP. It is important to consider that this value excludes the expenses taken to mitigate global warming and cost of research and preparation of disaster management relief efforts (Liverman and Glasmeir 2014).

Increased political conflicts: Due to degradation of natural resources such as water and land violence and terrorism get increase such as ISIS issue in Syria and effects on human security e.g. boarders of India-Pakistan is still unsecure due to intermittent clashes and environmental conflict due to depleting fish stock at Palk Strait between marine boarders of India and Sri Lanka.

Advantage of global warming

- Arctic, Antarctic, Siberia and other frozen regions of the earth experience more land for cultivation (opening of new lands) and more plant growth in favourable conditions.
- Northern Europe, Canada and Russia get benefited with increased harvest such as cereals, sugar beet, hay and potatoes.
- More sea transportation ways opens such as Canada's North West passage.
- Less energy and fuel requirement for warming up.
- Decrease in death due to freezing
- Longer the growing season could increase the agricultural production

(*Source*: Farhan 2015).

Carbon footprint

Total amount of carbon dioxide emission caused by individual, event, organization or product measured as carbon dioxide equivalent (CO_2e) or equivalent tons of carbon dioxide. Carbon footprint refers both direct e.g. fuel combustion in cooking and own vehicle transport (scope 1) and indirect emission of greenhouse gases e.g. purchasing thermal power (scope 2), purchasing food or products and employee's business travel (scope 3) (in Fig. 1)..

When we cook food, fuel contributes to carbon footprint by emission of $CO_{2.}$ When we buy food from the shop, food manufactured in one place and transported to another this causes CO_2 emission during transportation and packaging using electric machinery also contributes to carbon footprint (in case of thermal power generation), travelling in order to purchase food causes CO_2 emission from the transport, this also has to be considered (when we use private transport, its contribution is more than using public transport), and bag given with food also comprises its own greenhouse gas emission during its production and transportation thus cooking food contributes less to the carbon footprint than purchasing. In addition when food waste is disposed, waste is transported by municipal truck this causes CO_2 emission. During waste storage other greenhouse gas methane is generated due to anaerobic digestion and disposal methods such as incineration causes direct CO_2 emission to the atmosphere, open dumping also may leads to methane generation, thus every human activity is associated with greenhouse gas generation and contributes to the carbon foot print. ("Time for change," n. d.).

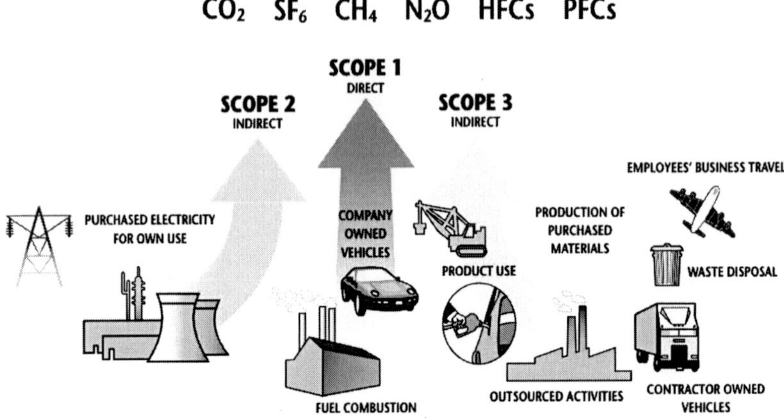

Fig. 1. Carbon footprint
Source: Bhatia and Ranganathan 2004

Mitigations for global warming

CO_2 mitigation

There are three basic ways suggested to lower the greenhouse effect. Firstly, stopping or reducing the emission of CO_2 into the atmosphere by ways such as using alternative green energy sources or renewable energy sources, upgrading the emission standards of the engine. Secondly, liquefying the CO_2 produced in the combustion and dump into the oceans, though it is a permanent disposal but it will result in ocean acidification which is currently becoming a major threat to aquatic life, thus underground injection or geologic sequestration and transportation/ storage of captured carbon in industries and power plants are quite safe methods. Thirdly, lowering the atmospheric CO_2 levels (post emission control) is done by increasing the sinks such as afforestation, reforestation and prevention of deforestation. Annually, about 2 billion tons of CO_2 ends up in oceanic organic deposits in the sea floor.

Air quality and emission trading: US EPA has proposed to reduce greenhouse gas emission, reduce emission from new vehicle such as up gradation to fuel efficient engines, reducing vehicular pollution via telecommuting and series of programs conducted by US EPA to reduce the vehicular emission.

Emission control during Beijing Olympics 2008, during the Olympic season 300,000 heavy emission vehicles (mostly trucks) were put away from the site, government encouraged public transport, rules allowed only some people to drive on certain days about 2 million vehicles were removed from roads, mobile data collection of CO_2 and soot in the atmosphere was done, as a result of the above action, presence of atmospheric carbon in that region went down by 33%.

Methods of carbon capture in power plants and industries

Post combustion capture (PCC)

This method involves separation of CO_2 from flue gas; solvent absorption using ammonia such as aqueous pure amines or blends of amines, in Alstom's Chilled Ammonia Process (ACAP), aqueous ammonium carbonate to bicarbonate reaction is used. Mono-ethanol-amine (MEA) in aqueous solution is used to capture CO_2 usually from boilers, Aker Clean Carbon is a mobile amine based facility, amino acid salt processes is the second generation method, amino acid salts has high absorption capacity than amines.

Adsorption methods are, using of a material where the CO_2 molecules get absorbed on to the solid surface e.g. 3X zeolites; this is comparatively advantages than liquid based absorption as regeneration energy is low, since the heat capacity of solid sorbent is lower than the aqueous solvents. Membranes are used to separate the CO_2 selectively, since CO_2 has high permeability than any other substances in the flue gas, however, it requires a pressure gradient for the separation; this is achieved by pressurizing flue gas on one side of the membrane and vacuuming the other side (Global CCS Institute 2012).

Pre combustion de-carbonization

This is achieved by providing 'synthesis gas' (mixture of H_2 and CO) for combustion where CO_2 is absorbed completely. Thus, the combustion occurs in the absence of CO_2. CO in the synthesis gas easily gets converted into CO_2 which is then captured using solvent. Here a hydrogen rich fuel is produced that facilitate the efficient burning in the turbine and minimizes the CO_2 emission.

Transportation of captured carbon dioxide can be done easily by regular transportation or shipment in a compressed cylinder (IEA Greenhouse Gas R and D Programme, n.d.).

Carbon sequestration

Carbon sequestration is a process providing long-term storage for captured carbon from industrial effluents, which helps to reduce the emission of carbon to atmosphere as CO_2. Captured compressed CO_2 can be injected underground using pipe line. Suitable geological formation for CO_2 sequestration are depleted oil and gas fields, solid, porous rock such as sandstone, shale, dolomite, basalt, or deep coal seams and saline formations. More precisely one or more layers below cap rock could be the ideal place which prevents the upward migration of CO_2 after being injected (Fig. 2) (EPA CCS 2015).

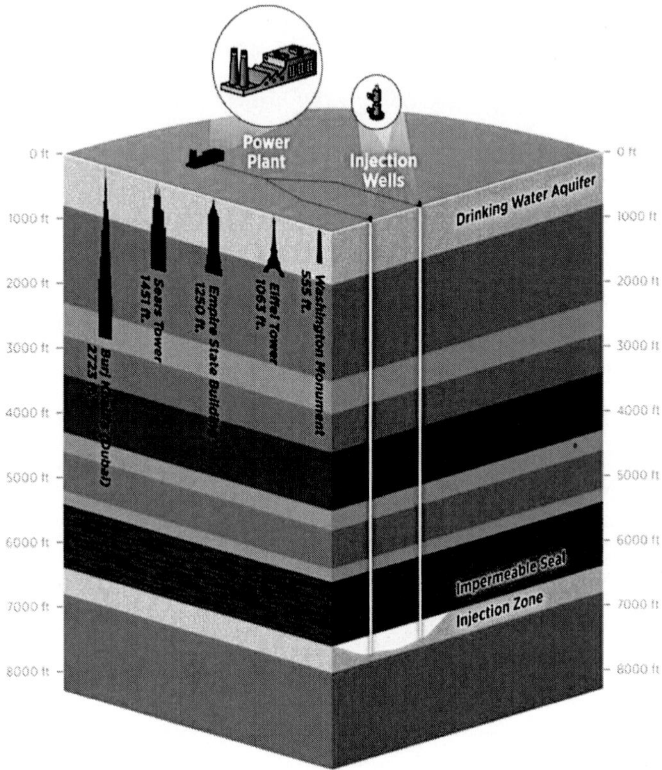

Fig. 2. Geographical location of carbon sequestration injection zone
Source: EPA CCS (2015)

NO_x mitigation

To reduce NO_x, methods such as Selective Catalytic Reduction process which has the NO_x reduction rate up to 80% where injection of reactive chemicals such as ammonia reacts with NO_x and convert into N_2 and O_2, changing air to fuel ratio and changing the combustion temperature. In automobile NO_x reduction, catalytic converters are used e.g. three way catalytic converters (1. conversion of NO_x into N_2 and O_2, 2. conversion of CO into CO_2 3. conversion of hydrocarbons into CO_2 and water) ('Reducing Acid Rain' US EPA 2012).

Absorption

Absorption is selectively isolating the pollutant; gaseous pollutants that can be dissolved within liquid scrubbers are coming under this category. In flue gas denitrification the mixing of nitrous oxides with water resulted with nitric acid compounds (which is a water and soil pollutant in liquid phase). In Selective Catalytic Reduction method ammonia is applied to the gas steam which reacts with the oxides of nitrogen at very high temperature (300 °C) in the presence of catalysts such as active Vanadium pentoxide and tungsten trioxide on a carrier of titanium which releases nitrogen and water.

Electrostatic precipitator

Negative corona is most preferred in industrial application as the industrial gases such as SO_2, CO_2 and H_2O have best ability to absorb free electrons and spark over voltage is higher in negative corona. However, negative corona generates higher level of Ozone; thus, it cannot be used in air conditioners.

Flare and thermo oxidizers

Flare stacks are used for burning off the flammable gas release. Generally used in petroleum refineries, natural gas processing plants and chemical plants. This also used to release the pressure of the equipment. Flares are designed for short-term combustion. To avoid most hazardous methane release during fermentation in beer factories flares are used to burn methane and releases in the form CO_2. Ground level flares are used in earth pits. Among thermal oxidizers regenerative thermal oxidizers are efficient up to 95%, the process is more simplified by the use of catalytic thermo oxidizers where the catalyst are used to reduce the ignition temperature and the reaction is employed in relatively low temperatures (reduction of 600 to 200 °C). There are ventilation air methane thermal oxidizer, thermal recuperative oxidizer and direct fired thermal oxidizer used for the relevant purposes ('Thermal oxidizer,' 2014).

Mitigation approaches for Global warming based on the sector of livelihood

1. Energy

(i) Increase energy efficiency in engines and boilers

(ii) Switching to low carbon fossil fuels such as natural gas

(iii) Introducing flue gas decarbonisation and carbon sequestration

(iv) Increasing the use of nuclear energy

(v) Increase the use of renewable energy sources

(vi) Conserve energy during the usage

2. Industry

(i) Reduce greenhouse gas emission such as methane

(ii) Reduce the material content of manufactured goods

(iii) Switch to energy efficient technology

(iv) Transferring and sharing technology mainly from developed to developing countries

(v) Recycle

3. Transport

(i) Improving energy efficiency of vehicles

(ii) Reducing vehicle emission

(iii) Reduce the vehicle weight and size to maximize the performance

(iv) Changing land use patterns and life styles to reduce transport requirements

(v) Integrate transport policies

(vi) Promote public transport option than personal vehicles

(vii) Promote greener vehicles such as electric cars

4. Agriculture

(i) Develop new management techniques to reduce tillage, recycling of crop residues, mixed cropping and avoid monoculture

(ii) Restoration of wetlands

(iii) Improve energy efficiency

(iv) Improve nutrition of ruminants and reduce methane generation

(v) Reduce biomass burning

(vi) Manage fertilizer use to reduce nitrous oxide production

5. Forestry

(i) Substitute burning of fuel wood for fossil fuels

(ii) Improve energy efficiency

(iii) Reduce biomass burning

(iv) Conserve CO_2 in living trees

(v) Afforestation and reforestation

6. Government

(i) Develop industrial land use plan to minimize energy consumption

(ii) Planning disposal of waste material to reduce production of methane and CO_2

(iii) Provide disincentives (tax) for excess energy consumption

(iv) Provide incentives for energy consumption and minimizing greenhouse gas emission such as reduce the taxes for electric and hybrid vehicles.

(v) Improve energy efficient, recycling and proper waste disposal

Source: Kemp 2014

Emission trading

It is an administrative approach of pollution control by giving economic incentives. Emission trading facilitates a market where parties can buy allowance or permits for emission of particular pollutant or credits are given for reduction of pollutants. There are several emission reduction projects under cap and trade scheme, here a cap (limit) values is defined for GHG emission.

Kyoto protocol 1997

This is an amendment to the U.N. Framework convention on climate change, parties are committed to bring down the emission of six greenhouse gases (Carbon dioxide (CO_2); Methane (CH_4); Nitrous oxide (N_2O); Hydrofluorocarbons (HFCs); Perfluorocarbons (PFCs); and Sulphur hexafluoride (SF_6) (UFCCC 2014) or reducing their production as the listed gases cause global warming, parties agreed to fund research on climate change and promoting alternative energy sources in both developed and developing nations, it also includes several international partnerships such as Asia- Pacific partnership on clean development and Climate. First commitment period was between 2008 and 2012 here 37 industrialized nations and the European community committed to reduce GHG emissions to an average of 5% against 1990 levels. Then Doha amendment was added in 2012, here parties committed to reduce GHG emissions by at least 18 % below 1990 levels in the period from 2013 to 2020.

Afforestation and reforestation

Forests are the natural sink for CO_2 and release oxygen. Maintaining or protection against forest degradation can be successful by planting, site preparation, tree improvement, fertilization, uneven aged stand management, thinning, pruning, weeding, cleaning, liberation cutting or other appropriate silviculture techniques, maintaining or increasing the landscape level carbon density using forest conservation strategies, longer forest rotations, fire management and protecting against insect pests (IPCC 2007).

Most popular afforestation and reforestation programs

Forest plantation in a land which does not have any forest in last 50 years of history is called afforestation, if it has an occurrence of forest within last five decades then it is reforestation.

- China annually increased its forest cover by 11,500 square miles, an area the size of Massachusetts, according to a report from the United Nations in 2011. China's Great Green Wall was designed to plant nearly 90 million acres of new forest (Jon 2012).
- Reforestation in Korea: Between 1961 and 1995, stocked forest land went up from 4 million ha. to 6.3 million ha. Total timber rose from 30.8 million cubic meters in 1954 to over 164.4 million cubic meters in 1984. By 2008, 11 billion trees had been planted in South Korea and about two-thirds of South Korea is now clothed with forest.
- Reforestation in Tanzania: the Kwimba Reforestation Project: During the nine year period of the project's run, over 6.4 million trees were planted.
- Reforestation in Mexico: Centre for Integral Small Farmer Development in the Mixteca region reforested with 1 million trees covered more than 1000 ha.
- Reforestation in the United States: the Appalachian Regional Reforestation Initiative (ARRI): 60 million trees have been planted on about 87,000 acres of active mine sites in Appalachia under ARRI's guidance.
- Reforestation in Colombia: Gaviotas villagers have successfully reforested about 20,000 acre, as a result rainfall has increased by 10%. (Sustainablog 2011).
- Japan after the World War II have done intensive reforestation from 1950-1970, during that period professional silviculture spread out in every Japanese village (Gerry 2005)

Forestry projects under the clean development mechanism (CDM) of the Kyoto Protocol

General features of this mechanism are reforestation of native forests, plantations for timber, agro forest or multipurpose tree plantations and healing barren lands.

Kyoto Protocol governs Land Use, Land Use, Change and Forestry (LULUCF) and modalities and procedures for CDM. Organizations such as International Tropical Timber Organization (ITTO) carried out the task according to the discussed strategies.

Role of international tropical timber Organization (ITTO)

International organizations such as ITTO, encourages conservation, sustainable development, use and trade of forest resources. It has 59 members represent about 80% of tropical forests and 90% tropical timber trade worldwide. ITTO collects analyses and circulates data on production and trade of timber and allocates funds since 1987. It has funded more than 750 reforestation and afforestation projects valued US$290 million. Donors are mostly Japan, Switzerland and the USA.

CDM projects

Timothy *et al.* (2006) reported following projects running under CDM.

- Pearl River Watershed Management, China: This project proposes to alleviate local poverty and reduce threats to forests by afforesting 4,000 hectares in the Guangxi Zhuang. Project also includes half of the Pearl River basin.
- Pico Bonito Forest Restoration, Honduras: This is a pilot project on agroforestry to support small scale farmers of 20 villages within the Pico Bonito National park buffer zone of 2,600 ha. Main roles of the project are introducing agroforestry for small scale farmers, reforestation to promote conservation, establishment of sustainable commercial grade plantation.
- San Nicolás Afforestation project: This project includes both forest and agroforestry plantation in an abandoned pasture land of 8,730 ha in San Nicolás, Colombia.

Agroforestry

According to U.S. Department of Agriculture (USDA) Agroforestry is a unique land management approach that intentionally blends agriculture (crop and animal farming) and forestry to enhance environmental, economic and social benefits. It is an intensive land management system that optimizes the benefits of biological interaction when trees, shrubs, crops and live stocks are combined. To satisfy the goals of agroforestry the management should satisfy the four "I"s. They are Intentional, Intensive, Integrated and Interactive. In the United States, there are five categories of agroforestry techniques in practice are silvopasture, riparian forest buffer, alley cropping, wind breaks and forest farming.

History of agroforestry

Agroforestry practices were first reported by O. F. Cook in a 1901 report of the U.S. Department of Agriculture mentioning that coffee plants were grown under shade trees (resembles alley cropping), he also justified the benefits of leguminous shade trees as they improves soil fertility by nitrogen fixing. According to Holdridge *et al.* (1977), Oelbermann and Smith (2011) *Alnus acuminate* K. planted in pasture high land of Costa Rica which could be considered as an initial effort of sivopasture. Budowski (1987) as cited in Oelbermann and Smith (2011) reported that use of *Cupressus lusitanica* L. windbreaks in highlands and the use of *Coediala lliondora* L. in lowland pastures in Costa Rica, in another work he also mentioned about secondary forest management in abandoned coffee plantation. According to Kass*et et al.* (1995), Oelbermann and Smith (2011) alley cropping was established in Costa Rica as new agroforestry practice in early 1980s as crops were grown between rows of trees. Similarly use of life-fences and increased biodiversity was mentioned in Harvey *et al.* (2005) as cited in Oelbermann and Smith (2011).

Classification of agroforestry in six structural categories

Type	Example
Crops under tree cover	Scattered trees in crop land, shade trees in plantation crops, park lands, crops in orchards, plantation crop combinations
Agroforest	Agroforestry home gardens, village forest gardens, mixed woodlots, agroforestry buffer zones
Agroforestry in linear arrangement	Wind breaks and shelterbelts, boundary planting, live hedges, living fences, soil conservation hedgerows, , woody strips, alley cropping, roadside planting
Animal agroforestry	Grazing or browsing in wooded or forest land, tree planting in rangeland, browse banks, animal feeding with collected browse
Sequential agroforestry	Shifting cultivation, tree-improved fallows, taungya
Minor agroforestry techniques	Sericulture, Lac production, apiculture with trees, tree-based aquaculture

Source: Torquebiau (200)

Agro forestry an overview

The agroforestry system provides continuum of options to land owners towards their own goals such as interplant crops, animal forage or trees. Agroforestry provides sustainable solutions to mitigate immediate consequences of global warming, food insecurity, rural poverty, struggle to develop new sources of economic growth and achieving sustainable development. It also provides means for sequestration of carbon in agro-soils, woody perennials and transferring carbon credit through market structures. Having trees along with annual crops increase the carbon sinking ability of the system as trees contribute more photosynthesis. Increasing human population increases the need for more land for agriculture. However, the annual crops are most vulnerable and often associated with immense irrigation practices that create more wetlands and subsequently more methane release to the atmosphere.

Together with perennial trees annual crops can find increased advantages such as less chance for pest out breaks as natural predators of pest species are abundant in microhabitats of large trees, less chance for spreading of diseases, more opportunities for natural means of reproduction or exposed to variety of pollinators such as insects and birds. Furthermore, shade provided by canopy reduces the loss of water from soil by evaporation; ultimately this reduces the irrigation time and also lessens the water requirements.

Trees can tolerate intra-annual weather changes of great magnitude and higher duration than annual plants. In addition, they provide more high valued products such as fruits, edible leaves, oil bearing nuts, medicinal species and feedstock for the production bio-fuels. This helps tremendously in rural economy as well as national export. Trees in farms also provide environmental services such as biodiversity conservation and watershed, they also mitigate environmental issues such as soil erosion, air pollution, odour, sedimentation in water bodies, etc., and the system also improves soil fertility by trapping soil nutrients and improving cycling of mineral, it protects downstream fishery by preventing sedimentation and riparian forest buffers further support in improving biodiversity in the fresh water aquatic systems (Pearce and Mourato 2004; Oelbermann and Smith 2011).

While satisfying the goals of Kyoto protocol and afforestation and reforestation projects such as CDM and voluntary carbon methods e.g. Chicago climate exchange, agroforestry also provide means for agriculture and support various other farm products. In case of Africa agricultural lands account for over double the area of forest lands. However, this shows significant reduction in natural carbon sequestration potential in the system, this clearly gives us a hint that the "future of the trees is depending on farms".

Agroforestry system provides a constant food supply or income as it gives variety of different products while maintaining regular employment. It arrests soil degradation and enhances effective cycling of nutrient flow, water and radiation. IPCC third Assessment report on Climate Change (IPCC 2001) stated that agroforestry can both sequester carbon and produce a range of economic, environmental and socioeconomic benefits. Microhabitats vastly increase the organism diversity and also enable the effective more closed cycling of nutrients by extraction of nutrients from deep soil horizon and continuous energy flow.

Agroforestry farms improve soil fertility through control of erosion, maintenance of soil organic matters, physical properties and increased nitrogen levels. Moderate microclimate in the system enables maximum genetic potential of many crop varieties that can only be produced when environmental parameters are close to optimum. Large trees in the system suppress the heat stress and bring favourable alterations in the micro climatic conditions by influencing radiation flux, air temperature, wind speed, saturation deficit if understory crops found, all these factors determine the rate and direction of photosynthesis and subsequent plant growth, transpiration and soil water usage.

Shade provided by trees support the growth of heat sensitive crops such as coffee, cocoa, ginger and cardamom. Wind breaks or shelter belts or parallel rows of trees of the landscape helps to reduce the evaporation and physical damages. Mulches reduce soil temperature and avoid nutrient depletion. And mixture of different crop trees maximize the resource utilization efficiency as the nutritional requirement of each species differ, this feature also maintain the soil minerals in a balance. As stated by Dewenter *et al.* (2007) and Kumar (2016) removal of shade trees increased soil surface temperature by 4 °C and relative humidity at 2 m above ground gets reduced by 12%. Agroforestry systems provide permanent cover to the soil while it consumes soil water and nutrients, system offer a promising income towards sustainable diversification of biochemical properties.

Agroforestry facilitates efficient use of rain water. World Agroforestry Centre has developed a system when 3 components that can be used in combination or separately 1) nitrogen fixing legume tree fallows 2) indigenous rocks phosphate in phosphor deficient soils 3) biomass transfer of leaves of nutrient accumulating swabs.

Agroforestry as CO_2 sinks

Trees in the agricultural land significantly sequester the atmospheric CO_2. Trees help to reduce the atmospheric CO_2 in following ways.

1. Conservation of existing carbon pools through practice such as avoiding deforestation and using alternative to slash and burn.
2. Sequester through improved fallows and integration with trees.
3. Substitution through biofuels and bioenergy plantations to replace fossil fuel use. (FAO, n.d.)

Studies revealed that agroforestry systems act as efficient carbon sinks (Alberecht and Kandji 2003). Assuming mean carbon content of above ground biomass of 50%, average carbon storage by the agroforestry system has been estimated to be 9, 21, 50 and 63 tons C ha^{-1} in semi-arid and sub humid, humid and temperate regions respectively. Worldwide it is estimated that 630 x 10^6 ha are suitable do the agro practices. Even in dry regions such as Sudan-Sahel zone of West Africa, technology could significantly contribute to curb land degradation and improving productivity. Studies of Smith and Wollenberg (2012) and IPCC (2000) revealed that agroforestry systems have 3 to 4 times more biomass than treeless cropping of the traditional agriculture. In a study held in India showed that the average carbon sequestration potential of the agroforestry system is 25t ha^{-1} over an area of 96 million ha. However there is substantial regional variability. It also showed that agroforestry systems increase the income, employment opportunities and livelihood security of the village community (Basu 2014).

A study Unrah *et al.* (1993) held in sub Saharan Africa where different types of agroforestry systems such as pasture, fruit, fuel wood, timber and shelter belt estimating total area of1548X 10^6 ha, where 4.5-19 t C ha^{-1} of total sequestration into biomass was recorded. Similarly, a study by Houghton *et al.* (1993) in the same region depicted average accumulation of 59 t C ha^{-1} over 888 × 10^6 ha area. A study on tropical region by Albrecht and Kandji (2003) revealed that carbon sequestration potential of agroforestry system estimated between 12 and 228M g ha^{-1}with a median of 95M g ha^{-1} it is also mentioned that suitable area for agroforestry in the earth is estimated as 585–1215×10^6 ha. Thus it has been suggested that the future of agriculture should find a way to mitigate the global warming and subsequent climate change by the effective implementation of agroforestry techniques.

Table 1. Potential carbon storage for agroforestry systems of different eco-regions of the world.

Region	Eco-region	System	t C ha^{-1}
Africa	Humid tropical high	Agrosilvicultural	29-59
South America	Humid tropical low	Agrosilvicultural	39-102 [a]
	Dry Low lands		39-195
Southeast Asia	Humid tropical	Agrosilvicultural	12-228
	Dry Low lands		68-81
Australia	Humid tropical low	Silvopastoral	28-51
North America	Humid tropical high	Silvopastoral	133-154
	Humid tropical low	Silvopastoral	104-198
	Dry Low lands	Silvopastoral	90-175
Nothern Asia	Humid tropical low	Silvopastoral	15-18

[a] Carbon storage values were standardized to 50 year rotation

Source: Dixon *et al.* 1993; Krankina and Dixon 1994; Schroeder 1993; Winjumet *et al.* 1992; Albrecht and Kandji 2003

In another study by Kim *et al.* (2016) showed that the average biomass carbon sequestration by trees and crop plant above ground is 13.9± 4.8t. CO_2 ha^{-1}y^{-1}

and average biomass carbon sequestration below ground by tree roots is 4.4 ± 1.5t. CO_2 $ha^{-1}y^{-1}$and average biomass carbon sequestration below ground by crop plants is 8.1 ± 4.4 t. CO_2 $ha^{-1}y^{-1}$and average biomass Methane sequestration below ground is -0.003 ± 0.03t CO_2 equivalents $ha^{-1}y^{-1}$.

However, it also accompanied by emission of other greenhouse gases such as N_2O emission -0.8 ± 3.2t CO_2 eq$ha^{-1}y^{-1}$ the net greenhouse gas sequestration by the biomass is 27.2 ±13.5 t CO_2 eq $ha^{-1}y^{-1}$ at least for the first 14 years after establishment. Agroforestry in humid tropics contain 50-75 t C ha^{-1} this is far less than forest but higher than annual crop plants (see graph 1)

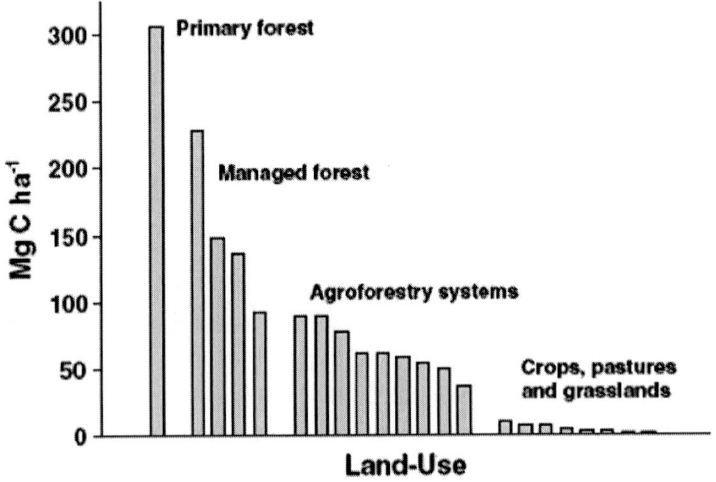

Graph 1. Carbon sequestration vs. Land use (Mg C ha^{-1}; Mg: Mega gram is equal to tons: t)
Source: Verchot *et al.* 2007; Torquebiau 2013

Unlike the more perennial system in the humid tropics, improved fallows are mostly short rotation, due to that reason they sequester much less carbon above ground. However, if the time average above ground carbon is considered they store substantial quantities of carbon compare to degraded land, croplands or pastures (Alberecht and Kandji 2003). In addition, agroforestry systems provide the unique opportunity to utilize the resources to achieve maximum return from unit of available soil, water, nutrition and sunlight in economically and ecologically sustainable manner. Highly productive agro forestry system including silver pasture systems contributes a major role in carbon sequestration in soils and woody biomass (above and underground). In Latin America traditional cattle management involves grass monocultures degrade 5 years establishment, release significant amount of CO_2.

Land management practices increase the uptake of CO_2 or reduce its emission decreases the atmospheric carbon significantly. If trees get harvested, this is followed by subsequent regeneration of the entire area. And sequestered carbon is locked through non-destructive (non-CO emitting) use of such wood. Even though afforestation and reforestation in degraded natural forest are useful options but agroforestry is much more beneficial and attractive.

1. It sequesters carbon vegetation & possibly in soils, depends on the pre-conversion of soil carbon.
2. More intensive use of land for agriculture production this lowers need for slash and burn of shifting cultivation which contributes deforestation.
3. Wood products produce under the system serve as a substitute for similar kind of products that are unsustainably harvested from natural forests.
4. Agro forestry increases the income of farmers and it reduces the incentives for further extraction from the natural forest for income augmentation.

It has been evidenced that agroforestry systems are promising management practices to increase aboveground soil carbon stocks to mitigate greenhouse gas emissions. (Umrani and Jain 2010).

Types of agroforestry practices

1. Silvopasture

Intentionally combining trees with forage and livestock production gives silvopasture system. Here saw logs are managed for high value; provide shade, shelter for livestock and forage. Trees also reduce stress and sometimes support forage production. Nutrients keep recycling in the system as leaf litter and animal manure bring the absorbed nutrients back to soil. Economically planting pines for pulp, hardwood for timber, nuts and fruit trees facilitate additional income form the same landscape. However, Silvopasture needs additional management skills. Nut crops can be another intermediate product (Agroforestry 2013).

Silvopasture system is created by either a) establishing trees into existing pasture or b) establishing forages in the woods.

a) *Establishing trees into existing pasture:* Success of the system depends on right tree choice which supports profitable livestock management and also beneficial as timber and forest products. Tree tubing and electric fencing protect young trees.

b) *Establishing forages in the woods:* This is to establishing sivopasture system by intensively manipulating forest environment. Factors such as light intensity, rotational grazing, and adjusting soil fertility are monitored and managed more frequently through tree shaping, thinning, pruning and more advanced agricultural and forestry techniques.

Trees: longleaf pine, shortleaf pine, loblolly pine, slash pine, ponderosa pine, Douglas fir, hardwoods, fine hardwoods and nut trees.

Forages: legumes and grasses

Cool season: grasses, legumes

Warm season: Bahia grass, Bermuda grass, Switch grass, Big bluestem, Little bluestem, Indian grass, legumes

Livestock: cattle, goats, sheep, horses, poultry, specialty

Source: the walden effect, n.d.)

Source: farm1, n.d.)

2. Riparian forest buffer

- Pasture grasses and legume plants helps to reduce soil erosion, produce quality forage in the understory and reduce unwanted trees.
- Natural recycling of elements such as Nitrogen, phosphorus and potassium from gazing herbivore to soil reduces the fertilizer requirements.
- Long term income goals can be met by techniques such as thinning of trees and pruning in order to increase the tree and forage growth.
- Production of additional products such as nuts and fodder
- Well distributed water within the system
- Grazing livestock reduces the maintenance cost as herbicide usage and mowing are not needed.

- Trees and livestock integrate and diversify entire farm system.
- Intensive management practices improve the growth of quality trees.
- Shading reduces stress and it improves animal productivity
- Enhance the habitats for wildlife and provisioning environmental conservation services.
- It improves scenic beauty of the area (aesthetic value).
- Improve plant vigor as faster and healthier tree growth, improved nutrient cycling and high quality forage.
- Lower animal stress as increased weight gain, milk yields, higher conception rate and lower the veterinarian cost.
- Reduce the risk of wildfire
- Products are meat, milk, hay, saw, timber, veneer logs, posts and poles, firewood, pulpwood, harvested game, organic mulches, nuts, fruit, ornamental flowers and greenery, mushrooms, etc.

Challenges of silvopasture

According to Zamora and Wyatt (2016) challenges of silvopasture are as follows:

- Distance and access to water competition for water between trees and pasture (seasonal irrigation may be needed for nuts and fruit trees).
- Challenges establishing young trees
- Challenges introducing forages to existing woodlands
- Maintain proper light intensity
- Fencing is quite labor intensive and costly

Riparian forest buffer

Here trees, shrubs, grasses and forbs are planted in strips or multiple rows along rivers, streams, lakes, wetlands and areas with groundwater discharge that can support woody vegetation as they are planted along the water bodies, they act as filter strips in order to prevent pollution form agro-runoff, improve habitats of wildlife and provide income for land owner from conservation, harvesting plants and timber. Width, layout and plant composition depend on flood plain characters, goals of land owner and requirements of the conservation programme. Zonation consists of three zone buffer system influenced by its functionality.

Figure Riparian Forest buffer (*Source*: "Forestasyst a", n.d.)

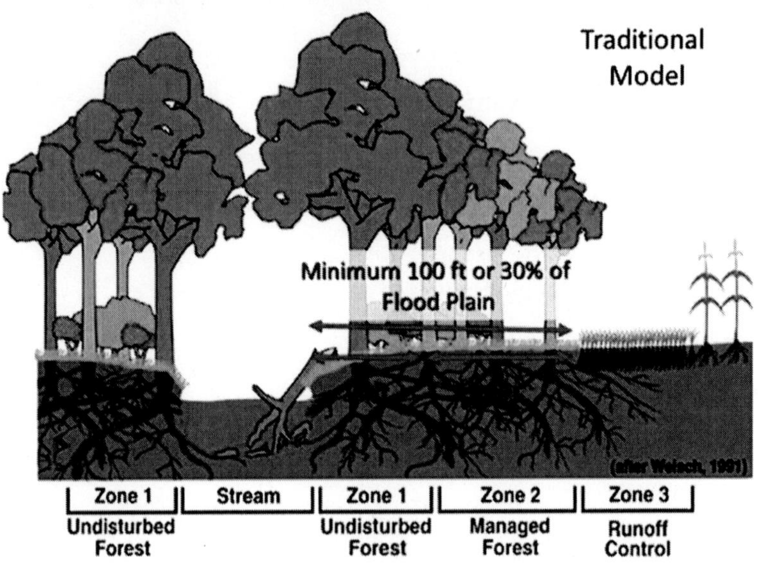

(*Source:* NAC 2016; Zamora and Wyatt 2016)

Fig. Riparian forest buffer

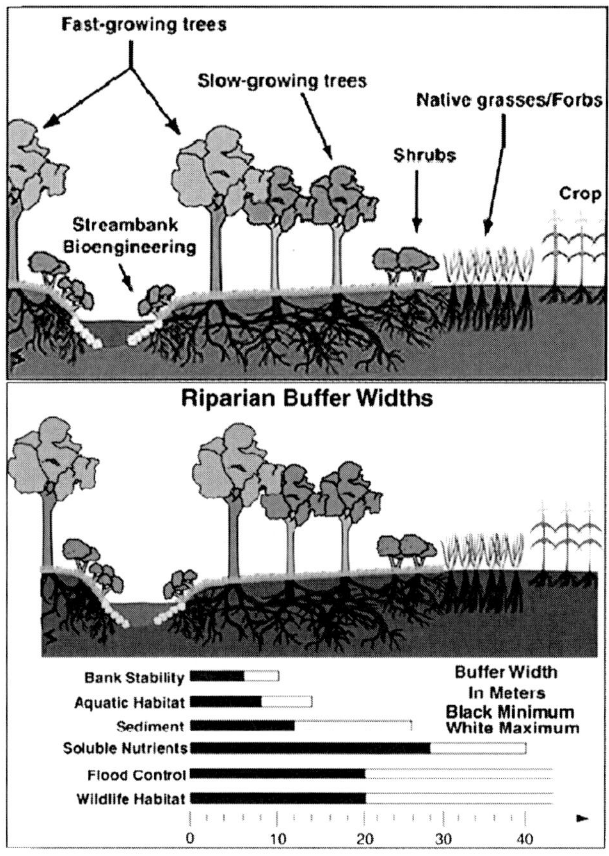

Riparian Forest Buffer, n.d.

Benefits of riparian forest buffer

- Maintain water quality by reducing the amount of sediments, excess nutrients, pesticides and other pollutants entering water bodies.
- Protect the ground water aquifers from the leaching pollutants.
- Flood plain protection: Slows flood water and decrease entire water volume of the stream
- Stabilizes stream banks and shoreline through root absorption
- Provide shade, shelter and food for fish and other aquatic biota.
- Provide detritus and large woody debris for aquatic and terrestrial fauna.
- Provide habitats and corridors for plants and animals

- Restore natural riparian communities
- Provide habit for pollinator species
- Prevent stream bank soil erosion
- Increase carbon sinks

Table 1. Benefits of riparian buffer or shelter belts based on type of flora

Benefits	Grass	Shrubs	Tree
Stabilize bank erosion	Low	High	High
Filter Sediment	High	Low	Low
Sediment	High	Low	Low
Aquatic habitat	Low	Medium	High
Range/pasture/prairie wildlife	High	Medium	Low
Forest wildlife	Low	Medium	High
Economic products	Low	Medium	High
Flood protection	Low	Medium	High

Source: Design of Riparian Forest buffer 2010

- Furnishes woodland recreational opportunities such as fishing, bird hunting, hunting, hiking and camping.
- Riparian forest buffers provides opportunities for additional earnings from timber, pulp, forage, firewood, specialty woodland products such as berries, nuts, mushrooms, greeneries, medicinal plants, and decorative floral materials e.g. flowers, boughs, stems and vines.
- Creates good stewardship mainly along recreational streams
- Straightens irregular fields, keep farm machinery away form stream banks and avoid the need to plant end rows where crop yields are over due to soil compaction
- Provides barrier against dust, noise, odor and light pollution
- They improve scenic beauty (aesthetic value).

Long term incomes of riparian buffer are

- Production of high value hardwoods such as Walnut, Oak and Maple
- Hybrid poplars grown for pulp wood saw logs or biomass
- Nuts and Berries from trees and woody shrubs (short term)

Short term income opportunities

- Decorative woody floral- ready in two years
- Barriers from woody shrubs

(*Source:* Brantly 2014)

Challenges of riparian forest buffer

- Trees and shrub species may provide alternate hosts to pest species
- Using multiple seed and seedling sources improves genetic diversity
- Herbicide tolerant varieties are considered as adjoining fields are subjected to herbicide application
- Layout, density and location of the buffer should mimic natural riparian forest
- Sites where the function of drains desired, penetration of woody root may eventually plug the underground structure (Riparian Forest Buffer 2015).

3. Alley Cropping

In alley cropping system, an agricultural crop is grown simultaneously with rows of long-term trees to provide annual income while the tree crop matures. Here trees and shrubs are planted in two or more sets of single or multiple rows with agronomic horticultural or forage crops cultivated in the alleys between the tree rows. Fine hardwoods such as walnut, oak, ash, and pecan are favorable species in alley cropping systems and can potentially give high-value lumber or veneer logs. Alley spacing depends on type of crop plants to be grown. Sun loving corn and soybean need more space between canopies. In between denser canopy shade tolerant varieties such as ornamental ferns and mushrooms can be harvested. Space also allows the equipment used for cultivation and harvest.

Source: Alley-cropping-corn-walnuts, n. d.)

Factors such as amount of rainfall, compatibility of trees and shrubs with the crop plant to ensure minimal competition for water, nutrients and light, spacing between and within rows, direction of sunlight and maintenance and use of equipment to be considered in the design prior to the establishment.

Benefits of alley-cropping

- Reduce soil erosion when established in sloping areas thereby improving water quality.
- Trees trap nutrients and chemicals leaching off site.
- Alley cropping improves crop performance or forage quality from the increased soil productivity from the added organic matter as well as from microclimate of crops and tree shades. Water consumption by plants increased, alley crops are protected from physical damages by extremes of wind and subsequent sand dust, improves pollinator foraging and nesting habitats.
- Reduce the use of chemical fertilizers and pesticides. Weeds are better controlled in the system, high nutritional usage and cycling and improved soil fertility without the use of fertilizers.
- Minimized nitrogen leaching improves water quality. Nitrogen leached beyond the cropping root zone is often captured by tree roots.
- Enhance biodiversity by increased microhabitats.
- Biological corridors allow wildlife to access habitats and water sources.
- Maximizes the land usage.
- Increased net carbon storage in soil and vegetation.

Challenges alley-cropping

- Need maintains for weeds and moisture
- May remove land from annual production, depends on the tree crop
- Complicate herbicide application
- Wrong alley spacing leads to failure of the crop
- Once trees are planted at defined spacing it cannot be changed for long term or the crop variety should suit with the design
- Alley spacing narrows the use of equipments or large plowing machinery cannot be applicable in between trees.

- Compete with crops for water, nutrients and light
- Requires marketing infrastructure for woody plant products

Potential trees: walnut, pecan, chestnut, pine and poplar

Tropical: coffee, coconut palm, Leucaena, Eucalyptus and papaya.

Potential shrubs hazelnut or filbert, (nuts) Willow, dogwood (decorative florals) choke cherry, high bush, cranberry, currant, elderberry, gooseberry, Saskatoon, sugar apple, pomegranate (fruits)

Potential alleyway crops row/cereal crops (corn, soybeans, milo and wheat)

Forage crops (legumes and grasses) specialty crops (vegetables, fruits, flowers, medicinal herbs)

Biomass (energy and feedstock)

4. Wind breaks

Wind breaks are trees, shrubs and grasses planted in single or multiple rows to reduce wind speed in an agricultural area for multiple purposes. Wind breaks are mainly aimed to avoid soil erosion in farm land. However they are valuable during extensive drought and soil erosion. It modifies air flow, odour plumes, sound waves and microclimate dynamics. It acts a filter by trapping airborne sediments, snow, nutrients, pesticides, pathogens and volatile organic compounds. Modification of the wind depends on six key factors are:

1. Height: height determines the distance of the downwind shelters (protection zone).
2. Density: high densities gives maximum wind reduction but short wind shadow, moderate densities give less wind reduction but longer wind shadow.
3. Orientation: directly influence the area protected and effects vary with critical weather periods and direction of wind.
4. Length: due to changing wind direction even to protect small area long wind break has to be established.
5. Width: density, wildlife and trapping efficiency and capacity influenced by the width of the wind break.
6. Continuity: gap in the wind break can result in damage or complication in downwind.

Adapted from Wind breaks (n.d.)

(*Source:* Forestasyst b, n.d.)

Benefits of windbreaks

- Reduce energy costs from heating and cooling in rural farmsteads and homes (Brandle *et al.* 2004; Agroforestry in Minnesota 2015).
- Mitigate odour around livestock operation (Brandle *et al.* 2009; Agroforestry in Minnesota 2015).
- Reduce wind stress on crops and protect the plants from drying wind, wind damage or damage by sand dust, increase crop microclimate and yield.
- Protect livestock from wind stress: livestock prefer grazing on sheltered environment (Brandle *et al.* 2009) animals grown in less downwind area are reportedly lack of wind stress (in winter), healthy, increased feeding efficiency and produce high yield.
- Manage snow around roads and farmsteads (Gullickson *et al.* 1999; Agroforestry in Minnesota 2015).
- Manage snow on croplands (Brandle *et al.* 2004; Agroforestry in Minnesota 2015).

- Reduce soil erosion
- Provide shelter for gazing animals
- Provide ecosystem services, wildlife habitats and recreational opportunities
- Multipurpose wind breaks: bioenergy, enhance food security, more habitats for wild life and provide valuable products such as timber, fruits, biomass and floral products (Streed and Walton 2001; Agroforestry in Minnesota 2015).
- Moderate noise or act as green belt and reduce the noise
- Increase aesthetic values
- It screen the views
- Reduce airborne chemical drift
- Improve efficiency of irrigation by lowering the down winds.
- Increase carbon storage and reduce the impacts of greenhouse effect

Challenges of windbreaks

- Need maintains for weeds and moisture
- Application of pre-emergent herbicides or mulch
- Compete with crops for water, nutrients and light

(Brandle *et al.* 2009; Agroforestry in Minnesota 2015)

5. Forest Farming

Forest farming is cultivation of high value specialty crops under the protection of a forest canopy that has been modified to provide the optimum level of shade required to produce optimum growth. It is cultivation of marketable non-timber forest products (NTFPs) in woodlands with relevant shade and site conditions. According to USDA National Agroforestry center forest faring is intentional manipulation, integration and intensive management of woodland that capitalizes specific plant interactions to produce non-timber products. This method diversifies the forest management and enhances the associated income opportunities. Forest farming improves forest composition and structure and long term health, quality and economic value. Active management of interaction between trees and understory crops are needed to maintain long term health and productivity. In local term it can be known as multistoried cropping, woodland gardening, farming the forest and home gardens which is done intentional and deliberative manner.

Two methods are used in forest farming such as woods cultivated and wild simulated method. Wood cultivated strategy is an intensive development of forest farming crops or farm products and this is a high cost method which involves farming in the forest. However wild simulated focusses on mimicking nature less input required thus lower the cost. Involvement a forestry specialist is essential to manage thinning, pruning, harvesting and even marketing the forest products and often labor intensive.

- Economically valuable crops with ready market are grown, understory comprises ground cover, herbs, shrubs and trees to produce products such as mushrooms, floral, greenery, herbs, fruits, vegetables, botanicals, landscaping, crafts, medicinal plants, Nuts and Pollen are adapted to forest type and climate. Hard wood and pine canopy gives shade required to maintain moisture by fruit plants.
- Medicinal products can be categorized under dietary supplements, herbal medicines and ethno botanicals e.g. Black cohosh, Goldenseal, Ginseng, May apple, Bloodroot, Witch-hazel, Pacific yew and Saw palmetto
- Foods are Fruits, Syrups, Mushrooms, Nuts, Vegetables and Honeys
- Decorative are either landscaping or florals e.g. Bittersweet, Red-twig, Sword fern, dogwood, Forsythia, pine straw, pine cones, Galax, Moss, Boughs and Salal.
- Crafts are derived from various parts of trees e.g. vines, branches, foliage, cones, bark, roots, burls and culls.

Fig. Mushroom cultivation in a forest on decaying logs

(*Source*: Forestfarming book n. d.)

Factors to be considered are soil pH, organic matter, mineral nutrients, drainage, land formation (slope, aspect, surface drainage, erosion), temperature, precipitation, over story canopy cover, existing forest vegetation, pests, pathogens and beneficial organisms such as pollinators, soil microorganisms and earthworms.

Benefits of forest farming

- Enhance forest health
- Improves composition of the forest as well as value and diversity of existing forest
- Improves timber quality
- Diversify income opportunities, this increases cash flow
- Products form forest farmed products
- Provide habitats for animals
- Enhance cultural and social relations. Knowledge from indigenous community can be used.

Challenges of forest farming

- Informal or immature market
- Variable yields
- More intensive management required more skills
- Considerable capital investment needed
- Limited available information on crop production
- Volatile markets for some products
- Poachers may attracted by crops
- Some plants may be endangered or subject to exploitation

Conclusion

Increasing human population increases the land usage for agriculture food production, this causes subsequent deforestation in order to find lands for crop production, and this ultimately reduced the natural carbon sinks as the forest cover shrinks and this situation elevates the CO_2 levels in the atmosphere and subsequently facilitates global warming. Global warming is an increasing environmental issue, earth's average temperature has warmed by 0.8°C, and annually 30 billion tons of CO_2 is being released to the atmosphere. It also affects the livelihood of human population by creating economical imbalance. Intergovernmental Panel Climate Change (IPCC) recently stated that even though it is difficult to measure the overall economic loss by the effects of climate change, roughly the aggregate loss of global economy have a more than 50% chance of being greater than 2% of Global GDP. Carbon capturing and sequestration methods are being widely used to minimize the levels of CO_2 in the atmosphere. Clean development mechanism (CDM) developed under Kyoto protocol promotes greenhouse gas emission reduction in developing world. Integrated Territorial Climate Plan (ITCP) implementation, making green certification as mandatory, assure the control of greenhouse gases, designing appropriate cap limits, spread the energy conserving techniques and relevant pollution control mitigation strategies and increase public awareness on all known effects of global warming.

Agroforestry is the intentional combining of agriculture and trees to create sustainable farming and ranching system. Here the same land used for agriculture is also providing habitat for forest trees which carry out considerable role in carbon storage and wildlife conservation. It also further reduces the atmospheric CO_2 levels as it provides biomass as fuel substitute for fossil fuels. In biomass CO_2 emissions from bio combustion is approximately considered to be used in a closed loop between growing and combustion of biomass thus the growth and combustion cycle nets to zero or carbon neutral. Similarly, methane released from livestock can be balanced by number trees or carbon sinks in the same land usage area. Agroforestry mitigate climate change due to its permanent tree cover and varied ecological niches, its diversified temporal and special management option assure the high resilience when disturbed. Agroforestry practices are associated with various other benefits to crop plants, live stocks, forest trees and wildlife. Being sustainable system it is more environmentally friendly as it prevents or mitigates global warming, soil erosion, flooding, drought, soil degradation, spreading of toxic chemicals, loss of soil nutrients, sedimentation in water bodies, loss of ground water quality or ground water contamination, odour plumes, animal stress, environmental stress on livestock, etc. Agroforestry increases the product quality and quantity; it also provides additional enterprises and diversified means of income to the landlords, products such as timber, mushrooms, floral, fruits, greenery, herbs, vegetables, landscaping, botanicals, crafts, medicinal plants, nuts, pollen, etc. Livelihood of village or rural as well as indigenous communities can improve in ways such as social, environment, economic, knowledge and health perspectives.

References

Agroforestry (2013) Retrieved on 05.09.2016 from http://www.usda.gov/wps/portal/ usda/ usdahome? contentidonly=true&contentid=agroforestry.html

Agroforestry in Minnesota (2015) Extension. Retrieved on 06.09.2016 from http:// www.extension.umn.edu/environment/agroforestry/docs/windbreak.pdf

Albrecht A, KandjiST(2003) Carbon sequestration in tropical agroforestry systems. Agriculture, Ecosystems, and Environment 99: 15-27. *Soil erodibility control and soil carbon losses under short term tree fallows in western Kenya.* Retrieved on 06.09.2016 from: https:// www.researchgate.net/publication/40438339_Soil_erodibility_control _and_ soil_carbon_losses_under_short_term_tree_fallows_in_western_Kenya

Alley cropping (n.d.) *USDA National Agroforestry Center (NAC), LincolnNE.* [ppt.] Retreived on 06.09.2016

Alley-cropping-corn-walnuts (n.d.) Retrieved on 05.09.2016 from https://upload.wikimedia.org/ wikipedia/commons/e/e9/Alley_cropping_corn_walnuts.jpg

Basu, J.P. Agroforestry, climate change mitigation and livelihood security in India Proceedings of the Third International Congress on Planted Forests New Zealand Journal of Forestry Science201444(Suppl 1):S11DOI: 10.1186/1179-5395-44-S1-S11.

Bhatia P,Ranganathan J (2004) The Greenhouse Gas Protocol: A Corporate Accounting and Reporting Standard, Revdedn., World Business Council for Sustainable Development (WBCSD).

Brandle JR, Hodges L, Tyndall J, Sudmeyer RA (2009) Windbreak Practices. In: Garett HE (ed) North American agroforestry: an integrated science and practice, 2nd ed. American Society of Agronomy, pp 75–104.

Brandle JR, Hodges L, Zhou XH (2004) Windbreaks in North American agricultural systems. Agroforestry Systems 61-62:65–78.

Brantly S (2014) Agroforestry notes: Forest Grazing, Silvopasture, and Turning Livestock into the Woods [pdf.] Retrieved on 04.09.2016 from http://nac.unl.edu/documents/ agroforestrynotes/an46si09.pdf

Budowski G(1987) The development of agroforestry in Central America. In: *Agroforestry:*

Design of Riparian Forest buffer (2010) Retrieved on 06.09.2016 from http:// www.extension.umn.edu/environment/agroforestry/riparian% 2Dforest% 2D buffers%2Dseries/design%2 Dof%2 Driparian%2Dforest%2Dbuffers/

Dewenter SI, Kessler M, Barkmann J, Bos M, Buchori D, Erasmi S (2007) Tradeoffs between income, biodiversity, and ecosystem functioning during tropical rainforest conversion and agroforestry intensification. Proc Natl AcadSci USA 104:4973–4978.

Dixon RK, Andrasko KJ, Sussman, FA, Lavinson MA,Trexler MC, Vinson TS(1993) Tropical forests: their past,present and potential future role in the terrestrial carbon budget.Water Air Soil Pollut 70:71–94.

Carbon sequestration in tropical agroforestry systems. Retrieved on 06.09.2016 from: https:// www.researchgate.net/publication/222668060_ Carbon_ sequestration _in_tropical_ agroforestry_systems

Earth observatory (n.d.) Retrieved on 2015 from http://www.earthobservatory.nasa.gov/

EPA CCS (2015) Carbon dioxide capture and Sequestration, Retrieved on 15.09.2015 from http://www.epa.gov/climatechange/ccs/.

FAO (n.d.) Three Opportunities and Constraints in Agoforestry sector Retrieved on 07.06.2016 from http://www.fao.org/docrep/w7745e/w7745e04.htm#TopOfPage

Farhan S. (2015) Global Warming Props& Cons, Retrieved on 07.05.2015 from http:// www.hamariweb.com/arcticles/arcticle.aspx?id=285

Farm 1(n.d.) Retrieved on 05.09.2016 from https://farm1.staticflickr.com/590/ 21305721669_702e991a72.jpg

Forest farming (n.d.) *USDA National Agroforestry Center (NAC), LincolnNE.* [ppt.] Retreived on 06.09.2016

Forestasysta(n.d.) Retrieved on 05.09.2016 from http://www.forestasyst.org/Agroforestry/0803-0224.jpg

Forestasyst b (n.d.) Retrieved on 05.09.2016 from http://www.forestasyst.org/Agroforestry/0903-0014.jpg

Forestfarmingbook (n.d.) on 05.09.2016 Retrieved from https://forestfarmingbook.files.wordpress.com/2013/04/shitakeoysterlogyard 660x 330.jpg? w=660&h=330&crop=1

Gerry M (2005) The Eco Tipping Points Project, Japan - How Japan Saved its Forests: The Birth of Silviculture and Community Forest Management, Retrieved on November.22.2014 from http://www.ecotippingpoints.org/our-stories/indepth/japan-community-forest-management-silviculture.html

Global Carbon Capture and Storage Institute (2012) Retrieved on 2015 from http://www.globalccsinstitute.com/

Gullickson D, Josiah SJ, Flynn P (1999) Catching the snow with living snow fences. University of Minnesota Extension Service, St. Paul, MN.

Harvey, C.A., Villanueva, C., Villacis, J., Chacon, M., Monuz, D., Lopez, M., Ibrahim, M., Gomez, R., Taylor, R., Martinez, J., Navasa, A., Saenz, J., Sanchez, D., Medina, A., Vilchez, S., Hernandez, B., Pereza, A., Ruiz, R., Lopez, F., Lang, I., Sinclair, F.L. (2005). Contribution of live fences to the ecological integrity of agricultural landscapes. *Agriculture, Ecosystems and Environment* 111:200-230, ISSN 0167-8809.

HoldridgeLR, Greneke WC, HathewayWH, Liang T, Tosi Jr JA (1977) *Forest Environments in Tropical Life Zones.* Pergamon Press, ISBN 0080163408, Oxford, U.K.

Houghton RA, Unruh JD, Lefebvre PA (1993) Current land use in the tropics and its potential for sequestering carbon. Global biogeochem. Cycles 7: 305-320

IEA Greenhouse Gas R&D Programme (n.d.) (pdf file) Retrieved on 15.09.2015 from http://www.ieaghg.org/docs/general_publications/3.pdf

IPCC reported (2007) as cited in National Geographic News (2014), Retrieved from http://news.nationalgeographic.com/news/2004/12/1206_041206_global_warming_2.html. Accessed on 10.11.2014.

IPCC. Land Use, Land Use Change and Forestry. WMO, UNEP (2000)

Unruh JD, Houghton RA,Lefebvre PA (1993) Carbon storage in agroforestry: an estimate for sub-Saharan Africa Climate. Research 3: 39–52.

Jon R. Luoma (2012) China's Reforestation Programs: Big Success or Just an Illusion?, Retrieved on November.22.2014 from http://e360.yale.edu/feature/chinas_ reforestation_ programs_ big_success_or_just_an_illusion/2484/.

KassDLC, Araya JS, Sanchez JO, Pinto LS, Rerreia P(1995) Ten years of experience with alley farming in Central America. In: Proceedings of the International Alley Farming Conference, IITA, Ibadan, Nigeria, pp. 393-402.

Kemp DD (2004) Exploring Environmental Issues, London; New York: Routledge.

Kim DG, KirschbaumMUF,Beedy TL (2016) Carbon Sequestration and net emissions of CH4 and N_2O under agroforestry: synthesizing available data and suggestions for future studies. Agriculture Ecosystems & Environment 226:65-78.

Krankina ON, Dixon RK(1994) Forest management optionsto conserve and sequester terrestrial carbon in the RussianFederation. World Resour. Rev. 6, 88–101.*Carbon sequestration in tropical agroforestry systems.* Retrieved on 06.09.2016 from: https://www.researchgate.net/publication/222668060_ Carbon_ sequestration_ in_tropical _agroforestry_systems

Kumar V(2016) Multifunctional Agroforestry Systems in Tropics Region. *In Researchgate* Retrieved on 07.09.2016 from https://www.researchgate.net/publication/289250935

Liverman D, Glasmeir A (2014) what are the Economic Consequence of Climatic Change, the Atlantic.com Retrieved on 2016 from http://www.theatlantic.com/business/archive/2014/04/the-economic-case-for-acting-on-climate-change/360995/

OelbermannMaren,SmithCarolyn E (2011) Climate Change Adaptation using Agroforestry Practices: A Case Study from Costa Rica, Global Warming Impacts - Case Studies on the Economy, Human Health, and on Urban and Natural Environments, Dr. Stefano Casalegno (Ed.), ISBN: 978-953-307-785-7

McMichael AJ (2003) Global Climate Change and Health: An Old Story Writ Large. Climate Change and Human Health: Risks and Responses. World Health Organization, Geneva.

NAC (2016) Retrieved on 06.09.2016 from http://nac.unl.edu/publications/agroforestrynotes.htm

NASA Global Climate Change,' (2015) Global Climate Change: How do we know? Retrieved on 05.05.2015 from http://climate.nasa.gov/evidence/

Osman KT(2014) Soil degradation, conservation and remediation, Springer Ed. 2014.

Pearce D, Mourato S(2004) The economic valuation of agroforestry's environmental services. In: Agroforestry ad Biodiversity Conservation in Tropical Landscapes, Schroth, G., da Fonseca, G.A.B., Harvey, C.A., Gascon, C., Vasconcelos, H.L., Izac, A.M.N., pp. 67-86. Island Press, ISBN 1559633565, Washington, USA.

Reducing Acid Rain' US EPA (2012) Retrieved on 05.05.2015 fromhttp://www.epa.gov/acidrain/reducing/

Riparian Forest Buffer (2015) *NRCS, AR* [pdf.]Retrieved on 04.09.2016 from https://efotg.sc.egov.usda.gov/references/public/AR/Riparian_Forest_Buffer.pdf

Riparian Forest Buffer (n.d.) *USDA National Agroforestry Center (NAC), LincolnNE.* [ppt.] Retreived on 06.09.2016

Royal Society (2005) Ocean acidification due to increasing atmospheric carbon dioxide. London, UK.

Sabine CL, Feely RA, Gruber N, Key RM, Lee K (2004) The oceanic sink for anthropogenic CO_2. Science 305: 367–371.

Sadek, T. (2010.Nov. 24-25) *Sustainable livelihood approach and climate change.* Paper presented at Expert Group Meeting on Promoting Best Practices On sustainable Rural Livelihoods in the ESCWARegio, Beirut.UN-ESCWA.

Schroeder P (1993) Agroforestry systems: integrated land use to store and conserve carbon. Climate Res 3:53–60.

SCOR – Scientific Committee on Ocean Research (2009) Report of the Ocean Acidification and Oxygen Working Group, International Council for Science's Scientific Committee on Ocean Research (SCOR) Biological Observatories Workshop.

Secretariat of the conservation on biological diversity (2003), Interlinkages between biological diversity and climate change. Advice on the integration of biodiversity considerations into the implementation of the United Nations framework conservation on climate change and its Kyoto protocol. Montreal, SCBD, 154p. (CBD Technical Series no. 10).

Shaftel H(2015) Sea Level, Global Climate change, retrieved on 05.05.2015 from http://climate.nasa.gov/vital-signs/sea-level/.

Silvopasture (n.d.) *USDA National Agroforestry Center (NAC), LincolnNE.* [ppt.] Retreived on 06.09.2016

Smith P,Wollenberg E (2012) Achieving mitigation through synergies with adaptation E. Wollenberg, A. Nihart, M.-L. Tapio-Boström, M. Grieg-Gran (Eds.), Climate Change Mitigation and Agriculture, ICRAF-CIAT, London-New York, pp. 50–57.

Steppler HA, Nair PKR (n.d.) A Decade of Development. International Council for Research in Agroforestry (ICRAF), ISBN 92 9059 036, Nairobi, Kenya, pp. 69-88

Streed E, Walton J (2001) Producing marketable products from living snow fences. University of Minnesota Extension Service, St. Paul, MN

Sustainablog (2011) 5 Successful Reforestation Projects, Retrieved on November.22.2014 from http://sustainablog.org/2011/07/reforestation-projects/.

Tech Ocean Science, (n.d.), Retrieved on May.05.2015 from http://www.teachoceanscience.net/teaching_resources/education_modules/coral_reefs_and_climate_change/how_does_climate_change_affect_coral_reefs/.

The Walden Effect (n.d.) Retrieved on 05.09.2016 from http://www.waldeneffect.org/20131103silvopasture.jpg

Thermal oxidizer, (2014) from Wikipedia, Retrieved on 07/12/2014 from http://en.wikipedia.org/wiki/Thermal_oxidizer.

Time for change (n.d.) *Timeforchange.org* Retrieved on 01.05.2016 from http://timeforchange.org/what-is-a-carbon-footprint-definition

Timothy P, Sarah W. Sandra B(2006) Guide book for the formulation of afforestation and reforestation projects under the clean development mechanism, *International Tropical Timber Organization, Technical Series 25* pdf.

Torquebiau E (2000). A renewed perspective on agroforestry concepts and classification. Comptesrendus de l'Académie des Sciences / Life Sciences 323: 1009-1017

Torquebiau E (2013) Agroforestry and climate change, *FAO Webinar* [ppt.]

Union of Concerned Scientists (2006) Retrieved on 2015 from http://www.ucsusa.org/global_warming#.V80jNvRhxIg

Verchot L *et al.* (2007) Climate change: linking adaptation and mitigation through agroforestry. Mitig Adapt Strat Glob Change 12: 901-918

Wind breaks (n.d.) *USDA National Agroforestry Center (NAC), LincolnNE.* [ppt.] Retreived on 06.09.2016

WinjumJK, Dixon RK, Schroeder PE (1992) Estimating the global potential of forest and agroforest management practices to sequester carbon. Water Air Soil Pollut. 64, 213–228. *Carbon sequestration in tropical agroforestry systems.* Retrieved on 06.09.2016 from: https://www.researchgate.net/publication/222668060_ Carbon_ sequestration _in_tropical_agroforestry_systems

Zamora and Wyatt (2016) Silvopasture Retrieved on 06.09.2016 from http://www.extension.umn.edu/environment/agroforestry/silvopasture/silvopasture.html

23

Potential of Agroforestry for Climate Change Mitigation

J.C. Tewari, Kamlesh Pareek, Shiran K and Mahesh Kumar Gaur

Central Arid Zone Research Institute, Jodhpur, Rajasthan, India

Introduction

Throughout the arid zones, there is no dearth of problems, but rapidly increasing desertification (some call it land degradation) is a problem of worldwide dimension. Water is a scare commodity in arid zones. Much of rainfall is lost by evapo-transpiration and as a result, ground water recharged only by seepage through soil profile. However, it is a common phenomenon in arid zones of the world that ground water is frequently used at the rate that exceeds recharge. The situation in arid tropics of India, which is spread over an area of 31.7 million ha, is no more different. Considering the role of agriculture in the social and economic progress of developing countries, the vulnerability of agricultural systems to the impacts of climate change has received considerable attention from the scientific community (Fischer *et al.* 2002; IISD 2003; Kurukulasuriya and Rosenthal 2003). Much of the available literature suggests that the overall impacts of climate change on agriculture especially in the tropics will be highly negative, although in a few areas there may be minor increases in crop yields in the short term (Maddison *et al.* 2007). Research on agroforestry as an adaptation to climate change and as a buffer against climate variability is in the process of evolving.

The Indian hot arid regions

The hot arid regions of India lie between 24° and 29° N latitude, and 70° and 76° E longitude, covering an area of 31.70 million hectares, and involving seven states: Rajasthan, Gujarat, Punjab, Haryana, Andhra Pradesh, Karnataka, and Maharashtra. An area-wise break up of hot arid regions is presented in

Table 1. In total, 11.8% of the country is under a hot arid environment. The arid regions of Rajasthan, Gujarat, Punjab, and Haryana together constitute the Great Indian Desert, better known as the Thar desert. As arid western Rajasthan accounts for 61% of hot arid region of the country, therefore it is considered principal hot arid region.

Table 1. Distribution of arid regions in different states of India

State	Area (million hectares)	Percent of total
Rajasthan	19.61	61.00
Gujarat	06.22	19.60
Punjab and Haryana	02.73	09.00
Andhra Pradesh	02.15	07.00
Karnataka	00.86	03.00
Maharashtra	00.13	00.4
Total	31.70	-

The production and life support systems in this part of the hot Indian arid zone are constrained by climatic limitations including: low annual precipitation (100–300 mm); very high temperature during the summer season (mean maximum temperature, 41°C) touching a maximum of 48° to 50°C; short (December to mid-February) cool and dry winters (the mean winter season temperature varies from 10° to 14°C); high wind speed (30–40 km hr^{-1}); high evapotranspiration; and general low humidity (an aridity index of 0.045–0.19) (Sharma and Tewari 2005).

Sand dunes are a dominant land formation of the region. More than 58% of the area is sandy and intensities of dunes vary from place to place. In general, soils contain 1.8–4.5% clay, 0.4–1.3% silt, 63.7–87.3% fine sand, and 11.3–30.3% coarse sand. They are poor in organic matter (0.04–0.12%), and low to medium in phosphorus content (0.05–0.10%). The nitrogen content is mostly low, ranging between 0.20 and 0.07% and infiltration rate is very high, at 7–15 cm hr^{-1} (Dhir 1997). Because of the complete absence of any aggregation, the soils are highly erodible.

Since the vast arid expanse of Ghaggar is blank on Chalcolithic and Iron Age maps of the Indian subcontinent, it seems that human occupation of the region was thin, and depended upon hunting and limited pastoralism (Dhir 1982). From the analysis of available authentic records, the population of the region registered an increase of 490% from 1901 to 1991. Decennial variations in population from 1901 to 1991 in the region showed a growth rate of 186% (between 1901 and 1971) as compared with 132% for the whole country. The population growth rate for the decades 1971–1981 and 1981–1991 were 36.7% and 30.7%, respectively. The changes that occurred in the density of population in the Thar between 1971 and 2011 were:

1971 48 persons km^{-2}
1981 69 persons km^{-2}
1991 89 persons km^{-2}
2001 101 persons km^{-2}
2011 127 persons km^{-2}

Above figure clearly indicates that the population density is quite high in principal of arid region of the country compared with the global average of 6–8 persons km^{-2} for arid zones. Similarly livestock population is also very high in this part of the country. Human livestock ratio in principal hot arid zone is 1.0: 1.5 in comparison to national average of 1.0: 0.5.

Agroforestry in arid region of western Rajasthan

The image of the desert of little but vegetation-less dense unbroken stretches of sand and inaccessible terrain conditions is shattered upon entering in arid western Rajasthan. There are sparsely distributed trees, and underneath growth of arable crops and or grasses, and other herbaceous flora (especially, during the kharif season, as agriculture is predominantly rainfed) in long stretches interspersed with distantly distributed village settlements and Dhanis (a unique settlement pattern of the arid western Rajasthan intended for the life of agriculturists families during the active cropping period away from the village, but nearby their fields) are features of the region (Tewari *et al.* 1999).

Traditional agroforestry systems

In general, rainfall appears to be governing factor for evolution of traditional agroforestry systems (Sharma and Tewari 2005). On the basis of rainfall, four types of major traditional agroforestry systems have been identified. In upper-transect extending from western border of district Jaipur to extreme west part of Ganganagar district, over 400 mm rainfall zone pre-dominated system is *Prosopis cineraria-Acacia nilotica* based; between 300 and 400 mm rainfall zone *P. cineraria* based; between 200-300 mm rainfall zone *Zizyphus* spp.-*P. cineraria* based; and in less than 200 mm rainfall zone *Zizyphus* spp.-*P. cineraria* – *Salvadora* spp. based (Narain and Tewari 2005). The tree/shrub density of the system forming species is also clearly governed by rainfall regime. Data indicated that with decrease in rainfall, the density of woody component of the system decreased substantially and as well as the productivity of arable crops and grasses were declined.

In upper transect of arid western Rajasthan rainfall seems to play determinant role in crop production. *P. cineraria – A. nilotica* based agroforestry system was found to be highly productive and *Zizyphus* spp. – *P. cineraria – Salvadora* spp. system was least productive. It was very interesting that yield of pearl millet, main cereal crop of the region, below the canopy of woody components in any system was not affected at all, rather in *Zizyphus* spp. – *P. cineraria* and *Zizyphus* spp. – *P. cineraria – Salvadora* spp. systems, it was substantially increased below the tree canopies in comparison of open spaces. However, legumes such as moth bean, cluster bean and moon beam exhibited slightly yield reduction below the tree canopies, 5.0 per cent (*Zizyphus* spp. – *P. cineraria* system) to 19.5 per cent (*P. cineraria* system); 12.5 per cent (*Zizyphus* spp. – *P. cineraria – Salvadora* spp. System) to 18.2 per cent (*Zizyphus* spp. – *P. cineraria* system); and 10.0 per cent (*P. cineraria* system) to 33.7 per cent (*P. cineraria – A. nilotica* system), respectively. Yield reduction under tree canopies for sesame ranged from 26.3 per cent (*P. cineraria – A. nilotica* system) to 46.1 per cent (*P. cineraria* system).

Foliage of trees constitutes nutritious components of animal feed in arid parts of Rajasthan. The leaf fodder of some tree species is as nutritious as leguminous fodder and is comparable with grass production from the pastures. Fuel wood and leaf fodder production potential of *P. cineraria – A. nilotica* system was maximum, while it was minimum for *Zizphus* spp. *A. cineraria – Salvadora* spp. system. Thus production from woody components of traditional agroforestry systems also follows the rainfall gradient like that of arable crops.

The lower transect in arid western Rajasthan encompassed Pali, Jalore, parts of Jodhpur and Barmer district in rainfall gradient of >500 to 200 mm from south-eastern margin to western part (Anonymous, 2008). Four distinct agroforestry systems have been identified in the study region. In south-eastern margin, where total annual rainfall was >500 mm, the pre-dominant agroforestry system was *Azadirachta indica - Acacia nilotica* var. *cupressiformis* based. This system extends from extreme eastern part of Pali district to western part of Jalore district. In the rainfall zone >400-500 mm, *Prosopis cineraria – Azadirachta indica* based system was prevalent. The density of woody component in the system was only 8.2 individual ha^{-1} with a very low total basal cover (0.6 m^2 ha^{-1}). In rainfall zone of <300-400 mm, the major traditional agroforestry system was *P. cineraria - Tecomella undulata - Salvadora* spp. based. The total basal cover contributed by system farming species was as high as 92 per cent thereby indicating that maximum biomass was accumulated on system farming species as biomass corresponds to basal cover. In western part of Barmer district (rainfall zone >200-300mm), the most prevalent system was *P. cineraria – Salvadora oleoides- T. undulata* based.

The major rain fed (kharif) crops grown in association of tree/shrubs in lower transect were similar to that of upper transect viz. pearl millet, mung bean, moth bean, cluster bean and sesame. The production pattern was also similar to that of upper transect.

Fuel wood and leaf fodder are main produce of woody component in the traditional agroforestry systems, which farmers collect by way of lopping and or/pollarding the trees and shrubs. The maximum fuel wood (4.9 t $ha^{-1}yr^{-1}$) and leaf fodder (1.78 t $ha^{-1}yr^{-1}$) production was recorded in *P. cineraria – S. oleoides-T. undulata* based agroforestry system, which was prevalent in western part of Barmer district (rainfall zone >200-300 mm). The fuel wood (1.2 t $ha^{-1}yr^{-1}$) and fodder (0.43t ha^{-1} yr^{-1}) production was minimum in *P. cineraria – A. indica* based system, which was dominant in 300-<400 mm rainfall zone. It appeared that with decrease in rainfall to a certain limit, the woody component in the system are protected by the farmers as they provide substantial amount of livelihood in environmentally more stressed areas.

Traditional agroforestry systems: crop production scenario

On an average, across all the villages studied in entire region, it was recorded that 60 per cent area was under pearl millet. The pearl millet grain yield exhibited clear declining trend with decrease in total annual rainfall. The grain yield of pearl millet was found to be comparable both in open field and under the tree canopy cover in all the systems. However, under the tree canopy cover yield of mung bean decreased 20.0-29.83 per cent, moth bean 15.69-32.43 per cent, cluster bean 20.0-36.61 per cent and sesame 21.05-23.53 per cent. The data clearly indicated that tall pearl millet crop production did not have any negative effect of tree canopy cover, however, production of grain in all short statured leguminous crops were adversely affected under the tree canopy cover.

Agroforestry: a climate resilient land use, suitable for climate change-adaption in the region

Considering the role of agriculture in the social and economic progress of developing countries, the vulnerability of agricultural systems to the impacts of climate change has received considerable attention from the scientific community (Fischer *et al.* 2002; IISD 2003; Kurukulasuriya and Rosenthal 2003). Much of the available literature suggests that the overall impacts of climate change on agriculture especially in the tropics will be highly negative, although in a few areas there may be minor increases in crop yields in the short term (Maddison *et al.* 2007). Table 2 presents some of the projected changes in climate and their potential impacts on agriculture as summarised in the IPCC report (Parry *et al.* 2007).

Table 2. Projected changes in climate and their impact on agriculture

Phenomena and direction of change	Likelihood of occurrence	Major projected impacts on agriculture
Warmer and fewer cold days and nights; warmer/more frequent hot days and nights over most land areas	Virtually certain (>99% chance)	Increased yields in colder environments decreased yields in warmer environments increased insect outbreaks
Warm spells/heat waves: frequency increases over most land areas	Very likely (90-99% chance)	Reduced yields in warmer regions due to heat stress; wild fire danger increase
Heavy precipitation events: frequency increases over most areas	Very likely	Damage to crops; soil erosion, inability to cultivate land due to water logging of soils
Area affected by drought: increases	Likely (66- 90% chance)	Land degradation, lower yields/crop damage and failure; increased livestock deaths; increased risk of wildfire
Intense tropical cyclone activity increases	Likely	Damage to crops; wind throw (uprooting) of trees; damage to coral reefs
Increased incidence of extreme high sea level (excludes tsunamis)	Likely	Salinisation of irrigation water, estuaries and freshwater systems

Carbon sequestration potential under different land use options are given in figure 1. Recognizing the ability of agroforestry systems to address multiple problems and deliver multiple benefits, the IPCC Third Assessment Report on Climate Change (IPCC 2001) states that "Agroforestry can both sequester carbon and produce a range of economic, environmental, and socioeconomic benefits. For example, trees in agroforestry farms improve soil fertility through

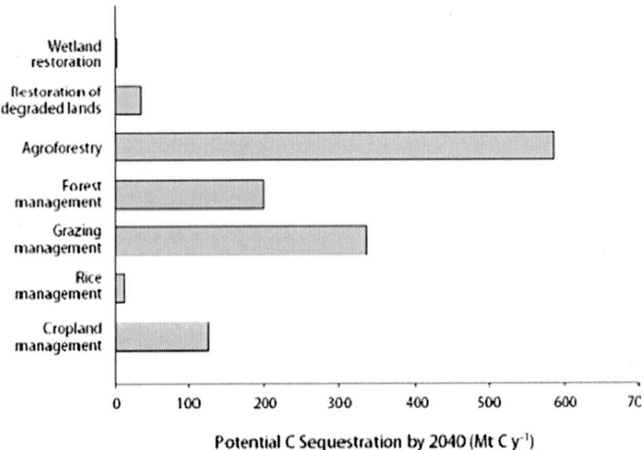

Fig. 1. Carbon sequestration of different land management options (adopted from IPCC 2000)

control of erosion, maintenance of soil organic matter and physical properties, increased N, extraction of nutrients from deep soil horizons, and promotion of more closed nutrient cycling. We believe that agroforestry interventions provide the best "no regrets" adaptation measures in making communities resilient to the impacts of climate change and do discuss the same in relation to the challenges posed by the changing and variable climate".

Agroforestry systems can be useful in maintaining production during drier years, a common phenomenon in arid regions of India. During drier years or in complete drought situations, deep root systems of trees are able to explore a larger volume for water and nutrients, which helps to maintain depleting soil moisture conditions to some extent. In drought prone environment of arid western Rajasthan, as a risk aversion and cooping strategy, the traditional agroforestry systems avoid long term vulnerability as trees act as an insurance against drought, insect-pests outbreaks and other threats, instead of a yield-maximising strategy aiming at short term monetary benefits (Rathore 2004).

Adaptation to climate change is now inevitable. Research on agroforestry as an adaptation to climate change and as a buffer against climate variability is in the process of evolving. Many pathways through which agroforestry qualify as an adaptation to climate change, especially in arid regions are discussed below.

Agroforestry systems: role in soil erosion control

There are several ways by which climate change manifests soil degradation. Higher temperatures and drier conditions lead to lower organic matter accumulation in the soil resulting in poor soil structure, reduction in infiltration of rain water and increase in runoff and erosion (Rao *et al.* 1998) while the expected increase in the occurrence of extreme rainfall events will adversely impact on the severity, frequency, and extent of erosion (WMO 2005). These changes will further exacerbate an already serious problem is being faced by arid regions of the country.

The woody component in agroforestry systems helps in reducing soil erosion which is the most harmful abiotic stress in arid region. It was observed if rows of trees are planted right angle to wind direction a tremendous amount of soil drop can be checked from agricultural fields. The shelterbelts on agricultural fields form a type of agroforestry practice (Gupta *et al.* 1984). Following data depict how effective are agroforestry practices for controlling the soil erosion in arid regions (Table 3).

Table3. Effect of different type of shelterbelts on soil erosion

Type of shelterbelt	Total amount of soil loss (kg ha^{-1})		
	Year I	Year II	Mean
Prosopis juliflora	93.2	609.3	351.2
Cassia siamea	91.5	277.1	184.2
Acacia tortilis	106.0	494.1	300.0
Agricultural field without trees	262.7	831.0	546.8

Agroforestry systems: inbuilt capacity to enhance the use efficiency of rain water

Water is already a scarce resource and climate change is expected to make the situation worse. Climate change has both direct and indirect impacts on water availability. The direct impacts include changes in precipitation patterns, while the indirect ones are increases in losses through runoff and evapo-transpiration.

There are several mechanisms whereby agroforestry may use available water more effectively than the annual crops. Firstly, unlike in annual systems where the land lies bare for extended periods, agroforestry systems with a perennial tree component can make use of the water remaining in the soil after harvest and the rainfall received outside the crop season. Secondly, agro-forests increase the productivity of rain water by capturing a larger proportion of the annual rainfall by reducing the runoff and by using the water stored in deep layers. Thirdly, the changes in microclimate (lower air temperature, wind speed and saturation deficit of crops) reduce the evaporative demand and make more water available for transpiration.

The tree canopies in agroforestry systems intercept the rain and reduce runoff (Khan *et al.*, 1995). In a study at CAZRI, it was found that in *Acacia tortilis* silvi-pasture stand canopy interception was 21.4%, whereas in *Colophospermum mopane* silvi-pasture it was 13.1%. Rainfall interception was positively related with canopy cover and negatively related with through fall. Average surface run off in *A. tortilis* based silvi-pasture stand was 53% higher than in *C. mopane* based silvi-pasture. This indicated that hydraulic response to rain is dominated by plant species character however; the percent annual runoff and soil erosion was very low in situations with trees on agricultural fields in comparison to bare soil condition. Thus, the enhanced use efficiency of rain water by woody species in agroforestry systems improves agricultural productivity.

Agroforestry systems: economically viable and environment friendly means to improve soil fertility

Nutrient mining from continuous cropping without adequately fertilizing or fallowing the land is often cited as the main constraint to increase in productivity in arid regions. It is observed that on average soils in Indian arid regions have poor nitrogen and phosphorus content. Organic carbon status in certain parts of arid regions has been depleted to an alarming state. While fertilisers offer an easy way to replenish the soil fertility, at the current prices it is very unlikely that there will be any change in the investments made by farmers of arid region in fertilizers. In this context, agroforestry systems have attracted considerable attention as an attractive and sustainable pathway to improve soil fertility. CAZRI, Jodhpur made substantial progress in the identification and promotion of agroforestry systems aimed at improving soil fertility.

Soil fertility under the tree species (*Prosopis cineraria* and *Acacia nilotica*) and in open field conditions was studied intensively (Singh and Lal 1969). In general, organic carbon content decreases with depth but the same was slightly higher under the canopy of both the tree species than in open field condition. The total nitrogen, available P and K are maximum under the canopy of *P. cineraria* followed by *A. nilotica*. The values for the same were minimum in open field condition.

The soil fertility build up under the fourteen year old stands of *P. cineraria* and *P. juliflora* were found to be of higher order as compared to open field condition. Similarly, status of micro-nutrients was also higher under tree canopies than in the open field (Table 4).

Table 4. Available macro and micro-nutrients under different tree based agroforestry systems

Tree species	Macronutrients (kg ha^{-1})			Micronutrients (ppm)			
	N	P	K	Zn	Mn	Cu	Fe
Prosopis juliflora	231	7	333	0.89	9.3	0.58	3.3
Prosopis cineraria	221	11	479	1.44	10.8	0..89	2.8
Open field	199	6	3.2	0.19	7.0	0.38	3.5

Agroforestry systems: moderating the micro-climate in the arid regions

The full genetic potential of many crops and verities can only be realized when environmental conditions are close to optimum. Trees on farm bring about favourable changes in the microclimatic conditions by influencing radiation flux, air temperature, wind speed, saturation deficit of understorey crops all of which will have a significant impact on modifying the rate and duration of photosynthesis and subsequent plant growth, transpiration, and soil water use (Monteith *et al.* 1991). Following examples present microclimatic moderation effect in harsh climatic conditions of arid regions of western Rajasthan.

Air temperature exhibited appreciable variation under the canopy of *A. tortilis* during monsoon period (Tewari *et al.*, 1989). A decline of 0.1°C to 0.7°C was evident beneath the canopy cover of *A. tortilis* than that of surrounding open areas at 07 hrs. Around 14 hrs, this decline was of the order of 0.6°C to 2.0°C. Similarly, it was found during monsoon season that the soil temperature just beneath the tree cover was lower by as much as 10°C to 16°C in top soil zone and 4°C to 5°C at 30 cm depth when compared to open field conditions, thereby indicating better soil-thermal regime. Seven year old *A. tortilis* trees provided more humidity on cluster bean grown with the trees. This finding proves the efficiency of agroforestry practices for maintaining favourable moisture status on arable lands (Ramakrishna and Shastri 1977).

Agroforestry systems: improve biodiversity

Continued deforestation is major challenge for forests and livelihoods. Although agroforestry may not entirely reduce deforestations, particularly in arid regions, it may acts as an effective buffer to deforestation. Trees in agro-ecosystems in Rajasthan have been found to support threatened cavity-nesting birds and offer forage and habitat to many other bird species. Biodiversity conservation may not be a primary goal of agroforestry systems. Nevertheless in some cases traditional agroforestry systems in arid western Rajasthan found to support very good species diversity and also act as a buffer to parks and protected areas (Pandey 2007).

In a study at Govardhanpura and Gokulpura cluster of villages in Bundi district having relatively little better rainfall regime, where a highly degraded silvi-pasture (45 ha) was re-vegetated and fenced to minimize abiotic and biotic disturbance exhibited marked difference in biodiversity of plant species (Dixit *et al.* 2005). The area was divided into 6 different blocks. Though initially only five woody and some grass species were introduced with some water conservation structures in the area, after five years the fortified area with physical and social fencing exhibited the woody species richness to the tune of 20 species with a Beta (between habitat) diversity of 2.2. Maximum number of woody species (25%) belonged to family Leguminosae/Mimosoideae. Rest of woody species were distributed among 11 other families/sub-families. The herbaceous vegetation richness increased to 36 (number of species) of which 80% were palatable. These 36 herbaceous species were distributed among 24 families. Cyperaceae family was represented by maximum percentage of the species. The Beta diversity of herbaceous species was 2.32 with index of general diversity of

3.437. The value of general diversity for woody and herbaceous species was very-very low in adjoining unprotected area.

Dominance-diversity (d-d) curves were developed for herbaceous vegetation on the basis of relative biomass values for all the blocks and also for open degraded land. With exception of open degraded land site, the d-d curves on all the blocks exhibited similar log normal distribution. This indicated that several herbaceous species were sharing relatively low importance values (biomass) or similar range of importance values. The log normal or curves approaching to log normal, in fact, indicate relatively stable population. The d-d curve of open degraded land site was closer to geometric series. Whittakar (1972) opined that communities having low diversity exhibit geometric series. In open degraded land ruthless exploitation of herbaceous vegetation has resulted in tremendous loss of biodiversity.

Conclusion

Climate change impacts on agriculture are of the greatest concern to most developing countries, particularly in tropics because of higher dependence on agriculture, subsistence level of operation, low adaptive capacity and limited institutional support. The scenario for arid tropics of India is no more different. In conclusion agroforestry systems in arid region are important option for climate change mitigation and sustainable development. Agroforestry systems act as sink for atmospheric carbon while helping to attain food security, increase farm income, improve soil health and discourage deforestation.

References

Anonymous (2008) Annual Progress Report. CAZRI (ICAR), Jodhpur, India, pp. 166.

Dhir RP(1982) The human factors in ecological history. In: Desertification and Development: Dryland Ecology in Social Perspective, Academic Press, New York, pp. 311-331.

Dhir R P (1997) Characteristics and behavioral aspects of arid and semi-arid zone soils. In: Yadav MS, Singh Manjit, Sharma SK, Tewari JC, Burman U (eds) Silvipastoral Systems in Arid and Semi-arid Ecosystems, CAZRI, Jodhpur, India, pp.39-46.

Dixit S, Tewari JC, Wani SP, Vineela C, Shaurasia AK, Panchal HB (2005) Participatory Biodiversity Assessment: Enabling Rural Poor for Better Natural Resource Management. ICRISAT, Patancheru, India, pp.16.

Fischer G, Shah M, van Velthuizen H (2002) Climate change and agricultural vulnerability. International Institute for Applied Systems Analysis. Report prepared under UN Institutional Contract Agreement 1113 for World Summit on Sustainable Development. Laxenburg, Austria

Gupta JP, Rao GGSN, Ramakrishna YS, Ramana Rao (1984) Role of shelterbelts in arid zone. Indian Farming 34(7):29-30.

IISD (2003) Livelihoods and climate change: Combining disaster risk reduction, natural resource management and climate change adaptation. In: New Approach to the Reduction of Vulnerability and Poverty, a conceptual framework paper prepared by the Task Force on Climate Change, Vulnerable Communities and Adaptation.

IPCC (2000) Special Report on Land Use, Land Use Change and Forestry. Summary for Policy Makers. Geneva, Switzerland. pp. 20.

IPCC (2001) Climate Change 2001: The Scientific Basis. Contribution of the Working Group 1 to the Third Assessment Report of the IPCC. Cambridge University Press,Cambridge.

Khan MA, Tewari JC, Issac VC (1995) Hydrology of small forested catchments in the arid region of Rajasthan. Annals of Arid Zone 34(4):259-262.

Kurukulasuriya P, Rosenthal S (2003) Climate change and agriculture: A review of impacts and adaptations. In: Climate Change Series Paper 91, World Bank, Washington, District of Columbia, pp.106.

Maddison D, Manley M, Kurukulasuriya P (2007) The Impact of Climate Change on African Agriculture: A Ricardian Approach. World Bank Policy Research Working Paper 4306.

Monteith JL, Ong CK, Corlett JE (1991) Microclimatic interactions in agroforestry systems. For Ecol Manage 45:31– 44.

Narain Pratap, Tewari JC(2005) Trees on agricultural fields: a unique basis of life support in Thar Desert. In: Tewari VP, Srivastava R L (eds) Multipurpose Trees in the Tropics:Management and Improvement Strategies. Arid Forest Research Institute, Jodhpur, pp. 516-523.

Pandey DN (2007) Multifunctional agroforestry systems in India. Current Science, 92(4): 455-463.

Parry ML, Canziani OF, Palutikof JP, Linden van der PJ, Hanson CE (2007) Climate change 2007: Impacts, adaptation and vulnerability. Contribution of Working Group II to the Fourth Assessment Report of the Intergovernmental Panel on Climate Change, Cambridge University Press, Cambridge, UK, pp. 1000.

Ramakrishna YS, Shastri ASRAS(1977) Microclimate under Acacia tortilis plantation. Annual Progress Report CAZRI, Jodhpur, pp. 69-70.

Rao KPC, Steenhuis TS, Cogle AL, Srinivasan ST, Yule DF, Smith GD(1998) Rainfall infiltration and runoff from an Alfisol in semi-arid tropical India. Soil and Tillage Research 48:61-69.

Rathore JS (2004) Drought and household coping strategies: A case of Rajasthan. Indian Journal of Agricultural Economics 59(4):689-708.

Sharma Arun K, Tewari JC(2005) Arid zone forestry with special reference to Indian hot arid zone. In: Forests and Forest Plants, Encyclopaedia of Life Support Systems (EOLSS), Developed under auspices of the UNESCO, EOLSS publishers, Oxford, UK.

Singh KS, Lal P(1969) Effect of *Prosopis cineraria* and *Acacia nilotica* on soil fertility and profile characteristics. Annals of Arid Zone 8:33-36.

Tewari JC, Bohra MD, Harsh LN (1999) Structure and production function of traditional extensive agroforestry system and scope of intensive agroforestry in Thar Desert. Indian Journal of Agroforestry 1(1):81-94.

Tewari JC, Harsh LN, Venkateswarlu J (1989) Agroforestry research in arid regions: a review. In: Singh RP, Ahlawat IPS, Saran G (eds.) Agroforestry Systems in India – Research and Development, Indian Society of Agronomy, New Delhi, pp. 3-17.

Whittakar RH (1972) Evolution and measurement of species diversity. Taxon 28:213-251.

Tewari JC, Tripathi D, Narain Pratap (2001) Jujube: A multipurpose tree crop for arid land farming systems. The Botanica 51:121-126.

WMO(2005) Climate and Land Degradation. World Meteorological Organization. WMO-No. 989.

24

Livestock as a Source of Livelihood Security in Arid Agroforestry System

B.K. Mathur

Central Arid Zone Research Institute, Jodhpur, Rajasthan, India

Introduction

Rain fed agroecosystem occupies 68% of India's cultivated area and supports 40% of the human beings and 65% of the livestock population. It produces 44% of food requirements, thus has and will continue to play a critical role in India's agriculture (Singh *et al.* 2004). Besides their well established role in agriculture, livestock have crucial role in food security and as risk aversion mechanism for sustaining family, whenever there is crop failure. Indian hot arid zone which is about 12% of total geographical area of the landmass of 0.32 million km^2 has maximum covering in western Rajasthan i.e. 61% of the total area whereas the other areas in arid region are available in the states of Gujarat, Punjab, Haryana, Andhra Pradesh and Karnataka accounting for 20, 5, 4, 7 and 3% of hot arid area whereas the cold arid area of 8.4 million ha lies in the state of Jammu & Kashmir covering the Leh and Ladakh region. The hot arid zone of Rajasthan is comprised of 12 districts of the state lying in the western part and these are Barmer, Bikaner, Churu, Hanumangarh, Jaisalmer, Jalore, Jhunjhunu, Jodhpur, Nagaur, Pali, Sriganganagar and Sikar and this arid region of Rajasthan has livestock population of 30.18 million which is about 52% (Livestock census 2012) of the total population of the state.

The harsh climatic conditions prevailing in the arid region, e.g., erratic rains and frequent droughts would suggest that, it is not very suitable for crop farming. Livestock farming has some in built superiority over crop farming as far as growth; stability and resource conservation are concerned. On an average, the region experiences 3 years of drought in every 10 years. The natural forces constituting the soil-climatic complex, which conspire to reduce the crop productivity and cause instability in agricultural production, have much less impact on livestock farming. This is due to differences in the nutritive value of natural vegetation, which mainly sustains livestock.

The superiority of livestock farming for development of arid region is further highlighted by the fact that this region is endowed with some of the best breeds of livestock and drought-hardy perennial grasses. The local breeds of livestock have acquired certain characters to withstand the arid climate, and the characters have been transmitted through generations to make the present hardy breeds of animals. However, due to lack of proper nutrition, genetic potentiality of these animals has not been expressed to its maximum level.

The major livestock breeds in the region are:

Cattle: Tharparkar, Rathi, Nagori and Kankrej
Sheep: Marwari, Jaisalmeri, Chokla, Nali,Magra, Pugal and Sonadi
Goat: Marwari, Kutchi and Parbatsari
Camel: Bikaneri and Jaisalmeri

For arid region of Rajasthan the general climatic conditions, topographical features and biotic factors do not encourage agricultural operations in the absence of extractive industry the peasantry has to fall upon animal husbandry as their main occupation. Rearing some of the finest breed of cattle, camel, sheep and goats known for their endurance making much use of the meagre feed resources which are grasses, herbs, shrubs, tree leaves and cultivated feed and fodder crops.

Table 1. Trend in livestock population (millions) of arid and non arid district of Rajasthan

S.No	Livestock	Area of Rajasthan	1997 Census % of Rajasthan	2003 Census % of Rajasthan	2007 Census % of Rajasthan	2012 Census % of Rajasthan
1	Cattle	Arid districts	4.96 (40.78)*	4.12 (37.97)	5.02 (40.46)	6.18 (46.38)
		Non Arid districts	7.20	6.73	7.39	7.14
		Total	12.16	10.85	12.41	13.32
2	Buffaloes	Arid districts	3.16 (32.37)	3.20 (30.65)	3.43 (29.77)	3.95 (30.41)
		Non Arid districts	6.60	7.24	8.11	9.03
		Total	9.76	10.44	11.54	12.98
3	Sheep	Arid districts	10.46 (73.30)	7.35 (73.5)	7.94 (70.37)	6.88 (75.75)
		Non Arid districts	3.82	2.65	3.34	2.20
		Total	14.31	10.00	11.28	9.08
4	Goat	Arid districts	9.53 (56.25)	8.36 (49.76)	11.77 (53.80)	12.79 (59.03)
		Non Arid districts	7.41	8.44	10.11	8.88
		Total	16.94	16.80	21.88	21.67
5	Camel	Arid districts	0.53 (70.66)	0.40 (80)	0.35 (82.14)	0.287 (85.19)
		Non Arid districts	0.21	0.10	0.08	0.048
		Total	0.75	0.5	0.43	0.325
6	Total Livestock (including, horse, pony, etc.)	Arid districts	28.57 (52.56)	27.5 (56.0)	29.08 (50.23)	30.18 (52.27)
		Non Arid districts	25.77	21.6	28.81	27.55
		Total	54.35	49.1 (10.12)**	57.89 (10.94)**	57.73 (8.87)**

*Data in parenthesis is percent population of Rajasthan state

**Data in parenthesis is percent population of India

Table 2. Common available feeding material from natural rangelands with crude protein percentage

S.No.	Botanical name	Local name	C.P.%
Grasses			
1.	*Aristida funiculate*	Lapra	4.38
2.	*Brachiaria ramosa*	Murat	11.37
3.	*Cenchrus biflorus*	Bhurat	15.31
4.	*Cenchrus ciliaris*	Dhaman	8-12
5.	*Cenchru s setigerus*	ModaDhaman	8.77
6.	*Cynodon dactylon*	Dob	10.10
7.	*Dactylo ctenium aegypticum*	Ganthia	10.10
8.	*Eluesine comprass*	Tantiya	13.99
9.	*Eragrostis tremula*	Chidi grass	6-10
10.	*Lasiurus sindicus*	Sewan	9.05
11.	*Panicum antidotale*	Gramna	7.1
12.	*Tetrapogon tenellus*	Kagiyo	13.36
13.	*Urochloa panicoides*	Kuri	12.02
Weeds			
1	*Achryanthes aspera*	Antipada	19.9
2	*Amaranthus viridis*	Chandelio	4.9
3	*Celosia argentia*	Makhamal	12.5
4	*Cyperus rotundus*	Motha	9.1
5	*Digeram uricata*	Lolru	12.07
6	*Indigofera cordifolia*	Bekario	18.0
7	*Trianthem aportulacastrum*	Sathi	14.31
8	*Tribulus terristris*	Kanti	18.18
Trees/shrubs			
1	*Acacia nilotica*	Desibabul	13.9
2	*Acacia senegal*	Kumat	17.48
3	*Acacia tortalis*	Israilibabool	14.6
4	*Azadirachita indica*	Neem	15.8
5	*Calotropis procera*	Aak	13.5
6	*Capparis deciduas*	Kair	2.05
7	*Cordia gharaj*	Gundi	15.3
8	*Euphorbia caducifolia*	Thor	5.61
9	*Prosopis cineraria*	Khejri	14.63
10	*Tecomella indica*	Rohida	12.2
11	*Prosopis juliflora (Pods only)*	Vilayati babul	13.8
12	*Salvadora persica*	Metha Jal	10.5
13	*Ziziphus nummularia*	Jharberi	14.1

Simultaneously, the grazing/browsing areas are progressively shrinking with encroachment by ever growing human population necessitating to evolve economical supplementary feeding (Mathur 1996). Cultivated feed and fodder crops: *Pennisetum typhodes* (Bajra), *Sorghum bicolor* (Jowar), *Phaseolus aconitifolius* (Moth), *Cyamopsis tetragonoloba* (Guar), *Sesamum indicum* (Til), *Phaseolus aureus* (Moong) etc.

Table 3. For stall feeding agro-by-products normally offered were:

S.No	Feeds and Fodder	Botanical name	C.P.%
1.	Moth charaor "Nira" (Fodderhavingstems, leaves crushed pods covering)	*Phaseeolus aconitifolius*	9.13
2.	Guar "phalgati" (Fodder having stems, leaves, crushed pods covering and also roots)	*Cyamopsis tetragonaloba*	9.95
3.	Ground nut chara (Fodder)	*Arachis hypogia*	12.93
4.	Missa Bhussa(Fodder)	*Cicer aritinum*	6.40
5.	Bajrakutter(Straw)	*Pennisetum typhoidium*	5.59
6.	Wheat straw.	*Triticum aestivum*	3.17

Along with fodder 1-2 kg concentrate mostly comprising of guar (*Cyamopsis tetragonaloba*) churi/moth (*Phaseeolus aconitifolius*) churi and also grounded bajra (*Pennisetum typhoidium*) were provided to cattle.

Acceptability and palatability of chaffed buffel grass v/s cluster bean and moth bean fodder

A feeding trial was conducted to evaluate the acceptability and palatability of chaffed buffel (*Cenchrus ciliaris*) grass v/s fodder of cluster (*Cyamopsis teragonalaba*) bean (guar) and moth (*Nigra aconfifolia*) bean in Marwari sheep. For the purpose, the adult Marwari rams of comparable live weight and of normal health were divided into 3 groups, each of 4 animals. Animals of different groups were kept in different enclosures and had free excess to water. The animals of group I, offered weighed *ad libitum* chaffed *C. ciliaris* straw, whereas, the animals of II and III, were offered *ad libitum* fodder of cluster and moth beans respectively. Their feed and water intake were recorded daily, and their live weights were recorded at weekly intervals. On an average the dry matter intake in sheep of I, II and III group were 2.76±0.92, 3.04+0.48 and 2.95+0.73 kg where as water intake was 7.53+1.23, 100kg body weight 1 day, 8.80+1.47 and 11.31+ 2.69 lit 100 kg body weight/day respectively. On an average, the sheep of I, II and III groups in 21 days feeding period gained 2.4%, 4.9% and 4.3% over their initial live weights. The results of the study indicated that amongst 3 feeds trial, the sheep has highest preference for cluster bean chaff, followed by moth bean, and had least preference for *C.ciliaris* straw. The trend in live weight gain was correlated with the higher water intake.

Table 4. Body weight gain, feed intake and water intake by different treatment group animals (data are average of 4 rams)

Animal group	Initial Av. B. Wt. (kg)	Final Av. B. wt. (kg)	B. wt. gain (kg)	Live B. wt. Gain (kg)	DMI 100 kg^{-1} B. wt. (kg)	Water intake 100 kg^{-1} B. wt. (kg)
Group I Grass as such *Cenchrus* grass	35.500	36.350	0.850	2.39	2.76±0.92	7.53±1.23
Group II Guar Chhara as such	36.800	38.600	1.800	4.89	3.04±0.48	8.80±1.47
Group III Moth Chhara as such	36.600	38.200	1.600	4.34	2.95±0.73	11.31+2.69

Utilization of tree fodders resources

These resources include leaves of various fodder trees, shrubs and pods. A large number of trees and shrubs have been identified which are being lopped for fodder purpose. Important tree fodders and utilization of their nutritive value have been reviewed by many authors (Patnayak *et al.* 1995; Patnayak 1992; Sharma *et al.* 1992; Singh 1981). Although a large number of fodder trees have been identified, only a very few species have been studied in detail and utilized in conventional feeding practice. There is need to identify and study in detail promising species of tree fodders in terms of forage production, animal acceptability and nutritive value. Although tree fodders are considered as emergency fodders, they serve as potential feed resources for small ruminants and provide about 50% of the total dry matter consumption. They are utilized as freshly loped leaves or by conserving as dry fodders. Dry matter in most tree varies from 12 to 20% on dry basis. Very high calcium and low phosphorus are a common feature in almost all tree fodders which when fed solely leads to a very wide Ca:P ratio. Dry matter intake from most of the tree leaves varies from 2.5 to 3.5 kg 100 kg^{-1} B W. Although they are rich in protein, the digestibility of crude protein is very low (25 to 30%) in many cases. In spite of higher intake of nitrogen and calcium, retention of nutrients in general is poor and phosphorus balance is negative or marginally positive. Tree fodders, therefore, should be fed in mixtures with other feeds, specially containing high phosphorus, such as wheat bran, rice bran, grains etc. Although tree dodders are consumed by all livestock, goats have special preference compared to others. The digestibility of nutrients in tree fodders in generally higher in goats with better growth rates than in sheep when fed dried tree leaves (Bohra 1980; Singh and Bhatia 1982). Low digestibility of protein is anti-nutritional factors such as mimosne, cyanogens, saponins, oxalates etc. The deleterious effects of various anti-nutritional factors and their possible amelioration have been reviewed by Kumar (1991).

Proper methods of collection, conservation and processing are important aspects of utilization of tree fodders such as *Ailanthus excelsa*, *Z. nummularia* and *P.cineraria* were used in complete feeds for meat lambs. (Bhatia and Patnayak 1982). The proximate analysis of some common top feeds particularly of desert region is given in table.

Nutritive value of some most common top feed: Leaves of *A. excelsa, Z. nummularia* and *P. cineraria*, the three most important top feeds, are also high in the TDN (total digestible nutrients) scale (64%, 50% and 41.0%, respectively), while the DCP (digestible crude protein) value of *A. excelsa* leaves in quite high (16%), whereas, that of the more palatable *Z. mummularia* or *P. cineraria* leaves, are considerably low (5.6% and 4.5%).

Table 5. Proximate components (% DM basis) of some desert top feeds

Scientific name	Local name	Crude protein	Ether extract	Crude fibre	Nitrogen free extract	Ash	Phosphorus	Calcium	Magnesium
Acacia nilotica	Babul	13.9	-	9.2	69.8	7.1	0.10	2.6	0.4
Acacia senegal	Kumut	10.3	-	9.5	65.7	16.4	0.05	6.9	0.6
Ailanthus excelsa	Ardu	19.6	3.7	13.5	47.7	15.5	0.20	2.3	-
Albizia lebbeck	Siris	16.8	4.0	31.5	36.2	11.5	0.15	2.57	-
Azadirachtaindica	Neem	12.4	3.5	11.4	66.6	9.4	0.10	3.4	0.6
Dalbergiasissoo	Shisham	16.2	4.2	24.6	46.2	9.2	0.21	2.3	-
Ficusbengalensis	Bargad	9.6	2.6	26.8	51.6	9.3	-	-	-
Ficusreligiosa	Pipal	9.7	2.7	26.9	45.8	-	-	-	-
Prosopis cineraria	Khejri	13.9	1.9	20.3	59.2	6.5	0.20	1.9	0.5
Prosopisjuliflora	Vilayati babul	21.4	-	20.8	50.0	7.7	0.20	1.5	0.5
Salvadoraoleides	Pieujal	9.6	-	9.3	40.2	40.8	0.10	11.9	0.7
Salvadorapersica	Kharajal	14.2	-	9.4	44.9	31.4	0.20	8.8	0.7
Tecomellaundulata	Rohida	12.2	-	15.8	63.0	8.9	0.20	3.4	0.9
Ziziphusnummularia	Bardi	11.7	3.0	16.0	65.0	7.2	0.20	1.6	0.3

Sources: Ganguli *et al.* 1964; Acharya and Patnayak 1977; Mathur *et al.* 2011.

Un-conventional resource of desert region used as feed

Tumba (*Citrullus colocynthis*) Seed Cake - a cheaper feed resource: Tumba plant, an annual creeper, grows naturally in abidance in hot arid areas of the country with minimum possible water availability. It grows and multiply very fast during monsoon season and its fruits; Tumba - are available in the month of October-November. Mature fruits are golden yellow in colour, ball like, 10-12 cm in diameter. The taste of the fruit is very bitter. Tumba seed are rich in fat, having up to 16% oil. The Tumba (*Citrullus colocynthis*) seed cake (TSC) a by product of the oil extraction industry is nutritionally rich as it contains 16 to 22% CP. Presently, TSC is available in abundance in arid regions and is being used as a fuel for furnaces in factories, and its thus wasted.

Feeding trials were conducted in Central Arid Zone Research Institute, Jodhpur since 1986 on cattle to identify and evaluate TSC as a source of livestock feed.

Tumba (*Citrullus colocynthis*) seed cake-a cheaper feed resource for livestock feeding in arid region

The studies clearly indicates that in cattle the conventional concentrates can safely be replaced by TSC to the extent of 25%, as a regular ingredient, which constituted guar (*Cymposis tetragonaloba*) korma and oil cakes mostly til (*Sesamum indicum*) and cotton (*Gossipuim spps.*) seed cake) and also pelleted cattle feeds.

It is observed that the TSC replacement did not affect the palatability and intake of feeds and fodder. There was no significant (1>0.5) difference in milk yield pattern of control and treatment groups, in terms of quantity and quality. TSC feeding to heifers up to calving and onwards did not show any ill effect on different reproductive parameters.

It can, therefore, be inferred that a simple practice of inclusion of TSC in animal feed will definitely lower the cost of animal feeding by 18% to 20% without having any adverse effect on the production, general health and reproductive performance. In addition, it would result in the utilization of locally available non-conventional cheaper protein source, abridging the gap between demand and supply of the scarce protein, thus, benefiting the marginal farmers appreciably (Mathur *et al.* 1989; Mathur 1996).

Vilayati babool (Prosopis juliflora) pods

Mesquite or Vilayatibabool (*Prosopis juliflora)* trees are extensively grown in desert areas, especially to check the shifting of sand dunes. *Prosopis juliflora* provides highly palatable and nutritious pods in large quantities and the flowers produce good quality nectar for honey. Ripe pods which have fallen on the ground are widely consumed by domestic stock and wild ungulates. These are rich in protein and free sugar, which gives them a sweet taste. In Mexico, ripe pods of some *Prosopis* species are ground into coarse flour (Pinot) for human consumption and baked after removing the seeds and pods are also fermented and brewed into a weak beer, because of high carbohydrates and protein contents. Studied pods palatability and its effect on dry matter intake (DMI), water intake, body weight changes, milk yield and blood constituents in goats, sheep and cattle.

Goats: Ten (10) goats in late lactation were divided into two group of 5 each i.e. group I control and group II treatment. Animals of each group were offered weighed quantity of roughage and concentrate on as such basis (consisting of 35% grinded bajara, 40% tumba seed cake and 25% groundnut cake). However, in treatment group 35% bajra in concentrate was replaced by *Prosopis juliflora* pod powder (PJPP) making ration near about isocaloric. Study showed that *Prosopis juliflora* pod powder can be used up to 35% in the concentrate of goats (Mathur *et al.* 2004). No significant effect was observed on blood parameters and milk yield of goats in late lactation, during extremes of summer in arid zone (Mathur *et al.* 2003).

Sheep: A feeding trial was conducted to evaluate the acceptability and palatability of ration comprised of *Prosopis juliflora* pod husk (PJPH) in Marwari sheep. The results of the study indicated that *P. juliflora* pod husk can be used up to 50 percent level in the concentrate along with tumba (*Citrullus colocynthis*) seed cake as low cost ration of the sheep without any adverse effect on animal health (Mathur *et al.* 2002).

Cattle: A cheaper and balanced concentrate feed mixture for arid region was tried by simply mixing the locally available ground feed ingredients including *Prosopis juliflora* pods, Tumba (*Citrullus colocynthis*) seed cake, mineral mixture etc. To reduce cost of cattle production initiated feeding trial on lactating Tharparkar cattle of this cheaper concentrate mixture for one year on Tharparkar cattle, at Research cum Demonstration Unit of Tharparkar cattle, KVK, CAZRI, Jodhpur.Farmer accepts this process technology very easily and is possible at livestock owner's doorstep. Since in India livestock keepers are feeding concentrate to their productive animals without engaging labour towards its preparation, which otherwise is practiced in developed countries having organized dairy farms. Since, with each process step of feed preparation cost/energy is involved and ultimately feed become highly costly and livestock owner in India do not follow it due to unaffordable labour cost involved and cost benefit ratio (B:C). During the present scenario when all the food and feed ingredients are at very high price, the cost of cheaper concentrate formulated is less by Rupees 200 per 100 kg concentrate.

Photo: Villayati Babool (*Prosopis juliflora*) Pods a Cheaper Source of Concentrate Mixture for Lactating Cattle

The acceptability and palatability of formulated concentrate mixture having *Prosopis juliflora* pods was high with no ill effect on health, increases milk and the cost of cattle production reduces significantly (Mathur *et al.* 2009).

Prosopis cineraria vs Colophospermum mopane leaves

It was observed that the palatability of dry *Colophospermum mopane* leaves was low and decreased with progress of feeding from 15 to 5% from 1st to 5th week and the traditional local *P. cineraria* leaves were better source of supplementation even in the dry form supporting the milk yield of goats (Mathur *et al.* 2006). Although sole feeding of fresh *C. mopane* leaves is not possible in goats but it can very well be fed as a part of complete rations at 35-40% of whole ration (Patil *et al.* 2007).

Lani (Salsola baryosma): Salty shrub Lani (*Salsola baryosma*) of arid region available at vegetative stage can be a good source of fodder to animals. Shoots consists of mainly fleshy stem, since leaves are very minute, consisting not more than 5-7% of the total biomass. The acceptability and palatability of fresh cut shoots showed that Lani is palatable (Mathur *et al.* 2007).

Lana (Haloxylon salicornicum) seeds: Lana (*Haloxylon salicornicum*) salty weed available in the arid region produces seeds having CP 18.60% found to be good source of protein in the concentrate part of lactating cows replacing about 25% of the conventional sesamum cake (Mathur *et al.* 2011).

Hardwickia binata as a supplement to goats: H. binata is available in plenty in the arid region having rainfall above 300 mm. the study showed that the fresh leaves are palatable to goats and sole feeding supports the body weight growth of growing kids (Patil *et al.* 2009).

Thornless cactus (Opuntia ficusindica): A new fodder source *Opuntia ficus*indica-a thornless cactus introduced in Indian arid region was found to have the fodder value as maintenance feed and was observed to reduce the water requirement if fed along with the dry roughages in goats, sheep and growing cattle. In addition its high mineral content may reduce the mineral requirement, as arid animals are suffering from mineral imbalances (Mathur *et al.* 2009).

Photo: Cactus (*Opuntia ficusindica*), thornless feed a mineral rich resource for cattle feeding

Cactus (*Opantia ficusindica*) cladodes were having DM 08.00%, with total ash 26.55% composed of AIA 02.50 and ASA 24. Cacti is rich in minerals particularly Calcium. The mineral rich cactus (*Opuntia ficusindica*) cladodes have good acceptability, palatability and dry matter intake with higher body weight gain in cattle.

Minimizing impact of extreme climatic conditions on production performance of grazing sheep in Thar desert

Thar Desert is characterized by high ambient temperature, low rainfall, high sun shine and wind velocity. An experiment was initiated in the month of February, 2012 at Chandan area of Jaisalmer district, maintaining sheep under continuous grazing system on fenced over mature sevan (*Lasiurus sindicus*) pasture as per the carrying capacity. The objective of study was to develop coping strategies for minimizing impact of extreme climatic conditions on livestock production of Thar Desert.The grazing time taken minutes hr^{-1} by sheep of supplemented group was higher than the non-supplemented group. Water consumed was higher in summer months in both the groups. Water consumption was significantly higher in the month of November 2012 than November 2013, may be due to changing ambient conditions during the period (Mathur *et al.* 2013). Supplementation of concentrate and health management results into increase in live body weights and wool yield. The study shows that maintaining grazing sheep under extreme climate condition exclusively on fenced over mature Sevan (*Lasiurus sindicus*) pasture causes animal production loss, which can be mitigated by adopting scientific feeding and health management practices(Mathur *et al.* 2016).

Feeding management of livestock during drought can be summarized as follow

1. Preparation of ration utilizing top feeds
2. Providing vitamin "A" doses
3. Cheaper and balanced concentrate utilizing local non-conventional feed resources
4. Addition of leguminous crop by products
5. Chopping of straw
6. Mineral mixture and common salt
7. Deworming

Health management of livestock

Broadly common diseases are divided in to four groups:-

1. Viral diseases: Rinder pest, (Cattle plague), Foot & Mouth disease, Rabies and Sheep pox.
2. Bacterial diseases: Anthrax, Tuberculosis, Mastitis, Haemorrhagic Septicaemia, Black-quarter and Pneumonia.
3. Parasites and Parasitic diseases: Ectoparasites (ticks and mites), Endoparasites (Protozoan, Nematodes, Trematodes and Cestods). Deworming: Animal should be dewormed with broad spectrum anthelmintic (Albendazole/Fenbendazole/ Closantel) thrice yearly. In case of ectoparasites: Spray/dusting should be done at regular intervals with ectoparasiticide (eg., deltamethrin). CLOSANTEL anthelmintic action is both endoparasiticide, and ectoparasiticide. These simple treatment will increase the availability of nutrients to livestock.

   4. Miscellaneous diseases & pathological conditions: Urinary calculi, milk fever, grass tetany, cobalt deficiency, PICA, vitamin A deficiency and paraplegia.

Strategies to mitigate drought

Immediate measures

- Utilization of crop residuals
- Utilization of non-conventional feed resources.
- Utilization of fodder trees

Long term measures

- Protection of natural rangeland for allowing restoration of grass cover, thereby increasing forage availability.
- Reseeding of natural rangeland with appropriate perennial grasses and legumes.
- Silvi- pastoral development- Two tier and three tier systems could increase forage yield.
- Agro-forestry system- Agro-forestry systems are ecological models similar to natural system and are sustainable.
- Horti-pasture and agro-horticulture systems.
- Social forestry system.
- Intercropping system- The system involves introduction of legume fodder in line sown cereal crops.
- Cultivated fodder production- Forage resources for small ruminants can also be increased by cultivation of fodder crops viz. Lucerne, berseem, oats in rabi season and guar, cowpea, field beans etc, in kharif season by following suitable agronomic practices.
- Animal feed storage: through establishment of fodder banks at village/ tehsil level and there monitoring.

Thus, Livestock have crucial role in food security providing milk, meat, egg, hide and skin, manure and also draught power in addition to employment, leading as risk aversion mechanism for sustaining healthy family.

References

Administrative Bulletin (2008-09) Directorate of Animal Husbandry, Govt. of Rajasthan Jaipur, Rajasthan.

Baba SH, Wani MH, Bilal AZ (2011) Dynamics and Sustainability of Livestock Sector in Jammu and Kashmir. Agricultural Economics Research Review 24: 119-132.

Census of Livestock (1997) Government of Rajasthan, Jaipur.

Census of Livestock (2003) Government of Rajasthan, Jaipur.

Census of Livestock (2007) Government of Rajasthan, Jaipur.

Census of Livestock (2012} Government of Rajasthan, Jaipur.

Singh Chakravarti, Rollefson IK (2005) Sheep pastoralism in Rajasthan, still a viable livelihood option? Workshop report, LPPS, Sadri, Pali, Rajasthan.

Dongre RA (2000) Haemato-biochemical Studies in Relation to Trace Mineral Status in Camel. M V.Sc. thesis, Rajasthan Agricultural University, Bikaner, Rajasthan.

Fakhruddin (1987) Mineral and vitamin deficiencies in livestock of western Rajasthan. In: National Symposium on Livestock Production vis-a-vis Fodder and Pasture Development in western Rajasthan, Abstract, Livestock Research Station, Rajasthan Agricultural University, Chandan, Jaisalmer, Rajasthan.

Gahlot AK (2005) Animal health management in drought prone area. In: Seminar on Sustainable livestock production in arid region, College of Veterinary and Animal Science, Bikaner, Rajasthan.

J and K Economic Survey 2012-13 (2015) Planning and Development Department, Govt. of Jammu and Kashmir. Livestock, pp. 241-260, Kashmir.

Kishore R (1998) Clinico-haemato-biochemical and Therapeutic Studies on Bovine Suffering from Allotriophagia. M.V.Sc. thesis, Rajasthan Agricultural University, Bikaner, Rajasthan.

Macala J, Sebolia B, Majinda RRT (1992) Proceedings of the Networks Workshop. In: Stares JES, Said AN, Kategile JA (eds) African Feeds Research Network, Gaborone, Botswana, 4-8 March 1991, International Livestock Centre for Africa, Addis Ababa. Ethopia, FAO, Rome.

Mathur BK (2013) *Juliflora* Based Animal Feed. A Compendium of Agro Technologies. Agri-Tech Investors Meet, New Delhi, National Agricultural Innovation Project, Krishi Anusandhan Bhawan –II, New Delhi, pp. pp. 122-123.

Mathur BK (2014) Nutritional constraints and health management of livestock: assistance to states for control of animal diseases (ASCAD). In: Seminar 2014 on New Horizons: Effective and Productive Livestock Development, Animal Husbandry Department, Government of Rajasthan, Ajmer, Rajasthan, pp. 31-41.

Mathur BK, Singh JP, Rathore VS, Singh NP, Beniwal RK (2011a) Utilization of hot desert shrub: Lana (*Haloxylon salicornicum*) seeds as feed resource in arid zone. Indian Journal of Small Ruminants 17(2):231-234.

Mathur BK (2015) Strategies of livestock feeding and health management in arid regions. Annals of Arid Zone 54 (3 & 4): 95-108.

Mathur BK, Singh JP, Ullha Ubed, Bohra RC (2016) Livelihood security through management of small ruminants in hot arid zone. In: Bhatt RK, Burman U, Painuli DK, , Singh DV, Sharma Ramavtar, Tanwar SPS (eds) Climate Change and Agriculture: Adaptation and Mitigation, Satish Serial Publishing House, Azadpur, Delhi, India, pp. pp.527-548.

Mathur BK, Bhati TK, Tewari JC (2006b) Comparative palatability of *Prosopis cineraria vs Colophospermum mopane* leaves in lactating Marwari goat in arid region. In: National Symposium on Livelihood Security and Diversified Farming Systems in Arid Region, Arid Zone Research Association of India, CAZRI, Jodhpur, No.3-2, pp.51-52.

Mathur BK, Mathur AC, Bohra HC, Sharma BK (2006a). Mineral status in anestrous cattle and buffaloes of arid region of Rajasthan. In: XII Animal Nutrition Conference on Technology Interventions in Animal Nutrition for Rural Prosperity, ANSI and ICAR, Anand Agricultural University, Anand, pp.78.

Mathur BK, Misra AK, Mathur AC, Singh JP (2011 b). Feeding strategies of livestock in arid Agro-ecosystem of India. In: Tiwari SP, Sanyal PK (eds) Veterinary Nutrition and Health, Satish Serial Publishing House, Commercial Complex, Azadpur, Delhi, India, pp.535-550.

Mathur BK, Patil NV, Mathur AC, Patel AK (2005) Impaction of Rumen of desert buck due to Polythene bags- a case study. Livestock International 10 (5):17-20.

Mathur BK (1996) Effect of Replacement of Cotton (*Gossypium* spp.) Seed Cake from Tumba (*Citrullus colocynthus*) Seed Cake and Sun Dried Poultry Droppings on the Performance of Sheep. Ph.D. thesis, Rajasthan Agriculture University, Bikaner, Rajasthan.

Mathur BK (2003) Livestock: human need for sustainability in arid environment. In: Narain, Pratap, Kathju S, Kar Amal, Singh MP, Kumar Praveen (EDS) Human Impact on Desert Environment, Arid Zone Research Association of India and Scientific Publishers, Jodhpur, India. pp. 506-514.

Mathur BK, Abichandani RK, Patel AK (1999) Higher incidences of twinning in Marwari goats by supplementary feeding in hot arid region. In: Proceedings IX Animal Nutrition Conference, Abstract, Acharya N.G. Ranga Agricultural University, Rajendra Nagar, Hyderabad, pp. 144.

Mathur BK, Mittal JP, Prasad S (1989) Effect of tumba cake (*Citrulluscolocynthis*) feeding on cattle production in arid region. Indian Journal of Animal Sciences 59: 1464-1465.

Mathur BK, Patel AK, Mathur AC, Abichandani KK, Kaushish SK (2000a) Wheat straw (untreated treated) consumption by tharparkar heifers during summer under stall fed conditions. In: Third Biennial Conference of Animal Nutrition Association on Livestock feeding strategies in the new millennium, CCSHAU, Hissar, pp. 206.

Mathur BK, Patel AK, Kaushish SK (2000b) Utilization of tumba (*Citrullus colocynthis*) seed cake of desert for goat production. 70 (4): 431-433.

Mathur BK, Patil NV, Mathur AC, Bohra HC, Bohra RC, Sharma KL (2009) Comparative mineral status of Jodhpur district villagers' cattle vs institute farm managed tharparkar cattle in hot arid zone. In: Proceedings of Animal Nutrition World Conference, New Delhi, India. pp. 61.

Mathur BK, Singh JP, BeniwalRK, SinghNP (2007) Utilization of salty shrub, Lani (*Salsola baryosma*) of arid region as drought feed for goats. FR-55. In: International Tropical Animal Nutrition Conference, NDRI, Karnal, pp. 33.

Mathur BK, Singh JP, Sani Nirmala, Pathak KML, RathoreVS, Kiraroo BD, Lukha Arjun, Beniwal RK (2009) Blood biochemical and mineral profile of camel maintained on different feeding regimen in hot arid region of India. In: Diversification of Animal Nutrition Research in the Changing Scenario, volume II, Abstract, 13th Biennial conference of Animal Nutrition Society of India, Banglore, pp. 80.

Mathur BK, Singh JP, Jat SR, Bhatt RK, Roy MM (2013a) Improving sheep productivity in the Thar desert through strategic supplementation and health care. In: National Seminar on Prospects in Improving Production, Marketing and Value Addition of Carpet Wool, Indian Society for Sheep and Goat Production and Utilization at Central Sheep and Wool Research Institute, Arid Region Campus, Beechwal, Bikaner, Rajasthan, India.

Mathur BK, Sirohi AS, Mishra AK, Patel AK, Mathur AC, Bohra RC (2013 b) Performance of goats fed on ardu (*Ailanthus excelsa*) and Neem (*Azadirachta indica*) leaves in arid region. Veterinary Practitioner 14 (2): 348-350.

Mathur BK, Siyak SR, Bohra HCn(2003) Feeding of mesquitel (*Prosopis juliflora*) pods powder to Marwari goats in arid region. Current Agriculture 27 (1-2): 57-59.

Mudgal VD, Pradhan K (1989) Proceedings of a consultation on Non-conventional Feed Resources and Fibrons Agricultural Residues - Strategies for Expanded Utilization. CCHAU, Hisar, pp.139.

Narain P, Sharma KD, Rao AS, Singh DV, Mathur BK, Ahuja Usha Rani (2000) Strategy to Combat Drought and Famine in the Indian Arid Zone. Central Arid Zone Research Institute, Jodhpur, pp. 65.

Patidar M, Mathur BK, Rajora MP, Mathur D(2008) Effect of grass – legume strip cropping and fertility levels on yield and quality of fodder in silvipastoral system under hot arid condition. Indian Journal of Agricultural Sciences 78: 394-398.

Patil NV, Mathur BK, Bohra HC, Patel AK (2005) Relative preference index for arid forages for goat and sheep. In: National Seminar on Conservation, Processing and Utilization of Monsoon Herbage for Augmenting Animal Production, Indian Society for Sheep and Goat Production and Utilization, RRS, CSWRI, Bikaner. pp.190.

Patil NV, Mathur BK, PatelAK, Bohra RC (2011) Nutritional evaluation of *Colophospermum mopane* as Fodder. Indian Veterinary Journal 88(1):87-88.

Radostits OM, Blood DC, Gay CC (1994) Veterinary Medicine, 8th ed., Bailliere Tindall. London (U.K.).

Sharma SN, Gahlot AK (1997) Veterinary, Jurissprudence, 4th ed., NBS Publishers, Bikaner, Rajasthan.

SinghAP,Gahlot AK (1999) Studies on sarcoptesscabei infestation in dromedary. In: 17thInternational Conference of the World association for the advancement of Veterinary Parasitology, Abstract, Denis Centre for Experimental Parasitology, The Royal Veterinary and Agricultural University, Copenhagen (Denmark).

Singh AP, Netra PR,Vashistha MS, Sharma SN (1994) Zinc deficiency in cattle. Indian J Animal Science 64:35-40.

Singh AP, Vyas SK,Tanwar RK, Sharma SN,Gahlot AK (1999) Studies on diagnosis and therapeutic aspects of sub clinical mastitis in cattle by using ampicillin and cloxacillin combination (AC-VET). Intas Polivet 1:13-19.

Tanwar RK, Kumar A, Sharma SN, Yadav JS, Mathur PD (1983) Clinical findings and therapy of polioencephalomalacia in goats. Indian J Vet Med 3:112- 114.

TanwarRK, Malik KS,Gahlot AK (1994) Polioencephalomalacia induced with amprolium in buffalo calves. Clinico-pathological findings. Journal of Veterinary Medicine 41: 396-404.

Venkatateswarlu J, Aantharam K, Purohit ML, Khan MS, Bohra HC, Singh KC, Mathur BK (1992) Forage 2000 AD The scenario for arid Rajasthan, CAZRI, Jodhpur pp. 32.

25

Molecular Characterization of Tree Species for their Genetic Improvement

S.K. Singh

Principal Scientist, Plant Pathology and In-charge Biotechnology Section Central Arid Zone Research Institute, Jodhpur, Rajasthan, India

Introduction

A number of researchers have used morphological descriptors such as growth habit, leaf, shoot, flower, fruit morphology to classify and discriminate different varieties/cultivars and or detect inter and intra-specific variations in different tree species of economic importance (Saran *et al.* 2006; Holland *et al.* 2009). Such morphological descriptors are quite homogenous and often found insufficient to distinguish varieties within morphological groups. The choice of suitable cultivars is of paramount importance for its success. The lack of breakthrough has been due to under utilization of genetic variability for superior quality and high yield potential. Most of the fruit tree species demonstrates a rich genetic diversity mainly through natural cross pollination due to self incompatibility (Godara 1980). The elite plant types with desirable traits have been released as varieties, mass multiplied and propagated through standard vegetative multiplication. However, the authenticity of cultivar identification remains unclear (Devanshi *et al.* 2007).

The morphological traits are influenced by environmental factors and developmental stages of plant, so the results do not provide true assessment of genetic diversity. Molecular markers offer several advantages over the conventional breeding tools for selection of diverse parents. Among various marker systems, the Randomly Amplified Polymorphic DNA (RAPD) is one of the most popular DNA based approaches (Williams *et al.* 1990). It allows the analysis of individual and large number of markers in relatively short time, as only a few primers allow the generation of sufficient data to obtain a robust estimate of diversity index and have allowed the resolution of complex taxonomic relationships. This technique has recently been used in a number of fruit trees *Mangifera indica* (Souza *et al.* 2011), Citrus species (El-Mouei *et al.* 2011), *and Punica granatum* (Sarkhosh *et al.* 2009; Narzary *et al.* 2009).

Recently Start Codon Targeted (SCoT) polymorphism has emerged as a new and promising marker technique for the genetic diversity assessment in plants (Collard and Mackill 2009) due to longer primer sequences and reproducibility. The SCoT marker is designed based on the conserved region surrounding the translation initiation codon, ATG (Sawant *et al.* 1999). It is used as a single primer for amplification of genomic DNA without prior genomic sequence information. Due to lesser recombination intensity between SCoT marker and the gene of interest, it can be used directly in marker-assisted breeding programs unlike random markers (Mulpuri *et al.* 2013). SCoT markers have been successfully used to assess genetic diversity and structure, identify cultivars, and for quantitative trait loci (QTL) mapping and DNA fingerprinting in different plant species including orange (Jiang *et al.* 2011), date palm (Al-Qurainy *et al.* 2015), *Pistacia* species (Amirbakhtiar and Sorkheh 2015), mango (Leo *et al.* 2011) and jojoba (Heikrujam *et al.* 2015).

Inter-simple sequence repeats (ISSR) are inter-tandem repeats of short DNA sequences and are present within the microsatellite repeats. They exhibit specificity of sequence-tagged-site markers showing the variation within unique regions of the genome at several loci. ISSR markers are inexpensive and readily adaptable technique for routine germplasm fingerprinting and evaluation of genetic relationship between accessions or genotypes (Agarwal *et al.* 2008) and construction of genetic linkage maps (Sahu *et al.* 2015). The technique does not require sequence information for primer synthesis having the advantage of random markers. The ISSR markers have great potential to detect genetic polymorphism, determine intra and inter-genomic diversity compared to other arbitrary primers and simultaneously reveal variation within unique regions of the genome. The primers used in ISSR analysis are based on di-, tri-, tetra- or penta-nucleotides motifs found at microsatellite loci and gives a wide array of possible amplification products (Al-Turki *et al.* 2015). The technique is useful in examining genetic diversity, phylogenetic studies, gene tagging, genome mapping and evolutionary biology studies in a wide range of crop species (Varshney *et al.* 2013). ISSR is a technique of choice for testing of genetic diversity in plant species for crop improvement (Pakseresht *et al.* 2013). Recently its utility has been demonstrated in genetic diversity analysis of *Eucalyptus* (Ballesta *et al.* 2015), chickpea (Bhagyawant *et al.* 2015), wild berry (Zoratti *et al.* 2015), *Ziziphus mauritiana* (Singh *et al.* 2007) and rice (Al-Turki *et al.* 2015).

Currently, nuclear ribosomal Internal Transcribed Spacer (ITS) intron regions encompassing highly conserved 5.8 rRNA gene is considered as one of the most useful phylogenetic marker due to less functional constrains and comparatively higher evolutionary changes (Alvarez and Wendel 2003). It evolves relatively quickly and can be useful in determining inter-specific (Singh *et al.* 2005) and sometimes intra-specific relationships (Kakani *et al.* 2011). The rate and patterns of ITS sequence mutation are typically appropriate for resolving relationships among species and genera (Saini *et al.* 2008). Although thousands of copies of the ITS exist in Angiosperm genomes, but are generally homogenized by concerted evolution, and thus can be treated as a single locus (Baldwin *et al.* 1995). The nuclear ribosomal RNA (r RNA) gene complex is a tandem repeat unit of one to several thousand copies. This complex has several domains that evolve at varying rates (Jorgenson and Cluster 1988) and thus have different phylogenetic utilities. The region of 5.8S rRNA gene was shown to contain considerable phylogenetic information (Mir *et al.* 2010). The ITS polymorphism might occur at a genus, species or individual levels, making it useful for phylogenetic, evolutionary and bio-geographical diversity studies (Carvalho *et al.* 2009). ITS sequences have been widely used in several species (Xiao *et al.* 2010) for phylogenetic inferences.

DNA extraction and quantification

The genomic DNA was extracted from 100 mg of fresh leaves of each of the tree sample, crushed with pestle and mortar in liquid nitrogen. The plant genomic DNA purification spin kit "Hi Pura" of Hi-media Company and protocols suggested by Birren and Lai (1993) and Sambrook *et al.* (1989) were followed for genomic DNA isolation. Finally, the genomic DNA was eluted with 200 μl of Tris-EDTA buffer (TE) for DNA fingerprinting. DNA was quantified with UV/VIS spectrophotometer by measuring OD_{260} and OD_{280}. The quantified DNA samples were diluted in TE buffer to make a final concentration of 50ng/ μl for PCR reactions.

PCR amplification of molecular markers

The decamer random primers of OPA, OPB and OPP series (Operon Technologies) were used for initial screening of varieties/cultivars/species. Based on the reproducibility of scorable bands, the multi-locus genotyping by RAPD was performed using selected decamer arbitrary primers. Each amplification was performed in a total volume of 25 ml containing: decamer primer, 1 μl (50 pmol/ μl); dNTP mix, 2 μl (2.5 mM/ μl Bangalore Genei); *Taq* DNA polymerase, 0.4 μl (5U/ μl, Sigma Chem); $MgCl_2$, 1 μl (25-mM, Sigma Chem); 10× PCR buffer, 2.5 μl (100 mM, Tris-HCl, pH 8.3, 15 mM $MgCl_2$, 250 mM KCl), 14.1 μl of dH_2O and 4 μl of genomic DNA (50 ng/μl). RAPD-PCR amplifications were performed in a gradient thermal cycler (Corbett Research, USA) with initial denaturation step of 94 °C for 3 min followed by 36 amplification cycles of 94 °C for 40 sec, 50 °C for 40 sec and 72 °C for 2 min and final elongation at 72 °C for 10 min. Amplicons were separated on to a 1.4% agarose gel pre-stained with ethidium bromide in 1× TAE buffer. The gel was run for 3 h at 50 V. The size of the amplified fragments was determined using 100 bp plus ladder (MBI Fermentas). Reproducibility of RAPD amplified DNA fragments were examined by repeating PCR reactions as well as running on gel twice.

A set of 36 Start Codon Targeted (SCoT) markers (Integrated DNA Technologies (IDT) were screened and 17 primers producing unambiguous and scorable fragments were selected for assessment of genetic diversity among 37 varieties of *Z. mauritiana*. Similarly, a set of 22 ISSR primers were initially tested and finally ISSR analysis was performed with 14 primers. Each PCR-amplification was performed in a total reaction volume of 25 μl containing primer, 50 pmol; dNTP mix, 0.25 mM, *Taq* DNA polymerase 1U (Sigma chem.); 10X PCR buffer, 5.0 mM (Tris-HCl, pH 8.3, 15 mM $MgCl_2$ 250 mM KCl), genomic DNA 50 ng in dH_2O. Amplicons were separated on 2 % agarose gel pre-stained with ethidium bromide solution using 1× TAE buffer. The gels were run for 2 h at 110 V. The size of the amplified fragments was determined using 100 bp plus ladder (MBI Fermentas). All the reactions were performed twice to test the reproducibility of the amplicon profiles generated by both the marker systems. SCoT–PCR amplification were performed in a gradient thermal cycler (Corbett Research, USA) with initial denaturation step of 95°C for 5 min followed by 40 amplification cycles of 95°C for 45 sec, annealing at 51°C for 45 sec and 72 °C for 1 min 30 secand final elongation at 72 °C for 10 min. Whereas, the reaction mixture and PCR amplification conditions of ISSR were same as that of SCoT amplification except the annealing temperature which was kept at 42 C for ISSR.

Molecular analysis of molecular markers

Reproducible bands of each locus of RAPD, SCoT and ISSR were scored as binary characters (present=1, absent=0). Combined data matrix obtained from all the 10 random primers was used to determine Jaccard's similarity coefficient with NTSYS-pc software (Rohlf 1997). The polymorphic information content (PIC) values for all the selected primers amplified by a particular primer pair was calculated for the RAPD markers to characterize the capacity of each primer to detect polymorphic loci using the formula derived by Smith *et al.* (1997). To perform analysis of molecular variance (AMOVA), germplasm was divided into populations based on the state from where they were originally developed. Principal Coordinate Analysis (PCA) via covariance matrix was calculated using GenALEx 6 software (Peakall and Smouse 2006). Whereas, diversity in the frequency of fragment size of RAPD patterns was apportioned within and among varieties using Shannon's information index (*i*) (Lewontin 1972) and gene diversity index (*h*) following Nei (1973), coefficient of genetic differentiation between populations (GST) and gene flow (Nm) using PopGen 32 programme.

PCR amplification of ITS region

The universal primers ITS-1 and ITS-4 developed by White *et al.* (1990) were used to amplify the ITS-1 and ITS-2 regions of ribosomal DNA, encompassing the 5.8S rRNA gene. Each PCR amplification was performed in a total volume of 50 ml containing: 1U Taq DNA polymerase (Bangalore Genei), 2.5 mM $MgCl_2$, 160 mM dNTP mix (MBI Fermentas, Germany), 50 pmol of each of ITS-1 and ITS-4 primers (Bangalore Genei), 50 ng genomic DNA in dH_2O. The reactions were performed in a thermal cycler (Corbett Research, USA) with following conditions: 1 min denaturation at 95 °C, 30 sec annealing at 50 °C, 80 sec elongation at 72 °C, for 34 cycles with a final elongation step of 72 °C for 10 min. Agarose gel was stained with ethidium bromide and photographed under UV light using Syngene gel documentation system.

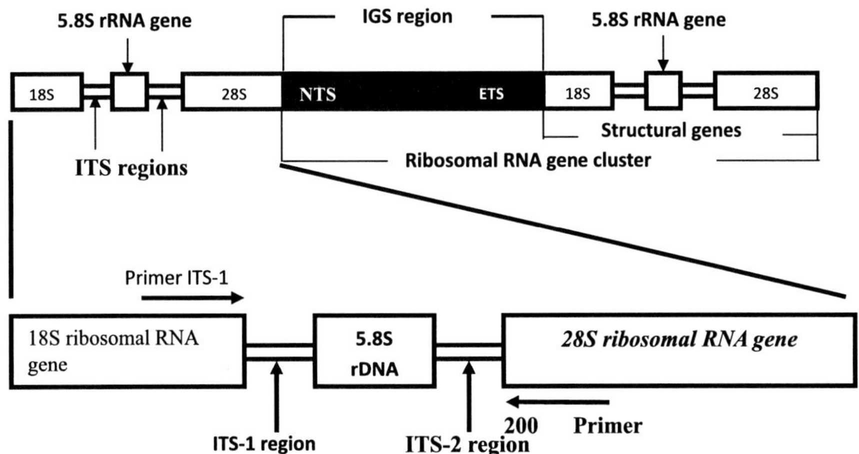

Sequence analysis

Amplified ITS regions were sequenced with an ABI Prism DNA sequencer (Applied Biosystems, Carlsbad, CA, USA) using ITS-1 and ITS-4 primers separately for labeling of each DNA by the BigDye terminator method (Applied Biosystems, Foster City, CA, USA). The sequenced data obtained from the ITS-4 primer were inversed using Gene Doc software and clubbed with the sequence data obtained with the ITS-1 primer, to obtain the complete sequence of the ITS region. Comparison of nucleotide sequences was performed using the Basic Local Alignment Search Tool (BLAST) network services of the National Centre for Biotechnology Information (NCBI) database (http://www.ncbi.nlm.nih.gov). Molecular characterization of tree species/varieties was done on the basis of similarity with the best aligned sequence of BLAST search. The phylogenetic relationships were established by multiple alignment of sequences using ClustalX 2.0.11 and generating phylogram depicting bootstrap values using NJ plot software based on Single Nucleotide Polymorphisms (SNPs), insertions/deletions (INDELS), and or length polymorphism in the ITS and 5.8S nuclear rDNA regions. To assess the robustness of phylogenetic relationships of pomegranate varieties, best aligned available reference sequences representing all the test genera from GenBank database were downloaded in FASTA format. A composite phylogenetic tree with bootstrap values showing grouping of varieties sequenced was generated to measure phylogenetic accuracy.

Genetic diversity in *Acacia senegal*

Acacia senegal is well adapted to arid environment of western Rajasthan and has a potential to restore soil fertility and sand dune stabilization. There is a scope of improvement by exploiting geographical genetic diversity. Seeds of five accessions from Niger (EC 87/ 7490), Mali (EC 87/ 7493, EC 87/ 7497 and EC 87/ 7499) and Senegal (EC 87/ 7500) were procured from CIFT Nogent Sur Marve. Seedlings of these accessions along with the accessions of Sudan (Sudan/87) and local collections from Rajasthan (Raj/87) were transplanted in 1988 at Central Research Farm of Central Arid Zone Research Institute (CAZRI), Jodhpur. The plants showing significantly high and low seed yields from each accession were selected for molecular diversity analysis (Table 1).

Table1. Details of *Acacia senegal* geographic populations used for molecular diversity analysis

S.N.	Tree No.	Accession Number	Seed yield	Origin	Longitude	Latitude	Altitude (m)	Precipitation (mm)
1	3	Sudan/87	Low	Sudan	30°00’E	15°00’E	400	161
2	35	Sudan/87	High	Sudan	30°00’E	15°00’E	400	161
3	68	EC87/7490	Low	Niger	8°00’ E	16°00’N	-	150
4	78	EC 87/7490	High	Niger	8°00’ E	16°00’N	-	150
5	113	EC87/7497	High	Mali	10°22’W	14°37’N	150	650
6	224	EC87/7493	High	Mali	11°35’W	14°57’N	100	560
7	235	EC87/7499	Low	Mali	07°10’W	15°08’N	270	450
8	244	EC87/7499	High	Mali	07°10’W	15°08’N	270	450
9	266	EC87/7493	Low	Mali	11°35’W	14°57’N	100	560
10	296	EC87/7500	Low	Senegal	14°49’ W	15°40’N	60	400
11	407	EC87/7500	High	Senegal	14°49’ W	15°40’N	60	400
12	411	Raj/87	Low	India	73°08’E	26°18’N	241	317
13	442	Raj/87	High	India	73°08’E	26°18’N	241	317

Thirteen selected plants showing significantly high and low seed yields from Rajasthan and exotic locations, viz. Niger, Mali, Senegal and Sudan, transplanted in 1988 at CAZRI were subjected to randomly amplified polymorphic DNA (RAPD) analysis. Six random primers generated a total of 86 scorable loci and exhibited 77.77 to 94.73% polymorphism. RAPD profiles showing polymorphism in banding patterns of *A. senegal* using OPP-9 random primer is shown in Fig.1. Dendrogram obtained from cluster analysis delineated all the 13 population samples representing 7 geographical populations and revealed existence of genetic diversity within and among geographical populations of *A. senegal*. The Indian population exhibited the maximum genetic diversity from rest of the African populations (Fig.2).

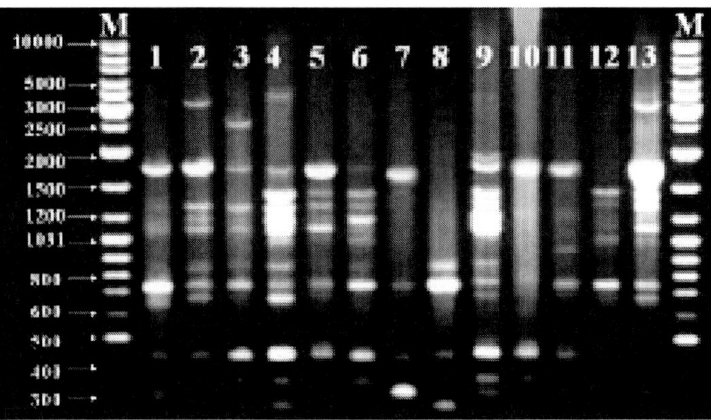

Fig.1. RAPD profiles showing polymorphism in banding patterns of *A. senegal* using OPP-9 random primer

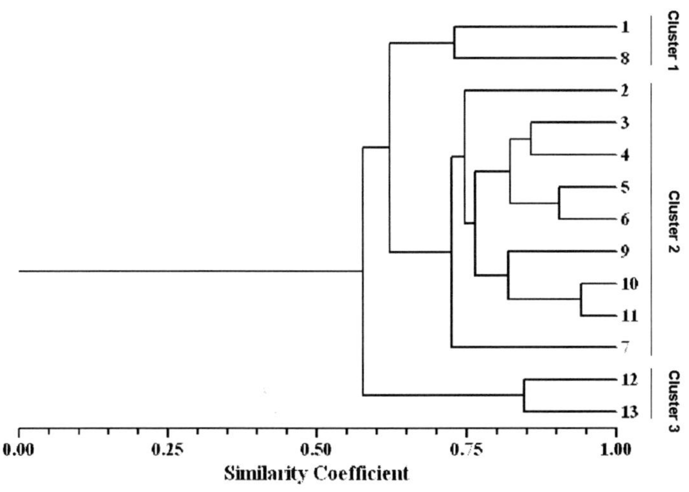

Fig.2. Dendrogram showing genetic diversity in geographical populations of *A. senegal*

All accessions from within and amongst different geographical locations exhibited 7 to 43% genetic diversity. The maximum genetic similarity of 93% was recorded between low and high seed yielding trees of accession EC 87/7500 from Senegal. The low seed yielding trees of accession Sudan/87 from Sudan showed about 74% genetic similarity with high seed yielding population of accession EC 87/7499 from Mali. The low and high seed yielding samples representing EC 87/7490, EC 87/7493, EC 87/7499 and EC 87/7500 accessions exhibited the maximum similarity within and amongst themselves and grouped together in cluster II. Whereas, the maximum genetically diverse accession Raj/87 representing Rajasthan, exhibited 43% genetic diversity from rest of the African populations of *A. Senegal* is attributed to the differences in the geographical conditions. The genetic similarity within Cluster II could be due to similar environmental ecology. Genetic similarity of Sudan/87 accession with EC 87/7499 of Mali (cluster I) and out grouping of Sudan/87 and EC 87/7499 (cluster II) is due to cross pollination and out crossing nature of *A. Senegal* whereby pollen from other populations also contribute to the gene pool of the seed (Morris *et al.* 2002). Simultaneously, separation of low seed yielding accession of Sudan/87 (cluster I) and out grouping of high seed yielding population of Sudan/87 within cluster II can be due to separation of Sudan from Niger, Mali and Senegal by Chad lake and significant differences in Longitude.

It is concluded that all high seed yielding exotic accessions had greater biomass than the Indian populations of *A. Senegal.* The RAPD dendrogram also exhibited substantial genetic diversity within and among exotic populations *vis-a-vis* Indian populations. The plus tree nos. 73, 78, 224, 244 and 296 may be used as one of the genetic diverse parents for cross pollination to induce heterosis with Indian populations to increase biomass and greater adaptability in arid environment of Rajasthan.

Inter-specific genetic diversity in selected species of *Prosopis*

Species of the genus *Prosopis* (Fabaceae Mimosoideae) are well adapted to grow in arid and semi-arid regions of the world. They are among the most important multipurpose leguminous trees and are used for re-vegetation, agroforestry, apiculture, fodder and fire wood. The interspecific genetic diversity in eight *Prosopis* speciesbelonging to Section Algarobia representing Series Pallidaeand Chilensis was determined using RAPD analysis (Table.2).

Table 2. Details of *Prosopis* species belonging to Section Algarobia

Accession no.	Prosopis species	Series	Morphological characters
Trujillo Prov. Proto Prac,	*P. pallida*	Pallidae	Leaflets 6-15 pairs / pinna, spines small, axillary, sometimes Peru EC 308211 absent, legumes fleshy, straight and sweet in taste
Introduced from Israel in 1968	*P. articulata*	Pallidae	Leaflets 6–20 pairs / pinna, spines axillary, legume dry, flattened, sweet in taste
Mexico EC 1208/83	*P. glandulosa*	Chilensis	Leaflets 6-17 pairs / pinna, spines axillary, uninodal, longand mostly solitary, legumes straight or moniliform, sweet in taste
Capayan, Catamarca, Argentina EC 308189	*P. chilensis*	Chilensis	Leaflets 10–29 pairs / pinna, spines axillary, uninodal, hard up to 6cm long. Legumes linear compressed with parallel margins, straight, falcate or sub falcate, sugary and edible
San martin, Catamarca,	*P. nigra*	Chilensis	Leaflets 20 -30 pairs / pinna, spines big, legumes straight or Argentina EC 308029 sub falcate, sub moniliform legumes, fleshy, compressed,rounded, edible and sweet
Anilaco, Catamarca, Argentina EC 308072	*P. flexuosa*	Chilensis	Leaflets 12–29 pairs / pinna, spines small, legumes straight to sub falcate, sweet and edible
Introduced from Mexico in 1917	*P. juliflora*	Chilensis	Leaflets 11-15 pairs / pinna, spines axillary, divergent, paired, legumes straight with incurved apex sometimes falcate, legumes bitter in taste
Neon Yacanta, Cardoba EC 308151	*P. alba*	Chilensis	Leaflets 25-50 pairs / pinna, spine scare and small only on strong shoots, legumes falcate to ring-shaped, compressed,

The combined dendrogram of the seven random primer data matrix delineated eight species into two clusters and three species as out groups (Fig. 3). Cluster I included *P. pallida*, *P.articulata* and *P. glandulosa*, cluster II included *P. chilensis* and *P. nigra*. Whereas three out grouped species *P. flexuosa*, *P. juliflora* and *P. alba* exhibited significant genetic diversity from cluster I and cluster II. *P. alba* showed the maximum genetic diversity of 60%, *P. juliflora* 55% and *P. flexuosa* about 49% from rest of the *Prosopis* species.

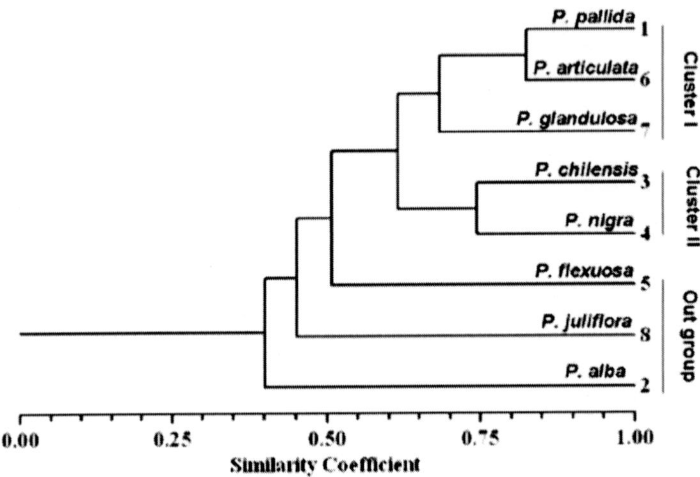

Fig. 3. Dendrogram of the seven random primer data matrix delineating *Prosopis* species

Our observation suggests that further gel extraction of species specific bands may provide unique primers that could help in resolving taxonomic chaos and address the problems of species-complex in the genus (Table 3).

Table 3. Species-specific diagnostic RAPD markers bands

Primer	Bands size (bp)	*Prosopis* species	Primer	Bands size (bp)	*Prosopis* species
OPA-9	850	*P. flexuosa*	OPP-7	1110	*P. articulata*
	575	*P. juliflora*		1031	*P. alba*
	500	*P. nigra*		500	*P. flexuosa*
OPA-10	1750	*P. juliflora*		475	*P. alba*
	450	*P. alba*	OPP-9	1031	*P. juliflora*
	375	*P. flexuosa*		950	*P. nigra*
	350	*P. glandulosa*		875	*P. alba*
OPA-13	1700	*P. alba*	OPP-16	2250	*P. juliflora*
	1200	*P. alba*		1600	*P. juliflora*
	700	*P. flexuosa*		1031	*P. Chilensis*
OPA-16	1300	*P. alba*		525	*P. pallid*
	775	*P. flexuosa*		425	*P. juliflora*

Genetic diversity in *Punica granatum*

Different cultivars of *Punica granatum* exist in the world vary widely in their morphological, agronomical and post harvest characteristics. The pomegranates have routinely been characterized using morphological and or biochemical markers (Ozgen *et al*. 2008; Tehranifara *et al*. 2010). These markers are of great significance but they exhibit limited information in distinguishing different varieties due to environmental plasticity. DNA markers reveal greater genetic diversity and relationships of horticultural plants and have been used to reconstruct phylogenetic relationships. Such information is of great importance for understanding the evolution and crop improvement in pomegranate. Thirteen *Punica granatum* cultivars were subjected to molecular characterization (Table.4)

Table 4. *Punica granatum* cultivars used for molecular characterization

Genotype	State
P-23	Maharashtra
P-26	Maharashtra
G-137	Maharashtra
Arakta	Maharashtra
Sinduri	Maharashtra
Jodhpur Red	Rajasthan
Dholka	Rajasthan
Gulsa Red	Maharashtra
GKVK	Karnataka
Basein Seedless	Karnataka
Ganesh	Maharashtra
Jalore Seedless	Rajasthan
Ganesh × Jalore Seedless	Maharashtra × Rajasthan Rajasthan

TenRAPD primers detected intra specific variations generating scorable amplicons reproducible patterns and generated 119 markers bands. Out of 119 marker bands, 110 markers were polymorphic amounting 92.44% polymorphism and exhibited 83.33 to 100% polymorphism in banding patterns. The number of PCR amplified products ranged from 8 (OPB-13) to 16 (OPA-02) with an average of about 12 bands per primer. The RAPD profile generated by OPB-04 exhibiting the maximum PIC is shown in fig.4. The dendrogram obtained from cumulative binary matrix analysis of 10 RAPD primers scorable fragments clearly delineated all the 13 varieties of *P. granatum* into three main clusters (Fig. 5). Cluster I contained four varieties i.e. P-23, G-137, Sindhuri and Ganesh. Cluster II included three varieties i.e. Dholka, Gulsa Red and GKVK. Whereas, the variety P-26 can be seen as distinct variety between cluster I and cluster II. Cluster III included three varieties namely, Jalore Seedless, Ganesh × Jalore Seedless and Basein Seedless. Jodhpur Red and Arakta were delineated as the most distinct varieties of pomegranate tested.

Upon multiple sequence alignment of all the 13 pomegranate sequences, we detected not only SNPs at number of places but also INDELS in both the ITS regions. ITS-1 region exhibited INDELS at three places and SNPs at four places. Whereas, 5.8S gene region was recorded with SNPs at 2 places one each in Arakta and Sindhuri. However, ITS-2 region exhibited INDELS at only one place and SNPs at 25 places by way of replacement of single nucleotide.The phylogram generated based on multiple sequences alignment showing delineation of pomegranate cultivars and two best aligned reference sequences downloaded from NCBI, Database is shown as fig.6. The pomegranate varieties which were shown to have clustering in RAPD dendrogram and were further delineated from each other along with their reference sequences with significant bootstrap values.

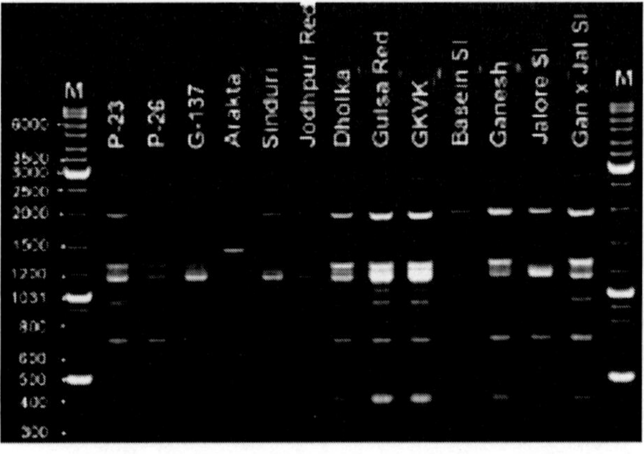

Fig. 4. The RAPD profile generated by OPB-04

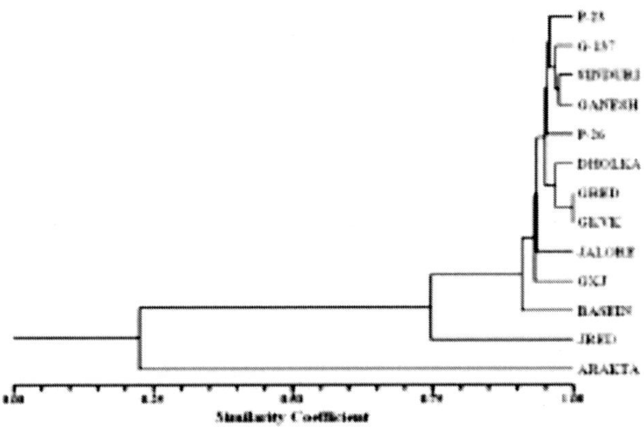

Fig. 5. The dendrogram of 10 RAPD primers

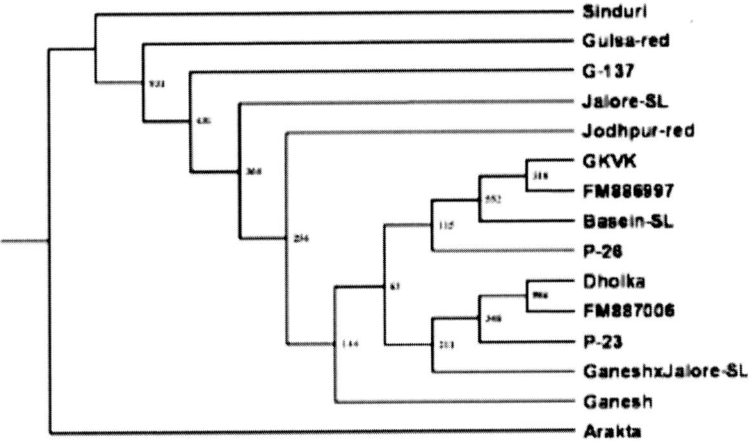

Fig. 6. phylogram generated based on multiple sequences alignment showing delineation of pomegranate cultivars

ITS sequence data determined the genetic diversity among putative Indian pomegranate varieties studied with high boot strap values. All the test varieties could be clearly distinguished by generating phylogram using NJ Plot. The morpho-physiological characterization proved insufficient to distinguish pomegranate varieties studied. This study validates the utility of ITS rDNA region as a reliable indicator of phylogenetic interrelationships, especially ITS-2 as probable DNA barcode at higher levels and can serve as an additional approach for identification and genetic cataloguing of pomegranate germplasm for crop improvement. The high genetic variability in this set of Indian pomegranate genotypes suggest that they might have originated from genetically divergent parents or has a long history of adaptation to their respective micro-climatic regions and could be of significance to contribute to pomegranate breeding programmes.

Genetic diversity of *Ziziphus mauritiana* cultivars using SCoT, ISSR and rDNA markers

Thirty-seven cultivars ofIndian jujube representing seven Indian states namely Rajasthan, Maharashtra, Punjab, Haryana, Gujarat, Uttar Pradesh and West Bengalare being maintained in randomized block design in horticulture block under field Gene Bank programme at Central Arid Zone Research Institute (CAZRI), Jodhpur, Rajasthan, India (Table 5).

Table 5. List and origin of 37 *Ziziphus mauritiana* cultivars

SN	Variety	State	SN	Variety	State
1	BC-1	Rajasthan	20	Kali	Rajasthan
2	Umran	Maharashtra	21	Jogia	Rajasthan
3	Seb	Rajasthan	22	Kaithli	Haryana
4	Illaichi	Punjab	23	Chhuhara	Maharashtra
5	Tikadi	Rajasthan	24	Dandan	Punjab
6	Gola	Haryana	25	Gola Gurgaon	Haryana
7	Reshmi	Haryana	26	Chonchal	Haryana
8	CAZRI-Gola	Rajasthan	27	Popular Gola	Haryana
9	Banarasi Karaka	Uttar Pradesh	28	Akrota	Haryana
10	Aliganj	Uttar Pradesh	29	Laddu	Uttar Pradesh
11	Katha	Rajasthan	30	Thar Bhubraj	Rajasthan
12	Z-G-3	Punjab	31	Ponda	Haryana
13	Mundia	Haryana	32	Wilayati	Punjab
14	Bagwadi	Punjab	33	Thar Sevika	Rajasthan
15	Maharwali	Rajasthan	34	Narikeli	West Bengal
16	Banarasi Pebandi	Uttar Pradesh	35	Sua	Haryana
17	S x K hybrid	Rajasthan	36	Vikas	Gujarat
18	Sanaur-5	Haryana	37	Babu	Gujarat
19	Thornless	Punjab			

Among 37 *Ziziphus mauritiana*cultivars, 17 SCoT primers amplified a total of 125 clear and bright DNA fragments with an average of 7.35 fragments per primer. Of 125 amplified DNA fragments, 77 (61.6 %) were found to be polymorphic. Except SCoT-23, all other primers exhibited higher polymorphic information content (PIC) ranging from 63.1 (SCoT-28) to 90.4 % (SCoT-31). The profile generated by SCoT-13 are shown in Fig.7 & 8.Among the cultivars, 14 ISSR primers yielded a total of 100 clear and bright DNA fragments and their sizes ranged between 275 bp and 3000 bp. The number of amplified fragments varied from 2 (ISSR-42) to 10 (ISSR-12, 30 and 39) with an average of 7.14 amplicons per primer. Of 100 fragments, 61 (61.0%) were polymorphic. The detected polymorphism per primer among the tested cultivars ranged from 33.3 (ISSR-40) to 100 % (ISSR-3 and ISSR-50). All the primers exhibited higher PIC ranging from 47.3 % (ISSR-42) to 88.8 % (ISSR-5, ISSR-30). The profile generated by ISSR-50 exhibiting 100% polymorphism is shown in Figs. 9 and10.

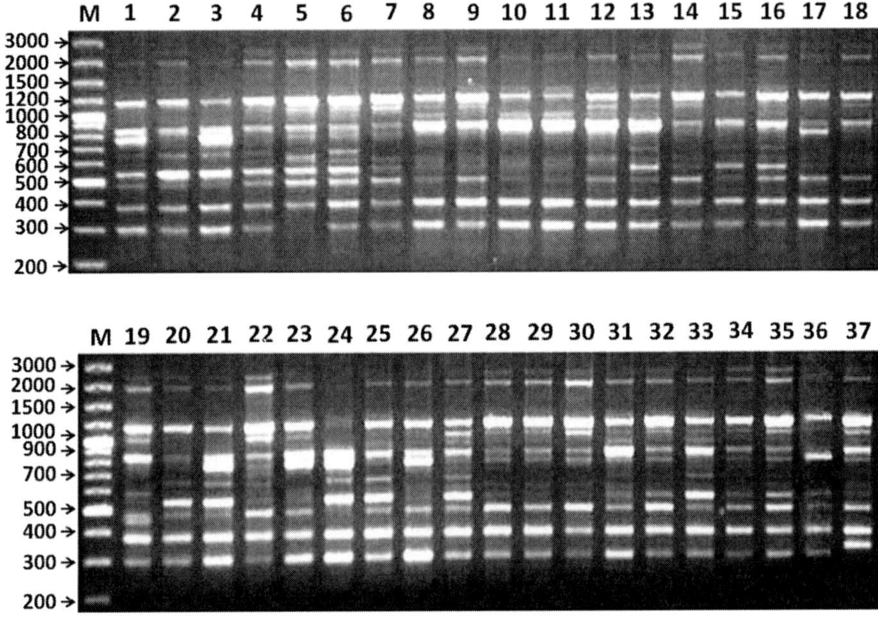

Figs.7 and 8. The profile generated by SCoT-13

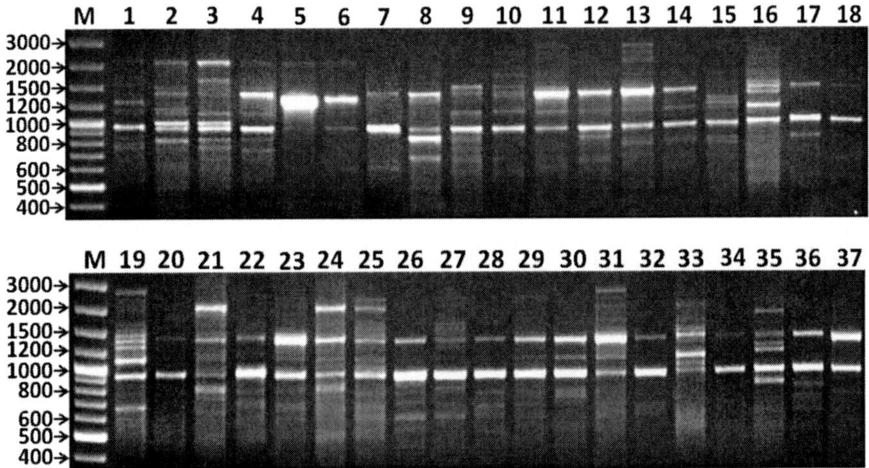

Figs 9 and 10. ISSR-50 exhibiting 100% polymorphism

The multiple sequence alignment of nucleotide sequences of all the 37cultivars together,in addition to total length polymorphismswe detected single nucleotide polymorphisms(SNPs) and insertion/deletions (INDELS)at several positions in ITS-1, 5.8S rRNA gene and ITS-2 region. Accordingly, the multiple sequence alignment of all the varieties of Z. *mauritiana* generated a phylogram using clustal X 2.0.22 and NJ plot (Fig.11).

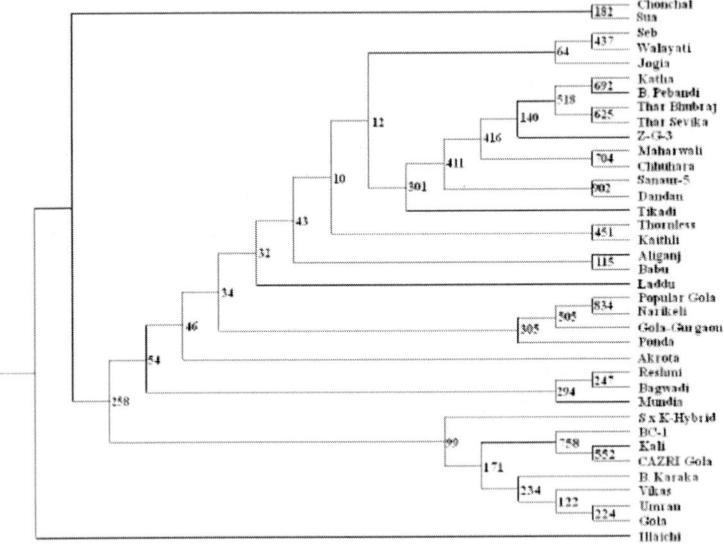

Fig.11. Phylogram showing the multiple sequence alignment of 37 varieties of Z. *mauritiana*

The perusal of phylogram revealed further delineated Z. *mauritiana* varieties into eight main clusters with four cultivars Tikadi, Laddu, Akrota and Illaichi as quite distinct out groups. A high degree of nucleotide sequence variation exhibited substantial intra-specific genetic diversity due to variations in the nucleotide sequences by way of SNPs, INDELS and ITS length polymorphism with significant boot strap values and allowed separation of all the 37 varieties of Z. *mauritiana* in the present study. Bootstrap values are dependable measures of phylogenetic accuracy and higher values are likely to indicate reliable group (Hillis and Bull 1993).

Identification of intraspecific hybrids in *Ziziphus mauritiana*

Breeding in *Ziziphus mauritiana* through hybridization is limited by its small sized flowers, cross incompatibility, low fruit set and poor retention. Correct identification of a hybrid is often difficult when it resembles more with one parent or when new morphological combination of characters arise from recombination of distinct genotypes. Under present study three varieties of Z.

mauritiana viz., Tikadi, Seb and Katha were found cross compatible and resulted in improved hybrids viz., F_1 (Seb X Tikadi), F_1 X Seb and Seb X Katha. The RAPD profiles of all the three parents and hybrids are shown in fig. 12.

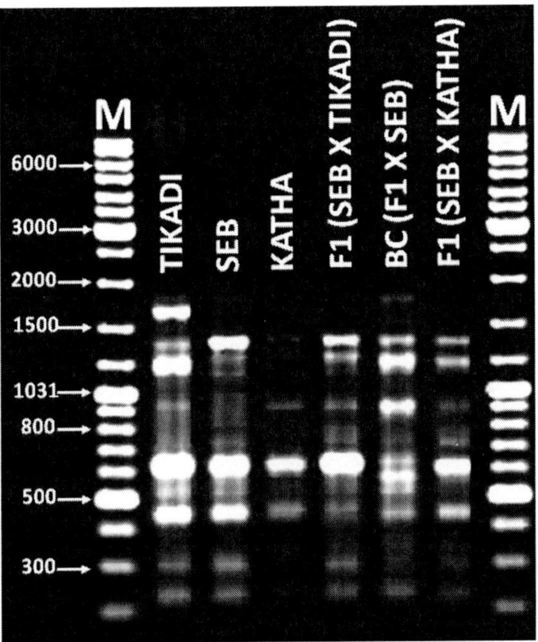

Fig. 12. RAPD profiles of parents and hybrids generated by OPA-16

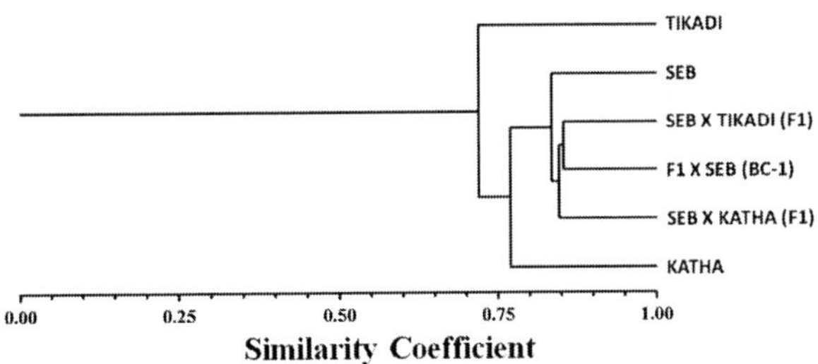

Fig. 13. Dendrogram clearly delineating variety Tikadi and Katha from Seb

The dendrogram clearly delineated variety Tikadi and Katha from Seb (Fig.13). All the three hybrids formed a single cluster with the variety Seb. The 5.8S gene region was found to be highly conserved (99.39%) followed by ITS-2 (97.77%). and ITS-1 (90.6 to 90.8%). Heterozygous positions in ITS sequences clearly show its hybrid nature and could perform as highly polymorphic molecular markers. Despite SNPs, INDELS and total length polymorphism among parents and their hybrids, all were molecularly identified as *Z. mauritiana* and did not cause any phylogenetic error. The nucleotide polymorphism in ITS-1 region can serve as a bar code to detect and address genetic diversity within *Z. mauritiana*.

References

Agarwal M, Shrivastava N, Padh H (2008) Advances in molecular marker techniques and their applications in plant sciences. Plant Cell Rep 27(4):617-31.

Al-Qurainy F, Khan S, Nadeem M, Tarroum M (2015) SCoT marker for the assessment of genetic diversity in saudi arabian date palm cultivars. Pak J Bot 47(2):637-643.

Al-Turki TA, Basahi MA (2015) Assessment of ISSR based molecular genetic diversity of Hassawi rice in Saudi Arabia. Saudi J Biol Sci 22:591–599.

Alvarez I, Wendel JF (2003) Ribosomal ITS sequences and plant phylogenetic inference. Mol Phylogenet Evol 29:435-455.

Amirbakhtiar N, Sorkheh K (2015) Analysis of diversity and relationships among wild *Pistacia* species using Start Codon Targeted (SCoT) markers. In: 1ST International and 9th International Congress of Islamic Republic of Iran, Shahid Beheshti University, Tehran, Iran pp. 1-4.

Baldwin BG, Sanderson MJ, Porter JM, Wojciechowski MF, Campbell CS, Donoghue MJ (1995) The ITS region of nuclear ribosomal DNA: a valuable source of evidence on angiosperm phylogeny. Ann Mol Bot Gard 82:247-277.

Ballesta P, Mora F, Contreras-Soto RI, Ruiz E, Perret S (2015) Analysis of the genetic diversity of *Eucalyptus cladocalyx* (sugar gum) using ISSR markers. Acta Scient 37:133-14.

Bhagyawant SS, Gupta N, Gautam A, Chaturvedi SK, Shrivastava N (2015) Molecular Diversity Assessment in Chickpea through RAPD and ISSR Markers. World J Agric Res 3(6):192-197.

Birren B, Lai E (1993) Pulsed field gel electrophoresis: A practical guide. Academic Press, San Diego, CA, pp. 253.

Carvalho A, Guedes-Pinto H, Lima BJ (2009) Genetic diversity among old Portuguese bread wheat cultivars and botanical varieties evaluated by ITS rDNA PCR-RFLP markers. J Genet 88(3):363-367.

Collard BCY, Mackill DJ (2009) Start codon targeted (SCoT) polymorphism: a simple, novel DNA marker technique for generating gene-targeted markers in plants. Plant Mol Biol Rep 27(1):86–93.

Devanshi, Singh AK, Sharma P, Singh B, Singh R, Singh NK (2007) Molecular profiling and genetic relationship among ber (*Ziziphus* sp.) genotypes using RAPD markers. Indian J Genet 67(2):121-127.

El-Mouei R, Choumane W, Dway F (2011) Characterization and estimation of genetic diversity in Citrus rootstocks. Intl J Agric Biol 13:571-575.

Godara NR (1980) Studies on Ffloral Biology and Compatibility Behavior in Ber (*Ziziphus mauritiana* Lamk.) Ph.D. thesis, Haryana Agricultural University, Hisar, India.

Heikrujam M, Kumar J, Agrawal V (2015) Genetic diversity analysis among male and female Jojoba genotypes employing gene targeted molecular markers, start codon targeted (SCoT) polymorphism and CAAT box-derived polymorphism (CBDP) markers. Meta Gene 5:90–97.

Hillis DM, Bull JJ (1993) An empirical test of bootstrapping as a method for assessing confidence in phylogenetic analyses. Syst Biol 42: 182–192.

Holland D, Hatib K, Bar-Yaakov I (2009) Pomegranate: botany, horticulture, breeding. Hortic Rev 35:127-191.

Jiang QQ, Long GY, Li WW, Deng ZN (2011) Identification of genetic variation in *Citrus sinensis* from hunan based on start codon targeted polymorphism. Agric Sci Technol 12(11):1594-1599.

Jorgenson RD, Cluster PD (1988) Modes and tempos in the evolution of nuclear ribosomal DNA: new characters of evolutionary studies and new markers for genetic and population studies. Ann Missouri Bot Garden 75:1238-1247.

Kakani RK, Singh SK, Pancholy A, Meena RS, Pathak R, Raturi A (2011) Assessment of genetic diversity in *Trigonella foenumgraecum* based on nuclear ribosomal DNA, internal transcribed spacer and RAPD analysis. Pl Mol Bio Rep 29:315-323.

Leo C, He XH, Chen H, Ou SJ, Gao MP, Brown JS, Tondo CT, Schnell RJ (2011) Genetic diversity of mango cultivars estimated using SCoT and ISSR markers. Biochem Syst Ecol 39:676-684.

Lewontin RC (1972) The apportionment of human diversity. Evol Biol 6:391-398.

Mir BA, Koul S, Kumar A, Koul MK, Soodan AS, Raina SN (2010). Intraspecific variation in the internal transcribed spacer (ITS) regions of rDNA in *Withania somnifera* (Linn.) Dunal. Indian J Biotechnol 9:325-328.

Morris AB, Baucom RS and Cruzan MB (2002) Stratified analysis of the soil seed bank in the cedar glade endemic *Astragalus bibullatus*: evidence of historical changes in genetic structure. American Journal of Botany89: 29-36.

Mulpuri S, Muddanuru T, Francis G (2013) Start codon targeted (SCoT) polymorphism in toxic and non-toxic accessions of *Jatropha curcas* L. and development of a codominant SCAR marker. Plant Sci 207:117–127.

Narzary D, Mahar KS, Rana TS, Ranade SA (2009) Analysis of genetic diversity among wild pomegranates in Western Himalayas, using PCR methods. Sci Hort 121:237–242.

Nei M, (1973) Analysis of gene diversity in subdivided populations. Proc Natl Aca Sci USA 70:3321-3323.

Ozgen M, Durgac C, Serce S, Kaya C (2008) Chemical and antioxidant properties of pomegranate cultivars grown in Mediterranean region of Turkey. Food Chem 111:703-706.

Pakseresht F, Talebi R, Karami E (2013) Comparative assessment of ISSR, DAMD and SCoT markers for evaluation of genetic diversity and conservation of landrace Chickpea (*Cicer arietinum* L.) genotypes collected from North-West of Iran. Physiol Mol Biol Plants 19(4):563-574.

Peakall R, Smouse PE (2006) GENALEX 6: genetic analysis in Excel. Population software for teaching and research. Mol Ecol Notes 6:288-295.

Rohlf FJ (1997) NTSYS pc: Numerical Taxonomy and Multivariate Analysis System Version 2.02h. Exeter software, New York.

Sahu AR, Mishra RR, Rath SC, Panigrah J (2015) Construction of interspecific genetic linkage map of pigeon pea using SCoT, RAPD, ISSR markers and simple inherited trait loci.Nucleus58(1):23-31.

Saini A, Reddy SK, Jawali N (2008) Intra individual and intra species heterogeneity in nuclear rDNA ITS region of *Vigna* species from subgenus *Ceratotropis*. Genet Res 90:299-316.

Sambrook J, Fritsch EF, Maniatis T (1989) Molecular cloning: A laboratory Manual, 2nd edition, Cold Spring Harbor Laboratory Press, Plainview, New York, USA.

Saran PL, Godara AK, Sehrawat SK (2006) Characterization of ber (*Ziziphus mauritiana* Lamk.) genotypes. Haryana J Hort Sci 35(3/4):215-218.

Sarkhosh A, Zamani Z, Fatahi R, Ranjbar H (2009) Evaluation of genetic diversity among Iranian soft-seed pomegranate accessions by fruit characteristics and RAPD markers. Sci Hort 121:313-319.

Sawant SV, Singh PK, Gupta SK, Madnala R, Tuli R (1999) Conserved nucleotide sequences in highly expressed genes in plants. J Genet 78:123–131.

Singh AK, Devanshi, Sharma P, Singh R, Singh B, Koundal KR, Singh NK (2007) Assessment of genetic diversity in *Ziziphus mauritiana* using inter-simple sequence repeat markers. J Plant Biochem Biotechnol 16(1):35-40.

Singh SK, Tiwari M, Kamal S, Yadav MC (2005) Morel phylogeny and diagnostics based on restriction fragment length polymorphism analysis of ITS region of 5.8S ribosomal DNA. J Biochem Biotechnol 14:179-183.

Smith JSC, Chin ECL, Shu H, Smith OS, Wall SJ, Senior ML, Mitchell SE, Kresovich S, Zeigle J (1997) An evaluation of the utility of SSR loci as molecular markers in maize (*Zea mays* L.): comparison with data from RFLPs and pedigree. Theor Appl Genet 95: 163-173.

Souza IGB, Valente SES, Britto FB, de SouzaVAB, Lima PSC (2011) RAPD analysis of the genetic diversity of mango (*Mangifera indica*) germplasm in Brazil. Genet Mol Res 10 (4):3080-3089.

Tehranifara A, Zareia M, Nematia Z, Esfandiyaria B, Vazifeshenasb MR (2010) Investigation of physico-chemical properties and antioxidant activity of twenty Iranian pomegranate (*Punica granatum* L.) cultivars. Sci Hort 126:180-185.

Varshney RK, Song C, Saxena RK, Azam S, Yu S, Sharpe AG, Cannon S, Baek J, Rosen BD, Tar'an B (2013) Draft genome sequence of chickpea (*Cicer arietinum*) provides a resource trait improvement. Nat Biotechno l31(3):240-246.

White TJ, Bruns SL, Taylor J (1990) *Amplification and direct sequencing of fungal ribosomal RNA genes for phylogenetics. In: Innes MA, Gelfand DH, Sninsky JJ, White TJ (eds) PCR Protocols, a Guide to Methods and Applications. San Diego, Academic Press, pp 315-322.*

Williams JGK, Kubelik AR, Livak KJ, Rafalski JA, Tingey SV (1990) DNA polymorphism amplified by arbitrary primers are useful as genetic markers. Nucleic Acids Res 18:6531-6535.

Xiao LQ, Moller M, Zhu H (2010) High nrDNA ITS polymorphism in the ancient extant seed plant cycas: incomplete concerted evolution and the origin of pseudogenes. Mol Phylogenet Evol 55:168-177.

Zoratti L, Palmieri L, Jaakola L, Häggman H (2015) Genetic diversity and population structure of an important wild berry crop. AoB Plants 7: 117.

26

Management of Salt-Affected Soils in Arid and Semi-Arid Ecosystem

Mahesh Kumar and N. R. Panwar

Central Arid Zone Research Institute, Jodhpur, Rajasthan, India

Introduction

The area under salt-affected soils in India is estimated to be 6.73 Mha spread over a number of states across the country. This 6.73 million ha are lying barren or produce very uneconomical yields of crops due to excessive accumulation of salts. These soils are agriculturally unproductive because of the high content of soluble salts which had harmful effects on plant growth due to moisture stress and toxicity/deficiency of some of the ions and the exchangeable sodium ions in clay complex lead to the unfavorable physical conditions of for the plant growth and water movement (Bhumla 1977). Saline and sodic soils have excess concentration of either soluble or exchangeable sodium, calcium and magnesium primarily in the form of chlorides and sulphates. Salt affected areas an expected to increase with spread of waterlogging and salinity due to increase in canal irrigation, and intensive exploitation of poor quality ground waters for agriculture in non canal commands. Certain states like Rajasthan and Haryana located in the western part of the country are endowed with 84 and 62% of poor quality ground waters, respectively. Continuous use of such waters for irrigation to agricultural crops is bound to increase the problem of salinity and sodicity in India (Yadav 1989; Minhas and Bajwa 2001). Introduction of irrigation without making proper provision for drainage is the major cause for the development of salinity in canal commands.It is severe on irrigated lands of the dry zone. It reduces crop yield and in severe cases causes complete abandonment of agriculture (Joshi and Dhir 1989; Joshi *et al.* 2000; Mahesh Kumar *et al.* 2016). The present paper deals with extent, causes, characteristics and management options of these soils for sustainability of agriculture in the area.

Extent of salt affected soils in India

National Remote Sensing Agency (NRSA), Hyderabad in association with other National and State level organizations like Central Soil Salinity Research Institute, Karnal; National Bureau of Soil Survey and Land Use Planning, Nagpur; All India Soil Survey and Land Use, Delhi; and State Government Agencies conducted survey and uses remote sensing data to prepare the maps of salt affected soils of India in 1996. The Landsat satellite images were used in mapping salt affected soils at 1:250,000 scale. Satellite images were interpreted for broad categorization of different types of salt-affected soils, sample areas for field verification were identified and surveyed for soil sampling and characterization. The salt affected soils were classified according to norms for pH, electrical conductivity (EC) and exchangeable sodium percentage (ESP). The state wise extent of salt affected soils in India is given in Table 1. It shows that maximum area of salt affected soils occur in Gujarat followed by Uttar Pradesh and Maharashtra which account for about 62.4 per cent. Due to the limitation of small scale some very small and isolated patches of salt affected soils occurring in the states of Delhi and Himachal Pradesh could not be detected. The salt affected soils accounts for 6.727 m ha equivalent to 2.1 per cent of the geographical area of the country.

Out of the total 6.727 million ha of salt affected soils, 2.956 million ha are saline and the rest 3.771 million ha are sodic. Out of the total 2.347 million ha salt affected soils in the Indo-Gangetic Plains, 0.56 million ha are saline and 1.787 million ha are sodic.

Table 1. Extent of state wise salt-affected soils India ('000 ha)

	Saline	Sodic	Total
Andhra Pradesh	77.598	196.609	274.207
Andaman and Nicobar Island	77.000	0	77.000
Bihar	47.301	105.852	153.153
Gujarat	1680.570	541.430	2222.000
Haryana	49.157	183.399	232.556
Karnataka	1.893	148.136	150.029
Kerala	20.000	0	20.000
Madhya Pradesh	0	139.720	139.720
Maharashtra	184.089	422.670	606.759
Orissa	147.138	0	147.138
Punjab	0	151.717	151.717
Rajasthan	195.571	179.371	374.942
Tamil Nadu	13.231	354.784	368.015
Uttar Pradesh	21.989	1346.971	1368.960
West Bengal	441.272	0	441.272
Total	2956.809	3770.659	6727.468

Natural or primary salinity

Salinity, primarily results from the accumulation of salts over long period of time, in the soil or groundwater, which is generally caused by two natural processes (Kolarkar *et al.* 1980; Singh *et al.* 1994).

- Weathering of parent materials breaks down rocks and release soluble salts of various types, mainly chlorides of sodium, calcium and magnesium, and to a lesser extent, sulphates and carbonates. With sodium chloride is the predominant soluble salt.
- The deposition of oceanic salt carried in wind and rain forms the second cause.
- Rainwater contains from 6 to 50 mg kg^{-1} of salt, the concentration of salts decreasing with distance from the coast to the inland areas.
- The amount of salt stored in the soil varies with the soil type, being low for sandy soils and high for soils contain a high percentage of clay minerals. It also varies inversely with average annual rainfall.

Secondary or human-induced salinity

Salinity occurs through natural or human-induced processes that result in accumulation of dissolved salts in the soil water to an extent that inhibits plant growth. Secondary salinisation results from human activities (anthropogenic) that change the hydrologic balance of the soil between water applied (irrigation or rainfall) and water used by crops (transpiration). The important causes for secondary salinisation are:

(i) land clearing and the replacement of perennial vegetation with annual crops;

(ii) use of salt-rich irrigation water; and

(iii) lands having insufficient drainage.

Sources and causes of accumulation of salts

The main causes of salt accumulation include

- Capillary rise from subsoil salt beds or from shallow brackish ground water;
- Indiscriminate use of irrigation waters of different qualities
- Weathering of rocks and the salts brought down from the upstream to the plains by rivers and subsequent deposition along with alluvial materials

- Ingress of sea water along the coast
- Salt-laden sand blown by sea winds

Characteristics of salt affected soils

Saline soils

These soils will have electrical conductivity of the saturation extract more than 4 dS m^{-1}, the exchangeable sodium percentage less than 15 and the pH is less than 8.5. With adequate drainage, the excessive salts present in these soils may be removed by leaching thus bringing them to normalcy. Saline soils are often recognized by the presence of white crusts of salts on the surface. The important soluble salts in these soils are cations sodium, calcium and magnesium with low amounts of potassium and anions, chloride, sulphate and some times nitrate. Owing to the presence of excess salts and the absence of significant amounts of exchangeable sodium, saline soils generally are flocculated and as a consequence the permeability is equal to or higher than that of similar non saline soils, which adversely affects the crop growth (Singh 2009). While the physical properties and water permeability in saline soils is comparable to their normal counterparts (Singh 2009), plant growth in sodic soils is hampered due to poor physical environment which adversely affects water and air flux, water holding capacity, root penetration and seedling emergence (Murtaza *et al.* 2006).

Saline-alkali soils

These soils will have electrical conductivity of the saturation extract more than 4 dS m^{-1}, the exchangeable sodium percentage greater than 15 and the pH is seldom higher than 8.5. These soils form as a result of combined process of salinisation and alkalisation. As long as excess soluble salts are present, these soils exhibit the properties of saline soils. Leaching of excess soluble salts downward, the properties of these soils will become like that of non-saline alkali soils. On leaching of excess soluble salts, the soil may become strongly alkaline (pH reading above 8.5), the particles disperse and the soil becomes unfavourable for the entry and movement of water and for tillage.

Alkali soils

These soils will have their exchangeable sodium percentage greater than 15, the electrical conductivity less than 4 dS m^{-1} and the pH range between 8.5 and 10. The exchangeable sodium content influences significantly the physical and chemical properties of these soils. As the ESP tends to increase, the soil tends to become more dispersed. In addition to the parameters proposed by the USDA, Indian scientists considered the nature of soluble salts. Further the pH value of

8.5 is too high, as isoelectric pH for precipitation of $CaCO_3$ at which sodification starts is 8.2 and mostly the pH is associated with the ESP of 15 or more.

Table 2. Properties of saline, saline-alkali and non-saline-alkali soils properties

	Saline soils	Saline alkali soils	Alkali soils
Electrical conductivity (dS m^{-1})	> 4.0	> 4.0	< 4.0
pH	< 8.5	> 8.5	> 8.5
Exchangeable Sodium Per cent (ESP)	< 15	> 15	> 15

Formations

Saline soils

When the soil contains excess of sodium salt while the clay complex still contains a pre-ponderance of exchangeable calcium, the soil is known as saline soil. The process of accumulation of salts leading to the formation of saline soil is known as salinization. The nature and abundance of salts presents depend on the composition of the rock and nature of weathering processes. The salts usually present in saline soils are the chlorides, sulphates, bicarbonates and sometimes nitrate of sodium, together with these of Ca and Mg in lesser proportions. Among the anions, chlorides and sulphate are present in greater proportions than bicarbonates and nitrates. The presence of chlorides and sulphates of sodium gives a white color to the encrustation formed on the surface from which the soil is known as white alkali soils when nitrates are in excess they give a brown color to the soil; such soil is known as brown alkali soil. These soils are formed in arid and semi arid regions, which have very low rainfall and high evaporation.

Saline-alkali soils

These soils form as a result of combined processes of salinization and alkalinization. When soluble salt accumulate in a soil over a prolonged period, sodium becomes the predominate cation in soil solution. Whereas, calcium and magnesium and potassium are absorbed by plants and removed from soil solution, much of the sodium is left behind, as it is not essential nutrient except for few crops. The concentration of sodium is further increases as Ca and Mg are precipitated as carbonates or Ca as $CaSO_4$, due to evaporation of moisture from soil or its absorption by plants. When the concentration of sodium is more than 50% of total cations the sodium ions replace the exchangeable Ca and Mg ions of clay complex by virtue of their high concentrations and thus normal Ca clay converted to Na clay.

$$\text{Ca-clay} + \text{NaD} \leftrightarrows \text{salt D Na-clay} + \text{Ca-salts}$$

Sodium now becomes the dominant exchangeable cation. Though the reaction is reversible and Ca salts are removed in drainage water as soon as they are formed. If the reaction proceeds in one direction from left to right, only. The process where by a normal soil is converted in to alkali soil is known as alkalization. Here two possibilities are now likely to arise during further pedogenic processes depending upon the movement of salts. In one case soil contains Na-clay as well as excess salts, and in other only Na-clay no soluble salts. If soluble salts are not leached out due to insufficiency of rain, they remain in the soil. The soil thus contain Na-clay and excess soluble salts in solution such soils are known as saline sodic soils.

Alkali soils

The soil colloids adsorb and retain cations on their surfaces. Cation adsorption occurs or consequence of the electrical charges at the surface of the soil colloids. While adsorbed cations are combined chemically with the soil colloids, they may be replaced other cations that occur in the soil solution. Calcium and magnesium are the principal cations found in the soil solution and on the exchange complex of the normal soils in arid regions. When excess soluble salts accumulate in these soils, sodium frequently becomes the dominant cation in soil solution. In arid region as the soil solution becomes concentrated through evaporation or water absorption by plants. The solubility limit of $CaSO_4$, $CaCO_3$ and $MgCO_3$ are often exceeds, in which case they are precipitated with a corresponding increase in sodium concentration under such conditions a part of original exchangeable Ca^{2+} and Mg^{2+} is replaced by sodium resulting alkali or sodic soils.

$$\text{Ca [clay + 2Na]} \leftrightarrows {}^{\text{Na}}_{\text{Na}}\text{[clay + Ca]}$$

Reclamation of saline soils

Salt leaching

Depending on the salt content of soil, Climatic condition etc. crop growth may be reduced or entirely prevented. Excess salts must be removed from root to improve crop growth. Reclamation means methods used to remove soluble salts from root zone and those commonly adopted are the following:

Scrapping

Salt crust from the soil surface can be removed mechanically. This method might temporarily improve crop growth.

Flushing

Surface salts can sometimes be washed away by horizontal flushing provided a drain or some low-lying wastelands are available nearby.

Leaching

This is the process of dissolving and transporting soluble salts from root zone by the down ward movement of water through the soil. This is accomplished by pounding fresh water on the soil surface and allowing it to infiltrate leaching is effective when salty drainage water is carried away through sub surface drains or when there is insufficient natural drainage and the water table is quite deep.

$$LR = \frac{D_{dw} \times 100}{D_{iw}} = \frac{EC_{iw} \times 100}{EC_{dw}}$$ Where, LR= Leaching requirement

EC_{dw} = EC of Drainage water

Ec_{iw} = EC of irrigation water

D_{dw}= Depth of drainage water

D_{iw}= Depth of irrigation water

Reclamation of sodic soils

Chemical reaction involving reclamation of sodic soils

1. Gypsum

When gypsum is applied to ameliorate salt affected soils the following reactions will take place and loss of exchangeable sodium occurs and calcium will take place of sodium on exchange complex as follows:

$Na_2CO_3 + CaSO_4 \leftrightarrows CaCO_3 + NaSO_4\downarrow$ Leachable

$Clay\ ]^{Na}_{Na} + CaSO_4 \leftrightarrows$ Ca clay + $Na_2SO_4\downarrow$ Leachable

2. Suphur

$2S + 3O_2 = 2SO_3$ (by sulphur oxidizing bacteria in soil)

$SO_3 + H_2O = H_2SO_4$

$H_2SO_4 + Na_2CO_3 \leftrightarrows CO_2 + H_2O + Na_2SO_4\downarrow$ leachable

On calcareous alkali soils

$H_2SO_4 + CaCO_3 \leftrightarrows CO_2 + H_2O + CaSO_4$

$CaSO_4 + {}^{Na}_{Na}[clay \leftrightarrows Ca\ [clay + Na_2SO_4\downarrow$ leachable

3. Iron sulphate

$FeSO_4 + H_2O \leftrightarrows H_2 + SO_4 + FeO$

In calcareous soils

$H_2SO_4 + CaCO_3 \leftrightarrows CaSO_4 + CO_2 + H_2O$

$CaSO_4$ +DCa [Clay $\leftrightarrows Na_2SO_4 \downarrow$ leachable

In non-calcareous soils

$H_2SO_4 \ {}^{Na}_{Na}$[clay $\leftrightarrows {}^{H}_{H}$[clay+$Na_2SO_4\downarrow$ leachable

4. Iron pyrite (FeS_2)

$FeS_2 + 3½O_2 + H_2O \leftrightarrows FeSO_4 + H_2SO_4$

FeSO4 + $2H_2O \leftrightarrows H_2SO_4 + Fe(OH)_2$

$H_2SO_4 + CaCO_3 \leftrightarrows CaSO_4 + H_2O + CO_2$

$CaSO_4 + \leftrightarrows$ Ca [Clay + $Na_2SO_4\downarrow$ leachable

5. Lime sulphur (CaS_2)

$CaS_2 + 8O_2 + 4H_2 O \leftrightarrows CaSO_4 + 4H_2SO_4$

$H_2SO_4 + CaCO_3 \leftrightarrows CaSO_4 + CO_2 + H_2O$

$CaSO_4 + {}^{Na}_{Na}$[clay $\leftrightarrows$ Ca [Clay + $Na_2SO_4\downarrow$ leachable

Other methods

1. Organic manures: Organic manures are known to facilitate the reclamation of sodic soils. The beneficial effect of organic manures are:

   a. The decomposition of organic matter to produce CO_2 and certain organic acids.

   b. Lowering of pH and the release of cations by solubilization of $CaCO_3$ and

   c. Replacement of exchangeable sodium by Ca thereby lowering the ESP.

2. Growing salt tolerant crops like wheat, barley, alfalfa, tomato, beets, etc. and grasses such as Karnal grass, para grass, Bermuda grass Dhaincha and sugar beet.

Management of soil irrigated with sodic water-a case study at Jodhpur Rajasthan

Demonstration of technology for the amelioration and management of soils irrigated with sodic water was conducted at two sites each in Barmer (Village-Budiwara) and Jodhpur (Village-Dhudhara) district during *rabi* season-2008-09 and at Eight sites in jodhpur in the Dhundharia village in year 2009-10, 2010-11under FPARP. The pH of the soils of experimental sites varied from 8.8 to 9.7 and that of electrical conductivity from 0.208 to 0.875 dS m^{-1}. The gypsum requirement (GR) of the soils varied from 4.3 to 10.7 tonnes per hectare. Irrigation water used for irrigation were sodic to slightly saline sodic in nature. The Residual sodium carbonate (RSC) of the irrigation water varied from 2.8 to 6.1 meL^{-1}. Based on the initial soil and water analytical data, experiments at farmer's field were laid down in four treatments. Treatments are comprised of three levels of gypsum i.e. 25, 50 and 100% of gypsum requirement of the soils along with a control plot (without gypsum) with recommended dose ofnitrogen and phosphorus through urea and DAP. Wheat (Variety- Raj 3077) was grown on the farmer's field on all the experimental sites in both the districts following all agronomic and other management practices followed in the region. Seven irrigations were given to crop by all the farmers. Gypsum application @ 50% of GR to the soil was found best among all the treatment at all the experimental sites. Growth attributes such as plant height, plants /$meter^2$, no. of effective tillers per plant, spike length and no. of leaves per plant were significantly higher when gypsum applied @50 % of GR of the soil in wheat in comparison to control and gypsum application @25 % of GR. However there was no significant difference in growth attributes of wheat crop between gypsum application @ 50 and 100 % of GR, but these were slightly higher in later (100% GR). Grain and straw yield of wheat was increased by 20-35 % over control and the grain yield was highest as 4.8 t/ha at Dhudharia village on field of Babu Singh and 4.4 t/ha on the field of Gobra Ram at Budiwara. The grain yield of wheat was also significantly higher in treatment receiving gypsum @100 % of GR over control in Budiwara and Dhundhara villages, respectively. However there was no significant difference in grain yield of wheat with gypsum application @ 50 and 100 % of GR, but these were slightly higher when gypsum was applied @ 50% of GR. The benefit cost ratio for all treatment at each location was also calculated and it was found highest as 2.45:1 and 2.37:1 when gypsum was applied @ 50 % of GR followed by 1.97:1 and 2.18:1 with gypsum application @100 % of GR at the field of Babu Singh at Dhundhara and Gobra Ram at Budiwara, respectively. BC ration was comparatively low in the plots receiving gypsum @25 % of GR at all the locations.

Table 3. Effect of gypsum application on soils irrigated with sodic water, yield of wheat and benefit / cost ratio

Name of farmer	*Treatment	Soil pH (before sowing)	Gypsum application (t ha^{-1})	RSC (meq L^{-1}) of irrigation water	Grain yield (t ha^{-1})	Straw yield (t ha^{-1})	Benefit /cost ratio
Gobra Ram	G3	9.1	5.4	4.7	4.4	5.6	1.97:1
	G2		2.7		4.2	5.2	2.37:1
	G1		1.4		2.9	4.0	1.50:1
	G0		0		2.2	3.2	---
Dhudha Ram	G3	9.5	4.8	2.8	3.5	4.2	1.57:1
	G2		2.4		3.1	3.7	1.72:1
	G1		1.2		2.5	2.9	1.18:1
	G0		0		2.2	2.7	--
Babu Singh	G3	9.7	7.8	5.7	4.8	5.6	2.45:1
	G2		3.9		4.4	4.6	2.18:1
	G1		1.9		3.4	4.5	1.50:1
	G0		0		3.2	4.0	---
Jetha Ram	G3	9.5	7.5	6.1	3.4	4.6	1.57:1
	G2		3.8		3.2	4.4	1.78:1
	G1		1.9		2.9	4.0	1.25:1
	G0		0		2.8	4.0	---

*G3-Gypsum application @100 % of GR, G2-Gypsum application @50 % of GR

G1-Gypsum application @50 % of GR, G0-Contro (Without Gypsum)

Reference

Bhumla DR (1977) Salt affected soils and their management. A position paper.

Joshi DC (2000) Sustainable management of brackish water irrigated soils for crop production. In: Assessment and Refinement Through Institute village Linkage Programme. Arid Agro-Ecosystem, NATP, CAZRI, Jodhpur. pp.1-35.

Joshi DC, Dhir RP (1989) Management of loamy sand soils irrigated with high RSCwater. In: Abstract-Part-11 International Symposium on Management of sandy soils, CAZRI, Jodhpur.

Kolarkar AS, Dhir RP, Singh N (1980) Characteristics and management of salt affected soils on south eastern arid Rajasthan. Journal of the Indian Society of Remote Sensing 8:85-93.

Mahesh Kumar, Amal Kar, Raj Singh (2016) Interventions of high residual sodium carbonate water-degraded soils amelioration technology in Indian Thar desert and farmers' response.Natl AcadSci Lett. 39 (4): 245-249

Minhas PS, Bajwa MS (2001) Use and management of poor quality water in rice wheat production system. J crop production 4: 273-306.

Murtaza G *et al.* (2006) Irrigation and soil management strategies for using saline-sodic water in a cotton-wheat rotation. Agricultural Water Management 81:98-114.

Singh N, Kolarkar AS, Bohra PC (1994) Quality of ground water and its effect on soil properties in Samdari – Siwana –Balotra area of western Rajasthan. Annals of Arid Zone 33:287-293.

Singh G (2009) Salinity-related desertification and management strategies: Indian experience. Land Degradation and Development 20:367-385.

Yadav JSP (1989) Irrigation induced soil salinity and sodicity. In: Proceeding of World Food Day Symposium on Environmental Problems Affecting Agriculture in The Asiaand Pacific Region held at FAO, Bangkok, Thailand. pp.47-62.

27

Wind Erosion Control Through Vegetative Measures to Combat Desertification

Priyabrata Santra, Suresh Kumar and P.C. Moharana

ICAR-Central Arid Zone Research Institute, Jodhpur, Rajasthan, India

Introduction

Desertification is affecting the livelihoods of millions of people, mainly poor in the dry lands, which occupy nearly 41% of the Earth's land area and are the residence of more than 2 billion people of the world. In India, the western part of Rajasthan is mostly affected by desertification process dominated by wind erosion. Kar *et al.* (2009) reported that 15.2 mha land in western Rajasthan is affected by wind erosion out of total 22.96 mha area under desertification. A brief account on issues and priorities related to wind erosion processes in Thar desert was reported in Santra *et al.* (2006). Plenty of incident solar energy in the region makes the desert surface very hot especially during summer months. Wind speed also remains very high in the region for most of the time in a year. Therefore, the loss of top fertile soils through wind erosion process is very active in the region. Eroded dust particles during severe wind erosion events commonly known as dust storms not only create the soils poor but also pose several environmental threats through generating aerosols in atmosphere. In severe dust storm events, the suspended particles may be transported by air over hundreds of kilometers and form a blanket of dust haze over the Indo-Gangetic plains and surrounding area. For example, average soil loss rate has been reported as 17 kg ha^{-1} min^{-1} during dust storm events and 25 kg ha^{-1} day^{-1} during local wind erosion events at Jaisalmer and the concentration of PM_{10} particles was found higher (30%) in eroded soil than in top soil (7%) (Santra *et al.* 2010). Average C and N content in eroded soils from Jaisalmer region of the

Indian Thar desert has been found 4 g C kg^{-1} and 0.77 g N kg^{-1}, respectively (Mertia and Santra 2012). In this paper, the basic wind erosion process is discussed in detail with different causative factors followed by different control measures. A Field experimental result from Indian Thar desert is also discussed as examples to understand the basic processes and how different factors affect it.

Wind erosion processes

Wind erosion is a set of processes by which soil particles are lifted from land surface, transported and deposited at another place. The conducive conditions for wind erosion process are i) presence of loose, dry and finely granulated particles on surface, ii) smooth soil surface with negligible amount of vegetation or any other surface cover, and iii) strong and turbulent wind regime to move soil particles. In the Indian Thar desert, most specifically at western boundary of Rajasthan covering areas of Barmer, Jaisalmer, and Bikaner, all these conducive conditions are highly favourable during summer months and thus wind erosion is found as a severe land degradation process in the area. A comprehensive summary on movement of soil particles by wind erosion processes was reported by Bagnold (1943) for desert sands and by Chepil and Woodruff (1963) for agricultural lands. In this paper, a brief of wind erosion process consisting initiation, transport, sorting, and abrasion are discussed.

Initiation of soil movement is started when the wind speed is reached to a level that starts the most erodible grains at surface in motion. This wind speed is also known as the threshold wind speed or threshold frictional velocity (u_{*t}). Bagnold (1943) defined threshold friction velocity as $(\tau_0/\rho)^{1/2}$ where τ_0 is the shear stress at the boundary and ρ is the air density and described it to vary as the square root of the product of equivalent diameter of soil particles and density relationship of fluid and grain as follows:

$$u_{*t} = A\sqrt{\frac{(\rho_s - \rho)}{\rho} gd} \qquad (1)$$

Where A is experimental coefficient, ρ_s is particle density, ρ is air density, g is gravitational constant and d is particle diameter. Several experiments have reported the value of A ranged from 0.08 to 0.12, which is largely dependent on Reynolds number (R). Frictional velocity is generally dependent on surface conditions and is usually obtained from data on wind velocity-profile as follows:

$$u_* = \frac{\bar{u}_z k}{\ln\left(\frac{z - D}{z_0}\right)} \qquad (2)$$

Where $\overline{u}_z$ is the mean wind speed at height z, k is von Karman's constant (0.4), D is effective roughness height, z_0 is roughness parameter.

Transport of soil particles, which are initially dislodged from surface by wind speed, is commonly described by three distinct modes: suspension, saltation, and surface creep. In the suspension mode, aggregates or particles that are removed from local source area is transported to high altitudes and over long distances, depending on their size, shape, and density. Chepil (1945) reported that 3 to 38% of total transport is occurred through suspension mode. Size of suspended aggregates/particles ranges from 2 to 100 μm with a mass median diameter of about 50 μ in an actively eroding field. This size range excludes the different size fractions of sand particles and aggregates of corresponding sizes, which remain in local area. Suspended particles generated through wind erosion process is generally considered as nutrient rich particles as plant nutrients and organic carbon are mostly associated with finer soil fractions. The enrichment ration even increases with the increase of coarse sand content in bulk soil. Consequently, suspension indirectly impacts soil productivity by removing nutrient rich fractions of soil or by leaving behind the less-fertile soil constituents. In the saltation mode, eroded particles are jumped or hopped from one place to another place. Salating particles leave the surface with wind force but are generally large enough to be suspended and thus returned back to surface. On returning to surface, saltating particles initiate the movement of another particle. Size of saltating particles/aggregates ranges from 100 to 500 μm. The bulk of transport during wind erosion, roughly 50 to 80% occurs through saltation mode. Majority of saltating particles/aggregates rises up to a height of 30 cm and in some cases up to a height of 120 cm. In the surface creep mode, coarse sand sized mineral particles having diameter 500-1000 μm are pushed and rolled on soil surface. About 7-25% of total mass transport occurs through surface creep mode (Lyles *et al.* 1985).

Transport of wind eroded particles/aggregates generally sorts the materials present on surface. Finer and lighter particles generally move faster than the coarser and denser ones. Erosive winds separate the soil into several distinct grades as follows (Chepil and Woodruf 1953): i) residual soil materials representing non-erodible clods and massive rock materials that remain in place, ii) large sands, large gravels, and large soil aggregates constituting semi-erodible grains that have been moved primarily by surface creep, iii) sand and clay dunes in the form of accumulated highly erodible grains that have been moved primarily in saltation, and iv) loess, which are mainly composed of dust lifted off the ground by saltation and carried high in the air and deposited in uniform layers both near and far from dunes. Distinct demarcations of size between various grades of wind-sorted materials do not exist and in most cases the size range of one grade overlaps with other grades. Non-selective removal of surface soil by wind is associated with loess, which was already sorted and deposited from the atmosphere during past. In case of selective removal of materials, winds tend to remove silt and clay particles leaving behind the sands and gravels on surface.

Abrasion by impacts of particles transported along the surface by wind is an important phase of the wind erosion process on all soils (Chepil and Woodruff 1953). Soils in arid areas usually are covered with a thin crust, which may be depositional crust or biological crust and are somewhat resistant to wind erosion. Abrasion of moving particles by wind disintegrates this thin crust and exposes more highly erodible soil underneath it. Abrasive force of saltating particles also gradually breaks the non-erodible particles. The materials detached from clods and surface crust by abrasion accumulate on the leeward side of fields or, if they are fine, are carried far through the atmosphere.

Management and control of wind erosion

Wind erosion can be controlled in two major ways: either by decreasing the erodibility of soil or by reducing the erosive energy of wind by erecting barriers. Basic principles to control wind erosion was discussed in detail by Chepil and woodruff (1963) and Tibke (1988). Here, four major control measures are discussed: i) Surface cover ii) wind breaks and shelterbelt, iii) tillage and iv) crop management. All these four control practices are designed to either take some of the wind force or to trap the eroded soil. For control of wind erosion in the Indian Thar desert several research efforts were made in past (Muthana 1982; Gupta *et al.* 1984; Kaul 1985; Mann 1985; Mertia 1992; Kar and Joshi 1995; Venkateswarlu and Kar 1996; Narain and Kar 2007), which were mostly reported as the sand dune stabilization techniques through vegetation covers in dune areas in checkerboard method, which became a popular wind erosion control technology in the region. Few literatures among these also reported the effect of shelterbelts on wind erosion control in the Indian Thar desert. Here, four major wind erosion control measures as outlined above are discussed in detail.

Surface cover

Use of surface cover to control wind erosion may be of two types, vegetative and non vegetative. Protection of land surface through vegetative surface cover is the one generally applicable method of permanent and effective control. Grass cover or crop covers are perhaps the most effective, easiest and most economical. In addition to the standing vegetation, crop residues often are place artificially on the soil to provide temporary cover until permanent vegetation cannot be established. It was further reported that in case of residue application, better control of wind erosion may be obtained if the residues are well anchored in the surface. Other than vegetative covers, various surface films such as resin-in-water emulsion (petroleum origin), rapid curing cutback asphalt, asphalt-in-water emulsion, starch compounds, latex in water emulsion (elastomeric polymer emulsion), by-products of the paper pulp industry and wood cellulose fibre were used to control wind erosion, which mainly aims to decrease soil erodibility. Sand or pebble (>2 mm) mulch are often used to control erosion. However, these types of materials as surface cover are mainly used in non agricultural lands where it is not feasible to obtain cover by growing and managing vegetation. The effectiveness of vegetative or non vegetative covers depends on type of covers and other several factors. Permanent grass cover in rangelands are thought to be most effective than crop covers, which exists in the field for short period. Among crops, dense row crops and creeping crops are highly favorable. After the plants complete their crop growth, residues become the primary cover. The decay of leguminous reside is faster than cereal or other crops and thus is less durable in field to control wind erosion. Moreover, more erect and finer and the denser is the residue, the smaller is the amount of wind erosion. Maintenance of grass cover in rangelands is very important to control wind erosion. It is always better to adopt control grazing practices in rangelands so as to maintain the primary productivity as well as to provide sufficient protective grass covers. From a field experiment at two grazing situations in Jaisalmer region of the Indian Thar desert had revealed that the aeolian mass transport rate was almost three times higher at the overgrazed site than at the controlled grazing site during mid of June to mid of July. Notably, the only difference between these two sites was the level of grazing pressure, which resulted in huge difference in mass transport rates. This clearly indicates that simple control of grazing in arid rangelands of the Indian Thar Desert may significantly reduce the active land degradation process due to wind erosion.

Wind break and shelterbelt

Shelterbelts are barriers of trees or shrubs that are planted to reduce wind speed and, as a result, prevent wind erosion; they frequently provide direct benefits to agricultural crops, resulting in higher yields, and provide shelter to livestock, grazing lands, and farms. Shelterbelts have been consistently reported in literature as barrier to prevent wind erosion and wind damage (Skidmore and Hagen 1977; Kort 1988). For establishment of shelterbelt, the design aspect was reported in detail by Mohammed *et al.* (1996) and Cornelis and Gabriel (2005). Shelterbelt when planted across and on the margins of agricultural fields effectively protect crops and control sand drifting (Ganguli and Kaul 1969).

The effectiveness of shelterbelt in reducing wind speed depends on wind speed itself, direction, shape, width, tree height and density etc.

All wind barriers provide maximum percentage reductions in wind velocity at leeward locations near the barrier with a gradual decrease downwind. In case of rigid barriers, the percentage reduction remains constant for different wind speeds. However, in case of flexible barriers i.e. tree shelterbelts, the degree of wind erosion control will be greater for low velocity winds than for high velocity winds. The direction of wind influences both the size and location of the leeward protected area. The area of protection is greater for wind blowing at right angles to the barrier length and is smallest or almost nil for wind blowing parallel with the barrier direction. The shape of windbreaks characterizes the outer perimeter or outer surface, which is in contact with the airstream. Abrupt vertical height of the barrier is less effective than the sloped triangular outer surface. Therefore, tree shelterbelt with pyramidal cross section will be more effective and such cross section may be obtained by planting tall trees in middle central rows followed by shrub of 2 rows at the outer ends with decreasing height. Porosity is other important factor influencing the effectiveness of wind breaks. Dense barriers provide large reduction of wind speed but for a short leeward distances whereas a porous barrier provide smaller reduction in wind speed but for a extended leeward distances. Therefore, a certain level of porosity is always desirable; however, large openings in barrier may be avoided because of air jetting through large opening may cause serious erosion in the leeward zone. Height of the barrier is another important factor influencing the effectiveness of barrier. Expressed in multiple of barrier height (H), the influence of a wind barrier may be up to 40 H to 50 H at leeward direction. For complete control of wind erosion in field wind breaks or tree shelterbelts need to be placed at close intervals as per the barrier height.

Trees for shelterbelt are planted at right angles to the prevailing wind direction. If the wind direction changes frequently a checkerboard pattern of plantings is required. Otherwise only parallel lines are needed. From 2-5 rows of fast growing trees of different heights should be planted to prevent any possible breaks in single rows, which would create a tunneling action with high and dangerous wind velocities. Tree shelterbelts with different species composition suitable for Indian Thar desert and their effect on wind speed reduction are reported in Mertia *et al.* (2006). The shelterbelt technology is adopted well in areas of Indian Thar desert where water resource is available either through tubewell command area (e.g. Lathi series at Jaisalmer) or IGNP command area. For arid conditions in the Indian Thar desert, Kaul (1969) suggested five row shelterbelt with a pyramidal shape having one row of tall trees (e.g. *Acacia tortilis*, *Tamarix articulate*, *Azadirachta indica*) followed by two rows of smaller trees (e.g. *Acacia senegal*, *Prosopis juliflora*, *Parkinsonia acculeata* etc.) and then followed by two rows of shrubs (e.g. *Aerva tometosa*, *Zizyphus spinachristi* and *Calligonum polygonoides*) at the edge in flank rows. Tree species suitable for different purposes in the Indian Thar desert is mentioned in Table 1.

Table 1: Design and species for shelterbelt plantation

Purpose	Design	Suitable species
Road side	3 to 5 4 staggered rows	*Acacia tortilis, Albizia lebbek, Acacia indica, Dalbergia sissoo, P. juliflora, Tamarix articulate*
Railway side	6 rows	*Prosopis acculeata, P. juliflora, Tecomella articulata*
Canal side	6 rows	*Acacia nilotica, Eucalyptus* spp., *T. undulata, A. tortilis, P. juliflora, D. sissoo, Prosopis cineraria, Acacia nubica*
Farm boundary (rainfed)	1/2/3 rows	*Acacia tortilis, A. lebbek, A. indica, D. sissoo, P. acculeata, P. juliflora*
Farm boundary (irrigated)	2 rows	*Acacia tortilis, A. lebbek, Dicrostachys cineraria, P. juliflora*

(*Source*: Mertia *et al.* 2006)

Mechanical wind barrier in the form of rock piles are often used by farmers in the Indian Thar desert. Other than shelterbelt, various non-vegetative structures were also tested as wind barrier throughout the world. Date-fronds mat fence was tested as mechanical barrier to control wind erosion in Arabian desert (Murai *et al.* 1990; Al-Afifi *et al.* 1990). Technical textiles were tested as wind screen for reducing the wind speed in Belgium (Dierickx *et al.* 2001). Gravel sand barrier was designed and tested in Tibet by Zhang *et al.* (2007) and pointed out that for most economical and scientific system to combat wind blown sand, engineering structures should be implemented at initial stage until the tree shelterbelts mature. More, recently, porous fence made of stainless steel was also tested to shelter the saltating sand particles inside wind tunnel in China (Zhang *et al.* 2010).

Conclusion

Wind erosion is a severe land degradation process in most arid parts of the world whether it is under hot or cold climate situation. The mechanism of wind erosion process mainly includes four basic steps: initiation of movement of soil particles by wind force, transport of moving particles basically in three modes (suspension, saltation and surface creep), sorting of transported particles, and the abrasion by moving soil particles, which further induce the erosion process.

All the particles initially carried out by wind are deposited in a distant places and thus affecting the source area by depleting the soil fertility and creating environmental hazards in deposited areas. Sometimes, the deposition of nutrient enriched eroded soil particles in places away from the source region increased the fertility. There are main five factors affecting the wind erosion process, which are climatic factor, soil erodibility factor, roughness factor, field length factor and vegetative factor. Considering these causative factors different control measures have been formulated, which mainly aims either to decrease the erosive energy of wind or to decrease the erodibility of soil by altering the surface soil characteristics or surface cover or roughness. Among different control measures, well anchored surface vegetative cover are the most applicable method to control wind erosion, which may be achieved through maintaining permanent grass cover in rangelands. Tree shelter belt is also a suitable options to control wind erosion in those places of desert, where water resource is available for their initial establishment.

References

Al-Afifi MA, Haffar I, Murai H, Itani S, Yokota H (1990) 1. Use of date-fronds mat fence as a barrier for wind erosion control. 2. Effect of barrier density on microclimate and vegetation. Agriculture, Ecosystems and Environment 33:47-55.

Bagnold RA (1943) The Physics of Blown Sand and Desert Dunes. William Morrow and Co., pp.265.

Chepil WS (1945) Dynamics of wind erosion: Nature of movement of soil by wind. Soil Science 60(4):305-320.

Chepil WS, Woodruff NP (1963) The physics of wind erosion and its control. Advances in Agronomy 15:211-302.

Cornelis WM, Gabriel D (2005) Optimal windbreak design for wind-erosion control. Journal of Arid Environments 61:315-332.

Ganguli JK, Kaul, RN (1969) Wind erosion control. ICAR, New Delhi. Techn Bull (Agric.) 20, pp.1-53.

Gupta JP, Rao GGSN, Ramakrishna YS, Rao BVR (1984) Role of shelterbelts in arid zone. Indian Farming 29-30.

Kar A, Joshi DC (1995) Sand movement and control of aeolinhazard. In: Sen AK, Kar A (eds) Land Degradation and Desertification in Asia and the Pacific region, Scientific Publishers, Jodhpur, pp. 19-40.

Kar A, Moharana PC, Raina P, Kumar M, Soni ML, Santra P, Ajai, Arya AS, Dhinwa PS (2009) Desertification and its control measures, In: Kar A, Garg BK, Singh MP, Kathju S (eds)Trends in Arid Zone Researches in India, Central Arid Zone Research Institute, Jodhpur, India, pp. 1-47.

Kaul RN (1985) Afforestation of dune areas. In: Sand Dune Stabilization, Shelterbelts and Afforestation in Dry Zones, FAO Conservation Guide 10. FAO, Rome, pp.75-85.

Kort J (1988) Benefits of windbreaks to field and forage crops. Agriculture, Ecosystem and Environment 22/23:165-190.

Lyles L, Cole GW, Hagen LJ (1985) Wind erosion: processes and prediction. In: Follett RF, Stewart, BA (eds) Soil Erosion and Crop Productivity, ASA-CSSA-SSSA, 677 South Segoe Road, Madison, WI 53711, USA.

Mann HS (1985) Wind erosion and its control. In: FAO Conservation Guide, Sand Dune Stabilization, Shelterbelts and Afforestation in Dry Zones. Food and Agriculture organization of the United Nations, Rome, pp.125–132.

Mertia RS (1992) Shelterbelt research in arid zone. J Tropical Forestry 8(3):196-200.

Mertia RS, Prasad R, Gajja BL, SamraJS, Narain, P (2006) Impact of shelterbelts in arid region of western Rajasthan. Central Arid Zone Research Institute, Jodhpur, India, pp. 1-76.

Mertia RS, Santra P (2012) Grazing practices in the rangelands of the Indian Thar desert and its impact on ecosystem and environment. In: Ramón J (ed) Grazing Ecology: Vegetation and Soil Impact NOVA Publication, NY, pp. 7-26.

Mohammed AE, Stigter CJ, Adam HS (1996) On shelterbelt design for combating sand invasion. Agriculture, Ecosystem and Environment 57: 81-90.

Murai H, Al-Afifi MA, Haffar I, Yoshizaki S (1990) Use of date-fronds mat fence as a barrier for wind erosion control. 2. Effect of barrier density on sand movement stabilization. Agriculture, Ecosystems and Environment 32:273-282.

Muthana KD (1982) A review of sand dune stabilization and afforestation. In: Proceedings of the workshop on the Problems of the Deserts in India. Geological Survey of India, Calcutta. Miscellaneous Publication 49, pp. 363-368.

Narain P, Kar A (2007) Desertification and its control in India. In: El-Beltagy A, Saxena MC, Wang T (eds)Human and Nature-working together for sustainable development of drylands. International Centre for Agricultural Research in the Dry Areas (ICARDA), Aleppo, Syria.pp. 84-94.

Santra P, Mertia RS, Narain P (2006) Land degradation through wind erosion in Thar Desert–issues and research priorities. Indian Journal of soil conservation 34 (3):214-220.

Santra P, Mertia RS, Kushawa HL (2010) A new wind erosion sampler for monitoring dust storm events in the Indian Thar desert. Current Science 99 (8):1061-1067.

Skidmore EL, Hagen LJ (1977) Reducing wind erosion with barriers. Transaction of ASAE 20:911-915.

Tibke G (1988) Basic principles of wind erosion control. Agriculture, Ecosystem and Environment 22/23:103-122.

Venkateswarlu J, Kar A (1996) Wind erosion and its control in arid north-west India. Annals of Arid Zone 35:85-99.

Woodruff NP, Siddoway FH (1965) A wind erosion equation. Soil Science Society of America Proceedings 29:502-608.

Zhang CL, Zou XY, Cheng H, Yang S, Pan XH, Liu YZ, Dong GR (2007) Engineering measures to control windblown sand in Shiquanhe town, Tibet. Journal of Wind Engineering and Industrial Aerodynamics 95:53-70.

Zhang N, Kang JH, Lee SJ (2010) Wind tunnel observation on the effect of a porous wind fence on shelter of saltating sand particles. Geomorphology 120:224-232.

28

Micropropagation: A Method for Mass Multiplication of Elite Trees

Rajwant K. Kalia, Sidhika Chhajer and Sanjay Kalia*

ICAR-Central Arid Zone Research Institute, Jodhpur- 342003, Rajasthan, India
**Department of Biotechnology, CGO Complex, New Delhi - 110053*

Introduction

The world demand for wood is continuously increasing due to increase in population. The ever increasing population has led to depletion of existing timber reserves and the forestland base has been reduced by increased allocation of forest areas to urban, agricultural, recreational and other uses. The growing awareness towards development of plantations has generated a voluminous increase in the demand for quality planting material. Conventionally, propagation is achieved through sexual (seeds) or vegetative methods (rooting of cutting, grafting, layering, etc). Propagation through seeds is routinely used for production of planting material but it generates immense genetic variability which is not desirable while propagating perennial tree species. In order to propagate selected trees with desirable features vegetative propagation methods like rooting of cuttings, grafting, layering etc have been developed. These methods can produce genetically identical off springs thereby preserving the advantageous traits; requires only one parent so eliminates the need for special pollination mechanisms also. Propagation of seedless varieties of fruits and species not producing viable seeds like bananas, sugarcane, potato, etc is routinely done using vegetative methods only. However, these methods lead to loss of genetic variation making the crops more prone to species specific diseases which can result in the destruction of entire crop. Season specificity, limited availability of propagules and loss of rooting potential with increase in mother tree age are additional limitations.

Tissue culture, an important area of biotechnology can be used to improve the productivity of planting material through enhanced availability of identified planting stock with desired traits. Plant tissue culture is a general term used for the cultivation of plant parts; cells, tissues, organs or even cells without cell walls (protoplasts) in synthetic medium under controlled, aseptic and *in vitro* conditions. German botanist Haberlandt (1902) was the first to conceive the idea of plant tissue culture, based on his prediction of 'cellular totipotency'. The genetic potential of a cell to regenerate a whole plant has been termed as 'totipotency'. Plant tissues when cultured *in vitro* exhibit a high degree of plasticity, allowing one type of tissue to initiate another. However, the extent of plasticity varies depending on the specific medium components used. The major applications of plant tissue culture include - clonal propagation, synthetic seeds, protoplast culture, in vitro breeding/ fertilization/ hybridization, embryo rescue, haploid breeding, development of disease free plants and somaclonal variants, screening for biotic an abiotic stress tolerance, production of secondary metabolites (suspension cultures), conservation of germplasm *in vitro* and through cryo-preservation in gene banks, international exchange of germplasm and development of tailored genotypes using genetic engineering. Successful and efficient production of improved transgenic plants relies heavily, if not exclusively, on the ability to regenerate whole plants from those cells, tissues, or organs in which foreign DNA has been inserted and expressed (Kalia 2002).

Clonal propagation through tissue culture is popularly called micropropagation, in other words micropropagation is vegetative propagation under *in vitro* conditions wherein plants are propagated using miniature plant tissues called explants grown aseptically in test tube or other containers. Micropropagation offers a rapid means for producing planting stock on a mass scale from a single nodal explant or seed, or callus raised from explants. Micropropagation has the following distinct advantages over conventional methods of vegetative propagation - small space requirement, high multiplication rate, freedom from seasonal influences and freedom from microbes. In micropropagation, generally, nodal shoot segments or shoot tips are exploited to form multiple shoots on appropriate nutrient medium. In majority of forest species, plantlets are produced from young juvenile material as rejuvenation of a mature tree is difficult because of 'aging effect', though it has been achieved in several systems. In many species the trees are rejuvenated by stumping to the base. The growth controls that operate in an intact plant can be broken down or eliminated under *in vitro* conditions, leading to profuse production of shoots from a single initial explant. The shoots can be separated and rooted to give rise to entire plantlet. Generally, micropropagation is approached in three ways - enhanced axillary bud break, adventitious bud differentiation and somatic embryogenesis. Stimulation of axillary and adventitious meristem development is accomplished by hormonal treatment of explant. Both methods are two step procedures involving shoot formation followed by rooting. Axillary shoots generally arise from pre-existing quiescent meristems located adaxial to juvenile branches and leaves in mature trees, and in the axils of cotyledons and leaves in the seedlings (Fig. 1). Regeneration from axillary meristems has been achieved from juvenile as well as mature tissues in many species. Propagation via adventitious meristems involves the formation of meristematic center or meristemoids, at places where it do not exist naturally, which develop into bud primordia and finally adventitious shoots with apical domes and leaf primordia (Fig. 1). These unipolar shoots are excised and cultured on rooting medium for induction of root meristem.

Axillary bud proliferation in chir pine (*Pinus roxburghii*)

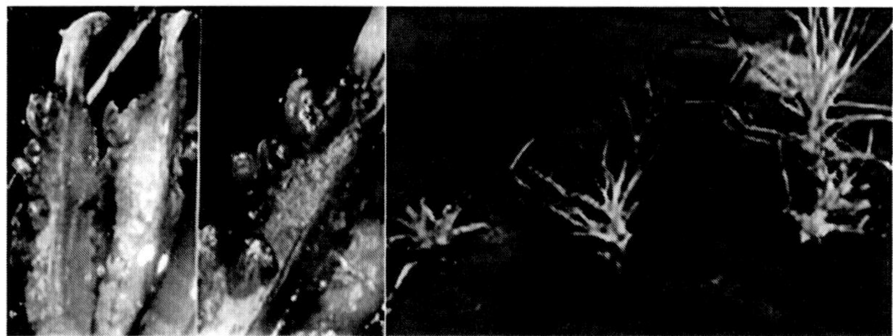

Adventitious bud differentiation on cotyledons of chir pine

Fig. 1. Shoot proliferation through axillary bud proliferation and adventitious bud differentiation
Source: Kalia 2002

Somatic embryogenesis is the process of formation of bipolar embryos from somatic tissue of the explants in a single step. The somatic embryos may directly be produced from the explant or indirectly from the callus or cell suspension cultures (Fig. 2). In this process, somatic cells behave like a zygote under the influence of growth regulators and develop into a bipolar structure with root and shoot primordia. Somatic embryos are being employed for producing synthetic seeds following encapsulation in sodium alginate beads.

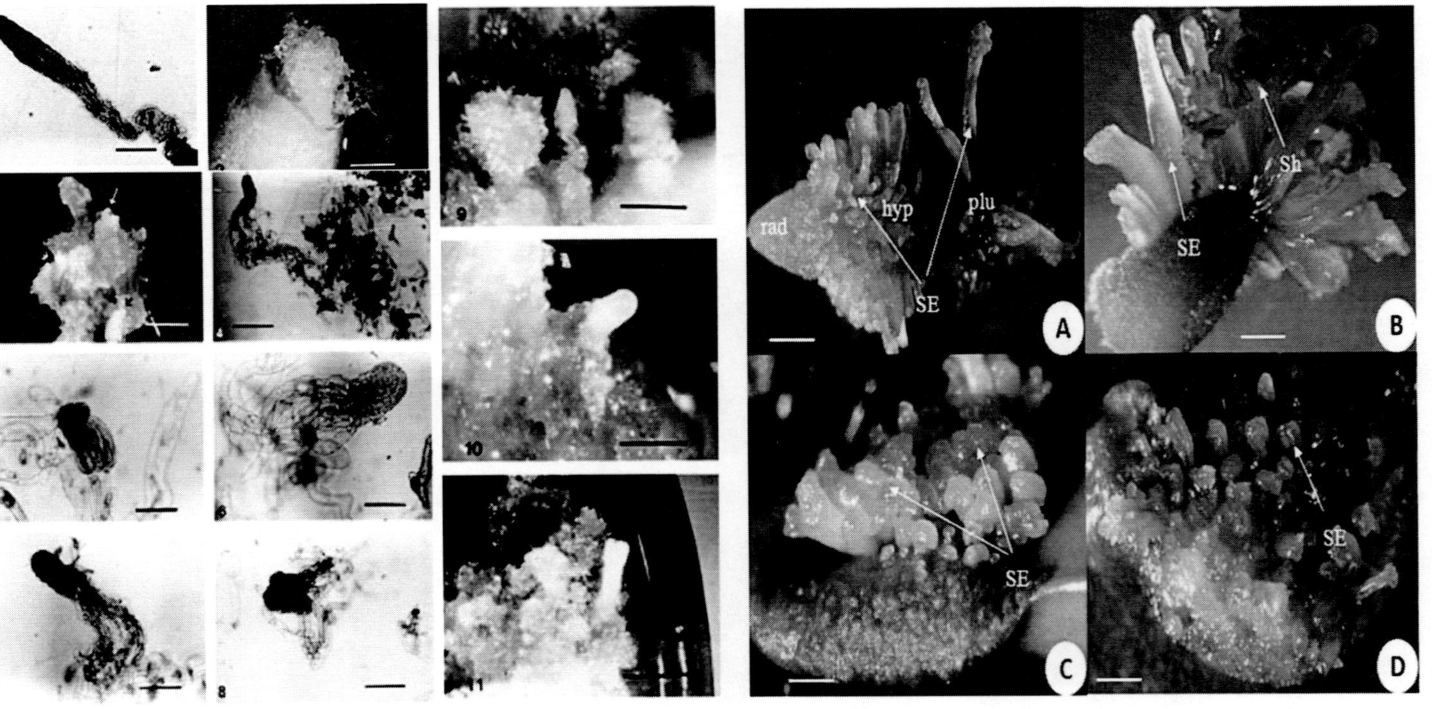

Fig. 2. Legend below B/W figure - Indirect somatic embryogenesis in Pinus roxburghii induced from immature zygotic embryos

Legend below coloured figure - Direct somatic embryogenesis in Dalbergia sissoo induced from zygotic embryos

Source: Arya *et al.* Kalia (2003)

Micropropagation

Stimulation of axillary meristems into shoots is the most commonly used method for mass propagation of plant species. This process involves four major steps namely induction/ establishment, multiplication of shoots, rooting of shoots, and hardening and acclimatization of plantlets.

Establishment is the most important and crucial step determining the success of a protocol. Usually nodal segments containing quiescent axillary buds are collected from young actively growing branches followed by sterilization and culture on synthetic nutrient medium. Explants from trees of different age groups exhibit difference in their *in vitro* response. Living plant materials from the environment are naturally contaminated on their surfaces (and sometimes interiors) with microorganisms, so surface sterilization of explants is done using chemical solutions (sodium or calcium hypochlorite, mercuric chloride). In case of severe contaminations, seeds can be raised aseptically or if possible, the mother plants can be transferred to green house atleast 2 months before explant collection.Shoot morphogenesis is greatly influenced by the season of explant collection (Fig. 3). This effect on explant behavior *in vitro* has been considered to result from the physiological status of the donor plant and the hormone levels at time of excision. Chhajer and Kalia (2017) reported that explants of *Tecomella undulata* collected during the spring season (March - April), which marks the end of dormancy and commencement of period of increased vegetative growth in nature; showed lesser contamination, early shoot initiation and better percent bud break with more shoots per explant (Fig. 3). The winter months (Dec - Jan) were highly unsuitable for *in vitro* establishment of nodal explants due to dormant growth. Maximum contamination was recorded in rainy season (July-Aug) when high moisture supports luxuriant growth of microbes on branches and in shoot axils. Similar reports of seasonal effects on contamination percentage and bud break in explants collected from mature trees are also available in many other plant species. Explants collected during active growth period are more responsive due to production of more auxins in young buds which stimulate cell division in the cambium thus supporting increased cell division and growth (Funada *et al.* 2001).

The sterilized explants are inoculated on semi-solid agar gelled/ liquid medium containing organic and inorganic nutrients, carbon source, growth regulators, etc. In addition to providing nutrients the medium provides access to atmosphere for gas exchange; dumping ground for plant metabolites; support for erect growth and maintains osmotic potential. Many nutrient media have been formulated by various research groups while working on different plant systems in the past. Many plant species have shown differential preference for specific medium during explant establishment and growth.To name a few important ones are:

Murashige and Skoog's (MS, Murashige and Skoog 1962) medium, Gamborg's (B5, Gamborg *et al.* 1968) medium, Nitsch and Nitsch (NN, Nitsch and Nitsch 1969) medium, Woody Plant medium (WPM, Lloyd and McCown 1980), Schenk and Hilderbrandt (SH, Schenk and Hildebrandt 1972) medium, etc (Table 1). Singh *et al.* (2012a) reported better response of *Dendrocalamus asper* nodal explants on MS medium compared to B5, SH or NN while Chhajer and Kalia (2017) reported better growth of *Tecomella* shoots on SH medium compared to MS, B5, WPM or NN.

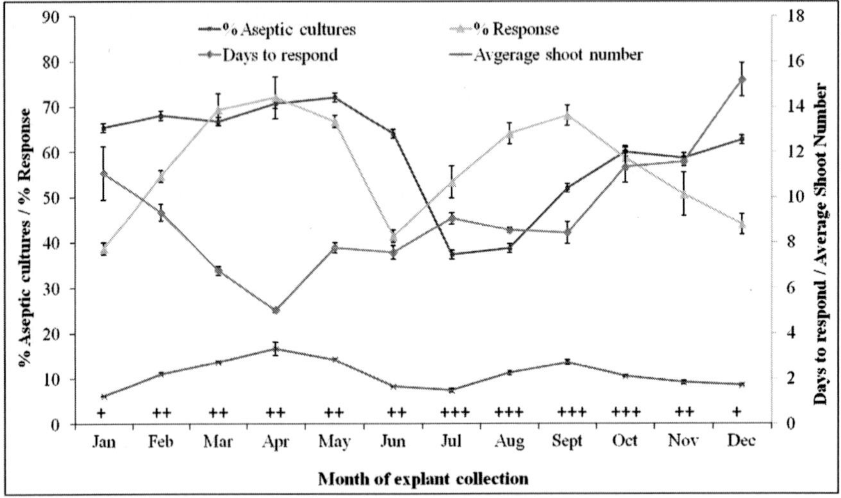

Fig. 3. Effect of month of explant collection on *in vitro* establishment and response of mature nodal explants of *Tecomella undulata* cultured on MS + 10µM BAP. Extent of callusing is depicted by +. Data (recorded after 3 weeks of inoculation) are mean ± SE
Source: Chhajer and Kalia (2017)

Table 1: Composition of various basal media

Salt*	Formula	MS	NN	SH	B5	WPM
Macroelements						
Potassium nitrate	KNO_3	1900	950	2500	2500	-
Ammonium nitrate	NH_4NO_3	1650	720	-	-	400
Magnesium sulphate	$MgSO_4.7H_2O$	370	186	400	250	180.69
Calcium chloride	$CaCl_2.2H_2O$	440	166	200	150	96
Potassium dihydrogen phosphate	KH_2PO_4	170	68	-	-	170
Sodium dihydrogen phosphate	$NaH_2PO_4.H_2O$	-	-	300	150	-
Ammonium sulphate	$(NH_4)_2SO_4$	-	-	-	134	-
Potassiumsulphate	K_2SO_4	-	-	-	-	990
Calcium Nitrate	$CaCl_2NO_3$	-	-	-	-	386.34

(*Contd.*)

Microelements						
Manganese sulphate	$MnSO_4 . 4H_2O$	22.6	25	10	10	22.3
Zinc sulphate	$ZnSO_4 . 4H_2O$	8.6	10	1	2	8.6
Boric acid	H_3BO_3	6.2	10	5	3	6.2
Copper sulphate	$CuSO_4 . 5H_2O$	0.025	0.025	0.2	0.025	0.25
Cobalt chloride	$CoCl_2 . 6H_2O$	0.025	-	0.1	0.025	-
Sodium molybdate	$NaMoO_4 . 2H_2O$	0.25	0.25	0.1	0.25	0.25
Potassium iodide	KI	0.83	-	1	0.75	-
Iron						
Ferrous sulphate	$FeSO_4 . 7H_2O$	27.8	27.8	15	27.8	27.8
Sodium EDTA	Na_2EDTA	37.3	37.25	20	37.3	37.3
Organics						
Thiamine Hydrochloride		0.1	0.5	5	10	1
Nicotinic acid		0.5	5	5	1	0.5
Pyridoxine Hydrochloride		0.5	0.5	0.5	1	0.5
Myo-inositol		100	100	1000	100	100
Biotin		-	0.05	-	-	-
Glycine		2	2	-	-	2

*Concentrations in mgl^{-1}

Weaker salt formulations like WPM, NN etc. are reported to promote axillary bud development in forest trees (McCown and Sellmer 1987). Various major and minor inorganic nutrients play diverse roles in plant growth and development (Table 2).

Table 2: Deficiency symptoms of major and minor inorganic salts in plants

Taken up as	Part of	Deficiency symptoms
Inorganic salts: major elements		
Nitrogen(NO_2^-,NO_3,NH_4^+)	Amino acids, chlorophyll, nucleic acids, proteins	Chlorosis of older parts first, Stunted growth
Phosphorus($H_2PO_4^-$,HPO_4^{-3})	Nucleic acids, phospho-lipids, ATP, NADP.	Stunted growth, disturbed metabolism, Necrotic spots on old leaves, delayed maturity
Potassium (K^+)	Membrane transport, stomatal opening, coenzyme	Mottling of margins, chlorosis in interveinal areas
Sulphur (SO_4^{--})	Amino acids-cysteine & methionine, thiamine, biotin	Dieback, Chlorosis of younger leaves
Magnesium (Mg^{2+})	Chlorophyll, ribosomes	Interveinal chlorosis of mature leaves
Calcium (Ca^{2+})	Middle lamella, Ca-pectate	Stunted growth, necrosis of meristematic areas
Iron (Fe^{3+})	Cytochromes, ferredoxin	Interveinal chlorosis of young leaves first

(Contd.)

Inorganic salts: minor elements		
Boron (BO_3^{-},$B_4O_7^{-}$)	Pollen germination, phloem transport	Brown heart or heart rot, cracks in stem
Zinc (Zn^{2+})	Auxin synthesis, part of dehydrogenases	"Rossette plants"Little leaf" and "White buds"
Copper (Cu^{2+})	Nitrite reductase, plastocyanin, polyphenol oxidase	Dieback disease, exanthema
Molybdenm (MoO_2^{++})	Nitrate reductase, also nitrogenase	Whiptail disease, inhibition of flowering
Manganese (Mn^{2+})	Activates carboxylases	Interveinal chlorosis, grey spots on leaves
Chlorine (Cl^{-})	Ion balance, O_2evolution in photosynthesis	Bronze leaves, wilting, flower abscission

The organic compounds include vitamins, amino acids and sometimes some undefined organic supplements are also used. Vitamins are organic molecules required in the diet in small amounts. Thiamine (Vit B1) essential for cultures is added as thiamine-HCl, it is required in 0.1 to 30 mgl^{-1}. Nicotinic acid and pyridoxine (Vit B6) stimulate growth of cultures while P-aminobenzoicacid, ascorbic acid, biotin, cyanocobalamine (Vit B 12), folic acid and riboflavin (Vit B 2) are also required sometimes.Except glycine amino acids are usually not added to the media. Natural forms are L-amino acids and some have been reported to have direct affects: L-tyrosine can contribute to shoot initiation; L-arginine facilitates rooting; L-serine helps in haploid embryo production; L-glutamine and L-asparagine enhance somatic embryogenesis. Sometimes complex organic supplements such as casein hydrozylate, yeast extract, malt extract, coconut water, orange juice, banana fruit extract, fish emulsion, skimmed milk etc are also used. Kalia (2002) regularly used casein hydrolysate (500 mg l^{-1}) for multiplication of embryogenic cultures of *Pinus roxburghii*. Agarwal *et al.* (2015) reported that medium containing adenine sulphate, ascorbic acid, citric acid and arginine improved shoot induction, multiplication and growth in *Alhagi maurorum*.

Several authors have suggested solutions to minimize the lethal browning or blackening of explants caused by phenolic compounds in plant tissue culture. These include addition of polyphenol adsorbents like activated charcoal (Arditti and Ernst 1993), antioxidants such as cysteine (Sanyal *et al.* 2005), ascorbic acid (Arditti and Ernst 1993), polyvinylpyrollidone (PVP, Laine and David 1994) or silver nitrate (Sanyal *et al.* 2005) to the culture medium. Singh *et al.* (2012b) reported a significant control of browning with enhanced shoots multiplication in *Dendrocalamus hamiltonii* when ascorbic acid (28 to 84 μM) was incorporated in the culture medium. Adsorbents such as polyvinylpyrollidone (PVP) and charcoal are used if there is excessive discoloration of the medium.

Saxena and Dhawan (1999) used PVP for improving shoot health in the cultures of *Dendrocalamus strictus*. Activated charcoal stimulates growth of cultures, because it adsorbs inorganic and organic molecules. It thus regulates the supply of some endogenous growth regulators and also removes secondary products secreted by cultures, which may be inhibitory.

Usually 2 or 3% sucrose or glucose is added as carbon source; when autoclaved sucrose results in sucrose + D-glucose + D-fructose. Fructose and starch are also used sometimes while myo-inositol at 100 mgl^{-1} is added routinely as a growth factor. Water potential difference between cell sap and media governs the uptake of water. Agar, carbon source and components of media are enough to build up osmotic but sometimes weekly metabolized sugars: mannitol or sorbitol, polyethylene glycol may be added. Kalia (2003) increased myo-inositol concentration to 200 mg/l for controlling tip burning and leaf fall in *in vitro* cultures of *Dalbergia sissoo*. Singh *et al.* (2012a, b) used table sugar in *D. asper* and *D. hamiltonii* respectively to make the protocol more cost effective without hampering shoot health and multiplication rates.

The pH of the media should be between 5.5 to 5.8 for best growth of cultures. Below pH 5.5, agar will not gel properly; above 6.0 it will be too firm. pH is affected by autoclaving, explant, utilization of organic acids and nitrogen etc. Stationary support to cultures may be provided by filter paper, cotton, cheese cloth, vermiculite etc. Gelling support is provided by 0.6 to 1.0% agar as substratum. Gelrite a transparent polysaccharide: a fermentation product of *Pseudomonas* species is also sometimes used to provide stationary support. Kalia *et al.* (2007) reported that softer (0.6% agar gelled) medium supported better rooting in *Pinusroxburghii* compared to harder medium (0.7 and 0.8% agar gelled) due to better root penetration and lesser radial growth of callus on medium surface. The shoots cultured on 0.6% agar gelled medium for 1 week were transferred to liquid medium with filter paper bridges to allow elongation of induced root primordial. Similarly, Mohammed and Vidaver (1988) reported that impediment of gaseous exchange in agar gelled medium led to poor root quality.

The hormones or the growth regulators are the most important components of the media determining the fate of the cultured cells and tissues. Hormones are the "chemical messengers" that control growth, development and several other functions in plants. Plant hormones are natural organic compounds which affect physiological processes at concentrations far below the level of nutrients and vitamins; their occurrence is wide-spread but effect is specific. The five classical plant hormones include auxins, cytokinins, gibberellins, abscisic acid and ethylene. However, now more classes of chemicals like polyamines, brassinisteroids, salicylic acid and jasmonates are also used as growth regulators. nterestingly, among all the classical naturally occurring PGRs, chemical identification of only abscisic acid was made from plant tissue. Identification of most others came from the extracts that produced hormone-like effects in plants. For example, auxin was identified from urine and the fungal cultures of *Rhizopus*, gibberellins from culture filtrates of the fungus *Gibberella*, cytokinins from autoclaved herring sperm DNA, and ethylene from illuminating gas. Though Haberlandt failed, the subsequent investigators (Gautheret, White and Nobecourt independently in 1930s) successfully established tissue culture technique for a number of plants. Credit for their success mainly goes to the discoveries of auxin by Went (1926), and Kogl and Haagen-Smit (1931). Later, the discovery of cytokinins by Miller *et al.* (1955) provided impetus to the advances in plant tissue culture.

Skoog and Miller (1957) were the first to point out that balance between the concentrations of auxin and cytokinin plays important role in initiation of shoot and root organogenesis. In plant tissue and organ culture, auxin and cytokinin play the most important roles of regulating growth and morphogenesis. A number of synthetic auxins and cytokinins with even better biological activity than that of the equivalent natural growth regulators have been discovered and are being used in plant tissue culture. Auxins [Indole-3 acetic acid (IAA), Indole-3 butyric acid (IBA), 2,4-dichlorophenoxyacetic acid (2,4-D), 2,4,5-trichlorophenoxyacetic acid (2,4,5-T), Picloram: 4-amino 3,5,6-trichloropicolinic acid, p-chlorophenoxyacetic acid (CPA), 1-naphthaleneacetic acid (NAA) and Dicamba: 3,6-dichloro-o-anisic acid] promotes cell division, cell elongation; inhibits lateral growth; adventitious root formation (at high conc.); adventitious shoot formation (at low conc.); induces somatic embryos (2, 4-D) and callus formation. While, cytokinins [Zeatin (Z), Isopentenyladenine (iP), Isopentenyladenosine (iPA), 6-bezylaminopurine/ benzyladenine (BAP), Kinetin, Thidiazuron (TDZ), N-(2 chloro-4-pyridyl)-N'-phenylurea (CPPU) and Metatopolin] cause adventitious shoot formation (at high conc.); inhibit root formation; cause cell division, callus formation and growth. Gibberellins promote shoot elongation; release dormancy in seeds, somatic embryos, apical buds; inhibits adventitious root formation; promote tuber, corm and bulb formation and facilitates acclimatization while abscisic acid leads to maturation of somatic embryos, facilitates acclimatization, bulb and tuber formation and promotes the development of dormancy in somatic embryos. Use of ethylene in tissue cultures is generally undesirable but inhibitors of ethylene production such as silver nitrate, silver sulfate, cobalt or nickel chloride, AVG (aminoethoxy vinyl glycine) or salicylic acid increase shoot regeneration and somatic embryo production.The role of conventional, non-conventional growth regulators (Fig. 4) and inhibitors of growth regulators have been reviewed at length by Kumar *et al.* (2016), Kalia *et al.* (2016) and Singh *et al.* (2016).

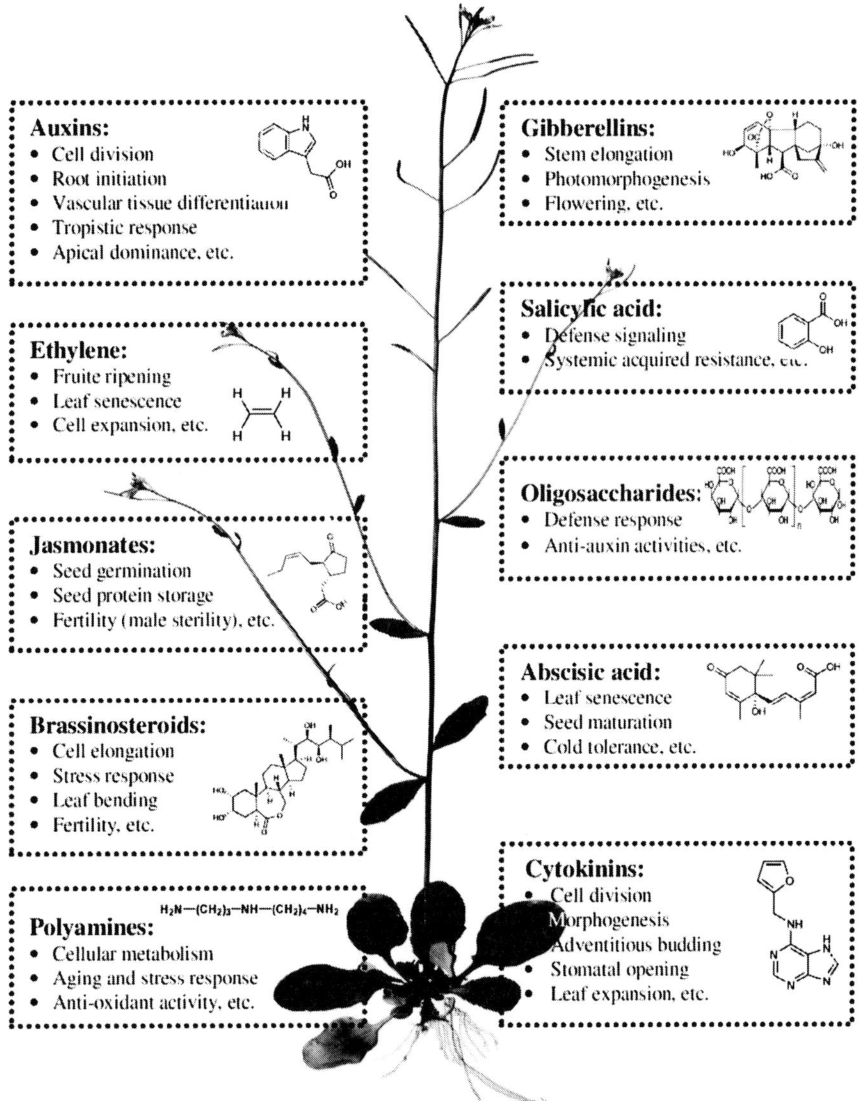

Fig. 4. Growth regulators control various aspects of plant's growth and development. Some of the classical and newly discovered PGRs with their major functions (structure of the representative compound is depicted)

Source: Kumar *et al.* (2016)

Auxins and cytokinins remain the most used growth regulators in tissue culture while gibberellins and ethylene are seldom used. Both auxins and cytokinins promote cell division in tissue culture, but probably effects at different steps hence complement each other. Manipulation of cytokinin and auxin ratio affects organogenesis: when cytokinin level is higher compared to auxins, shoots are formed; when ratios are about the same, callus is formed, and when auxins are more relative to cytokinins, roots are formed. Stage 1 (establishment /initiation) - requires both cytokinins and auxins for stimulation of buds and shoot primordia initiation; stage 2 (shoot multiplication) requires cytokinins while stage 3 (rooting) requires auxins. Callus initiation is achieved through high concentrations of 2, 4-D (4 to 8 mg/l) and auxins levels are lowered or eliminated for further development of somatic embryos or adventitious roots. Similarly, somatic embryogenesis requires induction of an embryogenic state by use of 2, 4-D or NAA. The embryogenic mass shows further development into embryos by removal or drastic reduction in the level of these auxins and ABA is added to promote somatic embryo maturation. Response of explants to growth regulators supplemented in the medium is also dependent on genotype and endogenous level of different growth regulators at the time of culture. Kalia *et al.* (2004) reported that responsiveness of *in vitro* cultures towards added NAA was genotype dependent, six clones of *Dalbergia sissoo* (from Rajasthan and Haryana), showed enhanced growth while another six clones (from Uttranchal and Uttar Pradesh) showed decline in response when NAA was added to the multiplication medium.

The ability of plant tissues to form adventitious roots depends on interactions of many different endogenous and exogenous factors like basal medium used and its strength, auxin type and concentration, and additives incorporated. The role of auxins in root development is well-established. Usually natural (IAA) and synthetic auxins (NAA and IBA) are used for rooting. During rhizogenesis, a diluted mineral fraction (50 to 75 percent) is used. Half strength SH medium proved better for *in vitro* root induction in *Tecomella undulata* compared to slow and delayed rooting response on 1/4x and 1x SH medium (Chhajer and Kalia 2017). Methods used to induce rooting in *in vitro* raised shoots include - (a) continuous culture on auxin supplemented (5-25mM) medium, (b) culturing shoots on high IBA/ NAA (50ppm) supplemented medium for 3 days followed by transfer to basal medium, and (c) treatment of shoot base with commercial rooting powder or IBA/ NAA (500ppm) for 15-30 min before transferring them to sterile soilrite for rooting. Sometimes the shoots are transferred to basal medium for 7 days before shifting to rooting medium to reduce the carry over effect of cytokinins. Sometimes additives like ferulic acid, silver nitrate, ascorbic acid, choline chloride or activated charcoal are added to rooting media to improve rooting response. Variable rooting percentages (10-100%) have been reported improve the rooting for different plant species cultured *in vitro*.

Plants cultivated *in vitro* are different from field grown plants. Plantlets developed *in vitro* grow heterotrophically on a medium containing ample sugar and nutrients, lower light intensity and under high humidity levels within the culture vessels. The micropropagated plantlets having a weak root system and poorly developed cuticle usually show high mortality when transferred to *ex vitro* conditions without proper hardening and acclimatization. The micropropagated plantlets have non functional stomata, weak root system and poorly developed cuticle. Gradual hardening and acclimatization helps the plants to develop phototrophic growth. During acclimatization to *ex vitro* conditions, leaf thickness generally increases, stomatal density decreases and stomatal form changes from circular to elliptical. Development ofcuticle, epicuticular waxes, and effective stomatalregulation of transpiration occurs leading to stabilizationof water potential of field transferred plantlets. The culture containers could be kept in the greenhouse with loose lids. Micropropagated plantlets can be left in shade for 3–6 days under diffused natural light to make them adjust to the conditions of new environment. This helps in semi-hardening of plants and leads to shoot elongation.

Fig. 5. Hardening and acclimatization of *in vitro* raised plants of *Tecomella undulata.*(A) Rooting on 1/2x SH + 10 μM IBA, (B) Hardening in 1/2x SH liquid medium with filter paper bridge, (C) *In vitro* raised plantlets. Plantlets growing in (D) sterile soilrite and pots containing soil and FYM in 1:1 ratio after 1 month and 3 months respectively and (E) 2-month-old field transferred plant

Large scale demonstrations of tissue culture raised bamboos were raised in various parts of the country under DBT funded multi-institutional projects so that more and more consumers can take advantage of this clonal propagation technique. Similarly, a field orchard of tissue culture raised date palm plants has been established at CAZRI, Jodhpur (Fig. 6) under ICAR funded project.

Fig. 6. Tissue culture raised date palm plants growing at CAZRI, Jodhpur

Somaclonal variations may appear due to cell cycle disturbances caused by exogenously applied growth regulators. The primary factors controlling induction of somaclonal variation *in vitro* include - culture method, explants source, ploidy level and *in vitro* culture age; high concentrations of growth regulators used; cell cycle disturbances caused by exogenously applied growth regulators; increased mutation rate per cell-generation over time, and accumulation of mutations over a period of time; alteration in DNA methylation patterns, DNA damage and mutation, and alteration of cell's ability to repair damaged and mutated DNA (Chhajer and Kalia 2016). The variability generated is a combined effect of genetic and epigenetic variations occurring during the course of culture and the genetic heterogeneity of the cells of explants. It is therefore extremely important to ascertain the suitability of a particular micropropagation protocol developed for a particular species, where commercial success in micropropagation depends solely on the maintenance of clonal uniformity. *In vitro* regeneration achieved through axillary bud proliferation remains the most trusted method for clonal propagation with minimum risk of generation of somaclonal variations as organized (pre-existing) meristems develop into shoots without involving dedifferentiation or redifferentiation of cells or tissues into meristems (*de novo* organogenesis). Commercial application of tissue culture to perennial crops must await adequate quality checks and field testing with proper controls. Therefore, it is important to devise methods of quality control at all the tissue culture stages to minimise the potential danger of such defects at later stages. Attempts have been made in the past to judge the genetic fidelity of *in vitro* raised plants using morphological and physiological traits, however, these methods require extensive observations until maturity. Further, the morphological and physiological differences may disappear over the growing seasons or the initially uniform looking plants may behave differently during flowering/ fruiting stages due to genetic aberrations. Therefore, more efficient detection tools like DNA markers are being used nowadays to ascertain the genetic fidelity of *in vitro* raised plants as these markers are not influenced by the age or tissue of the plant, growth stage or the prevailing environmental conditions (Fig.7).

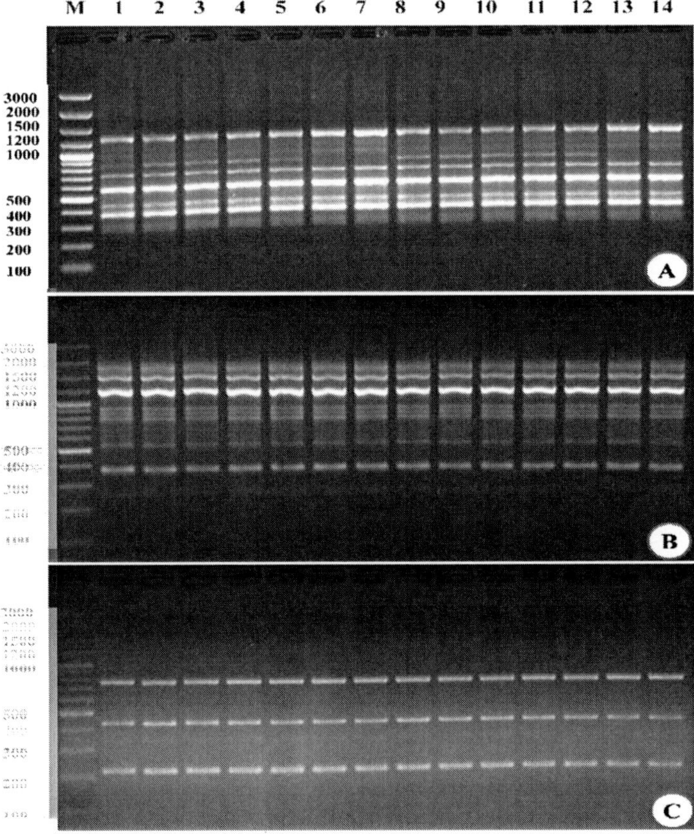

Fig.7. Genetic fidelity testing of *in vitro* raised plants of *Dendrocalamus asper*. DNA amplification with (A) RAPD, (B) ISSR and (C) SSR primers. Lanes: M-100bp plus DNA ladder, 1- mother plant, 2–11 *in vitro* cultures from 3rd to 30th passage after every 3rd sub-culture cycle, 12- plant at greenhouse/polyhouse stage, 13- plant at field stage after 1 year, and 14- plant at field stage after 2 years. *Source*: Singh *et al.* (2013a)

In last two decades, many DNA-based markers like RAPD, ISSR, SCoT, AFLP, SAMPL and SSR have been successfully used for assessment of genetic homogeneity of tissue culture raised plants. It has been suggested to use a combination of two or more marker types for genetic fidelity testing of plants so as to target a larger part of the genome under study (Singh *et al.* 2013a). Singh *et al.* (2013a, b) used RAPD, ISSR, SSR and AFLP markers for genetic fidelity testing in *Dendrocalamus asper* and *D. hamiltonii* respectively while Chhajer and Kalia (2016, 2017) used RAPD, ISSR, SSR and SCoT markers in *Tecomella undulata*. Genetic homogeneity of tissue culture raised plants was confirmed by generation of monomorphic DNA fragments with all the marker systems used.

Commercial plant tissue culture

Commercial plant tissue culture labs have been established around the globe. Most research institutes have plant tissue culture labs as a part of their routine activities. Over a thousand plant species have been cultured *in vitro*. NV Thomas & Co. in Kerala established the first commercial tissue culture unit in India in 1987 for large scale production of cardamom based on the protocol developed by NCL Pune, India. These pioneering efforts were followed by several other commercial companies. Indo-American Hybrid Seed Company developed the second tissue culture unit in the country at Bangalore in 1988. There are several companies in the field or micropropagation in the country now viz. Hindustan Lever, Tata Tea, Unicorn Biotech, Nath seeds, RPG Enterprises, Indian tobacco, Hindustan Agri Genetics limited, etc (Deptt. of Agriculture, Maharashtra). Most of these tissue culture units are located in Maharashtra, Andhra Pradesh, Karnataka and Kerala. These companies have been so far largely concentrating on exploiting the international markets.

Department of Biotechnology (DBT) identified plant tissue culture technology as a priority area revolutionizing the commercial agriculture sector therefore initiated a number of programs for integrated development and commercialization of this technology. Ministry of Science and Technology supported about 150 projects for research and development in commercial plant tissue culture in the country. As a result more than 50 commercial units were set up between 1987 and 1995 with a total installed capacity of about 210 million plants per annum (Singh and Shetty 2011).Various central and state government departments are encouraging the development of tissue culture industry through several schemes and have announced incentives also – Department of Biotechnology (DBT) supports research and development projects for standardization of tissue culture protocols for important plants at universities and research institutions. Funding is also provided for expansion of existing tissue culture units under the scheme called Small Business Innovation Research Initiative (SBIRI). DBT also established two micropropagation technology parks (MTPs) to provide service packages and training for generating skilled manpower. Ministry of Agriculture and Farmers' welfare (through DAC and Small Farmers Agri-business Consortium, SFAC), National Horticulture Board (NHB) and the state governments of Karnataka, Gujarat, Maharashtra and Andhra Pradesh provides financial assistance in the form of subsidies or loans for setting up tissue culture units, greenhouse and climate controlled polyhouse/shade houses while the Ministry of Commerce and Industry (through APEDA) has developed state-of-the-art air freight trans-shipment centres for tissue culture plants at New Delhi, Bombay and Bangalore airports (Singh and Shetty 2011).

As per the report of BCIL (2005) micropropagation protocols have been perfected for large scale propagation of many plant species (Table 3) and 46 established commercial tissue culture units with production capacity ranging between 1-5 million plants per annum are commercially producing plantlets through tissue culture.

Table 3. List of species perfected for large scale propagation *in vitro*

Plant category	Plants
Fruits	Banana, grapes, pineapple, strawberry, sapota
Cash crops	Sugarcane, potato
Spices	Turmeric, ginger, vanilla, large cardamom, small cardamom
Medicinal plants	*Aloevera*, geranium, stevia, patchouli, neem
Ornamentals	Gerbera, carnation, anthurium, lily, syngonium, cymbidium
Trees	Teak, white teak, bamboo, *Eucalyptus, Populus*

Source: BCIL 2005

Two tissue culture pilot plant facilities which have now been converted to micropropagation technology parks (MTPs) were developed at The Energy Research Institute (TERI), New Delhi and National Chemical Laboratory (NCL), Pune to provide an effective platform for transfer of proven technologies to entrepreneurs. These MTPs having production capacity of 1 million plantlets per annum have optimized a large number of technologies and also transferred them to industry. National facilities for virus diagnosis and quality control of tissue cultured plants (TCPs) have also been set up at many institutions in the country. Hardening units have also been established in different agro-climatic regions. These pilot plants have produced more than 5 million tissue cultured plantlets (BCIL 2005).

As per the survey report of BCIL (2005) the major consumers of tissue culture plants (TCPs) are the State Agriculture Department, Agri Export Zones (AEZs), sugar industry and private farmers. The paper industry, medicinal plant industry and State Forest Departments are using TCPs in a limited scale. The Spices Board, Cochin, has also brought large area under cultivation of TCPs particularly for small cardamom, vanilla and large cardamom through involvement of progressive farmers and is an active consumer of TCPs. There is an increasing awareness about the benefits of the TCPs over conventional plants among the sugar factories also. In addition, a number of progressive farmers and nurseries in the states of Andhra Pradesh, Maharashtra, West Bengal, Karnataka, Tamil Nadu etc., are the major consumers of TCPs particularly for flowers, banana, sugarcane and medicinal plants. The plants prioritized for tissue culture propagation by the above consumer segments are banana, grapes, pineapple, strawberry, sugarcane, potato, turmeric, ginger, large and small cardamom, vanilla, *Aloevera*, geranium, stevia, patchouli, gerbera, carnations, anthuriums, syngonium, lily and for few tree species namely teak,white teak, bamboo, eucalyptus and popular. As per the report of BCIL (2005), the priority plants of State Agriculture Departments include fruits (banana, papaya, strawberry, grapes, apple, sapota, mandarin orange, passion fruit, cherry, walnut, almond, pecan nut, pineapple, fig), spices (vanilla, ginger, turmeric, pepper, large cardamom), medicinal and aromatic plants (aloe, patchouli, gloriosa, senna, aswagandha, nightshade (*S. khasianum*), phyllanthus (*P. niruri*), dioscorea, neem, geranium) and ornamental plants (orchids, gerbera, anthurium, chrysanthemum, rose, dendrobium); the priority plants for state forest departments include sandal wood, *Bambusa* sp., *Ailanthus, Melocanna* sp., *Sapindus* sp. in Tamil Nadu; bamboo in Mizoram; teak, eucalyptus, bamboo, *Albizzia* and cane in Kerala; poplar, eucalyptus, teak and sissoo in Uttaranchal; amla and *Gmelina* sp. in Madhya Pradesh; eucalyptus, teak, *Nypa, Gmelina* sp., *Dalbergia* sp., *Pterocarpus, Xylocarpus* sp., *Machilus* sp. and Calamus sp. in West Bengal, and teak, bamboo, *Gmelina* and *Thyrostachys* in Tripura. The requirement of plant based industries is as follows: Paper industry (eucalyptus, subabul, *Acacia, Casuarina*, bamboo) and medicinal plant industry (aloe, patchouli, safedmuseli, *Taxus*).

The consumption of tissue culture raised plants for 2002-03 was approximately 44 million plants with banana, sugarcane, ornamentals, spices and medicinal plants constituting 41, 31, 14, 6 and 4% share respectively. The overall market for TCPs is expected to grow by at least 20 to 25% from 72 million in 2003-04 to 144 million over the next five years (BCIL 2005). The demand for tissue culture raised banana has increased by 25-30% and a similar trend has been observed for sugarcane for production of ethanol blended petrol. Foliages and ornamentals have a great potential in the international market. Major potplants and landscaping ornamentals like Ficus, Spathiphyllums, Syngoniums, Philodendrons, Nerium, Alpenia, Yucca, Cordylines, Pulcherrima, Sansevieria,

Gerbera, Anthuriums, Rose, Lilies, Alstromeria etc. are routinely produced by various plant tissue culture laboratories in India. About 212.5 million plants including 157 million ornamental plants amounting to 78% of the total production are reported. It may be pointed out that tissue culture laboratory can also be used to produce biofertilisers like rhizobium, azotobacter, azospirillum, phosphate solubilising bacteria culture as well as mushroom spawn culture that indirectly contribute to the agricultural sector (Singh and Shetty 2011).

The current progress of the tissue culture industry is encouraging but the expected rapid growth can be achieved byincreasing awareness level amongst the people about the application of this technology. Although within the country the research groups have put lots of effort in standardizing protocols for several plant species, the benefits have not been sufficiently demonstrated to the farmers at the field level. Micropropagation technology definitely holds the potential to improve the socio-economic status of Indian farmers and industries if amply used.

References

Agarwal T, Gupta AK, Patel AK, Shekhawat NS (2015) Micropropagation and validation of genetic homogeneity of *Alhagi maurorum* using SCoT, ISSR and RAPD markers. Plant Cell Tiss Organ Cult 120:313–323.

Arditti J, Ernst R (1993) Micropropagation of orchids. John Wiley & Sons, New York.

Arya S, Kalia RK, Arya ID (2000) Induction of somatic embryogenesis in *Pinus roxburghii* Sarg. Plant Cell Reports 19:775-780.

BCIL (2005) Summary report on market survey on tissue cultured plants prepared by Biotech Consortium India Limited for Department of Biotechnology and Small Farmers' Agri-Business Consortium.

Chhajer S, Kalia RK (2016) Evaluation of genetic homogeneity of *in vitro*-raised plants of *Tecomella undulata* (Sm.) Seem. using molecular markers. Tree Genetics and Genomes 12:100.

Chhajer S, Kalia RK (2017) Seasonal and micro-environmental factors controlling clonal propagation of mature trees of marwar teak [*Tecomella undulata* (Sm.) Seem]. Acta Physiol Plant 39:60.

Deptt of Agriculture, Maharashtra - http://www.mahaagri.gov.in/Technology/Tissue Culture.html

Funada R, Kubo T, Tabuchi M, Sugiyama T, Fushitani M (2001) Seasonal variations in endogenous indole-3-acetic acid and abscisic acid in the cambial region of *Pinus densiflora* Sieb. etZucc. stems in relation to earlywood/latewood transition and cessation of tracheid production. Holzforschung 55:128-134.

Gamborg OL, Miller RA, Ojima L (1968) Nutrient requirement of suspension culture of Soybean. Exp Cell Res 50:151-158.

Kalia RK (2002) Studies on standardization of micropropagation of *Pinusroxburghii*Sarg. Ph.D Thesis. Forest Research Institute, Dehradun.

Kalia RK, Arya S, Kalia S, Arya ID (2007) Plantlet regeneration from fascicular buds on seedling explants of *Pinus roxburghii*. Biol Plant 51:653-659.

Kalia RK, Singh R, Kalia S, Sharma SK, Kumar S (2016) Recent advances in understanding the role of growth regulators in plant growth and development *in vitro* - II. Non -conventional growth regulators.The Indian Forester 142 (6):524-535.

Kalia S, Kalia RK, Sharma SK (2004) Evaluation of clonal variability in shoot coppicing ability and *in vitro* responses of *Dalbergia sissoo*Roxb. SilvaeGenetica 53:212-220.

Kalia, S (2003) *In vitro* studies on morphogenesis and micropropagation of *Dalbergia sissoo* and *D. latifolia*. PhD Thesis, Forest Research Institute, Dehradun

Kogl F, Haagen-Smit AJ (1931) Uber die Chemie des Wuchsstoffs K. Akad. Wetenschap. Amsterdam. Proc Sect Sci 34:1411-1416.

Kumar S, Singh R, Kalia S, Sharma SK, Kalia RK (2016) Recent advances in understanding the role of growth regulators in plant growth and development *in vitro* - I. Conventional growth regulators. The Indian Forester 142 (5):459-470.

Laine E, David A (1994) Regeneration of plants from leaf explants of micropropagated clonal *Eucalyptus grandis*. Plant Cell Rep 13:473-476.

Lloyd G, McCown BH (1980) Commercially feasible micropropagation of mountain laurel (*Kalmia latifolia*) by use of shoot tip culture. Int Plant Prop Soc Comb Proc 30:421-427.

McCown BH, Sellmer JC (1987) General media and vessels suitable for woody plant cultures. In: Bonga JM, Durzan DJ (Eds) Tissue culture in forestry - General principles and biotechnology, vol. 2. Martinus Nijhoff Publication, Dordrecht, Boston, pp. 4-6.

Miller C, Skoog F, Von Saltza MH, Strong FM (1955) Kinetin, a cell division factor from deoxyribonucleic acid. J Am Chem Soc 77:1392.

Mohammed GH, Vidaver WE (1988) Root production and plantlet regeneration in tissue cultured conifers. Plant Cell Tissue Org Cult 14:137-160.

Murashige T, Skoog F (1962) A revised medium for rapid growth and bio assay with tobacco tissue culture. Physiol Plant 15:473-97.

Nitsch JP, Nitsch C (1969) Haploid plants from pollen grains. Science 163:85-87.

Sanyal I, Singh AK, Kaushik M, Amla DV (2005) *Agrobacterium* mediated transformation of chickpea (*Cicer arietinum* L.) with *Bacillus thuringiensis* cry1Ac gene for resistance against pod borer insect *Helicoverpa armigera*. Plant Sci168:1135-1146.

Saxena S, Dhawan B (1999) Regeneration and large-scale propagation of bamboo (*Dendrocalamus strictus* Nees) through somatic embryogenesis. Plant Cell Rep18:438-443.

Schenk RU, Hildebrandt AC (1972) Medium and techniques for induction and growth of monocotyledonous and dicotyledonous plant cell cultures. Can J Bot 50:199-204.

Singh G and Shetty S (2011) Impact of tissue culture on agriculture in India. Invited Review, Biotechnology. Bioinformatics and Bioeng Society for Applied Biotechnology, India. 1(2):279-288.

Singh R, Kumar S, Kalia S, Sharma SK, Kalia RK (2016) Recent advances in understanding the role of growth regulators in plant growth and development *in vitro* - III. Inhibitors of growth regulators. The Indian Forester 142 (11):1065-1072.

Singh SR, Dalal S, Singh R, Dhawan AK,KaliaRK (2012a) Seasonal influences on *in vitro* bud break in *Dendrocalamus hamiltonii* Arn. ex Munro nodal explants and effect of culture microenvironment on large scale shoot multiplication and plantlet regeneration. Indian J Plant Physiol 17:9-21.

Singh SR, Dalal S, Singh R, Dhawan AK, Kalia RK (2012b) Micropropagation of *Dendrocalamus asper* {Schult. &Schult. F.} Backer ex K. Heyne): an exotic edible bamboo. J Plant BiochemBiotechnol.

Singh SR, Dalal S, Singh R, Dhawan AK, Kalia RK (2013b) Ascertaining clonal fidelity of micropropagated plants of *Dendrocalamus hamiltonii* Neeset Arn. ex Munro using molecular markers. In Vitro Cell Dev Biol Plant 49: 572–583.

Singh SR, Singh R, Dalal S, Dhawan AK,KaliaRK (2013a) Evaluation of genetic fidelity of *in vitro* raised plants of *Dendrocalamus asper* (Schult. & Schult. F.) Backer ex K. Heyne using DNA-based markers. Acta Physiol Plant 35:419–430.

Skoog F, Miller C (1957) Chemical regulation of growth and organ formation in plant tissues cultured *in vitro.* Symp Soc Exp Biol 11:118-140.

Went FW (1926) On growth accelerating substances in the coleoptile of *Avena sativa.* Proc K Ned Akad Wet Ser C 30:10.

29

Potential Indigenous Agroforestry Systems of Northeastern Region of India

U. C. Sharma

Centre for Natural Resources Management, V. P. O. Tarore, Samba – 181133 J&K, India

Introduction

Trees play an important role in ecosystem in all terrestrials and provide a range of products and services to rural and urban people (Gogoi 2015). As natural vegetation is cut for agriculture and other types of development, the benefits that trees provide are best sustained by integrating trees into agricultural system - a practice known as agroforestry. Agroforestry focuses on the wide range of trees grown on farms and other rural areas. Among these are fertilizer trees for land regeneration, soil health and food security; fruit trees for nutrition; fodder trees for livestock; timber and energy trees for shelter and fuel wood; medicinal trees to cure diseases and trees for minor products *viz.* gums, resins or latex products. Many of these trees are multipurpose, providing a range of benefits. Although effective and valuable tools have been introduced to evaluate the risk to construction in areas prone to geo-hydrological hazards, fewer steps have been made in the field of ordinary land use planning at different scales. The region is predominantly hilly and mismanagement of rain water causes heavy loss of soil in the hills and silting of river beds and floods in the plains (Sharma and Prasad 1995; Sharma 1998). The prevalence of shifting cultivation in 3689 km^2, also results in heavy soil erosion, deforestation and water resources degradation (Sharma and Sharma 2004a). Social sanctions and belief system maintained a balance between resource potential and their utilization for a long time in the region, but due to the increase in the demographic pressure and indiscriminate use of natural resources, imbalance has been created. The fast growing population in the region has pressurized the food production base and to satisfy their needs, the people have mismanaged and misused water resources (Sharma 2003; Sharma and Sharma 2004a; Sharma and Sharma 2004b). There is annual loss of 83.3 million tonnes of soil and 10.65, 0.37 and 6.05 thousand tonnes of available N, P_2O_5 and K_2O, respectively due to shifting cultivation alone (Sharma and Prasad 1995). Eroded soil is deposited in the foothills of slopes and hollows and soil crusting, erosion rills and eroded soils make crop production methods more difficult (Sommer and Lindstrom 1998). Under this situation, the agroforestry would be the most appropriate for resource conservation and food and fodder security. This paper discusses the new and prevalent agroforestry systems in the region.

Study area and methodology

The place of study is the northeastern region of India, comprising seven states and having an area of 255 090 km[2] (Fig. 1). The region is predominantly hilly, with elevation above mean sea level

varying from 73 m to more than 7000 m. The agriculture is mostly done on hill slopes. Shifting cultivation is the predominant mode of cultivation. The practice was acceptable when the population was low and the shifting cycle was 25 to 30 years. The period was sufficient for the rejuvenation of vegetation but, with increase in demographic pressure, the shifting cycle has come down to 2 to 5 years, rendering the practice as most uneconomical and resource eroding

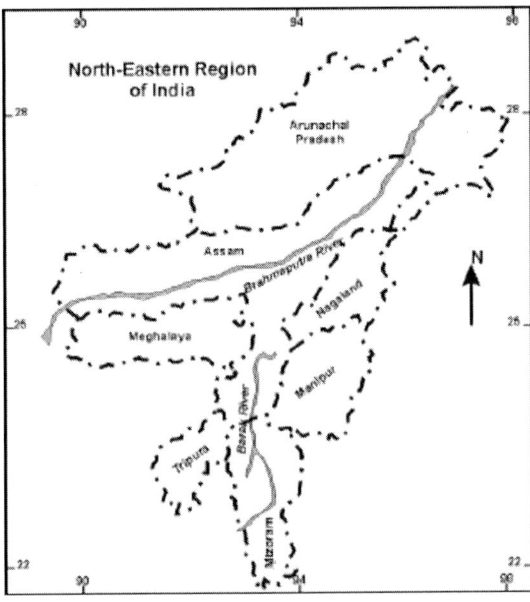

Fig.1. Northeastern region of India

Agroforestry models in north-eastern states of India

About 80% of the people of north-east India lives in the rural areas and almost all of them are directly or indirectly concerned with agriculture. Farmers are generally marginal and agroforestry practices can result an increase in their income without endangering the fragile ecosystem. The following agroforestry systems are generally practiced.

(i) Agri-Silviculture: (crops + trees)

(ii) Agri-Horticulture: (crops + fruit trees)

(iii) Silvi-Pasture: (trees + fodder crops)

(iv) Horti- Pasture: (fruit trees + fodder crops)

(v) Agri-Horti-Silviculture: (crops + fruit trees + trees)

(vi) Homestead Agroforestry : (may be the mixtures of crops, vegetables, fruit trees, fodder crops and trees)

The agroforestry in north- east India holds a great potential to make a positive and significant contribution to agricultural output, besides raising fuel-wood, timber, fodder, milk and meat production in one way and conserving the soil and water in other. The following agroforestry models have been identified:

Agroforestry with forest/ tree dominant

Agroforestry with tree dominant system generally occurs in natural conditions or artificially forested areas such as orchards or plantations. Agar tree (*Aquilaria malaccensis*) has been identified as a potential agroforestry species, especially for Assam due to their abundance in north- eastern region of India. The essential oil from agarwood is valued in high class perfumery. Some medicinal plants like patchouli (*Pogostemon cablin*), sarpagandha (*Rouvolfia serpentine*), Brahmi (*Brahmi indica*),Mosundary (*Houttynia cordata*), Mohavingaraj (*Wedelia calendulaceae*), Narasingha (*Musraya koenigii*) etc. are selected for intercropping with Agar trees in Assam. Tea, rubber, cardamom and coffee prefers diffuse sunlight and are most important commercial plantation crops in north- east India. These plantation crops are planted with a number of nitrogen fixing MPTS and some compatible crops such as black pepper, betel vine, arecanut etc. Such high density multistoried plantation system is found to be profitable and acceptable by the farmers, suiting the environment to the north-eastern region of India. Agricultural crops are also cultivated along with the economically most important forest tree species such as Gomari (*Gmelina Arborea*), Sissoo (*Dilbergia sissoo*), Kadam (*Michelia champaca*), Teak (*Tectona grandis*), Neem (*Azadirachta indica*), Acacia (*Acaciamangium*),

Arjun (*Terminalia arjuna*) etc. in the plain areas of NE region of India. The practice of growing agricultural crops along the forest tree species such as *Alnus nepalensis* (Hring), *Exbuclandia populania* can be observed in most of the hilly areas of north-eastern states of India.

Agroforestry with dominant crops

The agricultural crops like rapeseed (*Brassica campestris*), niger (*Guizotiaabyssinica*), buckwheat (*Fagopyrum esculentum* L.), sesamum (*Sesamum indicum*) etc. crops and some vegetables are cultivated in between the fruit trees such as jackfruit (*Artocarpus heterophylius* L.), citrus (*Citrus reticulata, Citruslimon* L.), guava (*Psidium guajava* L.), mango (*Mangiferaindica* L.), coconut (*Cosos nucifera* L.) etc. In between the fruit trees, growing of spices such as black pepper (*Piper nigrum* L.), ginger (*Zingiber officinale* L.), turmeric (*Curcuma longa* L.) etc. has been a common practice in north-easternregion of India. Tea plantation grown with shade trees in a two or three tier system of planting plays a pivotal role in maintaining ecosystem sustainability. On the shade trees of tea garden, cultivation of black pepper (*Piper nigrum* L.) and betel vine (*Piper betle* L.) has becoming a most profitable business in Assam and other NE states of India. Some medicinal plants like patchouli (*Pogostemon cablin*), sarpagandha (*Rouvolfia serpentine*), Brahmi (*Brahmi indica*),Mosundary (*Houttynia cordata*), Mohavingaraj (*Wedelia calendulaceae*), Narasingha (*Musraya koenigii*) etc. are selected for intercropping with Agar trees in Assam. Bamboo and other suitable MPTS along with some grasses like *Stylosanthes hameta, Cenchrus ciliaris, Panicummaximum, Cenchrus ciliaris etc.* can also be intercropped successfully for optimizing the overall system efficiency.

Agroforestry with fisheries dominant

Trees can be planted alongside rivers, ponds or other water bodies. Such fishery dominated agroforestry model ensures dams protection against water waves, decrease soil erosion. Tree species like *Psidium guajava* L. (Guava), *Cosos nucifera* L. (coconut), *Emblica officinalis* (Amala), *Aquilaria malaccensis*(Agar trees), *Terminalia cebula* (Silikha), *Mangifera indica*(Mango), *Citrus* spp., *Litchi chinensis* (Litchi), *Tamarindusindica, Sesbania grandiflora* are found to be suitable and common for such fishery dominated agroforestry model in NE India. Practices of arecanut and banana cultivation are alsowidespread on the bank of the fishery. To stabilize the embankment some perennial grasses are grown.

Bamboo based agroforestry

In north- east India, bamboos form an important component of homesteads, unproductive lands and are of common occurrence in the natural forests. Bamboo can be grown under all forms of agroforestry, involving many kinds of trees, shrubs, grasses, annual crops etc. Plantation of bamboos (*Bambusa bambos/ B. tulda/B. nutans*) in rows (at 5m × 5m spacing) with Gomari (*Gmelina arborea)* + turmeric (*Curcuma longa)* has already been reported to be practically feasible and economically viable, besides improving the physic-chemical properties of soil (Gogoi, 2011). In north-eastern region of India, the boundaries of paddy fields are often planted with different bamboo species such as *B. tulda, B. bambos, B. balcooa, D.hamiltonii* etc. This practice not only conserves soil, water and fertility, but also meets immediate requirement of the farmers *viz.* small timber, fuel and fodder.

Traditional agroforestry systems

Shifting cultivationis the most primitive and popular agroforestry system practiced in NE region. It is a the region is a complex system with wide variation that depends upon the ecology and cultural diversity among various tribal clans. However, the basic cropping practice is quite similar. Usually all the essential crops such as paddy, maize, tapioca, colocasia, millets, sweet potato, ginger etc. are grown on the same piece of land as mixed crop. In earlier 25-30 year cycle, the soils get enough time to rejuvenate and restore their health and productive capacity. However, with increase in population pressure on land resources, the Jhum cycle is getting reduced very fast. This makes the system unstable and lead to severe land degradation as a result of soil erosion and associated factors such as reduction in soil organic matter, nutrients etc. There is decline of forest cover due to shifting cultivation in the NEH region although the degree varies from one state to the other. In the region, trees are deliberately integrated with the crop and livestock production system. A number of crops like maize, ginger, pineapple, coffee, and vegetables are grown with tree species such as *Pinus kesiya*, *Alnus nepalensis*, *Schima wallichii*, *Pyrus communis*, *Prunus domestica*, *Areca catechu* etc. The choice of a particular tree species and intercrop depends upon the climatic conditions of the area and economic importance of the species.

Historical aspect of agroforestry in northeastern region

The tribal population in northeastern India remained more or less static for a very long period until inter-tribal warfare, head hunting, and pestilence and epidemics got effectively controlled. Gowswami (1981) mentioned that one Mr. Peal, who was sent on exploratory expedition in early nineteenth century by the British Government observed in his notes "from internal evidence, the tribal population seems to have been stable for a very long period, perhaps centuries. The checks are all positive, such as constant warfare, want of food, inducing diseases etc. As a consequence of head hunting and its isolating influence, we thus see a community of some hundred people perched on a hill and depending exclusively on their own resources, constantly fighting others similarly isolated on all sides, yet thoroughly able to maintain themselves. Perhaps in no other part of the world can so complete a tribal isolation seen. The every 40 or 50 square miles, there are about 4 villages of perhaps, one hundred families each, yet no more than an eighth or tenth of the land available can be cultivated at a time, and the population would seems to have reached its maximum". Thus an equilibrium was maintained for centuries between the population and available land on long cycles of 30 to 35 years for shifting cultivation. The equilibrium, however, got upset when the population started increasing paradoxically because of stopping of warfare amongst the tribes, health care and general awareness. Shortening of shifting cultivation cycle became inevitable as no migration could be contemplated till it had come to so short a period as 3 to 5 years. The factor that sustained crop yields in shifting cultivation of 30 to 35 years cycle disappeared when it came down to 3 to 5 years. Poor crop yields in shifting cultivation, the higher population and economic and social goals were no longer compatible with shifting cultivation as a way of life. The practice of shifting cultivation, which was prevalent in northeastern region of India since 7000 B.C. (Sharma 1976) did not keep pace with the food, fodder, fuel and fibre needs of growing tribal population. The homosapiens were under the awe of natural objects such as oceans, rivers, forest, springs etc. from the pre-historic age. Population was small in number, primitive in technology and their needs were extremely limited. With decrease in shifting cultivation cycle as a result of increase in population, the crop productivity decreased considerably, thereby necessitating the tribal people to migrate from one place to another for an easy access to food and water. Due to their ingenuity and skill, some tribes developed excellent agroforestry systems which ensured food and environment security.

(i) Zabo system

The tribal people of northeastern region of India have some social taboos and superstitions, which are to be followed religiously. One such tribe, known as Chakhesang, also migrated from the areas bordering Burma (now Myanmar) and settled in an area, now called Kikruma village in Phek district of Nagaland State of India. The village is located at an altitude of 1270m above mean sea level. There is a legendary story attached with the migration of the tribe. According to the village elders, the inhabitants of kikruma village have possibly migrated during nomadic period. During migration, the people were accompanying their leader, the chief of the tribe, for settlement at some suitable place, which should have fertile land for crop production and pastures for grazing of the domestic animals. The leader was carrying a big stone on his head and a cock in his hands. The belief of the tribe was that they would settle at a place where the stone must fall from the leader's head and the cock must crow. Only then that place will be free from diseases, good from crop production and overall welfare of the tribe. During the movement, the tribe moved through many fertile lands, but neither stone fell down from leader's head and not the cock crowed. This was not a good omen, the leader thought. The tribe kept on moving and reached a place where the stone fell down from the leaders head and the cock also crowed. This was thought to be God's will for settling down of the tribe. The place was, however, was not good for crop production and devoid of water source nearby. This is the place where now kikruma, village is situated (Sharma, 1998). The village is surrounded by two streams i.e. Seidzu and Khuzha, a few hundred meters down the settlement of the tribe on the hill. The tribe grew crops but the yields were poor in absence of irrigation. These was no drinking water around except to bring it from the streams after walking a couple of miles. This was the time when the people of Chakhesang tribe realized the importance of water for crop production, and for human and animal needs. The situation might have compelled the people to develop their own water resources due to their ingenuity and skill. The realization of importance of water and development of water resources transformed the life of the people of this tribe to a much better quality compared to others. This system of water harvesting and management is called Zabo system (Sharma 1997 1998; Sonowal *et al.* 1989).The environment of a place shapes and determines the habits, the mode of life and progress of civilization. The practice of shifting cultivation was evolved during Neolithic age and tribal people are involved with this socio-culturally. Shifting cultivation is not only a set of agricultural practices but implies the whole nexus of people's belief, attitude, self-image and tribal identity. The Zabo system, not only enhanced the crop productivity but helped in water resources development and sound resource base, improved the quality of life of the people and environment around as the deforestation was totally checked and soil erosion is negligible. The whole ecology of the region is in peril due to mismanagement of rain water as high intensity of rainfall causes soil erosion, especially from hill slopes.

Zabo means impounding of water in local language. The system has forest land on the top of the hill, water harvesting tanks (silt retention ponds and water storage tank) next down the hill slope, livestock enclosures and then terraced rice fields at the foot hills (Fig.2). The system is a unique combination of forest, agriculture, livestock and fisheries with a water and soil conservation base, which encourages purification of environment besides increasing crop productivity (Prasad and Sharma 1997; Sonowal *et al.* 1994).The system is followed on the land belonging to the farmers, a group of 10 to 15 farmers join together to practice Zabo system.

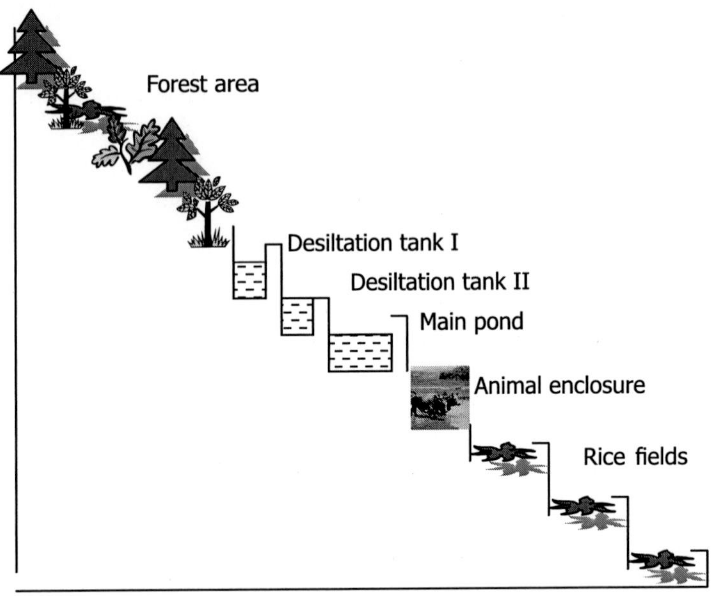

Fig. 2. Outline of Zabo system of water management

Forest land

The area on the hill top serves as catchment for rain water and is kept under natural vegetation. This area is normally not disturbed by cutting forest trees. The slope of this area is up to 100% or even more. Being permanent under vegetation, there is hardly any loss of soil through erosion.

Water harvesting system

Water harvesting ponds are dug-out with formation of earthen embankment below the catchment area. First two small ponds serve as silt retention tanks followed by a big tank down the slope. The size of these tanks depends on the catchment area but usual capacity is 300 to 600 m^3 with a depth varying from 1.5 to 2.5m. Silt retention tanks are constructed in between forest land and the main water storage tank to store run-off water from the catchment area for few days in these tanks so that soil, humus and organic matter is retained here before passing on the water to the main tank. Soil retention tanks are desilted annually and the soil containing organic matter is put in the rice fields for enhancing fertility. The ponds are rammed and compacted from inner sides so as to avoid water loss through seepage. The main water-harvesting tank is plastered on inner surface with mud mixed with chopped paddy straw to minimize water losses due to seepage. The bottom of the tank is rammed and animals are let loose in the beginning to make the bottom surface hard to reduce water losses.

Cattle enclosure

Enclosures are made with wood and bamboo for cattle and are managed by a group of farmers by stocking cattle on rotation basis. The enclosure is constructed on lower side of the pond. Buffaloes, cows, and pigs are the common animals with the farmers and about 30 to 50 of them are kept in one enclosure for a period of 10-15 days. During irrigation the water from the main tank is taken to the paddy fields by passing through the cattle enclosure and it carries dung and urine of the animals making the fields highly fertile.

Agricultural land

Paddy and other crop fields are located at lower elevation. Application of cow dung and run-off through open cattle yard are the common methods of manuring. Besides, leaves and succulent branches of *Alnus nepalensis* and *Albizia* lebbeck are also added to the fields and decomposed for soil fertility. No inorganic source of nutrients is added, making the whole system as organic farming. There is no pollution of water and environment due to fertilizer use. Some farmers practise paddy-cum-fish culture and about 50 to 60 kg of fish is obtained per ha which is an additional source of income to the farmers.

Impact on society

The Zabo system of water resource development and management is a unique system of farming which has helped the society in improving their quality of life and environment around. The farmers in other iso-agroclimatic condition are adopting the system to their advantage. The paddy yields in Zabo system are more than double compared to the average yields of the state and three-folds than shifting cultivation. The higher productivity not only fulfill food requirement of the tribal people and reduce their dependence on outside sources but also it is a source of income to be utilized for other rituals, festivities, marriages etc. The community lives under unpolluted environment free of frequent occurrence of common ailments. The system has added sustainability to agricultural productivity and healthy environment as it makes use of only locally available resources.

Apatani system

Water management

The Apatani tribe has developed a remarkable system of water management and irrigation. Every stream rising from the surrounding hills is tapped soon after it emerges from the forest, channelized at the rim of the valley and diverted by a network of primary, secondary and tertiary channels (Chaudhary *et al.* 1994; Prasad *et al.* 1994; Sharma 1997). At a short distance above the terraces, occurs the first diversion from the stream. Usually only a little water is allowed to flow in the first feeder channel while the stream continues its course. The feeder channel branches off at angles which lead water through the series of terraces, so that by blocking or opening the connecting ducts (Huburs), any field can be flooded or drained, as required (Fig. 3). The cross section of main channels ranges from 1.0-1.7 m in width and 0.45-0.65m in depth, while that of sub-channels from 0.75-0.96 m in width and 0.38-0.65 m in depth. These channels are generally pitched with boulders at the entry, which checks erosion of channels due to high flow of water. The most important aspect of right water management in low land rice fields is to keep water layer on the soil surface at the permissible depth. The Apatani farmers drain off the water from the rice fields twice during tillering, once during flowering and finally at maturity of rice crop. A 10cm water table is maintained in the plots by adjusting the height of outlet pipes. The water from the terraces is finally drained into the "Kale" river which flows through the middle of the valley. The discharge of this river recorded during rainy season was 4.8, 13.9 and 32.2 m^3 sec^{-1}, at the entrance, middle and end of the plateau, respectively.

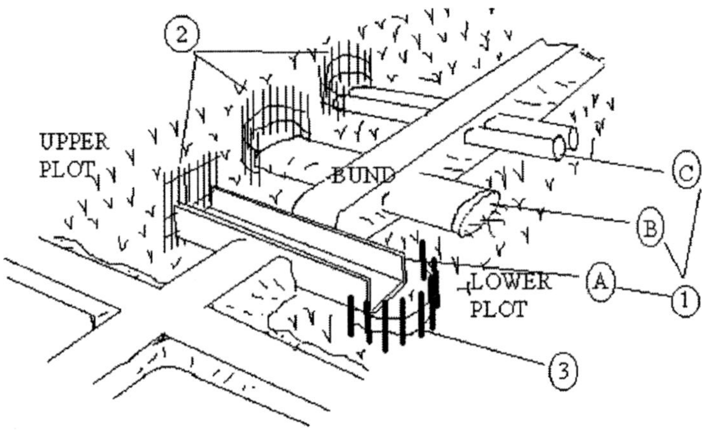

Fig. 3. Structures used for inter-plot irrigation cum drainage system in Apatani plateau. 1. Hubur (A, Made of plank; B, Made of pine tree trunk; C, Made of bamboo stem); 2. Obstruction for fish- Bamboo net; 3. Erosion control at water outlets – wood sticks

Nutrient management

Nutrient and fertility management of the terraces is done mainly through the recycling of agricultural wastes. Paddy straw, approximately 10.0 t ha^{-1}, is allowed to decompose in the wet terraces and finally incorporated at the time of land preparation. Burning of un-decomposed straw during January-February is also in practice especially on dry terraces. Pig and poultry manure, rice husk, kitchen wastes, ash and weeds removed during weeding are also recycled in the terraces every year for improving the soil fertility. After the rice crop is harvested, cattle are allowed free grazing in the fields from December to February and thus the cow dung is also recycled. Approximately, 20,000 cattle graze in the surrounding upper ridges of the plateau and discharge around 26,600 tonnes of dung every year. Since entire rain water of the surrounding hills is tapped for irrigation, nutrients from dung and even forest humus is also partly utilized through irrigation water. No inorganic fertilizers are used in the plateau and the yields of crops are 2 to 3 times higher in Apatani plateau than average of the state. Higher income from selling the produce after the family use has enhanced the quality of life of the people of Apatani tribe on sustainable basis.

Paddy-cum-fish culture

Fish channels of about 30 cm x 150 cm with 30 cm depth are dug at various locations in the rice fields. Fish fingerlings are put in these channels at the start of the cropping season and the fields are flooded. When the rice fields are drained before the harvest of the crop, the water remains in the fish channels. Average weight attained by the fish in four months varies between 130 and 400g. Normally, a loss of 50% fingerlings, either due to escape or mortality, is observed. In general, 185 to 250 kg fish is harvested per ha. Even after the family consumption, some fish is left for disposal. It is a subsidiary source of income for the farmers.

Community participation

In Apatani society, many big and hard operations, like small dams for diversion of water from the streams, making of primary, secondary and tertiary channels for carrying water to the fields, maintenance of the whole system to keep it always operational, fencing, maintenance of main irrigation channel and many unforeseen tasks are performed by the community as a whole. Men folk generally take care of hard works such as building terraces, irrigation channels, fencing, removing earth and planting trees while women folk look after nurseries, transplanting seedlings, weeding, fish management, harvesting, threshing, drying and storage. The periods of different operations and manpower requirement are given in Table 3.

Table 3. Period and manpower requirement of different operations

Operations required	Period	Mandays	Remarks
Fencing	February - March	4	Community operation
Maintenance of channels	January - April, July	3	- do -
Maintenance of risers	January - February	31	Individual operation
Nursery raising	February - March	18	-do-
Manuring	February - April	15	- do -
Land preparation	April - May	28	- do -
Transplanting	April - May	40	- do -
Weeding	Transplanting to harvest	41	- do -
Fish culture	May - September	7	- do -
Bird scaring and supervision	September - October	15	- do -
Harvesting, threshing and transporting	October - November	115	- do -
Drying and storage	October - November	11	- do -
Total		328	

Environment preservation

Apatani farmers are well aware and extremely cautious of their environment and ecology and have conserved their forests as sacred objects for improving the quality of environment and their health. Nobody can cut forest trees and vegetation without the permission of the community. The community decisions are binding and have to be honoured. Apart from conserving the soil from erosion, farmers have taken up the plantation of *Terminalis myrinalia, Altingia excelsa*, *Michelia* spp., *Magnolia* spp., Pines and bamboos. Thus, the entire hills surrounding the valley and uplands surrounding the villages are fully kept conserved as forest. This helps in conserving of natural resources, maintenance of ecological balance and continuous flow of streams. Thus, soil erosion, silting in rivers, drying of water sources, loss of nutrients, loss of flora, fauna and forest resources are negligible in this plateau and ecology has been properly taken care of.

Use of local resources

Use of local resources has made the system more sustainable. Apart from the utilization of rain water for irrigation and recycling of household wastes and by-products for fertility management, most of the important tools used in the Apartani farming viz. Sampya, Hilita, Damous and Mideing hiita etc. are made of wood. The plot to plot outlets in rice fields are made with wood or bamboos. Fishing nets and fish catching baskets are also made of bamboos. Local baskets and containers called Yagli, Yopho and Rajhu are made of bamboos too. Cattle and other animal thrive on forest resources while pigs are partly fed from the forest flora and partly from cultivated crops. Use of modern tools, implements and fertilizers is still not known to the Apatani farmers. However, the Department of Agriculture, Arunachal Pradesh is making all out efforts to make the system more productive and remunerative by introducing new techniques and improving the rice cultivation system of Apatanis.

Bamboo drip irrigation

Bamboo drip irrigation system is a unique example of efficient management of soil and water and reflects the ingenuity and skill of the tribal farmers. This is practiced by the Khasi and Jaintia tribes of the Meghalaya state of the region. The systems is based on the gravity flow and has been fully exploited by the farmers. The water is brought from upper reaches of hills to lower level through gravitational flow. The water is carried from the source to the place of use with the help of bamboo pipes cut into different shapes and forms (Singh 1989) (Figs. 4, 5). The flow of water can be controlled as per requirement by blocking and opening of bamboo slits at diversion points. The water is allowed to trickle down in drops at the plant site by having small holes at appropriate points. Once laid out, the system works round the clock, if desired. The farmers have the necessary skill to lay-out the bamboo network with proficiency and the whole system works efficiently and perfectly (Singh 1989). Depending on the plant positions where the water is to be applied, the network of bamboo channels is made with diversion and water control devices. Bamboos of large diameter are used at the start and the size is subsequently reduced as per requirement. Irrigation by this system is generally provided to plantation crops. Natural perennial as well as seasonal streams are the source of water for domestic and irrigation needs of the tribes of the region. Since the topography of the region is undulating, the cost of constructing of concrete channels for carrying the water to the desired sites is very high and the area for irrigation is very relatively low. A low cost technique can only be a suitable measure for carrying water through gravitational flow in such situations and the tribal farmers devised the bamboo drip irrigation system to satisfy their needs for water. Locally available bamboo is used for fabricating and laying-out the system, which can work as desired. The fabricating cost and maintenance costs of the whole system are nominal. The tribal community takes care of the system together. Use of local resources has made these systems more sustainable.

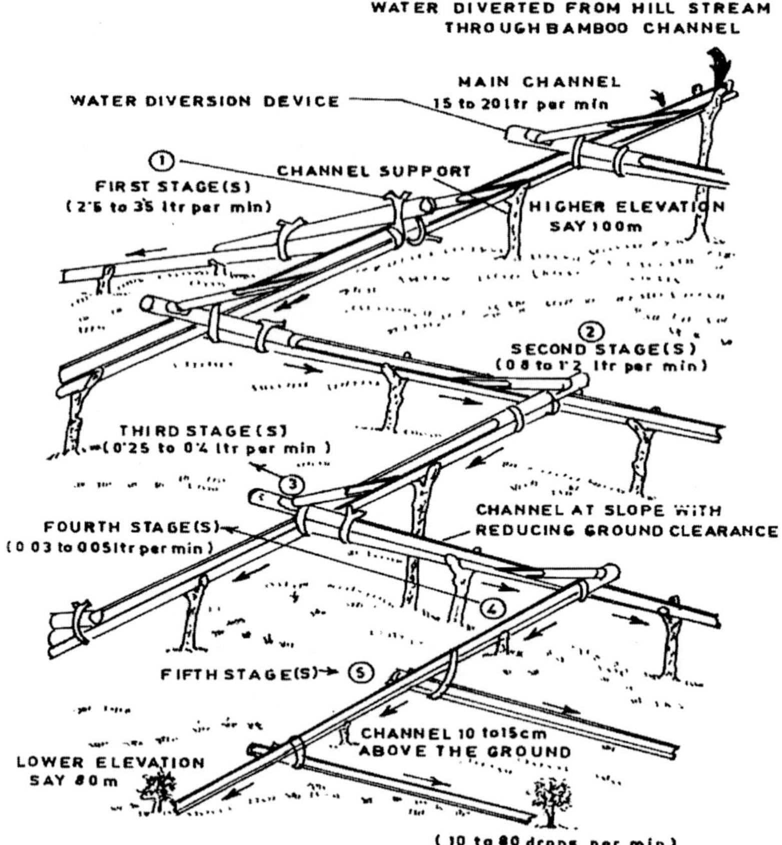

Fig. 4. Bamboo drip irrigation system

Fig. 5. Bamboo drip irrigation system laid-out in the hilly terrain

Agriculture with alder (Alnus nepalensis)

An organic land use system 'agriculture with alder' is being practiced by the Angami tribe of Nagaland state on terraced hill slopes, which has remained sustainable and highly productive for centuries. The study was conducted in Kohima district of Nagaland state on farmers' fields (Sharma and Singh 1994). The alder tree becomes ready for pollarding in about 6 to 8 years of planting and is pollarded at a height of 2m to 3m above the ground level. The trunk then sprouts shoots called coppices. One alder tree gives 100 to 200 coppices after pollarding. All coppices are cut, except 5 to 6 on the top of the main trunk. After pollarding, the fresh cut is covered with a stone to protect it from the frost injury.

The alder is a non-leguminous tree, fixing atmospheric nitrogen. Even 200 hundred year old alder trees can be seen in the fields. Annual addition of N to soil varied from 48.3 to 184.8 kg ha^{-2} with alder population varying from 60 to 625 trees ha^{-1}. The dry litter (leaves and succulent twigs) yield decreased significantly at less than 7m x 7m spacing per tree. Though the N fixed was more at higher alder density but a population of 250 trees ha^{-1} was found optimum for good crop productivity. Besides N, the organic matter content increased by 0.168% to 0.678%, annually, which helped in enhancing soil fertility and more *in-situ* retention of rain water (May to September). More *in-situ* retention of rain water provided sufficient moisture for winter (dry season) crops as well as reduced soil erosion through runoff, considerably. The cropping intensity in alder growing areas is 1.50% to 1.65% compared to 1.18% in other areas. The soil erosion from 'agriculture with alder' areas was 1.5 to 3.2 t ha^{-1} as against about 30 t ha^{-1} from general agriculture and 60 t ha^{-1} from shifting cultivation areas. The alder produced a nodule biomass of 150 to 307 kg ha^{-1}. The ash content of alder was found 3.85 % as against 2.95 % in *Eurya japonica* and 1.19 % in *Quercus lemellosa.* The alder leaves curl up after drying and act as natural mulch, thereby retaining more soil moisture. The system has helped in maintaining prefect soil health, resources conservation and improved environmental quality as unlike shifting cultivation, there is no deforestation for slash and burn. The crop productivity is two to three folds higher in alder land use compared to the average productivity of the state. No inorganic source of crop nutrition and other plant protection chemicals are used in the system and, so, there is no ground and surface water contamination as well as environmental pollution. The produce from these areas has more market value per unit weight compared to the produce from the areas where chemical fertilizers are used. The system is highly sustainable and need to be replicated in larger areas.

References

Choudhary RG, Kochhar S, Datta KK (1994) Rice based farming system of Apatani Plateau. In: Prasad RN, Sharma UC (eds) Potential Indigenous Farming systems of northeastern region, Indian Council of Agricultural Research complex, Northeastern region Meghalaya, India, pp. 1-12.

Gogoi Bhabesh (2011) Rejuvenation of degraded land through bamboos: A biological approach. In: Singh, Sanjay Singh, Das Rameshwar (eds) Productivity Enhancement and Value Addition of Bamboos by Excel India Publishers.

Gogoi Bhabesh (2015) Soil productivity management and socio-economic development through agroforestry in north-east India. Asian Journal of Science and Technology 6:2048-2053.

Gowswami PC (1981) Socio-economic aspects of soil conservation activities with special reference to northeast India. In: Socio-economic Aspects of Soil Conservation, Soil Conservation Department, Government of Meghalaya, Shillong, India, pp. 7-12.

Hutton JH (1928) The significance of head – hunting in Assam. J Royal Anthropological Inst 58:12-32.

Hutton JH (1969) The Angami Nagas. Oxford University Press, Bombay, India.

Prasad RN, Sharma UC (1994) Potential Indigenous Farming System of Northeastern Region. Indian council of Agricultural Research complex for northeastern region, Meghalaya, India.

Sharma TC (1976) The pre-historic background of shifting cultivation. North East India Council for Social Sciences Research, Shillong, Meghalaya, India, pp.2-12

Sharma UC (1997) Some indigenous farming systems of northeastern India. Asian Agri-Histroy 1:267-273.

Sharma UC (1998) Water management by the tribes of northeastern region: The traditional wisdom and future strategies. Fertilizer News: 43:43-49.

Sharma UC, Prasad RN (1995) Socio-economic aspects of acid soil management and alternate land use systems for northeastern states of India. In: DateRA, GrundonNJ, RaymentGE, Probert ME (eds) Plant-Soil Interactions at Low, pp. 689-696.

Sharma UC, Sonowal DK (1991) Zabo farming system of Nagaland. In: Proc. National Seminar on Shifting Cultivation, pp.110-126.

SharmaUC, Sharma Vikas (2004a) Implications of nutrient and soil transfer with runoff in the northeastern region of India. In: GolosovV, Belyaev V, Walling DE (eds) Sediment Transfer through the Fluvial System. IAHS Publ. No. 288, IAHS Press, Wallingford, U.K. pp. 488-493.

Sharma UC, Sharma Vikas (2004b) The 'Zabo' water management and conservation system: tribal beliefs in the development of water resources. In: Rodda John C, Ubertini Lucio (eds) Basis of Civilization-Water Science? pp. 184-191.

Sharma UC, Singh K (1994) Agriculture with alder (*Alnus nepalensis*) – a potential indigenous farming system of Nagaland. In: Proceedins Diamond Jubilee Seminar, Indian Society of Soil Science, New Delhi, India, pp.570-571.

Singh A (1989)Bamboo Drip Irrigation System. Indian Council of Agricultural Research Complex for Northeastern Hills Region, Barapani, Meghalaya, India.

Sonowal DK, Sharma UC, Tripathi AK (1994) Zabo farming system. In: Prasad RN, Sharma UC (eds) Potential Indigenous Farming Systems of Northeastern Hills Region, Indian Council of Agriculture Research Complex for Northeastern Hills Region, Shillong, Meghalaya, India, pp. 5-12.

Sonowal DK, Tripathi AK, Kirekha V (1989) Zabo, an indigenous framing system of Nagaland. Indian J Hill Res 2:1-8.

30

The Role of Agroforestry in Carbon Sequestration

U.C. Sharma

Centre for Natural Resources Management, Tarore, Jammu-181133, J & K, India

Introduction

There is a growing interest in the role of different types of land use systems in stabilizing the atmospheric CO_2 concentration and reducing the CO_2 emissions or on increasing the carbon sink of forestry and agroforestry systems. Forestry has been recognized as a means to reduce CO_2 emissions as well as enhancing carbon sinks. There is considerable interest to increase the carbon storage capacity of terrestrial vegetation through land-use practices such as afforestation, reforestation, and natural regeneration of forests, silvicultural systems and agroforestry (Brown 1996; Canadell and Raupach 2008). Agroforestry systems are very important given the area currently under agriculture, the number of people who depend on land for their livelihoods, and the need for integrating food production with environmental services (Garrity 2004; Makundi and Sathaye 2004).

The emphasis of land use systems that have higher carbon content than existing plant community can help achieve net gains in carbon, specifically and significant increases in carbon storage can be achieved by moving from lower biomass land uses [e.g. grasslands, crop fallows, etc. to tree based systems such as forests, plantation forests and agroforestry (Roshetko *et al.* 2007). Agroforestry provides a unique opportunity to combine the twin objectives of climate change adaptation and mitigation. Two major environmental issues of the world today are climate change and dwindling biodiversity. Increase in atmospheric concentrations of greenhouse gasses (GHG), of which the most common is carbon dioxide (CO_2), is considered to be the primary cause of global warming and estimates show that the current greenhouse gases (GHGs) concentrations are 30% more than the pre-industrial level. Carbon is accumulating in the atmosphere at a rate of 3.5 Pg (Pg = 1015 g or billion tons) per annum, the largest proportion of which resulting from the burning of fossil fuels and the conversion of tropical forests to agricultural production (Paustian *et al.* 2000). Scientific evidence suggests that increased atmospheric CO_2 could have some positive effects such as improved plant productivity. However, negative changes in the global climate (rising temperatures, higher frequency of droughts and floods) are often the most consequential processes possibly associated with an increased concentration of CO_2 in the atmosphere. Current terrestrial (plant and soil) C is estimated at 2000 ± 500 Pg, which represents 25% of global C stocks (DOE 1999). The sink option for CO2 mitigation is based on the assumption that this figure can be significantly increased if various biomes are judiciously managed and/or manipulated. In this connection, agricultural lands have the potential to remove and store between 42 and 90 Pg of C from the atmosphere over the next 50–100 years.

Agroforestry in acid soils

Agroforestry is a collective name for land use systems and technologies involving trees combined with crops and/ or animals on the same land management unit. It is a system that combines agricultural and tree crops of varying longevity, arranged either temporally or spatially, to maximize and sustain agricultural yield and minimize degradation of soil and water resources (Lal 1990). Phosphorus is one of the keys to agricultural sustainability for acid and highly weathered soils in the tropics. There is need to make the P deficient soils productive by raising their P availability for an ideal sustainable land use. However, a major constraint to this change from non-productive to productive soils is the strong resistance or buffer to changes in their actual P status (Novais and Barros 1997). Agroforestry production on acid soils is constrained by mineral deficiency or water or both, water and mineral deficiency. Agro-forestry, growing of multipurpose trees along with agricultural crops and rearing of animals, has been an important soil conservation practice. Sharma and Sharma (2009) reported that agroforestry can improve soil physical condition, increase in P status and ameliorate soil for optimum crop and forest productivity. With population increase, more and more slope lands have been turned into crop fields for grain production, resulting in serious environmental risks such as soil erosion and depletion of soil fertility (Sharma 1999; Sharma and Sharma 2004).

Potential C storage in agroforestry systems

Agroecosystems play a central role in the global C cycle and contain approximately 12% of the world terrestrial C. The terrestrial (plant and soil) C is estimated at 2000 ± 500 Pg, which represents 25% of global C stocks (DOE 1999). The sink option for CO_2 mitigation is based on the assumption that this figure can be significantly increased if various biomes are judiciously managed or manipulated. It is clear that forests have tremendous potential for C sequestration so as to reduce GHG concentrations in the atmosphere (Schoeder 1994). In this connection, AFS will have a great impact on the flux and long-term storage of C in the terrestrial biosphere (Dixon 1995) as the area of the world under AFS will increase substantially in the near future. The amount of C sequestered largely depends on the AFS put in place, the structure and function of which are, to a great extent, determined by environmental and socio-economic factors. Other factors influencing carbon storage in AFS include nature and kind of species of tree and management of system. About 20–25% of the total living biomass of trees prevail in the roots and there is constant addition of organic matter to the soil through decaying dead leaves and roots, which leads to improvements in the C status of the soil (Balkrishnan and Toky 1993).

Potential of agroforestry for carbon sequestration

Agroforestry, the practice of introducing trees in farming has played a significant role in enhancing land productivity and improving livelihoods in both developed and developing countries. Although carbon sequestration through afforestation and reforestation of degraded natural forests has long been considered useful in climate change mitigation, agroforestry offers some distinct advantages. The planting of trees along with crops improves soil fertility, controls and prevents soil erosion, controls water logging, checks acidification and eutrophication of streams and rivers, increases local biodiversity, decreases pressure on natural forests for fuel and provides fodder for livestock (Makundi and Sathaye 2004). The carbon storage capacity in agroforestry varies across species and geography (Newaj and Dhyani 2008). Further, the amount of carbon in any agroforestry system depends on the structure and function of different components within the systems put into practice.

Carbon (C) sequestration in agroforestry systems has recently generated international interest because of its potential impact on, and benefits for, agriculture and climate change (Schoeneberger 2008). Agroforestry systems store large quantities of C in vegetation and the soil (Sileshi *et al.* 2007), exchange C with the atmosphere (Schoeneberger 2008), protect forests, and improve soil productivity (Kaonga 2005). Recent predictions of global climate change have re-kindled the interest in agroforestry systems as potential C sinks. However, data on C dynamics in agroforestry systems are scarce because previous agroforestry research and development activities focused mainly on agricultural production. Improved fallows, in which leguminous trees and shrubs are grown in association with crops, have the potential to mitigate climate change by sequestering C in soils and biota, protecting existing natural forests, and conserving soil productivity (Kaonga 2005; Sileshi *et al.* 2007; Schoeneberger 2008).

With adequate management of trees in cultivated lands and pastures, a significant fraction of the atmospheric C could be captured and stored in plant biomass and in soils. However, increasing C stocks in a given period of time is just one step; the fate of those stocks is what ultimately determines sequestration. In agroforestry systems, C sequestration is a dynamic process and can be divided into phases for the sake of understanding. At establishment, many systems are likely to be sources of GHGs (loss of C and N from vegetation and soil). Then follow a quick accumulation phase and a maturation period when tons of C are stored in the boles, stems, roots of trees and in the soil. At the end of the rotation period, when the trees are harvested and the land returned to cropping (sequential systems), part of the C gets released back to the atmosphere (Dixon 1995). Therefore, effective sequestration can only be considered if there is a positive net C balance from an initial stock after a few decades. The demand and supply of different tree produce in India is given in Table 1.

Pasture management

On a global basis, grassland/grazing lands occupy 3460 Mha. Restoring degraded grazing lands and improving forage species is important to sequestering SOC and SIC. Furthermore, converting marginal croplands to pastures (by CRP and other set-aside provisions) can also sequester C. Similar to cropland, management options for improving pastures include judicious use of fertilizers, controlled grazing, sowing legumes and grasses or other species adapted to the environment, improvement of soil fauna and irrigation (Follett *et al.* 2001). Conant *et al.* (2001) reported rates of SOC sequestration through pasture improvement ranging from 0.11 to 3.04 Mg C ha^{-1} yr^{-1} with a mean of 0.54 Mg C ha^{-1} yr^{-1}.

Table 1: Demand and supply of different tree produce in India

2000	Timber	$m.m^3$	Fuel	Industrial	Green fodder	Dryfodder
Demand	64		201	28	593	178
	Forest	12	17			
Supply	Farm forest	31	98	12	145	482
	Total	43	115			
Deficit	-21		-86	-16	-448	+304

The primary source of the increase in atmospheric CO_2 concentrations is from the combustion of fossil fuels. Both past and future anthropogenic CO_2 emissions will continue to contribute to warming and sea level rise with grave implications globally. Developing countries are particularly at risk, as their infrastructures are most vulnerable to extreme events, and there is an expectation that climate change will worsen their food security, water availability and health, in addition to accelerating biodiversity losses (IPCC 2007). Important land uses and practices with the potential to sequester SOC include conversion of cropland to pastoral and forest lands, conventional tillage to conservation and no tillage, no manure use to regular addition of manure, and to soil specific fertilization rate (Huggins *et al.* 1998; Collins *et al.* 1999). India currently has one of the fastest growing economies in the world. Use of coal, gas and oil in large quantities, we are today, the fourth largest emitter of greenhouse gases worldwide, although our per-capita emissions are among the lowest in the world. The most recent IPCC report suggests that India will experience the greatest increase in energy and greenhouse gas emissions in the world if it sustains a high annual economic growth rate. The International energy Agency predicts that India will become the third largest emitter of greenhouse gases by as early as 2015. Burning of fossil fuels alone accounts for 83% of India's carbon dioxide emissions. Considering that just over half of India's current CO_2 emissions are from large point sources (IEA 2007), it may be that such current and future sources could be a suitable starting point for capturing emissions, transporting them, and then storing them in porous rock as a mitigation strategy against dangerous climate change. Fossil fuels such as coal, oil, and natural gas supply abundant energy at low cost. At present, 22 Gg of CO2 per year is emitted as a result of the use of fossil fuels (Gessinger 1997). Coal is the fuel most widely used for the generation of electricity worldwide because it is readily available, easily transportable, and relatively inexpensive. The increasing anthropogenic CO2 emission and global warming have challenged to find better ways to meet the world's increasing needs for energy while reducing greenhouse gases (Kane and Klein 1997). An effective strategy is needed to solve this increasingly urgent problem.

Climate Change Impact on Agriculture

With an economy closely tied to its natural-resource-base and climate-sensitive sectors such as agriculture, water, and forestry, India faces a major threat because of the projected changes in climate. Crucial sectors in India like agriculture, water resources, health, sanitation, and rural development are likely to be affected by climate change. India's large population primarily depends on climate-sensitive sectors like agriculture and forestry for livelihood. Climate change is the biggest global challenge affecting environmental, economical and social welfare throughout the world. The latest assessment report of the Intergovernmental Panel on Climate Change (IPCC 2007) projected that global average temperatures in 2100 will be between 1.8 and 4.0 °C higher than the

1980–2000 average. Precipitation is expect to be increased by 2.29% at the end of the century. On the other hand, sea levels are expected to increase by 0.59 meters by 2100 based on observed rates of ice flow from Greenland and Antarctica. Increased emissions of green house gas (GHG) to the atmosphere are the main causes for the climate change to happen (IPCC 2007).

Wheat production for the country as a whole may decline after 2020 and rice production may be adversely impacted in the eastern states. Studies by the Indian Agricultural Research Institute (IARI) indicate the possibility of a loss of 4–5 million tonnes in annual wheat production with every 1 ^{0}C rise in temperature. A study carried out by Indian Institute of Science assessed the impact of projected climate change on forest ecosystems in India. The main conclusion is that in 2085, between 68% and 77% of the forested grids in India are likely to experience shift in forest types depending upon projected climate change scenarios. Biodiversity is likely to be impacted under the projected climate scenarios due to the changes or shifts in forest or vegetation types in 57% to 60% of forests.

Agricultural systems and soil carbon

Increasing population pressure is increasing the demand on agricultural systems in many parts of the world and this has often led to the degradation of the soil resource. Soil carbon is a major determinant of sustainability of agricultural systems and changes can occur in both total and active, or labile, C pools (Blair *et al.* 1995). Soil carbon (both soil organic carbon, SOC and soil inorganic carbon, SIC) is important as it determines ecosystem and agro-ecosystem functions, influencing soil fertility, water-holding capacity and other soil parameters. It is also of global importance because of its role in the global carbon cycle and therefore, the part it plays in the mitigation of atmospheric levels of greenhouse gases (GHGs), with special reference to CO_2.

To reduce the emission of CO_2, carbon capture and storage (CCS) has been found to be an important option. The technique consists of three basic steps, viz. (i) capturing CO_2 at large and stationary point sources, (ii) transporting CO_2 from a source to sink, and (iii) injecting CO_2 in suited geological reservoirs or sinks. Among the other known sources to enhance CCS, the role of soils as an important natural resource, in capturing and storing carbon has not been adequately explained. The main issue of soil carbon management in India revolves around the fact that a few parts of the country have soils containing high amount of SOC and low amount of SIC, whereas other parts show a reverse trend. The most important fact is that soils act as a major sink and source of atmospheric CO_2 and therefore have a huge role to play in the CCS activity. The soils capture and store both organic (through photosynthesis of plants and then to soils as decomposed plant materials and roots) and inorganic carbon (through the formation of pedogenic calcium carbonates).

Soil carbon stocks

Carbon sequestration in different agroforestry systems occurs both belowground, in the form of enhancement of soil carbon plus root biomass and aboveground as carbon stored in standing biomass. Some of the earliest studies of potential carbon storage in agroforestry systems and alternative land use systems for India had estimated a sequestration potential of 68-228 Mg C ha^{-1} (Dixon *et al.* 1994). In a 6 year old *Gmelina arborea* based agri-silvicultural system 31.37 Mg C ha^{-1} was sequestered (Swamy *et al.* 2003). The first comprehensive report of SOC, SIC (soil inorganic carbon) and TC (total carbon) was carried out by Velayutham *et al.* (2000) and Bhattacharyya *et al.* (2000). Later these estimates were made useful for various mapping schemes (Bhattacharyya *et al.* 2008). Soil carbon stocks (SOC, SIC and TC) in seven different soil orders are shown in Table 2. Since carbon stock in soil is controlled by (mainly) content in carbon and the areal extent of soils, therefore, in soil if even carbon content is high the carbon stock will be less due to low area extent. It has been shown that level of SOC is determined among other factors, largely by climate (Jenny and Raychaudhuri 1960) and therefore, the density of soil carbon will vary in different climatic zones. Zonation of soil carbon stocks has been detailed earlier (Velayutham *et al.* 2000; Bhattacharyya *et al.* 2000, 2008). Soil organic carbon stock for India was first reported by Velayutham *et al.* (2000). Higher atmospheric temperature associated with low rainfall is responsible for high content of secondary carbonates. Calcium carbonate reported in the humid and subhumid region is considered mostly as inherited material in soils developed from strongly calcareous parent material, usually on young geomorphic surfaces. The SIC stock is relatively high in arid and semi-arid ecosystem (Bhattacharyya *et al.* 2000).

Carbon sequestration

Essentially, carbon sequestered is the difference between carbon 'gained' by photosynthesis and carbon 'lost' or 'released' by respiration of all components of the ecosystem, and this overall gain or loss of carbon is usually represented by net ecosystem productivity. Most carbon enters the ecosystem via photosynthesis in the leaves, and carbon accumulation is most obvious when it occurs in aboveground biomass. More than half of the assimilated carbon is eventually transported below ground via root growth, root exudates and litter deposition, and therefore soils contain the major stock of C in the ecosystem. Inevitably, practices that increase net primary productivity and/or return a greater portion of plant materials to the soil have the potential to increase carbon stocks.

Five principal C pools are; oceanic, storing about 38120 Pg of C; geological, storing about 5000 Pg of C; terrestrial carbon pool with about1560 Pg of organic and 700 Pg of inorganic carbon up to 1m depth; atmospheric C pool with 760 Pg of C and biotic pool with 560 Pg of carbon. The annual addition of C to the atmosphere is estimated to be 3.3 Pg.

Table 2. Carbon stocks in Indian soils

Soil order	Soil depth	Carbon stock (Pg)		
	range (m)	SOC	SIC	TC
Entisols	0-0.30-1.5	0.62 (6)2.56 (8)	0.89 (21)2.86 (8)	1.51 (11)5.42 (8)
Vertisols	0-0.30-1.5	2.59 (27)8.77 (29)	1.07 (26)6.14 (18)	3.66 (27)14.90 (23)
Inceptisols	0-0.30-1.5	2.17 (23)5.81 (19)	0.62 (15)7.04 (21)	2.79 (20)12.85 (20)
Aridisols	0-0.30-1.5	0.74 (8)2.02 (7)	1.40 (34)13.40 (39)	2.14 (16)15.42 (24)
Mollisols	0-0.30-1.5	0.09 (1)0.49 (2)	0.000.07 (0.2)	0.09 (1)0.56 (1)
Alfisols	0-0.30-1.5	3.14 (33)9.72 (32)	0.16 (4)4.48 (13)	3.30 (24)14.20 (22)
Ultisols	0-0.30-1.5	0.20 (2)0.55 (2)	0.000.00	0.20 (1)0.55 (1)
Total	0-0.30-1.5	9.5529.92	4.1433.98	13.6963.90

Parentheses show percentage of total SOC, SIC and TC

Source: Bhattacharyya *et al.* 2008, 2000b,

Agriculture to CO_2 emissions

Agricultural systems contribute to carbon emissions through several mechanisms:

(i) the direct use of fossil fuels in farm operations;

(ii) the indirect use of embodied energy in inputs that are energy-intensive to manufacture (particularly fertilizers); and

(iii) the cultivation of soils resulting in the loss of soil organic matter.

On the other hand, agriculture is also an accumulator of carbon, offsetting losses when organic matter is accumulated in the soil, or when above-ground woody biomass acts either as a permanent sink or is used as an energy source that substitutes for fossil fuels.

India's carbon dioxide emissions

India's carbon dioxide emissions due to consumption of energy were 1.1 Gt, of which emissions from combustion of coal were 0.7 Gt of carbon dioxide in 2004 (EIA 2007). Between 1990 and 2000, the overall increase in CO_2 emissions has been at a rate of 4.2% per annum and 5.1% between 2000 and 2005. On a per capita basis, India's emissions are remarkably low, even compared to other developing countries like Brazil, Mexico and China (Table 3).

Carbon sequestration potential of agro-forestry

Agricultural lands are major potential sink and could absorb large quantities of C if trees are introduced together with crops and/or animals. Agro-forestry is receiving wider recognition not only in terms of agricultural sustainability but also in issues related to climate change. The C sequestration potential of agro-forestry systems is estimated between 12 and 228 Mg ha^{-1}. Therefore, based on the earth's area that is suitable for the practice (585–1215 × 10^6 ha), 1.1–2.2 Pg C could be stored in the terrestrial ecosystems over the next 50 years. However, there are a number of shortcomings such as; the uncertainties related to future shifts in global climate, land-use and land cover, the poor performance of trees and crops on substandard soils and dry environments, pests and diseases.

Table 3. Carbon dioxide emission in top nine countries

S. No.	Country	Annual CO_2 emission (Pg)	% of global total	Emission percapita (tonnes)	Rank per capita
1.	China	7.031	23.3	5.3	78
2.	U.S.A.	5.461	18.1	17.5	12
3.	E. U. (27)	4.177	14.0	10.5	30
4.	India	1.742	5.8	1.2	145
5.	Russia	1.708	5.7	12.1	23
6.	Japan	1.208	4.0	9.5	38
7.	Germany	0.874	2.6	9.6	37
8	Canada	0.544	1.8	16.4	15
9.	Iran	0.538	1.7	7.3	54
	World	29.888	100.0	5.1	

Tree components in agroforestry systems can be significant sink of atmospheric carbon (C) due to their fast growth and high productivity. By including trees in agricultural production systems, agroforestry can, arguably, increase the amount of C stored in lands devoted to agriculture, while still allowing for the growing of food crops (Kursten 2000). In agroforestry systems, tree component is managed, often intensively by pruning for minimizing competition and maximize complementarity. The pruned materials are mostly non-timber products. Such materials often are returned to the soil. Besides, the amount of biomass and therefore C that is harvested and exported from the system is relatively low in relation to the productivity of the tree. Therefore, unlike in tree plantations and other monoculture systems, agroforestry seems to have unique advantage in terms of C sequestration.

The global contribution of agroforestry as a sink of C has been estimated in different eco-regions based on the tree growth rates and wood production, and assuming that the ratio of C in biomass is 50%, average C storage by agroforestry practices to be 9, 21, 50 and 63 Mg C ha^{-1} in semi-arid, sub-humid, humid and temperate regions (Schroeder 1994). At global scale, agroforestry could be implemented on 585 to 1275 x 106 ha and these lands could store 12 to 228 Mg C ha^{-1} under the prevalent climatic and edaphic conditions. With adequate management of trees in cultivated lands and pastures, a significant fraction of the atmospheric C could be captured and stored in plant biomass and in soils. However, increasing C stocks in a given period of time is just one step; the fate of those stocks is what ultimately determines sequestration. In agroforestry systems C sequestration is a dynamic process and can be divided into phases (Albrecht and Kandji 2003). At establishment, many systems are likely to be sources of GHGs (loss of C and N from vegetation and soil). Then follow a quick accumulation phase and a maturation period when tons of C are stored in the boles, stems, roots of trees and in the soil. At the end of the rotation period, when the trees are harvested and the land returned to cropping, part of the C will be released back to the atmosphere (Dixon 1995). Realistically, C storage in plant biomass is only feasible in the perennial agroforestry systems, which allow full tree growth and where the woody component represents an important part of the total biomass. Carbon storage can continue way beyond if boles, stems or branches are processed in any form of long-lasting products. The wood can serve as fuel, in which case an important part of the plant-stored C returns to the atmosphere. While C sequestration per se may be insignificant in the latter scenario, producing firewood from arable or grazed land may still present interesting opportunities in CO_2 mitigation through: (1) the protection of existing forests and other natural landscapes; (2) the conservation of soil productivity; and (3) the reduction of fossil energy consumption by using wood as energy sources (Kürsten 2000). Adequate understanding of these secondary effects of agroforestry with regards to CO_2 mitigation will require more research.

Impact of agroforestry on soil

Six years mean of annual sediment yield from different land use systems showed that it was within the critical limits in agroforestry land use compared to shifting cultivation, where it was much above the critical limit of 10 Mg ha^{-1} (Sharma 1999). The sediment yield was 36.21 Mg ha^{-1} in shifting cultivation followed by alder land use. The sediment yield was also impacted by the annual rainfall received at both the locations of study. The higher rainfall induced higher sediment load in the runoff. More than 90% of rainwater was retained *in-situ* in agroforestry land use systems compared to below 60% in shifting cultivation. More *in-situ* retention of rainwater helped in the availability of adequate moisture from the soil to the succeeding crops when the rainy season receded. Agroforestry systems help in nutrient build-up in the soil and maximum increase was found in agri-horti-silvo-pastoral land use where it increased from 2.7

to 21.5 mg P kg^{-1} of soil. In shifting cultivation land use, there was no significant change in available P status of the soil. The available K content of the soil increased from initial value of 105 mg kg^{-1} to 165 mg kg^{-1} of soil in agroforestry land use. Many tree species have soil ameliorative properties and help in increasing the pH in acid soils. With increase in pH, more fixed P becomes available. Integrating trees on the fields acted as natural source for nutrients from deeper layers of soil, added bio-fertilizer, conserved soil moisture and enhanced overall productivity of the system. Sanchez (1987) also reported that agroforestry systems are generally perceived to be sustainable and to enhance soil properties. Growing trees in conjunction with annual crops or pastures is believed to provide a more thorough plant cover to protect the soil from erosion and a deeper or more prolific root system to enhance nutrient cycling. The potential of using woody perennials has often been emphasized for conservation as well as production on hilly terrains (Young 1989; Kang *et al.* 1990). The role of woody perennials, both N-fixing and non-fixing, in improving the soil chemical, physical and biological properties has become the subject of investigations in the last decade (Jose *et al.* 2004). Plant growth in mineral acid soils is often restricted by many factors, including low soil pH and high amount of exchangeable Al. The increase in soil pH resulted in more availability of nutrients to the crops and, hence, higher productivity. This is supported by the studies of Keltjens (1995) who reported that malnutrition of plants on acid soils is mostly the result of limited soil nutrient availability, often supported by impaired uptake capability of the root and nutrient interaction at common root adsorption and absorption sites as interaction of Al and H^+ with P, K, Mg and Ca.

Conclusions

Agroforestry offers the possibility of improving food security by more effectively managing soil and water resources for the sustained production of annuals. The agroforestry systems, with suitable water and soil conservation measures, significantly reduced runoff from the watersheds on hill slopes and helped in more *in-situ* retention of rainwater, thereby reducing the soil and nutrient erosion. Practice of agroforestry, one-way or the other, served farmers as a source of food and their economic upliftment. Agroforestry can play a vital role in biological reclamation and sustaining soil fertility for growing crops and getting higher yields for livelihood security. There are suitable tree species which help in ameliorating the soils constraints. It will be worthwhile to develop and practise agroforestry systems involving fruit trees, plantation crops, spices, nitrogen fixing trees, legumes, crops and grasses and other multiple trees of economic importance. The trees in association with annuals could control runoff, improve soil quality, increase productivity and income of the farmers.

References

Albrecht A, Kandji ST (2003) Carbon sequestration in tropical agroforestry systems. Agriculture, Ecosystems and Environment 99:15–27.

Balkrishnan, Toky OP (1993) Significance of nitrogen fixing woody legume trees in forestry. Indian Forester 119: 126-134.

Bhattacharyya T, Pal DK, Chandran P, Ray SK, Mandal C, Telpande B (2008) Soil carbon storage capacity as a tool to prioritize areas for carbon sequestration. Curr Sci 95:482-494.

Bhattacharyya T, Pal DK, Velayutham M, Chandran P, Mandal C (2000) Total carbon stock in Indian soils: Issues, priorities and management. In: Special Publication of the International Seminar on Land Resource Management for Food, Employment and Environmental Security (ICLRM), New Delhi, pp. 1-46.

Blair GJ, Lefroy RDB, Leanne L (1995) Soil carbon fractions based on their degree of oxidation, and the development of a carbon management index for agricultural systems. Aust J Agric Res 46:1459-66.

Brown S (1996) Present and potential roles of forests in the global climate change debate. Unasylva 185: 3-10.

Canadell JG, Raupach MR (2008) Managing forests for climate change mitigation. Science 320:1456-1457.

Collins HP, Blevins RL, Bundy LG, Christenson DR, Dick WA, Huggins DR, Paul EA (1999) Soil carbon dynamics in corn-based agroecosystems: results from carbon-13 natural abundance. Soil Sci Soc Am J 63: 584-591.

Conant RT, Paustian K, Elliott ET (2001) Grassland management and conversion into grassland: effects on soil carbon. Applied Ecology 11: 343-355.

Department of Energy (1999) Carbon Sequestration: State of the Science. US DOE, Washington, DC.

Dixon RK (1995) Agroforestry system: source and sink of greenhouse gases? Agroforestry Systems 31:99–116.

Dixon RK, Brown S, Houghton RA, Solomon AM, Trexler MC (1994) Carbon pools and fluxes of global forest ecosystems. Science 263:185-190.

Follett RF, Kimble JM, Lal R (2001) The Potential of U.S. Grazing Lands to Sequester Carbon and Mitigate the Greenhouse Effect. CRC/ Lewis, Boca Raton, FL. pp. 442.

Garrity DP (2004) Agroforestry and the achievement of the millennium development goals. Agroforestry System 61:5-17.

Gessinger G (1997) Lower CO_2 emissions through better technology. Energy Convers Manage 38:25-30.

Huggins DR, Clapp CE, Allmaras RR, Lamb JA, Layese MF (1998) Carbon dynamics in corn-soybean sequences as estimated from natural carbon-13 abundance. Soil Sci Soc Am J 62:195-203.

IEA (2007) World Energy Outlook 2007: China and India Insights. International Energy Agency, Paris, France.

IPCC (2007) Climate change 2007: impacts, adaptation and vulnerability: contribution of working group II to the fourth assessment report of the IPCC. Cambridge University Press, Cambridge.

Jenny H, Raychaudhuri SP (1960) Effect of Climate and Cultivation on Nitrogen and Organic Matter Reserves in Indian Soils. ICAR, New Delhi, India, pp.126.

Jose S, Gillespie AR, Pallardy SG (2004) Inter specific interactions in temperate agroforestry. Agrofor Syst 61:237–255.

Kane R, Klein DE (1997) United States strategy for mitigating global climate change. Energy Convers Manage 38:13-18.

Kang BT, Reynolds L, Atta-Krah AN (1990) Alley farming. Advances in Agronomy 43: 315–359.

Kaonga ML (2005) Understanding Carbon Dynamics in Agroforestry Systems in Eastern Zambia. PhD Dissertation, Fitzwilliam College, University of Cambridge, UK.

Keltjens WG (1995) Mg uptake by Al-stressed maize plants with special emphasis on cation interactions at root exchange sites. Plant and Soil 171:141-146.

Kursten E (2000) Fuelwood production in agroforestry systems for sustainable land use and CO_2 mitigation. Ecological Engineering 16: 69–72.

Lal R (1990) Agroforestry systems to control erosion on arable tropical steep-lands. Proceedings of the Symposium "Research Needs and Applications to Reduce Erosion and Sedimentation in Tropical Steep-lands, IAHS-AISH Publ. No.192

Makundi WR, Sathaye JA (2004) GHG mitigation potential and cost in tropical forestry-relative role for agroforestry. Environment, Development and Sustainability 6:235-260.

Newaj R, Dhyani SK (2008) Agroforestry for carbon sequestration: Scope and present status. Indian Journal of Agroforestry 10:1-9.

Novais RF de, Barros NF de (1997) Sustainable agriculture and forestry production systems on acid soils: P as a case study. In: Munniz AC, Furlani AMC, Schaffert RE, Fageria NK, Rosolem CA, Cantarella H (eds) Plant-Soil Interactions at Low Ph: Sustainable Agriculture and Forestry Production, Brazilian Soil Science Society, pp. 39-51.

Paustian K, Six J, Elliott ET, Hunt HW (2000) Management options for reducing CO_2 emissions from agricultural soils. Biogeochemistry 48:147-163.

Roshetko JM, Lasco RD, Angeles MSD (2007) Small holder agroforestry systems for carbon storage. Mitigation and Adaptation Strategies for Global Change 12:219-242.

Sanchez PA (1987) Soil productivity and sustainability in agroforestry systems. In: Sanchez, PA, Steppler HA, Nair PKR (eds) Soil Productivity and Sustainability in Agroforestry Systems, International Council for Research in Agroforestry, Nairobi, Kenya, pp. 205-226.

Schoeder P (1994) Carbon storage benefits of agroforestry. Agroforestry Systems 27:89–97.

Schoeneberger MM (2008) Agroforestry: working trees for sequestering carbon in agroforestry systems.

Sharma UC, Sharma V (2004) Implications of nutrient and soil transfer with runoff in the northeastern region of India. In: Golosov V, Belyaev V, Walling DE (eds) Sediment Transfer through the Fluvial System, IAHS Publ. No. 288, IAHS Press, Wallingford, U.K., pp. 488-493.

Sharma UC, Sharma V (2009) Plant-livestock-soil-hydrology interactions in northeastern region of India. Biologia 64:460-464.

Sharma P, Rai SC (2007) Carbon sequestration with land-use cover change in a Himalayan watershed. Geoderma 139:371–378.

Sharma UC (1999) Food Security in the northeast: new paradigms. In: Sundariyal RC, Uma Shankar, Upreti TC (eds) Perspective for Planning and Development in North Eastern India. G.B. Pant Institute of Himalayan Environmental and Development, Almorah, India, pp. 197-212.

Sileshi G, Akinnifesi FK, Ajayi O, Chakaredza S, Kaonga M, Matakala P (2007) Contributions of agroforestry systems in miombo ecoregion of eastern and southern Africa. Environ Sci Tech 1(4):68–80.

Swamy SL, Puri S, Singh AK (2003) Growth, biomass, carbon storage and nutrient distribution in *Gmelina arborea* Roxb. stands on red lateritic soils in Central India. Bioresource Technology 90:109-126.

Velayutham M, Pal DK, Bhattacharyya T (2000) Organic carbon stock in soils of India. In: Lal R, Kimble JM, Stewart BA (eds) Global Climatic Change and Tropical Ecosystems Lewis Publishers, Boca Raton, F.L. pp.71-96.

Young A (1989) Agroforestry for Soil Conservation, CAB International, Walling Ford, Oxon, UK.

31

Agroforestry Rehabilitates Farmers through Clonal Plantations on Wastelands in Karnataka, India

S.K. Sharma

Captive Plantation Project, Raw Material Procurement Department West Coast Paper Mills Ltd., Dandeli (Uttara Kannada), Karnataka-581325 Society for Afforestation, Research and Allied Works, Bangur Nagar, Dandeli (Uttara Kannada)

Introduction

Agroforestry is a land-use management system where trees or shrubs are grown around or among crops or in pasture lands. It combines agricultural and forestry technologies to create more diverse, productive, profitable, healthy, and sustainable land use systems.

India has been at the forefront of agroforestry research ever since organized research in this area started worldwide about 25 years ago. Considering the country's unique land use, demographic, political and socio-cultural characteristics as well as its strong contributions to agricultural and forestry research, India's experience in agroforestry research has proved important to agroforestry development especially in developing countries.

The success stories of wasteland reclamation and Poplar-based agroforestry indicate that the wide adoption of these technologies rely on the understanding of the scientific principles behind it and the socioeconomic benefits they bring. An examination on the impact of agroforestry technology generation and adoption in different parts of the country highlights the major role of smallholders as agroforestry producers of the future. It is crucial that progressive legal and institutional policies are created to eschew the historical dichotomy between agriculture and forestry and encourage integrated land-use systems. Government policies hold the key to agroforestry adoption (Puri and Nair 2004).

Agroforestry in india

Agroforestry research was initiated in India about two decades ago and considerable progress has been achieved since then. Agroforestry practices, i.e., growing trees with food crops and grasses, are believed to have been practiced during the Vedic era (1000 BC). Although practiced since then, it was only until the establishment of the International Council for Agroforestry (ICRAF) in 1977, which was renamed to the International Centre for Research in Agroforestry in 1991, that agroforestry as a science was introduced (National Agroforestry Policy 2014).

India has been at the forefront of agroforestry research ever since organized research in this area started worldwide about 25 years ago. Considering the country's unique land use, demographic, political and socio-cultural characteristics as well as its strong contributions to agricultural and forestry research, India's experience in agroforestry research has proved important to agroforestry development especially in developing countries. It is crucial that progressive legal and institutional policies are created to eschew the historical dichotomy between agriculture and forestry and encourage integrated land-use systems. Government policies hold the key to agroforestry adoption.

Role of national agroforestry policy 2014

Current estimate show that about 65% of the country's timber requirement is met from the tree on farms. Agroforestry also generates significant opportunities. Recent development of National Agroforestry Policy 2014 is major breakthrough in this line. The National Agroforestry Policy is a path-breaker in making agroforestry an instrument for transforming lives of rural farming population, protecting ecosystem and ensuring food security through sustainable means. Among others, the policy will encourage farmers to practice agroforestry, enormously boost wood industry, and also help in biodiversity conservation (Tejwani 1994).

The role of the society for afforestation, research and allied works (SARA)

SARA was established as a nongovernmental organization in 2001 and supported by West Coast Paper Mills Limited, Dandeli, a leading paper manufacturing industry in India. It is a group of forestry, agricultural and financial experts and progressive farmers to help West Coast Paper Mills Limited procure wood raw materials as part of the Controlled Wood Policy created in 2001. The policy ensures that West Coast Paper Mills Limited can procure its wood raw materials from known sources to meet the annual production of 3.2 Lac metric tons of paper and paper products.

In the last eight years, SARA has been implementing contract farming under the Captive Plantation Project. Within this period, it has covered around 50, 000 acres of degraded, waste and marginal lands in and around 1900 villages across the various districts of Karnataka, Maharashtra, Andhra Pradesh States. It has also enjoined 8246 farmers to join SARA in this mega project (Table 1).

Table 1. Area and farmers covered under plantation at various locations since 2006 to 2016

S.No	Name of Area	State	Total plantation area (in acres)	No of farmers covered
1	Ram Nagar	Karnataka	2501	787
2	Khanapur	Karnataka	10844	2541
3	Kuluvalli	Karnataka	2461	34
4	Cheam manna Kittur	Karnataka	6303	780
5	Malgi-Sirsi	Karnataka	4343	601
6	Koppal	Karnataka	5987	1132
6	Chandgad	Maharashtra	5302	981
7	Radhanagri	Maharashtra	6218	1193
8	Gudur-Nellore	Andhra Pradesh	5466	197
	Total		49425	8246

Contract for farming- a model

Under the contract for farming agreement, the SARA-WCPM establish pulpwood plantations in the farmers' waste lands, barren and fallow lands with the support of SARA. The wood are then sold to SARA after five years.

All plantations under the project of SARA-WCPM certified under the Forest Stewardship Council-Forest Management by the Scientific Certification System since 2011 as Well-Managed Group Plantations. The Society ensures that all plantations comply with the certification objectives of promoting responsible management of forest; protecting and maintaining natural communities and high concentration of value forests; respecting the rights of communities and indigenous people; and building markets, adopting the best value and creating equal access to benefits. Amidst continuous mismatch between demand and supply for wood, SARA's commitment to ensure the steady supply of raw materials to West Coast Paper Mills Limited has resulted to over 1.7 million metric tons of pulpwood from SARA-affiliated plantations. To date, plantations under SARA's projects have yielded an estimated 25-30 metric tons per acre after every five-year harvesting cycle.

Clonal production of important pulpwood species

As part of contract farming, SARA is also actively engaged in producing quality superior clones of *Eucalyptus, Acacia, Casuarina* and Subabul. It has set up an advanced clonal nursery with an area of 3000 m^2 and with five mist chambers. This clonal nursery produces 50-60 lacs of clonal saplings which are transplanted to the farmers' fields (Fig. 1).

Fig. A -D : Clonal propagation of *Acacia hybrid* (*A. mangium, A. auriculoformis*) and field plantation

Fig. E -H : Clonal propagation of *Eucalyptus hybrid* and field plantation

Fig. I : Casuarina nursery. J : Intercroping with peanut K : Casuarina Clone L : Casuarina field Plantation

Fig 1:

SARA and West Coast Paper Mills Limited have adopted vegetative means of propagating techniques for *Eucalyptus* under its tree improvement program. The program aims to improve yield per unit area. As the demand for raw materials drastically increases daily so does the need to meet high demand. It is thus essential to grow superior clones of Eucalyptus on a large scale to increase bulk yield.

Clonal plantation for implementing agroforestry

Technology based farm-forestry plantations with genetically improved, high yielding and fast growing clonal planting stock of species like Eucalypts have tremendous potential for diversification of agriculture and meeting growing shortages of industrial timber on sustainable basis. *Eucalyptus* is one of the trees can be grown with crops. *Eucalyptus* is the most popular choice to be planted along the edges, or bunds, of agricultural fields, and appears to be well incorporated and accepted in agro-forestry in India. Silvicultural properties including straightness, narrow crown, self-pruning, high growth rates, adaptability to a wide range of soils and climates, coppicing ability, a tendency not to spread as a weed and wide utility of wood are some of the main features of *Eucalyptus* clones making it popular among the farmers for raising as block plantations (Sharma *et al.* 2009).

In addition to providing quality planting materials of *Eucalyptus*, Subabul, *Casuarina* and *Acacia* to the farmer-partners for contract farming, SARA also makes available the same quality planting materials at a subsidized price to farmers who are practicing agroforestry. Simultaneous with the distribution of planting materials is the provision of technical assistance to guide the farmers on how to improve yield. These initiatives are part of the Farm Forestry Program.

Under the program, SARA guides the farmers on how to intercrop *Eucalyptus* with agricultural crops (Sharma and Chopra 2014) (Fig. 2). *Eucalyptus* is recognized as a commercial crop in India and widely cultivated at a large-scale on farm lands and waste land by the farmers of Karnataka, Maharashtra and Andhra Pradesh. *Eucalyptus* is widely used as source of raw material for pulp, boards and furniture and also used as fuel wood. Integrating *Eucalyptus* in the farm lands has proven to be a big source of income to the farmers. In support of agroforestry farm establishment, SARA provides fodder to the livestock such as *Stylosanthes scabra* and *S. hamata*. These are also integrated in the agroforestry farms. In some areas, the farmers are earning extra income by intercropping maize, cashew, peanut and ginger. These are also integrated in the agroforestry farms. In some areas, the farmers are earning extra income by intercropping maize, cashew, peanut and ginger (Tewari 1998).

Fig 1:

Clonal plantation

SARA raised clonal plantations in its vicinity of around 250-300 kms to ensure a sustained supply to mills for its use. SARA has raised plantations in all types of soils and environmental conditions on degraded lands of farmers with a harvesting period of five years in various locations in districts of Karnataka, Maharashtra and Andhra Pradesh. The section wise physical and climatological data are given in Table 2.

The plantations were raised in dry area having rain fall of 650 mm to high rainfall area 2500-3000 mm. Site specific clones suited to a particular site were used. Clonal no. 7, 2135, 316, C-4, IFGTB, 103 were planted in block plantation considering a spacing of 1.5m x 3.0m and 888 plants were planted in one acre. Superior Clonal planting material of above clones was provided by Hi-tech clonal nurseries at Dandeli. Ripping lines were developed through Rippers with a depth of 1-1.5 fit.

Growth assessment and monitoring of plantation

SARA has also conducted monitoring and growth assement for its plantations of different years at various locations to ensure the available pulpwood from the plantation for its future use. Such study would give a concrete idea to Society and farmers as well to know the volume of standing crop. Expert team of SARA collected the field data in terms of survival percentage, total height (m) and girth (cm) for all the standing plantations. Data were compiled using statistical tools and recorded properly. Collected data showed overall superiority in terms of survival, height and girth. Field plantation of 5 years old matured clonal plantation on degraded land is shown in Fig. 3.

Fig. 3.

Table 2. Section wise physical & climatologically details

S.No	Parameters	Anawatti/ Anandpuram	Radhanagri	Chandgad	Khanapur	Ram nagar	Chennamana Kittur and Kulwalli	**Malgi**	**Koppal**	**Nellore**
1	District/State	Shimoga/ Karnataka	Kolhapur/ Maharashtra	Kolhapur/ Maharashtra	Belgaum/ Karnataka	Uttar Kannada/ Karnataka	Belgaum/ Karnataka	Uttar Kannada/ Karnataka	Koppal / Karnataka	Nellore/ Andhra Pradesh
2	Mean Temperature	20^0C-35^0C	20^0C-30^0C	20^0C-3520^0C	25^0C-35^0C	25^0C-35^0C	25^0C-35^0C	20^0C-30^0C	20^0C-30^0C	35^0C-40^0C
3	Latitude and Longitude	N 14^0 33' 27.54" E 75^0 09' 15.79"	N 16^0 24' 33.48" E 73^0 57' 28.67"	N 15^0 56' 33.95" E 73^0 59' 27.09"	N 15^0 42' 58.39" E 74^0 22' 52.02"	N 15^0 24' 57.51" E 74^0 29' 52.02"	N 15^0 35' 29.90" E 74^0 27' 50.51"	N 14^0 34' 58.84" E 75^0 00' 39.65"	N 15^0 35' 07.08" E 76^0 15' 54.31"	N 14^0 27' E 80^0 02'
4	Mean Annual Rainfall (mm)	1000-1200	2500	2500-3000	2000-2500	2000-2500	1000-1250	1200-1500	500-750	750-1000
5	Altitude	550-600m above MSL	550-600m above MSL	300-600m above MSL	650-750m above MSL	600-700m above MSL	680-750m above MSL	580-650m above MSL	350-450m above MSL	18-50 m above MSL
6	No. of Wet days	95	95	95	90	93	70	80	45	65
7	Soil Type	Red Muram + Black Cotton	Red Lateritic Soil	Red Lateritic Soil	Red Lateritic Soil	Sandy loam + Red Lateritic Soil	Red + Black Cotton Soil	Partly Red Muram+ Partly Black Cotton	Red Muram + Black Cotton	Red Muram + Black Cotton
8	Soil pH	7-8	7-8	7-8	7-8	6-7	6-7	7-8	7-8	7-8

Pulp wood harvesting

SARA also assists farmers in engaging expert contactors to harvest the trees that have completed the five-year harvesting cycle. The first rotation crop of 2006 to 2011 has been harvested and farmers were paid for their pulpwood. The actual harvesting, however, is carried out by employing local residents. This way, contract farming also provides employment to the whole farming community.

Second rotation crop

After harvesting of first crop farmer showed their keen interest to continue the agreement under Coppice Crop Agreement for another 5 years to take another advantage of their crop. SARA has covered around 3000 acres of coppice plantation in various plantations section on the basis of its suitability and last yield record. It has reduced the cost of land development and plantations cost which would certainly reduced the cost of pulpwood to WCPM. Also it would ensure the yield guarantee from second rotation crop on the basis of first yield record. All the maintenance work done by SARA-WCPM again on its own cost as per terms and conditions of the agreement. Such initiation would be helpful for farmers as well for society in a win-win situation.

Social, environmental and economical contributions

The initiatives of SARA in contract farming, clonal propagation of pulpwood species, pulpwood harvesting and agroforestry implementation not only provide numerous benefits to the paper industry but also to the farmer and his family. Overall, these initiatives help conserve and enrich our natural resources thereby reducing the pressure to source wood from natural forests. At the same time, saline lands are rehabilitated, soil erosion is reduced and carbon is effectively sequestered from the atmosphere. To date, there is still a big gap between pulpwood demand and supply. These initiatives will help increase available resources to meet pulpwood requirements of the paper industry and at the same time open alternative options to secure raw materials in the future.

Local employment will continue to thrive where the plantations are located. About 324 person-days per hectare per year are spent on land development, transplanting, and plantation maintenance, while 460 person-days per hectare per year are spent on harvesting, debarking and loading. Long-term income to the land owners is also secured with SARA's contract farming project.

To overcome the financial burden of small farmers they can adopt agroforestry models which allow growing of intercrops along with the trees crops which can fetch some immediate returns in the first two years.

Coclusion

Agroforestry always a topic of discussion for each and every person in such a way that it provides a helpful tool to take advantage for farm communities. Various agroforestry models provided a line to farmer so that he may adopt the best practise using particular species which suits to his land. On commercial level we need to promote agroforestry. Many crops and plant species are widely useful to promote agroforestry and are giving benefits to farmers so that he can earn good amount of his crop as well as from plantation also using best models. Many species like *Eucalyptus*, Poplar, Shisham, *Melia*, Teak which have been proved best selection for farmers now a days and farmer promoting these species on their lands with intercrop of wheat, sugarcane, groundnut, maize, cotton and others.

It would not be out of place to mention that in India, *Eucalyptus* is in trend and it is also recognized as a commercial agriculture crop. It has received much attention among commercial growers and agriculturists and they are going for its plantations on mega scale either alone or with intercrop of various other agriculture crop like wheat, sugarcane etc to get maximum return from the land. *Eucalyptus* plantations promoted by the private companies receive generous incentives such as technical know-how for establishing the trees on the farmer's land and contracts with the farmer to buy some or all of the first harvest for an agreed upon price at the time of harvesting, sales tax exemptions on the pulpwood, procurement through Agricultural Market Committees, no middle man involvement hence farmer can sell his produce directly to the end users and various tax holidays and tax exemptions for extended periods apart from element of subsidies on the *Eucalyptus* clones given by the company. These incentives put *Eucalyptus* at an advantage compared to other agricultural crops (including perennials), which receive no such promotion. Beside *Eucalyptus*, promotion of other commercial species and their plantations are required more attention so that farmer can use them as an alternative looking to their site suitability and available market. Government should also come forward and launch beneficial policies and schemes for farm communities with a fare commitment of financial and technical supports. Agriculture banks and institutions can also play a major role in this direction by providing them financial out puts through their schemes i.e. Agriculture loans. Crop Insurance of such commercial crop is also a need of the hour. In case something goes wrong at any stage then farmers could remain in a safeguard position. Further, market for such produce from plantations is required without any interference of middle man. Farmer should come directly and disposed of their crop and gain the right value for the crop. This step is very important to promote agroforestry on big scale because ultimate aim of the farmer is to get the right value of their produce at right time in a defined manner. Such efforts are immediately required by the Centre and State Government especially Agriculture department. In this Role of Forest department may not ignore. Further to this Forest department can also play a vital role to give relaxation and easy permission on such crops exempted already from any transit permit or pass. Farmer will declare the produce with crop name and if he will get the support from Forest department in easy manner and further receiving the permission timely he can sell his crop and cut whenever he required money for his need.

In last, to overcome the financial burden of small farmers they can adopt agroforestry models which allow growing of intercrops along with the trees crops which can fetch some immediate returns in the first two years. Looking to the overall scenario of available sources of wood it is now must to go for major plantations.

Acknowledgment

The author would like to acknowledge the contributions of Sh. Rajendra Jain, Executive Director, of West Coast Paper Mills Limited, Dandeli (UK), Karnataka.

References

Puri S, Nair PKR (2004) Agroforestry research for development in India: 25 years of experience of a national program. Agroforestry System 61(1-3):437-452.

National Agroforestry Policy (2014)

Tejwani KG (1994) Agroforestry in India. Oxford and IBH, New Delhi. pp. 10-47.

Sharma SK, Chauhan S, Kaur B, Arya ID (2009) Opportunities and major constraints in agroforestry system of western U.P. : A vital role of Star Paper Mills, Saharanpur (UP)-India. Agric Biol J N Am 1(3):343-349.

Sharma SK, Chopra RK (2014) Agroforestry Initiatives for Capacity Building and Social Security through Captive Plantations on Degraded Lands of Farmers in State Karnataka. India Jr Agri Sci Tech B (4):816-822.

Tewari SK (1998) Agroforestry, Pantnagar. pp. 1-58.

32

Economics and Yield Performance of Gamhar (*Gmelina arborea* roxb.) Under Agri-silvicultural System in East Singhbhum District in Jharkhand, India

[1]Kumar, S., [1]Malik, M.S. and [2]Vikas Kumar

[1]Faculty of Forestry, Birsa Agricultural University (BAU), Kanke, Ranchi, Jharkhand – 8340060, India
[2]College of Forestry, Kerala Agricultural University (KAU), Thrissur, Kerala – 680656, India

Introduction

The literatures and researcher data showed that there is negative impacts of agricultural expansion, landscape modification and deforestation on biodiversity, ecosystem services, alteringthe species composition and their ecological functions (Tscharntke *et al.* 2005; Priess *et al.* 2007; Flynn *et al.* 2009; Senior *et al.* 2013; Deguines *et al.* 2014; Kumar, 2016) which can lead to considerable changes in critical ecosystem processes. On the other hand, an increasingly industrialized global economy, rapid population growth, land degradation, land use pattern and role of various human activities have led to dramatically increased the pressure on the natural resources such as the available land for sustaining the livelihoods, and with over exploitation and extraction of the natural resources the ecosystems are becoming unsustainable and fragile since last century (Kumar 2017). Standardization of cultural practices is one of the primary objectives to make the system ecologically, sustainable and economically viable (Gill *et al.* 2009). To avoid these circumstances, tree growing in combination to agriculture (agroforestry systems) as well as numerous vegetation management regimes in cultural landscape (ethnoforestry systems) may improve soil fertility, carbon storage, provide fodder, produce tree fruits, expand fuel wood supplies and producea variety of wood products for farmers own use and sale without demanding additional land (Kumar 2015; Kumar 2016). Agroforestry system reducing vulnerability, increasing resilience of farming systems and buffering households against climate related risk in addition to providing livelihood security (NRCAR 2013). For example, agri-silvicultural systems, where trees and crops are grown together, are net sinks while agri-silvipastoral systems are possible sources of GHGs (Kandji *et al.* 2006). Agri-silvicultural systems are generally characterized by higher productivity on account of the vertical stratification of the shoot and roots systems of different components (Mathew *et al.* 1992; Kumar *et al.* 2001), they areextremely dynamic with available resources and environmental conditions changing over time. In addition, trees in managed species mixtures have a great potential to bring about 'micro-site enrichment' through processes such as efficient cycling of plant nutrients and nutrient pumping (Haines and DeBell 1979). Tree-crop compatibility trials in pursuit of identifying ideal combinations for understorey productivity improvements were conducted at several locations in India. Some of the reported combinations include rubber-banana (Rodrigo *et al.* 2001), *Hardwickia binata*-based agri-silvicultural system involving sorghum-pearl millet, pigeon pea, soybean and cotton (Khadse and Bharad 1996), coconut-based crop combinations for humid tropics such as coffee-banana, banana with ginger turmeric pineapple, papaya-pineapple, coffee-MPTs-pepper (Nair 1983; Lyiange *et al.* 1985).

Gamhar (*Gmelina arborea* Roxb) (Family- Verbenaceae) is fast growing species, itsintrinsic disease, fire and drought tolerance, as well asthe quality of its wood which is suitable for different types of uses such as paper pulping, plywood or particle board industry and furniture. It also known as white teak, a tropical deciduous tree native from moist tropicalforests of Asia (Rosero *et al.* 2011). It grows best on deep, well drained, base-rich soils with pH between 5.0 and 8.0 with wide range of conditions from sea level to1200 m elevation and annual rainfall from 750 to 5000 mm. In addition, Gamhar is considered a pioneer plant species capable of rapidly colonizing eroded or low nutritional quality lands, even-grained, soft, light and strong demanded for paneling, carriages, furniture, boxes and carpentry of all kinds which makes it interesting for reforestationor landscape restoration programs. The plant is widely used in Ayurveda, one of the ingredients of Dashamoola rishtada (Sharma *et al.* 2001; Khare 2004). Despite its ecological and increasing economic importance, very little is known about the biology of this species and its remarkable field behavior such as drought tolerance, atthe genetic, molecular and biochemical levels.

Planting trees and crops in association can also produce direct financial benefits (Nissen *et al.* 2001). In spite of the above advantages, there is substantial evidence that competition effects in intercropping systems may reduce or override overall productivity gains and financial returns compared with tree monocultures. Therefore, the choice of intercrop also depends on characteristics of particular tree speciese.g. root system, canopy, allelopathic effect of litter, agroclimatic and edaphic conditions (Batish *et al.* 2007).There is a great need to identify the suitable agricultural and horticultural crops, which can grow well along with tree species with limited solar energy available underneath the trees. In the present investigation on Gamhar woody perennial crop with the intention of growing agricultural crops viz., Gram (*Cicer arietinum*), Pea (*Pisum sativum*), Indian mustard (*Brassica nigra*) and horticultural crops such as Mango (*Magnifera indica*), Aonla (*Emblica officinalis*), and Papaya (*Carica papaya*) were intercropped. These crops were selected based on their adaptation, growing habit, production and requirement.

Study area

This experiment was conducted at field of Zonal Research Station, Darisai of Birsa Agricultural University, east Singhbhum District in Jharkhand. The altitude of site is about 622m above mean sea level. Geographically the site is located at

Fig. 1. Map of study area in East Singhbhum district in Jharkhand

21^{0}58' to 23^{0}36' N latitude and 85^{0}40' to 86^{0}54' E longitude in Chhotanagpur plateau. Darisai is situated on Zone VI (South Eastern Plateau Zone). The topographically the site is almost plain basic igneous rocks and granite in small pockets are the main geological sequence on which parent material and soil have developed.

Climate

The climate is classified as sub-humid with mean daily temperature of about 22.8^{0}C. The mean temperature of the coldest month (January) was 7^{0}C and the hottest month (May) was 38^{0}C. The mean relative humidity is about 70.88% in the area. The monsoon breaks out in the middle of June and last till mid-October. The annual rainfall in the area varies from 900-1500 mm. The mean wind velocity and evaporation varies from 3.61km/hr to 4.39 km hr^{-1} and 130 mm to 140mm, respectively.

Characteristics of soil

The soil of the field site is sandy loam to gravelly loam in texture, shallow soil depth (25-50cm), 0.3 percent and 3-5 percent slopes and slightly too moderately eroded, well drained with low water holding capacity and poor consistency. The soil colour varied from 7.5 YR 5/4 to 2.5YR 6/4 (brown to light yellowish brown). Medium in nutrient content, neutral to slightly acidic in reaction (pH 6.6).

Experimental design

The experiment was conducted in the field have boundary plantation of one tree species namely *Gmelina arborea* (Gamhar) and fruit crops namely Mango (*Mangifera indica*), Anola (*Emblica officinalis*), Papaya (*Carica papaya*). The agroforestry experiment along with the intention of growing agricultural crops namely Gram (*Cicer arietinum*), Pea (*Pisumsativum),* Indian mustard *(Brassica nigra)* and fruit crops namely Mango (*Magniferaindica*), Aonla (*Emblica officinalis*), Papaya (*Carica papaya*). During the experimental period, standard cultural practices like weeding, hoeing and watering were done at weekly interval *i.e.* after 7 days (weeding) or when it is required (watering) to maintain the proper growth of agricultural crops.

A field experiment involving six treatments and three replication but the stand was established in June, 1999 while agricultural crop was set up on December, 2011 and horticultural crops was planted in June, 2009. The size of plot was 16.0 X 16.0 m, spacing of tree species was 2.0 m from the line of farm boundaries, the spacing of agricultural crops was 0.30 X 0.30 m; spacing of horticultural crops was 5.0 X 5.0 m.

The different growth parameters such as height, stem girth, crown width, and yield of horticulture crops and germination, crop height, and yield of agricultural crops were recorded for each treatment. The fruit yield of agricultural crops by harvesting mature fruit and taking their weight after the final harvesting.

The data recorded on various parameters were subjected to statistical analysis for statistical validity of the results and interpretations. The main objective of t-test was to examine if there is any significant difference between the mean values of different parameter for each provenances/seed source. The significance of different sources of variation was tested by variance ratio of mean sum of square (F-test) at probability level of 5%. Standard of variance (ANOVA), standard error of mean (SE), coefficient of variation, critical difference, *etc.* were calculated for each parameter for the purpose of interpretation. For statistical calculation excel package of M.S. Office was used in a computer.

Observations

Agroforestry technologies vary from region to region. Adoption and practice of these technologies depends on the edaphic-climatic, socioeconomic status and needs of peasants. These attributes lead to variation in the structure and composition of recommended technologies and existing agrarianism (Panchal *et al.* 2017).

Height

It indicated that height of *Gmelina arborea* in the sole plantation was 10.18m which was lowest among agri-silvicultural system. Among the agri-silvicultural system, the Gamhar has recorded different height with different agriculture crops which ranged from 10.82 to 11.36m. The height of Gamhar is found highest in case of Gamhar+Gram which is 11.36m whereas the minimum highest was recoded in Gamhar + Pea (10.82m) (Table 1). Similar to the present study Malik *et al.* (2005) and Malik and Surendran (2000) have reported better growth of tree in Agri-silvicultural system of *E. globules* and potato and also Gill and Roy (1992) reported better height growth in agroforestry than in sole tree planting at National Research Centre on Agroforestry (NRCAF), Jhansi by taking 12 multipurpose tree species.

Diameter

It indicated that diameter of *G. arborea* is maximum (10.81cm) in agri-silviculture system where as minimum diameter (10.14cm) found in sole plantation. Among the agri-silvicultural system, the Gamhar has recorded different diameter with different agriculture crops which ranged from 10.75 to

10.81cm (Table 1). The ascending order of diameter was recorded in Gamhar>Gamhar+Gram>Gamhar+Indian mustard>Gamhar+Pea. Similar result has found by Gill and Roy (1992) reported better diameter and height growth in agroforestry than in sole tree planting at NRCAF, Jhansi. Similarly, better growth of tree species have been reported by different tree species in agroforestry system (Tree+Agricultural crop), *i.e. Dalbergia sissoo* intercropped with wheat and paddy (Sharma 1987).

Crown diameter

The observation ndicated that crown diameter of *G. arborea* is maximum (3.65cm) in agri-silviculture system where as minimum diameter (2.12cm) found in sole plantation. Among the agri-silvicultural system, the Gamhar has recorded different crown diameter with different agriculture crops which ranged from 3.34 to 3.65cm. The ascending order of crown diameter was observed in Gamhar>Gamhar+Gram>Gamhar+ Indian mustard >Gamhar+Pea (Table 1). Our result was also matched with Kaushik *et al.* (2002) was found crown spread of *Dalbergia sissoo* (40.10m^2), *A. indica* (24 m^2) and *M. alba* (10.90 m^2) in Agri-silvi-horticultural system in Arid India.

Basal area

The data showed that basal area of *G. arborea* (0.0092 m^2 plant^{-1}) in the agri-silviculture system was highest. Among the agri-silvicultural system, the Gamharwas recorded different basal area with different agriculture crops which ranged from 0.0091to 0.0092 m^2 plant^{-1} (Table 1). The basal area of Gamharwas found highest in Gamhar+Indian mustard and Gamhar+Pea (0.0092 m^2 plant^{-1}) whereas the minimum basal area was recorded in Gamhar+Gram (0.0091 m^2 plant^{-1}).

Volume

The observation was indicated that volume of *G. arborea* (0.103 m^3 pant^{-1}) in the agri-silviculture system is highest. Among the agri-silvicultural system, the Gamhar has recorded different volume with different agriculture crops which ranged from 0.099 to 0.103 m^3 pant^{-1}. The volume of Gamhar is found highest in case of Gamhar + Gram which was 0.103cu.m/plant, whereas the minimum volume was obtained in Gamhar + Pea (0.099 m^3 pant^{-1}). Further there is an increase of 16.5 % in tree volume in the combination of Gamhar + Gram over a period of one year.The ascending order of increment was Gamhar + Indian mustard = Gamhar + Gram > Gamhar + Pea > Gamhar (Table 1).

Table 1: Growth attributes of woody perennial species of Gamharin agri-silvicultural system

Treatments	Height (m)		Diameter (cm)		Crown diameter (m)		Basal area (m^2 $plant^{-1}$)		Volume (m^3 $plant^{-1}$)		Increment (m)
	2011	2012	2011	2012	2011	2012	2011	2012	2011	2012	
Gamhar only	10.18	10.45	10.14	10.37	2.12	2.22	0.0081	0.0085	0.082	0.088	0.27
Gamhar + Pea	10.82	11.16	10.81	11.13	3.65	3.72	0.0092	0.0097	0.099	0.109	0.34
Gamhar + Gram	11.36	11.79	10.75	11.39	3.34	3.43	0.0091	0.0102	0.103	0.120	0.43
Gamhar + Indian mustard	11.08	11.51	10.81	11.44	3.59	3.86	0.0092	0.0103	0.102	0.118	0.43
Mean	10.86	11.23	10.63	11.08	3.18	3.31	0.0089	0.0097	0.097	0.109	0.37
S.Em	0.319	0.312	0.174	0.184	0.172	0.143	0.001	0.001	0.003	0.003	0.016
CV %	13.479	12.751	7.520	7.588	17.568	14.045	15.091	15.142	14.620	13.429	19.672
C.D. at 5%	1.007	0.984	0.549	0.578	0.384	0.319	0.001	0.001	0.010	0.010	0.050

Growth attributes of horticultural and agricultural crops

The data was indicated that in horti-silvicultural system, the height of Mango (2.06 m), Aonla (3.12 m) and Papaya (1.82 m) were statistically significantly higher than Mango (1.92 m), Aonla (2.05 m) and Papaya (1.46 m) planting in sole cropping (Table 2). The data indicated that height in horti-silvicultural system was more than in sole planting. Similar observation had been observed by Gill (1992) regards higher growth of height and less of mango in horticultural crops based intercropped agroforestry system. The stem girth of Mango (24.65 cm), Aonla (23.66 cm) and Papaya (46.20 cm) were statistically significantly higher than Mango (22.76 cm), Aonla (21.66 cm) and Papaya (44.18 cm) planting in sole cropping (Table 2). Stem girth of horti-silvicultural system has more than in sole planting and similar observation has been recorded by Gill and Roy (1992). The crown width of Mango (109.60 cm), Aonla (279.64 cm) and Papaya (121.54 cm) were statistically significantly higher than Mango (103.30 cm), Aonla (264.12 cm) and Papaya (121.17 cm) planting in sole cropping. Crown width of horti-silvicultural system has more than in sole planting.

Yield of horti-silvicultural system, the yield of Papaya (553.11 q ha^{-1}) were statistically significantly higher than Papaya (476.53 q ha^{-1}) planting in sole cropping (Table 2). Yield in horti-silvicultural system was more than in sole planting. The production of Mango and Aonla fruit has not started till date (Table 2).

Germination

The germination (%) of all agricultural crops are maximum in agri-silviculture system where as minimum germination (%) found in sole plantation. Among the agri-silvicultural system, the Pea, Gram and Indian mustard were recorded different germination (%) which is 91, 90.67 and 87.67 per cent respectively (Table 3).

Crop Height

The data was indicted that in agri-silvicultural system, the height of Pea (84.67 cm), Gram (66.58 cm) and Indian mustard (124.01cm) were statistically significantly higher than Pea (75.66 cm), Gram (56.80 cm) and Indian mustard (113.38 cm) planting in sole cropping (Table 3). The data indicated that height in agri-silvicultural system was more than in sole planting.

Seed yield

Similarly, seed yield of Pea (7.32 q ha^{-1}), Gram (11.25 q ha^{-1}) and Indian mustard (4.44 q ha^{-1}) were statistically significantly higher than Pea (7.03 q ha^{-1}), Gram

Table 2: Growth attributes of horticultural crops species in hort-silvicultural system

Treatments	2011			2012			2011			2012			2011			2012			PapayaYield (q/ha)
	Height (m)						Girth (cm)						Crown width (m)						
	Mango	Aonla	Papaya	Mango	Aonla	Papaya	Mango	Aonla	Papaya	Mango	Aonla	Papaya	Mango	Aonla	Papaya	Mango	Aonla	Papaya	
Fruit Tree only	1.92	2.05	1.46	2.06	3.12	1.82	22.76	21.66	44.18	24.65	23.66	46.20	103.30	264.12	121.17	109.60	279.64	121.54	476.53
Gamhar + Fruit Tree	2.11	2.17	1.54	2.20	2.17	1.89	23.42	23.19	45.98	25.28	24.44	47.29	104.61	265.43	123.05	110.54	283.17	124.05	553.11
Mean	2.01	2.11	1.50	2.13	2.65	1.86	23.09	22.42	45.08	24.97	24.05	46.75	103.96	264.78	122.11	110.07	281.41	122.80	514.82
S.Em	0.005	0.005	0.011	0.005	0.011	0.011	0.128	0.524	0.515	0.358	0.259	0.395	0.480	0.437	0.419	0.069	1.253	0.385	21.434
CV %	0.566	0.574	1.785	0.589	1.030	1.489	1.354	5.721	2.800	3.510	2.640	2.069	1.131	0.404	0.841	0.154	1.091	0.768	10.198
C.D. at 5%	0.015	0.016	0.034	0.016	0.035	0.036	0.402	1.650	1.623	1.127	0.817	1.244	1.512	1.376	1.322	0.218	3.949	1.213	67.535

(9.29 q ha^{-1}) and Indian mustard (4.17 q ha^{-1}) planting in sole cropping. Statistically revealed that seed yield in agri-silviculture system was more than in sole planting. Similarly, Dhyani and Chauhan (1989) studied the performance of agricultural crops gave higher yield in partial shade as compared to open conditions. Singh and Pradhan (1989) observed that various intercrops with Mandrin (*Citrcus reticulata*) orchards can yield 60.9 kg ha^{-1} with Ginger, 149 kg ha^{-1} with maize and 59kg/ha along with finger millet in agri-silvicultural system. Similarly, Anon (1989) reported that out of various treatment evaluated, Coriander + Subabul produced significantly higher yield (10.3 q ha^{-1}) than (8.5 q ha^{-1}) with Coriander alone. Findings of Varadaranganatha and Madiwalar (2010) have reported six prominent agroforestry systems practiced in the three distinct agro-ecological situations. In all the three situations, bund planting was the most prominent agroforestry practiced by farmers, followed by horti-silviculture system and less prominent practice was block plantation. Mango was found as dominant fruit tree species.

Table 3. Germination, crop height and seed yield of agricultural crops at maturity in sole and in agri-silvicultural system

Treatments	Germination (%)			Crop Height (cm)			Seed yield (q ha^{-1})		
	Pea	Gram	Indian mustard	Pea	Gram	Indian mustard	Pea	Gram	Indian mustard
Agricultural Crops only	89.33	88.00	85.33	75.66	56.80	113.38	7.03	9.29	4.17
Gamhar + Agricultural Crops	91.00	90.67	87.67	84.67	66.58	124.01	7.32	11.25	4.44
Mean	90.17	89.33	86.50	80.17	61.69	118.70	7.17	10.27	4.31
S.Em	0.236	1.027	0.850	0.120	0.325	0.742	1.110	0.747	0.104
CV %	0.640	2.817	2.407	0.368	1.290	1.532	3.792	17.826	5.886
C.D. at 5%	0.743	3.237	2.678	0.379	1.024	2.339	0.350	2.355	0.326

Financial yield of fuel wood, horticultural and agricultural crops

Productivity in terms of money, i.e. financial yield was calculated for tree, fruit and crop on the basis of market rate (Gamhar Rs. 600 ft^{-3}, Papaya Rs. 17 kg^{-1}, Pea Rs. 40 kg^{-1}; Gram Rs. 55 kg^{-1} and Indian mustard Rs. 35 kg^{-1}) (Table 4). The result indicated that in all cases the agroforestry system gave more income as compared to sole plantation. In agroforestry system, highest financial yield of Gamhar (Rs. 72,972 of total boundary plants), Papaya (Rs. 9,40,287 ha^{-1}), Pea (Rs. 29,280 ha^{-1}), Gram (Rs. 61,875 ha^{-1}), and Indian mustard (Rs. 15,540 ha^{-1}) whereas in sole plantation, the financial yield of Gamhar (Rs. 21,406 of total boundary plants), Papaya (Rs. 8,10,101 ha^{-1}), Pea (Rs. 28,120 ha^{-1}), Gram (Rs. 51,095 ha^{-1}), and Indian mustard (Rs. 14,595 ha^{-1}). Kumar *et al.* (2004)

noticed that agricultural fields are one of the potential areas, where large scale planting of trees can be taken up along with the agricultural crops. Agroforestry models adopted by the farmers in Haryana and Uttaranchal states of India are highly lucrative, therefore, attracting farmers in a big way. Net Present Value (NPV) for different models on six years rotation varies from Rs. 26,626 to 72,705 $ha^{-1}yr^{-1}$, whereas B/C ratio and IRR vary from 2.35 to 3.73 and 94 to 389 per cent, respectively.

Table 4: Economic of Gamhar, horticultural crop species and agricultural crops in horti-silvicultural and agri-silvicultural system

Income	Tree species (Rs. ha^{-1})	Fruit tree species (Rs. ha^{-1})			Agricultural crops (Rs. ha^{-1})		
	Gamhar	Mango	Aonla	Papaya	Pea	Gram	Indian mustard
Tree/Fruits/ Crops only	21,406	—	—	8,10,101	28,120	51,095	14,595
Gamhar+ fruits/crops	72,972	—	—	9,40,287	29,280	61,875	15,540
Gross income	94,378	—	—	17,50,388	57,400	1,12,970	30,135
Net income from agroforestry system	51,566	—	—	1,30,186	1,160	10,780	945

Conclusion

i. Growth performance of horticultural and agricultural crop was found maximum under Gamhar tree as compare to sole plantation.

ii. Growth performance of Gamhar was also found better in agroforestry system.

iii. The yield of agriculture crop of Pea, Gram and Indian mustard was found maximum in agri-silvicultural system as compare to sole plantation.

iv. The overall performance of Gamhar, horticultural and agricultural crops was better under agroforestry system as compare to grown in sole system.

References

Batish DR, Kohli RK, Jose S, Singh HP (2007) Ecological Basis of Agroforestry. CRC Press, New York.

Deguines N, Jono C, Baude M, Julliard R, Fontaine C (2014) Largescale trade-off between agricultural intensification and pollination. Front Ecol Environ 12:212-217.

Dhyani SK, Chauhan DS (1989) Evaluation of crops in relation to shade in densities of Khasi pine *(Pinuskhasya)*. Progress Report Agroforestry Division, ICAR Research Complex for N.E.H. Region, Shillong.

Flynn DF, Gogol-Prokurat M, Nogeire T, Molinari N, Richers BT, Lin BB, Simpson N, Mayfield MM, DeClerck F (2009) Loss of functional diversity under land use intensification across multiple taxa. Ecol Lett 12: 22-33.

Gill AS, Roy RD (1992) Tree growth and crop production under agrisilvicultural systems. Range management and Agroforestry 12(1): 69-78.

Gill RIS, Singh B, Kaur N (2009) Productivity and nutrient uptake of newly released wheat varieties at different sowing times under poplar plantation in north-western India. Agroforest Syst 76:579-590.

Haines SC, DeBell DS (1979) Use of N-Fixing plants to improve and maintain productivity of forest soils. In: Proc Impact of intensive harvesting on forest nutrients cycling. School of Forestry, Syracuse, NY. USA.

Kandji ST, Verchot LV, Mackensen J *et al.* (2006) Opportunities for linking climate change adaptation and mitigation through agroforestry systems. In: Garrity D, Okono A, Grayson M, Parrott S (eds) World Agroforestry into the Future. World Agroforesty Centre, Nairobi. pp. 113-122.

Kaushik N, Kasashik RA, Saini RS, Deswal RPS (2002) Performance of agri. silvi. horticultural system in arid India. Indian Journal of Agroforestry 4(1):31-34.

KhadseVM, Bharath GM (1996) Performance of annual crops under canopy of *Hardwickiabinnata* in agroforestry systems. J Soil and Crops 6:151-153.

Khare CP (2004) Indian herbal remedies: rational western therapy, ayurvedic, and other traditional usage, botany. New York: Springer Science & Business Media, pp. 523.

Kumar BM, George SJ, Suresh TK (2001) Fodder grass productivity and soil fertility changes under four grass+tree associations in Kerala, India. Agroforest Syst 52:91-106.

Kumar BM, Nair, PKR (2004) The enigma of tropical homegardens. Agroforestry Systems 61: 135-152.

Kumar V (2015) Estimation of carbon sequestration of agroforestry systems. Van Sangyan 2(5):17-23.

Kumar V (2016) Multifunctional agroforestry systems in tropics region. Nature Environment and Pollution Technology 15(2):365-376.

Kumar V (2017) Agrobiodiversity, Structural Compositions and Species Utilization of Homegardens in Humid Tropics, Kerala, India. PLOSone (In press).

Lyiange MDS, Tejawani KG, Nair PKR (1985) Intercropping under coconut in Srilanka. Agrofor Syst 2:215-228.

Malik MS, Surendran C (2000) Intercropping with multipurpose tree species (MPTS) based Industrial plantation and its beneficial effects. In: Bisaria Solank, Honda (eds) Multipurpose Tree Species Research, Agrobios (India), pp. 321-327.

Malik MS, Surendran C, Kailaasham K (2005) Predicting growth of Eucalyptus globules under Agroforestry plantation. Indian Journal of Agroforestry 7(1):57-61.

Mathew T, Kumar BM, Babu KVS, Umamaheswaran K (1992) Comparative performance of four multipurpose trees associated with four grass species in the humid regions of southern India. Agroforest Syst 17: 205-218.

Nair PKR (1983) Soil Productivity of Aspects of Agroforestry. International council for research on agroforestry. Nairobi, Kenya, pp. 85.

Nissen TM, Midmore DJ, Keeler AG (2001) Biophysical and economic tradeoffs of intercropping timber with food crops in the Philippine uplands. Agricul Syst 67(1):49-69.

NRCAF (2013) NRCAF Vision 2050. National Research Centre for Agroforestry, Jhansi, pp.03.

Panchal JS, Thakur NS, Jha SK, Kumar V (2017) Productivity and carbon sequestration under prevalent agroforestry systems in Navsari district, Gujarat, India. The Ecoscan (In press).

Priess JA, Mimler M, Klein AM, Schwarze S, Tscharntke T, Steffan- Dewenter I (2007) Linking deforestation scenarios to pollination services and economic returns in coffee agroforestry systems. Ecol Appl 17:407-417.

Rodrigo VHL, Stirling CM, Teklehaimanot Z, Nugawela A (2001) Intercropping with banana to improve fractional interception and radiation use efficiency of immature rubber plantations. Field Crops Research 69: 237-249.

Rosero C, Argout X, Ruiz M, Teran W (2011) A drought stress transcriptome profiling as the first genomic resource for white teak-Gamhar-(*Gmelinaarborea*Roxb) and related species. In: Rosero *et al.* (eds) BMC Proceedings, 5 (Suppl 7): pp.178.

Senior MJ, Hamer KC, Bottrell S, Edwards DP, Fayle TM, Lucey JM, Mayhew PJ, Newton R, Peh KSH, Sheldon FH, Stewart C, Styring AR, Thom MDF, Woodcock P, Hill JK (2013) Trait dependent declines of species following conversion of rain forest to oil palm plantations. Bio divers Conserv 22:253-268.

Sharma KK (1987) Effect of trees on agricultural crops. Institutional seminar. FRI, Dehradun.

Sharma PC, Yelne MB, Dennis TJ (2001) Database on medicinal plants used in ayurveda. New Delhi: Central Council for Research in Ayurveda and Siddha, Government of India.

Tscharntke T, Klein AM, Kruess A, Steffan-Dewenter I, Thies C (2005) Landscape perspectives on agricultural intensification and biodiversity-ecosystem service management. Ecol Lett 8:857-874.

Varadaranganatha GH, Madiwalar SL (2010) Studies on species richness, diversity and density of tree/shrub species in agroforestry systems. Karnataka J Agriculture Science 23 (3): 452-456.

33

Poplar (*Populus deltoides*) Based Agroforestry Systems: An Economically Viable Livelihood Option for the Farmers of North India

N. V. Saresh[1], Archana Verma[2], D. S. Rana[1], Ravi Bhardwaj[3] B. Hulikatti Mahantesh[4], S. M. Raghavendra[5] and M.S. Sankanur[6]

[1]*Regional Research Station (Punjab Agricultural University), Ballowal Saunkhri SBS Nagar, Punjab, Inida*
[2]*Central Arid Zone Research Institute, Jodhpur, India*
[3]*College of Horticulture and Forestry Neri, Hamirpur, Dr YSP UHF Solan, H.P. India*
[4]*Union Bank of India, Gudgeri Branch, Dharwad, Karnataka, India*
[5]*Department of Forestry and Environmental Science, UAS, Bengaluru Karnataka India*
[6]*Department of Forest Biology and Tree Improvement, College of Forestry, NAU Navsari, Maharashtra, Inida*

Introduction

Forest cover of India has been estimated to be 7,01,673 km^2 (70.17 million ha) which is 21.34 per cent of the total geographical area of the country. The tree cover of the country is estimated to be 92,572 km^2 which is 2.82 per cent of geographical area. The total forest and tree cover is 7,94,245 km^2 which is 24.16 per cent of the total geographical area of the country. Out of this, 9.14per cent forest cover is in the form of open forests and another 1.26 per cent is scrub forests (FSI 2015). According to National Forest Policy 1988, one-third (33.33%) of the land area should be under the forest cover for sound ecological balance. It means that we have to bring another 11.66 % area under forest cover and at the same time improve quality of the degraded forests. However the horizontal expansion of land under tree cover is not possible. In order to increase the tree cover and fulfill the requirements of the people and industries, social forestry programs should be launched. As the country has a net sown area of 40.25 per cent, it is one of the most important potential areas for tree growing along with the agriculture crop. In order to attract farmers toward agroforestry, we should have viable agroforestry models, which can provide attractive financial returns to the farmers (Kumar *et al.* 2004). There are different agroforestry systems prevalent in many parts of India, which is having a high potential to support our agriculture in a sustainable manner.

India is an agriculture intensive country where traditional rice-wheat cropping system is contributing towards food security of the world. Presently farmers are going for diversification of the system of agriculture by incorporating tree component in their cropping system. Agroforestry has emerged as one of a viable alternative for crop diversification in the existing agriculture systems. Different tree species are being grown in the various parts of our country either as a block or boundary plantation. Poplar (*Populus deltoides*) is one of the most adopted species on the agriculture fields (Puri *et al.* 2001). Poplar as an agroforestry component helps in regulating nutrient cycling and adds organic matter in the form of leaf litter which helps in maintaining soil health. With proper management of poplar based agroforestry system, farmers are producing an average of 100 q of wood/acre/year which is yielding an income ranging from Rs. 70,000 – Rs. 80,000 acre^{-1} yr^{-1} (Gill *et al.* 2015). The poplar based agroforestry system is providing high economic returns and has helped in improving the livelihood status of the farmers in Northern India.

Why Poplar based agroforestry system?

The poplar based Agroforestry system has gained popularity because of the following reasons:

1. They have features favourable for Agroforestry system, i.e. fast growth, adaptability with agricultural crops.
2. Readily available market.
3. Encouragement from the private firms due to its multiple uses.
4. Huge demand in wood based industries due to easy availability.
5. Use of poplars for Carbon sequestration.
6. Use of poplars for Phytoremediation.
7. Absorption of nitrate pollutants from farms.
8. To mitigate salinity (Armitage 1985).

Features of poplars suitable for agroforestry

1. Fast growth, 20-25 $m^3 ha^{-1} yr^{-1}$ (Singh *et al.* 2001)
2. Straight clean bole.
3. Leaflessness during winter.
4. Multiple uses-Pulpwood, packing cases, poles, board and plywood, etc.
5. Compatibility with agricultural crops.
6. High economic returns of Rs. 88,749 $acre^{-1}$, @12 percent interest in seven years (Dillon *et al.* 2001)
7. Short rotation period (6-8 years).
8. High CO_2 exchange rate (Nelson 1984).
9. High water use efficiency (Dickmann and Keathley 1996).

Silviculture of poplars

Nursery Practice: Nursery techniques for raising *Populus* spp. have been standardized at FRI. The same technique is applicable to all *Populus species* planted in the country.

Preparation of cuttings

Poplars are generally raised by vegetative means by using cuttings. Cuttings are derived from one year old shoots from lower two-third portion or from nursery grown one year old plants, during the dormant season. Cuttings from the shoots are prepared using sharp cutters like secateurs or *gandasa*. The optimum diameter of cuttings (22 cm long) varies from 1-3 cm. Cuttings can safely be drawn at any time from the middle of January. Both ends of the cuttings should be sealed by wax, as protection against moisture loss. The cuttings must be submerged in fresh water immediately after preparation and kept for 28 hours prior to planting, cuttings should be drenched with Aldrin (30 EC thoroughly mixed in 100 liter water) emulsion. Thereafter the cuttings are treated with Emisan – an organo-mercurial fungicide (250 gm Emisan in 1000 litre of water).

Planting of cuttings

The cuttings are inserted vertically in well prepared nursery beds. The usual spacing between cuttings is generally kept at 50 cm or 60 cm and between rows 60 to 80 cm. The entire length of cuttings should be inserted into the soil keeping one bud above ground level. The soil around each cutting should be compacted gently but firmly without injuring the bark. The nursery raised plants called Entire transplants (ETP's) attain a height of about 3 to 4 meters in one growing season. These are utilized for planting in the field.

Distribution

World

Poplars in their natural range occur interspersed throughout the forests of temperate regions in the Northern Hemisphere between the southern limit of around latitude 30^0 N and northern limit of latitude 45^0 N. Several species occur naturally in the land mass. There are 35 species of poplars currently recognized in the world.

Table 1: Important countries undertaking poplar plantations

Countries	
France	Canada
Japan	India
Italy	Pakistan
Korea	Turkey
Hungary	Syria
Yugoslavia	Iran
Australia	Iraq
Romania	Afghanistan
Germany	New Zealand
Netherlands	Morocco
Belgium	Tunisia
U.S.A.	Algeria

Table 2: Countries reporting Poplar stand area

Country	Natural Poplar stands area (1000 ha)	Planted Poplar stands (1000 ha)
Canada	28,300	-
Russian Federation	21,900	-
United states	17,700	-
China	2,100	4900
Germany	100	-
Finland	67	-
France	40	236
India	10	1000
Italy	7	119
Turkey	-	130
Argentina	-	64
Total	70,224	6,449

Source: Ball *et al*. (2005)

Globally, 91 per cent of poplars grow in natural forests, 6 per cent in plantations and 3 per cent in AF systems. China (76%) and India (15.5%) are the major countries having higher planted area (Ball *et al.* 2005).

Table 3: Area trends in different countries

Country	Poplars		
	Natural forest species	Planted	Agroforestry/trees outside forests
Argentina		▲	
Belgium	▼	▼	▼
Bulgaria	▼	●	▲
Canada	●	▲	▲
Chile		●	●
China	▲	▲	▲
Croatia	▲	▼	
Finland	●	▲	
France		▲	
Germany	▼	▼	▲
India	●	●	●
Italy		▲	
Russian Federation	▲	▼	●
Serbia and Montenegro	▼	●	▲
Spain	●	▲	▲
Sweden			
Turkey	●	▼	▼
United Kingdom	●	▲	
United States	▼	▲	▲

The reports by IPC members suggest that the area of planted poplars appears to be increasing globally. Regionally, the area is decreasing or stable in Europe, increasing or stable in Asia, increasing in North America and increasing or stable in South America. The planting of poplars in smallholder woodlots and in agroforestry systems is increasingly enhancing land use in Asia (especially China and India) and South America.

India

Populus is widely planted above 28°N latitude in Jammu and Kashmir, Punjab, Haryana, Uttar Pradesh, North Bengal, Himachal Pradesh, Sikkim and Arunachal Pradesh, along roads, canals, in agriculture fields, towns, parks, orchards and home gardens.

Indigenous and exotic *Populus* spp.

Indigenous Populus *species*

There are six *Populus* spp. indigenous to India growing along water courses in the higher hills, in valleys and also on hill sides exposed due to landslides etc.

Most of the indigenous *Populus* spp. grow in areas having mean minimum and maximum temperature of 6° C and 10° C respectively. They grow well on low lying and moist areas preferring loamy soils but may be planted on river beds with sandy soils and in areas with clayey loams in forest soils. The soil pH best suited to poplars ranges between 5.0 to 6.5. Poplars are known to occur naturally in sub-tropical broadleaved hill forests, wet temperate, moist temperate deciduous and dry temperate forests. Their distribution in nature is as follows:

Table 4 : Indigenous *Populus* species distribution

Species	Distribution
Populus ciliata	Temperate and Sub-temperate region of Himalayas; altitude: 1200-3500m.
P. laurifolia	North-west Himalayas; altitude: 2400-4000m.
P. gamblei	Eastern Himalayas (Sikkim, North Bengal, Arunachal Pradesh); altitude: 400-1300m.
P. euphratica	North-west Himalayas; altitude: up to 400m.
P. alba	Western Himalayas; altitude: 1200-1300m.
P. jaquemontiana var. glauca	Eastern Himalayas (Sikkim, North Bengal); altitude: 1500-3200m.

Source: Tewari (1993)

Among the six species, *P. sanoria* and *P. gamblei* are fast growing and offer a potential to meet the increasing demand of wood for packing cases and for other industrial uses. *P. euphratica* is seen to be thriving well in the cold desert region of higher Himalayas. Other species grow slow but yield good quality wood (Tewari 1993).

Exotic *Populus* species

Since poplars prefer longer hours of day light, the natural zone of poplars lies upwards from 31^0 N latitude. Therefore exotic poplars were initially introduced in India in areas lying between 28^0 N to 31^0 N. Individual plants of the genus *Populus* are either male or female therefore their seeds produce a large number of hybrids. This character offers excellent opportunities for improvement and selection. To produce true to type plants they must invariably be propagated as a vegetative.

Research trials on poplars at Forest Research Institute (FRI), Dehradun and State Forest Departments helped in identifying suitable clones of *Populus species* to suit the different agro-climatic conditions in the country. The nursery raising technology developed by the FRI scientists have helped popularize the species amongst Forest Departments and farmers within a short span of less than two decades. After initial screening, the State Forest Departments and companies like WIMCO (Western India Match Company Limited) are provided with the germplasm of the successful clones, which they multiply and distribute to the farmers.

In India large number of exotic species/clones have been tested, screened and recommended for raising large scale commercial plantations above 28° N latitude. Several states have raised commercial plantations of exotic poplars. Suitable clones per species of *Populus* adapted to the different regions/states are as follows:

Table 5: Suitable exotic species for different states

State	Suitable exotic *Populus* species
Uttar Pradesh	*Populus yunnanensis, P. robusta, P. deltoides* 'G-3', 'G-48', 'D-121'.
Jammu Kashmir	*P. nigra* var. italica, *P. euramericana*'I-488', *P. eugenei, P. robusta, P. deltoides*'IC'.
Himachal Pradesh	*P. euamericana* 'I-488', 'I-65', 'I-15', *P.Rubrapoiret, P.deltoides* 'IC'.
Punjab	*P. deltoides* 'G-3', 'G-48', 'D-121'.
Haryana	*P. deltoides* 'G-3'.
Arunachal Pradesh	*P. deltoides* 'G-3', 'IC'.
Maharastra	*P. deltoides* 'D-121', 'G-3', 'G-48'.

Source: Tewari 1993

Some of the new successful clones recently introduced in the farming systems of North India are 'ST-67' 'S7C4', 'S7C8'. Clones of 'L' series developed at Haldwani (U. P.) from open pollinated seeds of 'G-48' and 'D-121' have shown promising results and are expected to replace the existing clones in the near future.

Selection of crops under poplar based agroforestry systems

All *rabi* and *Kharif,* except paddy can be successfully grown under block plantations of poplar during the initial three years.

Table 6: Crops grown at different stages under poplar in different cropping seasons

Year	*Kharif*	*Rabi*	Annual
1	Mentha, Mung, Maize, Sorghum and *Arvi*	Wheat, Mustard, Potato, Berseem and Marigold	Turmeric and sugarcane
2	Mentha, Mung, Bajra, Sorghum, Cowpea and *Arvi*	Wheat, Mustard, Potato, Berseem, oats and Marigold	Turmeric and sugarcane
3	Bajra, Cowpea and *Arvi*	Wheat, Mustard, Potato, Berseem, oats and Marigold	Turmeric
4-6	Not economical to grow crops	Wheat, Mustard, Potato, Berseem and oats	-

Source: Gill *et al.* 2015

Economics of poplar based agroforestry system

The economic efficiency measures used for analysis of agroforestry systems are:

(i) Net Present Value (NPV)

It is the present value of net benefits that the project had generated over and above that was available, when the amount proposed to be invested in the projects was invested at the current rate of interest elsewhere.

$$NPV = \sum_{i=1}^{n} \frac{(Bi - Ci)}{(1+r)^i}$$

Where, Bi = Benefits in the ith year

Ci = Costs in i^{th} year

i = Year: 1, 2, 3……….n

n = No. of years

r = Rate of interest

(ii) Benefit – Cost Ratio (BCR)

It is the ratio of sum of discounted benefit to the sum of discounted costs.

$$BCR = \frac{\sum_{i=1}^{n} \frac{Bi}{(1+r)^i}}{\sum_{i=1}^{n} \frac{Ci}{(1+r)^i}}$$

(iii) Internal Rate of Return (IRR)

It is defined as that rate of discount, which equalized the present value of stream of net benefits with the initial investment outlay.

$$Co = \sum_{i=1}^{n} \frac{(Bi - Ci)}{(1+r)^i}$$

Where, Co = Initial investment outlay

I = Year: 1,2,3,…….n

r' = Internal rate of return

Bi = Benefits in i^{th} year

Ci = Costs in i^{th} year

Poplar on farm lands

Performance of intercrops in agroforestry system: the case of poplar (*Populus deltoides*) in Uttar Pradesh, India

Jain and Singh (1999) examined the costs and returns of intercrops with *Populus deltoides* in Uttar Pradesh. The data was collected through stratified sampling techniques from Bhjavalkhera, Daddrail, Kant, Puwayan, Tilher, and Nigohi blocks of Shahjahanpur district of Uttar Pradesh where the Wimco agroforestry project has been under implementation since 1986. This district was selected considering the number of Poplar based agroforestry adopters and the area under Poplar plantation. The costs and returns of intercrops along with Poplar declined with the age of the trees in comparison to sole crops. However, the net returns of intercrops were higher than those of sole crops in the early stage of poplar intercropping.

The net income of the wheat intercrop over the total cost increased till the fourth year in comparison to that of the sole crop because of exclusion of the rental value from cost of the intercrop. The benefit-cost ratios (B / C ratio) for sole crop (1.24) was over ceded by the intercropping system and it ranged between 1.29 to 1.75 for the rotation period of poplar. The income from sugarcane was also higher. The benefit-cost ratios of sugarcane intercrop were higher than that of the sole crop (1.80) in the first (2.30) and second year (1.95) but were lower in the third years (1.67). Thus wheat and sugarcane with Poplar can provide high supplementary income in the early stages and wheat can also give some income in the later stages also.

Bio-economics of cropping systems combining medicinal and aromatic herbs with commercial timber tree species

A study by Dutt and Thakur (2004) in Himachal Pradesh attempted to explore the prospects of successfully intercropping high value medicinal and aromatic herbs with commercial timber tree species with the aim to boost economy of farmers. The study was undertaken for two consecutive experimental years in the field consisting of six-year old timber tree species (*Populus* hybrid, G-48) planted in East-West direction with three spacings of 8x3m, 6x4m, 5x5 m and 4x6m.The rotation period for the timber species is 8 years. Uniform and healthy nursery grown seedlings of medicinal plant species *Ocimum sanctum* was transplanted at the onset of rainy season in the first week of July and planted with 40cm plant x plant and row x row spacings in the plots between the rows of poplar trees. Sole plots of medicinal herbs without trees was maintained in the same field. The highest gross returns (Rs. 78810.0, 96750.0 ha^{-1}) were observed for sole crop, followed by plots under 8 x 3m poplar spacing (Rs. 58590.0 ha^{-1}) for *Ocimum sanctum* as intercrop during the first and second year of experimentation when calculated on Panchang basis. The observed trend was as open control > 8x3m > 6x4m > 5x5m > 4x6m.

The growing of medicinal and aromatic plant species with poplar was observed to be highly remunerative, since profit was significantly higher under agroforestry systems compared to sole cropping of these medicinal and aromatic herbs. The maximum net returns (Rs. ha yr^{1}) from *Ocimum sanctum* based agroforestry system was obtained when grown with poplar at 8X3 m spacing and sold as Panchang (Rs. 70798 and 86186). The higher net returns under agroforestry intervention is due to the substantial additional contribution by timber trees, while this extra benefit is absent under sole cropping.

The combination of medicinal and aromatic herbs with commercial timber tree species on the same farmland ensures higher profit to the farmers. The high value medicinal and aromatic herb namely, *Ocimum sanctum* can be grown successfully as intercrops. Poplar trees planted at 8x3m and 6x4mspacings may be adopted as suitable tree spacings to cultivate medicinal herbs for diversification and boosting economy of farmers under rainfed conditions.

Performance of poplar (*Populus deltoides*) based agroforestry system using aromatic crops

Chauhan (2000) conducted an experiment at CIMAP, Uttar Pradesh to know the economics of Poplar based agroforestry using aromatic plants. The experiment consisting of sole crop of poplar, Poplar + Lemon grass, Poplar + Citronella java, Poplar + Palmrosa and Poplar + Japanese mint. Maximum net profit was received from agroforestry system using aromatic crops (i.e. Rs. 1,81,850 to 2,17,958) as compared to sole poplar plantation (Rs.1,28,450). Highest cost of cultivation for Japanese mint with poplar trees was observed (Rs. 21,625 $ha^{-1} yr^{-1}$) followed by citronella java (Rs. 14,050 $ha^{-1}yr^{-1}$) Palmrosa (Rs. 12,840 $ha^{-1} yr^{-1}$) and lowest for lemongrass (Rs. 8,520 $ha^{-1} yr^{-1}$). Highest cost of cultivation was of Japanese mint due to planting done ever year and it required more cultural operation, fertilizer and irrigation than perennial aromatic grasses. Maximum net return (Rs. $ha^{-1} yr^{-1}$) was obtained from lemongrass (Rs. 43,590), followed by citronella java (Rs. 40,160) Palmrosa (Rs. 39,670) and minimum for Japanese mint (Rs. 36,370) inter-cropped under poplar plantation. It was concluded that farmers can raise their net income amounting to Rs. 43,590 $ha^{-1}yr^{-1}$ by growing lemongrass and Rs 36,370 $ha^{-1} yr^{-1}$ by growing Japanese mint with poplar during a period of five years.

Poplar in blocks

Viable agroforestry models and their economics in Yamunangar district of Haryana and Haridwar district of Uttarakhand

Kumar *et al.* (2004) studied viable agroforestry models and their economics in Yamunangar district of Haryana and Haridwar district of Uttarakhand. Reconnaissance field surveys of agroforestry were done in the Yamunanagar District of Haryana state and Haridwar District of Uttarakhand in 2001-2002 to select the study area. These two districts were chosen as the agroforestry is well developed in these two areas. Results of the study are based on the Participatory Rural Appraisal (PRA) technique followed to study agroforestry in the villages. While computing the economic returns from poplars, 10% mortality was presumed in all plantations, which was generally observed on account of damage by wind, insects and pests.

Discounted values of costs for various agricultural crops have been calculated on the presumption that entire cost is invested in the beginning i.e. at the time of sowing. Similarly, in case of poplar, it has been presumed that entire maintenance cost for a particular year is incurred in the beginning of the year. Rental value of the land has not been incorporated in the cost while calculating B:C ration, NPV and IRR because land owners are growing forestry and agricultural crops themselves. Depending on the geometry of poplar plantation and utilization of the land under poplar for agriculture, various viable agroforestry models are as follows:

Model 1: Poplar – sugarcane – turmeric block plantation model

In the first year, one-year-old ETPs of poplar and cuttings of sugarcane are planted in the field in the first week of February. Sugarcane is harvested in December. From the left over clump, new culms of sugarcane come up which are harvested in the second year in November. After this potato is sown in November itself and harvested in April of the third year. Then turmeric is sown in April and harvested in March of the fourth year. Similarly two more crops of turmeric are harvested. In the sixth year after harvesting turmeric Chari fodder crop is sown in May and harvested during August-September. Poplar is harvested at the end of sixth year. For this model, Net Present Value (NPV) at 12, 9 and 6 % discount rates come to Rs. 53,685; 62,349; and 72,705 $ha^{-1}yr^{-1}$ with Benefit / Cost (B : C) ratio as 2.87, 3.06 and 3.27, respectively and Internal Rate of Return (IRR) as 97 %.

Model 2: Poplar – sugarcane – wheat-chari block plantation model.

As described in the model-1, poplar and sugarcane are planted in the first year and for the first two years two, crops of sugarcane are harvested. In the second year wheat is sown in December, which is harvested in April of the third year. Then in May, chari fodder is sown, which is harvested during August-September. Similarly chari and wheat crops are taken in alternation till sixth year. Poplar crop is harvested after six year of age. For this model, NPV at 12, 9 and 6 per cent, discount rates come to Rs. 46,126; 53,733 and 62,956 ha^{-1} yr^{-1} with Benefit /Cost (B:C) ratio as 3.23, 3.47, 3.73, respectively and IRR as 94 %.

Model3: Poplar-sugarcane-wheat-chari-potato-maize-bajra block plantation model

In this model also, operations in the first two years are same as model-1. In the third year after harvesting of wheat, chari fodder is sown in May and harvested during August-September. Then potato is sown in November and harvested in February of the fourth year. Then maize fodder crop is sown in March and harvested in June. Bajra crop is sown in July and harvested in October. Similarly, potato, maize fodder and bajra crops are taken in a sequence till sixth year.At the six years of age, poplar trees are also harvested.For this model NPV at 12, 9 and 6 per cent, discount rates come to Rs. 49,280; 57,317 and 66,942 $ha^{-1}yr^{-1}$ with Benefit / Cost (B : C) ratio as 2.46, 2.58, 2.71, respectively and IRR as 96 %.

Model 4: Poplar – sugarcane – potato – barseem -chari block plantation model

In this model also, operations in the first two years are same as model-1 as far as poplar and sugarcane are concerned. However, in the second year, immediately after the harvesting of sugarcane in November, potato crop is sown in the field, which is harvested in February of the third year. Then chari crop is sown in the month of May and harvested during August-September. Then barseem fodder is sown in October and harvested four times from January to April in the fourth year. Similar operation is repeated till sixth year for taking chari and barseem fodder crops. At the end of sixth year, poplar is harvested. For this model, NPV at 12, 9 and 6 per cent discount rates come to Rs. 49,348; 57,338 and 66,902 ha^{-1} yr^{-1} with Benefit /Cost (B:C) ratio as 2.83, 3.01, 3.22, respectively and IRR as 97 %.

Table 7: Different poplar block plantation models

Model No.	Agroforestry models	NPV (12, 9 and 6% discount rate) and ranking		B:C Ratio and ranking	IIR (%) and ranking
		Total (Rs./ha)	Yearly (Rs. /ha)		
1.	Poplar	322112	53685 I	2.87 II	97 I
	Sugarcane-	374096	162349 I	3.06 II	
	Turmeric	436232	172705 I	3.27 II	
2.	Poplar-	276758	46126 IV	3.23 I	94 III
	Sugarcane-	322400	53733 IV	3.47 I	
	Wheat-chari	377134	62856 IV	3.73 I	
3.	Poplar	295679	49280 III	2.46 IV	96 II
	Sugarcane-	343902	57317 III	2.58 IV	
	Wheat-chari-	401652	66942 II	2.71 IV	
	Potato-maize-				
	Bajra				
4.	Poplar-	296088	49348 II	2.83 III	97 I
	Sugarcane-	344028	57338 II	3.01 III	
	Potato-	401409	66902 III	3.22 III	
	Barseem-chari				

Source: Kumar *et al*. (2005)

Poplar on boundary

Farmers with marginal and small land holdings have adopted Boundary Plantation Models. In general, these farmers are financially weak and unable to sacrifice nominal agricultural production in lieu of much higher return from block plantation of poplar at the end of six years. For boundary plantations, poplar is now getting popularized with the farmers. Gross income from the boundary plantation comes to Rs. 91,448 ha^{-1} (Sharma and Dadhwal 1996).

Model 5: Poplar – paddy-wheat boundary plantation model

Under this model, poplar is planted on the raised bunds along the field boundary in the first week of February. In the November itself wheat crop is sown, which is harvested in April of the second year. Then paddy and wheat crops are taken alternately till sixth year. At the end of sixth year poplar crop is also harvested. For this model, NPV at 12, 9 and 6 per cent discount rates come to Rs. 26,626; 29,738; 29,738 and 33,392 ha^{1} yr^{-1} with Benefit / Cost (B : C) ratio as 2.35, 2.42, 2.49, respectively and IRR as 389 %. IRR is high due to short rotation of agricultural crops and relatively less return from the poplars (as compared to return from the block plantation) after six years of rotation.

Model 6: Poplar – sugarcane – wheat-paddy boundary plantation model

Under this model also, poplar is planted on the raised bunds along the field boundary and sugarcane is sown in the field in the first week of February. Sugarcane was harvested in December. Next sugarcane crop is harvested in the November month of second year. Then wheat crop sown in November itself and harvested in April month of the third year. Then in July paddy is sown in the field and harvested in November. Like this, crops of wheat and paddy are taken alternately till sixth year. At the end of sixth year poplar trees are also harvested. For this model, NPV at 12%, 9% 6% discount rates come to Rs. 31,519; 34,946; 38,937 ha^{-1} yr^{-1} with Benefit /Cost (B:C) ratio as 2.67, 2.73, 2.81, respectively and IRR as 216 %. In this model, IRR is lesser as compared to model-5 because relatively longer rotation agricultural crop of sugarcane was taken in the first two years.

Table 8: Different poplar boundary plantation models

Model No.	Agroforestry models	NPV (12, 9 and 6% discount rate) and ranking		B:C Ratio and ranking	IIR (%) and ranking
		Total (Rs./ha)	Yearly (Rs. /ha)		
5	Poplar- Paddy-wheat	159754 178429 200349	26626 II 29738 II 33392 II	2.35 II 2.42 II 2.49 II	389 I
6	Poplar- Sugarcane- Wheat-paddy	189115 209677 233623	31519 I 34946 I 38937 I	2.67 I 2.73 I 2.81 I	216 II

Source: Kumar *et al.* (2004)

Rankings of the agroforestry models based on NPV, B: C ration and IRR have been shown (Table 7 and 8). Among the block plantation models, models1 and 2 rank first and fourth respectively, while models 3 and 4 have almost same NPV. Although model 2 ranks fourth based on NPV but has the highest B: C ration. IRR of the models vary from 94% to 97%. Among the boundary plantation models, model-6 was found better based on NPV and B: C ratio but model -5 has a much higher IRR as compared to model-6.

The Agroforestry models adopted by the farmers in Haryana and Uttaranchal states of India are highly lucrative, therefore, attracting farmers in a big way. NPV for different models on six years rotation varies from Rs. 26,626 to Rs. 72,705 ha^{-1} yr^{-1} whereas benefit / cost ratio and IRR vary from 2.35 to 3.73 and 94% to 389%, respectively. Thus Agroforestry has not only uplifted socioeconomic status of the farmers but also contributed towards overall development of the region.

Conclusion

Poplar based agroforestry systems play an important role in the socio-economic development of people in North India. They help in meeting the growing demand for timber, fuelwood and fodder. Both indigenous as well as exotic species are cultivated for meeting the demands. Poplars have the features which are suitable for agroforestry systems. Agricultural crops are more compatible with poplar in the initial years. Poplar based agroforestry systems provide about Rs. 70,000 to 80,000 yr^{-1} $acre^{-1}$, which is approximately three times more than the rice-wheat rotation in central plains of Punjab. This has boosted the economy and improved the livelihood pattern of the poplar growers, and helped in the development of plywood industry in the state. This poplar based agroforestry system is economically more viable and profitable livelihood option than many of the crop rotations. It is also capable of providing continuous employment on farms and preserving ecological system as well.

References

Armitage FB (1985) Irrigation Forestry in Arid and Semiarid Lands. IDRC, Ottawa, Canada. pp.160.

Ball J, Carle J, Del Luno A (2005) Contribution of poplars and willows to sustainable forestry and rural development. Unasylva 56 (221):8-9.

Chauhan HS (2000) Performance of Poplar (*Populus deltoides*) based agroforestry system using aromatic crops. Indian Journal of Agroforestry 2:17-21.

Dhillon A, Sangwan V, Malik DP,Luhach MS (2001)An economic analysis of poplar cultivation. Indian forester 127(1): 86-90.

Dickmann DI, Keathley DE (1996) Linkage physiology and molecular genetics on *Populus* ideotype. In: Stettler RF, Bradshaw Jr HD, Heilman, Hinckley TM (eds) Biology of *Populus*, NRC Research Press, Ottawa, Canada. pp. 468-490.

Dutt V, Thakur PS (2004) Bio-economics of cropping systems combining medicinal and aromatic herbs with commercial timber tree species. Indian Journal of Agroforestry 6(1):1-7.

FSI (2015) State of Forest Report, Forest Survey of India, GoI, Dehra Dun. pp. 16.

Gill RIS, Singh B, Kaur N, Sangha KS (2015) Agroforestry- A viable option for crop diversification in Punjab. Deptt of Forest and wildlife preservation, Govt. of Punjab, pp.24.

Jain SK, Singh P (1999) Performance of intercrops in agroforestry system: the case of poplar (*Populusdeltoides*) in Uttar Pradesh. Indian Forester 125 (2):195-205.

Kumar R, Gupta PK, Gulati Ajay (2004) Viable agroforestry models and their economics in Yamunangar district of Haryana and Haridwar district of Uttaranchal. Indian Forester 130(2):131-148.

Nelson ND (1984) Woody plants are not inherently low in photosynthetic capacity. Photosynthesis 18:600-605.

Puri S, Swamy SL, Jaiswal AK (2001) The potential of poplar (*P. deltoides*) in the sub-humid tropics of central India: survival, growth and productivity. Indian Forester 127(2): 182-185.

Sharma NK, Dadhwal KS (1996) Prominent agroforestry practices in the Saharanpur plains of Uttar Pradesh – A case Study. Indian Journal of Soil Conservation 24: 172-173.

Singh NB, Kumar Dinesh, Gupta Rajesh K, Singh Kadam, Khan GH (2001) Role of poplar in sustainable development of India. Indian Journal of Agroforestry 3(2): 92-99.

Tewari DN (1993) Poplar. Surya Publication, Dehradun. pp.30.

34

Tree Diversity and Ecosystem Services of Rural and Urban Homegardens Agroforestry of Kottayam District Kerala, India

Vikas Kumar, Abhijith R. and T. K. Kunhamu

Department of Silviculture and Agroforestry, College of Forestry, Kerala Agricultural University, Thrissur, Kerala, India

Introduction

Homegardens are traditional agricultural systems, particularly common in the tropics characterized by a high plant diversity (Kumar and Nair 2004 2006; Scales and Marsden 2008; Galluzzi *et al.* 2010), maintain high level of productivity, stability, sustainability and equitability (Soemarwoto and Conway 1992), serve as a refuge for wildlife (Perfecto and Vandermeer 2008), contributing significantly to the household diet by producing a large variety of food commodities (Kumar and Nair 2004; Wiersum 2006; Pulido *et al.* 2008; Kortright and Wakefield 2011; Kumar 2017; Kumar and Tripathi 2017), nutritional supplies (Mendez *et al.* 2001; Perrault and Coomes 2008; Vlkova *et al.* 2010; Balooni *et al.* 2014), supplementary cash income for owner (Huai *et al.* 2011; Arifin *et al.* 2012; Yang *et al.* 2014) and important spaces for transmission of cultural heritage (Galluzzi *et al.* 2010). Many native or endangered species can be found in homegardens (Albuquerque *et al.* 2005; Akinnifesi *et al.* 2010; Milow *et al.* 2013), and they are niches for domesticating semi-wild species (Akinnifesi *et al.* 2010). Plants grown, intentionally or unintentionally, in homegardens are important for household subsistence and economy. Homegardens, with their diversified agricultural crops and trees, are of vital importance to the subsistence economy of many areas in the tropics (Nair 1993; High and Shackleton 2000). Plants in homegardens contributeto carbon sequestration and mitigate effects of climate change (Balooni *et al.* 2014). Human population density is usually high in these areas; and the average size of landholding is less than one hectare (Nair and Sreedharan 1986). Homegardens of Kerala are primary models of such integrated farming system (Geetha *et al.* 2015). The most conspicuous characteristics of all homegardens are their layered canopy arrangements and admixture of compatible species, with each component occupying a specific place and function (Nair 1993).

Most homestead systems consist of an herbaceous layer near the ground and a tree layer at higher levels. However, a closer look at the vegetation shows that plants seem to be organized in different patterns or layers. Thus, homegarden is also called as "Multi-tier system" or "Multi-tier-cropping" as it consists of different canopy strata (Fig. 1). Homegardens, woody perennial crops are the main component because they provide sustainable products and ecosystem services (Kindt *et al.* 2004; Kehlenbeck 2007; Kabir and Webb 2008). For instance, five canopy strata have been identified in the homegardens of Kerala. The first layer lies within 2 m height from the ground and is constituted by vegetables: *Cajanus cajan* (pigeon pea), *Arachis hypogaea* (peanuts), Phaseolus, Psophocarpus and Vigna species (beans and other legumes); tuber crops: *Colocasia esculenta* (taro), *Dioscorea alata* (greater yam), *Dioscorea esculenta* (sweet yam), *Ipomoea batatas* (sweet potato), *Manihot esculenta* (cassava), *Xanthosoma* species (tannia or cocoyam); grasses: *Cymbopogon citratus* (lemon grass); spices: *Zingiber officinale* (ginger), *Kasthuri manjal* (kasthuri), *Piper methysticum* (kava), *Curcuma longa* (turmeric), *Cinnamomum zeylanicum* (cinnamon), *Areca catechu* (betel nut), *Piper betle* (betel vine) and other herbaceous plants; *Ananas como*sus (pineapple),

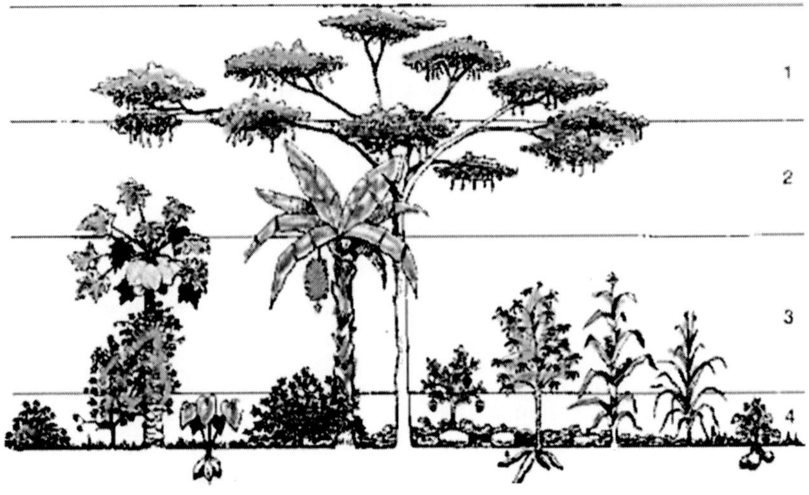

Fig. 1. Overall view of Kerala homegardens in humid tropics

Passiflora edulis (passion fruit), *Saccharum officinarum* (sugarcane), *Zea mays* (corn or maize) (Sankar and Chandrashekara 2002; Kumar *et al.* 2016; Kumar and Tripathi 2017).

The second and third layers (within 2 to 10m height) are almost continuous and overlapping each other. Some of the common constituents of these layers in Kerala are banana, nutmeg, papaya, mango, cocoa, young coconut palms and saplings of trees. The upper most canopy layer is formed by *Anacardium occidentale* (cahewnut), *Artocarpus heterophyllus* (jackfruit), *Citrus* species (lemon, lime, orange and tangerin), *Annona* species (soursop and sweetsop), *Swietenia macrophylla* (mahogany), *Ailanthus triphysa* (tree of heaven), *Averrhoa carambola* (carambola), *Artocarpus altilis* (breadfruit), *Carica papaya* (papaya), *Psidium guajava* (guava), *Mangifera indica* (mango), *Azadirachta indica* (neem), *Musa* species (bananas and plantains), *Persea americana* (avocado), *Cocus nucifera* (coconut), *Spondias dulcis*, *Syzygium malaccense* (malay apple), *Tamarindus indica* (tamarind), *Hevea brasiliensis* (rubber), and other tall trees of 10 to 25m height (Kumar *et al.* 2016; Kumar and Tripathi 2017). However, the choice species is determined by the agro-climatic and farmers' socio-economic conditions. Analysis of area occupied by canopy cover of different constituents indicated that in tradition homegardens, which are not less than 40-45 years old, the crown to land ratio ranged from 210 to 88% (Sankar and Chandrashekara 2002; Kumar *et al.* 2016). Several studies in Asia indicated that with commercialization, often a gradual change from subsistence to commercial crops occurs in homegardens, while the crop diversity decreases (Kumar and Nair 2004; Das and Das 2005; Peyre *et al.* 2006; Abdoellah *et al.* 2006; Kumar 2011; Kunhamu *et al.* 2015). Moreover, species diversity, size, shape and plant diversity also vary from place to place depending on cultural (Rico-Gray *et al.* 1990), ecological and socio-economic factors (Soemarwoto 1987).

George and Chattopadhyay (2001) observed that such shifts in land-use may have profound implications for the food security of the state, which already depends on 'outside supplies'to meet more than half of its food grain requirements. Cultivation of commercial crops is the major threat in farming at rubber is the main invader in Kerala fields; peoples are more concentrated on this commercial crop cultivation on economical basis. Nevertheless no comprehensive reports characterizing species composition and structure of rural and urban homegardens of Kerala are available. Urban homegardens are important to enhance community's aesthetic value (Balooni *et al.* 2014), improve the community's microclimate, cultural identity, urban biodiversity (Taylor and Lovell 2014). Hence an attempt was made (a) to undertake a qualitative and quantitative analysis of rural and urban homsteads; (b) to characterize the structural and functional attributes of selected homegardens by analyzing the species composition; (c) to understand the present status of homesteads especially in the urban sector consequent to the fast changing socio-economic equations of the state, and (d) to devise strategies for the improvement of the productivity and product diversity of the urban and rural homegardens.

Study area

The study was conducted at Kottayam district, Kerala, two representative's site. (i) Karoorpanchayat, Lalam block of total area is 36.84 sq.km, and (ii) Pala municipal town (away from 28 km east) and total area of 15.93 sq.km on the banks of the Meenachil river of Kottayam.

The field work involved conducting detailed 60 house-hold surveys of individual farmers belonging to the small, medium and large holding size categories in the Karoor Panchayat and Pala municipality of Kottayam district, Kerala. First of all, we collected the information of farmers name and land holding of given study area from Panchayat officer as well as Agriculture officer (Table 1). The survey included besides gathering general information on specific questionnaires, crop and livestock production enterprises, structure (species composition), arrangement of components in space and time (vertical and horizontal stratification), cropping pattern, cropping intensity, enumeration (by measuring the height and girth at breast height (above the 1.37 m ground level) of all scattered trees on the homestead and border trees except palms and also recorded on management and financial aspects of the crops were collected from the owners of the homegardens by method of questionnaire based survey. Profile diagram of the plants across the land was drawn by examining the vertical stratification of homegardens.

Land holdings

Rural homesteads

Total 60 homegadens were surveyed with farmer ranged from 0.10 to 2.5 acres, an average of homegardens area was 0.67 acres and total covered area was 48.58 acres. On the basis of land available for cultivation, homegardens were categorized in to three; small homegardens (d" 0.5acre), medium homegardens (0.5-1 acre) and large homegardens (>1 acres) in such a way that each categories class was selected 20 homestead. The total area of all the surveyed small homegardens was recorded 5.16 acres with the average area 0.50 acres of each homegardens. Similarly, medium homegardens of the total areas was 13.91 acres with the average area per homegarden 0.69 acres and the larger homegardens of the total area was recorded 29.51 acres with the average land available per homegarden was 1.48 acres.

Urban homesteads

Total 60 homegadens were surveyed with farmer ranged from 0.04 to 0.70 acres, an average of homegardens area was 0.25 acres and total covered area was 15.26 acres. On the basis of land available for cultivation, homegardens were categorized in to three; small homegardens (≤ 0.2 acre), medium homegardens (0.2-0.5 acre) and large homegardens (>0.5 acres) in such a way that each categories class was selected 20 homestead. The total area was covered 1.48, 4.58 and 9.21 acre with respect to small, medium and large homestead.

Assessment of floristic wealth

Plants were identified in respect of taxonomic position with the help of the standard flora (Hooker 1872; Gamble 1915; Manju *et al.* 2008). Common names and families of trees were traced to the publication of the Sasidharan (2004; 2012). For assessing the plant species diversity, Simpson's diversity index, Shannon-Wiener diversity functions and Equitability index were computed (Simpson 1949; Shannon and Wiener 1963).

Observation

Components

It was observed that a variety of trees, shrubs, herbs and livestock were grown, by the farmers in their homegardens for their daily consumption and for economic benefits.

Species frequency

In general there was wide variability in crop composition among the homesteads visited in Karoor Panchayat and Pala Municipality of Kottayam district. Cash crops, fruit crops and multipurpose trees (MPTs) were found in all homegardens. Majority of the areas are habituated by cash crops such as rubber (*Hevea braziliensis*). Presence of food crops and timber trees are also noticed in most of the homegardens (Fig. 2). Figure shows the less frequency was observed in urban homegardens compare to rural homegardens viz., livestock, spices, fodder crops, vegetables and other miscellaneous species. The area lies in the rubber belt of Kerala and majority of the people present here are rubber planters, of which some owns huge areas of land.

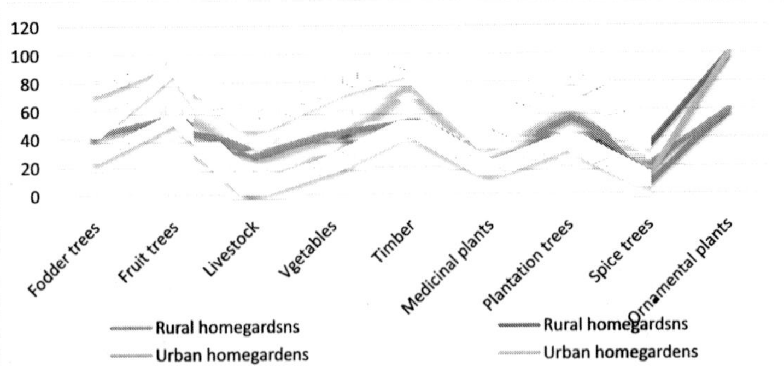

Fig. 2: Frequency of species and number in different (small, medium & large) categories of rural and urban homegardens

Species diversity

It was found that the cash crops comprising mainly of rubber and coconut occupy a large share in the homegardens. Density of fruit crops, timber species, livestock and other miscellaneous species are higher in small homegardens. The density of plantation trees were higher in medium homegardens and larger homegardens. The numbers of plants belonging to various crop components were calculated on per acre basis and are represented (Fig. 3). The species diversity of the various categories vary according to the necessity to which the farmer put.

The total availability was more in medium scale gardens when compared to the counter-parts. The intensity of cash crops coming into the situation was higher in large homesteads. In the coming parts the comparison present among each tree and each components were brought into the view (Fig. 3).

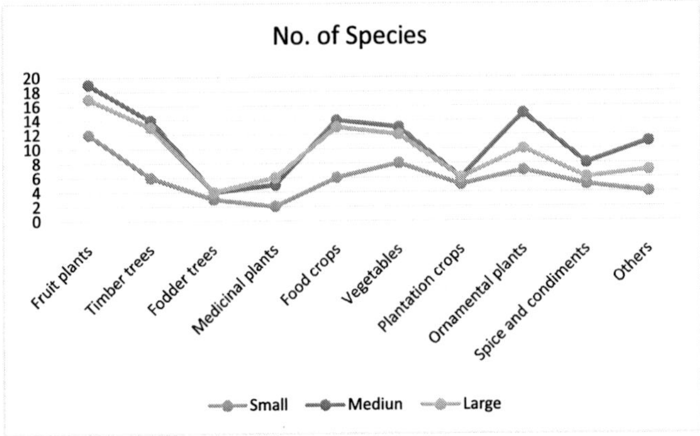

Fig. 3. Total number of species in different (small, medium & large) categories of rural and urban homegardens

Table 1. Details of the rural and urban homesteads farmers and land capacity details

Sl. No.	Name and address (Karoor Grama Panchayath)	Landholdings (acres)	Sl. No.	Name and address (Pala municipality)	Landholdings (acres)
	1. Small homegardens			**1. Small homegardens**	
i.	Tomy Thomas, Thundathil	0.38	i.	Chacko Joseph, Konnath	0.08
ii.	Gopalakrishnan. K, Kalappurayil	0.5	ii.	Baby Joseph, Chakkalackel	0.09
iii.	Isahmanaver, Perumballoor	0.45	iii.	C.Chacko, Niranathadathil	0.08
iv.	Manaverjoseph, Puthenpurackel	0.47	iv.	Mathai M, Vedikkunnel	0.07
v.	C.C Antony, Akkiyil	0.25	v.	Scaria Joseph, Padannanniyil	0.07
vi.	ThommanDevassy, Puthenpurayil	0.3	vi.	Jameesan T., Pannikkottuthadathil	0.04
vii.	Jose Thomas, Aanappara	0.15	vii.	Tomy Francis, Nellikkel	0.08
viii.	Vakkachan.C, Chalayil	0.1	viii.	Sebastian, Elaniyel	0.1
ix.	B. Sasidharan, Vallomkuzhiyil	0.18	ix.	James Kurian, Njavelikunnel	0.1
x.	N.Mohanan, Konnakkal	0.4	x.	GimmyCyriac, Kandathil	0.07
xi.	OusephOuseph, Kizhakkacherukunnel	0.2	xi.	Baby John, Kunnappillel	0.05
xii.	Binu Philip, Mundakalil	0.21	xii.	Joy Joseph, Thottumkal	0.09
xiii.	K.M.Mathew, Kulathinal	0.37	xiii.	K.T.Thomas, Kathalikkunnel	0.09
xiv.	Pailo Mathew Kozhipallil	0.25	xiv.	N.B Scaria, Naikkazhakunnel	0.08
xv.	C.J.Augustin, Chempanikkel	0.1	xv.	Benny Augustin, Nellikkunnel	0.09
xvi.	Zakaria, Palaniyil	0.18	xvi.	George Thomas, Muthirenikkel	0.05
xvii.	P.V.Thambi, Puthenpurackel	0.15	xvii.	Thomas Joseph, Moolamkuzhikel	0.06
xviii.	M.V.Thankappan, Mundathilparambil	0.08	xviii.	Chandrasekharan Nair, Manayil	0.06
xix.	Justin Jose, Njezhukumkattil	0.13	xix.	Benny Scaria, Vattamthottil	0.07
xx.	Augustin Joseph, Vellimoozhayil	0.31	xx.	Domis Thomas, Kannattucheruvil	0.05
	2. Medium homegardens			**2. Medium homegardens**	
i.	Mathew thomas,pathiyankal	0.7	i.	K.S. Cheriyan, Kattakkayam	0.13
ii.	Abraham Devasya, Shrathazh	0.6	ii.	Josukutty, Kizhathadiyoor	0.15
iii.	Manoj Madhavan, Kuzhikulam	0.6	iii.	JoyJohn, Kuriathanom	0.25

(Contd.)

iv.	Devasyachan,Uppoothil	0.92	iv.	Appachan J., Annachirakkunnel	0.3
v.	AneesChakko,Kurishummoottil	0.85	v.	Mathai Joseph, Manjankal	0.3
vi.	George, T, Vettukunnel	0.8	vi.	Mathukutty.C, Thakadiyel	0.38
vii.	James Varkey,Vellamkunnel	0.72	vii.	Thomas.T, Mathaikulath	0.35
viii.	P N Madhumohan	0.62	viii.	Sumeshkesavan, Thundathil	0.4
ix.	Jose Mathew, Kulapurath	0.52	ix.	Jony Antony, Kadaplakkal	0.45
x.	Augustin, Champanikel	0.7	x.	Justin Antony, Thakadiyel	0.42
xi.	Siby Joseph, Panniyanikkel	0.64	xi.	TomichanIssac, Mannoor	0.2
xii.	Joseph Mathew,Pazhayidam	0.5	xii.	E.K.Kumaran, Nirappel	0.12
xiii.	ScariaC.,Cholakku	0.53	xiii.	George Mathew,Kappil	0.12
xiv.	ChandiPothen. Mundakkalil	0.84	xiv.	P.C.Kuriakose,Vysyanparambil	0.15
xv.	Rajamma,Sankarasadanam	0.81	xv.	P.T.Joseph,Puzhakkattukarayil	0.1
xvi.	Robin Alex, Ellimoottil	0.62	xvi.	Prince Francis,Palakkattukunnel	0.14
xvii	Joseph,Puthenpurayil	0.89	xvii.	D.A.Sebastian,Moonnanikkal	0.12
xvii	Joseph Mathai,Madavath	0.68	xviii.	AugustinThomas,Mundankel	0.17
xix.	Manuvel Mathew, Mundathanath	0.65	xix.	Cheriyamma Abraham, Anithottam	0.18
xx.	Ajesh Thankachan, Mappalackel	0.72	xx.	Saju Thomas, Karikukkunnel	0.15
	3. Large homegardens			**3. Large homegardens**	
i.	George Thomas,Varacheriyil	1.4	i.	Chandrasekharan G, Mullavellil	0.55
ii.	Vinukumar, Chazhikattu	1.3	ii.	Joseph.Thomas, Thakadiyel	0.5
iii.	Naveen Thomas,Kalkunnel	1.35	iii.	Mons Antony, Nellamthadathil	0.52
iv.	Tomy Joseph, Puthenpurayil	1.56	iv.	Sunny Thomas,Poovathanikunnel	0.55
v.	Abraham Thomas, Vallokuzhiyil	1.6	v.	Dinesh.Damodaran, Poovathumkal	0.6
vi.	Sunny Paul, Parayaruthottathil	1.1	vi.	Sabu.Kuttappan, Puthanparambil	0.6
vii.	George Mathai,Chalil	1.3	vii.	Manoj. R., Thekkenellanikal	0.68
viii.	K. Sivadas, Vallokuzhiyil	1.7	viii.	Jojo Antony, Nellamthadathil	0.7
ix.	Vijayakumar.R,Kalkunnel	1.7	ix.	Baby Thomas, Poovathanikunnel	0.71
x.	Tomy, Parayil	1.25	x.	ThomasJoseph, Kunnel	0.75
xi.	Neelkandan. P,Nelithazhathu	1.7	xi.	Mathew.M., Illimoottil	0.22
xii.	Ruby Augustin.Chembottickel	2	xii.	Biju, Koottamakal	0.24

(Contd.)

xiii.	G.SanthoshKumar,Puthusseril	1	xiii.	JoseAbraham, Vellaniyil	0.23
xiv.	SivaramanKutty, Pothanickel	1.5	xiv.	P.J,antony, Puthiyidom	0.42
xv.	Anil kumar, Kaithakkal	1.5	xv.	Alex Antony, Chakkankel	0.3
xvi.	Radhakrishnan P,Parvathybhavan	1.25	xvi.	Anu Thomas, Thottumakal	0.3
xvii.	George, Puthiyidam	2.5	xvii.	M.B.George, Maniyanchira	0.62
xviii.	Madhu N, Kandathumkarayil	1.1	xviii.	Jose Thomas, Thottathil	0.23
xix.	T.NPrathap,Cherolikel	1.2	xix.	K. Suresh, Kathirinal	0.25
xx.	N.Suresh, Naduviledut hu	1.5	xx.	Thomas George, Mundamattom	0.24

Tree diversity

Fruit trees

Barring *Mangifera indica* and *Artocarpus heterophyllus* which were included under Multi-Purpose trees (MPTs), major fruit trees dominant in the area was *Psidium gujava, Citrus limon, Carica papaya, Artocarpus* species, etc. in small homegardens. Its density is more in small farms followed by medium and large. In general, farmer's preference for fruit trees was very less in the homegardens. The highest number of species are those that have more than one uses e.g. *Mangifera indica, Artocarpus heterophyllus, Artocarpus hirsutus* etc. In the rural homesteads people usually have some fruit trees which the can add to their nutrient intake (Fig. 4).

Fig. 4. Density of fruit trees in different (small, medium and large) categories of rural homesteads

In the urban scenario, majority of the fruit trees are present in the large homesteads (Fig. 5). The cases of small land holdings, occupation of the land owners etc. may be the reasons for it. Here also the major fruit species are the multi-purpose trees due to their various uses.

Fig. 5. Density of fruit trees in different (small, medium & large) categories of urban homesteads

Timber trees

Teak was the dominant timber tree in the rural and urban contests. Teak predominates in all categories of homegardens and its density is more in small homegardens (Fig. 6). Farmers who own medium homegardens prefer mahogany compared to teak. As the area is dominated by cash crops, farmer's attitude towards timber trees is less. In the rural homesteads trees like *Tectona grandis, Swietenia macrophylla, Ailanthus tryphysa, Artocarpus* spp. etc. dominated in the overall space.

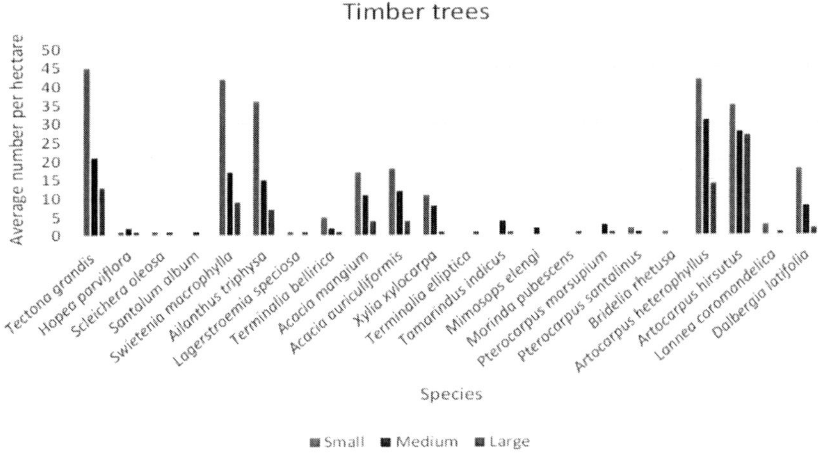

Fig. 6. Density of timber trees in different (small, medium and large) categories of rural homesteads

In the urban scenario, the diversity is less when compared, the highest occurrence are for species like *Tectona grandis, Swietenia macrophylla, Ailanthus tryphysa* etc (Fig. 7).

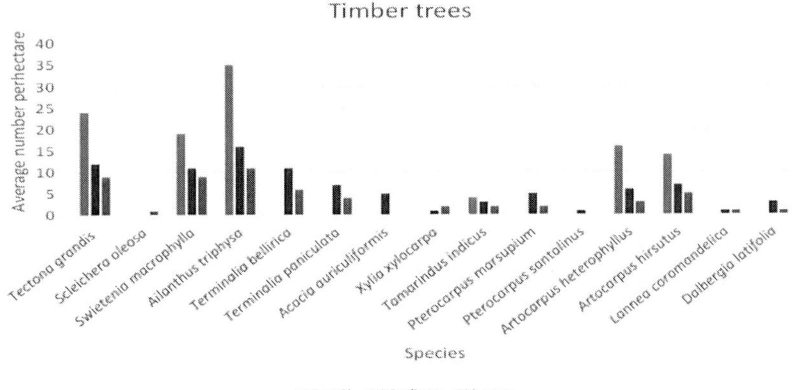

Fig.7. Density of timber trees in different (small, medium and large) categories of rural homesteads

Fodder species

The main fodder species includes *Mulberry, Glyricidia, Subabul* and *Artocarpus*. These are the main species that are usually seen in rural homegardens (Fig. 8).

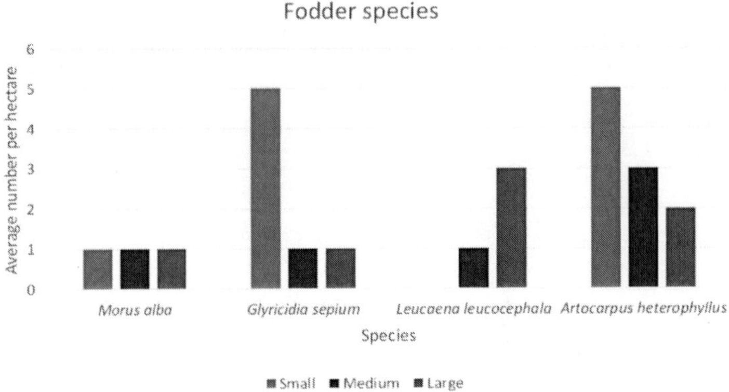

Fig. 8. Density of fodder trees in different (small, medium and large) categories of rural homesteads

In urban homegardens, mainly seen are *Gliricidia* and *Artocarpus*. In most of the cases *Gliricidia* are used as live fences and as support trees, but the population is higher in rural as the number of livestock are more in then (Fig. 9).

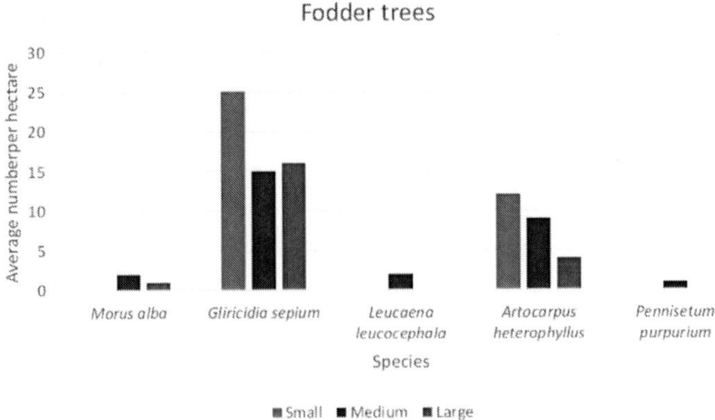

Fig. 9. Density of fodder trees in different (small, medium & large) categories of urban homesteads

Medicinal plants

In majority of the homesteads there are certain number of medicinal plants which are planted or of natural growth. In many cases the common species are Neem, Koovalam, Elenji, Nelli etc. in rural (Fig. 10) and urban homegardens (Fig. 11).

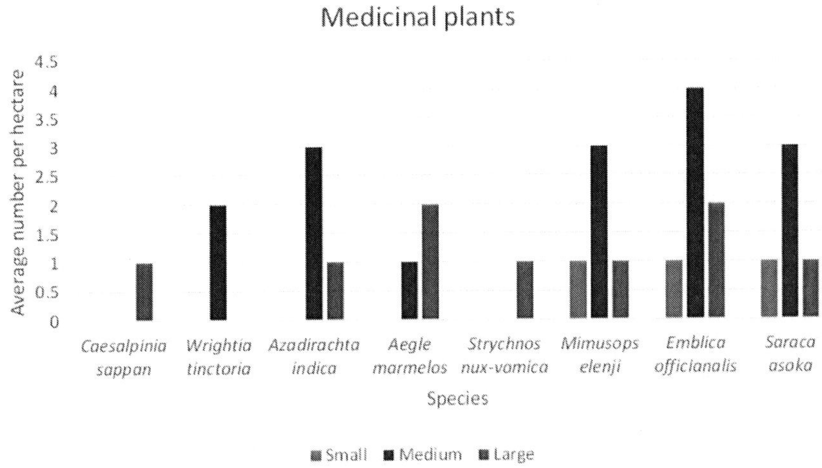

Fig. 10. Density of medicinal trees in different (small, medium and large) categories of rural homesteads

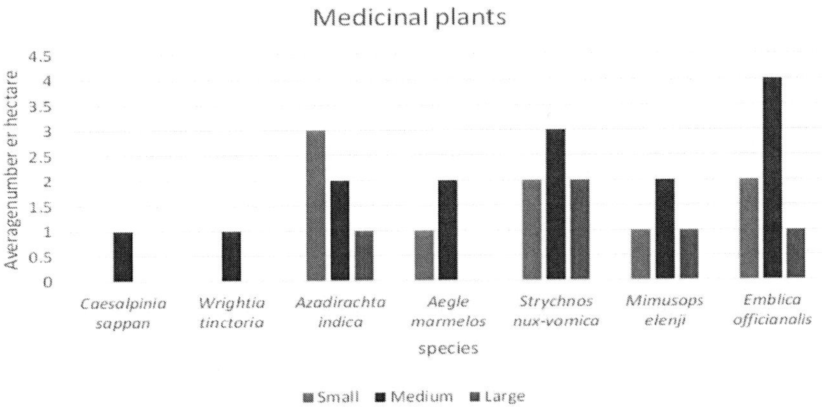

Fig. 11. Density of medicinal trees indifferent (small, medium and large) categories of urban homesteads

Horticultural plantation crops

There was huge availability of plantation crops mainly rubber and coconut. Majority of the farmers are growing rubber for finance in rural homegardens (Fig. 12). In these areas it can be mainly said as rubber based cropping system, it varies from rural to urban, in various aspects of holdings, scale of production etc.

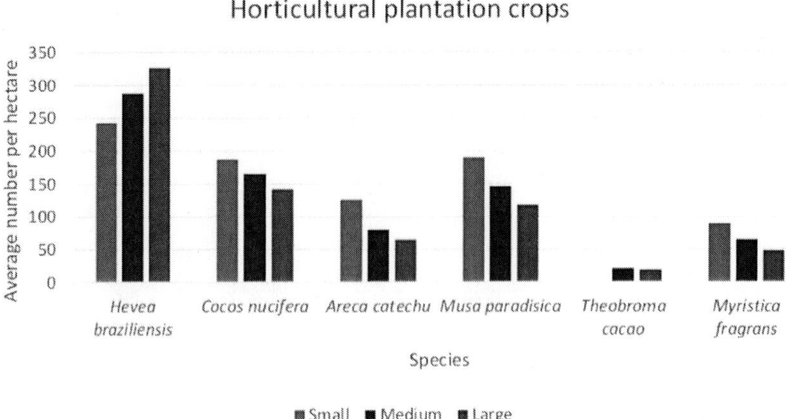

Fig. 12. Density of horticultural plantation crops indifferent (small, medium and large) categories of rural homesteads

The cultivation of these crops are increasing due to the shift from subsistence farming to cash crops in urban homegardens (Fig. 13). This lead to clearing of all other crops such as paddy, vegetables etc.

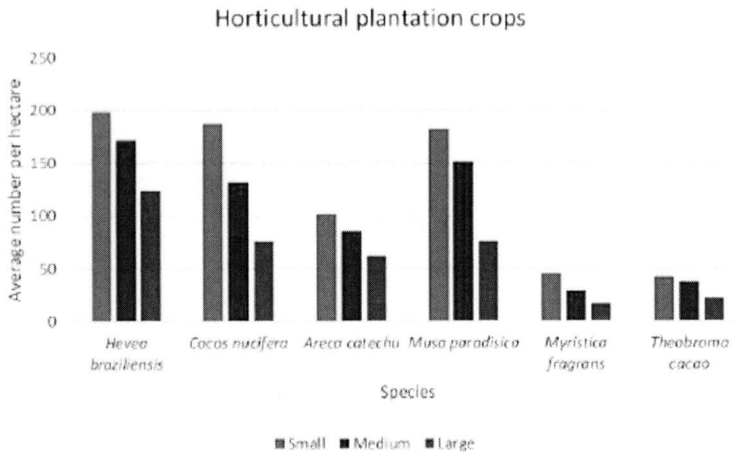

Fig. 13. Density of horticultural plantation crops indifferent (small, medium and large) categories of urban homesteads

Livestock

Livestock is found to be very less in all types of homegardens of the areas. Comparatively the density of most of the livestock is higher in small holdings in rural homegardens (Fig.14). The predominant livestock in this area is hen, poultry dominates.

Fig. 14. Density of livestock indifferent (small, medium and large) categories of rural homesteads

The reason for declining livestock population can be attributed to fodder scarcity and space limitations in urban homegardens (Fig. 15).

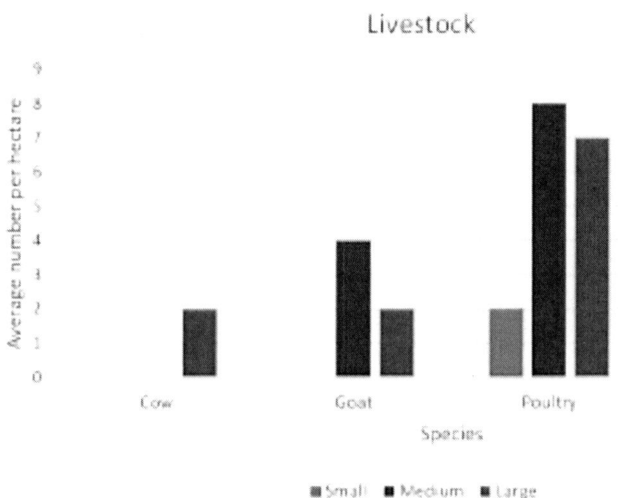

Fig. 15. Density of livestock indifferent (small, medium and large) categories of urban homesteads

Factors determining homegardens diversity

Homegardens with higher numbers of plant individuals seemed to have more species richness and diversity. A positive effect of woody plant abundance on species richness and diversity was also found in this study. The individual tree species with total volume of different (small, medium and large) categories of urban homesteads and rural homeagens were mentioned (Table 4). The comparison between rural and urban homegardens characteristics was mentioned (Table 5). The Diversity indices computed for tree species separately for different size categories suggest moderate changes. The Shannon index was higher for small (3.64),while lowest for large (2.873.23) holdings in rural homegardens while in same trend found in urban homegardens where the higher Shannon's diversity in small size (2.45) and the lowest for the large homegardens (1.97) (Table 6). The Simpson's dominance index showed modest variation among garden size classes (0.83 to 0.94 in rural homegrdens whereas, in urban homegardens varies from 0.82 to 0.92) (Table 7).

Arrangement of components

Spatial arrangement

Cash crops like coconut, arecanut, nutmeg and plantain followed a specified spacing for its proper growth and development. Coconut was planted at a spacing of 7 × 7m, while arecanut at a spacing of 2.4 × 2.4 m. Both coconut and arecanut were raised in rows. In nutmeg-coconut intercropping, coconut has a spacing of 7 × 7m and the quincunx pattern of planting was followed for nutmeg. Arecanut was planted at a spacing of 2.4 × 2.4m either in single or double row in between the rows of coconut. Tapioca is planted at a spacing of 0. 60 × 0.60 m in separate blocks. Elephant foot yam and colocasia followed a spacing of 0.90 × 0.90 cm, while 1.5 × 1.5 m spacing was followed for the monoculture of plantain. Turmeric and ginger are planted in blocks in raised earthen beds. In most of the homesteads, teak and mahogany are planted along the borders as wind breaks. Fruit trees and other miscellaneous trees were found scattered in the homegarden without any specific spacing. In the urban counterpart similar spacing are followed in the areas of land availability and in most of small and in some medium gardens they simply planted without any such rules according to the land availability. Most of the space will be occupied by the buildings and the remaining areas are planted accordingly.

Vertical stratification

A graphical representation of vertical and horizontal stratification of the homestead are shown in the profile diagram

- Upper storey: Upper storeys of small and medium homegardens were composed of coconut, arecanut, teak and jack. In addition to this species like teak, mahogany, *Casuarina*, jack fruit and tamarind were seen in the upper strata of large homegardens (Fig. 13). The MPTs and timber trees were found scattered in the homegarden without any specified planting geometry. These are more or less similar in case of rural and urban homesteads. In case of plantations the area will be completely covered by rubber canopy.

- Middle storey: *Mangifera indica, Myristica fragrans, Manilkara sapota, Annona squamosa, Moringa oliefera, Morinda tinctoria, Adathoda vasica and Tecoma stans* were found in the middle storey of small homegardens. *Nutmeg, Artocarpus communis, Mangifera indica, Michelia champaca and Azadirachta indica* were seen in the middle storey of medium homegardens. *Mangifera indica, Loobica, Gliricidia,* Sapota *and Annona squamosa* were found in the middle storey of large homegardens.
- Under storey: In the lower storey of small homegardens colocasia, cassava, turmeric and bilimbi were dominant. The medium sized homegardens consisted of plantain, citrus, vegetables and guava. The Large homesteads in addition to the above species consisted of vegetables, dioscorea, elephant foot yam etc. Napier grass and guinea grass were grown as cover crops in large homesteads mainly in coconut based inter crop systems.

Temporal arrangement

Majority of areas has been plantation land where there was only one component of rubber and in certain cases some understorey of shade loving species. In nutmeg-coconut association coincident arrangement of components can be seen. Nutmeg has been planted under the shade of coconut and arecanut. Pepper under coconut also followed the same arrangement. Annuals like ginger, turmeric, colocasia, elephant foot yam, and tapioca and plantain were intercropped under coconut and arecanut intermittently irrespective of the season.

Species preference

Farmers showed more preference for plantation crops like rubber, cocoa, coconut, plantain, arecanut and nutmeg. In general, crop composition varies considerably among the homesteads visited. Nutmeg and plantain intercropped with coconut was found in medium and large homesteads respectively. The area is mainly used for intercropping the coconut field with nutmeg, arecanut and plantain considering their economic advantages. In some cases there was vast areas of monocultures of rubber. There were species such as teak, vegetables etc. coming in various levels.

Table 4. Total volume of trees species in Karoor panchayat and Pala Municipality in each categories according to their species

Sl. No.	Species name (Rural homegardens)	Total Volume (m^3)	Sl. No.	Species name (Urban homegardens)	Total Volume (m^3)
	1. Small Homegardens			**1. Small Homegardens**	
i.	*Acacia mangium*	1.32	i.	*Ailanthus triphysa*	3.70
ii.	*Ailanthus triphysa*	2.26	ii.	*Annona reticulata*	2.15
iii.	*Artocarpus heterophyllus*	3.25	iii.	*Artocarpus heterophyllus*	6.69
iv.	*Casuarina equisetifolia*	1.21	iv.	*Casuarina equisetifolia*	2.46
v.	*Dalbergia latifolia*	0.58	v.	*Dalbergia latifolia*	7.51
vi.	*Mangifera indica*	2.31	vi.	*Garcinia gummi-gutta*	3.20
vii.	*Swietenia macrophylla*	0.89	vii.	Hevea braziliensis	10.66
viii.	*Syzygium cumini*	0.57	viii.	*Mangifera indica*	4.64
ix.	*Tectona grandis*	4.19	ix.	*Psidium gujava*	3.97
x.	*Terminalia bellerica*	0.54	x.	*Saraca asoka*	1.85
xi.	*Garcinia gummi-gutta*	0.42	xi.	*Swietenia macrophylla*	4.95
xii.	*Citrus limon*	0.21	xii.	*Syzygium cumini*	5.20
xiii.	*Annona reticulata*	0.28	xiii.	*Tectona grandis*	7.56
	2. Medium homegardens			**2. Medium homegardens**	
xiv.	*Terminalia bellerica*	7.21	xiv.	*Ailanthus triphysa*	3.97
xv.	*Annona squamosa*	0.24	xv.	*Ailanthus triphysa*	5.750
xvi.	*Artocarpus heterophyllus*	4.89	xvi.	*Annona squamosa*	0.72
xvii.	*Artocarpus hirsutus*	3.25	xvii.	*Artocarpus heterophyllus*	20.24
xviii.	*Briedelia retusa*	1.21	xviii.	*Artocarpus hirsutus*	8.64
xix.	*Dalbergia latifolia*	4.58	xix.	*Briedelia retusa*	2.65
xx.	*Delonix regia*	2.10	xx.	*Dalbergia latifolia*	5.30
xxi.	*Hopea parviflora*	4.3	xxi.	*Hevea braziliensis*	4.69
xxii.	*Hangifera indica*	12.60	xxii.	*Hopea parviflora*	9.32
xxiii.	*Morus alba*	2.10	xxiii.	*Mangifera indica*	7.16

(*Contd.*)

xxiv.	*Pouteria campechiana*	0.98	xxiv.	*Morus alba*	1.05
xxv.	*Swietenia macrophylla*	4.88	xxv.	*Pouteria campechiana*	2.51
xxvi.	*Syzygium aqua*	0.57	xxvi.	*Swietenia macrophylla*	17.26
xxvii.	*Tectona grandis*	8.90	xxvii.	*Syzygium aqua*	3.62
xxviii.	*Terminalia bellerica*	1.69	xxviii.	*Tectona grandis*	34.72
xxix.	*Xylia xylocarpa*	5.99	xxix.	*Terminalia arjuna*	7.85
xxx.	*Terminalia arjuna*	4.57	xxx.	*Terminalia bellerica*	11.21
xxxi.	*Aegle marmelos*	3.29	xxxi.	*Xylia xylocarpa*	4.52
	3. Large homegardens			**3. Large homegardens**	
xxxii.	*Ailanthus triphysa*	6.14	xxxii.	*Aegle marmelos*	2.36
xxxiii.	*Annona squamosa*	2.50	xxxiii.	*Ailanthus triphysa*	4.40
xxxiv.	*Artocarpus heterophyllus*	9.15	xxxiv.	*Annona squamosa*	0.95
xxxv.	*Artocarpus hirsutus*	8.49	xxxv.	*Artocarpus heterophyllus*	18.64
xxxvi.	*Briedelia retusa*	0.21	xxxvi.	*Artocarpus hirsutus*	10.31
xxxvii.	*Butea monosperma*	2.96	xxxvii.	*Briedelia retusa*	2.81
xxxviii.	*Dalbergia latifolia*	7.25	xxxviii.	*Butea monosperma*	1.23
xxxix.	*Delonix regia*	1.87	xxxix.	*Dalbergia latifolia*	3.58
xl.	*Hevea braziliensis*	18.24	xl.	*Delonix regia*	4.20
xli.	*Hopea parviflora*	5.30	xli.	*Hevea braziliensis*	7.57
xlii.	*Mallottus philippensis*	2.57	xlii.	*Hopea parviflora*	4.98
xliii.	*Mangifera indica*	9.53	xliii.	*Mallottus philippensis*	2.63
xliv.	*Morus alba*	2.14	xliv.	*Mangifera indica*	18.72
xlv.	*Pouteria campechiana*	0.98	xlv.	*Macranga peltata*	8.28
xlvi.	*Swietenia macrophylla*	8.02	xlvi.	*Pouteria campechiana*	1.02
xlvii.	*Syzygium aqua*	1.28	xlvii.	*Swietenia macrophylla*	21.22
xlviii.	*Tectona grandis*	18.89	xlviii.	*Syzygium aqua*	2.86
xlix.	*Terminalia bellerica*	9.87	xlix.	*Tectona grandis*	30.19
Xlx.	*Xylia xylocarpa*	8.75	Xlx.	*Terminalia bellerica*	3.44
Xlxi.	*Xylia xylocarpa*	4.52			

Table 5. Comparison between rural and urban homesteads in Kottayam district, Kerala

Characteristics	Rural homesteads	Urban homestead
Species diversity	Higher	Lower
Spatial arrangement	More dense	Less dense
Temporal arrangement	More well arranged	Less arranged
Income generation*	More	Less
Land holding	More	Less
Cost of land	High	Less
Family labours	More	Less

*In rural areas people are more dependent on these gardens for subsistence

Table 6. The Shannon's diversity index of timber species in the rural and urban homesteads of Kottayam district, Kerala

Categories	Rural homesteads	Urban homestead
Small	3.64	2.45
Medium	3.21	2.19

Discussion

The studied urban homegardens were important places for keeping plants used in daily life, especially for daily consumption. Edible species were one of the most important components in these homegardens. For this reason, many researchers across the world have reported on the abundance of food plants in homegardens (Abebe *et al.* 2006; Wezel and Ohl 2006).

Plant diversity of homegardens

Study carried out in the homegardens of Karoor panchayat (rural homegardens) revealed the presence of 84 plant species (excluding vegetables) with 7 food crops species, 19 fruit crops species, 6 cash crops species, 15 timber species, 13 MPTs, 7 spices and condiments species, 20 other miscellaneous species and 4 fodder species. In the Pala municipality was about 72 species with 5 food crops, 15 fruit trees, 5 cash crops, 15 timber species and many other miscellaneous species and 3 fodder species. Apart from these, livestock like cow, buffalo, goat, hens, duck, buffalo, apiculture and pisciculture. Nair and Sreedharan (1986) had reported 30 arboreal taxa from the homegardens of Kerala. Babu *et al.* (1992) however observed a total of 36 species of woody perennials from the homesteads of southern districts of Kerala. Eighty four species listed from rural homegardens and seventy two from urban homegardens show great diversity in their utility value which includes timber trees, MPTs, vegetables, spices and condiments, fruit trees, food crops, cash crops etc. Cash crops like coconut, rubber, arecanut and plantain dominates in most of the homegardens. The dominant timber trees include teak, mahagony and matti while the dominant fruit trees include tamarind, annona, garcinia, guava, citrus, etc. The density and diversity of spices and condiments were much less in the homegardens studied. Colocasia and cassava were the major food crops in the homegardens apart from rice which was not included in the study. Six species of animal components were also found during the study although their density was comparatively very less. This being the case, giving more attention to homegardens development could improve the food security of households in the future (Kebebew *et al.* 2011). Moreover, it should be noted that products were sustainably generated from these woody plant species (Kindt *et al.* 2004; Kehlenbeck *et al.* 2007). Sharing surplus fruit between kin and neighbors was common and important. This indicated that homegardens were also important food sources for the community, help into promote community food security (Kortright and Wakefield 2011).

Small homegardens

In small homegardens farmers are not interested in raising crops for income generation. They depend the homegardens mainly to satisfy their day to day needs. Even though the density of planting in small homegardens is very high due to space constraints they are interested in sustainable food production rather than economic benefits. Food crops and vegetables were more prominently cultivated in marginal homegardens. Cow dung is used as manure in the homegarden itself which helps to improve the productivity of the crops. The fruit and timber trees were less in marginal homegardens compared to bigger homegardens. The main timber tree seen in this type of homegarden is usually teak, that were planted along the boundaries of the land and not within the plot. Cash crops, rubber, coconut and arecanut are the only marketed items in most of the small homegardens.

Medium homegardens

The diversity is more in medium homegardens, compared to small homegardens. This is because in these types of homegardens, in addition to dietary needs the farmer is interested in generating income, such that they raise cash crops in all available areas. Teak, mahogany, and matti are the timber trees mainly seen in this type of homegardens. These species are good timber, but are also good wind breaks. Fruit trees were more prominent in this type of homegarden which include guava, mango, sapota, jack, garcinia etc. The increase in fruit trees is attributed to the reason that more area is available to the farmer for its cultivation and also because of the aesthetic value these trees and the fruits provides. The density of planting in medium homegardens is less compared to small home gardens.

Large homegardens

In large homegardens there were more number of species due to larger land holding, and also agricultural crops were cultivated extensively. Such types of homegardens are highly revenue oriented and cultivate multiple types of crops to generate income. In most of the homegardens cash crops like rubber, coconut, arecanut, and nutmeg were cultivated at a commercial basis. Less area was allocated for the cultivation of food crops. Cultivation of fodder grasses like Napier and guinea grasses fetches an addition income to the farmer. Good deal of land area is allocated for poultry farm in certain larger homegardens. It was also noted that the intensity of planting is very less in large homegardens as compared to small homegardens.

Structure of homegardens

Cropping pattern

Tree plantation-based intercropping systems are being popularized on a large scale around the globe especially in the tropical countries (Nair 1983). Smallholder farmers are inclined to intercrop young trees with intensively managed arable crops to confer nutrient and weeding benefits while gaining short-term returns. Tree-crop compatibility trials in pursuit of identifying ideal combinations for understorey productivity improvements were conducted at several locations in India. Some of the reported combinations include rubber-banana (Rodrigo *et al.* 2001), Leucenia leucocephala-pearlmillet (Bhatia and Kanaujia 2000), Hardwickia binata-based agrisilvicultural system involving sorgum-pearl millet, pigeon pea, soyabean and cotton (Khadse and Bharad 1996), coconut-based crop combinations for humid tropics such as coffee-banana, banana with ginger, turmeric,pineapple, papaya-pineapple, coffee-MPTs-pepper (Nair 1983; Lyiange *et al.* 1985). The different types of land use pattern that we observed are given below,

Rubber-based intercropping systems

Intercrops such as pineapple, banana and yam are cultivated during the immature phase of rubber whereas coffee, cocoa and medicinal plants are recommended for the mature phase (Siju *et al.* 2012; Rubber Board 2015).In smallholder rubber plantations world-wide, food crops such as rainfed rice, groundnut, cassava and plantains are grown between the rows of rubber trees during the initial years of thc plantations. In a study conducted by Lisha (2005), it was revealed that coffee and cocoa could be grown as intercrops in mature rubber without adverse effect on growth and yield of rubber. Though competition existed between the component crops for above ground resources the soil fertility was improved under the intercropped situation compared to monoculture rubber.

Medicinal plants such as *Piper longum*, *Sorbilanthus haenianus*, *Boerhaavia diffusa*, *Desmodium gangeticum* and *Pseudarthria viscid* are recommended for cultivation in the inter-space of rubber plantations in place of cover crops. Besides being ground cover their cultivation helps the farmer in generating subsidiary income. Being shade tolerant they can be cultivated even in mature rubber plantations. The introduction of medicinal plants in rubber plantations, as ground cover will enhance the biodiversity and sustain our herbal wealth (Mathewkutty 2002). The probability of adoption of intercropping was highest for three intercrops, banana cassava and pineapple (Rajasekharan and Veeraputhran 2002). Shade tolerant crops such as coffee and cocoa could be grown throughout the lifespan of rubber provided that temporary shades is given during the early stages (Rodrigo, 2002).A study under intercropping systems on the growth of *Curcuma longa* planted in one and three year-old *Hevea brasiliensis*, and four year-old *Citrus reticulata* plantations was conducted by Khaunkuab *et al.* (2008) in Thailand. Monocropping of *Curcuma longa* was also established. The results showed that the greatest length of leaves and number of plants/clump of para rubber, 72 and 216 after planting, were 7.64 cm, 12.21 cm, 20.05 cm, 38.33 cm, 1.01 and 3.43 chief/grove, respectively, and the size of rhizome, weight of rhizome and the average yield of 1-year-old para rubber were significantly higher than the data of mono-cropping systems.

Coconut-based intercropping systems

Multispecies and multi-storeyed cropping system ensures maximum resource capture and use, leading to higher yield per unit area of soil, water and light. Improvement in the soil properties and biological activities in the root region of coconut due to intercropping, results in the modification of soil environment for the benefit of the plant growth (Maheswarappa *et al.* 1998). Menon *et al.* (2013) observed leafy plants like *Boerhaavia diffusa*, *Cassia tora* and *Alternanthera sessilis* performed better in treeless open than coconut garden. *Boerhaavia diffusa, Cassia tora* and *Alternanthera sessilis* yielded 44, 94 and 57 % more in open than in shaded conditions respectively. Kumar *et al.* (2005) revealed that galangal yield was not affected by light interception levels by the upper-storey canopy up to 82 % of the open. Shade-tolerant and rhizomatic MAPs (Medicial Arometic Plants) can be grown on a longer-term basis in widely spaced plantations. Intercropping of medicinal plants in coconut (*Cocos nucifera*) stands is an age-old practice in India and other parts of south- and south-east Asia. These palms allow 30 % to 50 % of incident light to the underneath, which is ideal for some MAPs, including cardamom (*Elettaria cardamomum*). Galangal (*Kaempferia galanga*) is traditionally intercropped in mature coconut gardens in Kerala, India. Twelve year old coconut trees did not adversely affect the growth and yields of a number of medicinal species grown as intercrops compared to the yields in the open (Nair 1989).

Cashew-based intercropping systems

Performance of intercrops varies with age of tree and also the extent of canopy coverage. Major area under cashew is on undulated lands of hill slope. And hence, the number of intercrops cultivated are also comparatively lesser. Information on cashew based intercropping systems is very less. Adeyemi (1998) reported that weed suppression was better in intercropped than cashew alone. Weed suppression was best in plots carrying cashew + cassava and cashew + plantain + cassava mixtures with a 56 to 60 per cent reduction in the frequency of weeding per annum. Growth and yield of the annual food crops intercropped with newly planted cashew and two year old cashew were not significantly different to sole annual cropping as reported by Abeysinghe *et al.* (2003). Pawar and Sarwade (2006) reported soyabean sequenced with mustard intercropped in mango resulted the higher grain yield as compared to mango alone. Gill and Ajit (2006) recorded higher grain yield and straw yield in wheat intercropping with mango var. Amrapalli under Jhansi (Uttar Pradesh) conditions. Rathna and Swain (2006) recorded the maximum average fruit number obtained in mango based intercropping with french bean (80 fruits plant^{-1}) followed cowpea (75 fruits plant^{-1}) compared to the minimum obtained from sole mango (47 fruits plant^{-1}). Pawar *et al.* (2009) reported the higher growth and yield attributes in mango based intercropping systems with soya bean under Latur (Maharashtra) conditions.

Arecanut based intercropping systems

In these localities, arecanut based system had only subsidiary importance compared to coconut-based systems. Probably the lower returns from arecanut coupled with high harvesting cost and labour scarcity, deter the farmer from the large scale cultivation of arecanut. In certain cases fodder grasses were also raised as cover around the arecanuts, may be competitive interaction among the two crops less. Banana, nutmeg, black pepper, tapioca, vegetables, turmeric and ginger are grown successfully under arecanut. Farmers of Kerala prefer dioscorea, yams and tapioca as intercrops in arecanut plantation (Khader 1982). Although the practice of mixed cropping a perennial tree/shrub such as citrus, jack and coconut with arecanut is common, the cultivation of cacao is recent (Bhat 1979).

Other tree-based intercropping system with rhizomatous spice crops

Bavappa (1990), based on the trials conducted in India, Philippines, Sri Lanka, Malaysia and West Samoa reported that, tuber crops, rhizomatous spice and chilli are found to be more profitable crops under different agro climatic conditions in coconut plantation. Jayachandran *et al.* (1992) observed significant differences in the performance of turmeric grown under different shaded (25, 50 and 75 percent) conditions. The turmeric crop grown under 50 per cent shade recorded maximum plant height (44.30 cm) followed by 25 per cent (43.30 cm) and 75 per cent shade (42.08 cm) whereas crop grown under open condition (full light) recorded the lowest plant height (32.50 cm). Sujatha *et al.* (1994) recorded the maximum number of leaves (27.75) in ginger cv. PGS-10 when it was grown as intercrop with coconut. Hegde (1998) recorded maximum number of leaves per clump (13.75) in ginger variety Suprabha under arecanut based intercropping system when compared to sole crop in open area (44.30). Latha *et al.* (1995) reported that, plant height, number of leaves, leaf area index and fresh weight was higher in turmeric cultivar under *Leucaena*. Further, they indicated the need for standardising optimum light requirement for each cultivar for higher yield. Bandyopadhyay *et al.* (2003) recorded maximum plant height (157.3 cm), leaf production (11.4), leaf length (75.3 cm) and leaf breadth (19.1 cm) in turmeric cv. Sugandhum when intercropped with young arecanut plantation followed by ACC-360 and Roma. There was no adverse effect of turmeric on growth of arecanut plant. Kumar (2004) reported higher plant height and number of tillers was produced by turmeric (69.33 cm and 6.69, respectively) under intercropping compared to sole cropping (49.66 cm and 5.29, respectively). Similarly, in ginger higher plant height and number of tillers (11.33) in tamarind plantation compared to open area (26.50 cm height and 7.06 tillers). Significantly higher plant height (43.66 cm) and number of tillers per clump (6.06) were recorded by turmeric under intercropping compared to sole cropping (34.46 cm and 5.13, respectively) and in ginger higher plant height (47.10 cm) and number of tillers per clump (9.06) under tamarind based intercropping situation compared to open area (32.26 cm height and 7.46 tillers) as reported by Kumar (2005).

Economic benefits

Homegardens are economically and biologically sustainable systems. This study used inventories, survey information and market data to estimate the productivity of 60 homegardens in Kottayam. All homegardens except small homesteads were found to be economically profitable and also better economic utility. Profit value of gardens is based on size of farm land and the scale of products produced in homegardens significantly with cash crops and food crops. Homegardens can contribute to household income in several ways. The household may sell products in the homegarden that include fruits, vegetables, animal products and other valuable materials. Economic benefits were determined based on the information from farmers as well as inventory of the gardens. Market values were determined based on existing prices. The most important contributors to the economic profit were rubber, coconut, arecanut, nutmeg and plantain and the distribution of profit varied across garden sizes.

In addition to other benefits, the range of products in homegardens significantly improves the family's financial status. Homegardens can contribute to the household with cash crops as well as food crops (Hoogerbuugge and Fresco 1993). Homegardens can contribute to household income in several ways. The household may sell products in the homegarden including fruits, vegetables, animal products and other valuable materials. Coconuts are usually harvested for nuts every three months. Integrated farming with crop and livestock increased family income by more than 70% and reduced the cost of production by 50% as compared to monoculture. The rubber plantations are the most favoured ones of these areas. The structural and functional diversity of components in the model ensure a high level of resource use efficiency meeting the multiple demands of the farmer (Geetha 2004).

Sustainability of homegardens at threat

Homegardens are undergoing extensive changes that deteriorate their intrinsic characteristics. New government policies and population pressures leads to changes in agriculture scenario and the high market-orientation, do exert considerable pressures on homegardens (Kumar and Nair 2004). Arrival of high-input- intensive agricultural practices and the substitutes for the homegarden products resulted to face serious setbacks in content and quality of these traditional systems. There are plenty of reports that deals with alarming rates of conversion of homegardens to other systems more prominently to monoculture based commercial systems (Kumar and Nair 2004; Abdoellah *et al.* 2006). In a study, out of 30 homegardens studied from Kerala 50% still continues traditional features, whereas 33% adopted modern practices. The modernization process includes a gradual concentration on a limited number of cash crop species, a decrease of the tree/shrub diversity, an increase of ornamental plants, a gradual homogenization of homegarden structure and an increased use of external inputs (Peyre *et al.* 2006). Commercialization, switched cultivation systems more specialized and rural people turned employing in non-primary production activities. This results changes in many rural farming systems in general, and homegardens in particular. Loss of 27 varieties of mango during a 60-year period reported from West Java, Indonesia (Soemarwoto 1987) and decline in diversity due to fragmentation in Chagga homegardens of Tanzania (Rugalema *et al.* 1994) are examples. It has raised the question as to whether homegardens are becoming extinct (Kumar and Nair 2004). The uprising fears that the traditionally diverse and ecologically sustainable homegardens will gradually disperse into monospecific agricultural practices with undetermined sustainability are will be a setback to the earlier concepts of homegardens (Soemarwoto 1987).

Homegardens diversity

The Shannon index was higher for small (3.64),while lowest for large (2.873.23) holdings in rural homegardens while in same trend found in urban homegardens where the higher Shannon's diversity in small size (2.45) and the lowest for the large homegardens (1.97). A similar trend was observed for the Kerala homegardens (Kumar *et al.* 1994). Peri-urban homegardens of Neyyatinkara Municipality area, southern Kerala reported that Shannon's diversity index was 3.77, 3.23 and 3.87, respectively for large, medium and small homegardens and respective value for Simpsons Dominance Index 0.92, 0.89 and 0.81 (Kunhamu *et al.* 2015). Similar reports also have been observed from homegardens of North-East India (Sahoo *et al.* 2010) with a decrease in the Shannon index with increases in size of holdings (3.28, 2.59 and 2.98, respectively for small, medium and large land holdings). The reports from central Kerala suggest large (> 0.4 ha) and small(< 0.4 ha) homegardens having Shannon index value of 2.27and 2.38, respectively (Saha *et al.* 2009). Pandey *et al.* (2006) observed Shannon dominance index for homegarden (1.38) and home forest gardens (3.168) from South Andaman.

The Simpson's dominance index showed modest variation among garden size classes (0.83 to 0.94 in rural homegrdens whereas, in urban homegardens varies from 0.82 to 0.92). Other studies reported from Kerala suggested a lower Simpson index within the range of 0.251 to 0.739 for different thaluks of Kerala (Kumar *et al.* 1994). Probably, these lower values of diversity indices could be attributed to the fact that those studies considered only tree species for computing diversity indices while the present study included total plant diversity. A high value (0.834) was reported in different size categories in a village of Thiruvavanthapuram. Pandey *et al.* (2006) reported 0.40 and 0.69 in homegardens and home-forest-gardens, respectively. Kumar (2011) reported a reduction in Simpson index with increase in holdings. Central Kerala had a value of 0.64, 0.41 and 0.46, respectively for small, medium and small homegardens. Simpson's diversity index is a measure of diversity, which takes into account both richness and evenness.

References

Abdoellah OS, Hadikusumah HY, Takeuchi K, Okubo S, Parikesit (2006) Commercialization of homegardens in an Indonesian village: vegetation composition and functional changes. In: Kumar BM, Nair PKR (eds) Tropical homegardens: A time-tested example of sustainable agroforestry, Springer, Dordrecht, pp 233-250.

Abebe T, Wiersum KF, Bongers F, Sterck FJ(2006) Diversity and dynamics in homegardens of southern Ethiopia. In: Kumar BM, Nair PKR (eds) Tropical homegardens: time-tested example of sustainable agroforestry. Springer, Dordrecht, pp. 123-142.

Abeysinghe DC, Sangakkkara UR, Jayasekera SJBA (2003) Intercropping of young cashew (*Anacardium occidentale* L.) and its effects on crop productive and land utilization. Tropical Agric Res 15:10-19.

Adeyemi AA (1998) Effects of intercropping on weed incidence in cashew (*Anacardium occidentale*) plantations. Nigerian J Tree Crop Res 2(1):83-94.

Akinnifesi FK, Sileshi GW, Ajayi OC, Akinnifesi AI, De Moura EG, Linhares JF, Rodrigues I (2010) Biodiversity of the urban homegardens of São Luís city, Northeastern Brazil. Urban Ecosyst 13:129-146.

Albuquerquea UP, Andradeb LHC, Caballeroc J (2005) Structure and floristics of homegardens in Northeastern Brazil. Journal of Arid Environments 62:491-506.

Arifin HS, Munandar A, SchultinkG, Kaswanto RL (2012) The role and impacts of small-scale, homestead agroforestry systems ("pekarangan") on household prosperity: An analysis of agro-ecological zones of Java, Indonesia. International Journal of AgriScience 2(10): 896-914.

Babu KS, Jose D, Gokulapalan C (1982) Species diversity in Kerala homegarden. Agroforest Today 4(3):15.

Balooni K, Gangopadhyay K,Kumar BM (2014) Governance for private green spaces in a growing Indian city. Landscape and Urban Planning 123:21-29.

Bandyopadhyay A, Ghosh D, Hore JK (2003) Performance of different turmeric germplasm as intercrop in young arecanut plantation. The Orissa J Hort 31(1):22-25.

Bhat KS (ed). (1979) Multiple Cropping in Coconut and Arecanut Gardens. Tech Bull No 3, CPCRI, Kasaragod, India. pp. 11.

Bhatia KS, Kanaujia, VK (2000) Growth response of pearl millet and barley to *Leucaena leucocephala* based agroforestry system. Indian J Agrofor 2:33-36.

Das T, Das AK (2005) Inventorying plant biodiversity in homegardens: A case study in Barak Valley, Assam, North East India. Current Science 89(1):155-163.

Galluzzi G, Eyzaguirre P, Negri V (2010) Home gardens: neglected hotspots of agro-biodiversity and cultural diversity. Biodiversity and Conservation 19(13):3635-3654.

Gamble JS (1915) Flora of the Presidency of Madras- Bishen Singh and Mahendra Pal Singh, Dehradun, Vol I-III, pp. 2017.

Geetha V (2004) Shade response and nutrient requirement of common rainfed intercrops of coconut. M. Sc. (Ag.) thesis, Kerala Agricultural University, Thrissur, pp. 216.

Geetha VS, Appavoo MR, Vinuba AA, Jeeva S (2015) Species richness of edible plants grown in homegardens of the foothills of Western Ghats. Science Research Reporter 5(2): 187-191.

George PS, Chattopadhyay S (2001) Population and land use in Kerala. In: Growing Populations, Changing Landscapes: Studies from India, China and the United States. National Academy of Sciences, Washington DC, pp. 79-106.

Gill AS, Ajit(2006) Intercropping of wheat in mango (*Mangifera Indica* Linn.). Indian J Forestry 29(4):417-420.

Hegde NK (1998) Crop diversification studies in arecanut (*Areca catechu* L.) plantation. Ph. D. Thesis, Univ Agri Sci Dharwad, Karnataka.

High C, Shackleton CM (2000) The comparative value of wild and domestic plants in homegardens of a South African rural village. Agrofor Syst 48:141-156.

Hoogerbrugge I, Fresco LO (1993) Homegarden system: Agricultural characteristics and challenges. Intl Inst Environ Devepment Gaxekeeper Series pp.39.

Hooker JD (1872) Flora of British India. L Reeve, London 7: 1872- 1897.

Huai H, Xu G, Wen G, Bai W (2011)Comparison of the homegardens of eight cultural groups in Jinping Country, Southwest China. Economic Botany 65(4):345-355.

Jayachandran BK, Meerabai M, Mammen MK, Mathew KP(1992) Influence of shade on growth and productivity of turmeric. Spice India 3(4):2-9.

Kabir ME, Webb EL (2008) Floristics and structure of southwestern Bangladesh homegardens. International Journal of Biodiversity and Management 4:54-64.

Kebebew Z, Garedew W, Debela A (2011) Understanding homegarden in household food security strategy: Case study around Jimma, southwestern Ethiopia. Research Journal of Applied Sciences6(1):38-43.

Kehlenbeck K (2007) Rural homegardens in Central Sulawesi, Indonesia: An example for a sustainable agro-ecosystem? Georg-August-University.

Kehlenbeck K, Arifin HS, Maass B (2007) Plant diversity in homegardens in a socio-economic and agro-ecological context. In: TscharntkeT, LeuschnerC, ZellerM, GuhardjaE, BidinA (eds) The Stability of Tropical Rainforest Margins, Linking Ecological, Economic and Social Constraints of Land Use and Conservation. Springer Verlag, Berlin.pp.297-319.

Khaaunkaub N, Suwunnalert S, Chotipha R, Namphech L, Sinphai J, Chaiburi J (2008)Study on intercropping system between turmeric (*Curcuma longa* Linn.) in para rubber and fruit plantation on upper southern area. Proceedings of the 46th Kasetsart University Annual Conference, Kasetsart, pp. 373-379.

Khadse VM, Bharath GM (1996) Performance of annual crops under canopy *of Hardwickiabinnata* in agroforestry systems. J Soil and Crops 6:151-153.

Kindt R, Simons AJ, van Damme P (2004) Do farm characteristics explain differences in tree species diversity among Western Kenyan farms? Agroforestry Systems 63(1):63-74.

Kortright R, Wakefield S (2011) Edible backyards: A qualitative study of household food growing and its contributions to food security. Agriculture and Human Values 28(1): 39-53.

Kumar BM (2011)Species richness and aboveground carbon stocks in the homegardens of central Kerala, India. Agricultural, Ecosystems and Environment 140:430-440.

Kumar BM, Nair PKR (eds) (2006) Tropical homegardens: a time-tested example of sustainable agroforestry. Advances in Agroforestry, Netherlands, Springer.

Kumar BM, Nair PR (2004) The enigma of tropical homegardens. Agroforestry Systems 61(1-3): 135-52.

Kumar BM, George SJ, Chinnamani S (1994) Diversity structure and standing stock of wood in the homegardens of Kerala in peninsular India. Agroforestry Systems 25:243-262.

Kumar BM, Kumar SS, Fisher FF (2005) Galangal growth and productivity related to light transmission in single –strata, multistrata and no-over-canopy systems. J New Seeds7(2): 111-126.

Kumar RD (2005) Intercropping studies in tamarind (*Tamarindus indica* L.) plantation. M.Sc. thesis, Univ Agri Sci Dharwad, Karnataka.

Kumar V (2017) Importance of Homegardens Agroforestry System in Tropics Region. In: Prithwraj Jha (ed) Biodiversity, Conservation and Sustainable Development. New Academic Publishers, New Delhi. Vol. 2, Issues and Approaches (In press).

Kumar V, Tripathi, AM (2017) Vegetation composition and functional changes of tropical homegardens: Prospects and challenges. In: Gupta SK (ed), Agroforestry for increased production and livelihood security in India, published in NIPA, New Delhi. pp 475-505.

Kumar V, Niyas P, Unnithan SR (2016) Medicinal plants based on intercrops homegardens system. Van Sangyan 3(7):20-26.

Kunhamu TK, Ajeesh R, Kumar V (2015) Floristic analysis of peri-urban homegardens of southern Kerala, India. Indian Journal of Ecology 42(2):300-305.

Latha P, Giridharan MP, Naik BJ (1995) Performance of turmeric (*Curcuma longa* L.) cultivars in open and partially shaded conditions under coconut. J Spices and Aromatic Crops 4:139-144.

Lisha C (2005) Competitive interactions in a rubber based agroecosystem with coffee and cocoa as intercrops, Disseration, M.Sc. Centre for Plantation Development Studies, University of Calicut, Calicut, India, pp. 66.

Lyiange MDS, Thejawani KG, Nair PKR (1985) Intercropping under coconut in Srilanka. Agrofor Syst 2:215-228

Maheswarappa HP, Hegde MR, Dhanapal R, Biddappa CC (1998) Mixed farming in coconut garden - Its impact on soil physical, chemical properties coconut nutrition and yield. J Plantn Crops 26:139-143.

Manju CN, Rajesh KP, Madhusoodanan PV (2008) Checklist of the bryophytes of Kerala, India. Tropical Bryology Research Reports 7:1-24.

Mathewkutty T (2002) Medicinal plants as ground cover in rubber plantations In: Jacob C Kuruvilla (ed) Global competitiveness of Indian rubber plantations industry: Rubber Planters Conference. Rubber Research Institute of India, Kottayam, India, pp. 89-92.

Mendez VE, Lok R, Somarriba E (2001) Interdisciplinary analysis of homegardens in Nicargua: Micro-zonation, plant use and socioeconomic importance. Agroforestry Systems 51: 85-96.

Menon MV, Krishnankutty J, Pushpalatha PB (2013) Identification and evaluation of selected underutilized leafy plant species in the homesteads of Kerala, India. In: II International Symposium on Underutilized Plant Species: Crops for the Future - Beyond Food Securitypp. 625-630.

Milow P, Malek S, Mohammad N, Ong H (2013) Diversity of plants tended or cultivated in Orang Asli homegardens in Negeri Sembilan, Peninsular Malaysia. Human Ecology 41(2): 325-331.

Nair MA, Sreedharan C (1986) Agroforestry farming system in the homesteads of Kerala, Southern India. Agrofor Syst 4:339-363.

Nair PKR (1983)Soil Productivity of Aspects of Agroforestry. International Council for Research on Agroforestry. Nairobi, Kenya, pp. 85.

Nair PKR (1989) The role of trees in soil productivity and protection. In: Nair PKR (ed)Agroforestry systems in the Tropics. Kluwer Academic Publishers, Dordrecht, Netherlands, pp. 567-589.

Nair PKR (1993) An Introduction to Agroforestry. Kluwer, Dordrecht, pp.499.

Pandey CB, Lata K, Venkatesh A, Medhi RP (2006) Diversity and species structure of home gardens in South Andaman. Tropical Ecology 47:251-258.

Pawar HD, Sarwade SD, Jadhav KT, Aglawe BN(2009) Effect and performance of intercrops in mango orchards. J Maharashtra Agric Univ34 (2):237-238.

Pawar HD, Sarwade SD (2006) Evaluation of performance of intercrops in mango orchards. J Soils and Crops 16 (2)352-354.

Perfecto I, Vandermeer J (2008) Biodiversity conservation in tropical agroecosystems. Ann N Y Acad Sci 1134:173-200.

Perrault A, Coomes O (2008) Distribution of agrobiodiversity in home gardens along the Corrientes River, Peruvian Amazon. Economic Botany 62:109-126.

Peyre A, Guidal A, Wiersum KF, Bongers F (2006) Dynamics of homegarden structure and function in Kerala, India. Agroforestry Systems 66:101-115.

Pulido MT, Pagaza-Calderon EM, Martinez-Balleste A, Maldonado-Almanza B, Saynes A, Pacheco RM (2008) Home gardens as an alternative for sustainability; Challenges and perspectives in Latin America. In: Albuquerque UP, Alves-Ramos M. (eds) Current Topics in Ethnobotany. Research Signpost, India. pp. 55-79.

Rajasekharan P, Veeraputhran S (2002) Adoption of intercropping in rubber smallholdings in Kerala, India: A tobit analysis Agrofor Syst56(1):1-11.

Rathna S, Swain SC (2006) Performance of intercrops in mango orchard in Eastern Ghats Land Zone of Orissa. Indian J Dryland Agri Res and Development 21 (1):12-15.

Rico-Gray V, Garcia FJG, Chemas A, Puch A, Sima P (1990) Species composition, similarity, and structure of Mayan homegardens in Tixpeual and Tixcacaltuyub, Yucatan, Mexica. Economic Botany 44:470-487.

Rodrigo VHL (2002) Rubber-based intercrops for improved productivity: The best available today In: Jacob C Kuruvilla (ed)Global Competitiveness of Indian Rubber PlantationsIndustry:Rubber Planters Conference. Rubber Research Institute of India, Kottayam, India. pp. 24-25.

Rodrigo VHL, Stirling CM, Teklehaimanot Z, Nugawela A (2001) Intercropping with banana to improve fractional interception and radiation use efficiency of immature rubber plantations. Field Crops Research69:237-249.

Rubber Board (2015) Available: http://www.rubberboard.org.in.

Rugalema GH, Johnsen FH, Rugambisa J (1994) The homegarden agroforestry system of Bukoba district, North-Western Tanzania. Agroforest Syst26:205-214.

Saha KS, Nair PKR, Nair DV, Kumar BM (2009) Soil carbon stock in relation to plant diversity of homegardens in Kerala, India. Agroforestry Systems 76:53-65.

Sahoo UK, Rocky P, Vanlalhriatpuia K, Upadhyaya K (2010) Structural diversity and functional dynamism of traditional home gardens of North-East India. The Bioscan 1:159-171.

Sankar S, Chandrashekara UM (2002) Development and testing of sustainable agroforestry models in different agroclimatic zone of Kerala with emphasis on socio-cultural, economic, technical and institutional factors affecting the Sector. KFRI Research Report 234. Kerala Forest Research Institute, Peechi, Kerala.

Sasidharan N (2004) Biodiversity documentation for Kerala Part 6. Flowering Plants. Kerala Forest Research Institute, Peechi, Kerala.

Sasidharan N (2012) A Digital database of Flowering plants of Kerala Ver. 2. Kerala Forest Research Institute, Peechi, India

Scales BR, Marsden SJ (2008) Biodiversity in small-scale tropical agroforests: a review of species richness and abundance shifts and the factors influencing them. Environ Conserv 35:160-172.

Shannon CE, Wiener W (1963) The Mathematical Theory of Communication. Urban University Press, IL, USA.

Siju T, George KT, Lakshmanan R (2012)Changing dimensions of intercropping in the immature phase of natural rubber cultivation: a case study of pineapple intercropping in central Kerala. Rubber Science 25(2):164-72.

Simpson EH (1949) Measurement of diversity. Nature163:688.

Soemarwoto O (1987) Homegardens: a traditional agroforestry system with a promising future. In: Stepper HA, Nair PKR (eds) Agroforestry: A decade of development, ICRAF, Nairobi.pp. 157-170.

Soemarwoto O, Conway GR (1992) The Javanese homegarden. J Farming Syst Res Exten 2(3): 95-118.

Sujatha K, Latha P, Nair NK (1994) Evaluation of ginger varieties as an intercrop in coconut gardens. South Indian Hort 42 (5):339.

Taylor JR, Lovell ST (2014) Urban home gardens in the Global North: A mixed methods study of ethnic and migrant home gardens in Chicago, Illinois. Renewable Agriculture and Food Systems First View:1-11.

Vlkova M, Polesny Z, Verner V, Banout J, Dvorak M, Havlik J, Lojka B, Ehl P, Krausova J (2010) Ethnobotanical knowledge and agrobiodiversity in subsistence farming: Case study of home gardens in Phong My commune, central Vietnam. Genetic Resources and Crop Evolution 58:629-644.

Wezel A, Ohl J (2006) Homegarden plant diversity in relation to remoteness from urban centers: A case study from the Peruvian Amazon region. In: Kumar BM,Nair PKR (eds). Tropical Homegardens: A Timetested Example of Sustainable Agroforestry. Springer, Netherlands.pp. 143–158,

Wiersum KF (2006) Diversity and change in homegarden cultivation in Indonesia. In: Kumar BM, Nair PKR (eds). Tropical Homegardens: A Time-tested Example of Sustainable Agroforestry. Springer, Netherlands.pp. 13-24.

Yang L, Ahmed S, Steppp JR, Mi K, Zhao Y, Ma J, Liang C, Pei S, Huai H, Xu G, Hamilton AC, Yang Z, Xue D (2014) Comparative homegarden medical ethnobotany of Naxi healers and farmers in Northwestern Yunnan, China. Journal of Ethnobiology and Ethnomedicine 10(1):6.

35

Agroforestry: A Multipurpose Multi-Storeyed Renewable Plant Treasure with Medicinal Values Giving Multi-Returns for Livelihood and Benefits over Climate Change

[1]S. Panda, [2]P. K. Dhara, [3]S. Sarkar and [4]N. C. Das*

[1]Jhargram, BCKV, PaschimMedinipur – 721 507, West Bengal, India
[2]Department of Soil Water Conservation, F/Ag, Bidhan Chandra Krishi Viswavidyalaya, Mohanpur, Nadia – 741 252, West Bengal, India
[3]Jhargram, BCKV, PaschimMedinipur – 721 507, West Bengal, India
[4] Bidhan Chandra KrishiViswavidyalaya, Mohanpur, Nadia – 741 252 West Bengal, India

Introduction

Agriculture expanded and forest dwindled. That eventually occurred with the expansion of civilization. Clearing of forests, granted in past, continued in modern times. Cutting of forests reached to such an alarming level that thinking of people turned back to create more forests. Sometimes over extraction of forest resources made forests unstable to yield perpetually according to necessity in future for which forests are turned unsupportive for livelihood and people tried for innovative means to enhance forest productivity in future (Nair 1993). Also with time agriculture faced the similar unfortunate condition towards intensive cultivation. So, peoples' thinking turned back in agriculture how to make it stable and more supportive for livelihood year after year. Culturally human societies are connected both with agriculture and as well as forests from the time immemorial. It is a universal truth that Forest is mother of all livelihoods. As agriculture and civilisation expand, community people retain their primitive close association with various plant and trees in their homestead garden, around temples, houses, community lands for their regular cultural practices including various rituals. So, association of agriculture and forests is deep seated in human feelings (Shiva and Mathur 1996; Blatter and Millard 1954; Talbot 1984). Thus, forests are still associated with modern farming community and civilisation as we see planting of trees by the road sides and along the boundary of agricultural fields also. As the agricultural fields extend around a tree, sometimes that is left out to grow bigger. Though except fruit trees farmers hardly prefer to plant trees within their agricultural plots (Dhara *et al.* 2016). In present days, agroforestry system can resolve the dilemma of intensification of agriculture as well as increasing areas under forests. Agroforestry is deliberate use of woody perennials like trees, shrubs, palms and bamboos on the same land management unit as agricultural crops, pastures and /or animals. This may involve both vertical and horizontal mixed spatial arrangements at the same time or sequentially over time on the same piece of land (FAO 2015). The field of scientific agroforestry has rapidly developed since the early 1980s. Presently both scientific and development communities are quite aware about it. Such developments in agroforestry in recent years have resulted in establishing agroforestry systems in a concerted scientific and technical manner for land development. Agroforestry has come to lime light recently for concern over sustainability of agricultural development in the context of rapid depletion of natural resource base including soil fertility and increase in global warming and equally addressing food security (Nair 1993). Agroforestry is creation of forests outside forest areas including practice of agriculture is a quiet push to grow crops on the inter-space of trees. So, this system has multipurpose values to give return in the form of agricultural produce as well as wood production from multi-storeyed plants and trees grown on same piece of land. As all plants have their medicinal values with some plants and trees, having especially enriched with such qualities, extraction of medicinal resources from agroforestry plants and trees can add to more return from same sort of agroforestry systems. Growing trees can sequester carbon from atmosphere which is of much interest to resolve global warming. Again all those are naturally renewable. In this way agroforestry including its medicinal resources can act as multi-returns support for livelihood in mitigating climate change.

Study area and experimental details

Since 1983 various agroforestry systems had been tried under All India Coordinated Research Project (AICRP) on Agroforestry, Jhargram, Regional Research Station (Red and Laterite Zone) under Bidhan Chandra Krishi Viswavidyalaya (BCKV), West Bengal. Jhargram (22.4370°N, 86.9924°E) is located in the district of Paschim Medinipur in West Bengal at an elevation of 78.77 m above mean sea level under humid and sub-humid climate. Those field trials were done to identify some profitable agroforestry systems (Dhara *et al.* 2016; Dhara 2016).

Additionally those systems would be the source of many other important subsidiary beneficial materials including medicinal resources. Possibilities of harnessing medicinal resources from those agroforestry systems have been examined from literatures (Kirtikar *et al.* 1918; WAC 2009; NMPB 2016) to find out other possible economic returns which would be of further benefits for livelihood support from same agroforestry systems. Such possible activities would be able to yield more benefits over the sole tree or crop culture.

Apart from those benefits raising trees on crop fields could sequester atmospheric carbon. All those approaches are quantified from the results of different field trials and available market price of different promising produces from each agroforestry system including possible medicinal materials out of those systems. Estimation of carbon sequestration was done on the basis of biomass produced from tree (Chaturvedi and Khanna 1982; Chaturvedi and Khanna 2000) and crops. Above ground biomass of those trees and alley crops were estimated using standard procedures. 50 per cent of biomass was estimated to be carbon sequestered from atmosphere (Birdsey 1992; Paladinic *et al.* 2009).

Soil testing for physical and chemical parameters was done following standard laboratory methods (Piper 1966; Jackson 1962; Ghosh *et al.* 1983; INM Division 2011; USDA 1993). Interactions of various agroforestry parameters were analysed following standard procedures (Kumar and Nair 2011; Batish *et al.* 2008).

Observations

Out of the field trials eighteen different agroforestry base systems had been identified and were in the process of standardisation for other locations under humid and sub-humid climate in West Bengal. Those agroforestry systems were found to be economically beneficial (Dhara *et al.* 2016; Dhara 2016).

Soils under field trials were degraded with low water holding capacity and highly erodible, poor inorganic carbon and N, P and K status. Inconsistent annual monsoon rainfall through years makes the agriculture naturally unstable (Dhara *et al.* 2016).

Suitability studies of different agroforestry systems

Classification of agroforestry systems can be initially done according to (a) the components present like (i) Agri-silviculture (i.e. tree based alley cropping), (2)

Silvi-pasture (i.e. trees and animals), (3) Agri-silvipasture (i.e. trees with crops and animals). Next level classification is based on arrangements of components in space and time. In spatial arrangements tree and crop and / or animal components are grown as mixtures, with trees planted over the whole land unit. As an alternative to that spatial arrangement trees are systematically planted in rows or as elements like field boundaries or fodder reserves.

Based on such spatial arrangements eighteen different Agroforestry System Bases were identified (Table 1) on degraded red and lateritic soil under humid and sub-humid climate in West Bengal. Along with those system bases various alley crops could be successfully cultivated with economic profitability (Table 2). Considering those system bases to be grown with a fruit tree (Mango/ Guava/Sweet orange/Ber/ Aonla) along with at least one alley $^{18}C_3$ combinations, i.e. 816 Agroforestry Systems could be successfully formulated which would be economically viable and those can be confidently innovated among farming community according to their choice for their assured livelihood from available land holdings. Out of those systems eight agroforestry systems had been identified for extension in another area like New Alluvial Zone (NAZ) under humid and sub-humid climate in West Bengal (Dhara *et al.* 2016; Dhara 2016).

Table 1. Different Agroforestry Systems under Field Trials at Jhargram, BCKV, Humid and Sub-humid Climate in West Bengal

S. N.	Agroforestry Systems	Alley Crops
1.	*Eucalyptustereticornis*+ Mango – Arable Crops	Rice, Mustard, Pigeon Pea, Elephant Foot Yam, Turmeric, Kalai, Cow Pea, Ground Nut, Lady's Finger, Bottle Gourd, Maize, Olitorius Jute(Various combinations of those crops with the trees could produce many other profitable agroforestry systems.)
2.	*Eucalyptustereticornis*+Guava - Arable Crops	
3.	Gamhar(*Gmelinaarborea*) + Mango - Arable Crops	
4.	Gamhar + Guava- Arable Crops	
5.	Gamhar +Sweet orange - Arable Crops	
6.	Akashmoni (*Acacia auriculiformis*) + Guava	
7.	*Akashmoni* +Sweet orange + Arable Crops	
8.	Lamboo + Mango (*Mangiferaindica*) + Arable Crops	
9.	Kadam (*Anthocephaluscadamba*) + Mango + Arable Crops	
10.	Ber - Arable Crops +Lamboo(Disoxylumbinectariferum) as boundary plantation	
11.	Mahua (*Madhucalongifolia*) +Sweet orange – Arable Crops	
12.	Kapok (*Ceibapentandra*) - Arable Crops	
13.	Karanja (*Pongamiapinnata*) - Arable Crops	
14.	Sal (*Shorearobusta*) - Arable Crops	
15.	Neem (*Azadirachtaindica*) - Arable Crops	
16.	*Eucalyptus tereticornis* - Arable Crops	
17.	*Bambusabalcooa* - Arable Crops	
18.	*Bambusatulda* - Arable Crops	

Table 2. Benefit: Cost ratio of different fruit based agroforestry systems in cv. *Amprapalli* Plantation at 7th year at Jhargram, BCKV, Humid and Sub-humid Climate in West Bengal

Treatments	Fruit- based agroforestry systems	*Benefit: cost ratio*	Parameters Considered
T_1	Mango + Eucalyptus - Pigeon pea	3.2	1. Yield of Mango
T_2	Mango + Eucalyptus - Black gram followed by Mustard	2.8	2. Cost of Cultivation
T_3	Mango + Eucalyptus – Bottle gourd followed by Mustard	3.4	3. Gross return from Mango, Crop and Eucalyptus
T_4	Mango + Eucalyptus – Lady's Finger followed by Mustard	3.5	
T_5	Mango + Eucalyptus	1.5	
T_6	Eucalyptus (Sole)	1.2	
T_7	Mango (Sole)	1.2	
SEm(±)		0.399	
CD(P=0.05)		0.284	

Fruit – basedagri-horti-silvi agroforestry system – most preferred among farmers

Among so many numbers of choices agroforestry systems being taken to farmers for their possession for favourable practice, only fruit-based agroforestry systems had been found to be mostly chosen by the farmers. Other Silvi – trees were chosen mostly as boundary plantation. Among those choices following were found most successful under humid and sub-humid climate in West Bengal (Dhara 2016):

1. Mango + Eucalyptus (*E. Tereticornis*) based agri-horti-silvi system in Red and Laterite Zone.
2. Mango + Gamhar (*Gmelina arborea*) based agri-horti-silvi system in Red and Laterite Zone.
3. Mango + Lamboo (*Dysoxylumbinectariferum*) as boundary plantation based agri-horti-silvi system in New Alluvial Zone.
4. Guava based agri-horti system in New Alluvial Zone.
5. Ber based agri-horti system in New Alluvial Zone.
6. Homestead Agroforestry.

Boundary plantations–confident innovative through fares

Practising fruit based agroforestry systems was proved to be a successful land use method for vertically enhancing productivity and coverage of risk in relation to unstable weather even on a degraded land and soil conditions. Still innovating agroforestry systems, in general, in farmers plot was found to be a hard nut to crack. In that context boundary plantation was successful through fares for taking agroforestry as a usual component in normally followed farming systems. With the growth of boundary trees and seeing the effect of shades of those trees farmers would be easily convinced for introducing agroforestry trees in their regular farming plots.

Homestead agroforestry – needs more attention

Naturally some trees are grown by the farming communities around their homes. Those trees are helpful for supply of timbers for their own use or as cash for to meet up unavoidable and emergent expenditures for family members. Those homestead trees are also of regular utilities for using leaves, twigs, branches, etc. towards performing rituals and social occasions. So, such agroforestry systems need more attention for growing more trees out of forest areas.

Social agroforestry – a concept need extension

As we are aware about social forestry for growing trees on vested and government lands and by the village paths and roads and also in towns and cities. Within that forestry systems agriculture as a component is needed to be introduced for attracting more people and, thus, creating a facility for needy people for provision of their minimum necessity for timber and fruits. Such agroforestry will be more beneficial also for nearby buildings due to regular pruning of trees. In that way social forestry would be able grow more trees on human settlements.

Various benefits from agroforestry systems as envisaged from those field trials can be elucidated as follows:

1. *Sustainable soil fertility*: Agroforestry could ensure viable crop culture on inter-spaces of trees grown. Apart from this, agroforestry system can create congenial micro-environment in agroecology for more beneficial economic return from fruit trees and / or forest trees and crops raised on same piece of land which in turn would be beneficial for societal enlightenment. Thus, agroforestry integrates elements of both agriculture and forestry in a sustainable production. From long term field trial it has been revealed that productivity of alley crops was maintained at a certain level year after year under agroforestry systems. Such stable rate of productivity is an exquisite requirement of a sustainable agriculture and that might be possible due to sustenance of soil fertility under agroforestry.
2. *Improvement in soil pH, organic carbon and N, P and K status:* From various field trials it was revealed that soil pH of acid soil could be improved to some extent, while status of soil organic carbon, N, P and K increased substantially under agroforestry system in Red and Lateritic

Soil areas. Such observations were reported year after year, though soil pH could not be corrected through agroforestry which might be due to inherent acidic nature of red and lateritic soil. On the other hand organic carbon reserve in that soil could not be built up due to high air temperature during summer, being responsible for degradation of organic matter formed from tree and crop culture. Still tree cultivation acted as a stable carbon sink in towards continual enhancement of organic carbon reserve in soil.

3. *Sustainable soil moisture control:* As scientific studies on agroforestry systems was started from the early nineteen eighties, early works were involved in studies on spacing of trees, moisture holding capacity of soil under root of agroforestry trees, etc. From such works (Roy *et al.* 1989) it was revealed that agroforestry system could maintain soil moisture at different depths under trees in even dry seasons on degraded red and lateritic soils.

4. *Synergistic effects of alley crops on wood volume of trees:* From different field studies it was found that productivity of alley crops were sustainably maintained year after years while wood volumes were recorded to be higher for tree crops grown with alley crops than sole tree/ sole tree + fruit tree.

5. *A new approach on possibility of extracting medicinal resources:* Trees and crops grown under agroforestry systems could be a good source of medicinal resources. Examples can be cited from trees and alley crops cultivated under field trials like bark and root of Gamhar; root of Akashmoni; bark, wood and leaf of Eucalyptus; fruit of Mango; seed of Lady's Finger; fruit and leaf of Bottle gourd; leaf, fruit and seed of Pigeon pea; beans of Black gram; fruit of Sweet orange; leaf of Guava; young shoots and leaf of Bamboo; bark and root of Kadam; bark and seed coat oil of Lamboo; seed of Sesame; grain of Rice; leaf, fruit and leaf of Pointed gourd; etc. including well known medicinal values of neem and karanja. An economic analysis of benefits of medicinal resources (i.e. tannin from bark and wood and oil from leaf) of *Eucalyptus tereticornis* showed an extra income (Table 3) of over three lakh rupees more than traditionally followed benefit: cost analysis (Table 2) of agroforestry systems. For maintaining quality of such medicinal resources agroforestry might be followed organically.

6. *Other possible renewable returns:* Apart from wood and crop produce, medicinal resources of agroforestry systems can be treasure of many other returns. Mango based Eucalyptus – Alley crops agroforestry systems, for example, can be employed for apiculture; and pruned twigs

Table 3: Estimation of Extra Return obtainable from Medicinal Resources (Tanin + Leaf Oil) from Eucalyptus

Wood Production ($m^3ha^{-1}yr^{-1}$)	Wood Weight ($kg^3ha^{-1}yr^{-1}$)	Bark Weight ($kg^3ha^{-1}yr^{-1}$)	Bark obtainable ($kg^3ha^{-1}yr^{-1}$)	Wood obtainable ($kg^3ha^{-1}yr^{-1}$)	Tanin Weight ($kg^3ha^{-1}yr^{-1}$)	Return from Tanin @ Rs 2000 /kg (Rs)	Leaf obtainable (tree to tree spacing 10m X 10m) ($kg^3ha^{-1}yr^{-1}$)	Eucalypt Leaf Oil ($ml^3ha^{-1}yr^{-1}$)	Return from Eucalypt Leaf Oil @ Rs 600/ 1000 ml (Rs)	Extra Return from Tanin + Leaf Oil (Rs)
Column 1	2	3	4 #	5 #	6	7 ®	8 #	9	10 ®	11
	(1*0.65* 10*100)	(2*21.36%) 312.390	(3*50%) 1462.500	(2*50%) 159.740	9% (4+5)		(0.42 kg/ plant *121*50%)	(8*1.15%)		(7 + 10)
4.5	2925	624.780	312.390	1462.500	159.740	3,19,480	25.41	29.222	17,533	**3,37,013**

#considering 50% permissible pruning of trees/year
® Prices as obtained from India Mart Website:
www.indiamart.com

and numerous plant materials could be the sources of making toys and materials for entertainment and interior decoration and walking by that agroforest would be itself of entertaining and those would be the sources of more returns. Thus, agroforestry can be regarded as multi- purpose multi-storeyed plant treasure having multi-return capabilities on renewable basis.

7. *Mitigation of climate change:* As it was found that cultivation of alley crops could help the tree to grow more wood volume; agroforestry could give more wood than sole tree and/or fruit tree cultivation. It was estimated that 2.925 Mg wood (Table 3) could be obtained from one hectare of Eucalyptus from Mango based Eucalyptus - alley crops agroforestry in a year and that can be estimated to be 1.4625 Mg ha^{-1} yr^{-1} of carbon sequestered from atmosphere. Apart from that extra amount of carbon could be sequestered through fruit trees and alley crop biomass production. Thus, alley crops could act as an active component for mitigation of climate change through sequestering of such huge amount of carbon from atmosphere.

Conclusion

1. Proper planning of agroforestry could help to identify suitable and economically viable agroforestry systems aimed at mitigation of climate change.
2. Growing Silvi component along with fruit tree and alley crops could enhance economic benefits helpful for poverty alleviation including food and fodder security. Thus, agroforestry system acts as multistoreyed plant treasure.
3. Agroforestry system could maintain soil moisture even in dry season and helped to maintain organic carbon and N, P and K fertility status of soil and, thus, it could maintain Soil Health.
4. An approach for harnessing medicinal resources from grown trees and crops would make same unit of agroforestry system able to earn more return. Along with those resources other possible resources from apiculture and possibilities of making toys and interior decoration materials from discarded numerous tree and crop materials would make agroforestry amultipurposemultireturn plant treasure.
5. For procuring medicinal resources from agroforestry systems, trees and crops would needed to be grown organically.

6. Lastly, but not least it might be concluded that Agroforestry System can mitigate climate change by sequestering huge amount of carbon through biomass production by trees and crops and continuous enrichment of soil for organic carbon.

References

Nair PKR (1993) An Introduction to Agroforestry. Kluwer Academic Publishers, Dordrecht.

Shiva MP, Mathur RB (1996) Management of Minor Forest Produce for Sustainability. Oxford &IBH Publishing Co. Pvt. Ltd, N. Delhi.

BlatterE, Millard WS (1954) Some Beautiful Indian Trees.2nd Edition:Stearn, WT(Ed.): Bombay Natural History Society, Bombay.

Talbot WA (1984) Systematic List of the Trees, Shrubs and Woody –Climbers of the Bombay Presidency. Government Central Press, Bombay.

Dhara PK, Panda S, Sarkar S, Das N (2016) Agroforestry systems practiced in new alluvial zone of West Bengal. In: Proc Intl. Conf. Agriculture, Food Sci., Nat. Resource Management and Environmental Dynamics: The Technology, People and Sustainable Development, 13-14 August, Kalyani, India. pp. 162-164.

FAO (2015) Agroforestry–Definition. http://www.fao.org/corp/copyright/en/.2016

DharaPK (2016) Humid and sub-humid region-BCKV, Jhargram (West Bengal). In: Chaturvedi OP *et al.* (ed) Agroforestry Technologies for Different Agro-Climatic Zones of the Country. ICAR-Central Agroforestry Research Institute, Jhansi. pp. 51– 64.

Kirtikar KR, Basu BD, An ICS (1918) Indian medicinal plants. In: Blatter E, Caius JF, Mhaskar KS (eds) 2nd Edition in Four volumes. M/s Bishen Singh Mahendra Pal Singh, New Cannaught Place, Dehradun. 1918 (Reprint 1975).

Orwa C *et al.* Agroforestree Database: a tree reference and selectionguide version 4.0. World Agroforestry Centre, Kenya. http://www.worldagroforestry.org/resources/databases/agroforestree and www.feedipedia.org/node/1650

World Agroforestry Centre 2009 *Eucalyptus tereticornis* Smith–Mysore gum.www.worldagroforestry.org/treedb2/AFTPDFS/Eucalyptus_tereticornis.pdf

National Medicinal Plants Board. New Delhi. Monthly Market Price. May and June 2016.

Chaturvedi AN, Khanna LS (1982) Forest Mensuration.International Book Distributors, 9/3, Rajpur Road, Dehradun, India. pp.98-99.

Chaturvedi AN, Khanna LS (2000) Forest Mensuration and Biometrics.3rd Ed, Khanna Bandhu, 7 Tilak Road, Dehradun, India.

Birdsey RA (1992) Carbon Storage Accumulation in United States Forest Ecosystems, General Technique Report WO – 59. USDA Forest Service, North Eastern Forest Expt. Station, Radnor, PA.

Paladinic E *et al.* (2009) Forest biomass and sequestered carbon estimation according to main tree components on forest stand scale. Periodicum Biologorum, UDC 57:61, CODENPDBIAD, ISSN 0031 – 5362. III (4):459-466.

Piper CS (1966) Soil and Plant Analysis. Hans Publishers, Bombay.

Jackson ML (1962) Soil Chemical Analysis. Prentice Hall of India Pvt. Ltd., N. Delhi.

Ghosh AB *et al.* (1983) Soil and Water Testing Methods: A Laboratory Manual, Division of Soil Sc. and Agril. Chem., IARI, N. Delhi.

INM Division, GOI. Methods Manual - Soil Testing in India. Dept. of Agri. & Coop., Ministry of Agriculture, Govt. of India. January 2011.

USDA Handbook (1993) Soil Survey Division Staff. Soil Survey Manual. Soil Conservation Service.

Kumar BM, Nair PKR (2011) Carbon Sequestration Potential of Agroforestry Systems: Opportunities and Challenges. Springer, Dordrecht.

Batish, DR *et al.*(ed) (2008) Ecological Basis of Agroforestry. CRC Press, USA.

Roy GB, Majhi SK, Ghosh RK, Panda S (1989) Study of soil moisture status in different planting methods of Akashmoni (*Acacia auriculiformis* A. Cunn.) trees in an agroforestry system of cultivation on laterite soil of West Bengal. Ind J Landscape Systems and Ecological Studies 12(2):31-34.

36

Entomofauna in Agroforestry Systems Friends or Foes

Nisha Patel

Central Arid Zone Research Institute, Jodhpur – 342 003, India

In contrast to the earlier and existing non-sustainable land use practices, agro forestry presents a more dynamic, productive and sustainable solution to the increasing demand of food, fodder, timber and other agricultural and tree goods. According to the United Nations,the world population was more than 7 billion in 2011 which is expected to go up to 9.3 billion by the mid-century. Consequently food production needs to be increased by over 60% by the year 2050. Agroforestry, which is a multifunctional, ecologically based, natural resource management system can help in dealing with such environmental, economic or social issues.

Plants and insects have coexisted, interacted and developed a variety of associations with each other for more than 350 million years. There is enormous species diversity in both these groups, and the interactions have played vital role in the development of this diversity. Plants in the form of crops,shrubs or trees provide insects with food,shelter and oviposition sites (Panda and Khush, 1995). Insects which constitute almost half of the total faunal biodiversity (Speight *et al.* 1999) provide valuable major ecosystem services such as pollination, decomposition, recycling, herbivory and serving as important components of the food chain, etc (Hein, 2009; Hillstrom and Lindroth, 2008). Many of the ecosystem services provided by insects such as nutrient cycling, pest regulation and pollination – directly contribute to agricultural production (Losey and Vaughan 2006). The healthy functioning of these ecosystem services ensures the sustainability of agricultural systems. On the other hand insects as pests are the most important group of organisms causing injury to plants in agroforestry systems.

Interactions among the insects and plants in agroforestry system can be either neutral, beneficial or harmful. When the pest problems in an agroforestry system do not exceed those in monoculture of crop or tree species the interaction can be termed neutral. However, they will be considered harmful when insect pest problems increase in the agroforestry system as compared with a monoculture plantations. If the insect pest problems decrease in the components of agroforestry, it indicates a positive interaction from the insect-pest management point of view. The insect pests in an agroforestry system are basically the pests of its components: the agricultural crops and the perennial components. The composition and complexity of agroforestry systems affects the diversity and population of insect pests and their natural enemies. The entomofauna of the system will play both a beneficial as well as harmful role and the net outcome will depend on the type of agroforestry system type and composition of the plant species in agriculture or tree component and the types and relative population of insects.

Insect Pests in Agroforestry

Pests have often been named as one of the factors reducing the benefits from tropical agroforestry (Mchowa and Ngugi 1994; Rao *et al.* 2000). Day and Murphy (1998) and Rao *et al.* (2000) have discussed chiefly about insect pests affecting agroforestry trees and their management. Schroth and coworkers (2000) dealt with insect pests and diseases in agroforestry systems of the humid tropics. The insect-pest intensity in agroforestry is affected by various factors. Floral diversity in the system is an important factor. Monocultures suffer from increased pest problems as these are easily recognised and colonised by insects whereas fields with diverse vegetation have lesser number of potential host plants resulting in reduced colonization on the crop.The plants which belong to the same or very close taxonomic groups usually suffer the attack of common pests. Many insects pests are polyphagous and if the agroforestry system has as its component the plant species which are devoured by such insects, the population of these may increase. Many leguminous plants are components of agroforestry systems. If crops such as pulses or some oilseeds are grown with leguminous woody perennials,the pest populations common to both components may increase. Bruchid species *Caryedon serratus* infests many legume tree seeds and also groundnut. The combination of such plants will ensure the availability and thus build up of population pest insects throughout the year. The damage caused by the pod borer, *Etiella zinckelia*, in peas has been reported to be increased by the presence of Acacia trees in the vicinity, as reported by Szeoke and Takacs (1984). The selection of plant species for agroforestry, should therefore include plants from different taxonomic groups to avoid the possibility of common pests.

The food of larval stages of insects is often different than that of its adult stage. Sometimes the perennial components provide shelter and food to the adult insect whose larval stage may damage the crop component in agroforestry system. In arid zones crops such as pearl millet, cowpea, green gram, moth bean and sesame are often co cultivated with tree species as *Prosopis cineraria, Azadirachta indica* and *Zizyphus* spp. The leaves of these tree species are the food for adult scarabaeid beetles (*Lachnosterna* spp., *Holotrichia* spp., *Anomala* spp., *Adoretus* spp. etc) whose larvae cause heavy losses to rainy-season crops in some areas. The presence of such host plants in the crop fields may sometimes increase pest problem in crops. The adults survive for a few days and lay eggs near roots of crops. The eggs hatch into white grubs, which infest the crops, causing heavy losses.

The insect vectors of plant diseases often survive on a variety of host plants that are not taxonomically close. Sucking pests such as white flies and some aphid and jassid species are vectors of viral, bacterial and other plant diseases. The presence of refuge host plants in an agroforestry system may provide a source of infection for the crops if the plant harbours vectors that spread these disease.

Agroforestry Systems and Pest Management

Well-designed agroforestry systems can reduce the need for pesticides in several ways. The larval or young stages of many beneficial insects feed on other insects but the adults of most of the predators and parasitoids of pest insects require nectar and pollen at certain stages of their life for growth and reproduction. A diversified ecosystem with many types of trees, shrubs and small plants provides microhabitats, food sources (prey, nectar, pollen), alternative hosts and shelter for natural enemies and thus encourages colonization and population build up of natural enemies. Diversified cropping systems, such as agroforestry systems are considered to be more stable and more resource conserving (Vandermeer 1995). Such systems encourage abundance and efficacy of natural enemies insects due to their inherent plant diversity thereby regulating herbivore populations (Altieri and Letourneau 1984). Diversified ecosystems such as agroforestry have the presence of flowering plants that provide pollen and nectar which serve as alternative food for natural enemies which increase their fecundity and longevity. Higher parasitism levels have been observed in annual crops and orchards with rich floral undergrowth (Leius 1967).Cultivated ground cover plants and weeds provide ecological niches of various species in the community and variety of resources for predators and parasitoids, including shelter, food, and information on the location of their herbivorous prey (Bugg and Waddington 1994).Flowering plants can also increase populations of non-pestiferous herbivores (neutral insects) in crop fields.

Such insects serve as alternative hosts or prey to entomophagous insects thus improving the survival and reproduction of these beneficial insects in the agroecosystem. Many researchers have reported that the presence of neutral insects on flowering plants within or near crop fields increase predation and parasitism of specific crop pests (Altieri and Whitcomb 1979).

Plant diversity around crop fields is also important in determining the diversity and abundance of natural enemies within agroecosystems. Enhancement of natural enemies and more effective biological control has been observed where wild vegetation remains at field edges in close association with crops (Altieri, 1994). Under arid conditions ber and henna in the organic farm provided pollen and nectar to variety of pollinators, predators and parasitoids besides providing shelter for insects like lacewing.Many lacewing eggs were seen on henna (*Lawsonia inermis*) and ber (*Zizyphus mauritiana*) plants. Calotropis shrubs encourage beneficial insects like coccinellids and syrphids. Moreover,Calotropis and henna also support populations of prey insects viz., aphids and whiteflies which provide food to predators when the crops are not in field. *Prosopis cineraria* and *Acacia senegal* trees are also good sources of pollen and nectar and attract many types of pollinators and parasites on the farm.

Insects for Recycling of Nutrients

Insects play an important role in degradation processes. Many insects viz., termites, dung beetles, flesh flies, flies, etc are primary or secondary decomposers. Insect decomposers are crucial for the breakdown of organic matter of both animal and plant origin material (both leaf litter and woody material)which facilitates its return to the soil as nutrients. Their action is often necessary before other groups of decomposers (fungi, bacteria) can take advantage of the material. Without the help of insects, break down and disposal of wastes, dead animals and plants would be impossible and this waste would accumulate in our environment. Termites are major decomposers of plant material in the deserts and contribute to nutrient cycling, soil fertility, soil aeration stability etc.

References

Altieri, M.A. (1994). Biodiversity and pest management in agroecosystems. Haworth Press, New York, 185pp.

Altieri, M.A. and Whitcomb,W.H. (1979). The potential use of weeds in the manipulation of beneficial insects. *Hortscience* 14:12-18.

Altieri, M.A. and Letourneau, D.K. (1984). Vegetation diversity and insect pest outbreaks. *CRC Critical Reviewsin Plant Sciences,* 2: 131-169.

Bugg, R.L. and Waddington, C. (1994). Using cover crops to manage arthropod pests of orchard: a review. *Agriculture, Ecosystem and Environment,* 50:11–28.

Day, R., Murphy, S.T. (1998). Pest management in tropical agroforestry: prevention rather than cure. *Agroforestry Forum,* 9:11–14.

Hein Lars (2009). The economic value of the pollination service, a review across scales. *The Open Ecology Journal,* 2: 74-82.

Hillstrom, M.L., Lindroth, R.L. (2008). Elevated atmospheric carbon dioxide and ozone alter forest insect abundance and community composition. *Insect Conservation and Diversity,* 1: 233-241.

Leius, K. (1967). Influence of wild flowers on parasitism of tent caterpillar and codling moth. *Canadian Entomologist,* 99: 444-446.

Losey, J.E., Vaughan, M. (2006). The economic value of ecological services provided by insects. *Bioscience,* 56 (4): 331–323.

Mchowa, J.W. and Ngugi, D.N. (1994). Pest complex in agroforestry systems: the Malawi experience. *Forest Ecology and Management,* 64:277–284.

Panda, N., Khush, G.S. (1995). Host plant resistance to insects. CAB International, Wallingford.

Rao, M.R., Singh, M.P. and Day, R. (2000). Insect pest problems in tropical Agroforestry systems: contributory factors and strategies for management. *Agroforestry Systems,* 50:243–277.

Schroth, G., Krauss, U.,Gasparotto, L., Duarte Aguilar, J.A. and Vohland, K. (2000). Pests and diseases in agroforestry systems of the humid tropics. *Agroforestry Systems,* 50:199–241.

Speight, M.R., Hunter, M.D., Watt, A.D. (1999). Ecology of Insects – concepts and applications. Oxford, Blackwell Science, 340p.

Szeoke, K. and Takacs, L. (1984). Damage caused by lima bean borer *(Etiella zinckella Tr)* in peas. 1. Study of habits. *Novenyvedelem,* 20(10):433- 438.

Vandermeer, J. (1995). The ecological basis of alternative agriculture, *Annu. Rev. Ecol. Syst.* 26: 201-224.

www.scientificamerican.com/article/insects-provide-billions/Insects provide billions in free service –Scientific American

www.xerces.org/pollinator-conservation/The Xerces Society » Pollinator Conservation

About the Editors

Dr. C. B. Pandey is a well-known agro-forester with an experience of more than 25 years working in the field of agroforestry. He has conducted extensive studies on different aspects of agroforestry that include, soil-tree-crop interactions,nutrient cycling, carbon sequestration and N-transformations. He discovered nitrogen conserving mechanism, and phosphorus availability mechanism in iron rich soils in the humid tropics. He has authored more than 100 papers in the journals of national and international of repute and edited several books.

Dr. Pandey has completed his undergraduate (1982) and post graduate (1984) studies from Gorakhpur University and Ph.D. from Banaras Hindu University (BHU) in 1990 and pursued post-doctoral study from 1990 to 1993 in North Eastern Hill University, Shillong and BHU, Varanasi.

He served Indira Gandhi Agricultural University as an Assistant Professor from 1993 to 2000, Senior Scientist from 2000 to 2008 at Central Agricultural Research Institute, Port Blair, Andaman and Nicobar Islands, Principal Scientist from 2008 to 2012 at Central Soil Salinity Research Institute, Karnal (Haryana) and is presently heading the Natural Resources and Environment division in Central Arid Zone Research Institute, Jodhpur. He is a Fulbright Scholar and a fellow of National Academy of Agricultural Sciences, and is a recipient of Fakhruddin Ali Ahmed and Dr. K.G. Tejwani awards for his outstanding works in the field of agroforestry.

Dr. Mahesh Kumar Gaur is working as Senior Scientist at ICAR-Central Arid Zone Research Institute, Jodhpur, India since September 2012. He specializes in Arid lands Geography and application of satellite remote sensing, GIS and digital image processing for natural resources mapping, management and assessment; drought, desertification and land degradation. He has worked as Associate Professor for more than 13 years with the Department of Higher Education, Government of Rajasthan state, India. He hasconducted various sponsored studies of ISRO, SAC, UGC, various Ministries of Government of India, ICARDA, etc. He has published four books and is author of a number of research papers and book chapters that has appeared in national and international research journals.

He is associated with Global Geography Education, Association of American Geographers, USA; Associate Editor, *RIGEO*, Turkey; *Journal of Arid Environments*, the Netherlands; Society for Conservation Biology, Arilington, USA; Member, Scientific Board of International *Journal of Ecosystems & Ecography*, USA; etc. He was awarded prestigious *Associate ships* by the UGC of India and Scientific Assembly of the International Committee on Space Research (COSPAR). He has been awarded *Government Citizen Karmveer Award*, 2011 by iCONGO for working in the field of *Higher Education, Environment and Technology Applications for Community Upliftment* and *Millennium Award* for working in the field of Environment. He has been to South Africa, Namibia and Costa Rica on academic assignments. He is member of a number of national and international scientific organization and Editorial Boards of journals. He is Secretary of INCA, Jodhpur Chapter and Member, Government of India constituted *task force on preparation of mountain zone regulations & policies*.

Dr. R. K. Goyal, B.E. in Agricultural Engineering (1987) from College of Technology and Engineering, Udaipur; M. Tech in Water Resources Engineering (1989) from Indian Institute of Technology, New Delhi and Ph.D. in Water Resources Engineering from M.B.M. Engineering College, Jodhpur, is presently working as Principal Scientist (Land and Water Management Engineering) at ICAR-Central Arid Zone Research Institute, Jodhpur. He possesses expertise of more than 23 years in arid lands' hydrology; water/watershed management and climate change modelling.

Dr. Goyal is recipient of certificate of appreciation "SAVING THE DRYLANDS" from United Nations Environment Program (UNEP) for outstanding contribution in combating desertification and controlling land degradation in dryland environments (1996). Dr. Goyal has published more than 125 research articles in national and international journals. He has visited The University of Queensland, Brisbane, Australia (1996) under AusAID fellowship of Government of Australia.

NIPA
81.